Preface

This is the third manuscript I have prepared on the same general subject of noise effects. The first, entitled "The Effects of Noise on Man," was a monograph for the Office of Naval Research in 1950 and was published by the American Speech and Hearing Association. The second, of the same title, was sponsored by the Office of Naval Research, the Surgeon General of the Army, and the National Aeronautics and Space Administration, and was published by Academic Press as a book in 1970. The present volume, except for chapter 12, was prepared for the National Aeronautics and Space Administration with support from the U.S. Department of Transportation and from the U.S. Environmental Protection Agency. Chapter 12 was originally prepared for the City of Santa Monica, California. All these manuscripts were intended to be critical reviews and interpretations of the relevant original source literature for the measurement of noise in terms of its effects on people. Except for a relatively few papers, only English language publications were included.

In addition to providing major findings of published research, an attempt is made in the present volume to integrate, where possible, the findings into some theoretical framework. In this process, using data from a number of sources, I undertook analyses and modeling of certain topics. This was done particularly with respect to presbycusis, sociocusis, and nosocusis (chapter 7), damage to hearing (chapter 8), and reactions to community noise (chapter 11). Because of possible interest to researchers in fields beyond "noise effects," chapter 6, in part, has been published in the Journal of the Acoustical Society of America (vol. 73, pp. 1897-1919, and vol. 74, pp. 1907-1909, 1983).

There has been a particularly significant increase since 1970 in the store of knowledge about noise and its effects, and today many of the seeming conflicts among research findings of the past are found to be more apparent than real. However, the state of the art on certain topics (approximately 20 percent of the book) remains best represented by findings published prior to 1970. Although a number of research questions remain, objective methods now exist for measuring noise environments that predict, with considerable accuracy, the effects of noise on people and communities.

The first four chapters are concerned with some definitions of terms and research on the fundamentals of noise, hearing, and auditory perception. The last eight chapters are concerned with research on noise effects on more complex human behavior and the application of this research to the regulation of noise in work and living areas.

I gratefully acknowledge the authors and publishers of journals and books for permission to reproduce their figures. I owe many thanks to Joyce Garbutt for her skills and patience during innumerable typings of the manuscript, to James R. Young for his help with respect to formulas in chapter 2, and to members of the staff of NASA Langley Research Center for their valuable editorial suggestions and comments.

Needless to say, I am deeply grateful to the organizations cited above for their support over the years. I can only hope this book provides some justification for that support.

Contents

Chapter 1
Definitions of Noise and Noise Pollution

Introduction

The adverse effects of noise, or unwanted sound, have been the subject of extensive research for many years in the fields of psychoacoustics and physiological acoustics. This research grew out of the desire for scientific understanding of these effects on people, especially because of social problems created by the steady increase in the intensity and prevalence of noise found in living and work environments since the start of the industrial revolution.

Definition of Noise

In the fields of electronics, neurophysiology, and communication theory, "noise" means signals that bear no information and whose intensities usually vary randomly over time. The word "noise" is sometimes used in that sense in acoustics, but for our purposes, noise will be defined as an audible acoustic energy that adversely affects the physiological or psychological well-being of people. This is consistent with the usual definition of noise as being "unwanted sound."

It is not possible to define noise on the basis of physical aspects of sounds. Indeed, sounds can be wanted at one moment, such as when they contain information that is either of aesthetic or practical interest to a person, whereas in other contexts or for other people, the same sounds may be considered as noise because they interfere with the hearing of other sounds that are wanted. For example, the sounds of an engine can convey useful information to a mechanic, whereas the sound is noise to another person attempting to talk in its presence. Likewise, music may be considered as pleasing sound under most circumstances but as noise when it interferes with conversation or sleep or when it is not pleasing aesthetically.

Another factor complicating the definition of noise is that a given sound can sometimes be wanted psychologically and yet be physiologically unwanted or damaging. Very intense electronically amplified "rock" music is an oft-cited example. Although enjoyed by the musicians, the intense music can result in permanent damage to the neural receptors in their ears after a certain length of exposure time.

Also, some confusion about what is "noise" stems from the fact that certain sounds are sometimes called noise because the sound indicates a source that can be harmful. The noise from an airplane or the buzz of a mosquito may be considered as very "noisy" or unwanted, when actually the danger from, or anxiety about, the source of the sound itself is what is so objectionable. However, the evaluation and control of noise must be predicated on more direct psychoacoustic and physiological acoustic effects to be of practical use.

Noise Pollution

By and large, the types of noise of concern to society and government have the same meanings to most people and come from sources that normally represent no direct physical threat to people. Heavy machinery in industry, transportation vehicles, and appliances around the home make sounds about which there is little confusion as to their unwantedness and to which large segments of our society are exposed daily.

As the extent of noise and its effects on people have become measurable, private and governmental efforts have increased environmental noise control to protect those people who are exposed. However, establishing broad-based and effective noise-control programs has been a difficult and contentious matter. This has been partly because of the relatively small amount of previous research data concerning the effects and measurement of noise, and partly because of economic conflicts created, or threatened to be created, by proposed noise control programs and regulations.

Conflicts of interest inhibiting the establishment of noise controls are exemplified by the following cases: (1) some weaving mills would have to close down if forced to pay the costs involved in reducing the noise to levels that do not damage the auditory system of the workers; (2) labor unions and builders have objected to noise zoning that would prevent residential areas along noisy highways on the grounds that work would be taken away from the local areas; and (3) landowners do not wish to have their land use restricted by noise zoning.

Another conflict over noise control in communities involves noises that do not, at any one time, necessarily reach a large number of people and are more private than public. These noises are from neighbors, from dogs barking, from gatherings of people, and from frequent temporary use of noisy tools at or near homes or offices. The fact that the public interest is generally not served by the noise is a force towards prohibiting such noises.

Under some circumstances, however, there are reasons for permitting, or even promoting, the presence of noise. For example, the exterior noise generated by machinery and equipment that is operated for or by the public, especially those of transportation such as aircraft, buses, automobiles, trucks, and trains. The conflict is between the benefits from the transportation versus the "costs" to those exposed to the exterior noise. This problem is complicated by the fact that although the operation of the vehicles is of economic and social benefit to most of the public, only a relatively small portion of the general population is exposed daily to the adverse effects, or costs, of the intense exterior noise generated by these vehicles. In brief, the personal well-being of a small minority of the people is in conflict with the economic interests of the population in general, including the minority exposed to the noise.

Indeed, local governments cannot enforce laws that attempt to control noise in a way that substantially interferes with interstate commerce. The reason is that the commerce, even if not of local interest, is a matter of interest to people living in areas not impacted by the noise. At the same time, if a noise environment (*e.g.*, that from commercial aircraft) is protected from local government control because of interstate commerce or other reasons, the citizens whose health is affected or whose property values are reduced because of the noise may be entitled to monetary compensation. The issue then becomes who should be liable for paying any compensation that might be justified—the local government who permitted improper land use, the airport-aircraft operator, or the people engaged in the commerce. There are, however, practical problems to the description of noise pollution in economic terms. Specifically, translating psychological, physiological, or other possible damage effects of the noise on individuals into "market-place" economic equivalents is obviously a complex matter.

The control of environmental noise pollution and the resolution of conflicts of interest over noise pollution presumably depend upon executive, legislative, and, sometimes, judicial governmental actions. In short, noise pollution is as much a political problem as it is a scientific matter. However, proper resolution of the conflicts of interest created by noise pollution rests upon the proper use and interpretation of the scientific research findings and concepts pertaining to the effects of noise.

Chapter 2
Physical Measures of Sound and Noise

Introduction

Before it is anything else, noise is sound, and to understand the effects of noise on people, it is helpful to consider the nature of sound and the auditory system. Discussed at the beginning of this chapter are some units of sound and noise measurements that are essential to the interpretations and use of the research findings and concepts developed in later chapters.

Definitions of Sound

Frequency

Imagine, if you will, the action of the diaphragm of a loudspeaker as it moves back and forth because of an electrical signal applied to a magnetic coil in the speaker. The pushing out of the diaphragm causes the air particles at that point to be pushed together, and the pulling back leaves a rarefaction of the air particles. This positive air pressure (compared with normal) followed by a reduced air pressure (compared with normal) travels through the air like a wave and pushes back and forth the eardrum upon which it falls, setting into action the auditory system. The sound can, of course, be generated by any vibrating physical object, such as a violin string, the parts of a truck, the column of air puffing from the exhaust of a jet engine, the muzzle of a gun, or the mouth of a person talking. The number of times per second the air pressure increases above, then decreases below, and then returns to normal pressure is defined as the frequency in hertz (Hz) (also referred to as cycles per second (cps)).

For the human listener, sound is defined as acoustic energy between 2 Hz and 20 000 Hz, the typical frequency limits of the ear. The lowest frequency of sound that has a pitch-like quality is about 20 Hz. There is further discussion of sound frequencies following definitions of the intensity variables of sound.

Sound pressure and energy

The degree to which the particles of air at a point in space (*e.g.*, at the eardrum of a person) are compressed and rarefied from their normal ambient state is called the pressure of sound at an instant in time p_i. (See fig. 2.1.) The integral $\int_{t_i}^{t_n} p_i^2\, dt$ is proportional to sound energy. The act of squaring p_i reflects the fact that the work required to rarefy the air (the negative pressure) is the same as that to compress the air (the positive pressure) by a like amount.

The mean, or average, of the squared pressures P^2 at a number of instants in a longer period of time is related to the sound energy in the given period of time as follows:

$$P^2 = \frac{1}{n} \sum_{i=1}^{n} p_i^2$$

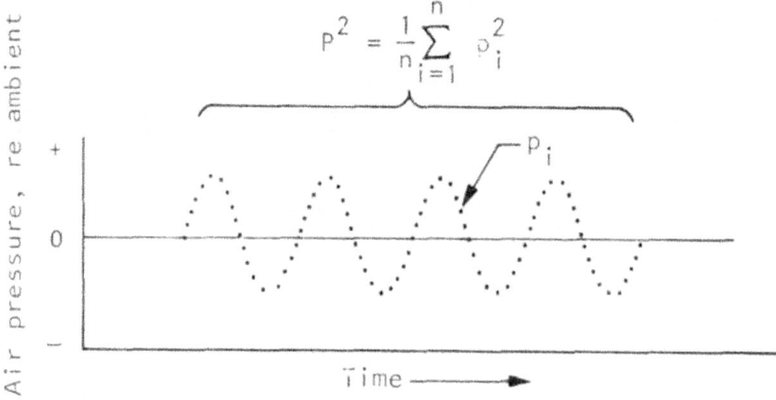

FIGURE 2.1. Representation of instantaneous sound pressure p_i. Summing and averaging the sound pressures squared over time give the average instantaneous sound pressure P.

where n is the number of instantaneous pressures sampled or measured over 1 sec. In order to obtain accurate measures of varying sound pressures, the sampling rate must be at least 2 times the highest frequency in the sound signal being measured.

For our purposes, the air is presumed to act as a pure resistance to sound waves of different frequencies. For reasons to be discussed later, the basic period of time used for most measurements of brief sounds and noise in terms of how they affect people is 1 sec. When briefer or longer periods are used, they are specified for the benefit of the reader.

It is common practice to express sound in a unit called the decibel (dB). In acoustics the decibel is 10 times the common logarithm of the ratio of 2 entities proportional to power (*e.g.*, instantaneous sound pressure). For sound pressure, these entities are the measured pressure $p_{i,\text{meas}}$ and a reference pressure $p_{i,\text{ref}}$ (20 μPa for SPL) by the following:

$$\text{Sound pressure level, dB} = 10 \log_{10} (p^2_{i,\text{meas}} / p^2_{i,\text{ref}})$$

Measures of sound

The following three general measures of sound have been developed for the purpose of relating the intensity of sound to its effects on people:

1. One-second sound pressure level (SPL) in decibels. As mentioned previously, the measured SPL is 10 times the common logarithm of the measured instantaneous sound pressures $p^2_{i,\text{meas}}$ over the reference amount $p^2_{i,\text{ref}}$. This measure is estimated for continuous sounds by using a standard sound level meter (SLM). By international standards, pressure is measured in newtons per square meter, or pascals. The reference pressure used with sound level meters is 20 μPa.

Mathematically, the 1-sec SPL (also designated as L) in decibels is

$$\text{SPL} = L = 10 \log_{10}\left(\frac{1}{n} \sum_{i=1}^{n} \frac{p^2_{i,\text{meas}}}{p^2_{i,\text{ref}}}\right)$$

where n is the number of instantaneous pressures measured. Again, unless otherwise noted, the period over which p_i is averaged is 1 sec. The relations between SPL in decibels and various units in which P can be expressed are shown in figure 2.2. (See also Harris, ref. 1.)

It is often the maximum 1-sec level that occurs during a noise event that is of interest. This is designated as SPL_{max} (or L_{max}) in decibels.

2. Event exposure level L_{ex} in decibels. As we shall see later, 1-sec SPL is an appropriate measure for relating the physical energy in sounds of 1 sec or shorter duration to auditory system responses to sounds. However, most sounds or noises are of longer durations. The event exposure level L_{ex} is used in this latter context. Although not the only method used in actual sound-level-measurement

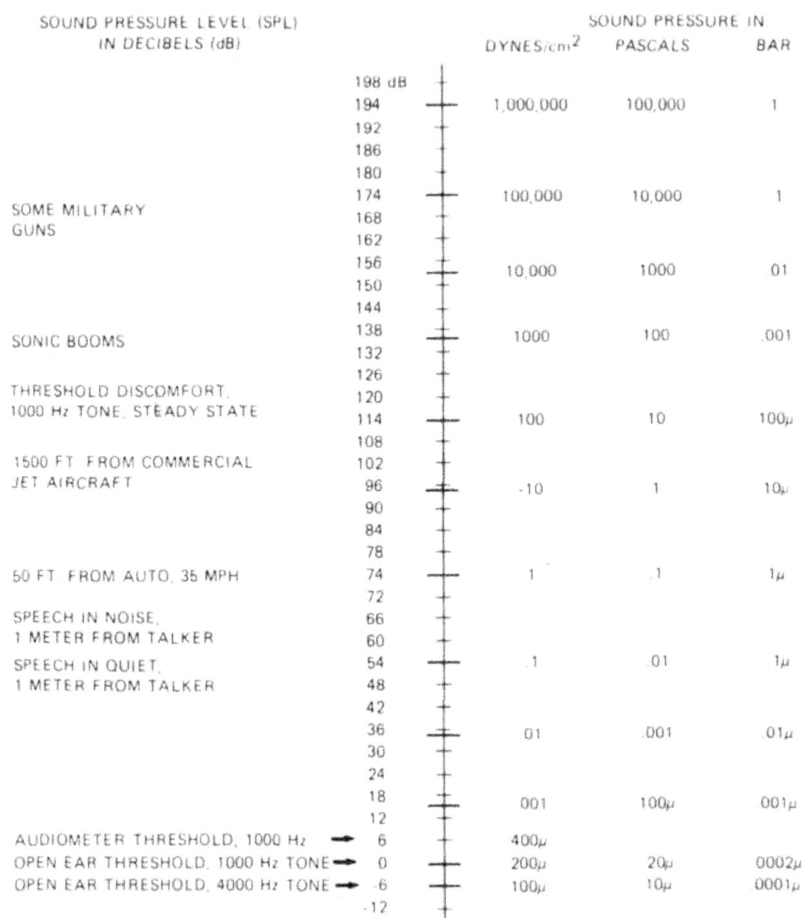

NOTE

1. SOUND PRESSURE LEVEL IN DECIBELS (dB) = $10 \log_{10} \left(\frac{P}{20\mu Pa}\right)^2$
2. NORMAL ATMOSPHERE (0° C, SEA LEVEL) 1 BAR = 14.7 LBS/FT² = 2117 LBS/IN² = 10^5 PASCALS = 10^6 DYNES/cm² = 194 dB re 20μ Pa
3. EVERY DOUBLING OF THE SOUND PRESSURE CAUSES AN INCREASE OF 6 dB
4. A 1.41 INCREASE IN SOUND PRESSURE CAUSES AN INCREASE OF 3 dB IN SOUND PRESSURE LEVEL AND REPRESENTS A DOUBLING OF SOUND ENERGY
5. THE SPL's GIVEN HERE ARE FOR LINEAR FREQUENCY WEIGHTING

FIGURE 2.2. Sound pressure level in decibels as a function of various units of sound pressure. Some approximate SPL's for certain sounds and noises are indicated.

procedures, it is convenient and appropriate to consider the amount of energy in the total, identified sound or noise event (*e.g.*, the sound of an airplane flying overhead).

Mathematically, L_{ex} in decibels is

$$L_{ex} = 10 \log_{10} \sum_{i=1}^{n} 10^{L_i/10}$$

where $i = 1$ is the first second of the sound event, $i = n$ is the last second of its presence, and L_i is the SPL of each 1-sec interval over the specified period of time. For practical reasons related to the logarithmic character of decibels, as well as other reasons discussed in chapter 6, it is sometimes practically sufficient to sum only SPL's between the times the sound is within 10 dB of the maximum or peak SPL reached by the sound.

Multiple-event sound equivalent level L_{eq} in decibels

In real life, the cumulative effects of the energy in repetitions of the same or different sounds over days and years of exposure are usually of primary concern in the assessment of noise pollution. For these purposes, a quantity labeled L_{eq} has been developed. This measure of noise exposure is proportional to the sum of the energy of the 1-sec SPL's over a fixed, specified long period of time (*e.g.*, 1 hour, 1 day, 1 year, 50 years, *etc*). However, unlike L_{ex}, for which the measured pressures in successive 1-sec intervals are merely summed, in L_{eq} the sum of the measured 1-sec SPL's is averaged over the total duration, in seconds, of the fixed, specified period of time. Accordingly, L_{eq} is proportional to the average 1-sec SPL. If it were present for each and every second during the fixed, specified period of time, the SPL would add up to, or be equivalent to, the amount of sound energy actually present during the entire period of time.

Mathematically, L_{eq} in decibels can be expressed in the following three ways:

$$L_{eq} = 10 \log_{10} \frac{1}{n} \left(\sum_{i=1}^{n} 10^{L_i/10} \right)$$

where $i = 1$ is the first second, and $i = n$ is the last second, and L_i is the SPL of each 1-sec interval during a specified period of time. When appropriate, L_{eq} is

$$L_{eq} = L_{ex_i} + 10 \log_{10} N_i + 10 \log_{10} \frac{1}{T}$$

where L_{ex_i} is the value of a particular type or level of noise events occurring during the specified period of time T in seconds and N_i is the number of such occurrences. The third way L_{eq} can be expressed is

$$L_{eq} = 10 \log_{10}\left(\frac{1}{n} \sum_{i=1}^{n} 10^{L_{ex_i}/10}\right)$$

where L_{ex_i} is the first value and L_{ex_n} is the last value for each L_{ex} present during the specified period of time.

The reason for using the decibel for quantifying the above measures of sound energy is that the range of times (1 sec to years of exposure) and of sound pressures (from less than 20 μPa to millions of micropascals) requires a means of compressing the number scales involved. The use of the decibel achieves this end. In this document all logarithms are to the base 10.

Modifications to Measures of Sound Energy

Up to this point the physical quantities described have been based on relatively simple physical assumptions that the ear responds to sound as an energy sensing device with a given frequency bandwidth and uniform sensitivity to frequencies within that bandwidth. However, it is necessary to introduce some reference to the shaping and modification of these quantities that experience and research have shown to be needed in order to maximize the correlation between measures of the physical world of sound and some basic psychological-physiological responses in people to that sound.

Impulsive and steady-state sound

In the intensity-time dimensions, sound may be labeled as being either "impulsive" or "nonimpulsive." Impulsive sound is herein defined as a change in SPL, above certain values of SPL (to be specified later), of more than 10 dB in 1 sec or less; all other 1-sec intervals of sound are nonimpulsive. Sound is also defined here to be steady state when the SPL remains relatively constant (within ±2.5 dB) for successive periods of 1 sec. Sound, unless shorter than 1 sec in total duration, can go from impulsive to nonimpulsive and vice versa during its existence. This definition is related to the fact that, as discussed later, changes in sound level as rapid as described have psychological effects that differ from those of other intervals of sound.

I would like to point out the use in my earlier book (ref. 2) of 0.5 sec rather than 1 sec for this period. For various reasons shown later, it is in keeping with available community noise assessment procedures, as well as being in reasonable agreement with the operating characteristics of the human ear, to use the 1-sec period.

These definitions of impulsive and steady-state sound have the effect of detaching any meaning or identification the sounds or noises may have to the listener. As discussed elsewhere, the practical assessment of noise pollution is not concerned with the contributions to annoyance from sounds or noise attributable to any idiosyncratic meanings to particular individuals or groups.

Broadband noise

When the sound pressure varies in a very regular way and at the same unvarying frequency, a "pure tone" is heard. When more than one tone or frequency is present, the sound is said to be complex, that is, to have a band of frequencies. Most everyday noises consist of broadband sounds such as those that commonly occur from the operation of machinery, the firing of guns, the operation of airplanes, and so forth. By this we mean that the sound contains low-, middle-, and high-frequency components; some of these components may be stronger than others, giving a somewhat tonal character to the overall sound.

A way of explaining this frequency band, or spectrum as it is called, of a sound or noise is to consider the kind of sound that would come from a piano when all the keys were simultaneously struck, each key causing a different tonal frequency to be emitted. The noise of an airplane engine, for example, caused by many different vibrating parts and air exhausts, will usually have different and varying amounts of energy at more frequencies in its spectrum than the sound of the piano (with all strings struck simultaneously); however, both emit a broadband-frequency spectrum of sound. Likewise, the human voice, when a person is speaking, has a broadband-frequency spectrum. This is not to say that a single frequency or tone is not, at times, considered as a noise, even a very objectionable one, but that by and large, most sounds and noises are broadband in nature.

Overall frequency weighting method

The ear has the ability to analyze out individual parts of broadband sounds and also to respond to the sound over all frequencies; that is, the noise from an automobile is identifiable as a total, single noise. However, the ear is more sensitive to some frequencies or parts of the spectrum than to others. Figure 2.3 shows the results of experiments (refs. 3 and 4) in which subjects were presented, one at a time, with individual 1/3-octave bands of sound frequencies (shown by the small horizontal bars) and asked to adjust its intensity level until it sounded as noisy (solid curve) or as loud (dashed curve) as the 1/3-octave band centered at 1000 Hz. The higher frequency bands, those above 1000 Hz, are physically less intense (lower SPL in decibels) than the lower frequency bands in order to sound equal in noisiness and loudness to the band centered at 1000 Hz. The frequencies included in each 1/3-octave band are given in the left-hand column of table 2.1.

Sound level meters (SLM's) are supposedly built to integrate the energy in a noise over all its frequencies simultaneously. Depending on its mode of operation, more weight can be given to the higher than to the lower frequencies as in

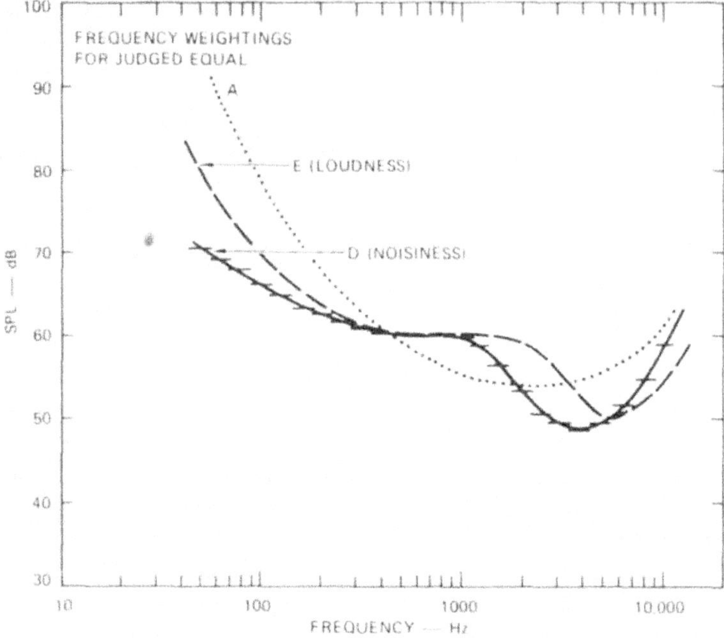

FIGURE 2.3. Relative frequency weighting for loudness (ref. 3), perceived noisiness (ref. 4), and A-weighting (ref. 5).

the human ear. One such frequency weighting, called A-weighting, is also shown in figure 2.3. As shown in figure 2.3, the frequency-weighted curve is about the same shape as those curves found when the subjects are asked to judge how noisy or how loud are the different 1/3-octave frequency bands.

Because the A-weighting follows somewhat the shape of the equal-loudness and equal-noisiness contours and is by national and international standards (ref. 5) built into most SLM's, it has become the most widely used method of measuring broadband sounds to predict how loud or noisy they will sound to the average person. Accordingly, two noises that have the same A-weighted level will presumably sound equally loud or noisy to the average listener. Some of the limitations to this generalization are mentioned in a few paragraphs below and in later chapters.

More accurate frequency weightings for estimating how loud or noisy different sounds will be are shaped more like the equal-loudness and equal-noisiness contours in figure 2.3. These other frequency weightings, called D- and E-weightings, are not used to any great extent, however, because they are not standardized parts of most SLM's. The relative values of a number of actual or proposed frequency weightings are given in table 2.1. The B- and C-weightings are

TABLE 2.1.

Cut-Off Frequencies and Center Frequencies of Preferred 1/3-Octave-Band Filters and Frequency Weightings for Sound Level Meters

Cut-off frequencies, Hz	Center frequencies, Hz	A-weighting, dB	B-weighting, dB	C-weighting, dB	D_1-weighting, dB	D_2-weighting, dB
45 to 56	50	−30.2	−11.7	−1.3	−12	−19
56 to 71	63	−26.1	−9.4	−.8	−11	−17
71 to 90	80	−22.3	−7.4	−.5	−9	−14
90 to 112	100	−19.1	−5.7	−.3	−7	−11
112 to 140	125	−16.2	−4.3	−.2	−6	−9
140 to 180	160	−13.2	−3.0	−.1	−5	−7
180 to 224	200	−10.8	−2.1	0	−3	−5
224 to 280	250	−8.0	−1.4	0	−2	−3
280 to 355	315	−6.5	−.9	0	−1	−2
355 to 450	400	−4.8	−.6	0	0	0
450 to 560	500	−3.3	−.3	0	0	0
560 to 710	630	−1.9	−.2	0	0	0
710 to 900	800	−.8	−.1	0	0	0
900 to 1 120	1 000	0	0	0	0	0
1 120 to 1 400	1 250	.5	−.1	−.1	2	2
1 400 to 1 800	1 600	1.0	−.1	−.1	6	6
1 800 to 2 240	2 000	1.2	−.2	−.2	8	8
2 240 to 2 800	2 500	1.2	−.3	−.3	10	10
2 800 to 3 550	3 150	1.2	−.5	−.5	11	11
3 550 to 4 500	4 000	1.0	−.8	−.8	11	11
4 500 to 5 600	5 000	.5	−1.3	−1.3	10	10
5 600 to 7 100	6 300	−.2	−2.0	−2.0	9	9
7 100 to 9 000	8 000	−1.1	−3.0	−3.0	6	6
9 000 to 11 020	10 000	−2.5	−4.3	−4.3	3	3

incorporated in standard SLM's, but they are considered obsolete as loudness weighting functions and are not used for most noise-assessment purposes.

In order to avoid the need to identify the quantity referred to in some of these variations in frequency weightings every time a decibel value is cited, the frequency-weighting letter designation is often given as follows: SPL_A, or simply L_A, in decibels; SPL_C, or L_C, in decibels; and so forth. An alternate means of designating the frequency weighting used is to attach the weighting letter to the unit, that is, dBA, dBC, and so forth. This practice is a convenience when decibels, sometimes representing two or more different quantities, are presented in the same graphs, tables, or discussion. A further means of ensuring unambiguous shorthand designation is to attach, when appropriate, "max" to the unit (*e.g.*, max dBA or dBA_{max}) when referring to the maximum of peak sound pressure level that occurs during a noise event.

1/3-octave-band method

A less common procedure for predicting the level of loudness or noisiness of a broadband sound than that of measuring energy over all sound frequencies with a frequency-weighting meter is the following: (a) measure the SPL in decibels in each 1/3-octave band; (b) adjust the SPL for each band in accordance with the A-weighting value pertaining to each frequency band; and (c) sum, on an energy basis, these 1/3-octave-band SPL's. The result is the A-weighted SPL in decibels over all frequencies, as would also be measured on an SLM with A-weighting. The individual band values give the noise-control engineer or equipment designer more insight into what changes in the noise spectrum can best be made to reduce its overall loudness or noisiness.

Another procedure is to weight the individual 1/3-octave-band levels in accordance with the shape of the equal-noisiness contours of figure 2.3 and to sum these values, using certain prescribed procedures that take into account the relative bandwidth of a noise, to arrive at the perceived noise (noisiness) level measured in decibels (PNdB). A variation of this unit, effective perceived noise level in decibels (EPNdB), is primarily used for the engineering specifications of the noise from aircraft and makes allowances for pure-tone factors as well as durations that can influence how noisy a sound is judged to be (refs. 2 and 6).

The calculation of the PNdB level of a noise at a given moment of time is a two-step process. The number of "noy" units (a perceived, subjective quantity) for each of twenty-four 1/3-octave-band SPL's is found from table 2.2. (This table of noy values represents an approximation of the contours of fig. 5.11.) These noy values are summed in accordance with the following formula:

$$N = n(k) + 0.15\left[\left(\sum_{1}^{24} n\right) - n(k)\right]$$

and

$$\text{PNdB} = 40.0 + 33.3 \log N$$

where N is the perceived noisiness in noys, where n is the number of noys in a 1/3-octave band (see table 2.2), and $n(k)$ is the largest of the $24n$ values.

References 2 and 6 contain detailed descriptions of formulas and procedures for calculating modifications to PNdB to take into account temporal, pure-tone, and impulse corrections. Newly developed impulse-correction procedures are presented in chapter 5.

Temporal factors

To predict the role of frequency spectrum in the undesirability of sounds, no real distinction is made for general noise-assessment purposes between judged loudness and noisiness (*i.e.*, a dBA level is used to predict either judgment). But to predict the effects of temporal factors, loudness is not an adequate attribute of sound to use because the loudness of a noise at a level of 80 dBA lasting, say, 10 sec will appear to be about as loud as the same noise lasting, for example, 20 sec. As further discussed later, the noisiness, or the undesirability of a sound, tends to increase as its duration is increased or, in some cases, as its duration is decreased to the degree it becomes impulsive.

Multiple occurrences

Realistically, a single occurrence per day of a noise that is an expected part of one's environment is usually not sufficiently annoying or bothersome to be a matter of practical concern. Rather, the typical total daily dosage of a number of occurrences of a noise or noises that are a regular, recurring part of the living environment appears to be that to which people respond in rating the acceptability or unacceptability of their noise environment. As mentioned previously, the effect of the "meaning" of a sound or noise on its judged acceptability is not considered in the physical measurement of noise dosage.

Sound measurements aimed at assessing this aspect of the noise environment also follow, for the most part, the equal-energy principal used for equating individual noise occurrences of different levels and/or durations. For example, the average resident in a neighborhood will rate the noise environment to be about equally acceptable with 100 daily occurrences of a 10-sec-duration noise at a level of 95 dBA as with but 50 average daily occurrences of the same duration at a level of 98 dBA. The total daily sound energy for these two conditions is the same. The summed duration in seconds of our example of 50 noise events would, of course, be one-half the summed duration in seconds of the 100 events. However, doubling the sound pressure squared (which is represented by an increase of 3 dBA) doubles the sound energy present per second each noise is present and compensates, in our example, for the loss in energy due to halving the number of noise events.

In the United States, the practice is to add to this equal-energy method of measuring environmental noises an adjustment for the time of day at which regularly occurring noises are present. The simplest practice is to add a 10-dB "penalty" to the actual A-weighted levels occurring at nighttime (10 p.m. to 7 a.m.) compared with the same levels when they are present during the day (7 a.m. to 10 p.m.). The result is called day-night level L_{dn} (or sometimes DNL). An additional weighting, or penalty, is used in a quantity called the community noise equivalent level (CNEL), wherein a 5-dB penalty is also applied to noises occurring during the hours of 7 p.m. to 10 p.m.

In the United States, these presently used units of noise measurement (L_{dn} and CNEL) evolved from the original "composite (summed energy) noise level" procedure for expressing in a single number the daily noise dosage a variety of highly related units—composite noise rating (CNR, refs. 1 and 7), noise ex-

TABLE 2.2. Antilog (Base 10) of SPL/10 and Noys as a Function of SPL

[From ref. 2]

SPL-dB	LOG_{10}^{-1} (SPL/10) (EXPON NUMERIC FORM EXAMPLE 7.94E 02 = 7.94 × 10^2)	BAND CENTER FREQUENCY																							
		50HZ	63HZ	80HZ	100HZ	125HZ	160HZ	200HZ	250HZ	315HZ	400HZ	500HZ	630HZ	800HZ	1KHZ	1.2KHZ	1.6KHZ	2KHZ	2.5KHZ	3.1KHZ	4KHZ	5KHZ	6.3KHZ	8KHZ	10KHZ
29	7.94E 02																			1.0	1.0				
30	1.00E 03																		1.0	1.1	1.1				
31	1.26E 03																		1.1	1.1	1.1	1.0			
32	1.58E 03																	1.0	1.1	1.2	1.2	1.1			
33	2.00E 03																	1.1	1.2	1.3	1.3	1.2	1.0		
34	2.51E 03																1.0	1.2	1.3	1.4	1.4	1.3	1.1		
35	3.16E 03																1.1	1.3	1.4	1.5	1.5	1.4	1.2		
36	3.98E 03																1.2	1.3	1.5	1.6	1.6	1.5	1.4		
37	5.01E 03																1.3	1.4	1.6	1.8	1.8	1.6	1.5	1.0	
38	6.31E 03															1.0	1.3	1.5	1.8	1.9	1.9	1.7	1.6	1.1	
39	7.94E 03															1.1	1.4	1.6	1.9	2.0	2.0	1.9	1.8	1.2	
40	1.00E 04										1.0	1.0	1.0	1.0	1.0	1.2	1.5	1.7	2.0	2.2	2.2	2.0	1.9	1.4	
41	1.26E 04										1.1	1.1	1.1	1.1	1.1	1.3	1.6	1.8	2.2	2.4	2.4	2.2	2.0	1.5	1.0
42	1.58E 04									1.0	1.1	1.1	1.1	1.1	1.1	1.3	1.7	2.0	2.4	2.6	2.6	2.4	2.2	1.7	1.1
43	2.00E 04									1.1	1.2	1.2	1.2	1.2	1.2	1.4	1.8	2.2	2.6	2.8	2.8	2.6	2.4	1.8	1.2
44	2.51E 04								1.0	1.1	1.3	1.3	1.3	1.3	1.3	1.5	2.0	2.4	2.8	3.0	3.0	2.8	2.6	2.0	1.4
45	3.16E 04								1.1	1.2	1.4	1.4	1.4	1.4	1.4	1.6	2.1	2.6	3.0	3.2	3.2	3.0	2.8	2.2	1.5
46	3.98E 04							1.0	1.2	1.3	1.5	1.5	1.5	1.5	1.5	1.7	2.3	2.8	3.2	3.4	3.4	3.2	3.0	2.4	1.7
47	5.01E 04							1.1	1.3	1.4	1.6	1.6	1.6	1.6	1.6	1.8	2.4	3.0	3.4	3.6	3.6	3.4	3.2	2.6	1.8
48	6.31E 04						1.0	1.2	1.4	1.5	1.7	1.7	1.7	1.7	1.7	2.0	2.6	3.2	3.6	3.9	3.9	3.6	3.4	2.8	2.0
49	7.94E 04						1.1	1.3	1.5	1.6	1.9	1.9	1.9	1.9	1.9	2.1	2.8	3.4	3.9	4.1	4.1	3.9	3.6	3.0	2.2
50	1.00E 05						1.2	1.4	1.6	1.7	2.0	2.0	2.0	2.0	2.0	2.3	3.0	3.6	4.1	4.4	4.4	4.1	3.9	3.2	2.4
51	1.26E 05					1.0	1.3	1.5	1.7	1.9	2.1	2.1	2.1	2.1	2.1	2.4	3.2	3.9	4.4	4.7	4.7	4.4	4.1	3.4	2.6
52	1.58E 05					1.1	1.4	1.6	1.9	2.0	2.3	2.3	2.3	2.3	2.3	2.6	3.5	4.1	4.7	5.0	5.0	4.7	4.4	3.6	2.8
53	2.00E 05				1.0	1.2	1.5	1.7	2.0	2.1	2.5	2.5	2.5	2.5	2.5	2.8	3.7	4.4	5.0	5.3	5.3	5.0	4.7	3.9	3.0
54	2.51E 05				1.1	1.3	1.6	1.9	2.1	2.3	2.6	2.6	2.6	2.6	2.6	3.0	4.0	4.7	5.3	5.7	5.7	5.3	5.0	4.1	3.2
55	3.16E 05				1.2	1.4	1.7	2.0	2.3	2.4	2.8	2.8	2.8	2.8	2.8	3.2	4.3	5.0	5.7	6.1	6.1	5.7	5.3	4.4	3.5
56	3.98E 05			1.0	1.3	1.5	1.9	2.2	2.4	2.6	3.0	3.0	3.0	3.0	3.0	3.5	4.6	5.3	6.1	6.5	6.5	6.1	5.7	4.7	3.7
57	5.01E 05			1.1	1.4	1.7	2.0	2.4	2.6	2.8	3.2	3.2	3.2	3.2	3.2	3.7	5.0	5.7	6.5	7.0	7.0	6.5	6.1	5.0	4.0
58	6.31E 05			1.2	1.5	1.8	2.2	2.6	2.8	3.0	3.5	3.5	3.5	3.5	3.5	4.0	5.3	6.1	7.0	7.5	7.5	7.0	6.5	5.3	4.3
59	7.94E 05			1.3	1.7	2.0	2.4	2.8	3.0	3.2	3.7	3.7	3.7	3.7	3.7	4.3	5.7	6.5	7.5	8.0	8.0	7.5	7.0	5.7	4.6
60	1.00E 06		1.0	1.4	1.8	2.2	2.6	3.0	3.2	3.5	4.0	4.0	4.0	4.0	4.0	4.6	6.1	7.0	8.0	8.7	8.7	8.0	7.5	6.1	5.0
61	1.26E 06		1.1	1.5	2.0	2.4	2.8	3.2	3.5	3.7	4.3	4.3	4.3	4.3	4.3	5.0	6.5	7.5	8.7	9.3	9.3	8.7	8.0	6.5	5.3
62	1.58E 06		1.2	1.7	2.2	2.6	3.0	3.5	3.7	4.0	4.6	4.6	4.6	4.6	4.6	5.3	7.0	8.0	9.3	10.0	10.0	9.3	8.7	7.0	5.7
63	2.00E 06		1.3	1.8	2.4	2.8	3.2	3.7	4.0	4.3	4.9	4.9	4.9	4.9	4.9	5.7	7.5	8.7	10.0	11.0	11.0	10.0	9.3	7.5	6.1
64	2.51E 06	1.0	1.5	2.0	2.6	3.0	3.5	4.0	4.3	4.6	5.3	5.3	5.3	5.3	5.3	6.1	8.0	9.3	11.0	11.0	11.0	11.0	10.0	8.0	6.5
65	3.16E 06	1.1	1.6	2.2	2.8	3.2	3.7	4.3	4.6	5.0	5.7	5.7	5.7	5.7	5.7	6.5	8.7	10.0	11.0	12.0	12.0	11.0	11.0	8.7	7.0
66	3.98E 06	1.2	1.8	2.4	3.0	3.5	4.0	4.6	5.0	5.4	6.1	6.1	6.1	6.1	6.1	7.0	9.3	11.0	12.0	13.0	13.0	12.0	11.0	9.3	7.5
67	5.01E 06	1.4	2.0	2.6	3.3	3.7	4.3	5.0	5.4	5.9	6.5	6.5	6.5	6.5	6.5	7.5	10.0	11.0	13.0	14.0	14.0	13.0	12.0	10.0	8.0
68	6.31E 06	1.6	2.2	2.8	3.6	4.0	4.6	5.4	5.9	6.4	7.0	7.0	7.0	7.0	7.0	8.0	11.0	12.0	14.0	15.0	15.0	14.0	13.0	11.0	8.7
69	7.94E 06	1.8	2.3	3.0	3.9	4.3	5.0	5.9	6.4	6.9	7.5	7.5	7.5	7.5	7.5	8.7	11.0	13.0	15.0	16.0	16.0	15.0	14.0	11.0	9.3
70	1.00E 07	2.0	2.5	3.3	4.2	4.6	5.4	6.4	6.9	7.5	8.0	8.0	8.0	8.0	8.0	9.3	12.0	14.0	16.0	17.0	17.0	16.0	15.0	12.0	10.0
71	1.26E 07	2.2	2.8	3.6	4.6	5.0	5.9	6.9	7.5	8.0	8.6	8.6	8.6	8.6	8.6	10.0	13.0	15.0	17.0	19.0	19.0	17.0	16.0	13.0	11.0
72	1.58E 07	2.3	3.0	3.9	5.0	5.4	6.4	7.5	8.0	8.7	9.2	9.2	9.2	9.2	9.2	11.0	14.0	16.0	19.0	20.0	20.0	19.0	17.0	14.0	11.0
73	2.00E 07	2.5	3.3	4.2	5.4	5.9	6.9	8.0	8.7	9.3	9.8	9.8	9.8	9.8	9.8	11.0	15.0	17.0	20.0	21.0	21.0	20.0	19.0	15.0	12.0
74	2.51E 07	2.8	3.7	4.6	5.9	6.4	7.5	8.7	9.3	10.0	10.6	10.6	10.6	10.6	10.6	12.0	16.0	19.0	21.0	23.0	23.0	21.0	20.0	16.0	13.0
75	3.16E 07	3.0	4.1	5.0	6.4	6.9	8.0	9.3	10.0	11.0	11.3	11.3	11.3	11.3	11.3	13.0	17.0	20.0	23.0	24.0	24.0	23.0	21.0	17.0	14.0
76	3.98E 07	3.3	4.5	5.4	6.9	7.5	8.7	10.0	11.0	11.0	12.0	12.0	12.0	12.0	12.0	14.0	19.0	21.0	24.0	26.0	26.0	24.0	23.0	19.0	15.0
77	5.01E 07	3.7	5.0	5.9	7.5	8.3	9.3	11.0	11.0	12.0	13.0	13.0	13.0	13.0	13.0	15.0	20.0	23.0	26.0	28.0	28.0	26.0	24.0	20.0	16.0
78	6.31E 07	4.1	5.4	6.4	8.3	9.1	10.0	11.0	12.0	13.0	14.0	14.0	14.0	14.0	14.0	16.0	21.0	24.0	28.0	30.0	30.0	28.0	26.0	21.0	17.0
79	7.94E 07	4.5	5.9	6.9	9.1	10.0	11.0	12.0	13.0	14.0	15.0	15.0	15.0	15.0	15.0	17.0	23.0	26.0	30.0	32.0	32.0	30.0	28.0	23.0	19.0
80	1.00E 08	5.0	6.4	7.5	10.0	11.0	11.0	13.0	14.0	15.0	16.0	16.0	16.0	16.0	16.0	19.0	24.0	28.0	32.0	35.0	35.0	32.0	30.0	24.0	20.0
81	1.26E 08	5.5	6.9	8.3	11.0	11.0	12.0	14.0	15.0	16.0	18.0	18.0	18.0	18.0	18.0	20.0	26.0	30.0	35.0	37.0	37.0	35.0	32.0	26.0	21.0
82	1.58E 08	6.1	7.5	9.1	11.0	12.0	13.0	15.0	16.0	17.0	19.0	19.0	19.0	19.0	19.0	21.0	28.0	32.0	37.0	40.0	40.0	37.0	35.0	28.0	23.0
83	2.00E 08	6.8	8.3	10.0	12.0	13.0	14.0	16.0	17.0	19.0	20.0	20.0	20.0	20.0	20.0	23.0	30.0	35.0	40.0	42.0	42.0	40.0	37.0	30.0	24.0
84	2.51E 08	7.5	9.1	12.0	13.0	14.0	15.0	17.0	19.0	20.0	21.0	21.0	21.0	21.0	21.0	24.0	32.0	37.0	42.0	45.0	45.0	42.0	40.0	32.0	26.0
85	3.16E 08	8.3	10.0	13.0	14.0	15.0	16.0	19.0	20.0	21.0	23.0	23.0	23.0	23.0	23.0	26.0	35.0	40.0	45.0	47.0	47.0	45.0	42.0	35.0	28.0
86	3.98E 08	9.1	12.0	13.0	15.0	16.0	17.0	20.0	21.0	23.0	24.0	24.0	24.0	24.0	24.0	28.0	37.0	42.0	47.0	50.0	50.0	47.0	45.0	37.0	30.0
87	5.01E 08	10.0	13.0	14.0	16.0	17.0	19.0	21.0	23.0	24.0	26.0	26.0	26.0	26.0	26.0	30.0	40.0	45.0	50.0	55.0	55.0	50.0	47.0	40.0	32.0
88	6.31E 08	11.0	13.0	15.0	17.0	19.0	20.0	23.0	24.0	26.0	28.0	28.0	28.0	28.0	28.0	32.0	42.0	47.0	55.0	60.0	60.0	55.0	50.0	42.0	35.0

BAND CENTER FREQUENCY

SPL-dB	LOG_{10}^{-1} (SPL/10) (EXPON NUMERIC FORM EXAMPLE 7.94E 02 = 7.94 x 10^2)	50HZ	63HZ	80HZ	100HZ	125HZ	160HZ	200HZ	250HZ	315HZ	400HZ	500HZ	630HZ	800HZ	1KHZ	1.2KHZ	1.6KHZ	2KHZ	2.5KHZ	3.1KHZ	4KHZ	5KHZ	6.3KHZ	8KHZ	10KHZ
89	7.94E 08	12	14	16	19	20	21	24	26	28	30	30	30	30	30	35	45	50	60	63	63	60	55	45	37
90	1.00E 09	14	15	17	20	21	23	26	28	30	32	32	32	32	32	37	47	55	63	67	67	63	60	47	40
91	1.26E 09	15	16	19	21	23	24	28	30	32	34	34	34	34	34	40	50	60	67	71	71	67	63	50	42
92	1.58E 09	16	17	20	23	24	26	30	32	35	37	37	37	37	37	42	55	63	71	75	75	71	67	55	45
93	2.00E 09	17	19	21	24	26	28	32	35	37	39	39	39	39	39	45	60	67	75	80	80	75	71	60	47
94	2.51E 09	19	20	23	26	28	30	35	37	40	42	42	42	42	42	47	63	71	80	86	86	80	75	63	50
95	3.16E 09	20	21	24	28	30	32	37	40	42	45	45	45	45	45	50	67	75	86	93	93	86	80	67	55
96	3.98E 09	21	23	26	30	32	35	40	42	45	49	49	49	49	49	55	71	80	93	100	100	93	86	71	60
97	5.01E 09	23	24	28	32	35	37	42	45	47	52	52	52	52	52	60	75	86	100	108	108	100	93	75	63
98	6.31E 09	24	26	30	35	37	40	45	47	50	56	56	56	56	56	64	80	93	108	116	116	108	100	80	67
99	7.94E 09	26	28	32	37	40	42	47	50	55	60	60	60	60	60	69	86	100	116	125	125	116	108	86	71
100	1.00E 10	28	30	35	40	42	45	50	55	60	64	64	64	64	64	74	93	108	125	133	133	125	116	93	75
101	1.26E 10	30	32	37	42	45	47	55	60	64	69	69	69	69	69	79	100	116	133	142	142	133	125	100	80
102	1.58E 10	32	35	40	45	47	50	60	64	69	74	74	74	74	74	84	108	125	142	150	150	142	133	108	86
103	2.00E 10	35	37	42	47	50	55	64	69	74	79	79	79	79	79	91	116	133	150	162	162	150	142	116	93
104	2.51E 10	37	40	45	50	55	60	69	74	79	84	84	84	84	84	97	125	142	162	173	173	162	150	125	100
105	3.16E 10	40	42	47	55	60	64	74	79	84	91	91	91	91	91	104	133	150	173	186	186	173	162	133	108
106	3.98E 10	42	45	50	60	64	69	79	84	91	97	97	97	97	97	111	142	162	186	200	200	186	173	142	116
107	5.01E 10	45	47	55	64	69	74	84	91	97	104	104	104	104	104	119	150	173	200	215	215	200	186	150	125
108	6.31E 10	47	50	60	69	74	79	91	97	104	111	111	111	111	111	128	162	186	215	232	232	215	200	162	133
109	7.94E 10	50	55	64	74	79	84	97	104	111	119	119	119	119	119	137	173	208	232	250	250	232	215	173	142
110	1.00E 11	55	60	69	79	84	91	104	111	119	128	128	128	128	128	147	186	215	250	266	266	250	232	186	150
111	1.26E 11	60	64	74	84	91	97	111	119	128	137	137	137	137	137	158	200	232	266	284	284	266	250	200	162
112	1.58E 11	64	69	79	91	97	104	119	128	137	147	147	147	147	147	169	215	250	284	300	300	284	266	215	173
113	2.00E 11	69	74	84	97	104	111	128	137	147	158	158	158	158	158	181	232	266	300	324	324	300	284	232	186
114	2.51E 11	74	79	91	104	111	119	137	147	158	169	169	169	169	169	194	250	284	324	346	346	324	300	250	200
115	3.16E 11	79	84	97	111	119	128	147	158	169	181	181	181	181	181	208	266	300	346	372	372	346	324	266	215
116	3.98E 11	84	91	104	119	128	137	158	169	181	194	194	194	194	194	223	284	324	372	400	400	372	346	284	232
117	5.01E 11	91	97	111	128	137	147	169	181	194	208	208	208	208	208	239	300	346	400	430	430	400	372	300	250
118	6.31E 11	97	104	119	137	147	158	181	194	208	223	223	223	223	223	256	324	372	430	464	464	430	400	324	266
119	7.94E 11	104	111	128	147	158	169	194	208	223	239	239	239	239	239	274	346	400	464	500	500	464	430	346	284
120	1.00E 12	111	119	137	158	169	181	208	223	239	256	256	256	256	256	294	372	430	500	532	532	500	464	372	300
121	1.26E 12	119	128	147	169	181	194	223	239	256	274	274	274	274	274	315	400	464	532	568	568	532	500	400	324
122	1.58E 12	128	137	158	181	194	208	239	256	274	294	294	294	294	294	338	430	500	568	600	600	568	532	430	346
123	2.00E 12	137	147	169	194	208	223	256	274	294	315	315	315	315	315	362	464	532	600	648	648	600	568	464	372
124	2.51E 12	147	158	181	208	223	239	274	294	315	338	338	338	338	338	388	500	568	648	692	692	648	600	500	400
125	3.16E 12	158	169	194	223	239	256	294	315	338	362	362	362	362	362	416	532	600	692	744	744	692	648	532	430
126	3.98E 12	169	181	208	239	256	274	315	338	362	388	388	388	388	388	446	568	648	744	800	800	744	692	568	464
127	5.01E 12	181	194	223	256	274	294	338	362	388	416	416	416	416	416	478	600	692	800	860	860	800	744	600	500
128	6.31E 12	194	208	239	274	294	315	362	388	416	446	446	446	446	446	512	648	744	860	928	928	860	800	648	532
129	7.94E 12	208	223	256	294	315	338	388	416	446	478	478	478	478	478	549	692	800	928	1000	1000	928	860	692	568
130	1.00E 13	223	239	274	315	338	362	416	446	478	512	512	512	512	512	588	744	860	1000	1064	1064	1000	928	744	600
131	1.26E 13	239	256	294	338	362	388	446	478	512	549	549	549	549	549	630	800	928	1064	1136	1136	1064	1000	800	648
132	1.58E 13	256	274	315	362	388	416	478	512	549	588	588	588	588	588	676	860	1000	1136	1200	1200	1136	1064	860	692
133	2.00E 13	274	294	338	388	416	446	512	549	588	630	630	630	630	630	724	928	1064	1200	1296	1296	1200	1136	928	744
134	2.51E 13	294	315	362	416	446	478	549	588	630	676	676	676	676	676	776	1000	1136	1296	1384	1384	1296	1200	1000	800
135	3.16E 13	315	338	388	446	478	512	588	630	676	724	724	724	724	724	832	1064	1200	1384	1488	1488	1384	1296	1064	860
136	3.98E 13	338	362	416	478	512	549	630	676	724	776	776	776	776	776	891	1136	1296	1488	1600	1600	1488	1384	1136	928
137	5.01E 13	362	388	446	512	549	588	676	724	776	832	832	832	832	832	955	1200	1384	1600	1720	1720	1600	1488	1200	1000
138	6.31E 13	388	416	478	549	588	630	724	776	832	891	891	891	891	891	1024	1296	1488	1720	1856	1856	1720	1600	1296	1064
139	7.94E 13	416	446	512	588	630	676	776	832	891	955	955	955	955	955	1098	1384	1600	1856	2000	2000	1856	1720	1384	1136
140	1.00E 14	446	478	549	630	676	724	832	891	955	1024	1024	1024	1024	1024	1176	1488	1720	2000			2000	1856	1488	1200
141	1.26E 14	478	512	588	676	724	776	891	955	1024	1098	1098	1098	1098	1098	1261	1600	1856					2000	1600	1296
142	1.58E 14	512	549	630	724	776	832	955	1024	1098	1176	1176	1176	1176	1176	1351	1720	2000						1720	1384
143	2.00E 14	549	588	676	776	832	891	1024	1098	1176	1261	1261	1261	1261	1261	1448	1856							1856	1488
144	2.51E 14	588	630	724	832	891	955	1098	1176	1261	1351	1351	1351	1351	1351	1552	2000							2000	1600
145	3.16E 14	630	676	776	891	955	1024	1176	1261	1351	1448	1448	1448	1448	1448	1663									1720
146	3.98E 14	676	724	832	955	1024	1098	1261	1351	1448	1552	1552	1552	1552	1552	1783									1856
147	5.01E 14	724	776	891	1024	1098	1176	1351	1448	1552	1663	1663	1663	1663	1663	1911									2000
148	6.31E 14	776	832	955	1098	1176	1261	1448	1552	1663	1783	1783	1783	1783	1783	2048									

posure forecast (NEF, ref. 8), and, eventually, L_{dn} (refs. 7 and 9). These and somewhat similar units developed in other countries are discussed in detail in references 7 and 8.

As a matter of historical and some practical importance, it might be noted that the estimated 10-dB nighttime penalty was expressed, inadvertently it appears, as about 13 dB in the CNR and NEF formulas for calculating these units: CNR $= \overline{\text{PNdB}}_{max} + 10 \log N_f - 121$ and NEF $= \overline{\text{PNdB}}_{eq} + 10 \log N_f - 88$, where PNdB_{max} is the maximum perceived noise level in decibels, PNdB_{eq} is the equivalent perceived noise level in decibels, $N_f = N_d + 16.7\, N_n$, N_d and N_n are the number of daytime and nighttime events, and barred variables are the average of the energy levels of the variable. (Interestingly, the very earliest estimation procedure proposed for a CNR confused a 10-dB nighttime penalty with a 3-dB energy weighting for numbers of nighttime operations versus numbers of daytime operations. See table 4 of ref. 10.)

In reality, these formulas for CNR and NEF represent a solution to the question as to what weighting would be required to achieve CNR's or NEF's that differ by 10 dB when the hourly rate of nighttime aircraft operations is the same rate as for the daytime. This, of course, is a different proposition than the concept that a given aircraft noise occurrence at nighttime should be penalized (given 10 dB more weight) more than if it had occurred during the daytime, which is the avowed intention for CNR and NEF. The actual achievement L_{dn} formulation is the following:

$$L_{dn} = \overline{L_{ex}} + 10(\log N) - 49.4$$

where

$$N = N_d + 10N_n$$

Damage to hearing

Generally the average daily dose of environmental noise is related to the acceptability or unacceptability of the noise as a source of annoyance and stress, although the sum of noise dosages at work and elsewhere over days and years must be considered in assessing the potential permanent damage to hearing that can occur from exposure to noise. As discussed later, an equal-energy dose over a person's lifetime of daily doses of damage to hearing is used for predicting damage risk to hearing from noise exposure. However, the procedures for predicting damage risk to hearing from physical measures of noise are somewhat complicated by the fact that the equal-energy principle is, for many types of daily noise occurrences, inadequate as a predictor of the amount of damage to hearing occurring within a day's exposure. A model and a formula for predicting damage risk to hearing from physical measures of sound are developed and presented later. The result is called damage level (DL).

Measures of Noise Energy

Rules for labels

Complete standardization of the measures and labeling of sound energy for specifying their degree of undesirability, or noisiness, has not been reached yet. For the purposes of the present document the following "rules" will be used for the labels for some common aspects of noise energy that are related to its effects on people:

Level: Level L is the ratio, in decibels, between a measured and a reference amount of sound pressure that has been shaped or modified from a purely linear physical measure to achieve correlation with some unwanted effect, or attribute, of the sound on people. A letter indicative of the attribute is assigned as a prefix to L, for example, loudness level (LL), perceived noisiness level (PNL), or damage level (DL).

Frequency shaping: The designation of the frequency shaping that might be required is shown as a subscript suffix, for example, A-weighted loudness level would be LL_A; A-weighted perceived noise level is PNL_A, and so forth.

One-second energy: The designation of the basic unit of time for measuring sound energy, 1 sec (1s), is noted as a subscript following the frequency-weighting designation; for example, LL_{A1s}, PNL_{A1s}, and so forth.

Temporal shaping: Designations for temporal energy shaping beyond the 1-sec period are shown as subscript suffixes following the frequency-weighting and energy subscripts, for example, the A-weighted perceived noise level of a sound of some unspecified exposure (ex) duration is $PNL_{A1s,ex}$, or the equivalent (eq) A-weighted perceived noise level of a number of exposures to noises over some specified duration is $PNL_{A1s,eq\ 1\ year}$, or the A-weighted damage level over a 50-year career of exposure is $DL_{A1s,eq\ 50\ years}$, or A-weighted perceived noise level of noises over 1-year exposure with an "extra" penalty (*i.e.*, 10 dB) given to noises occurring during nighttime hours is $PNL_{A1s,eqdn,1\ year}$.

Labeling of measures of noise energy

Although following such rules in labeling quantities of noise should result in their clear, unambiguous identification, the alphabet labels become long and confusing in their own right. Further, the context in which they are being used often makes clear the quantities under discussion so that further abbreviations can be used.

Table 2.3 is an attempt to summarize the physical measures of sound energy with the shaping or modifications that make them most meaningful for the prediction of how people will be affected. Also given are the labels, and their abbreviations, that can be used to identify these modified measures of sound energy. It is helpful to consider these as either quantities of noise (not sound) or as quantities of sound transformed to predict their unwanted effects on people

TABLE 2.3.

Basic Sound Energy Measures and Their Frequency and Time Weightings

Label for sound energy level	Equation for sound energy level (a)	Weighting of sound required to correlate human responses with physical measures for— Frequency (b)	Time	Labels for frequency- and time-weighted SEL's correlated with human responses of— Loudness (L) (c) Complete	Loudness (L) (c) Abbreviated	Perceived noise level (PNL) (d) Complete	Perceived noise level (PNL) (d) Abbreviated	Hearing damage (D) (e) Complete	Hearing damage (D) (e) Abbreviated
Sound pressure level, SPL, dB	[f]SPL = $10 \log_{10} \frac{1}{n} \left(\sum_{i=1}^{n} \frac{p_{i,\text{meas}}^2}{p_{i,\text{ref}}^2} \right)$	A	1 sec (1s)	LL_{A1s}	LL_A	[g]PNL_{A1s} [g]$PNL_{A1s,max}$	L_A $L_{A,max}$		
		PNdB	1 sec (1s)			[f]PNL_{PNdB1s} [f]$PNL_{PNdB1s,max}$	L_{PNdB} $L_{PNdB,max}$		
Event exposure level, L_{ex}, dB	$L_{ex} = 10 \log_{10} \sum_{i=1}^{n} 10^{L_i/10}$	A	Event exposure, sec			[g, h]$PNL_{A1s,ex}N$	[i]L_{ex}		
		PNdB	Event exposure, sec			[g, h]$PNL_{PNdB1s,ex}N$	[i]L_{ex}		
Equivalent exposure level, L_{eq}, dB	$L_{eq} = 10 \log_{10} \left(\frac{1}{n} \sum_{i=1}^{n} 10^{L_i/10} \right)$	A	Total time T of a specified period			[g]$PNL_{A1s,eqT}$	L_{eqT}		
		PNdB	Total time T of a specified period			[g]$PNL_{PNdB1s,eqT}$	L_{eqT}		
		A	Day-night (dn) daily avg. for 1 year (y)			[g]$PNL_{A1s,eqdn(1y)}$	[i]L_{dn}		
		PNdB	Day-night (dn) daily avg. for 1 year (y)			[g]$PNL_{PNdB1s,eqdn(1y)}$	[i]L_{dn}		
		A	50 years					$DL_{A1s,eq50y}$	DL

[a] *See appropriate sections of text for definitions of symbols.*

[b] *Frequency weightings: A, C, D, E, phons, and PNdB; A-weighting is most commonly used.*

[c] *Loudness is perceived by average person as intensity of a sound at a given moment in time.*

[d] *Noisiness is the perceived undesirability or annoyance of a sound or multiple sounds over time independently of any meaning the sounds may have to listeners. Based on laboratory tests and attitude surveys in communities.*

[e] *Hearing damage is the permanent shift in the threshold of hearing of pure tones in specified percentages of an exposed population due to exposure to noise. Based on laboratory tests and surveys of hearing in communities, industry, and military.*

[f] $P^2_{\text{meas}} = \frac{1}{n} \sum_{i=1}^{n} \frac{P_i^2}{P^2_{i,\text{ref}}}$.

[g] *PNL is the integrated, frequency-weighted energy in 1-sec intervals of time plus, if required, a startle-related penalty for impulsive, 1-sec intervals of noise (i) (see fig. 5.20) and a 5-dB house-vibration penalty for 1-sec intervals of sound exceeding certain levels (v) (see fig. 5.29). The use of these corrections is to be indicated by the parenthetical subscripts (i, v).*

[h] *When required, classes or types of noise events are identified as N following their specified level, e.g.,* L_{ex}, *N1,* L_{ex}, *N2, etc.*

[i] *Single-event noise exposure level (SENEL) and single-event level (SEL) are often practically synonymous with* L_{ex}. *However, for aircraft flyover noise, SEL, SENEL, and* L_{ex} *are often estimated from the formula* $L_{ex} = L_{max} + 10 \log (t/2)$, *where t is the duration (in seconds) and the flyover noise is within 10 dB of the known* L_{max}. *EPNdB is also used for measuring a noise event. (See text.)*

[j] *Community noise equivalent level (CNEL) differs from* L_{dn} *solely in that CNEL includes an "evening" (7 p.m. to 10 p.m.) time weighting not used in* L_{dn}. L_{dn} *is also sometimes identified as day-night level (DNL). CNEL time weightings are 0 dB for day (7 a.m. to 7 p.m.), 5 dB for evening (7 p.m. to 10 p.m.), and 10 dB for night (10 p.m. to 7 a.m.);* L_{dn} *time weightings are 0 dB for day (7 a.m. to 7 p.m.) and 10 dB for night (10 p.m. to 7 a.m.). In the United States,* L_{dn} *is most commonly used quantity for predicting PNL of a living environment.*

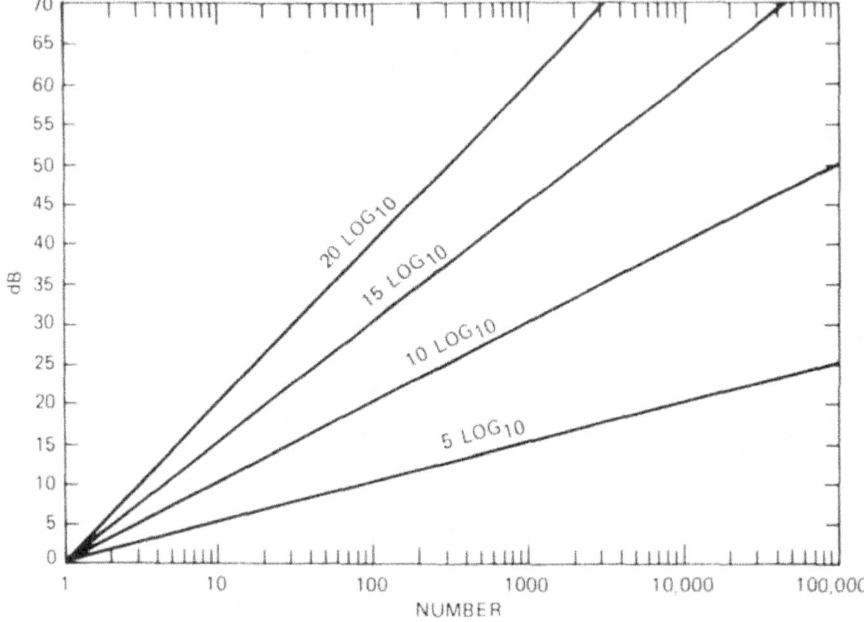

FIGURE 2.4. Graph for finding number of decibels of a number.

independently of any meaning in the sound. It is convenient that the units for expressing the magnitude of these quantities has been the common decibel, a logarithm of a ratio of the amount at hand to a reference amount. The conversion of various numbers to be discussed in later chapters to decibels and the summation of decibels representing different sounds or noises, as may be of interest, can be aided by the graphs shown in figures 2.4 and 2.5.

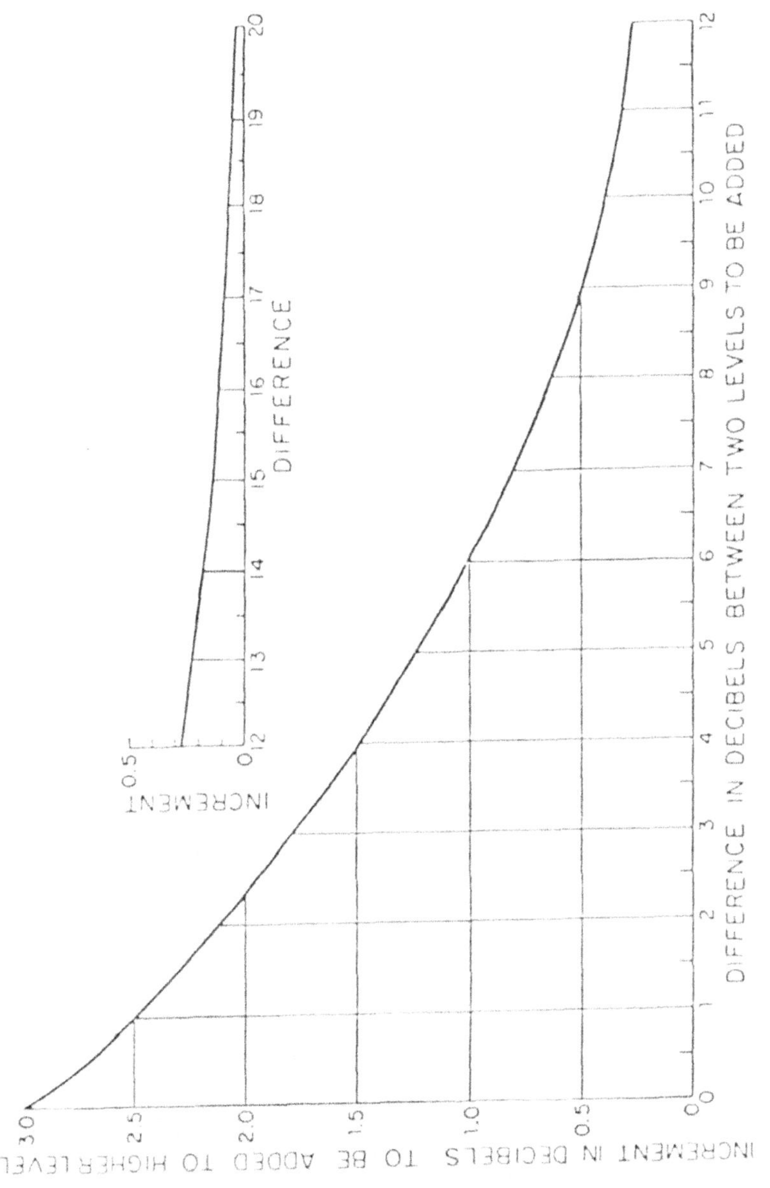

FIGURE 2.5. Chart for combining sound levels by "decibel addition."

(From ref. 9.)

References

1. Harris, Cyril M., ed.: Handbook of Noise Control, Second ed. McGraw-Hill Book Co., c.1979.
2. Kryter, Karl D.: The Effects of Noise on Man. Academic Press, Inc., 1970.
3. Stevens, S. S.: Perceived Level of Noise by Mark VII and Decibels (E). J. Acoust. Soc. America, vol. 51, no. 2, pt. 2, Feb. 1972, pp. 575-601.
4. Kryter, Karl D.; and Pearsons, Karl S.: Some Effects of Spectral Content and Duration on Perceived Noise Level. J. Acoust. Soc. America, vol. 35, no. 6, June 1963, pp. 866-883.
5. American National Standard Specification for Sound Level Meters. S1.4-1971, American Nat. Stand. Inst., Inc., Apr. 27, 1971.
6. Noise Standards: Aircraft Type Certification. Federal Aviation Regulations, vol. III, pt. 36, FAA, Dec. 1969, as amended.
7. Information on Levels of Environmental Noise Requisite To Protect Public Health and Welfare With an Adequate Margin of Safety. Rep. 550/9-74-004, U.S. Environ. Prot. Agency, Mar. 1974. (Available from NTIS as PB 239 429.)
8. Galloway, William J.; and Bishop, Dwight E.: Noise Exposure Forecasts: Evolution, Evaluation, Extensions, and Land Use Interpretations. Rep. FAA-NO-70-9, Aug. 1970.
9. Bolt Beranek and Newman, Inc.: Calculation of Day-Night Levels (L_{dn}) Resulting From Civil Aircraft Operations. EPA 550/9-77-450, U.S. Environ. Prot. Agency, Jan. 1977. (Available from NTIS as PB 266 165.)
10. Land Use Planning With Respect to Aircraft Noise. AFM 86-5, TM 5-365, NAVDOCKS P-98, U.S. Dep. Defense, Oct. 1, 1964. (Available from DTIC as AD 615 015.)

Chapter 3
Physiological Functioning of the Ear and Masking

Introduction

The peripheral parts of the auditory system are the outer ear (pinna and auditory canal), the middle ear (eardrum and ossicles), and the inner ear (the cochlea). The ear, possibly largely independent of processing by the central nervous system (ref. 1), is capable of analyzing the spectral content of complex, broadband sounds. This spectral analysis is primary and essential to the identification of and discrimination among different sounds and to the learning of their meanings by the higher brain centers.

Although not well understood by scientists, this spectral analysis capability can be either partly or totally disturbed by overexposure to intense sound or noise. Unfortunately, as is discussed subsequently in some detail, the noises and sounds of modern society take some toll on the ear before most lifetimes are complete. In addition, noise can have an immediate adverse effect by overloading the ear, causing some temporary loss in hearing ability.

Analysis of Sound by the Ear

Figure 3.1 is a schematic of the path and means by which sound enters the human ear, at which time it is transduced into motion in the fluid (perilymph) of the cochlea. This fluid action causes nerve fibers on the basilar membrane to send impulses to higher nerve centers, where the impulses are perceived or interpreted as sound.

In classical auditory theory, the analysis of sound, with respect to the physical dimensions of frequency and intensity, takes place in the cochlea. Except for special circumstances, phase information (related to the positive-negative pressure relations among different frequency components in a broadband sound) appears to be of little significance to the subjective response to sounds. The primary psychological aspects of frequency and intensity are pitch and loudness. Psychological dimensions other than pitch or loudness—for example, density volume (size) and perceived noisiness—have also been related to the frequency-intensity characteristics of sounds. (See ref. 2.)

The attribute of pitch has been ascribed to possible cues of place of stimulation on the basilar membrane of the cochlea and to the frequency of firing of peripheral neural units; the attribute of loudness has been ascribed to the number and rate at which neural impulses are generated in the cochlea. It is, however, the perception of the patterns of complex pitch and loudness in the flow of time that makes audition such a useful and pervasive part of man's consciousness.

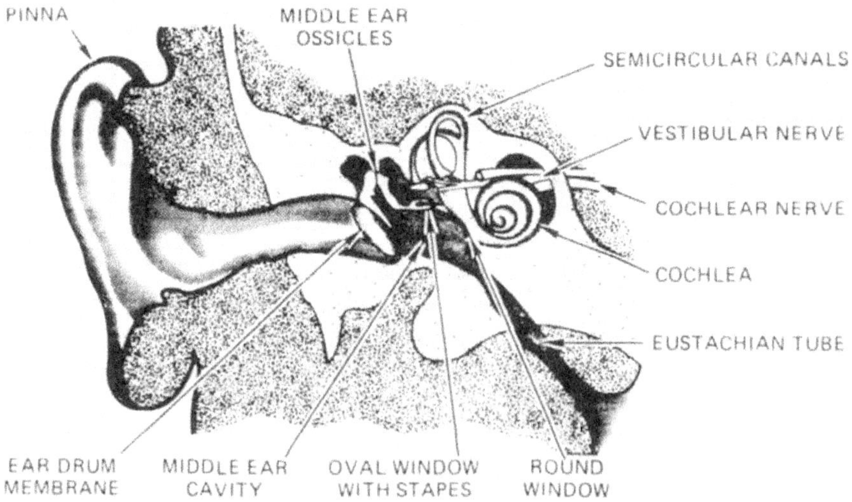

FIGURE 3.1. Parts of the human ear.

Critical bandwidth of the ear

The abilities of the ear to perceive pitch as a function of frequency and loudness as a function of intensity, and to detect small changes in these attributes, were well mapped out prior to 1950. (See refs. 3 and 4.) However, the ability of the ear to behave as a bandpass filter was not extensively delineated until after 1950.

As conceived by Fletcher (ref. 5), the concept of filters within the ear having what are called critical bandwidths has proved to be significant. This concept has furnished a basis for explaining some auditory behavior with respect to speech perception, auditory fatigue, loudness, pitch perception, and masking. Basically, the cochlea and its associated nerve nets often seem to behave as a very large set of overlapping bandpass filters connected in parallel. These filters, like most filters, have skirts that are not sharp. (See ref. 6.) Their bandwidths change as a function of frequency, and they become broader when the signal intensity is increased. In particular, it seems that at high intensity levels the upper skirt of the filter becomes much less steep than the lower skirt. The critical band concept indicates that increasing the bandwidth of a masking noise beyond a certain width does not increase the degree to which a pure tone located at the center of the band is masked. Only the energy at frequencies nearer the center frequency contribute to the analysis and masking of the tone at the center.

References 5 and 7 to 9 are experiments in which the intensity level of a pure tone, presented with a broadband random noise, was adjusted until the tone was just audible. This process was repeated with pure tones of different frequencies. Swets *et al.* (ref. 10) conducted a similar experiment in which a narrow band of noise was masked with a wider band. Masking is when a tone, or other signal is made nonaudible to a listener because of the presence of another sound or noise. The results of figure 3.2 show that the width of the critical band and critical ratio vary as a function of frequency. (Figs. 3.2 through 3.22 are from ref. 11.) For the two bottom curves in figure 3.2, critical band is defined as the ratio (called the critical ratio) between the spectrum level (level per Hz) of white noise and the pure tone at a masked threshold. That is, the critical band was defined as the band of noise around a pure tone, at center frequency, whose acoustic power equaled that of the pure tone when at masked threshold. References 12 to 20 show that when the critical band was measured directly in a variety of perceptual judgment tests, its bandwidth varied more or less as a function of frequency, as did the critical band when measured indirectly by the masking of a pure tone by a white noise. However, the width was about 2 1/2 times greater when measured directly. (See fig. 3.2.)

Shown in figure 3.3 and also in figure 3.4, at the points 6 dB down from the peaks, is the bandwidth of the resonance functions for the basilar membrane of the cochlea as measured in references 21 and 22. From an analysis of perceptual data on the critical band, and from Von Békésy's direct measurements of basilar membrane resonance, Greenwood (ref. 23) concludes that one critical bandwidth

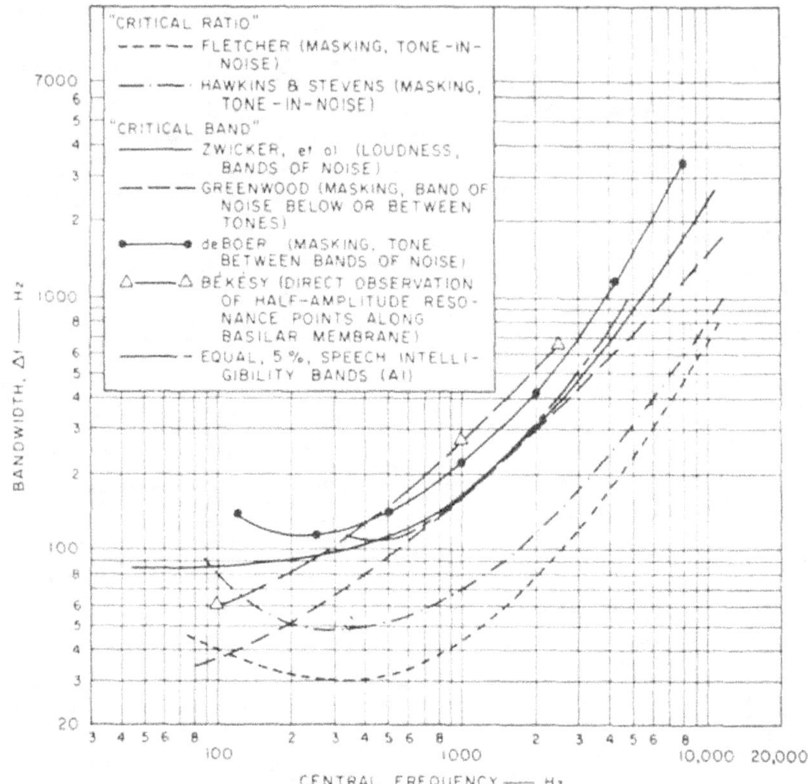

FIGURE 3.2. Bandwidth of sounds as function of band center frequency for various parameters. (Data from Fletcher (ref. 5), Hawkins and Stevens (ref. 7), Greenwood (ref. 16), De Boer (ref. 18), Von Békésy (ref. 21), and Zwicker *et al.* (ref. 14).)

(not ratio) extends about 1 mm along the basilar membrane in the frequency region from about 400 to 6000 Hz. (See right-hand vertical ordinate in fig. 3.2.)

It should be recognized that the auditory system is capable of perceiving pitch changes, trills, beats, *etc.*, from frequency changes in an acoustic stimulus that are much narrower than the critical band or the gross hydromechanical patterns that result from stimulation of the cochlea. It is probable that neural mechanisms present in the brain stem are responsible for some of this further sharpening action. (See ref. 24.)

The hydromechanical behavior of the inner ear is such that the low frequencies cause turbulence (the presumed stimulator, in some manner, of the receptor cells on the basilar membrane) toward the part of the cochlea, called the apex, farthest from the place where sound vibrations enter the inner ear. Also, this turbulence is asymmetrical, spread out toward the base of the cochlea and truncated

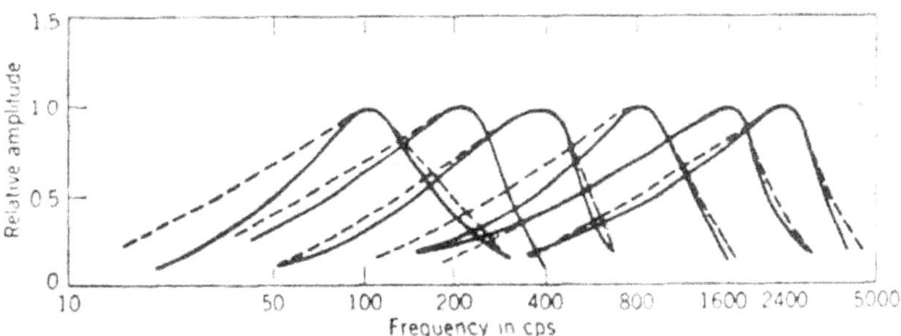

FIGURE 3.3. Resonance curves for six points on the cochlear partition. The solid curves are measured values, and the dashed curves are theoretical values. (From ref. 22.)

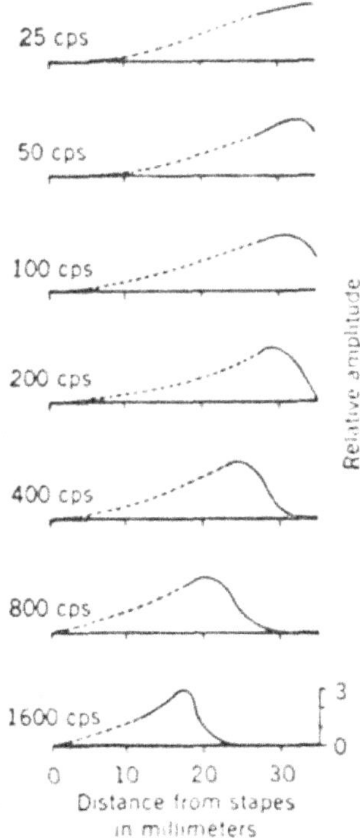

FIGURE 3.4. Displacement amplitudes along the cochlear partition for different frequencies. The stapes was driven at a constant amplitude, and the amplitude of vibration of the cochlear partition was measured. The maximum displacement amplitude moves toward the apex as the frequency is decreased. (From ref. 21.)

toward the apex. (See figs. 3.3 and 3.4.) These two facts contribute to a characteristic effect of sound, namely, that low-frequency sounds tend to stimulate many more receptor fibers on the basilar membrane than do high-frequency sounds.

Critical summation time of the ear

Just as in the frequency domain, where there is a critical bandwidth over which the ear processes loudness, there is a "critical time period" over which the ear summates loudness. This time period, sometimes called the "time constant" of the ear, is about 1 sec for detection of pure tones in the quiet (minimum loudness) and in the presence of a masking noise. The time constant for the detection of a change in the loudness of a sound when it is at suprathreshold levels (far above minimum loudness with no masking noise present) is about 0.3 sec. The auditory system apparently has a more difficult loudness discrimination task at absolute and masked thresholds than at suprathreshold.

The critical summation time is also shown by the fact that sound energy occurring outside a time frame of 0.3 sec at suprathreshold levels, and 1 sec at absolute or masked threshold levels, does not contribute to the loudness of a sound perceived for that period of time. This is, of course, analogous to the critical frequency bandwidth of the ear where sound energy present outside the critical bandwidth does not contribute to the loudness of the sound perceived in the critical band.

This critical time period is fairly independent of the frequency content of the sounds. However, it is important to note that it is the perceived loudness that is being processed or integrated, and not the sound energy *per se*. In short, this independence of time and frequency is revealed only when the spectral contributions of a sound to loudness are taken into account. Perceptual data pertaining to the critical summation time, or time constant, of the ear is discussed in some detail in chapter 5 of this report.

Model of the inner ear

The model of the inner ear subscribed to for the purposes of this report is as follows: (1) the time-pressure envelope of a sound is created and is related by hydromechanical means to the construction of the inner cochlea and the fluids it contains and varies with frequency along the basilar membrane within the cochlea; (2) the neural receptors on the basilar membrane can respond to the changes in pressure and turbulence of the cochlear fluid; (3) the auditory nervous system is capable of interpreting neural firings from the basilar membrane with respect to the number of neural firings, the place on the basilar membrane initiating the firings, and the periodicity of the firings; and (4) when the rms pressure of the sound stays steady for a period longer, within limits, than about 0.3 sec, the rate and/or periodicity of neural responses from the basilar membrane becomes stabilized. For reasons to be more fully discussed subsequently, it is proposed that this temporal interval be taken as 1 sec rather than 0.3 sec.

Outer and middle ear

The outer and middle ear appear to have the function not only of transmitting to the inner ear the pressure waveform of the sound, but also of protecting the inner ear from having to be exposed to sounds outside its capacity. In regard to the latter: (1) the middle ear can prevent the transmission to the inner ear of pressure waves with rise times longer than 200 msec by means of the action of the eustachian tube (or even by rupturing of the eardrum); (2) when presented with high-intensity pressure waves, small muscles in the middle ear can contract, which stiffens the ossicular chain and attenuates the transmission of at least some frequencies of the sound (with very intense sound, the ossicular chain appears to rotate from its normal axis in a way that limits or even reduces the pressure level reaching the inner ear), and (3) the mass and stiffness of the ossicular chain prevent transmission of a pressure wave with a rise time of less than 50 μsec. These time durations, 200 msec to 50 μsec, correspond to the period of the frequencies of 5 Hz to 20 000 Hz and match the frequency band limits of the inner ear which are set largely by the dimensions of the inner ear and by the nature of the basilar membrane. (See ref. 25.)

The acoustic resonance of the ear canal and of the outer ear contributes to the frequency response characteristics of the ear. It is significant that, relative to frequencies below 1000 Hz, the ear canal effectively amplifies higher sound frequencies, particularly around 4000 Hz. (See fig. 3.5.) Although this may at times contribute to overstimulation of the inner ear at the higher frequencies, this resonance may be a useful compensation for the nonlinear attenuation of these higher frequencies during the transmission of sound through air and most other media.

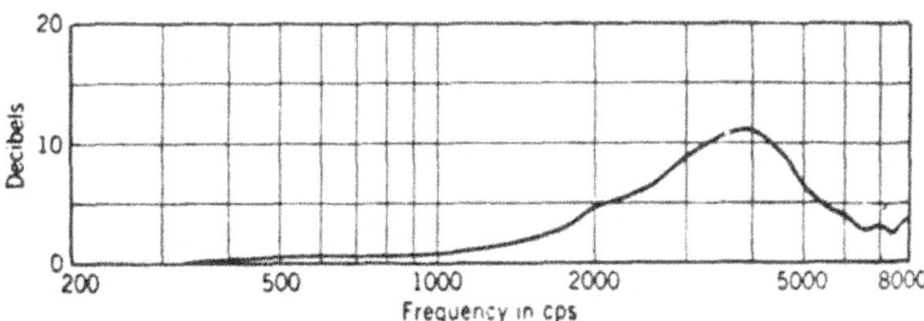

FIGURE 3.5. Effects of resonance in external meatus. Ordinate shows ratio in decibels between sound pressure at eardrum and sound pressure at entrance to auditory canal. (From ref. 26.)

Aural Reflex

The tympanic and stapedius, two small muscles in the middle ear, are attached to the small ossicular bones that connect the eardrum with the cochlea. They mediate the so-called aural reflex and thereby play a significant role in audition, particularly when noise is present. The aural reflex can influence the effects of noise with regard to masking, loudness, and auditory fatigue. The tympanic and stapedial muscles contract when the ear is exposed to a sound that is about 80 dB above threshold level. In humans, the reflex action is inferred from (1) various perceptual auditory tests, (2) physical measurements of changes in the volume of the external ear canal, and (3) changes in the acoustic impedance of the eardrum.

If sufficiently intense, a sound in one ear will activate the reflex in both ears, although the contraction in the sound-stimulated ear is somewhat stronger than that in the nonexposed ear. Up to a point, as the intensity of the sound increases the degree of contraction increases. It appears that the reflex is more responsive to broadband sounds than to pure tones, and is more responsive to lower frequencies than to higher frequencies.

The reflex appears to adapt or relax in the presence of continued stimulation after about 15 min of exposure to an intense steady-state noise. That the muscles are not fatigued can be shown by the fact that the reflex can be reactivated by changing the acoustic stimulus. (See ref. 27.) It is conceivable that the reflex gradually relaxes during continued stimulation in order to compensate for, or because of, a gradual decrease in the loudness of sounds with long-duration stimulation.

The aural reflex seems to be most readily activated and maintained by intermittent, intense impulses of noise. Latency of the reflex is about 35 to 150 msec, depending on the intensity of the stimulus, and relaxation time following an impulse of noise is reported to be as long as 2 to 3 sec for complete relaxation, with most of the recovery probably occurring within about 0.5 sec. The effect of this reflex upon auditory fatigue from gunfire is shown subsequently.

The reflex is involuntary and, except to the specially trained subject, its occurrence is not detectable. It can apparently be conditioned to light and other stimuli, and some people can cause it to contract voluntarily. Whether people can, by volition, cause the reflex to relax when active is another question, although it has been suggested as an explanation of some phenomena related to threshold measurements.

What mechanical changes take place in the ossicular chain of bones as the result of their contraction of the aural muscles are a matter of some conjecture. For one thing, the acoustic impedance of the eardrum is changed, but the relative position of the eardrum is probably unaltered with total reflex activity, because the two muscles appear to be antagonistic to each other. It is reasonable to think that the aural reflex merely serves to stiffen the eardrum and the bones of the middle ear so that they will not transmit sound as effectively as normal. Something like this must occur, but it affects the transmission of sounds mostly below about 2000 Hz.

An argument can be made that the aural reflex is designed to protect the ear from drastic changes in the velocity of movement of the eardrum-cochlear fluid system. It is conceivable that, when the velocity of movement of the cochlear fluid exceeds a certain critical value, the disturbance differs from the usual hydromechanical turbulence in the cochlea and spreads both downwards and upwards along the cochlear partition. The best way to protect the ear against this type of trauma is to reduce the transmission of frequencies below about 2000 Hz because of the following: (1) the middle ear is most compliant in the frequency region from 300 to 1500 Hz, and (2) acoustic stimuli that occur at levels in excess of about 150 dB are invariably impulsive and have their major energy in the frequency region below 2000 Hz.

Studies concerned primarily with the perceptual-psychological effects of the aural reflex, as distinct from their physiological correlates, fall into the following three general categories: (1) contralateral threshold shift, (2) loudness, and (3) auditory fatigue.

Contralateral threshold shift

One of the effects of the aural reflex is investigated in the following way. A sound at an intensity sufficient to elicit the aural reflex is presented to one ear and at the same time the listener is tracking his threshold for pure tones in the opposite ear. When activated, the reflex, being bilateral, will cause a rise in the threshold at some frequencies in the "quiet" ear. It is, of course, necessary that the frequency content of the sound used to elicit the reflex be sufficiently different in frequency from the tones being tracked, so as not to have present direct masking due to bone conduction of sound through the listener's head.

Figures 3.6 to 3.8 (from ref. 28), figure 3.9 (from ref. 29), and figure 3.10 (from ref. 30) reveal many of the salient facts about this contralateral threshold shift. First, it is shown in figure 3.6 that the greatest shift seems to occur for a 500-Hz tone regardless of the frequency content of the stimulus used to arouse the reflex. Figure 3.7 shows that the reflex is more responsive to lower than to higher frequency sounds. Figure 3.7 also shows that, up to a point, the contralateral threshold shift increases linearly with stimulus intensity. Figure 3.8 shows that with continued exposure the reflex adapts.

Loudness

Finding the threshold of hearing for a tone is, in a sense, finding its minimum loudness. Loeb and Riopelle (ref. 29) had subjects find not only the contralateral threshold shifts but also the loudness (relative to pre-reflex loudness) of the 500-Hz test tone presented at suprathreshold levels. (See fig. 3.9.) They found that when the contralateral-reflex activating signal (a tone of 2200 Hz) was pres-

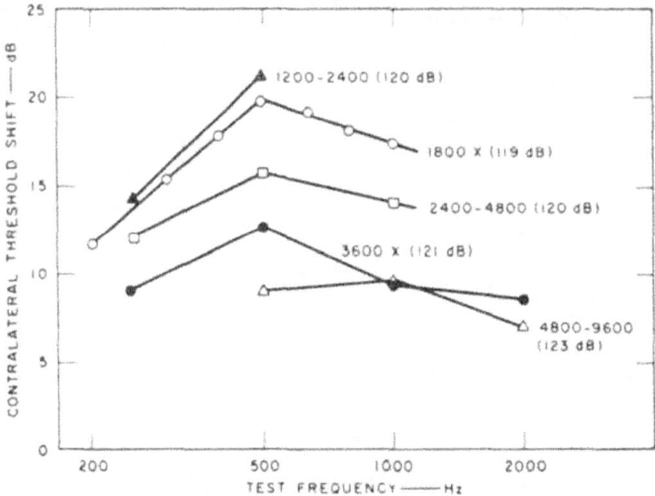

FIGURE 3.6. Contralateral threshold shift 30 sec after noise onset as a function of frequency. Adapting noise spectrum is the parameter; two noises with center frequencies of 1800 and 3600 Hz, marked "X," have very narrow bandwidths (less than 1/3 octave). (From ref. 28.)

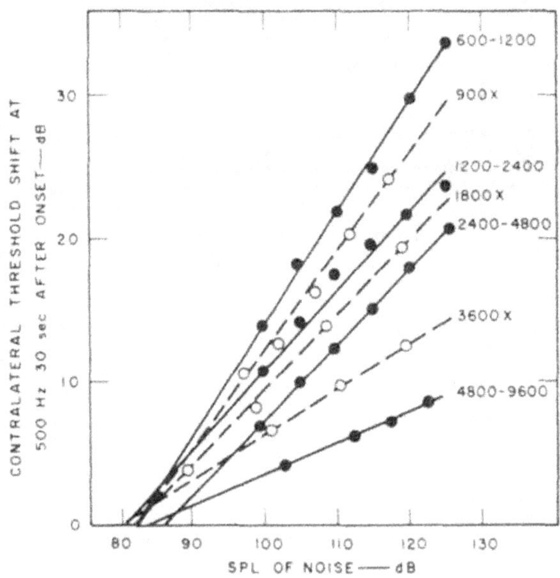

FIGURE 3.7. Growth of contralateral threshold shift at 500 Hz with level of arousal noise. Filter setting is the parameter; three noises with center frequencies of 900, 1800, and 3600 Hz, marked "X," have very narrow bandwidths (less than 1/3 octave). (From ref. 28.)

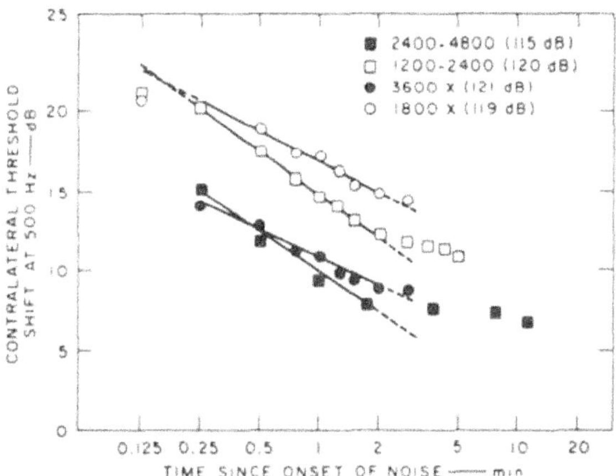

FIGURE 3.8. Adaptation of contralateral threshold shift. (From ref. 28.)

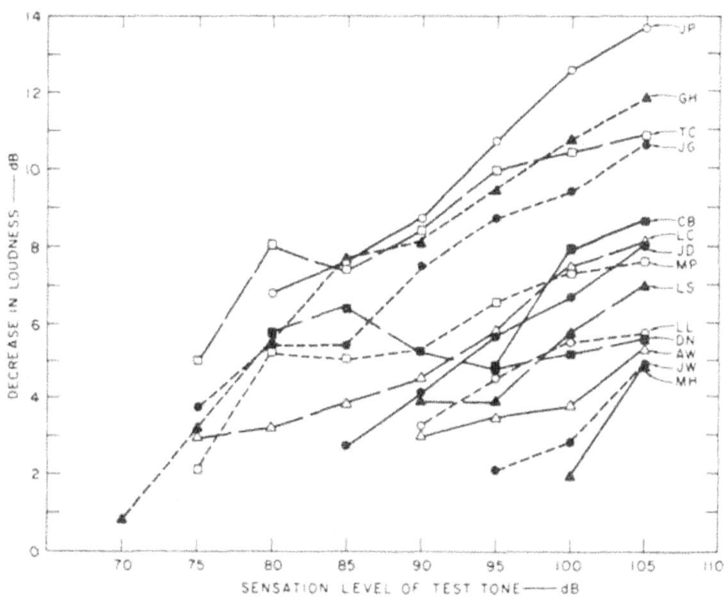

FIGURE 3.9. Decrease in perceived loudness as a function of sensation level of test tone. Each curve represents data from a different subject. (From ref. 29.)

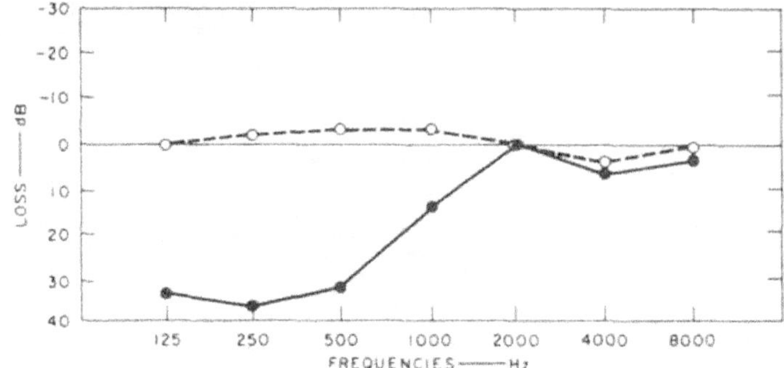

FIGURE 3.10. Dotted line shows average hearing level of eight selected ears. Solid line shows average hearing level of the same ears during maximum voluntary contraction of middle ear muscles. (From ref. 30.)

ent, the loudness of the 500-Hz test tone at a sensation level of 105 dB decreased relative to its loudness when the reflex activating signal was not present. The decrease in loudness for the test tone at 105 dB was equivalent to a decrease of 5 to 14 dB in intensity of the test tone; at a sensation level of only 20 dB, the decrease in loudness of the 500-Hz test tone with the reflex present was only 2 to 5 dB.

This result led to the conjecture that the reflex possibly had a nonlinear "snubber" action for intense signals but would not attenuate weak signals. This conclusion must remain in doubt in view of the fact that contralateral threshold shifts of 10 to 15 dB, rather than only 3 to 5 dB, have been obtained by others (ref. 31) for roughly comparable stimulus conditions.

Voluntary control of reflex

It appears that some people are able to voluntarily activate their aural reflex. (See ref. 30.) These people are apparently not only aware of a reduction in the loudness level of some sounds, but they also hear the sound made by the contraction and relaxation of the intra-aural muscles. It is estimated in reference 30 that 1 to 2 percent of people have this ability and that others can be trained. Figure 3.10 shows the threshold shift observed at various frequencies during maximum voluntary contraction of the aural reflex.

Auditory fatigue

The fact that the aural reflex (fig. 3.8) adapts or relaxes when a noise stimulus continues for longer than 15 min or so suggests that the reflex would not offer significant protection from hearing loss due to exposure to intense, more or less continuous noise. However, it is shown in references 32 and 33 that the aural reflex can protect the ear from auditory fatigue as the result of exposure to gunfire. Fletcher and Riopelle (ref. 32) elicited the reflex with a brief 1000-Hz tone at 98

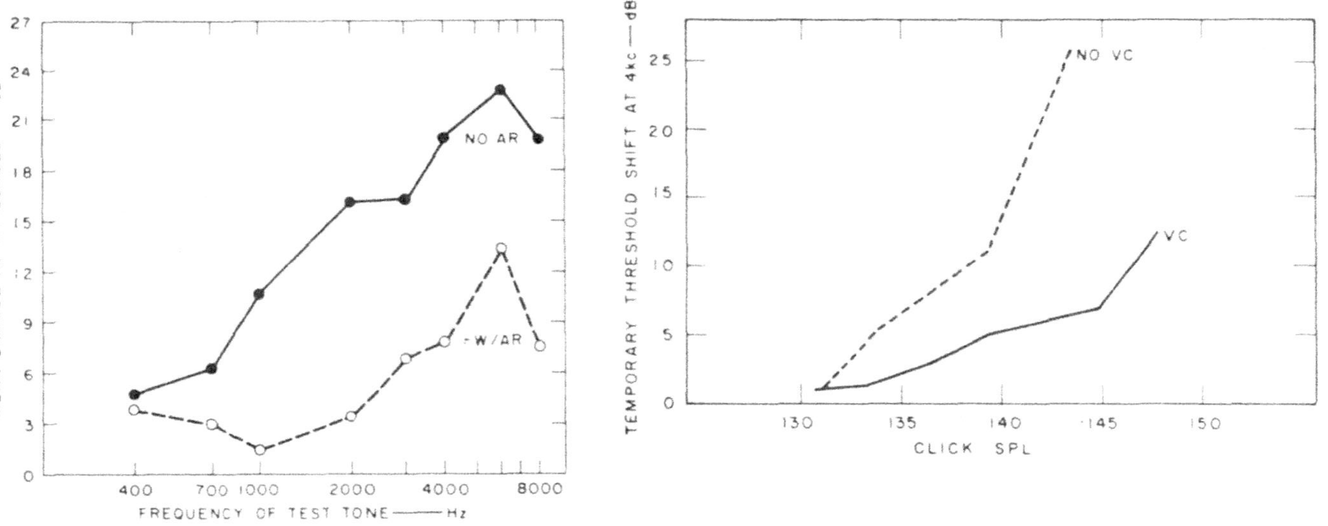

FIGURE 3.11. Left graph: mean temporary threshold shifts (TTS) at various frequencies for the two experimental conditions. (W/AR is with aural reflex; and no AR is no aural reflex.) (From ref. 32.) Right graph: average growth of TTS with successive exposures to increasing pulse levels with and without voluntary contraction (VC) of middle ear muscles. Attenuation produced by VC is given by amount by which function generated with VC is shifted to the right. (From ref. 34.)

dB in one experiment. In later experiments, they elicited the reflex with a click and a band of noise, presented 200 msec before exposing the ear to the impulse from the firing of a gun. They measured the threshold shifts, which were temporary, after 200 rounds of firing. A second experimental condition was the same as that just described except the tone, click, or band of noise was withheld prior to each round of firing. Presumably, the aural reflex was not active in the latter case when the gun noise reached the ear. Figure 3.11 shows that the aural reflex afforded as much as 15 dB of protection from temporary threshold shift (TTS).

Physiological studies have shown that the reflex relaxes following cessation of a stimulus that causes full contraction. Thus, whenever impulsive sounds are separated by more than a certain period, the reflex action presumably present as the result of each impulse provides no attenuation to the succeeding impulse. Germane to this is an experiment of Ward (ref. 35) in which he exposed subjects to acoustic impulses separated by 1-, 3-, 9-, and 30-sec intervals. As shown in fig-

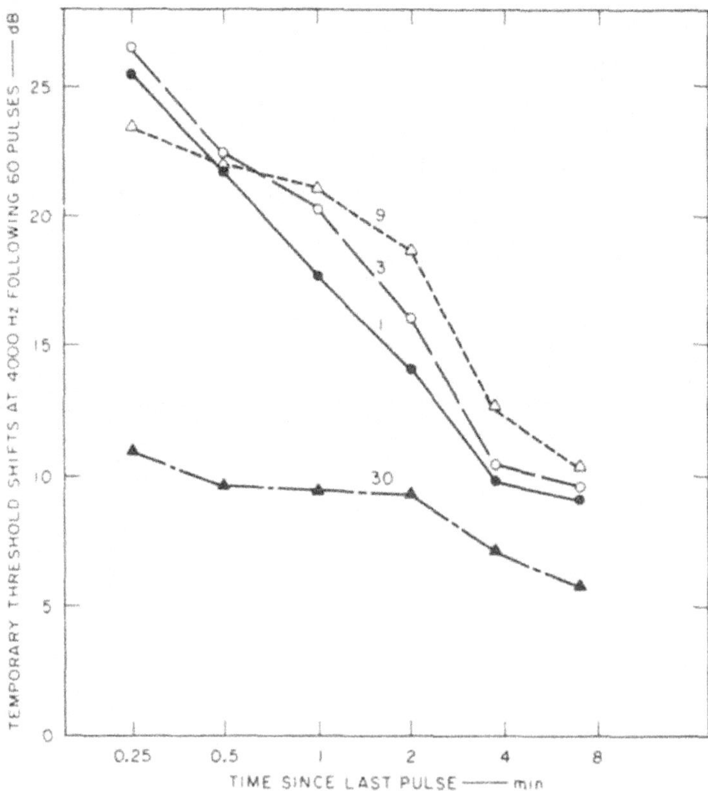

FIGURE 3.12. Recovery in time of TTS at 4000 Hz produced by 60 high-intensity pulses. Interpulse interval, in seconds, is the parameter. (From ref. 35.)

ure 3.12, the threshold shifts are roughly the same for intervals up to 9 sec. Ward concludes that the reflex was therefore inactive within 1 sec after the impulse, or the shift with the 1-sec intervals would have been less than with 3- or 9-sec intervals. Other experimental results from exposure to impulse noise suggest that there might be some residual reflex contraction up to 3 sec in some people.

Aural reflex in persons with hearing loss

The aural reflex is used as a means of diagnosing certain types of hearing disorders. (See ref. 36.) In particular, its absence indicates conductive difficulties in the middle ear.

Terkildsen (ref. 37) found that persons in industry with significant hearing losses as a result of being exposed to traumatic noise had somewhat weaker aural reflexes than persons with normal hearing. On the other hand, Hecker and Kryter (ref. 38) found that soldiers with large permanent hearing losses (presumably due to exposure to gunfire) showed greater aural reflexes than soldiers with normal hearing. These investigators also found that the men with greater reflex activity showed less TTS when exposed to gunfire than was exhibited by men with lesser aural reflexes; however, this difference may have been because of the decreased sensitivity (larger permanent hearing losses) of the former group relative to the latter, and not because of some increased protection afforded by the aural reflex, although this latter is a possibility.

The aural reflex is a subject of active research. (See refs. 39 to 44.) Also, the aural reflex is discussed in subsequent chapters of this report regarding hearing loss from noise and nonauditory system responses to noise.

Masking

A major function of the auditory system is the analysis of acoustical signals so that information-bearing components in a sound wave can be discriminated or separated from the unwanted or noisy parts. In a sense, noise is always present during the hearing process; in the limiting case of quiet, it is the internal noise floor of the auditory system, but usually noise is present in the acoustical signal along with the information-bearing components. The interference or masking of wanted signals by noise is, in a sense, the converse of the analysis process.

The masking of pure tones and bands of noise are of considerable interest and provide some basis for understanding the effects of noise on speech communications. The effects of noise on the process of speech communication *per se* is discussed subsequently in this chapter; in a subsequent chapter, the annoyance experienced by people in everyday living because of the effects of noise on speech communications is discussed.

The general method used for measuring masking of tones or bands of noise is as follows. Using a pure tone or narrowband noise generator, the threshold of audibility is determined at a number of frequencies for the listener in the quiet. Then, while a masking pure tone or a band of noise is presented, the listener redetermines his threshold of audibility by means of other (probe) tones or bands of noise. The increase in level required for the probe tone or band of noise to be audible at each frequency represents the amount of masking caused by the masking tone or band of noise.

Direct and frequency-spread masking

Direct masking occurs when the receptors in the cochlea that normally process a signal of a given frequency are stimulated by another signal of the same frequency or when the receptors are affected by the upward and downward spread of masking along the basilar membrane from another signal. Ehmer (ref. 45), Small (ref. 46), and Carter and Kryter (ref. 47) confirmed and extended the masking patterns for pure tones that had been found previously by Wegel and Lane (ref. 48). Egan and Hake (ref. 8), Ehmer (ref. 49), Zwicker (ref. 50), Saito and Watanabe (ref. 51), and Carter and Kryter (ref. 47) measured the masking pattern of narrow bands of noise.

Typical examples of the results obtained are shown in figure 3.13. It is shown in the figure that the masking pattern from narrow bands of noise is much smoother, particularly at the vicinity of the center frequency of the masker, than those found with pure tones; the latter masking functions are disturbed by audible beats that occur between the probe tone and the masking tone and its harmonics. These harmonics are introduced by nonlinear distortion in the ear.

The curves in figure 3.13 reveal the following interesting characteristics of direct masking:

(1) The band of noise causes more masking around its center than does the pure tone. Increased masking near the center frequency of the masker band of noise would, of course, be expected from the integrative action of the critical bandwidth of cochlear functioning. Also, as Egan and Hake and Ehmer suggest, the pure-tone masking may be only apparently lessened at the locus of the masking tone, because beats between the probe and masking tone cause a false measurement of the threshold of the tone.

(2) There is an asymmetrical upward spread of masking that becomes more severe at higher intensity levels. Von Békésy's observations of asymmetrical resonance patterns along the basilar membrane offer an apparent mechanism to explain the asymmetrical upward spread of masking.

Of particular interest are the masking patterns obtained in reference 52 with intense low-frequency tones ranging from 10 to 50 Hz. (See fig. 3.14.) The masking pattern for a tone as low as 25 Hz and an intensity level of 130 dB appears to extend from almost flat to as high as 4000 Hz. The masking effects of tones as low as 50 Hz on tones and upon speech have also been investigated; the findings, which are discussed subsequently, agree reasonably well with those of reference 52.

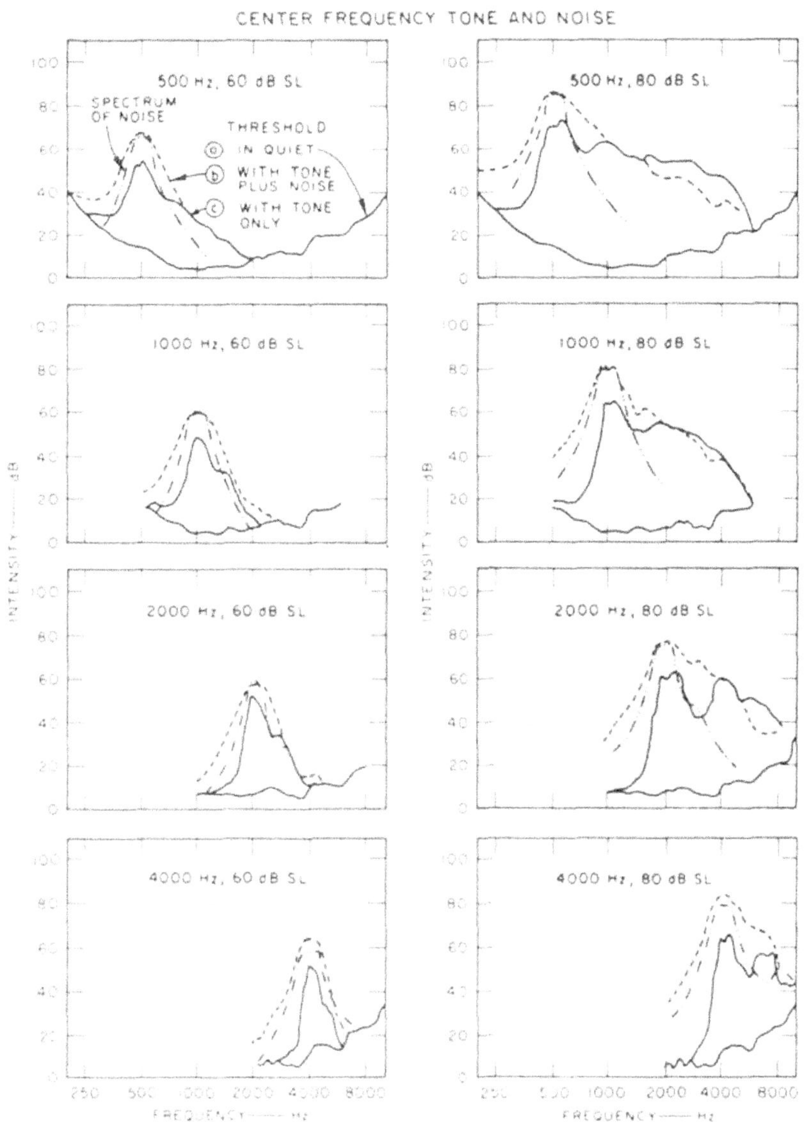

FIGURE 3.13. Masked thresholds for pure tones and narrow bands of noise. Center frequency and sensation level (SL) of the tone and noise are parameters. (From ref. 49.)

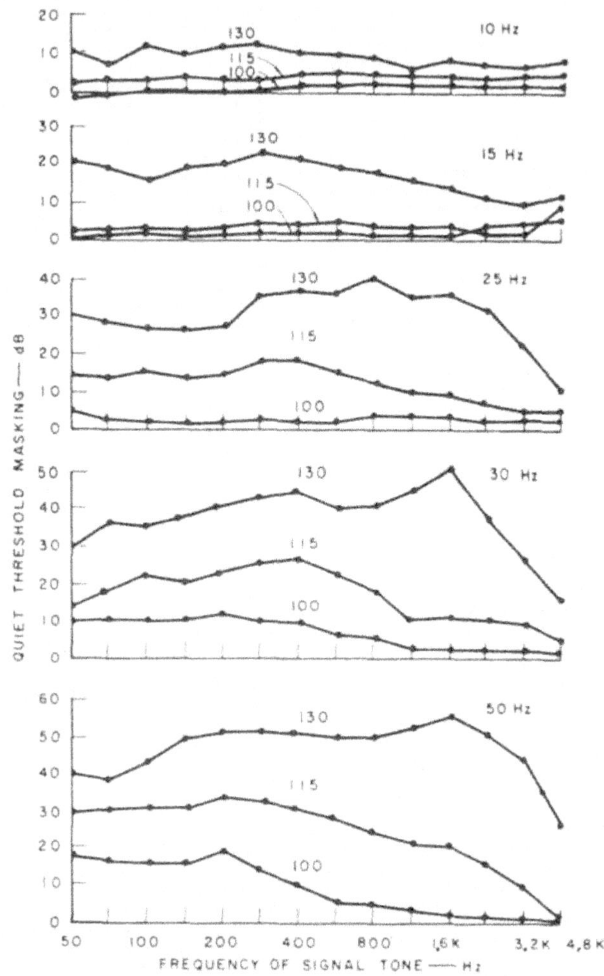

FIGURE 3.14. Masking with 5 low-frequency pure tones (10, 15, 25, 30, and 50 Hz) and 3 intensity levels (100, 115, and 130 dB SPL). Ordinate shows masking in dB relative to quiet threshold. Abscissa shows frequency of signal tone in Hz. Parameter is SPL of masking tone in dB, re 0.0002 dyne/cm^2. Each point is average masking experienced by 5 listeners. (From ref. 52.)

It is of some interest to consider what effect, if any, masking noise has upon loudness and on the ability of the ear to discriminate or detect changes in signal level. This detection of change is called the difference limen for intensity. (See ref. 53 for a review of research on the difference limen for sound intensity.) It is a generality, as is also shown in experiments on the intelligibility of speech in noise, that the signal-to-noise ratio, not the absolute level of the masking noise, up to about 110 dB, determines the detectability of changes in signal intensity. Figure 3.15 (from ref. 54) shows that the difference limen for intensity ΔI for an octave band of noise is essentially constant in sensation levels above 20 dB.

The general constancy of the difference limen is no doubt related to the well-established fact that the ear recruits loudness in the presence of noise so that a sound only a few decibels above a masking noise appears almost as loud as it would if the masking noise were not there. Figure 3.16 (from ref. 55) demon-

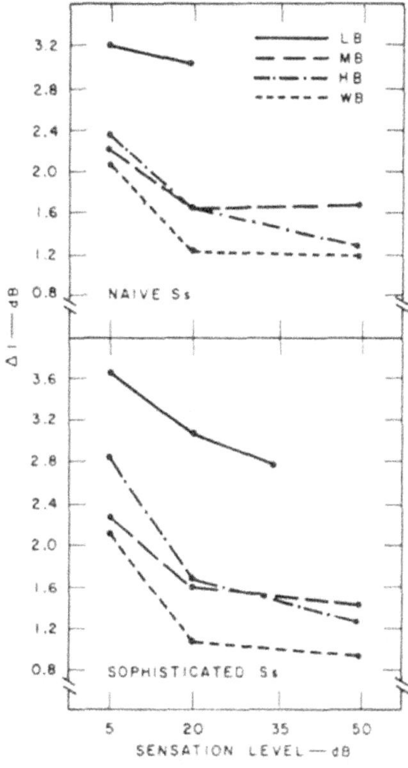

FIGURE 3.15. Mean increase in intensity ΔI for naive and sophisticated subjects (Ss) to detect change in intensity. Each data point represents 44 threshold determinations in upper panel and 24 in lower panel. WB = 127-8160 Hz; LB = 127-255 Hz (octave band); MB = 1040-2080 Hz (octave band); HB = 4080-8160 Hz. (From ref. 54.)

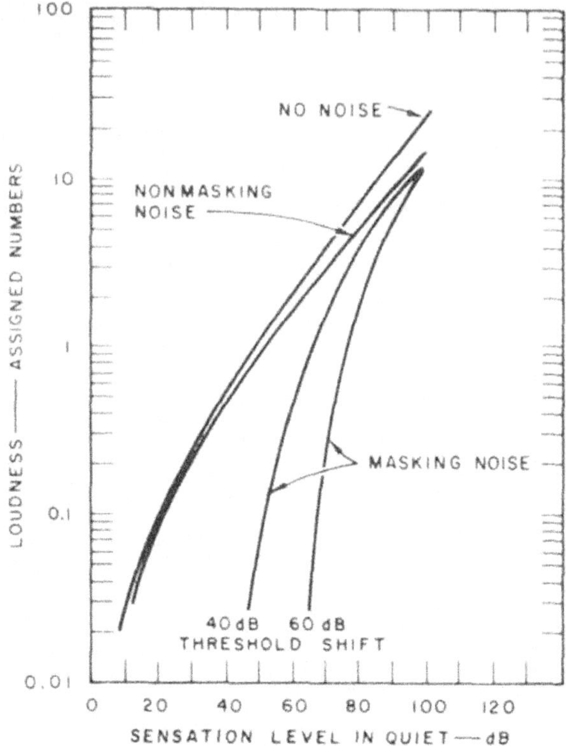

FIGURE 3.16. Masked monaural-loudness curves, obtained by method of numerical magnitude balance, compared with curves of numerical magnitude balance without masking and in the presence of a nonmasking noise. (From ref. 55.)

strates that the loudness of a sound grows much more rapidly above its threshold in noise than in quiet. A matter of both theoretical and practical importance is that loudness grows for intense signals in the ear with a sensorineural hearing loss, as does loudness in the normal ear when the normal ear is in the presence of a noise sufficient to cause a masked threshold shift comparable to the permanent shift in the ear with the hearing loss. (See fig. 3.17.) That is, a person with the hearing loss perceives sounds above an intense background noise as being about as loud as the sounds appear to be to the person with normal hearing.

Pitch changes with direct masking

A number of investigations (refs. 56 to 58) have reported that the pitch of a tone may change when heard in the presence of a band of noise. If the band of noise is of a higher frequency than the tone, the pitch decreases slightly; if the noise is of a lower frequency than the tone, the pitch increases. Both of these effects occur only when the loudness of the tone and the noise are about the same.

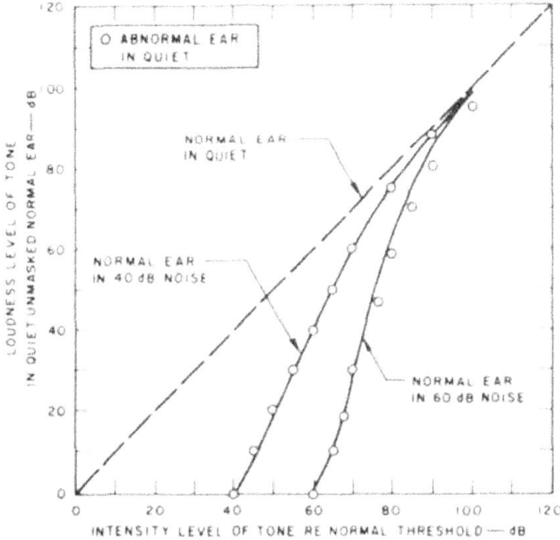

FIGURE 3.17. Loudness-level curves of a partially masked tone, obtained by method of adjustment, compared with loudness-balance data in ears with sensorineural hearing loss exhibiting loudness recruitment. (From ref. 55.)

Egan and Meyer (ref. 57) offer a convincing explanation of why these pitch changes may occur. The argument, put forth also in reference 18, is that the locus or central tendency of the area on the basilar membrane that has the highest signal-to-noise ratio determines what pitch is perceived. This concept, which is a general model for direct masking in the cochlea, is illustrated in figure 3.18. It is shown in this figure that the point on the frequency scale enjoying the maximum signal-to-noise ratio is not the center frequency of the pure tone, but is lower in frequency for the tone below the band of noise and higher for the tone above the band of noise.

Remote masking

Remote masking was discovered and named by Bilger and Hirsh. (See ref. 9.) Remote masking refers to the fact that a high-frequency band of noise, provided it is sufficiently intense, will elevate the audibility threshold for pure tones of low frequency. This is shown in figure 3.19 (from ref. 59). It is usually presumed that this masking is direct masking caused by the presence of low-frequency distortion products resulting from the amplitude distortion that occurs when the signal strength is sufficiently intense to overload the ear.

Bilger (ref. 60) demonstrated remote masking with subjects whose intra-aural muscles had been cut. This result seems to rule out masking as the result of attenuation of low-frequency sounds due to the reaction of the aural muscles to intense sound.

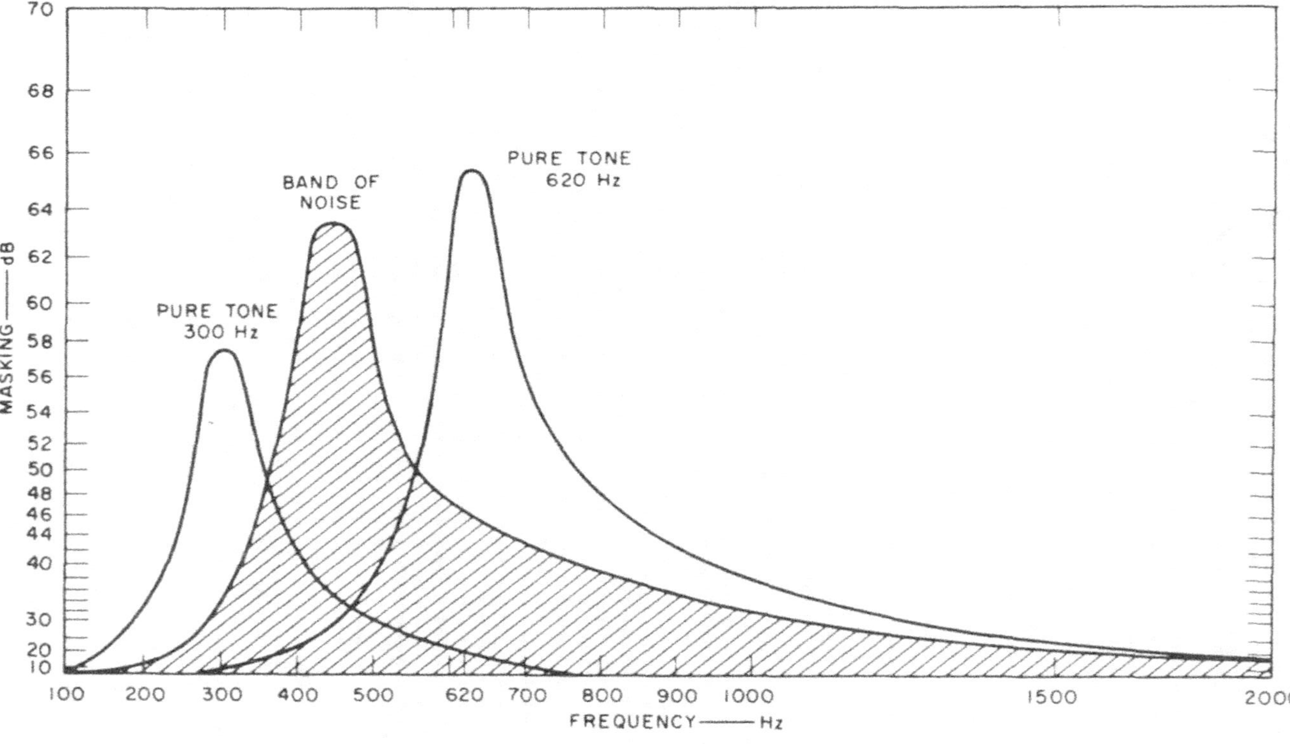

FIGURE 3.18. Loudness patterns. Coordinates are such that when the amount of masking produced by a stimulus is plotted as a function of frequency, area under resulting curve is proportional to loudness of that masking stimulus. (From ref. 57.)

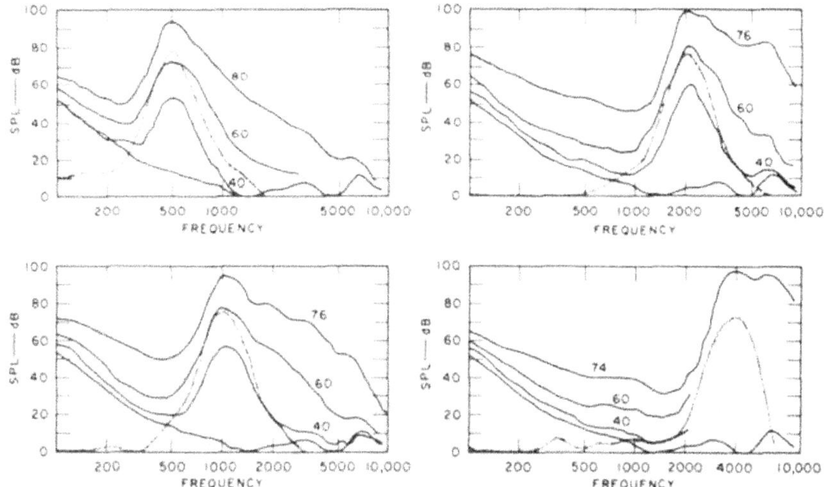

FIGURE 3.19. Solid curves represent average binaural pulsed pure-tone thresholds in quiet (bottom curves) and in presence of several levels of noise centered at 500, 1000, 2000, and 4000 Hz. The spectrum SPL (sound pressure level per Hz) of the noise producing each threshold curve is shown as the parameter in each figure. Hatched curve in each plot shows spectrum of noise at maximum level used. All data were obtained from the same five individuals. (From ref. 59.)

Central masking

Central masking occurs when sound presented to one ear raises the threshold of sound presented to the opposite ear in a way that cannot be attributed to contralateral direct masking, action of the aural reflex, or binaural phase interactions. Contralateral direct masking occurs when sound presented to one ear reaches the other ear. (This masking is usually small because, upon reaching the opposite ear, the sound presented to one ear is usually attenuated by about 50 dB due to transcranial conduction.) The aural reflex, which acts bilaterally, may cause a threshold shift at the lower frequency in the same ear and in the opposite ear—this phenomenon is sometimes called contralateral remote masking. Binaural phase interactions are also discussed subsequently in this report. In general, central masking is a phenomenon that is relatively small and unexplored.

Ward (ref. 28) summarized the several masking effects as shown in figure 3.20. The curves are labeled in accordance with the type of masking that was affecting the threshold change at 500 Hz in the right ear of listeners. Although direct ipsilateral masking is 30 to 100 dB more effective than the other types, these other types cannot be ignored; for example, ipsilateral remote masking contributes to the masking of speech by high-frequency noise. However, the amount of central masking is apparently rather small and cannot be separated from direct contralateral masking or aural reflex (AR) effects.

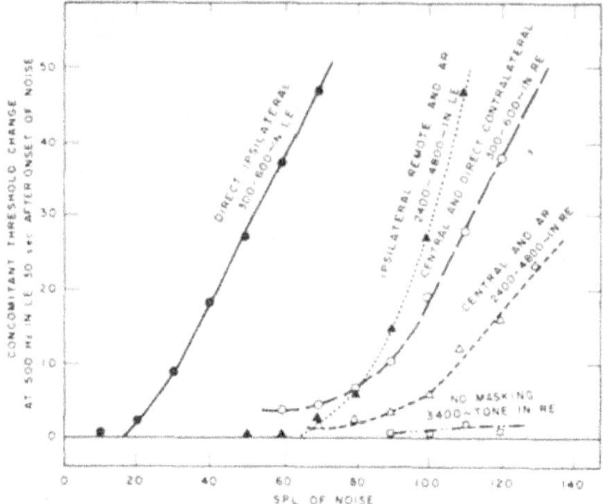

FIGURE 3.20. Growth of direct and remote masking. Threshold at 500 Hz in left ear was measured in presence of low-frequency (300–600 Hz) or high-frequency (2400–4800 Hz) noise, or a 3400-Hz pure tone, in the left ear (LE) or right ear (RE). (From ref. 28.)

Temporal masking

Since the pioneer work of Samojlova (ref. 61), Pickett (ref. 62), and Chistovich and Ivanova (ref. 63), considerable attention has been given to the temporal pattern of masking. In these investigations, a probe tone of very brief duration is presented both before and after a masking tone or noise. Figure 3.21 shows typical results; the small amount of masking for dichotic listening (probe tone in one ear, masking tone in opposite ear) indicates that temporal masking is primarily of a direct, or at least ipsilateral, sort.

Forward masking in time is not surprising—it could be a manifestation of temporary auditory fatigue or some sort of refractory period due to the previous stimulation. But how can a masker elevate the threshold of a sound preceding it in time? Wright (ref. 65) and Zwislocki (ref. 66) suggest that the effect is due to a restriction in the time available for the auditory system to summate the energy and loudness of the tone or a click preceding the masking noise. Apparently, a given length of time is required because of a stimulus-intensity-neural-response time factor, wherein, it is hypothesized, the neural impulses from the much more intense masking noise reach the brain sooner than the impulses resulting from the test tone or click at threshold. Presumably, the growth of the perception of a signal is the integral of the distribution of impulses in the various neural pathways from the cochlea to the higher centers. Since the weaker, preceding sound activates the slower pathways, its growth of loudness occurs at a slower rate than that of the later, more intense masking sound.

Temporal masking is obviously a factor in the detection of temporal order of two stimuli. Hirsh (ref. 67) found that a 10- to 20-msec separation is required be-

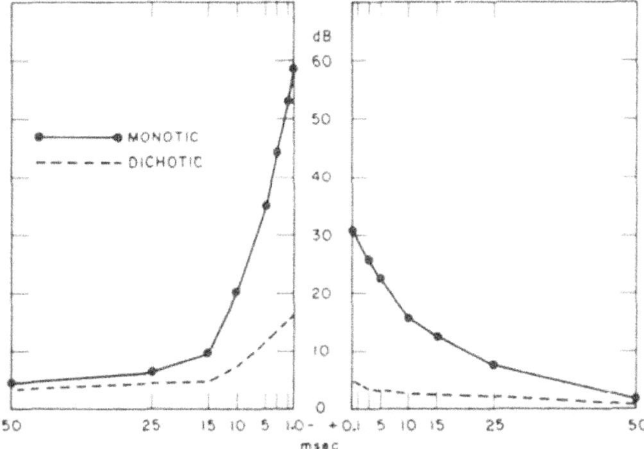

FIGURE 3.21. Backward and forward masking under conditions of 90-dB masking and 5-msec probe duration. Abscissa represents masking interval (maskers not present) with positive values of forward masking on the right and negative time values of backward masking on the left. Ordinate represents amount of masking in dB, that is, difference between masked threshold and unmasked threshold of probe. Data on monotic listening condition are shown by solid line and dotted line represents dichotic listening. (From ref. 64.)

tween two sounds for the human observer to correctly detect which of the two sounds came first. With only a 2- to 3-msec delay, a separation between two sounds was heard, but the order in which the two sounds came could not be identified.

Binaural effects

Binaural masking attributable to the aural reflex, contralateral transmission of sound, and central masking has been discussed. At this point, those effects that are due to variations in phase relations between the ears for either the signal or the masking noise are considered.

Jeffress and his colleagues (ref. 68), Pollack (ref. 69), and others following the work of Licklider (ref. 70) and Hirsh (ref. 71) have extended the knowledge of how the two ears work together in terms of signal detection in noise. For some unknown reason, if two signals are presented simultaneously to both ears, they mask each other by the minimal amount if one signal is in phase with respect to itself at the two ears and the other is out of phase with respect to itself at the two ears. However, if the phases are the same at the two ears for both signals (*i.e.*, both in or both out of phase), mutual masking is increased from 0 to 16 dB, depending upon the frequency spectra involved. Intermediate degrees of phase correlations cause intermediate effects. (See ref. 72.) Thus, certain obvious advantages may be gained in communication systems operated in noise fields when control of the signal and noise phase relations at the two ears is possible. This topic is discussed more thoroughly in chapter 4 of this report.

The conclusions described above are for the binaural presentation of the signal and the noise. Also of interest are the conditions under which the noise is presented to one ear or to both ears. In both of these cases, the signal is presented to only one ear. A startling finding, as shown in figure 3.22, is that adding noise to the ear opposite the ear receiving both the tone signal and the noise reduces the masking of the tone when the tone is at a sensation level of more than about 10 dB

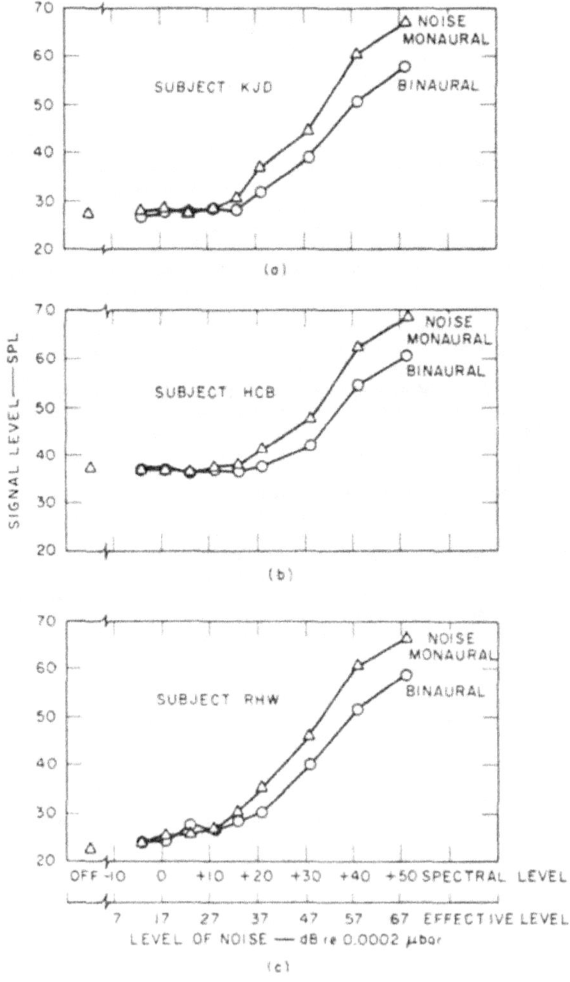

FIGURE 3.22. Masked thresholds (required level of signal to be audible in presence of noise) for one ear versus noise levels in that ear or in both ears. (From ref. 73).

above its threshold in the quiet (from ref. 73). That is, the addition of noise at the opposite ear improves the detection of the signal. This phenomenon has been labeled masking level difference (MLD), the difference in the level of the tone at the detection threshold when being masked by noise in the same ear and when masked by noise in both ears.

These binaural phase effects are clearly caused by an analysis process going on in the central nervous system—a process that is also consciously recognized as a locus of the signal and the noise in so-called phenomenal space. Phenomenal space is the locus, on introspection, of the source of a sound. For example, when a recording of a musical instrument is played via earphones with the signal in phase at the two ears, the impression is that the sound is centered in the middle of the head. Changing the phase relations between the two ears tends to externalize the source and place it to the side of the head where the lower frequency components in the signal lead inphase.

Finally, under binaural listening, it should be mentioned that in addition to phase differences for a signal or signals at the two ears, intensity differences also make important contributions to the detection and localization of sound in space. As shown in reference 74, phase differences between tonal signals in the presence of noise improve detection for frequencies only below 2000 Hz, and pressure level differences between the two ears increase detection for all frequencies above about 500 Hz. Effectiveness is increased at the higher frequencies.

Localization (perhaps a better term is lateralization) of the source of impulsive sound with respect to the listener is, apparently, based on at least two cues. The first of these is the well-known precedence effect (first investigated by Wallach *et al.* (ref. 75) but known as the "Hass effect" in architectural acoustics, see ref. 76), where the position of the source of sound is ascribed to the side of the person, or ear, first receiving the sound. The second cue consists of phase and intensity differences between the sound at the two ears. The precedence and intensity cues were investigated by Freedman and Pfaff (ref. 77). They found that a 25- to 45-msec (depending on the experimental method used) temporal difference for a click at the two ears was equivalent to about a 1-dB difference in dichotic intensity with respect to lateralization of the source of the clicks.

References

1. Zwislocki, Jozef J.: Sound Analysis in the Ear: A History of Discoveries. American Sci., vol. 60, 1981, pp. 184-192.
2. Stevens, Stanley Smith; and Davis, Hallowell: Hearing—Its Psychology and Physiology. John Wiley & Sons, Inc., 1938.
3. Shower, E. G.; and Biddulph, R.: Differential Pitch Sensitivity of the Ear. J. Acoust. Soc. America, vol. III, suppl. no. I, pt. 2, July 1931, pp. 275-287.
4. Riesz, R. R.: Differential Intensity Sensitivity of the Ear for Pure Tones. Phys. Rev., second ser., vol. 31, no. 5, May 1928, pp. 867-875.

5. Fletcher, Harvey: Auditory Patterns. Rev. Mod. Phys., vol. 12, no. 1, Jan. 1940, pp. 47–65.
6. Schafer, T. H.; Gales, R. S.; Shewmaker, C. A.; and Thompson, P. O.: The Frequency Selectivity of the Ear as Determined by Masking Experiments. J. Acoust. Soc. America, vol. 22, no. 4, July 1950, pp. 490–496.
7. Hawkins, J. E., Jr.; and Stevens, S. S.: The Masking of Pure Tones and of Speech by White Noise. J. Acoust. Soc. America, vol. 22, no. 1, Jan. 1950, pp. 6–13.
8. Egan, James P.; and Hake, Harold W.: On the Masking Pattern of a Simple Auditory Stimulus. J. Acoust. Soc. America, vol. 22, no. 5, Sept. 1950, pp. 622–630.
9. Bilger, R. C.; and Hirsh, I. J.: Masking of Tones by Bands of Noise. J. Acoust. Soc. America, vol. 28, no. 4, July 1956, pp. 623–630.
10. Swets, John A.; Green, David M.; and Tanner, Wilson P., Jr.: On the Width of Critical Bands. J. Acoust. Soc. America, vol. 34, no. 1, Jan. 1962, pp. 108–113.
11. Kryter, Karl D.: The Effects of Noise on Man. Academic Press, Inc., 1970.
12. Licklider, J. C. R.; and Guttman, Newman: Masking of Speech by Line-Spectrum Interference. J. Acoust. Soc. America, vol. 29, no. 2, Feb. 1957, pp. 287–296.
13. Gässler, Von G.: Über Die Hörschwelle Für Schallereignisse Mit Verschieden Breitem Frequenzspektrum (On the Threshold of Hearing for Sounds With Different Spectrum Width). Acustica, vol. 4, Heft 1, 1954, pp. 456–462.
14. Zwicker, E.; Flottorp, G.; and Stevens, S. S.: Critical Band Width in Loudness Summation. J. Acoust. Soc. America, vol. 29, no. 5, May 1957, pp. 548–557.
15. Hamilton, P. M.: Noise Masked Thresholds as a Function of Tonal Duration and Masking Noise Band Width. J. Acoust. Soc. America, vol. 29, no. 4, Apr. 1957, pp. 506–511.
16. Greenwood, Donald D.: Auditory Masking and the Critical Band. J. Acoust. Soc. America, vol. 33, no. 4, Apr. 1961, pp. 484–502.
17. Scharf, B.: Complex Sounds and Critical Bands. Psychol. Bull., vol. 58, 1961, pp. 705–717.
18. De Boer, E.: Note on the Critical Bandwidth. J. Acoust. Soc. America, vol. 34, no. 7, July 1962, pp. 985–986.
19. Plomp, R.; and Levelt, W. J. M.: Tonal Consonance With Critical Bandwidth. J. Acoust. Soc. America, vol. 38, no. 4, Oct. 1965, pp. 548–560.
20. Green, David M.: Detection of Multiple Component Signals in Noise. J. Acoust. Soc. America, vol. 30, no. 10, Oct. 1958, pp. 904–911.
21. Von Békésy, Georg: Uber die Resonanzkurve und Die Abklingzeit der Verschiedenen Stellen der Schneckentrennwand (On the Resonance Curve and the Decay Period at Various Points on the Cochlear Partition). Acus-

tica, vol. 8, 1943, pp. 66-76. (Available in English translation J. Acoust. Soc. America, vol. 21, no. 3, May 1949, pp. 245-254.)

22. Von Békésy, Georg; and Rosenblith, Walter A.: The Mechanical Properties of the Ear. Handbook of Experimental Psychology. S. S. Stevens, ed., John Wiley & Sons, Inc., 1960, pp. 1075-1115.
23. Greenwood, Donald D.: Critical Bandwidth and the Frequency Coordinates of the Basilar Membrane. J. Acoust. Soc. America, vol. 33, no. 10, Oct. 1961, pp. 1344-1356.
24. Licklider, J. C. R.: Three Auditory Theories. Psychology: A Study of a Science, S. Koch, ed., McGraw-Hill Book Co., 1959.
25. Von Békésy, G.: Über akustische Reizung des Vestibularapparates. Pfluegers Arch. Gesamte Physiol., vol. 236, 1935, pp. 59-76.
26. Wiener, Francis M.: On the Diffraction of a Progressive Sound Wave by the Human Head. J. Acoust. Soc. America, vol. 19, no. 1, Jan. 1947, pp. 143-146.
27. Wersall, R.: The Tympanic Muscles and Their Reflexes. Acta Oto-Laryngol. Suppl. 139, 1958.
28. Ward, W. Dixon: Studies of the Aural Reflex. I. Contralateral Remote Masking as an Indicator of Reflex Activity. J. Acoust. Soc. America, vol. 33, no. 8, Aug. 1961, pp. 1034-1045.
29. Loeb, M.; and Riopelle, A. J.: Influence of Loud Contralateral Stimulation in the Threshold and Perceived Loudness of Low-Frequency Tones. J. Acoust. Soc. America, vol. 32, no. 5, May 1960, pp. 602-610.
30. Reger, S. N.; Menzel, O. J.; Ickes, W. K.; and Steiner, S. J.: Changes in Air Conduction and Bone Conduction Sensitivity Associated With Voluntary Contraction of Middle Ear Musculature. Seminar on Middle Ear Function, J. L. Fletcher, ed., Rep. 576, U.S. Army Medical Res. Lab., 1963, pp. 171-180.
31. Ward, W. Dixon: Studies on the Aural Reflex. II. Reduction of Temporary Threshold Shift From Intermittent Noise by Reflex Activity; Implications for Damage-Risk Criteria. J. Acoust. Soc. America, vol. 34, no. 2, Feb. 1962, pp. 234-241.
32. Fletcher, John L.; and Riopelle, Arthur, J.: Protective Effect of the Acoustic Reflex for Impulsive Noises. J. Acoust. Soc. America, vol. 32, no. 3, Mar. 1960, pp. 401-404.
33. Fletcher, J. L.: TTS Following Prolonged Exposure to Acoustic Reflex Eliciting Stimuli. J. Aud. Res., vol. 1, 1961, pp. 242-246.
34. Fleer, R.: Protection Afforded Against Impulsive Noise by Voluntary Contraction of the Middle Ear Muscles. Seminar on Middle Ear Function, J. L. Fletcher, ed., Rep. 576, U.S. Army Medical Res. Lab., 1963.
35. Ward, W. Dixon: Effect of Temporal Spacing on Temporary Threshold Shift From Impulses. J. Acoust. Soc. America, vol. 34, no. 9, pt. 1, Sept. 1962, pp. 1230-1232.

36. Jepsen, O.: Middle Ear Muscle Reflexes in Man. Modern Developments in Audiology, J. Jerger, ed., Academic Press, Inc., 1963.

37. Terkildsen, K.: The Intra-Aural Muscle Reflexes in Normal Persons and in Workers Exposed to Intense Industrial Noise. Acta Oto-Laryngol., vol. 52, 1960, pp. 384-396.

38. Hecker, Michael H. L.; and Kryter, Karl D.: A Study of Auditory Fatigue Caused by High-Intensity Acoustic Transients. Rep. No. 1158 (Contract No. DA-49-007-MD-985), Bolt Beranek and Newman, Inc., Oct. 15, 1964. (Available from DTIC as AD 450 707.)

39. Moller, A. R.: The Acoustic Middle Ear Muscle Reflex. Handbook of Sensory Physiology, Springer-Verlag, 1974.

40. Richards, Alan M.; and Goodman, Allan Cooper: Threshold of the Human Acoustic Stapedius Reflex for Short-Duration Burst of Noise. J. Aud. Res., vol. 13, no. 3, July 1977, pp. 183-189.

41. Sesterhenn, G.; and Breuninger, H.: On the Influence of the Middle Ear Muscles Upon Changes in Sound Transmission. Arch. Oto-Rhino-Laryngol., vol. 221, no. 1, Aug. 31, 1978, pp. 47-60.

42. Borg, Erik; and Ödman, Bengt: Decay and Recovery of the Acoustic Stapedius Reflex in Humans. Acta Oto-Laryngol., vol. 87, no. 5-6, May-June 1979, pp. 421-428.

43. Zakrisson, J. E.: The Effect of the Stapedius Reflex on Attenuation and Post-stimulatory Auditory Fatigue at Different Frequencies. Acta Oto-Laryngol., Suppl. 360, 1979, pp. 118-121.

44. Brask, Torben: The Noise Protection Effect of the Stapedius Reflex. Acta Oto-Laryngol., Suppl. 360, 1979, pp. 116-117.

45. Ehmer, Richard H.: Masking Patterns of Tones. J. Acoust. Soc. America, vol. 31, no. 8, Aug. 1959, pp. 1115-1120.

46. Small, Arnold M., Jr.: Pure-Tone Masking. J. Acoust. Soc. America, vol. 31, no. 12, Dec. 1959, pp. 1619-1625.

47. Carter, N. L.; and Kryter, K. D.: Masking of Pure Tones and Speech. J. Aud. Res., vol. 2, 1962, pp. 68-98.

48. Wegel, R. L.; and Lane, C. E.: The Auditory Masking of One Pure Tone by Another and Its Probable Relation to the Dynamics of the Inner Ear. Phys. Rev., first ser., vol. 23, no. 2, Feb. 1924, pp. 266-285.

49. Ehmer, Richard H.: Masking by Tones vs. Noise Bands. J. Acoust. Soc. America, vol. 31, no. 9, Sept. 1959, pp. 1253-1256.

50. Zwicker, E.: Über psychologische und methodische Grundlagen der Lautheit. Akust. Beih., Heft 1, 1958, pp. 237-258.

51. Saito, Shuzo; and Watanabe, Shingo: Normalized Representation of Noise-Band Masking and Its Application to the Prediction of Speech Intelligibility. J. Acoust. Soc. America, vol. 33, no. 8, Aug. 1961, pp. 1013-1021.

52. Finck, Alfred: Low-Frequency Pure Tone Masking. J. Acoust. Soc. America, vol. 33, no. 8, Aug. 1961, pp. 1140-1141.

53. Harris, Charles S.; and Shoenberger, Richard W.: Human Performance During Vibration. AMRL-TR-65-204, U.S. Air Force, Nov. 1965. (Available from DTIC as AD 624 196.)
54. Small, Arnold M., Jr.; Bacon, W. Edward; and Fozard, James L.: Intensive Differential Thresholds for Octave-Band Noise. J. Acoust. Soc. America, vol. 31, no. 4, Apr. 1959, pp. 508-510.
55. Hellman, Rhona P.; and Zwislocki, J.: Loudness Function of a 1000-cps Tone in the Presence of a Masking Noise. J. Acoust. Soc. America, vol. 36, no. 9, Sept. 1964, pp. 1618-1627.
56. Corso, J. F.: Historical Note on the Thermal Masking Noise and Pure Tone Pitch Changes. J. Acoust. Soc. America, vol. 26, no. 6, Nov. 1954, p. 1078.
57. Egan, James P.; and Meyer, Donald R.: Changes in Pitch of Tones of Low Frequency as a Function of the Pattern of Excitation Produced by a Band of Noise. J. Acoust. Soc. America, vol. 22, no. 6, Nov. 1950, pp. 827-833.
58. Webster, J. C.; Miller, P. H.; Thompson, P. O.; and Davenport, E. W.: The Masking and Pitch Shifts of Pure Tones Near Abrupt Changes in a Thermal Noise Spectrum. J. Acoust. Soc. America, vol. 24, no. 2, Mar. 1952, pp. 147-152.
59. Spieth, Walter: Downward Spread of Masking. J. Acoust. Soc. America, vol. 29, no. 4, Apr. 1957, pp. 502-505.
60. Bilger, Robert C.: Remote Masking in the Absence of Intra-aural Muscles. J. Acoust. Soc. America, vol. 39, no. 1, Jan. 1966, pp. 103-108.
61. Samojlova, I. K.: The Masking Effect of Short Signals as a Function of the Time Between the Masked and Masking Sound. Biofizika, vol. 4, 1959, pp. 550-558.
62. Pickett, J. M.: Backward Masking. J. Acoust. Soc. America, vol. 31, no. 12, Dec. 1959, pp. 1613-1615.
63. Chistovich, L. A.; and Ivanova, V. A.: Backward Masking by Short Sound Pulses. Biofizika, vol. 4, 1959, pp. 170-180.
64. Elliott, Lois L.: Backward Masking: Monotic and Dichotic Conditions. J. Acoust. Soc. America, vol. 34, no. 8, Aug. 1962, pp. 1108-1115.
65. Wright, H. N.: Temporal Summation and Backward Masking. J. Acoust. Soc. America, vol. 36, no. 5, May 1964, pp. 927-932.
66. Zwislocki, J.: Theory of Temporal Auditory Summation. J. Acoust. Soc. America, vol. 32, no. 8, Aug. 1960, pp. 1046-1060.
67. Hirsh, Ira J.: Auditory Perception of Temporal Order. J. Acoust. Soc. America, vol. 31, no. 6, June 1959, pp. 759-767.
68. Jeffress, Lloyd A.: Masking and Binaural Phenomena. DRL-A-245 (Contract Nos. NObsr-72627, NE 051247-6, NE 051456-4, Nonr-3579(04), NR 142-190, and Fund Transfer R-129), Def. Res. Lab., Univ. of Texas, June 14, 1965. (Available as NASA CR-64362.)
69. Pollack, Irwin: Monaural and Binaural Threshold Sensitivity for Tones and for White Noise. J. Acoust. Soc. America, vol. 20, no. 1, Jan. 1948, pp. 52-57.

70. Licklider, J. C. R.: The Influence of Interaural Phase Relations Upon the Masking of Speech by White Noise. J. Acoust. Soc. America, vol. 20, no. 2, Mar. 1948, pp. 150–159.
71. Hirsh, Ira J.: The Influence of Interaural Phase on Interaural Summation and Inhibition. J. Acoust. Soc. America, vol. 20, no. 4, July 1948, pp. 536–544.
72. Jeffress, Lloyd A.; Blodgett, Hugh C.; and Deatherage, Bruce H.: The Masking of Tones by White Noise as a Function of the Interaural Phases of Both Components. I. 500 Cycles. J. Acoust. Soc. America, vol. 24, no. 5, Sept. 1952, pp. 523–527.
73. Blodgett, Hugh C.; Jeffress, Lloyd A.; and Whitworth, Randolph H.: Effect of Noise at One Ear on the Masked Threshold for Tone at the Other. J. Acoust. Soc. America, vol. 34, no. 7, July 1962, pp. 979–981.
74. Gardner, Mark B.: Binaural Detection of Single-Frequency Signals in the Presence of Noise. J. Acoust. Soc. America, vol. 34, no. 12, Dec. 1962, pp. 1824–1830.
75. Wallach, Hans; Newman, Edwin B.; and Rosenzweig, Mark R.: The Precedence Effect in Sound Localization. American J. Psychol., vol. 62, no. 3, July 1949, pp. 315–335.
76. Gardner, Mark B.: Historical Background of the Haas and/or Precedence Effect. J. Acoust. Soc. America, vol. 43, no. 6, June 1968, pp. 1243–1248.
77. Freedman, S. J.; and Pfaff, D. W.: Trading Relations Between Dichotic Time and Intensity Differences in Auditory Localization. J. Aud. Res., vol. 2, 1962, pp. 311–317.

Chapter 4
Speech Communications in Noise

Introduction

The most common complaint about noise is that it interferes with or masks speech signals. Indeed, the masking of speech by noise greatly reduces the performance of work that involves speech communications and is a cause of accidents in industry, in the office, and in the home. The masking of speech by noise appears to be the most harmful effect noise has upon people from the point of view of either practical, economic consequences or emotional reactions. Before presenting information on the effects of noise on speech communications, including the masking of speech, it is in order to first describe the general physical characteristics of speech and the methods used to measure speech masking.

The Speech Signal

Some physical characteristics of the speech signal

Considerable insight on the reception of speech in noise is gained from an examination of the speech spectra shown in figure 4.1 (from ref. 1). The root-mean-square (rms) level, measured over 1/8-sec intervals of the acoustic speech wave for speech uttered at a constant level of effort encompasses a range of nearly 30 dB.

Von Tarnóczy (ref. 2) measured the spectrum level of speech for six European languages and found only small variations among the different languages. The similarity between Von Tarnóczy's spectra and the long-term idealized spectrum shown in figure 4.1 indicates that the so-called articulation index, developed by French and Steinberg (ref. 3) and discussed later, is usable with most, if not all, European languages and possibly others. In figure 4.2 (ref. 4)

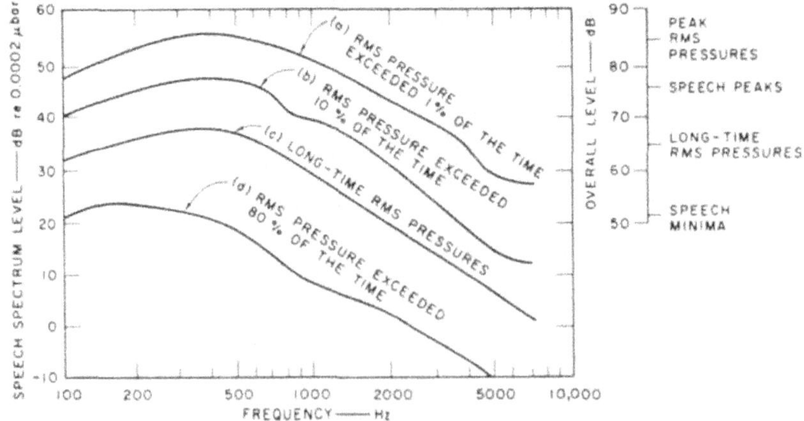

FIGURE 4.1. Spectrum level of male speech measured over 1/8-sec intervals. (From ref. 1.)

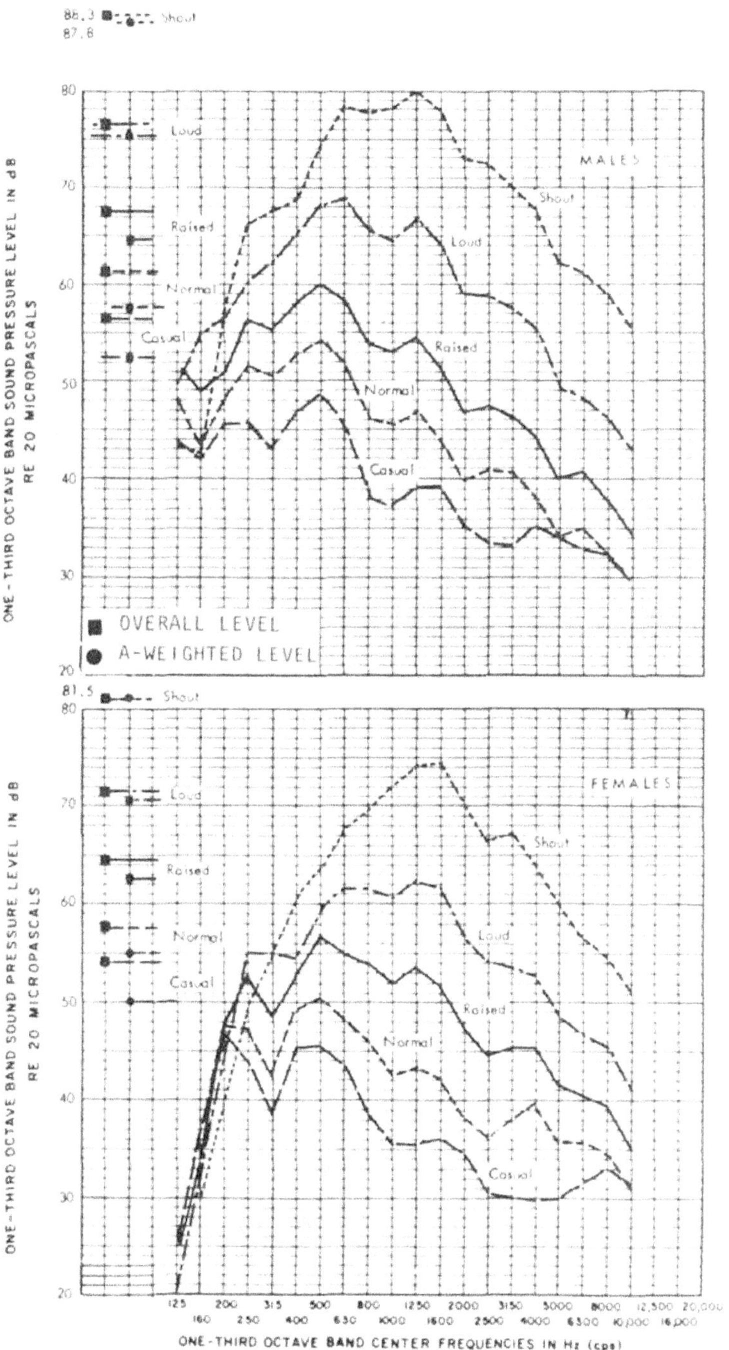

FIGURE 4.2. Average speech spectra for five vocal efforts. (From ref. 4.)

are shown the 1/3-octave-band levels of speech measured 1 m from male and female talkers using various vocal efforts.

Speech levels

Relating the speech levels as measured in the laboratory to more typical sound level meter (SLM) readings of speech in the field is a rather complex procedure. In the first place, the laboratory data are usually in terms of so-called L_{eq} (the long-term energy, unweighted in frequency, in 1 or 2 minutes of continuous speech, integrated and averaged to 1 second). Secondly, speech levels, being so variable from moment to moment (see fig. 4.1), are difficult to read on an SLM except when the meter is set on "slow" action.

Taking the arithmetic average of the peak levels reached by each word in typical sentences uttered at a conversational level of effort and measured on a sound level meter set on "slow" and A-weighted gives a level about equal to the true L_{eq} in decibels, unweighted. Accordingly, in discussions of speech level in dBA herein, reference is being made to the speech measured in decibels on a sound level meter set on A-weighting and slow meter or from actual frequency unweighted L_{eq} measures, on the assumption that the two values will be approximately the same. That this is a reasonable approximation, at least in so far as A-weighting is concerned and for speech in the normal range of intensity, is seen in figure 4.3 from Pearsons *et al.* (ref. 4).

Male and female speech levels

Table 4.1 shows the means and standard deviations (σ) of the distribution of speech levels at various vocal efforts by males, females, and children (under 13 years) when speaking in the quiet. The speech level of males is 2 to 3 dB greater than that of females at casual to raised levels of effort and 5 to 7 dB greater at higher levels of effort.

Noise and vocal effort

Most measures of variations in the intensity of speech have been taken under quiet, laboratory conditions, as were those shown in figures 4.2 and 4.3 and table 4.1. In such studies, the talkers were instructed to use different vocal efforts at different times. However, noise in the talker's environment will cause the talker involuntarily to increase somewhat his vocal effort (Korn (ref. 5), Webster and Klumpp (ref. 6), Kryter (ref. 7), Pickett (ref. 8), and Gardner (ref. 9)).

The investigations of Pearsons *et al.* (ref. 4) on this matter are particularly important because some of their measured speech levels were obtained in a variety of real-life circumstances. Some of their findings are shown in figure 4.4 (homes and schools) and figure 4.5 (television and various other environments). Figure 4.6 summarizes these data and also shows calculated and estimated regression lines to the data when segmented into three sections.

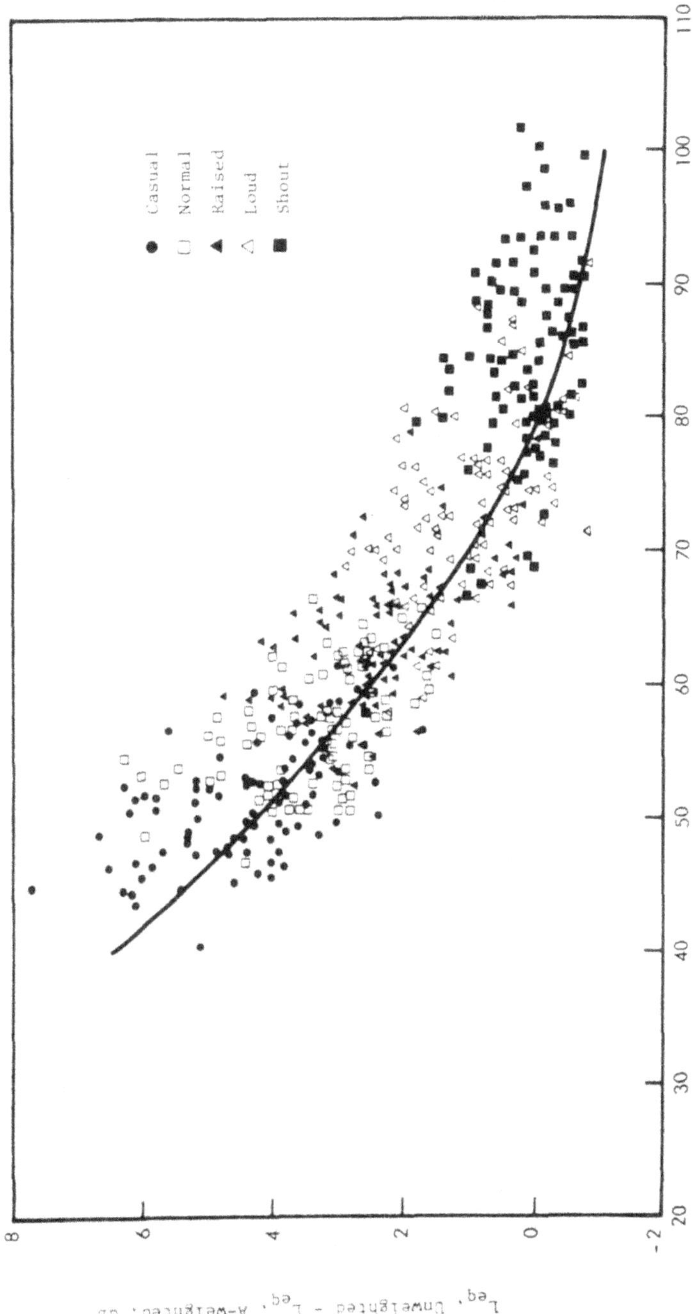

FIGURE 4.3. Differences between unweighted and A-weighted L_{eq} of speech. The A-weighted L_{eq} is calculated, not measured. (From ref. 4.)

TABLE 4.1

Mean Speech Levels at Various Vocal Efforts Measured in an Anechoic Chamber

[Background level L_{eq} = 16 dB; from ref. 4]

Vocal effort	Speech level, dB (a)						
	Males		Females		Children		Average
	L_{eq}	σ	L_{eq}	σ	L_{eq}	σ	L_{eq}
Casual	52.0	4.0	50.0	4.0	53.0	5.0	52.0
Normal	58.0	4.0	55.0	4.0	58.0	5.0	57.0
Raised	65.0	5.0	63.0	4.0	65.0	7.0	64.0
Loud	76.0	6.0	71.0	6.0	74.0	9.0	73.0
Shout	89.0	7.0	82.0	7.0	82.0	9.0	85.0

[a]*Results were rounded to the nearest decibel.*

It appears from figure 4.6 that for typical speech communication purposes involving sentences, a person will raise his voice about 0.5 dB (from a speech level of about 55 dBA) for each 1-dB increase in background noise. Background noise levels began at 45 to 50 dBA and were increased up to about 75 dBA. The speech level 1 m from the talker in 75-dBA noise usually is about 65 to 70 dBA.

Table 4.2 is a statistical summary of the data contained in figures 4.5 and 4.6. It is important to note in table 4.2 that in trains and aircraft, the speech levels at the listeners' ears were about equal to the background noise level, but in the other situations except the classroom, the speech levels were maintained somewhat above the background noise. In the classrooms, however, the teachers spoke at a level of effort comparable to that used in aircraft even though the background noise was low; probably because the classrooms involved were large (about 20 × 35 ft), and a higher than normal level of effort was required to have the speech reach all the students at a level sufficient to be properly heard. However, because of reverberation, room acoustics tend to keep the speech level from declining at distances farther than about 12 ft from the talker. (The approximate effect is shown later.)

Effect of noise on duration of conversations

It seems clear that over a certain range people will increase their vocal efforts in order to overcome, at least to some extent, the masking effects of noise. A practical and interesting question is that of determining how long people can converse in noise without feeling vocal strain.

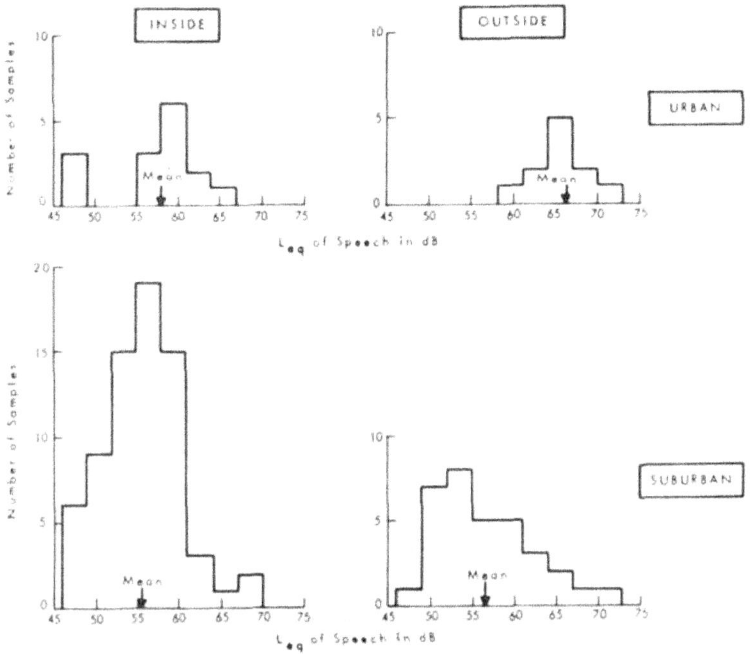

(a) Homes. Measured about 1 m from talker.

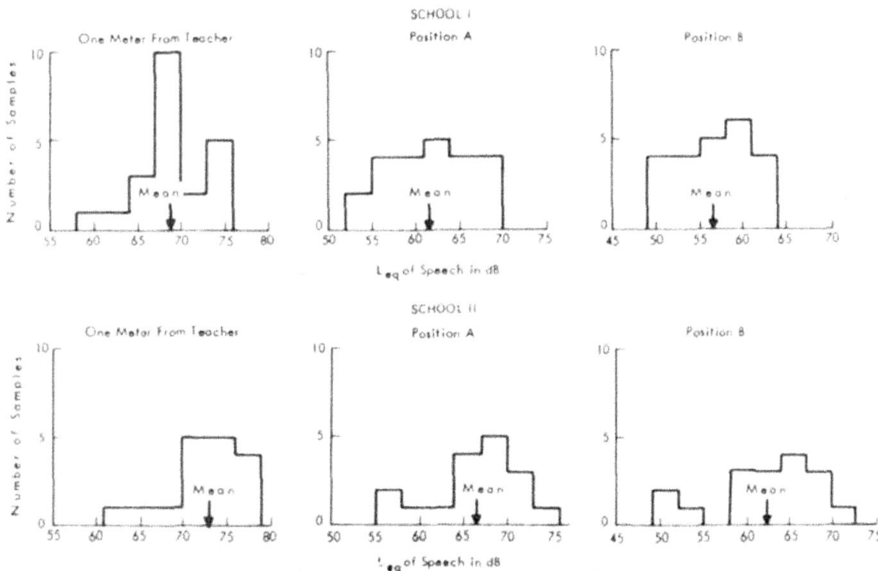

(b) Schools. Speech levels produced by teachers; positions A and B were about 2 m and 7 m, respectively, from the teacher.

FIGURE 4.4. Distribution of speech levels in homes and schools. (From ref. 4.)

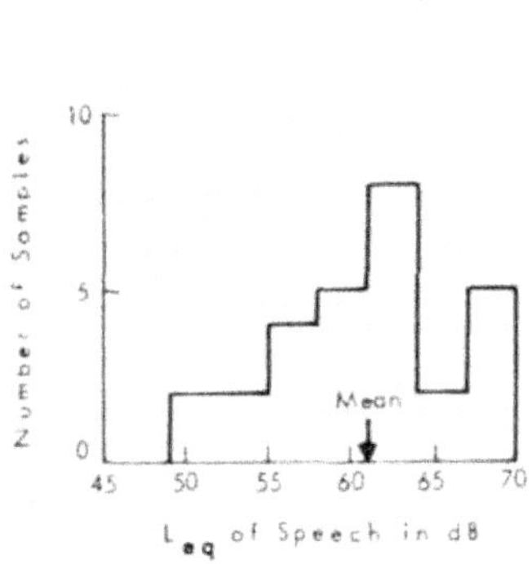

(a) Television speech levels. Measured about 3 m from TV.

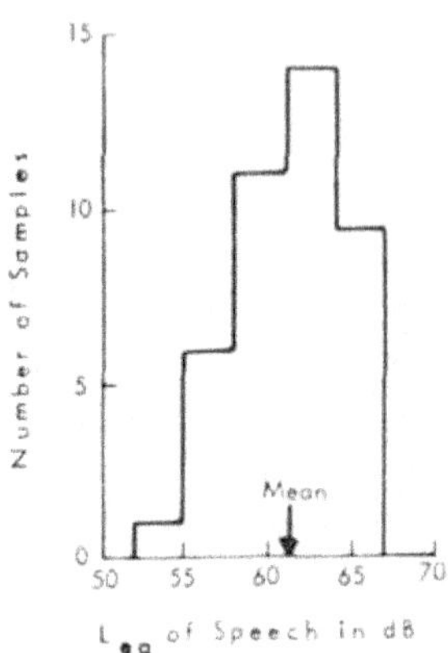

(b) Department stores. Measured about 1 m from talker.

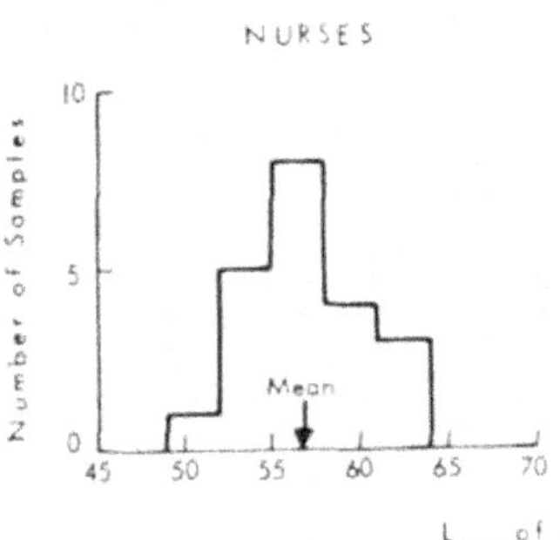

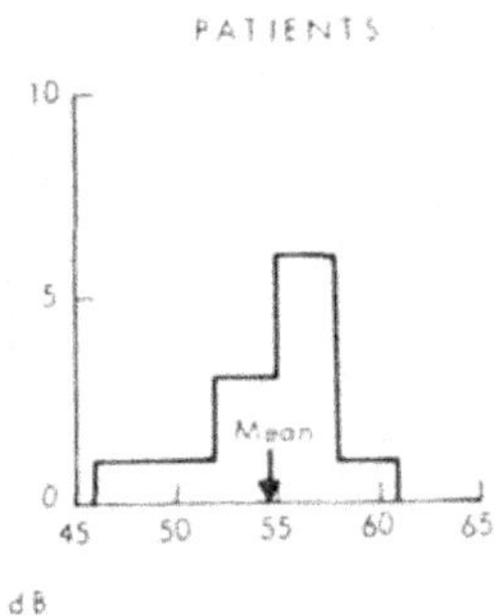

(c) Hospitals. Measured about 1 m from talker.

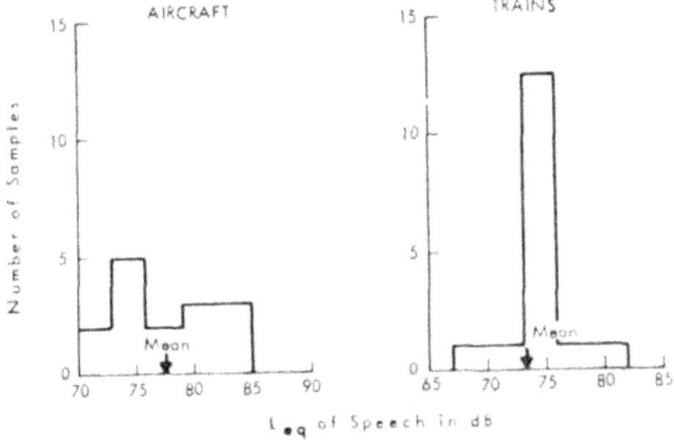

(d) Transportation vehicles. Measured about 0.4 m from talker.

FIGURE 4.5. Speech levels in various environments. (From ref. 4.)

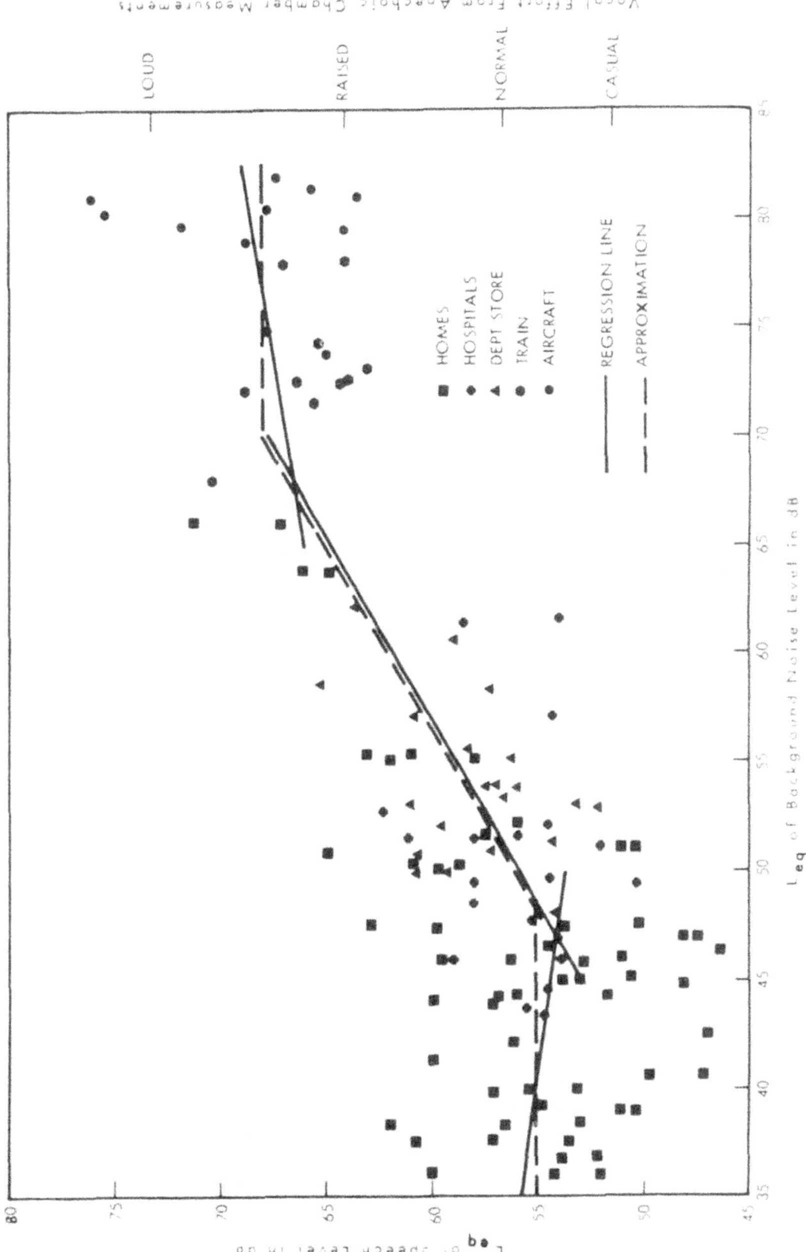

FIGURE 4.6. Conversing speech levels in several environments normalized to 1 m as a function of background noise. (From ref. 4.)

TABLE 4.2

Average Speech Levels in Various Environments

[Data from ref. 4]

Environment	Background level, dB		Distance from talker, m	Speech level, dB (a)	
	L_{eq}	σ		L_{eq}	σ
Schools					
I	[b]48.0	2.0	1	69.0	4.0
	↓	↓	2	62.0	5.0
			7	57.0	4.0
II	[b]51.0	3.0	1	73.0	4.0
			2	66.0	5.0
	↓	↓	7	62.0	6.0
Homes					
Outside, urban	61.0	5.0	≈ 1	66.0	4.0
	61.0	5.0	Corrected to 1	65.0	4.0
Outside, suburban	48.0	4.0	≈ 1	56.0	5.0
	48.0	4.0	Corrected to 1	55.0	5.0
Inside, urban	48.0	2.0	≈ 1	57.0	6.0
	48.0	2.0	Corrected to 1	57.0	6.0
Inside, suburban	41.0	3.0	≈ 1	55.0	5.0
	41.0	3.0	Corrected to 1	55.0	5.0
Hospitals					
Nurses	52.0	5.0	≈ 1	57.0	4.0
	52.0	5.0	Corrected to 1	56.0	3.0
Patients	45.0	2.0	≈ 1	55.0	1.0
	45.0	2.0	Corrected to 1	56.0	2.0
Dept. Stores	54.0	4.0	≈ 1	61.0	3.0
	54.0	4.0	Corrected to 1	58.0	3.0
Transportation vehicles					
Trains	74.0	3.0	≈ 0.4	73.0	3.0
	74.0	3.0	Corrected to 1	66.0	2.0
Aircraft	79.0	3.0	≈ 0.3	77.0	4.0
	79.0	3.0	Corrected to 1	68.0	4.0

[a]*Results were rounded to the nearest decibel.*

[b]*Measurements were made with typical student activity. Background values of classrooms during the phonetically balanced word test and other "quiet periods" were 47 dB for School I and 43 dB for School II.*

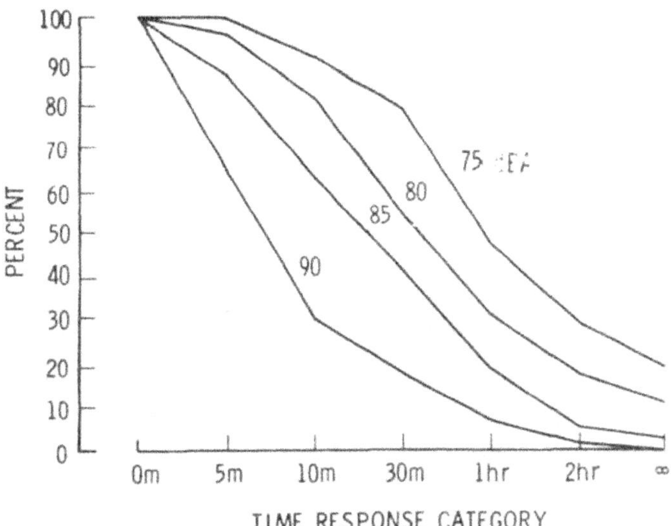

FIGURE 4.7. Cumulative distributions of subject estimates of how long an airplane trip could last without undue vocal strain from conversation in different levels of aircraft background noise. (From ref. 10.)

Rupf (ref. 10) studied this question in a simulated small aircraft situation (two persons seated side by side) with aircraft noise present for different tests and at different levels. The subjects were instructed to engage in conversation during a series of 5-minute segments of noise. Among other things, they were asked to assume the noises would be present during a trip (with continuous discourse not explicitly stated) and to estimate how long the airplane trip could last without undue strain on their voices. The results are shown in figure 4.7. Fifty percent of the subjects estimated that if the trip lasted more than about 1 hour and the noise level was about 75 dBA, they would feel undue vocal strain. However, other data collected from the subjects showed that with actual conversation, only 50 percent of the subjects considered a noise level of 75 dBA as being acceptable (without undue voice strain) for only 5 minutes with a distance of about 1 ft between the talker and listener.

Speech Intelligibility

Message set

In the discussion of the masking of speech by noise, masking effectiveness, unless otherwise specified, is in terms of the degradation in test scores of the understandability of speech in the presence of noise. These tests are variously called intelligibility or articulation tests; the distinction is usually made on the basis of how they are scored. If the sense of the meaning of the word, phrase, or sentence is of interest, the test is called an intelligibility test, whereas if communication performance is measured in terms of the individual phonemes or speech sounds in each word, the test is called an articulation test.

Miller *et al.* (ref. 11) demonstrated that the intelligibility or understandability of speech in noise is a strong function of the probability of occurrence of a given speech sound, word, or phrase. The larger the message set being used in a given communication system, the lower the probability of occurrence of a particular member of the message set, and the more susceptible is the communication process to interference from noise. As illustrated in figure 4.8, the understandability of the words is as much influenced by the message, or information, set size as the masking noise. The importance of this factor (size of message set) in speech communications is illustrated again later in this chapter.

Masking of speech by noise

As seen in figure 4.9, the masking effectiveness of different frequency bands of noise varies with signal-to-noise ratio. Clearly, the ratio of the long-term speech-to-noise sound pressure level alone is not an adequate indicator of the masking of speech by noise. We shall see later that by taking into joint account the nature of the speech spectrum, the critical bandwidth for speech, and direct (including upward and remote) masking, it is possible to make general statements about the direct masking of speech by noise and to predict fairly well the kinds of results shown in figure 4.9.

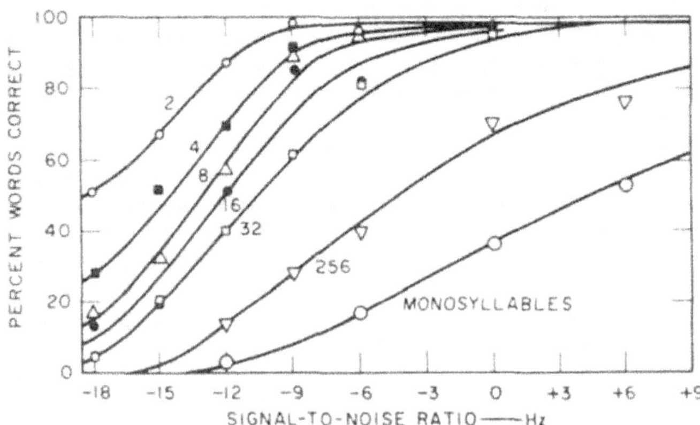

FIGURE 4.8. Word intelligibility scores obtained at various speech-to-noise ratios for test vocabularies containing different numbers of English monosyllables. The bottom curve was obtained with a vocabulary of approximately 1000 monosyllables. (From ref. 11.)

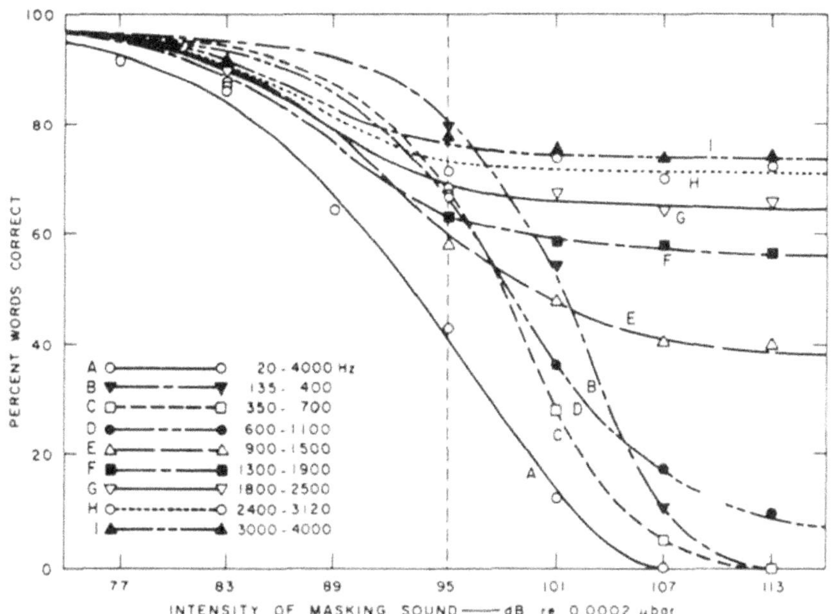

FIGURE 4.9. Percentage of words correct as a function of the intensity of narrow bands of masking noise. This speech was not filtered, and its level was held constant at 95 dB. (From ref. 11.)

The importance of considering the signal-to-noise ratio at a number of points along the frequency scale follows not only from the fact that different frequency regions of speech are somewhat more important than others to speech intelligibility, but also because the long-term speech spectrum is curved, and decreases at the rate of about 8 dB per octave above about 500 Hz. (See fig. 4.1.) The spectrum of the "instantaneous" peaks of speech is flatter than the rms pressure spectrum; however, it appears that the rms pressure spectrum is the effective spectrum with respect to the understanding of speech. For this reason, the lower speech frequencies are the last to be masked by noises whose spectra fall off less steeply than the speech spectrum as the signal-to-noise ratio is decreased. French and Steinberg (ref. 3) demonstrated this fact by progressively reducing the level of filtered speech until it was made inaudible (was masked) by the threshold of hearing, and Webster and Klumpp (refs. 12 and 13) found a similar pattern by measuring speech intelligibility under a variety of noise spectra and levels.

Licklider and Guttman (ref. 14) did a study of the masking of speech by pure tones and continuous spectra to show how best, for a given amount of noise power, one could mask or reduce the intelligibility of speech. They varied the number and relative amplitude of the masking components. The density of spac-

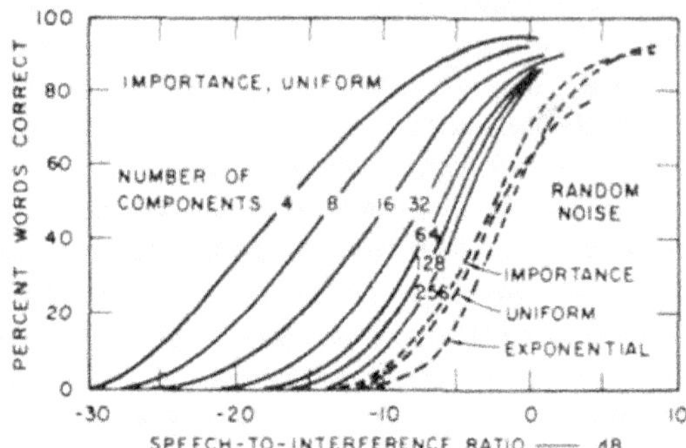

FIGURE 4.10. Masking of unfiltered speech by line-spectrum interference (solid curves) and by continuous-spectrum interference (dashed curves) in regular articulation tests. The density of spacing of the line components was governed by the importance function, and the lines were uniform in amplitude. The continuous spectra were shaped by filters. (From ref. 14.)

ing between components was varied in accordance with the critical bandwidth function of the ear over the range from about 200 to 6100 Hz, called the "importance function." (The critical bandwidth and the width of bands equally important to intelligibility are proportional to each other, see fig. 3.2.) Also these investigators masked the speech by means of random noise, with the amplitude as a function of frequency either uniform, negatively exponential, or proportional to the critical bandwidth of the ear. It appears from the results shown in figure 4.10 that:

1. Two hundred fifty-six components that are separated according to the relative importance function are 2 or 3 dB less efficient in terms of sound power required for masking, than a continuous spectrum random noise in the 200 to 6100 Hz band, which has a spectrum that declines as a function of frequency at the rate of 3 dB per octave (so called pink noise, or negative exponential).

2. The masking effectiveness of most components and of noises indicates that the upward spread of masking and remote masking, as found with pure tones, contributes significantly to speech masking. This effect was not found for components containing only a small number of tones.

A similar finding—the importance of the upward spread of masking—is demonstrated by the speech interference effects of pure tones of 50, 100, or 200 Hz. (See fig. 4.11, from ref. 15.) It is also seen in fig. 4.11 that pure tones of equal sensation level cause more speech masking than do tones of equal sound pressure level.

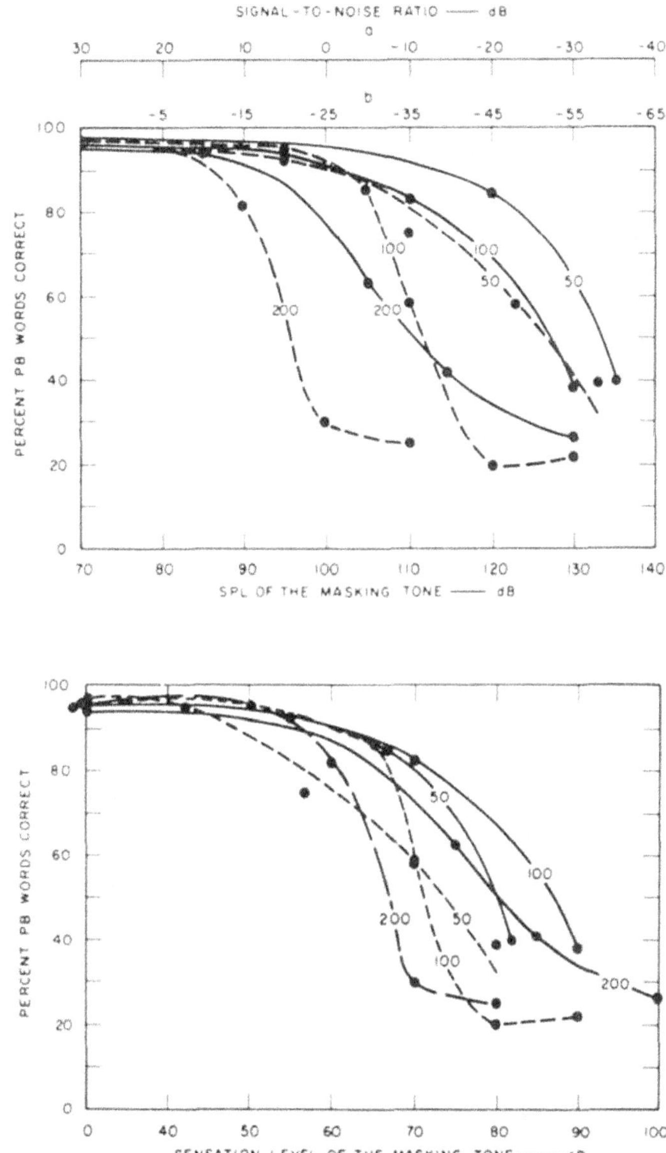

FIGURE 4.11. Percent PB words correct as a function of sound pressure level (upper graph) and sensation level of the masking tone (lower graph). Solid curves are for speech at an SPL of 100 dB (upper abscissa [a]). Dashed curves are for speech at an SPL of 75 dB (upper abscissa [b]). Parameter is frequency in hertz of masking tone. (From ref. 15.)

Effects of vocal effort on intelligibility

The effect of noise level upon vocal effort and speech level was discussed above. It is of interest to know what the effect of vocal effort in talking might have upon the intelligibility of the speech, with signal-to-noise ratio held constant. Pickett (ref. 8) conducted research on this problem and found the results shown in figure 4.12. Clearly, speech uttered with very weak or very high levels of effort is not as intelligible as speech in the range from about 50 to 80 dB (measured at 1 m from the talker) even though the speech-to-noise ratio is kept constant.

Effects of speech intensity on intelligibility

In many noisy environments, the speech signal is often speech that has been spoken into a microphone at a normal or near normal level of effort and then amplified by electronic means to make it audible above the noise. Of course, here the question is how intense can the speech be before it loses intelligibility, presumably because it is distorted due to overloading of the ear. Pollack and Pickett (ref. 16) found that in the quiet (signal-to-noise ratio of 55 dB), there was no loss in intelligibility even at speech levels of 130 dB. When there was some noise present, however, speech above about 85 dB declined in intelligibility with signal-to-noise ratio kept constant. (See fig. 4.13.)

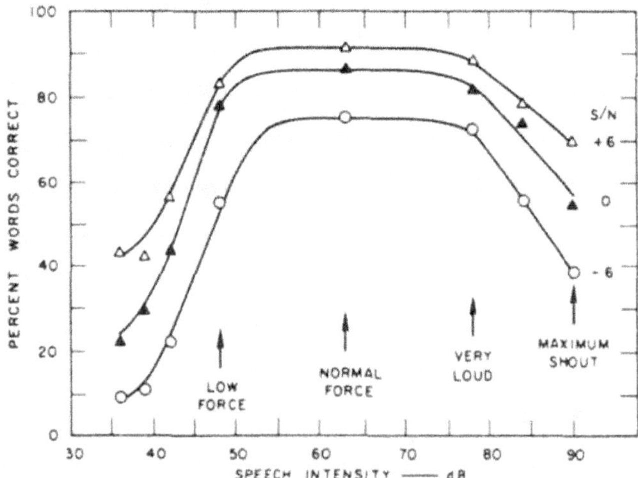

FIGURE 4.12. Relations between speech intelligibility in noise and vocal force. Noise, 70 dB, flat spectrum. The speech was uttered at these various levels, and then the level of microphone pick-up signal was adjusted prior to mixing with the constant level of 70-dB noise to provide the indicated S/N ratios to the listener. (From ref. 8.)

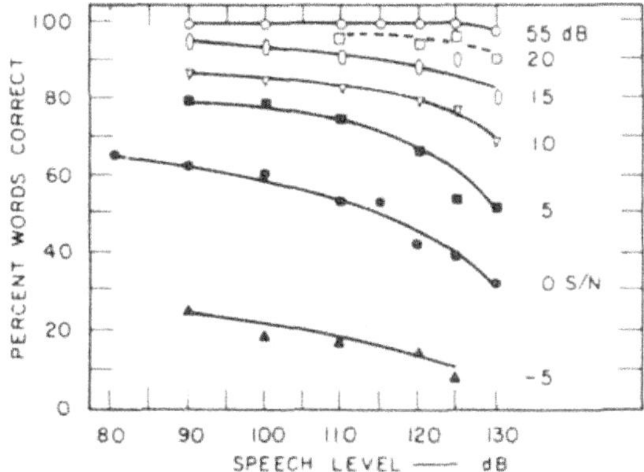

FIGURE 4.13. The deterioration of intelligibility at high levels. (From ref. 17.)

These data demonstrate the interesting fact that the intelligibility tests with nearly 100 percent of the words correct (the case of speech of 130 dB in the quiet) are somewhat insensitive indicators of the true fidelity or undistorted nature of a given speech signal. The potentially degrading effects of overloading the ear with very intense speech upon intelligibility are, no doubt, actually present when the speech is heard in the quiet, but can only be measured when the test scores have been lowered and made more sensitive by some additional stressful condition such as noise.

It was stated earlier that masking is usually not particularly affected by temporary auditory fatigue in the normal ear provided the signal-to-noise ratio remains constant, and the signal consists of pure tones. However, Pollack (ref. 17) found that the effective masking of speech did increase significantly during a 13-minute exposure to random broadband noise (flat, 100 to 5000 Hz) at levels above about 115 dB. This effect (see fig. 4.14) is presumably due to an inability of the fatigued ear to discriminate among the speech sounds as well as the unfatigued ear.

Interrupted noise

When the speech signal is masked, either partially or completely, by a burst of noise, its intelligibility changes in a rather complex manner, as shown in figure 4.15. These functions are explained by Miller and Licklider (ref. 18) as follows. At interruption rates of less than about 2 per sec (which, for a noise on-time of 50 percent would make the duration of each burst of noise at least 0.25 sec), whole words or syllables within a word tend to be masked. At interruption rates between about 2 and 30 per sec, the noise duration is so brief that the listener is able to hear a portion of each syllable or phoneme of the speech signal, and the amount of masking thereby tends to be reduced. When the interruption

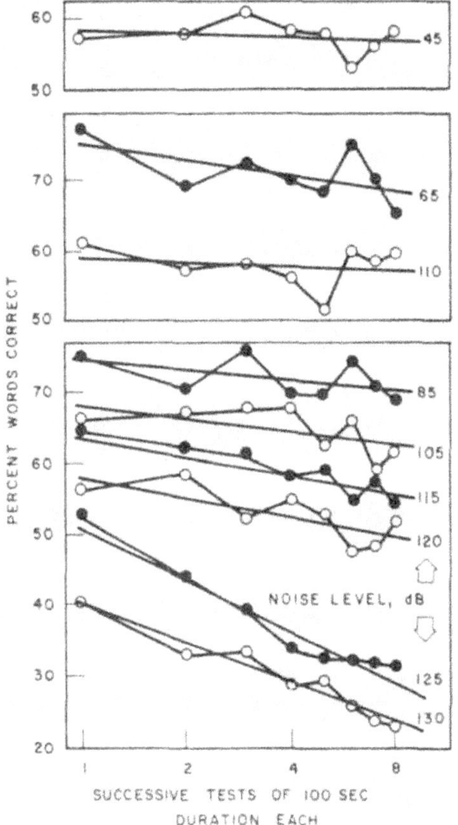

FIGURE 4.14. Speech intelligibility in noise as a function of continuous noise (and speech) exposure. The vertical axis has been broken to avoid overlap among conditions. S/N ratio is 0 dB. Each point represents the average of 500 determinations consisting of one 25-item test list read by each of four talkers to a testing crew of five listeners. (From ref. 17.)

rate is more frequent than 30 per sec, the spread of masking in time around the moment of occurrence of a burst of noise results in increased masking, until by 100 interruptions per sec there is effectively continuous masking. This is in good agreement with the temporal masking results shown in figure 3.21, where it was seen that appreciable masking occurs only for 5 to 10 msec before and after an intense 90-dB sound.

That the increased masking of speech is due to a spread in time, presumably both forward and backward, is demonstrated in figure 4.16, where the intelligibility of speech that is interrupted by turning it off and on in the quiet can be compared with that of speech that is turned off during noise bursts and on between noise bursts. The signal-to-noise ratio refers to the signal in the quiet

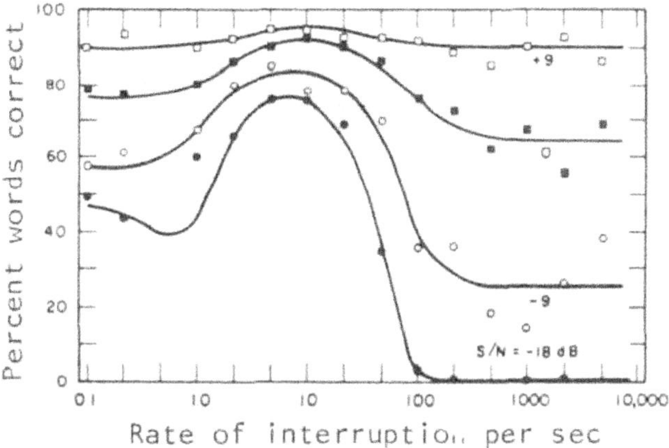

FIGURE 4.15. The masking of continuous speech by interrupted noise. Noise on-time is 50 percent. (From ref. 18.)

versus the noise alone. The temporal masking does not degrade the speech until the interruption rate exceeds 20 or 30 per sec.

Miller and Licklider (ref. 18) found that the above effects were the same for random or regularly spaced interruptions and that varying the speech on-time did not appreciably change the nature of the relationship between interruption rate and intelligibility. Pollack (ref. 19) found that, over relatively wide limits, varying the signal-to-noise ratio at rather slow rates provided intelligibility comparable to that observed with a steady-state signal-to-noise ratio.

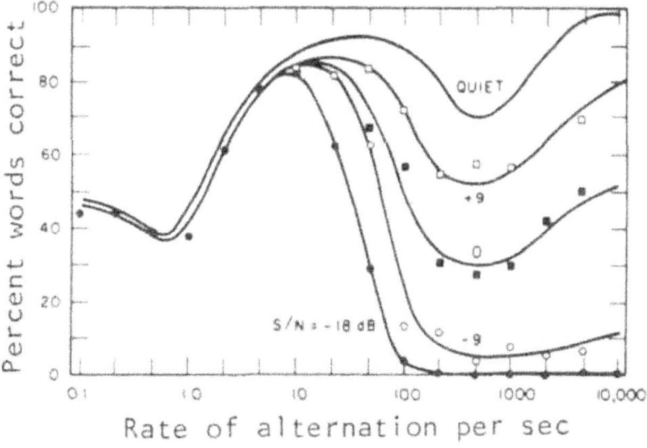

FIGURE 4.16. Word articulation as a function of the frequency of alternation between speech and noise, with signal-to-noise ratio in decibels as the parameter. (From ref. 18.)

One of the most common noises that masks speech is speech itself—the babble of other voices. Figure 4.17 (from ref. 20) shows how speech intelligibility is affected as a function of the number of competing voices. By the time eight voices are present, the "noise" spectrum is apparently continuously present.

Binaural factors in speech perception

As discussed previously, a person listening with both ears is able to some extent to separate, by some central nervous system mechanism, a signal from noise on the basis of relative phase and temporal relations of the signal and the noise reaching the two ears. If there are some temporal or frequency differences at the two ears between the signal and the noise, it appears that the listener may direct his attention to the sound he wishes to perceive without conscious regard to localization or phenomenal space. This is particularly noticeable when the noise consists of other speech signals—what has come to be called the "cocktail party" effect—and when the competing signals differ somewhat in spectra (refs. 21, 22, and 23).

A somewhat extreme situation for direct person-to-person communication, but one which demonstrates clearly the advantages of binaural listening as compared with monaural listening in the presence of masking sounds, was studied by Pollack and Pickett (ref. 22). They presented, *via* earphones, a speech signal in phase at the two ears against one background of speech presented to one ear and another background of speech to the other ear; some of the results are shown in figure 4.18. The control condition in figure 4.18 was achieved by merely disconnecting one of the listener's earphones. It is obvious that some of the direct masking of the speech that takes place with monaural listening is

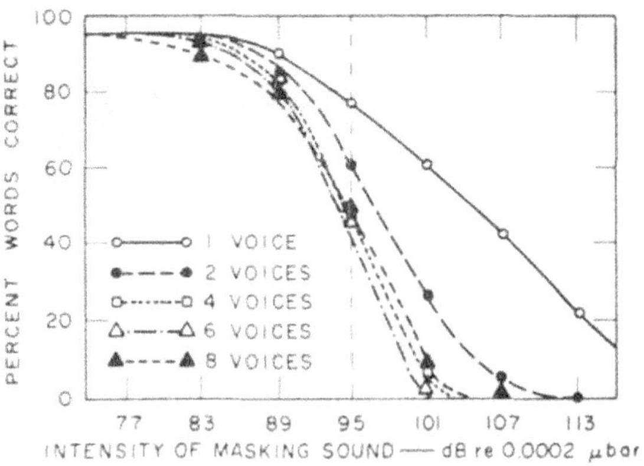

FIGURE 4.17. Word intelligibility as a function of intensity of different numbers of masking voices. Level of desired speech was held constant at 94 dB. (From ref. 20.)

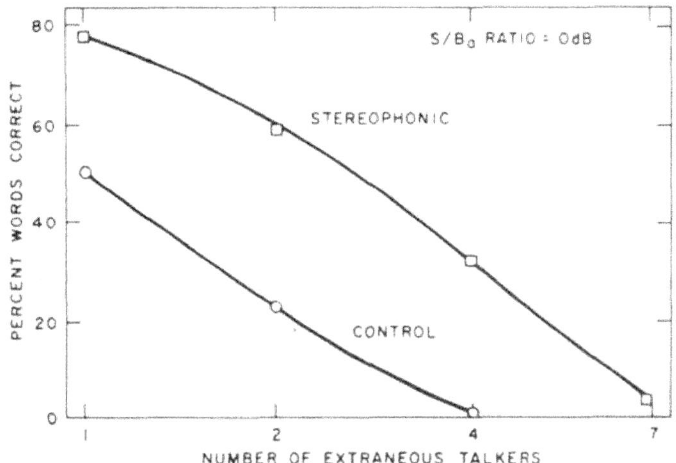

FIGURE 4.18. Comparison between average intelligibility scores of an adjusted speech-to-background noise for binaural-stereophonic listening condition and for monaural-control listening condition. (From ref. 22.)

TABLE 4.3

Monaural-Binaural Presentation and Interaural Phase Relations as Factors Influencing the Masking of Speech by White Noise[a]

[From ref. 21]

		Percent PB words correct				
		Binaural noise			Monaural noise	
		+	0	−	R	L
Binaural speech	+	18.0	27.4	35.4	98.0	99.0
	−	43.0	27.3	15.8	98.1	98.8
Monaural speech	R	30.3	13.2	20.1	16.6	98.7
	L	18.1	8.3	15.2	98.4	15.4

[a] *+-in phase; −-out of phase; 0-random phase; R-right ear; L-left ear.*

appreciably overcome on the basis of cues available with binaural-stereophonic listening.

Experiments conducted by Licklider (ref. 21) led to an understanding of binaural listening to speech in noise. Table 4.3 shows the effect on intelligibility of all combinations of monaural and binaural listening to speech and noise over earphones. It is seen in table 4.3 that for binaural noise and speech, speech intelligibility is highest when they are of opposite phase. For monaural noise and speech, speech intelligibility is at a minimum when the noise and speech are heard in the same ear. Weston *et al.* (ref. 23) demonstrated that localization cues available in free-field listening because of phase and intensity differences between the two ears were responsible for the increased intelligibility that occurred when noise and speech sources were separated in space.

Combating Noise Interference Effects

Maintaining satisfactory speech communications is obviously of great importance in many situations, especially in certain business and military operations. Over the years a number of techniques for overcoming the adverse effects of noise on speech communications have been developed.

Message set reduction

The importance of reducing alternatives (*i.e.*, reducing the size of the set of possible messages) to the intelligibility of speech was mentioned in the section on "Speech Intelligibility." This factor can be used to advantage in various ways in combating the masking of speech by noise.

A simple, and apparently the most beneficial, way of effecting this reduction of alternatives is to restrict the talkers to a limited number of specific words, phrases, or sentences that they can use when communicating by speech in a given situation. Such constraints have been found to be effective for many military operations. Moser (ref. 24) has contributed to the standardization of voice message procedures for the U.S. Air Force and international commercial aviation. Some benefit is, of course, also gained by prescribing the exact procedures—order of talking and how to talk. Standardizing the procedures and the messages to be used reduces the amount of information with which the listener must cope and thereby improves speech communications in noise (Pollack (ref. 25) and Frick and Sumby (ref. 26)).

Information redundancy

Another technique for overcoming the adverse effects of noise on speech communication is information redundancy; that is, having the talker repeat his words or messages. Thwing (ref. 27), for example, found that the intelligibility of single words increased by about 5 to 10 percent (equivalent to a reduction in the noise level of about 3 dB) when each word was repeated once. Further repetition caused little additional improvement.

Increasing the signal level

Increasing the level of the signal relative to that of the noise is the most effective way to avoid masking of speech. This may not be possible for a variety of reasons:

1. With direct person-to-person talking, the noise may be so intense (or the listener so far away) that the talker cannot effectively override it.

2. When a communication system such as a telephone or radio telephone is involved, the power available for amplification of the speech signal may be limited.

3. The masking noise may be mixed with the speech at the talker's microphone so that amplification of the signal increases the noise and leaves the signal-to-noise ratio and intelligibility relatively constant.

4. The masking noise may be so intense at the listener's ears that increasing the speech level by means of an electronic amplifier system is not practical because making the speech more intense would overload the ear and cause distortion and possible pain to the listener.

Methods of alleviating the masking effects of noise for each of the above-listed conditions have been investigated and will now be presented.

Megaphone

Pickett and Pollack (ref. 28) report that a small megaphone improved speech intelligibility relative to the unaided voice by an amount equivalent to a reduction of the noise level by 6.5 to 11.5 dB, depending on the noise spectrum. The least gain was found with a "flat" white noise; the greatest, with a noise having a −12 dB slope above 100 Hz.

Peak clipping

It is obvious from an examination of the amplitude waveform of a speech signal, as for example in figure 4.19 (from ref. 29) that those parts of the speech wave usually associated with a consonant are less intense than the parts present when vowels are uttered. Adding noise masks the consonant sounds at a lower level than that required to mask the vowels.

In order to have the consonants override the noise and yet not increase the peak power requirements of a transmission system, the level of the consonant sounds must be increased relative to that of the vowel sounds. This can be done simply by passing the speech through a so-called "peak clipper" and then re-amplifying the result to whatever peak power level is available. This process is illustrated in figure 4.19, where the peak-to-peak amplitude of the consonant "j" is made equal to the unclipped and clipped vowel "o" by peak clipping and amplifying the speech by 20 dB. Speech thusly clipped has a greater average speech power for a given peak power and is more intelligible in noise than is unclipped speech.

However, certain precautions must be kept in mind when peak clipping is to be used. First, speech peak clipped by more than about 6 dB sounds distorted and noisy due to the clipping of the vowel waveform when listened to in the quiet. (When heard in noise, peak-clipped speech sounds relatively undistorted because the distortion products from the speech signal tend to be masked by the noise.) Secondly, when there is noise mixed with the speech prior to peak clipping, the amount of clipping that is beneficial is limited. This latter fact is revealed through a comparison of the top and bottom graphs in figure 4.20 (from ref. 30).

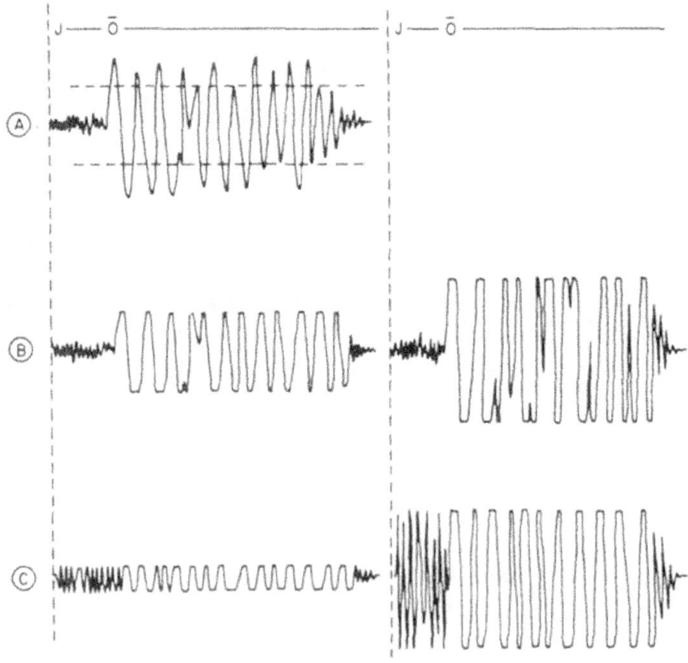

FIGURE 4.19. Schematic representations of word "Joe." A is undistorted; B is after 6-dB clipping; and C is after 20-dB clipping. Clipped signals in B and C are shown reamplified until their peak-to-peak amplitudes equal the peak-to-peak amplitude of A. (Data from ref. 29.)

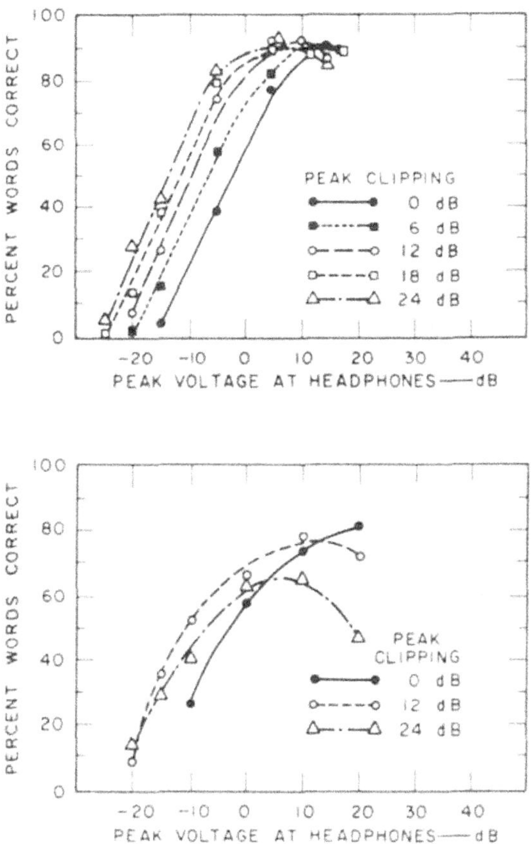

FIGURE 4.20. Results of intelligibility tests conducted with talkers in the quiet and listeners in ambient airplane noise (upper graph), and intelligibility tests conducted with both talkers and listeners in simulated airplane noise (lower graph). In the lower graph, note that when the microphone picks up noise, clipping is not so beneficial; it is even detrimental to speech intelligibility. A dynamic microphone (non-noise canceling) was used. (From ref. 30.)

Noise exclusion at the microphone

One way to keep noise out of a microphone is to attach the microphone directly to tissues of the throat and head so that it will not pick up airborne noise but will pick up the speech signal through the body tissues. Such microphones are reasonably effective in excluding noise when attached to the throat, ear, teeth, or forehead, but they tend to somewhat distort the speech signal. (See Moser *et al.* (ref. 31).)

Placing an air-activated microphone in a shield (usually a cup that forms a seal around the talker's mouth) will typically achieve noise exclusion, as shown in figure 4.21 (from ref. 32). A third method is to use a close-talking pressure gradient microphone. Here both surfaces of the active element of the microphone are exposed to air. Depending on frequency wavelength, random incidence sound waves (the noise) will impinge on both sides of the element more or less simultaneously. Thus, they tend to cancel each other; that is, the microphone element does not move. The speech signal, on the other hand, is highly directional when the microphone is held close to the lips and correctly oriented and, therefore, activates the moving element of the microphone. The amount of noise cancellation achieved is shown in figure 4.22 (from ref. 32) as a function of frequency.

Noise exclusion at the ear

Earplugs and muffs for over the ears have received considerable attention as a means of protecting the ear against auditory fatigue from exposure to intense noise. (See fig. 4.23 from refs. 33, 34, and 35.) It is to be noted in figure 4.23 that in practice worker-fitted earmuffs and especially earplugs give somewhat less protection (sound attenuation) than do these devices when carefully fitted by an experimenter-tester. Smoorenburg (ref. 36) and Edwards *et al.* (ref. 37) report about 13 to 34 dB (depending on pure-tone test frequency) less attenuation for the average or median military or industrial user of earplugs (self-fitted) when measured on the job than that afforded by the same devices when tested under laboratory conditions, as in the upper graph of figure 4.23.

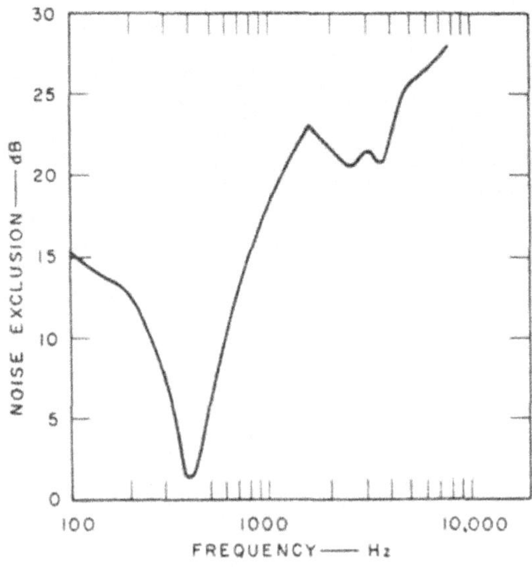

FIGURE 4.21. Noise exclusion with a noise shield. (From ref. 32.)

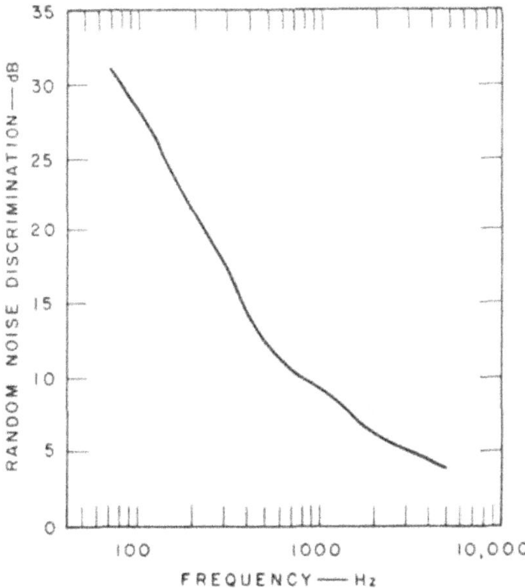

FIGURE 4.22. Random-noise discrimination of a noise-canceling microphone. (From ref. 32.)

However, ear protective devices interact in various fortunate and unfortunate ways with the reception of speech by users. In the first place, earplugs or muffs attenuate equally, at any one frequency, the speech signal and ambient noise passing through them. Since the signal-to-noise ratio at any one frequency remains constant at the listener's eardrum, speech intelligibility would be expected to be the same whether or not earplugs or muffs were worn. However, what happens is that: (a) in high-level noise, speech intelligibility is improved when earplugs or muffs are worn because the speech and noise are reduced to a level where the ear is not overloaded and, therefore, discriminates the speech from the noise somewhat better; and (b) in lower-level noise, on the other hand, speech intelligibility is decreased when earplugs or earmuffs are worn because the speech is reduced along with the noise to a level below the listener's threshold of hearing. (See fig. 4.24 from refs. 7 and 37.)

Thus, persons who are suffering some hearing loss will not benefit, in terms of speech communication, from wearing earplugs or earmuffs in a low noise level as much as will the person with normal hearing. (See Alberti *et al.* (ref. 35) and Rink (ref. 38).) The effect of earplugs and muffs upon speech communication (or any other signal detection) in noise can only be predicted from a knowledge of the hearing of the listener, the spectrum of the noise at the listener's ears, and the sound attenuation characteristics of the earplug or earmuff. Some of these interactions are illustrated in figure 4.25. Figure 4.25 shows that in moderate noise, persons with normal hearing and persons with some degree of hearing loss both

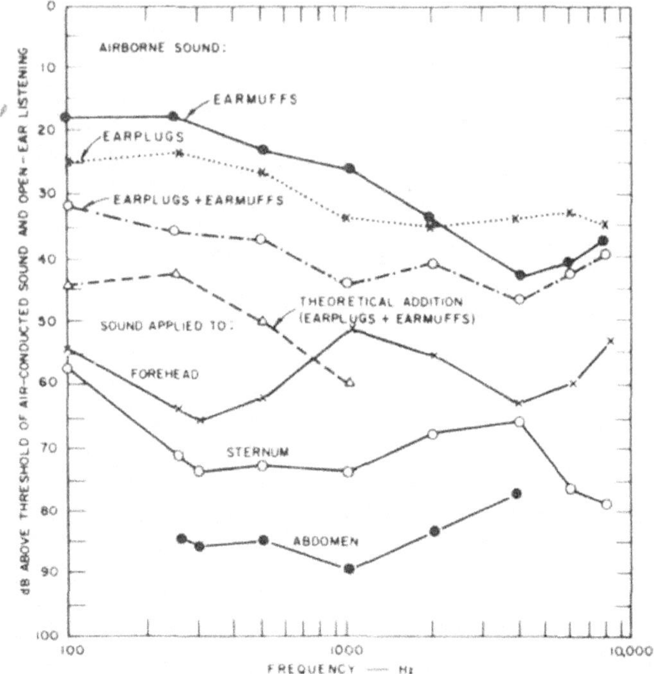

(a) The top four curves show measured attenuation curves for carefully fitted earplugs and earmuffs. (Data from ref. 33.) The bottom three curves show the audibility of sound when applied to different parts of the body and conducted to the ear through tissues of the body. (Data from ref. 34.)

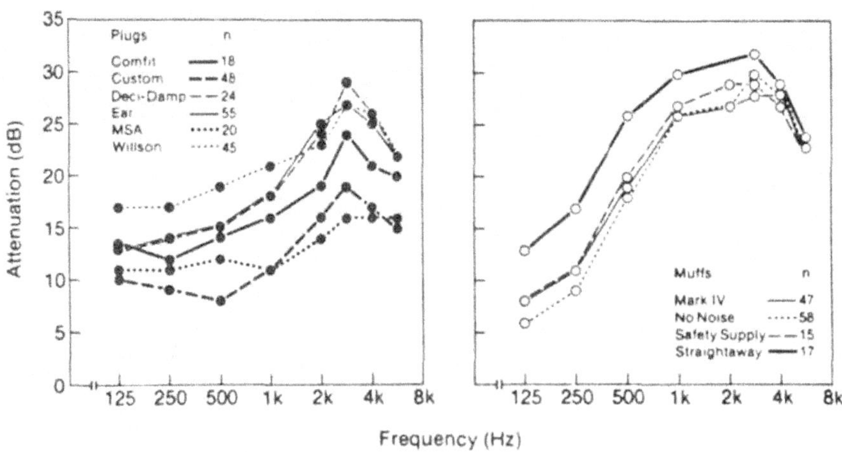

(b) Average attenuation for 10 types of earplugs and earmuffs when fitted by the worker with no adjustments by the tester. (From ref. 35.)

FIGURE 4.23. *Attenuation curves for earplugs and earmuffs.*

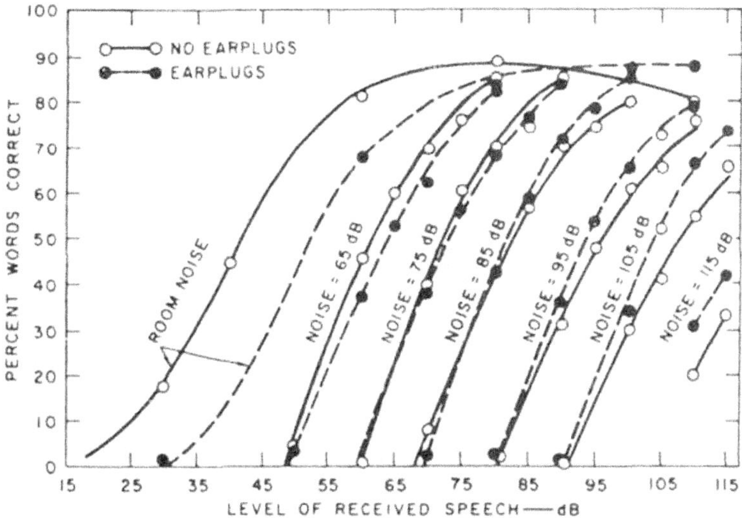

(a) Relationship between intelligibility and speech level with noise level as the parameter for listeners with normal hearing. (From ref. 7.)

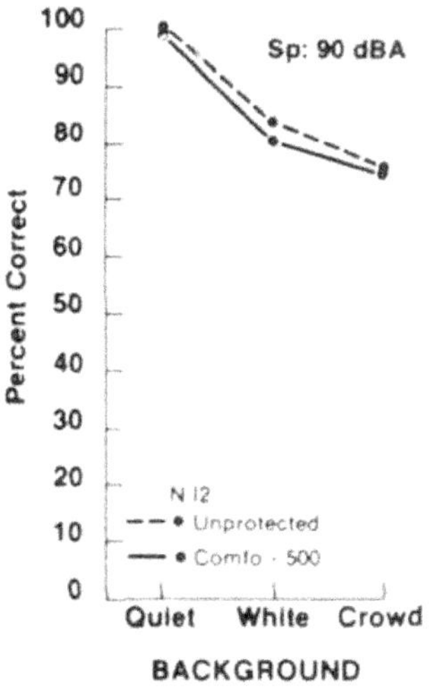

(b) Speech discrimination scores in quiet and with a background noise of 85 dB with and without hearing protectors. Normal hearing subjects, aged 30 to 35. (From ref. 37.)

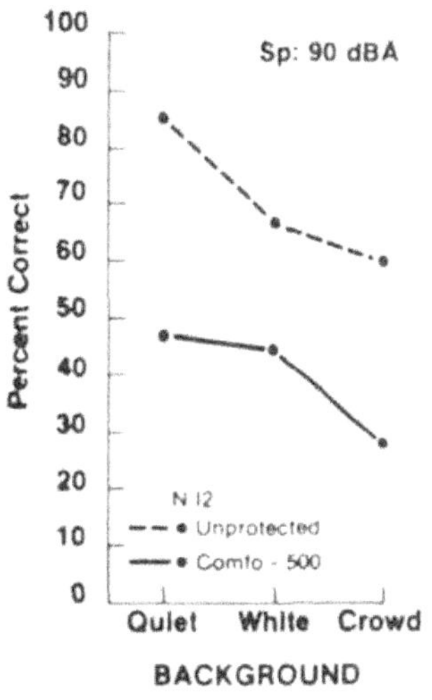

(c) Speech discrimination scores in quiet and with a background of white and crowd noise of 85 dB, with and without hearing protectors. Subjects with high-frequency hearing loss, aged 30 to 35. (From ref. 37.)

FIGURE 4.24. Effects of earplugs on the understanding of speech in noise.

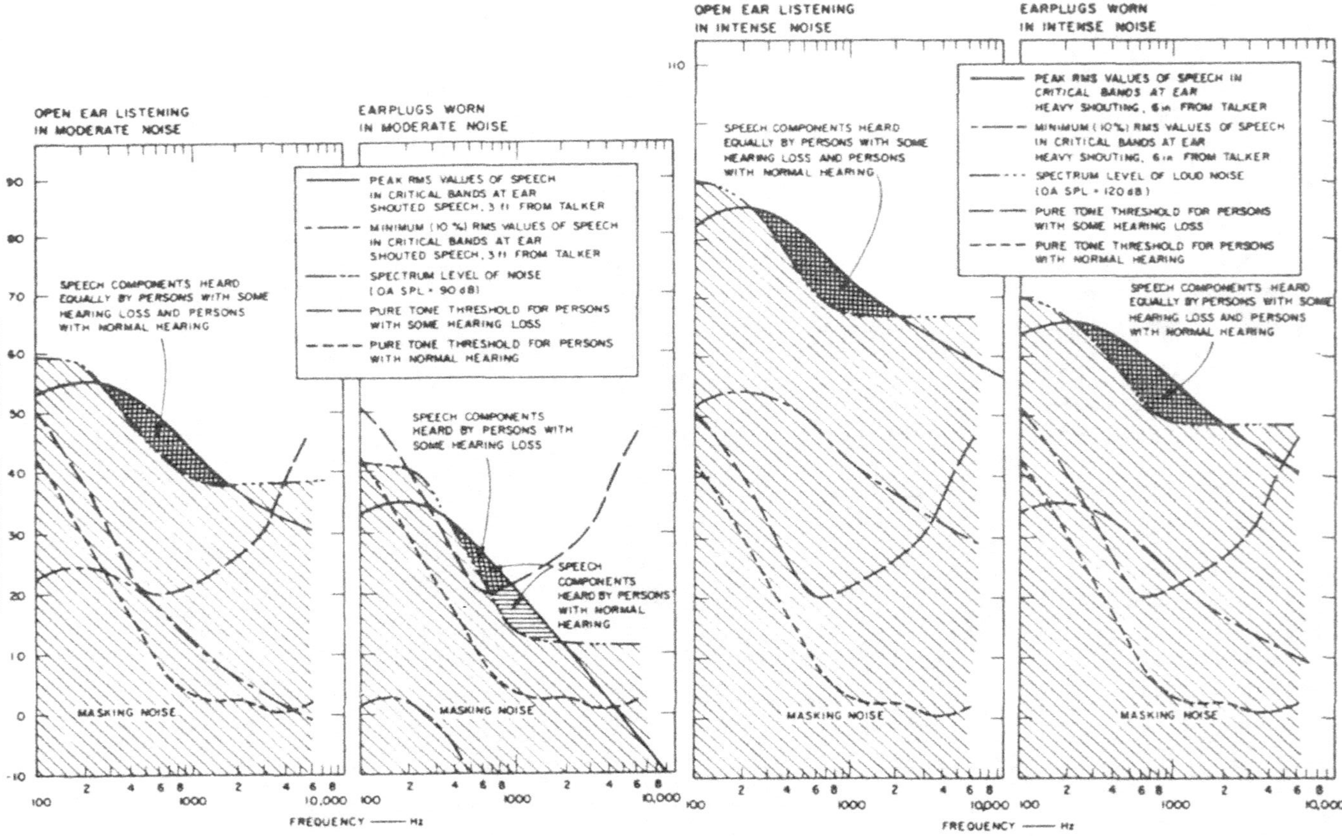

FIGURE 4.25. Amount of speech signal audible in presence of moderate and intense noise by persons with normal hearing and by persons with some degree of hearing loss.

hear equal amounts of speech with ears open, but when wearing earplugs, persons with hearing loss hear less speech than persons with normal hearing. In intense noise, however, both classes of listeners hear equal amounts of speech when listening with ears open or ears plugged.

It should also be noted that when one is in noise, plugging the ears results in a drop of 1 to 2 dB in voice level (ref. 7). Howell and Martin (ref. 39) also report a similar finding. Apparently the earplugs or ear coverings attenuate the ambient noise without lowering to as great an extent the speaker's own speech, which he hears both by tissue and bone conduction through his head and by airborne sound. When the speaker wears earplugs in the quiet, he raises his voice level by 3 to 4 dB, since his own voice now sounds weaker to him because of attenuation of the airborne components of the speech wave. (See fig. 4.26.)

Nonlinear earplugs

With regard to speech communication, earplugs or earmuffs appear to be contraindicated in the situation where they are probably needed—namely, in the presence of intermittent, impulsive noise, such as gunfire. Here, the wearer of earplugs or earmuffs cannot hear weak speech during the silent intervals between impulses. The ideal solution would be a nonlinear device that would let weak sounds through at full strength but would attenuate intense sounds.

Zwislocki (ref. 40) and Collins (ref. 41) have described the theoretical basis for such nonlinear devices and have built and tested models of them. The devices, which are essentially acoustic filters, operate on a frequency selective basis. They

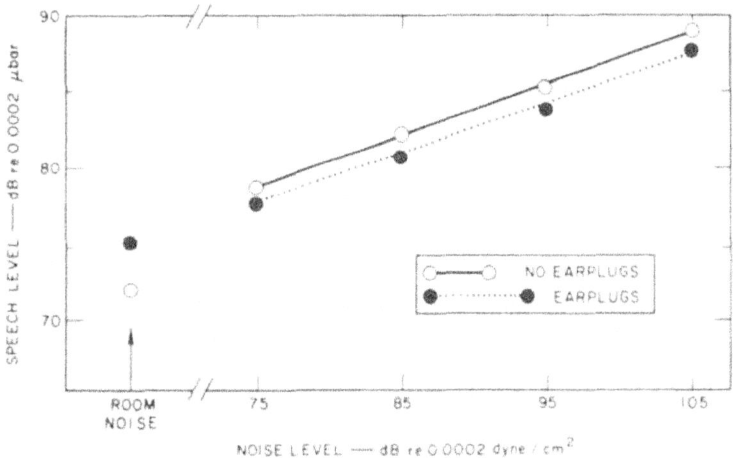

FIGURE 4.26. Effect of noise level on average speech intensity used by eight speakers with and without earplugs. Speech and noise levels were measured at listener's position (7 ft from speaker). (From ref. 7.)

afford significant attenuation at frequencies above 1500 Hz but offer little or no attenuation at frequencies below 1500 Hz, the region containing the strongest speech components. Also, the device can be made significantly nonlinear only after the overall sound pressure level exceeds a certain level.

Noise cancellation

Olson and May (ref. 42) developed a device for actively cancelling ambient noise. In one version of this system, a microphone mounted close to the listener's ear picks up environmental noise. The signal from this microphone is then amplified and fed into a loudspeaker (or to the listener's earphones) so that it is 180° out of phase with the noise signal and, therefore, acoustically cancels the noise at the listener's ears. This procedure is effective only for frequencies below about 200 Hz and, for this reason, will generally not help speech communication in noise.

This procedure was tried as a means of cancelling the hum of powerful industrial transformers that were annoyingly audible in residential areas near the transformers. Since the noise was primarily 60 Hz hum, considerable—about 20 dB—cancellation of the noise could be achieved. However, wind could cause "drifting" of the transformer noise and make cancellation of the noise unreliable.

"Electrical" stimulation of hearing

It has been observed a number of times in the past that an electrical signal that has been modulated by speech or other audible waveform and applied to the skin near the ear can be heard by the subject as though the stimulus had been applied acoustically to the ear. It has been suggested that the electrical stimulus is conducted by means of fluid in the tissue around the ear to the auditory nerve or to the auditory nerve fibers within the cochlea, and that the fibers are thereby stimulated. However, it is highly probable that the hearing that occurs as the result of application of an electrical signal to the external skin near the ear is the result of the electrical stimulation being transduced into an acoustic signal by some mechanical or electromechanical process external to the nerve endings. This acoustic signal is then transmitted to the cochlea either *via* the ossicular chain or by conduction through the bone surrounding the cochlea. The fact that the electrical signal can be a modulated radio-frequency carrier lends credence to the notion that a mechanical detection system is involved in this type of "electrical" stimulation of the ear.

Another form of electrical stimulation of the auditory system is achieved by electrical impulses from electrodes inserted into or near the auditory nerve fiber endings in the cochlea. Simmons (ref. 43) and others have demonstrated that such stimulation can elicit hearings of a rather crude sort.

Regardless of the precise mechanism involved, a practical application of "electrical stimulation of the ear," assuming an efficient transducer instrument can be developed, is that it would provide a possible means of avoiding the ambient noise that may be present in the listener's environment. In this case, the noise can perhaps be eliminated by the use of ear plugs or ear covering devices with the acoustic signal being applied electrically to the skin near the listener's ears.

Estimating Speech Intelligibility in Noise From Physical Measures

Articulation index (AI)

On the basis of data related to the intelligibility of filtered speech and certain assumptions regarding the equivalence of bandwidth and signal level (when both are measured in equivalent power), French and Steinberg (ref. 3) outlined a procedure whereby an index to the intelligibility of speech could be calculated from purely physical measurements. They called this the articulation index, or AI.

The AI concept holds that speech intelligibility is proportional to the average difference in dB between the masking level of noise and the long-term rms plus 12 dB level of the speech signal taken at the center frequency of 20 relatively narrow frequency bands. The masking spectrum of a noise may be different from the noise spectrum because of the spread of masking and remote masking. This proportionality holds provided the difference falls between 0 and 30 dB. These 20 bands, which were chosen because they were found to contribute equally to the understanding of speech, are proportional to the critical bandwidth of the ear as determined from studies of loudness and masking. (See fig. 3.2.)

The only significant modifications that have been made to the calculation procedures for AI as proposed by French and Steinberg dealt with the specification of exact procedures to be followed in converting noise spectra to noise-masking spectra and with methods for calculating AI from octave and 1/3-octave-band speech and noise spectra. The steps to be followed in the calculation of AI have been further developed (refs. 44 and 45) and are published as ANSI Standard S3.5-1969 (ref. 46). Figure 4.27 gives the work sheet used for calculating AI from 1/3-octave-band speech and noise spectra and an example of the calculation of an AI. Figure 4.28 shows the general relation between AI and various other measures of speech intelligibility.

Two points are made in Standard S3.5 that bear repetition here: (1) the AI can be applied properly only to communication systems and noise environments as specified in the subject document; and (2) there are types of communication systems and noise-masking situations that can only be evaluated by direct speech intelligibility or other performance tests. In particular, communication systems that process speech signals in various ways in order to achieve bandwidth compression cannot be evaluated validly by the AI procedure.

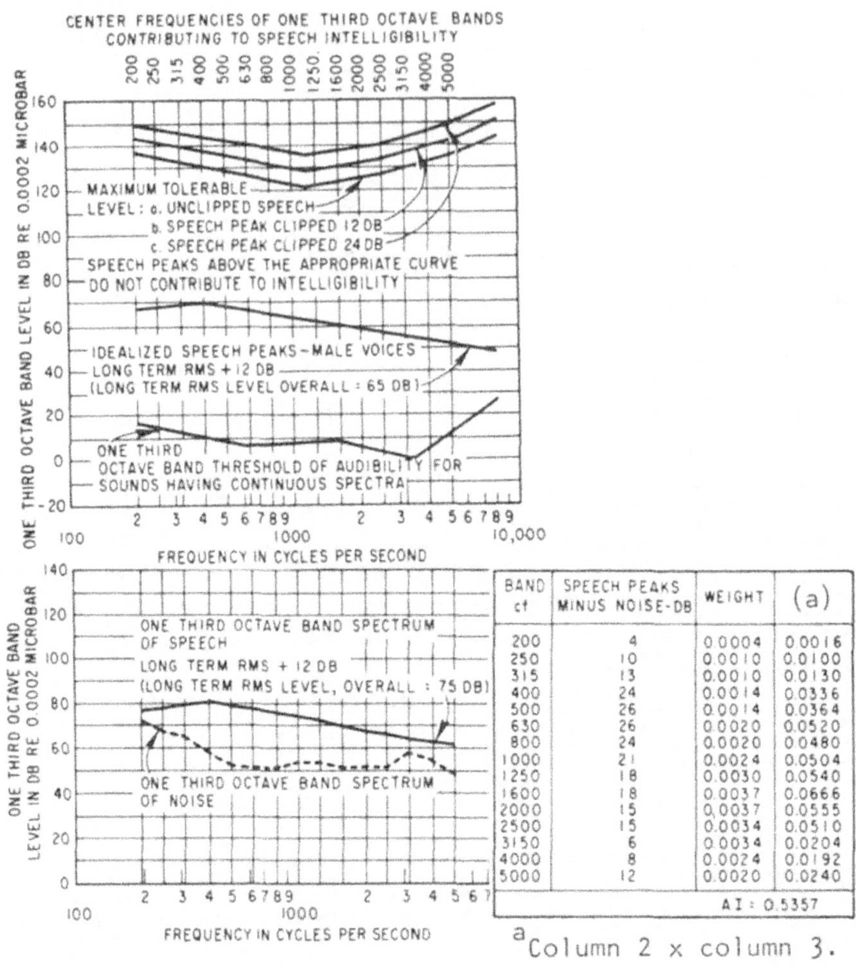

BAND cf	SPEECH PEAKS MINUS NOISE-DB	WEIGHT	(a)
200	4	0.0004	0.0016
250	10	0.0010	0.0100
315	13	0.0010	0.0130
400	24	0.0014	0.0336
500	26	0.0014	0.0364
630	26	0.0020	0.0520
800	24	0.0020	0.0480
1000	21	0.0024	0.0504
1250	18	0.0030	0.0540
1600	18	0.0037	0.0666
2000	15	0.0037	0.0555
2500	15	0.0034	0.0510
3150	6	0.0034	0.0204
4000	8	0.0024	0.0192
5000	12	0.0020	0.0240
		AI = 0.5357	

[a] Column 2 x column 3.

FIGURE 4.27. Upper graph: work sheet for AI, 1/3-octave-band method. Lower graphs: example of calculation of AI by 1/3-octave-band method. (From ref. 46.)

Suggestions are given in Standard S3.5 for refinements to AI to take into account such things as the vocal effort used by the talker, interruption in the noise, face-to-face talking, and reverberation present in the listening situation. In this regard, it should be noted that other procedures for making allowances for reverberation besides that used in S3.5 have been proposed. Bolt and MacDonald (ref. 47) suggested that reverberation effects could be properly accounted for by adding to the measured noise level an amount that depends upon the reverberation time. More recently, Janssen (ref. 48) recommended that the measured level of the speech signal be reduced to an effective level by an amount that depends

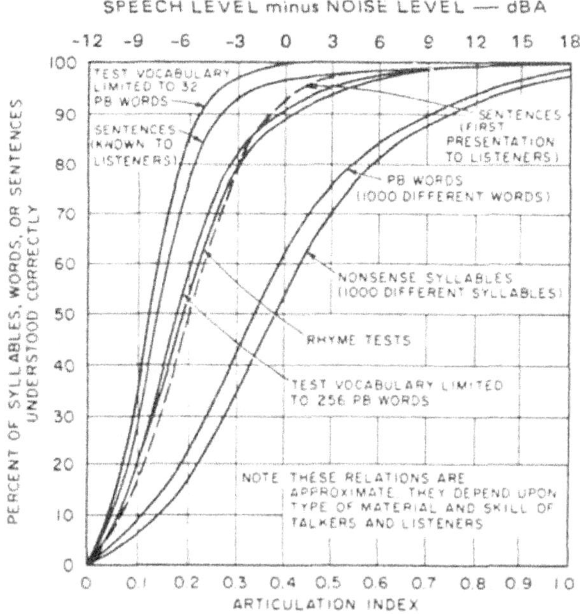

FIGURE 4.28. Relation between AI and various other measures of speech intelligibility. (Data from ref. 46.)

upon the reverberation time. Also, Levitt and Rabiner (ref. 49) proposed that the effects of binaural phase relations of speech and noise upon speech intelligibility, as discussed earlier, can be predicted by AI when certain adjustments are made to measured speech-to-noise ratios present in different frequency bands.

While not cast in terms of the AI *per se*, the findings of Lochner (ref. 50) are relevant to understanding the effects of reverberation on speech intelligibility. He found that placing sound-absorbing baffles between rows of metal-working machines did not reduce the noise level at the position of the worker at a given machine but did increase the perceived localization of the noise as coming from that machine. Speech communication was said to be improved as a result. This improvement could be related to binaural phase discrimination factors discussed in chapter 3 and earlier in this chapter.

Other AI procedures

Procedures similar to those used for finding AI have been proposed in several countries over the past 15 or 20 years (refs. 51, 52, 53, and 54). The variations are principally in terms of the width of the frequency bands in which the signal-to-noise ratios are determined.

Cavanaugh *et al.* (ref. 55) suggested the use of a graphical procedure for the estimation of AI. In their method, the spectrum of the noise is plotted on the same graph as the peak instantaneous levels reached by speech signals. The area between the noise spectrum and the speech peaks, adjusted for the relative importance assigned to different speech frequencies, is proportional to the AI for that speech and noise condition.

Validity of the AI procedure

Data collected by French and Steinberg (ref. 3), Miller (ref. 20), Egan and Wiener (ref. 56), and others with respect to masking of speech by noise of various bandwidths and spectra shapes provide a basis for demonstrating the ability of AI to predict the relative proficiency of given communication systems or conditions. Figure 4.29 shows some of these findings.

It should be emphasized that the values of the scores obtained on speech intelligibility tests are influenced by the proficiency and training of the talker and listening crew, as well as the difficulty of the speech material being used. Therefore, one cannot expect that a given communication system will provide identical test scores when tested in different laboratories and, particularly, with different groups of listeners and talkers, even though the AI of the system remains constant. In fact the inherent variability in speech intelligibility testing, while often not very large when similar test materials are used, is a recommendation for the use of AI whenever appropriate.

Criteria of acceptable noise levels for speech communications

Intelligibility tests and related calculation procedures are of paramount value in the evaluation, selection, and design of the components of speech communication systems and in the control of environmental noise conditions for the operation of such systems. However, the assessment of the masking effects of noise on speech, either by intelligibility testing or by calculation procedures, does not, of course, directly indicate how bothersome this masking will be in a given communications situation. A communications system, including the noise environment in which it is operated, that gives satisfactory performance when used with a special vocabulary and trained operators (for example, air traffic control by radio (ref. 26)) could be judged as completely unacceptable if used by untrained operators or in situations where flexible, nonstandardized speech communication is permitted.

Additional variables are the importance of the messages and the standards of the users. To my knowledge, no study has been reported which relates the scores of speech intelligibility tests to performance ratings of given communications systems for various classes of communications requirements. However, some laboratory tests of ratings of effort required to use a telephone system in free

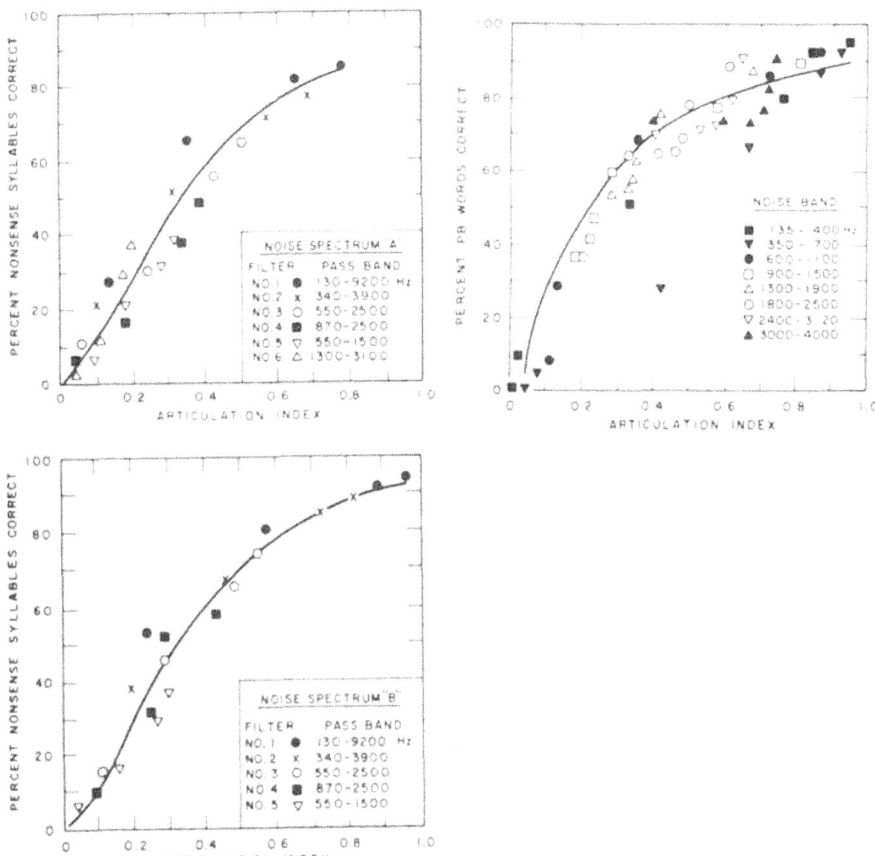

FIGURE 4.29. Left graphs: comparison of obtained and predicted test scores for speech passed through a bandpass filter and heard in the presence of a broadband, negatively sloped spectrum noise set at various intensity levels. Right graph: comparison of obtained and predicted test scores for broadband speech in the presence of narrow bands of noise set at various intensity levels. (From ref. 45.)

conversation and general satisfaction ratings of telephone communications have been obtained (ref. 57).

Nevertheless, some standards of expected satisfaction with speech communication systems have evolved. Beranek (ref. 58), for example, suggests that a communication system with an AI of less than 0.3 will usually be found unsatisfactory or only marginally satisfactory; one with an AI of at least 0.3 but less than 0.5 will generally be acceptable; one with an AI of 0.5 to 0.7 will be good; and a system with an AI higher than 0.7 will usually be considered very good to excellent.

Relations between AI, SIL, and other units

Beranek (ref. 59) proposed a simplified version of AI to be used in predicting the effectiveness of person-to-person speech communication in the presence of noise. Beranek estimated what the average speech level would be in the octave bands 600 to 1200, 1200 to 2400, and 2400 to 4800 Hz at various distances from a talker using various vocal efforts. Assuming the noise spectrum was a relatively continuous broadband noise, he estimated what noise levels would be required in these same octave bands to give an AI of about 0.5. The averages of the decibel levels in the three octave bands from 600 to 4800 Hz were tabulated for this condition, as shown in table 4.4. These averages are called SIL's (speech interference levels). The SIL's in table 4.4, presumably equivalent to an AI of 0.5, should allow sentence intelligibility scores of about 95 percent correct and PB word scores of about 75 percent correct. (See fig. 4.28.) As noted later, the noise present in various combinations of octave bands other than those between 600 and 4800 Hz has been proposed as a means of calculating,the SIL.

Various other ways of measuring sound exist that will provide reasonable estimates of or indices to AI, provided that the energy in a masking noise is concentrated predominantly in the frequency region covered by the normal speech spectrum. In addition to SIL, the following methods have been used: (a) sound level meter readings with either A or D frequency weightings, (b) loudness level in phons and perceived noise level in PNdB, and (c) the noise rating contour (NR, NC, or NCA) procedures. In the NR (noise rating), NC (noise criteria), and NCA (compromise noise criteria) procedures, the octave-band spectrum of a noise is plotted on special graphs. (See fig. 4.30 from refs. 60 and 61.) The highest NC (also called NR) or NCA contour touched by any octave band of the noise is assigned to that noise. Procedures for the calculation of loudness level as proposed by S. S. Stevens (phon(S)) and by E. Zwicker (phon(Z)) and perceived noise level in PNdB are discussed in detail in chapter 5.

While these various methods, particularly SIL, have been found to be a reasonably accurate method for evaluating speech communication in many noises, as will be shown below, unlike AI they should not be applied to noise spectra that have intense low- or high-frequency components. Other limitations of SIL and these other procedures, compared with AI, are that certain broad assumptions must be made in their use regarding the interactions between room acoustics, noise present, vocal effort used by the talker, and level of speech received by the listeners. The reader will find discussions of many of these matters in the proceedings of an international symposium on speech intelligibility (ref. 62).

As a practical matter, it is important to be able to measure and report a noise environment in units other than AI. Not only might the instrumentation and methods be more convenient, but there is also the possibility of establishing criteria of acceptability for noise with respect to more than one effect of noise (masking of speech, annoyance, damage to hearing, *etc.*) on the basis of a common single measurement unit. The next two sections present experimental test

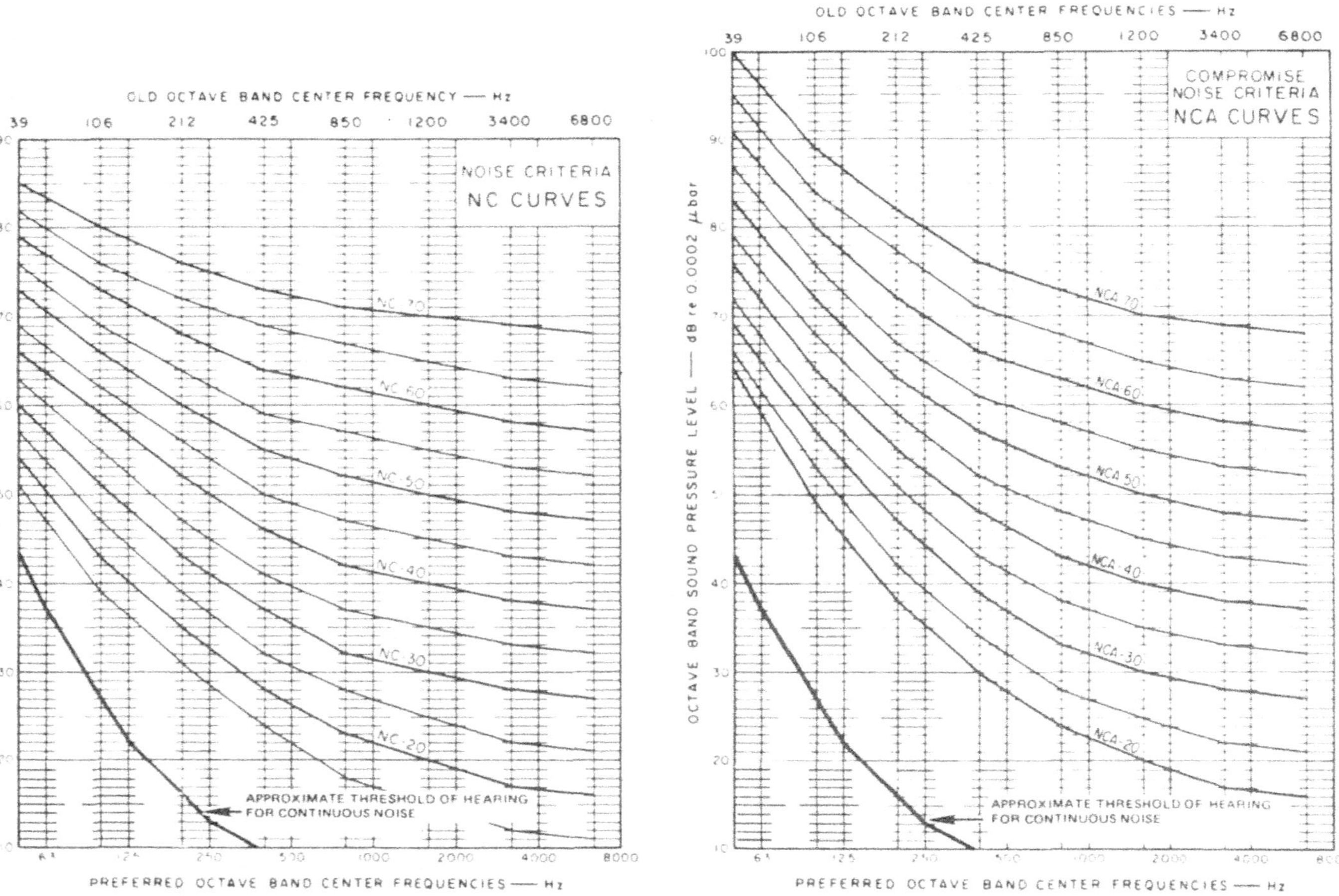

FIGURE 4.30. Left graph: noise criteria (NC curves) referred to old and preferred series of octave bands. Right graph: compromise noise criteria (NCA curves) referred to old and preferred series of octave bands. (From ref. 60; approximate threshold of hearing for continuous noise from ref. 61.)

TABLE 4.4

Speech Interference Levels That Permit Barely Reliable Conversation, or the Correct Hearing of Approximately 75 Percent of PB Words

[From ref. 59]

Distance between talker and listener, ft	Speech interference level, dB			
	Voice level			
	Normal	Raised	Very loud	Shouting
0.5	71	77	83	89
1	65	71	77	83
2	59	65	71	77
3	55	61	67	73
4	53	59	65	71
5	51	57	63	69
6	49	55	61	67
12	43	49	55	61

data that evaluate the relative effectiveness with which other noise measurement techniques estimate the effect of noise on the intelligibility of speech.

Test results

Klumpp and Webster (refs. 13 and 63) and Webster and Klumpp (ref. 12) report the band spectra of 16 noises they found to provide equal interference with speech (rhyme word test scores of about 50 percent correct), and overall values calculated or measured from these spectra for the following frequency and/or bandwidth weighting procedures: AI, SIL (3- and 4-octave-band methods covering frequencies from 355 to 4800 Hz), A-weighting, B-weighting, C-weighting, DIN 3 (a frequency weighting available on sound level meters made in Germany), NC, NCA, loudness level in phons (Stevens), and PNL. Their major findings are illustrated in figure 4.31.

Since the noise spectra reported are those present when the speech test scores were about equal (50 percent correct), the value obtained by a given measurement procedure should be the same for all 16 noises if that measure is to be considered a good index for 50 percent rhyme word intelligibility. The greater the deviations, given in terms of equivalent decibel units, of the values of each noise from the mean of all 16 noises, the less well is the test score of 50 percent predicted.

In the Klumpp and Webster study (ref. 63) SIL 355 to 2800 Hz, (the arithmetic average of the octave bands centered at 500, 1000, and 2000 Hz) or SIL 300 to 2400 Hz (octave-band-center frequencies of 425, 850, and 1700 Hz) gave reasonably good predictions, and SIL 600 to 4800 Hz (octave-band-center frequen-

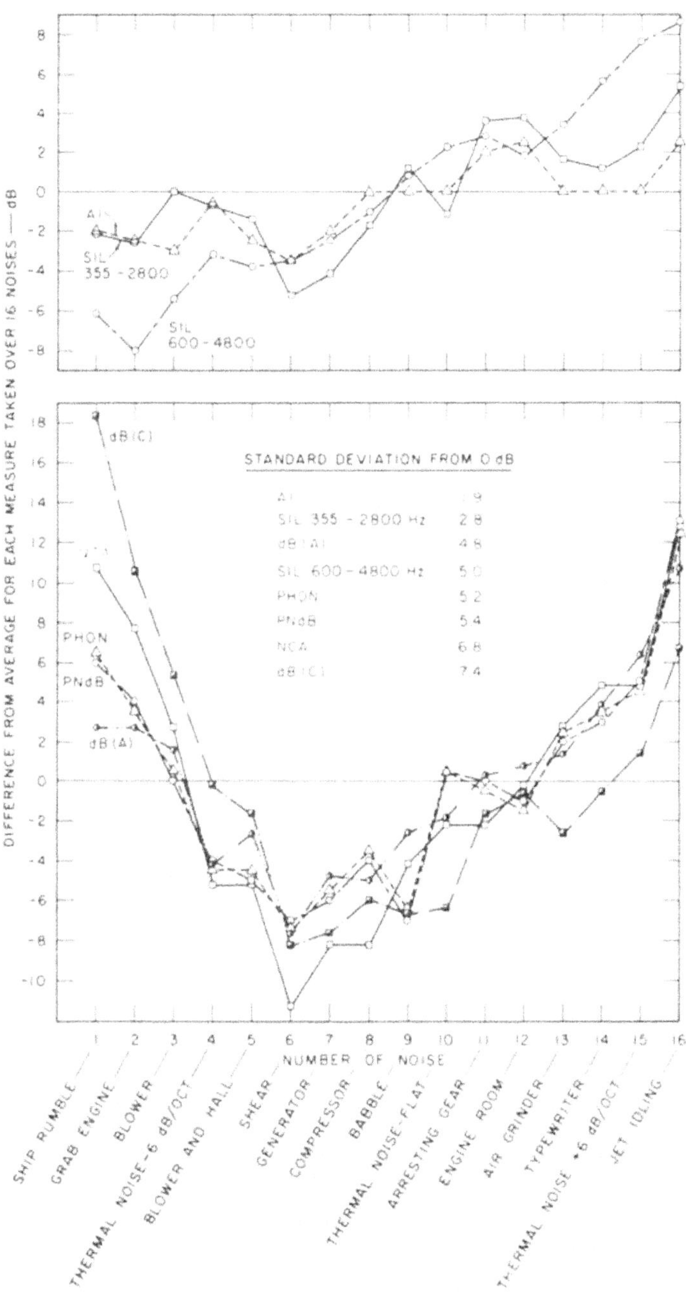

FIGURE 4.31. Difference between average for each measurement procedure taken over the 16 noises and the value for each of the noises when speech intelligibility equaled 50 percent words correct. (Data from refs. 12, 13, and 63.)

cies of 750, 1700, and 3400 Hz) gave generally poorer predictions of the speech interference effects of the noises. Most of the noises were recorded aboard naval vessels, and 10 of the 16 noises contained most of their energy in the frequency region below 1000 Hz. Webster and Klumpp (refs. 12 and 13) recommend SIL 355 to 2800 Hz over SIL 300 to 2400 Hz, even though the latter performed slightly better in their study, because the octave bands involved correspond to those now specified as preferred by various standardization groups (ref. 64).

On the other hand, in the Kryter and Williams (ref. 65) and Williams *et al.* (ref. 66) studies with aircraft noises, some of which had most of their energy at or above 1000 Hz, the reverse was true—SIL 600 to 4800 Hz appeared to be better correlated with the speech test scores than other SIL's. (Williams *et al.* did not, however, report values for SIL 300 to 2400 Hz.) Presumably this difference is partly attributable to the greater predominance of lower frequency noise in the Klumpp and Webster (ref. 63) study than in these aircraft noise studies.

In addition, it should be noted that a possible serious restriction to the generality of the Klumpp and Webster deductions lies in the use of 50 percent rhyme test scores as the measure of speech communication performance. As seen in figure 4.28, this represents a generally low level of communication proficiency (equivalent to an AI of about 0.2) and one which could be achieved only at noise levels where, particularly for their noises, the portions of the speech signals above about 2000 Hz were completely obliterated by the noise. At perhaps somewhat more realistic speech-to-noise ratios, the SIL's based on higher frequency bands than those recommended by Webster and Klumpp (refs. 12 and 13) would probably be more accurate predictors of speech interference.

In any event, the results of speech-masking experiments clearly demonstrate that in order to best predict the speech interference effects of a given noise (rather than the average of some type of noise spectrum), a measurement such as AI rather than SIL, A-weighting, PNL, or loudness level is required. It should be noted that NC (or NCA), which has been extensively used for noise evaluation, does not indicate very well the effects of noise on speech communication.

Calculated comparisons

In an attempt to provide a perhaps more general and realistic estimate of the accuracy with which AI will be estimated by the various measures of noise, seven noise spectra were chosen from the research literature. Five spectra represent different but relatively common sources, and two spectra were tailored from a random noise generator, as shown in table 4.5 (from refs. 63, 67, 68, and 69). Figure 4.32 shows how these seven noises are related to AI. The average deviations of the levels of the seven noises, as measured by each of the units cited when the noises were set to the same AI value, are given in the lower right-hand column of numbers. The column shows that when the AI values of the seven noises were all the same the standard deviations of the levels of the seven noises were: for dBA, 2.6 dB; for dBD, 2.2 dB; and for SIL 600 to 4800 Hz, 0.7 dB.

TABLE 4.5

Representative Octave-Band Spectra Used for the Evaluation of Various Procedures for Estimating Some Effects of Noise on Man

Noise description	SPL, dB, for octave bands, Hz, at—								OASPL, dB					
	53	106	212	425	850	1700	3400	6800	Un-weighted	A-weighted	C-weighted	D_1-weighted	D_2-weighted	D_3-weighted
Thermal noise (−6 dB/octave above 106 Hz)	89	93	87	81	75	69	63	57	95	82	95	89	87	80
Commercial jet landing, 610-ft altitude (ref. 67)	81	88	89	91	94	95	92	93	101	100	100	107	107	103
Planer (ref. 68)	82	84	85	87	88	88	87	85	95	94	95	101	101	96
Trolley buses (ref. 69)	68	72	74	73	69	64	58	52	79	73	79	78	77	72
Automobiles (ref. 69)	70	73	72	67	62	58	54	50	77	68	77	74	73	67

Noise description	SPL, dB, for octave bands, Hz, at—								OASPL, dB					
	63	125	250	500	1000	2000	4000	8000	Un-weighted	A-weighted	C-weighted	D_1-weighted	D_2-weighted	D_3-weighted
Thermal noise, "flat" (ref. 63)	60	64	68	69	71	78	75	72	82	82	81	90	90	85
Motor generator (ref. 63)	71	70	71	72	65	71	68	60	79	76	79	83	83	78

Also indicated in figure 4.32 are qualitative statements concerning the acceptability of speech communications systems having certain AI's, when the speech at the listener's ears is of a specified level. The horizontal lines are speech levels 1 m from the talker, or source. The speech levels can be converted to those present at other distances in a room (approximately 15 × 20 ft) by means of figure 4.33 (from ref. 70). These seven noise spectra must be used with some caution. For one

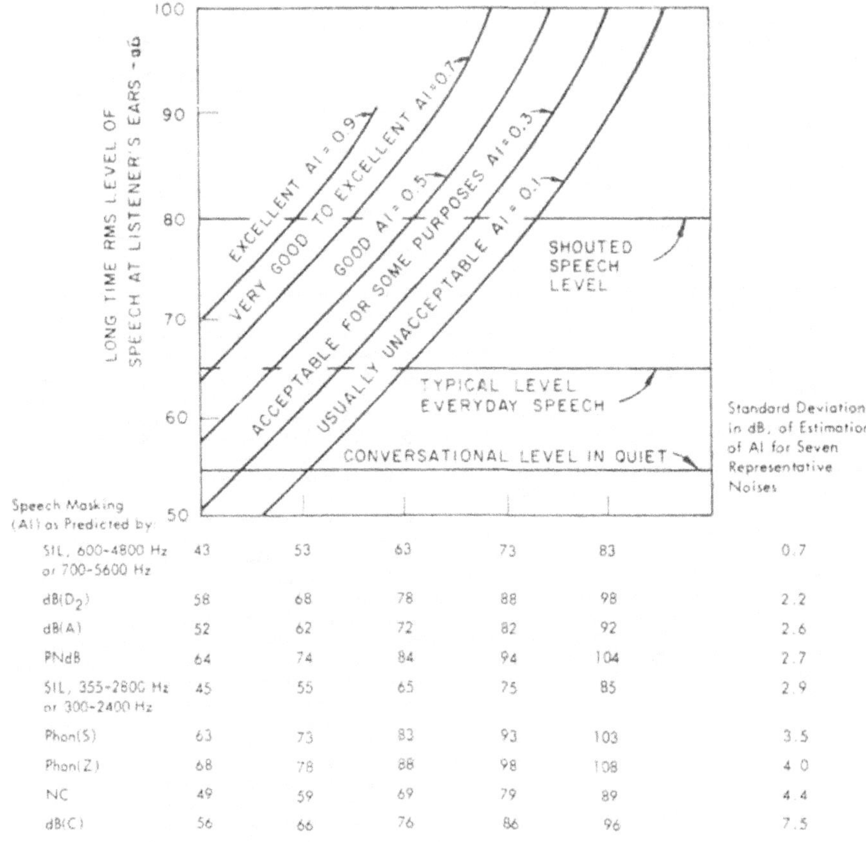

Speech Masking (AI) as Predicted by:						Standard Deviation in dB, of Estimation of AI for Seven Representative Noises
SIL, 600-4800 Hz or 700-5600 Hz	43	53	63	73	83	0.7
$dB(D_2)$	58	68	78	88	98	2.2
dB(A)	52	62	72	82	92	2.6
PNdB	64	74	84	94	104	2.7
SIL, 355-2800 Hz or 300-2400 Hz	45	55	65	75	85	2.9
Phon(S)	63	73	83	93	103	3.5
Phon(Z)	68	78	88	98	108	4.0
NC	49	59	69	79	89	4.4
dB(C)	56	66	76	86	96	7.5

NOTE: *Percent of word or sentence intelligibility scores for a given noise measurement (SIL, Phons, PNdB, etc.) and speech level approximately correct only when:*
(a) Noise has spectrum shaped like the average of the seven spectra given in Table 4.
(b) Speech is relatively undistorted

FIGURE 4.32. Average speech interference of the seven noises of table 4.5 measured by various units as predictors of AI, the parameter, as a function of the intensity of speech reaching the listener's ears. The overall SPL of each noise was adjusted until the AI for each noise had the same value. The average value was then found for each of the other measures of speech masking (SIL, PNdB, *etc.*): these averages and the standard deviations of the seven noises from the averages are tabulated on the abscissa. The deviations are approximately correct for AI values from about 0.15 to 0.75.

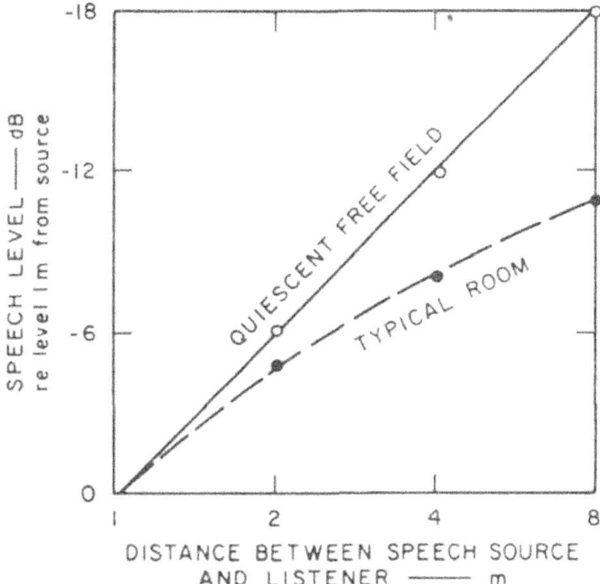

FIGURE 4.33. Approximate effect of acoustic environment and distance (in meters) between speech source and listener upon received speech level. (Data from ref. 70.)

thing, any small set of noise spectra may be unrepresentative, in kind or numbers, of the noise present in the real world. It is also true that noise control engineering is usually applied to given specific noise sources or environments one at a time. And, unless one is dealing with a homogeneous set of noises, it is somewhat risky, from an accuracy point of view, to make noise measurements that do not permit calculation of AI.

Webster (ref. 71) also plotted, figure 4.34, some of the general relations shown in figure 4.32 with distance between talker and listener as a variable and vocal effort as the parameter. Webster also showed some relations between AI and other physical units of noise measurement listed in descending order of statistical accuracy for evaluating the masking of speech. PSIL (the average octave-band level in the octaves centered at 500, 1000, and 2000 Hz), was the best measure and the one that varied least for predicting speech interference of Navy shipboard noises at levels higher than 70 dB. The next best measure was SIL averaged over the three octaves from 600 to 4800 Hz, followed by A-weighted sound level meter reading, PNL in PNdB, and finally, C-weighted sound level meter reading. Again the statistics shown in figure 4.34 are based on speech at a very low level of intelligibility in the presence of 16 shipboard noises, and the relative proficiency, except for AI, of the various units of noise measurements is not necessarily representative of their proficiency with other noises and at other levels of noise masking.

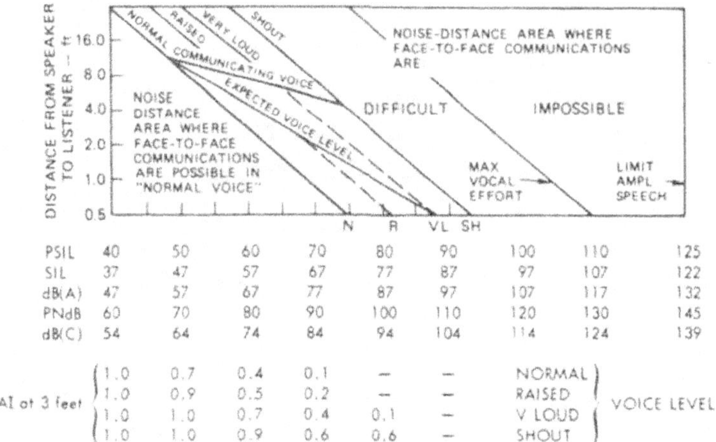

PSIL	40	50	60	70	80	90	100	110	125
SIL	37	47	57	67	77	87	97	107	122
dB(A)	47	57	67	77	87	97	107	117	132
PNdB	60	70	80	90	100	110	120	130	145
dB(C)	54	64	74	84	94	104	114	124	139

AI at 3 feet							Voice level
	1.0	0.7	0.4	0.1	—	—	NORMAL
	1.0	0.9	0.5	0.2	—	—	RAISED
	1.0	1.0	0.7	0.4	0.1	—	V LOUD
	1.0	1.0	0.9	0.6	0.6	—	SHOUT

FIGURE 4.34. Voice level and distance between talker and listener for satisfactory face-to-face speech communication as limited by ambient noise level. Under the graph are rank-ordered, from best to worst, various objective measures of noise level. (Data from ref. 71.)

Devices for Evaluating Speech Systems

The AI procedures were proposed as a means of evaluating the performance of a communication system without requiring the administration of time-consuming and costly speech intelligibility or articulation test procedures. But, it is often as impractical to make the physical measurements necessary to find the noise and speech spectrum at the listener's ears for the calculation of AI as it is to apply speech intelligibility test procedures.

Twenty tones

Tkachenko (ref. 72) proposed a very simple, yet excellent, method for obtaining the AI of a speech communication system. He developed an artificial speech signal that consisted of 20 pure tones spaced at the center frequency of the 20 bands found by French and Steinberg (ref. 3) to be equally important to the understanding of speech. Each tone is audited separately by the listener, who adjusts an attenuator controlling the level of the tone until it is just audible. The level of the tone is read from the attenuator, which is calibrated in terms of equivalent, normal speech level, and 12 dB is added to take into account the speech peak factor. This process is repeated for each of the 20 pure tones, and the average of the levels required for the tones to be just audible is proportional to AI.

Although this method of Tkachenko is not as automatic as some other methods, which are described below, it has much to recommend it. For one thing, it uses the human listener to determine the exact masking effect of the system noise, something which, short of conducting speech intelligibility tests, is only estimated in AI and related calculation procedures.

Automatic devices

Licklider *et al.* (ref. 73) developed an electronic device which when applied to a speech communication system will automatically provide a number that is usually proportional to AI. This machine is actually based on a somewhat different concept than that of AI, although it uses the frequency-weighting importance function and the signal-to-noise-ratio weighting function developed for AI.

The Licklider *et al.* device repeatedly plays a recorded brief (several seconds) sample of speech over the communication system to be evaluated. At any one moment, the output of the speech system being evaluated is simultaneously compared in a narrow frequency band with the recorded speech input. This process is repeated a number of times, each time the locus of the frequency band in which the input and output speech are to be compared is changed. Suitable integrating circuits average the measured correlations between the input and output signals.

As well as revealing the interference or disruptive effects of any noise present in a system, the average correlation calculated for a given system appropriately reflects the presence of distortions such as frequency shifts and rapid fluctuations in noise and speech levels. These latter distortions are usually ignored in the calculation of AI. On the other hand, this device processes only actual noise spectra present in a communication system and does not make allowances for upward and remote masking. Robertson and Stuckey (ref. 74) found this device predicted speech intelligibility test scores quite well for most, but not all, noise interference conditions tested.

Goldberg (ref. 75) described a machine called a "voice interference analysis set" that attempts to compute the AI for a communication system. This machine applies an amplitude-modulated tone (1000 Hz) to the system under test. The level of the received signal is compared with the noise level found at the receiver of the system under test in 10 bands that are proportional in width to the 20 AI bands. Certain corrections are electronically determined when the noise present would cause a significant upward spread of masking.

Kryter *et al.* (ref. 76) designed a speech communication index meter (SCIM) that is similar to the Goldberg machine in that it attempts more or less directly, to calculate AI from measured signal-to-noise ratios. SCIM transmits a broadband signal that simulates normal speech with respect to long-term spectrum shape and, to some extent, amplitude variations. The simulated signal-to-actual-noise ratios are found in nine frequency bands and appropriately averaged to arrive at an approximate AI value. This device takes into account the effects on the speech intelligibility of direct masking (including the upward spread), frequency shifting, frequency distortion, and amplitude limiting of the speech signal. Preliminary comparison of SCIM-measured AI with speech intelligibility scores indicates reasonable accuracy in the prediction of the test scores. SCIM has recently been modified to reflect the effectiveness of speech communication systems operating with a signal or noise level which varies with time (Hecker *et al.*, ref. 77).

Simplified measuring set

It appears that the monitoring of the background noise levels in commercial telephone circuits can be adequately accomplished with a device that, like a sound level meter, measures the noise energy over all frequencies (ref. 78). The measuring instrument, called Bell 3A, weights the frequencies in accordance with judgments made by listeners over a telephone of speech interfered by 14 different pure tones. It also has an integration time of 0.2 sec in order to simulate the growth of loudness as judged by listeners. It should be noted that this instrument is based *on judgments of how tones would interfere with speech on a standard telephone* circuit, not on speech intelligibility tests.

Aikens and Lewinski (ref. 78) found the interesting fact that telephone users will accept a 3 to 4 dB lower signal-to-noise ratio (as measured on the Bell 3A message circuit noise set) when the speech level is at a modest level than when the speech is at a level only 10 dB higher. It is possible that the 3A device evaluates the annoyance or loudness of circuit noise more than its masking effect on speech, since a 3 to 4 dB decrease in signal-to-noise ratio at the important speech frequencies should cause a noticeable reduction in speech intelligibility.

References

1. Dunn, H. K.; and White, S. D.: Statistical Measurements on Conversational Speech. J. Acoust. Soc. America, vol. 11, no. 3, Jan. 1940, pp. 278-288.
2. Von Tarnóczy, T.: Das Durchschnittliche Energie-Specktrum der Sprache (für Sechs Sprachen). Acustica, vol. 24, no. 2, Feb. 1971, pp. 57-74.
3. French, N. R.; and Steinberg, J. C.: Factors Governing the Intelligibility of Speech Sounds. J. Acoust. Soc. America, vol. 19, no. 1, Jan. 1947, pp. 90-119.
4. Pearsons, Karl S.; Bennett, Ricarda L.; and Fidell, Sanford: Speech Levels in Various Noise Environments. EPA-600/1-77-025, U.S. Environ. Prot. Agency, May 1977. (Available from NTIS as PB 270 053.)
5. Korn, T. S.: Effect of Psychological Feedback on Conversational Noise Reduction in Rooms. J. Acoust. Soc. America, vol. 26, no. 5, Sept. 1954, pp. 793-794.
6. Webster, John C.; and Klumpp, Roy G.: Effects of Ambient Noise and Nearby Talkers on a Face-to-Face Communication Task. J. Acoust. Soc. America, vol. 34, no. 7, July 1962, pp. 936-941.
7. Kryter, K. D.: Effects of Ear Protective Devices on the Intelligibility of Speech in Noise. J. Acoust. Soc. America, vol. 18, no. 2, Oct. 1946, pp. 413-417.

8. Pickett, J. M.: Limits of Direct Speech Communication in Noise. J. Acoust. Soc. America, vol. 30, no. 4, Apr. 1958, pp. 278-281.
9. Gardner, Mark B.: Effect of Noise on Listening Levels in Conference Telephony. J. Acoust. Soc. America, vol. 36, no. 12, Dec. 1964, pp. 2354-2362.
10. Rupf, John A.: Noise Effects on Passenger Communication in Light Aircraft. [Preprint] 770446. Soc. Automot. Eng., Mar.-Apr. 1977.
11. Miller, G. A.; Heise, G. A.; and Lichten, W.: The Intelligibility of Speech as a Function of the Context of the Test Materials. J. Exp. Psychol., vol. 41, 1951, pp. 329-335.
12. Webster, J. C.; and Klumpp, R. G.: Articulation Index and Average Curve-Fitting Methods of Predicting Speech Interference. J. Acoust. Soc. America, vol. 35, no. 9, Sept. 1963, pp. 1339-1344.
13. Klumpp, R. G.; and Webster, J. C.: Speech Interference Aspects of Navy Noises Research and Development Report, January 1961-December 1964. Rep. NEL-1314, U.S. Navy, Sept. 1965. (Available from DTIC as AD 625 262.)
14. Licklider, J. C. R.; and Guttman, Newman: Masking of Speech by Line-Spectrum Interference. J. Acoust. Soc. America, vol. 29, no. 2, Feb. 1957, pp. 287-296.
15. Carter, Norman L.; and Kryter, Karl D.: Masking of Pure Tones and Speech. J. Aud. Res., vol. 2, 1962, pp. 66-98.
16. Pollack, Irwin; and Pickett, J. M.: Intelligibility of Peak-Clipped Speech at High Noise Levels. J. Acoust. Soc. America, vol. 31, no. 1, Jan. 1959, pp. 14-16.
17. Pollack, Irwin: Speech Intelligibility at High Noise Levels: Effect of Short-Term Exposure. J. Acoust. Soc. America, vol. 30, no. 4, Apr. 1958, pp. 282-285.
18. Miller, George A.; and Licklider, J. C. R.: The Intelligibility of Interrupted Speech. J. Acoust. Soc. America, vol. 22, no. 2, Mar. 1950, pp. 167-173.
19. Pollack, Irwin: Speech Communications at High Noise Level: The Roles of a Noise-Operated Automatic Gain Control System and Hearing Protection. J. Acoust. Soc. America, vol. 29, no. 12, Dec. 1957, pp. 1324-1327.
20. Miller, G. A.: The Masking of Speech. Psychol. Bull., vol. 44, 1947, pp. 105-129.
21. Licklider, J. C. R.: The Influence of Interaural Phase Relations Upon the Masking of Speech by White Noise. J. Acoust. Soc. America, vol. 20, no. 2, Mar. 1948, pp. 150-159.
22. Pollack, I.; and Pickett, J. M.: Stereophonic Listening and Speech Intelligibility Against Voice Babble. J. Acoust. Soc. America, vol. 30, no. 2, Feb. 1958, pp. 131-133.
23. Weston, Peter B.; Miller, James D.; and Hirsh, Ira J.: Release From Masking for Speech. J. Acoust. Soc. America, vol. 38, no. 6, Dec. 1965, pp. 1053-1054.

24. Moser, H. H.: Research Investigations on Voice-Communication in Noise. Rep. CSD-TDR-62-5 (Contract AF 19(604)-6179), Ohio State Univ. Res. Found., 1961. (Available from DTIC as AD 279 870.)
25. Pollack, Irwin: Message Procedures for Unfavorable Communication Conditions. J. Acoust. Soc. America, vol. 30, no. 3, Mar. 1958, pp. 196-201.
26. Frick, F. C.; and Sumby, W. H.: Control Tower Language. J. Acoust. Soc. America, vol. 24, no. 6, Nov. 1952, pp. 595-596.
27. Thwing, Edward J.: Effect of Repetition on Articulation Scores for PB Words. J. Acoust. Soc. America, vol. 28, no. 2, Mar. 1956, pp. 302-303.
28. Pickett, J. M.; and Pollack, Irwin: Prediction of Speech Intelligibility at High Noise Levels. J. Acoust. Soc. America, vol. 30, no. 10, Oct. 1958, pp. 955-963.
29. Licklider, J. C. R.: Effects of Amplitude Distortion Upon the Intelligibility of Speech. J. Acoust. Soc. America, vol. 18, no. 2, Oct. 1946, pp. 429-434.
30. Kryter, K. D.; Licklider, J. C. R.; and Stevens, S. S.: Premodulation Clipping in AM Voice Communication. J. Acoust. Soc. America, vol. 19, no. 1, Jan. 1947, pp. 125-131.
31. Moser, H. H.; Dreher, J. J.; O'Neill, J. J.; and Oyer, H. J.: Comparison of Mouth, Ear, and Contact Microphones. Tech. Rep. 37, AFCRC TN-56-68, U.S. Air Force, 1956. (Rev. 1958.)
32. Hawley, Mones, E.; and Kryter, Karl D.: Effects of Noise on Speech. Handbook of Noise Control, Cyril M. Harris, ed., McGraw-Hill Book Co., c.1957, pp. 9-1 - 9-26.
33. Zwislocki, J.: Ear Protectors. Handbook of Noise Control, Cyril M. Harris, ed., McGraw-Hill Book Co., c.1957, pp. 8-1 - 8-27.
34. Von Gierke, H. E.: Personal Protection. NOISE Contr., vol. 2, no. 1, Jan. 1956, pp. 37-44.
35. Alberti, P. W.; Abel, S. M.; and Riko, K.: Practical Aspects of Hearing Protector Use. New Perspectives on Noise-Induced Hearing Loss, Roger P. Hamernik, Donald Henderson, and Richard Salvi, eds., Raven Press, 1982, pp. 461-469.
36. Smoorenburg, Guido F.: Damage Risk Criteria for Impulse Noise. New Perspectives on Noise-Induced Hearing Loss, Roger P. Hamernik, Donald Henderson, and Richard Salvi, eds., Raven Press, 1982, pp. 471-490.
37. Edwards, R. G.; Hauser, W. P.; Moiseev, N. A.; Broderson, A. B.; and Green, W. W.: Effectiveness of Earplugs as Worn in the Workplace. Sound & Vib., vol. 12, no. 1, Jan. 1978, pp. 12-18, 20, 22.
38. Rink, Timothy L.: Hearing Protection and Speech Discrimination in Hearing-Impaired Persons. Sound & Vib., vol. 13, no. 1, Jan. 1979, pp. 22-25.
39. Howell, K.; and Martin, A. M.: An Investigation of the Effects of Hearing Protectors on Vocal Communication in Noise. J. Sound & Vib., vol. 41, no. 2, July 22, 1975, pp. 181-196.
40. Zwislocki, J.: Acoustic Filters as Ear Defenders. J. Acoust. Soc. America, vol. 23, no. 1, Jan. 1951, pp. 36-40.

41. Collins, G.: The Design of Frequency Selective Earplugs. A. R. L./D/R12, Admiralty Res. Lab. (British), Feb. 1964.
42. Olson, Harry F.; and May, Everett G.: Electronic Sound Absorber. J. Acoust. Soc. America, vol. 25, no. 6, Nov. 1953, pp. 1130-1136.
43. Simmons, F. B.: Electrical Stimulation of the Auditory Nerve in Man. Arch. Otolaryngol., vol. 78, 1966, pp. 24-54.
44. Kryter, Karl D.: Methods for the Calculation and Use of the Articulation Index. J. Acoust. Soc. America, vol. 34, no. 11, Nov. 1962, pp. 1689-1697.
45. Kryter, Karl D.: Validation of the Articulation Index. J. Acoust. Soc. America, vol. 34, no. 11, Nov. 1962, pp. 1698-1702.
46. American National Standard Methods for the Calculation of the Articulation Index. ANSI S3.5-1969, American Nat. Standards Inst., Inc., Jan. 16, 1969.
47. Bolt, R. H.; and MacDonald, A. D.: Theory of Speech Masking by Reverberation. J. Acoust. Soc. America, vol. 21, no. 6, Nov. 1949, pp. 577-580.
48. Janssen, J. H.: A Method for the Calculation of the Speech Intelligibility Under Conditions of Reverberation and Noise. Acustica, vol. 7, no. 5, 1957, pp. 305-310.
49. Levitt, H.; and Rabiner, L. R.: Predicting Binaural Gain in Intelligibility and Release From Masking for Speech. J. Acoust. Soc. America, vol. 42, no. 4, Oct. 1967, pp. 820-829.
50. Lochner, J. P. A.: The Influence of Noise on the Intelligibility of Speech. Univ. Port Elizabeth, South Africa, 1978.
51. Golikov, E. E.: Calculating the Articulation in Noisy Rooms. Soviet Phys.—Acoust., vol. 6, no. 3, Jan.-Mar. 1961, pp. 407-408.
52. Jeffress, Lloyd A.: Masking and Binaural Phenomena. DRL-A-245 (Contract Nos. NObsr-72627, NE 051247-6, NE 051456-4, Nonr-3579(04), NR 142-190, and Fund Transfer R-129), Def. Res. Lab., Univ. of Texas, June 14, 1965. (Available as NASA CR-64362.)
53. Richards, D. L.: A Development of the Collard Principle of Articulation Calculation. Proc. Inst. Electr. Eng. (London), vol. 103, 1956, pp. 679-691.
54. Rozhanskaya, E. V.: On the Question of the Mathematical Foundation of the Theory of Intelligibility. Akad. Nauk SSSR Komissiya Akustika Tr., vol. 7, 1953, pp. 53-60. (Available from NTIS as Translation 62-11581.)
55. Cavanaugh, W. J.; Farrell, W. R.; Hirtle, P. W.; and Watters, B. G.: Speech Privacy in Buildings. J. Acoust. Soc. America, vol. 34, no. 4, Apr. 1962, pp. 475-492.
56. Egan, James P.; and Wiener, Francis M.: On the Intelligibility of Bands of Speech in Noise. J. Acoust. Soc. America, vol. 18, no. 2, Oct. 1946, pp. 435-441.
57. Richards, D. L.; and Swaffield, J.: Assessment of Speech Communication Links. Proc. Inst. Electr. Eng. (London), vol. 106B, 1959, pp. 77-89, 90-92.

58. Beranek, L. L.: The Design of Speech Communication Systems. Proc. Inst. Radio Eng., vol. 35, 1947, pp. 880-890.
59. Beranek, L. L.: Noise Control in Office and Factory Spaces. Ind. Hyg. Found. America, Trans. Bull. 18, 1950, pp. 26-33.
60. Schultz, T. J.: Noise-Criterion Curves for Use With the USASI Preferred Frequencies. J. Acoust. Soc. America, vol. 43, no. 3, Mar. 1968, pp. 637-638.
61. Robinson, D. W.; and Whittle, L. S.: The Loudness of Octave-Bands of Noise. Acustica, vol. 14, no. 1, 1964, pp. 24-35.
62. Intelligibilité de la Parole (Speech Intelligibility). Fédérations des Sociétés Européennes d'Acoustique, Nov. 1973.
63. Klumpp, R. G.; and Webster, J. C.: Physical Measurements of Equally Speech-Interfering Navy Noises. J. Acoust. Soc. America, vol. 35, no. 9, Sept. 1963, pp. 1328-1338.
64. American National Standard—Preferred Reference Quantities for Acoustical Levels. ANSI S1.8-1969, American Nat. Stand. Inst., Inc., Feb. 24, 1969.
65. Kryter, K. D.; and Williams C. E.: Some Factors Influencing Human Response to Aircraft Noise: Masking of Speech and Variability of Subjective Judgments. FAA-ADS-42, June 1965.
66. Williams, C. E.; Stevens, K. N.; Hecker, M. H. L.; and Pearsons, K. S.: The Speech Interference Effects of Aircraft Noise. FAA-DS-67-19, Sept. 1967.
67. Kryter, K. D.; and Williams, Carl E.: Masking of Speech by Aircraft Noise. J. Acoust. Soc. America, vol. 39, no. 1, Jan. 1966, pp. 138-150.
68. Karplus, H. B.; and Bonvallet, G. L.: A Noise Survey of Manufacturing Industries. American Ind. Hyg. Assoc. Q., vol. 14, 1953, pp. 235-263.
69. Bonvallet, G. L.: Levels and Spectra of Transportation Vehicle Noises. J. Acoust. Soc. America, vol. 22, no. 2, Mar. 1950, pp. 201-205.
70. Beranek, Leo L.: Acoustics. McGraw-Hill Book Co., c.1954.
71. Webster, John C.: Effects of Noise on Speech Intelligibility. Noise as a Public Health Hazard, W. Dixon Ward and James E. Fricke, eds., ASHA Rep. 4, American Speech and Hearing Assoc., Feb. 1969, pp. 49-73.
72. Tkachenko, A. D.: Tonal Method for Determining the Intelligibility of Speech Transmitted by Communication Channels. Soviet Phys.—Acoust., vol. 1, nos. 1 & 2, 1955, pp. 182-191.
73. Licklider, J. C. R.; Bisberg, A.; and Schwartzlander, H.: An Electronic Device To Measure the Intelligibility of Speech. Proceedings of the National Electronics Conference, vol. 15, 1959, pp. 1-6.
74. Robertson, D. W.; and Stuckey, C. W.: Investigation and Evaluation of the Gel Speech System Test Set. Rep. RADC TR 61-88, Rome Air Development Center, 1961.
75. Goldberg, J. M.: The Voice Interference Analysis Set. An Instrument for

Determining the Degradation of Signal Quality of a Voice Communication Channel. J. Audio Eng. Soc., vol. 11, 1963, pp. 115-120.

76. Kryter, K. D.; Ball, J. H.; and Stuntz, S. E.: SCIM—A Meter for Measuring the Performance of Speech Communication Systems. Paper 19, IEEE Transactions of the Third Canadian Symposium on Communications, Sept. 1964.

77. Hecker, Michael H. L.; Von Bismarck, Gottfried; and Williams, Carl E.: Automatic Evaluation of Time-Varying Communication Systems. IEEE Trans. Audio & Electroacoust., vol. AU-16, no. 1, Mar. 1968, pp. 100-106.

78. Aikens, A. J.; and Lewinski, D. A.: Evaluation of Message Circuit Noise. Bell Syst. Tech. J., vol. 39, no. 4, July 1960, pp. 879-909.

Chapter 5
Loudness, Noisiness, and Vibration Effects

Introduction

As determined in the laboratory, the relations between physical measures of the spectral-temporal character of sounds and judgments of their loudness and/or noisiness are presumed to predict the relative annoyance to be expected from noises heard in real life. In addition to the relative contribution of different physical aspects of sound to these judgments (*e.g.*, higher frequencies are judged to be louder and noisier than lower frequencies), laboratory tests have also been conducted on the more absolute question of the acceptability of noises in and around homes.

The acceptability-unacceptability of noises in real life depends considerably on the degree to which they interfere with sleep and with speech and other auditory communication. Accordingly, the usefulness of laboratory loudness and noisiness judgment data rests upon how well these data correlate with such things as the speech and sleep interference effects of noise in real life.

As shown in the present and later chapters, there are sufficient consistencies between laboratory data and data on annoyance from noise in real-life situations to allow definition of simple physical measures of noise that predict psychological and behavioral effects of noise in real life. A number of research questions remain, of course, for further investigation.

This chapter presents the findings, rationale, and arguments of loudness and noisiness judgment research, as involved in the development of these physical measurement procedures. Also discussed are some factors related to the prediction of real-life annoyance when noise-induced house vibrations are involved.

Loudness

Loudness is defined as the subjective intensity of sound, independent of any meaning the sound might have. It is generally believed that the louder a sound is, the more unacceptable, or noisy, it is. While this is generally true, it does not follow that measuring the physical energy in a sound is sufficient for predicting the subjective noisiness or unwantedness of different sounds (even though a frequency weighting is used to adjust the measurement in accordance with the subjective loudness of different parts of the sound spectrum). As mentioned before, the temporal aspects of sounds also influence their subjective noisiness.

Loudness is somewhat analogous to the brightness of a light; for example, a 100-W light bulb is perceived to have a certain brightness that does not appreciably change with time. However, the continuing presence of a bothersome light (perhaps when one is trying to go to sleep) increases its perceived unwantedness and cause-of-annoyance value, but not its brightness. Similarly, increasing the duration of an unwanted sound of a given intensity increases its perceived unwantedness, or noisiness, and cause-of-annoyance value, but not its loudness.

Stevens (ref. 1) proposed melding the results of judgment tests of loudness and noisiness as a function of the frequency of pure tones or narrow bands of noise. He suggested that the resulting contours be called perceived level or magnitude functions. The author's opinion, based on information and definitions given below, is that this nomenclature is not sufficiently definitive to be useful in quantifying noise.

Nevertheless, the concept of loudness and the methods developed for measuring sounds to estimate their loudnesses are highly relevant to the subject at hand. Indeed, the basic frequency-weighting schemes for loudness are used as is, or with slight modification, as one of the steps required to measure sounds for purposes of predicting their subjective noisiness.

There are four basic psychological-physical relations that depict the perceptual attribute of sound called loudness:

1. The intensity levels required to make each tone or critical band of frequencies in the audible frequency range appear subjectively to be equally loud (discussed in chapter 3; see fig. 3.2)

2. The duration required to achieve a stable perception of the loudness of a sound (the critical summation time, about 1 sec, for the human ear, discussed in chapter 3)

3. The growth of loudness as the bandwidth of the sound spectrum widens

4. The growth of loudness as the physical intensity of a given sound increases

Dependence of loudness on frequency

Fletcher and Steinberg (ref. 2) and Fletcher and Munson (ref. 3) appear to have made the first major attempts to measure loudness. Fletcher and Munson conjectured that loudness was proportional to the number of impulses leaving the cochlea per second. Fletcher and Munson specified a 1000-Hz tone as the standard sound against which other tones would be judged for loudness. Stevens (ref. 4) suggested that the unit of loudness be called the sone and that 1 sone be ascribed to a 1000-Hz tone set at a sound pressure level of 40 dB. On the sone scale, a sound twice as loud as a sound of 1 sone is given a value of 2 sones; four times as loud, 4 sones; *etc.*

Since those of Fletcher and Munson's, a number of studies have been conducted in which the loudness of pure tones and narrow frequency bands of noise has been judged relative to a tone or band of noise centered at 1000 Hz. For all intents and purposes, the D- and E-weighting contours shown in figure 2.3, and their respective counterparts, PNL and Mark VII (see fig. 5.1) fairly represent the findings. Figure 5.1 compares the shapes of various equal-loudness and equal-noisiness contours, separated to improve legibility (refs. 1, 5-7, and 8).

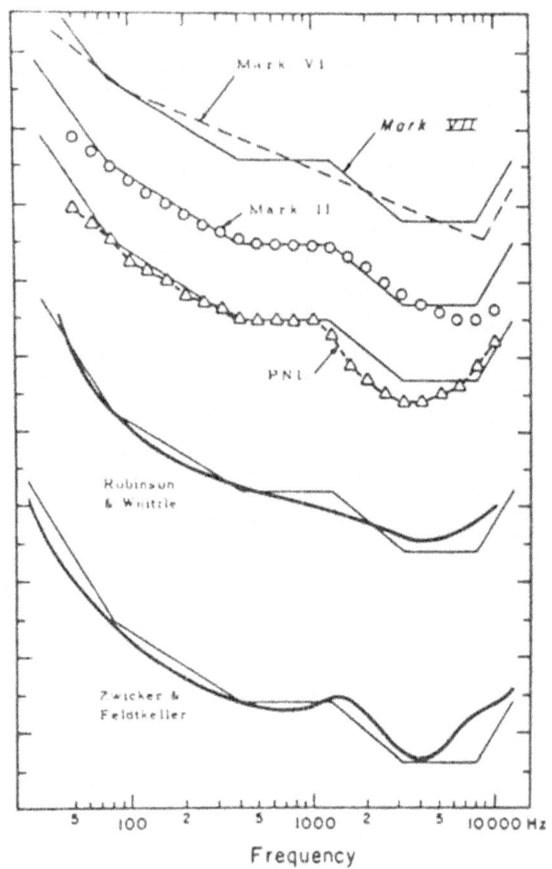

FIGURE 5.1 Comparison of various equal-loudness and equal-noisiness (PNL in PNdB) contours: Mark II, VI, and VII (proposed by Stevens); Robinson and Whittle's contour (ref. 6); Zwicker and Feldtkeller's contour (ref. 7); (PNL Kryter and Pearsons, refs. 5 and 8). Level of 1000-Hz comparison band was 80 dB for all the contours, here separated to improve legibility. (From ref. 1.)

Effect of bandwidth—Stevens' methods

Equal-loudness contours, whether for pure tones or bands of noise, are of somewhat academic interest unless they can somehow be used for evaluating the loudness of complex noises and sounds found in real life. Fletcher and Munson proposed a procedure for calculating from physical measurements the loudness of a complex sound consisting of a number of tones. However, their method was not used much because of its complexity.

Churcher and King (ref. 9) and later Beranek *et al.* (ref. 10) proposed that a simple summation in sones of the loudness of octave bands of sound would give a reasonable approximation to the perceived loudness of a complex sound consisting of one or more octave bands of random noise. For these purposes, an octave band of random noise having the same overall SPL as a pure tone of the same center frequency was assumed to be equally loud. In addition to the equal-loudness contours for octave bands of random noise, Stevens (refs. 1, 11, 12, and 13) also published new procedures for evaluating the total loudness of sounds with broad, continuous spectra. Stevens demonstrated that his method was more accurate in predicting the judged loudness of complex sounds consisting of bands of random noise than the method of simply adding together the sone values of individual octave bands.

Stevens' general formula is to add to the sone value of the loudest band a fractional portion of the sum of the sone values of the remainder of the bands:

$$\text{Loudness} = S_m + f(\Sigma S - S_m)$$

where ΣS is the sum of the sone values in all bands, S_m is maximum number of sones in any one band, and f is a fraction dependent on bandwidth. Stevens derived the fraction to be applied when the spectrum of the sound was measured in either full-octave ($f = 0.3$), one-half-octave ($f = 0.2$), or one-third-octave ($f = 0.15$) bands.

However, instead of expressing loudness in terms of sones, the loudness of a given sound is expressed in practice in terms of the sound pressure level, in decibels, of a reference sound when it is as loud as the given sound. The result is called loudness level in phons. The unit phon can be calculated from psychological units, sones, but not directly from physical measurements of sound pressure because the relation between sound pressure level and loudness varies with frequency differently at different levels of intensity.

Stevens (ref. 13) slightly modified his earlier method (ref. 12) of calculating loudness and named this new method Mark VI. Mark VI has been adopted by the American National Standards Institute as the procedure to be used for the calculation of loudness of noise measured in either octave, one-half-octave, or one-third-octave bands (ref. 14). The International Standardization Organization (ISO) (ref. 15) has recommended Mark VI as the method to be used for calculating the loudness of sounds measured with octave band filters, and Zwicker's method (refs. 16 and 17) described below, when the sounds are measured with one-third-octave band filters.

Effect of bandwidth—Zwicker's method

As previously mentioned, Fletcher and Munson suggested that loudness is proportional to the number of nerve impulses per second reaching the brain from the auditory nerve fibers. Further, they noted that two tones competing for the attention of a single nerve fiber would interfere with simple loudness summation and therefore all components within a certain frequency band should be grouped together and treated as a single component. The width of these frequency bands was estimated by Fletcher and Munson to be 100 Hz for frequencies below 2000 Hz, 200 Hz for frequencies between 2000 and 4000 Hz, and 400 Hz for frequencies between 4000 and 8000 Hz. From subjective tests of loudness and masking, Zwicker *et al.* (ref. 18) also determined the frequency groupings, "frequenzgruppen," that take place in the cochlea of the ear (see fig. 3.2). Frequenzgruppen are sometimes referred to as critical bands.

Zwicker (ref. 17) determined the spread of masking for narrow bands of noise, the threshold of audibility of pure tones, and the change in level of a 1000-Hz tone to obtain a doubling (or halving) of loudness. His results on the growth of loudness are similar to those found by Stevens (ref. 12) and Robinson (ref. 19). His published results for spread of masking for narrow bands of noise are more or less like the spread of masking data obtained by Egan and Meyer (ref. 20), Ehmer (ref. 21), and Carter and Kryter (ref. 22). Zwicker's assumption that there is a functional correspondence between masking and loudness is well substantiated by data on the critical bandwidth of the ear.

On the basis of these concepts, Zwicker (ref. 16) developed a graphic method for depicting and calculating the loudness of a complex sound. For calculation purposes, he prepared 10 graphs (covering both diffuse and free-field conditions, see fig. 5.2 for an example) in which the horizontal ordinates are marked off in equal frequenzgruppen (approximated for practical purposes by one-third-octave steps above 280 Hz), and the vertical divisions for each frequenzgruppen, in phons, are proportional to sones. The short-dashed curves show the area covered by the upward spread of masking. Plotting a sound spectrum on Zwicker's graph and drawing in the lines for spread of masking are supposed to show what proportion of available "nerve impulse units" are made operative as the result of exposure of the ear to a given sound. Accordingly, this area on the graph is presumably proportional to total loudness.

Dependence of loudness on intensity (growth of loudness)

Scaling the growth of loudness of a sound as a function of changes in its intensity into steps that are subjectively equal in size has been a somewhat controversial problem. Reviews of the work in this area have been made by Stevens (ref.

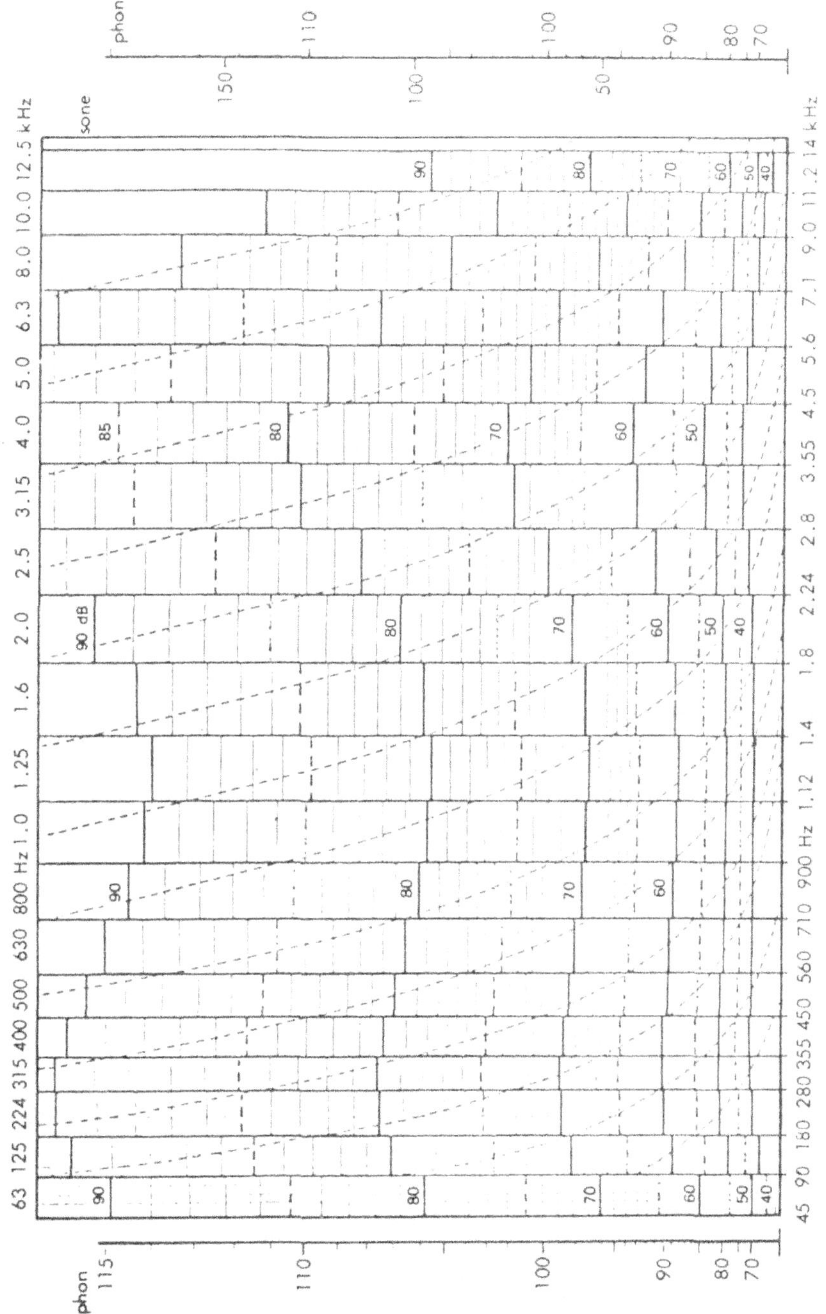

FIGURE 5.2 Example of loudness computation graph. (From ref. 15.)

23) and Gzhesik *et al.* (ref. 24). Three general methods have been used for scaling the growth of loudness of a sound, usually a 1000-Hz tone, as a function of changes in sound pressure level: (1) monaural vs. binaural loudness, (2) magnitude and ratio estimation, and (3) equal section or equal interval.

Monaural versus binaural loudness

The monaural versus binaural loudness argument used in reference 3 by Fletcher and Munson (which followed from their assumption that loudness was proportional to the number of auditory nerve impulses reaching the brain per second) was that a given sound when presented to two ears should appear to be twice as loud as when presented only to one ear. Fletcher and Munson found that the level of the monaurally presented tone had to be set about 10 dB higher in level than the level of an equally loud binaurally presented tone. Thus, they concluded that over at least the middle range of loudness levels, loudness about doubles for each 10 dB increase in the sound pressure level of a sound.

Reynolds and Stevens (ref. 25) found that the loudness scale for monaural listening was somewhat different from the loudness scale for binaural listening and thus that Fletcher and Munson's assumption about the summation of loudness from the two ears appeared less than perfect, at least at some intensity levels. However, Hellman and Zwislocki (ref. 26) later found nearly perfect, within experimental error, interaural summation of loudness, as shown in figure 5.3.

Magnitude and ratio estimation

The monaural versus binaural equal-loudness scale is very similar to the average of those developed on the basis of magnitude estimations of the loudness of sound presented only monaurally or only binaurally. In this method, the subjects assign a number, say 100, to a tone at a particular SPL, say, 100 dB; they are then asked to assign the number 50 to the tone when it sounds half as loud as it did at 100 dB. Another method is that of estimating loudness ratios or fractions; here the subjects may adjust the level of a tone until it is one-half, one-tenth, *etc.*, as loud as a standard or reference level.

Results of studies by various investigators using the magnitude estimation and ratio judgment methods differ rather widely (Kryter, ref. 27). Garner (refs. 28 and 29) believes that the differences among these results are due in part to context effects. That is, subjects' judgments about what appears half as loud are different when they know the total range of levels available to them for judgment and when they do not.

In reviewing loudness scaling procedures, Stevens (ref. 23), makes the point that although obtaining a loudness scale from a listener is a difficult problem, this scale must be determined if the concept of loudness is to have any practical utility. Stevens suggests that the best method is "magnitude production" in which each subject is allowed to choose a number scheme and then results are

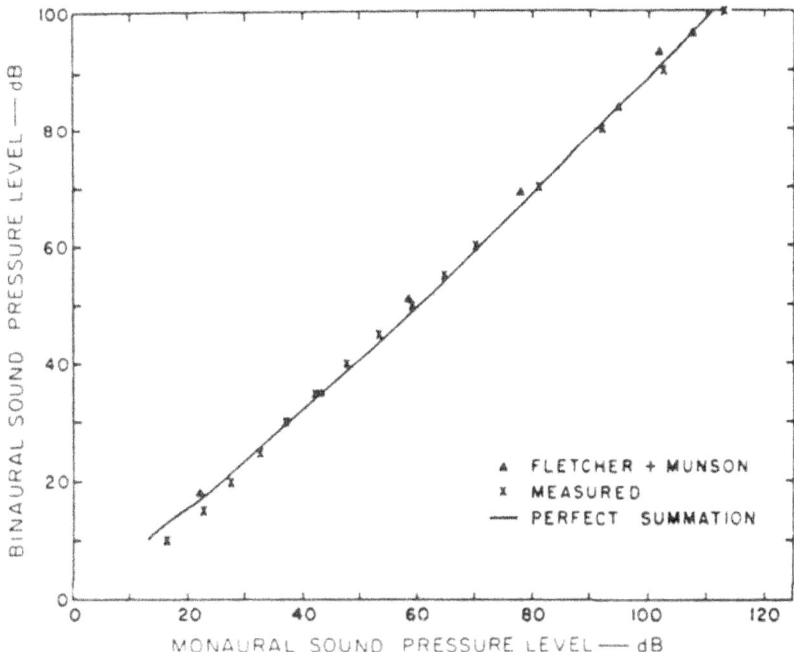

FIGURE 5.3. Binaural sound pressure level as a function of monaural sound pressure level at equal loudness. (From ref. 26.)

averaged across subjects, after normalizing the results for individual differences in the choice of numbers used.

Equisection loudness scale (equal intervals)

In addition to the monaural versus binaural and the methods of magnitude and ratio estimation, a method of equal intervals, or equisections, has been suggested as a suitable method for deriving a scale of loudness. In this method, the subjects hear a tone presented at, in the simplest case, two different levels of intensity; they are then told to adjust the third level of the same tone so that the difference in loudness between the second and third levels is equal to that between the first and second levels. Using this method, Wolsk (ref. 30), Kwiek (ref. 31), and Garner (ref. 32) measured equal intervals over various ranges of intensity of a 1000-Hz tone.

Unlike the magnitude and ratio estimation methods, the results obtained by various investigators using the method of equal intervals agree closely with each other. However, no real knowledge is obtained from the equal interval method as to what changes in level are required in order for the listener to report a subjective sensation of, for example, the doubling or halving of the loudness of a sound.

Since the loudness scale derived by the equal-interval method is so different from the scales derived by other methods (see fig. 5.4), we must choose one or the other for a practical use. It would seem reasonable to choose the form of loudness function on the basis of how the loudness scale is to be used. If, for example, we intend to say that sound A is twice (or some portion) as loud as sound B, then we are obliged to use a loudness scale based on ratio or magnitude judgments. On the other hand, if we want to decide whether the difference in loudness between sounds A and B is equal to the difference in loudness between B and C, then the Garner-Kwiek equal-interval loudness scale would be more meaningful. The general interest in loudness judgments in real-life situations is probably more in terms of apparent magnitude or relative loudness than in terms of equal intervals, and thus the loudness scale based on magnitude estimation seems the more appropriate for general use.

Critical summation time for loudness

As mentioned in chapter 3, the ear utilizes a 1-sec integration period of frequency-weighted (for loudness) sound energy to fully perceive loudness level when a sound is near its masked threshold. These findings, based on data from Garner and Miller (ref. 33), are shown in figure 5.5 (from ref. 34). Sensation level, the abscissa, refers to the difference, in decibels, between the intensity of a sound and its intensity at the threshold of audibility in quiet.

When the sounds are at suprathreshold levels, this critical summation period is about 0.3 to 0.5 sec. These findings, from judgments of brief bursts of narrow-band noise impulses, are shown from studies of the noisiness of suprathreshold impulses by Fidell *et al.* (ref. 35) in figure 5.6, and of the loudness of suprathreshold bursts of a 1000-Hz tone by Pedersen *et al.* (ref. 36) in figure 5.7. Judged loudness and noisiness follow the same temporal summation course when the durations are less than 1 sec. As mentioned before, the ear probably requires a somewhat longer loudness integration time for sounds at masked threshold (fig. 5.5) than at suprathreshold levels (1 sec compared with about 0.3 sec) because of the greater physical uncertainty in the signals being processed and the more difficult discriminations required at threshold than at suprathreshold conditions.

Perceived Noisiness

The following assumption appears to be implicit to most of the quantitative approaches that have been made to the evaluation of annoyance due to environmental noise: Since individuals in a community live in somewhat similar houses and daily repetitive ways, the average amount of annoyance due to noise interfering with auditory communications, sleep, and rest and to house vibration would be predictable, in a statistical sense, from certain physical measures of the environmental noise. Without this assumption, concepts such as loudness and perceived

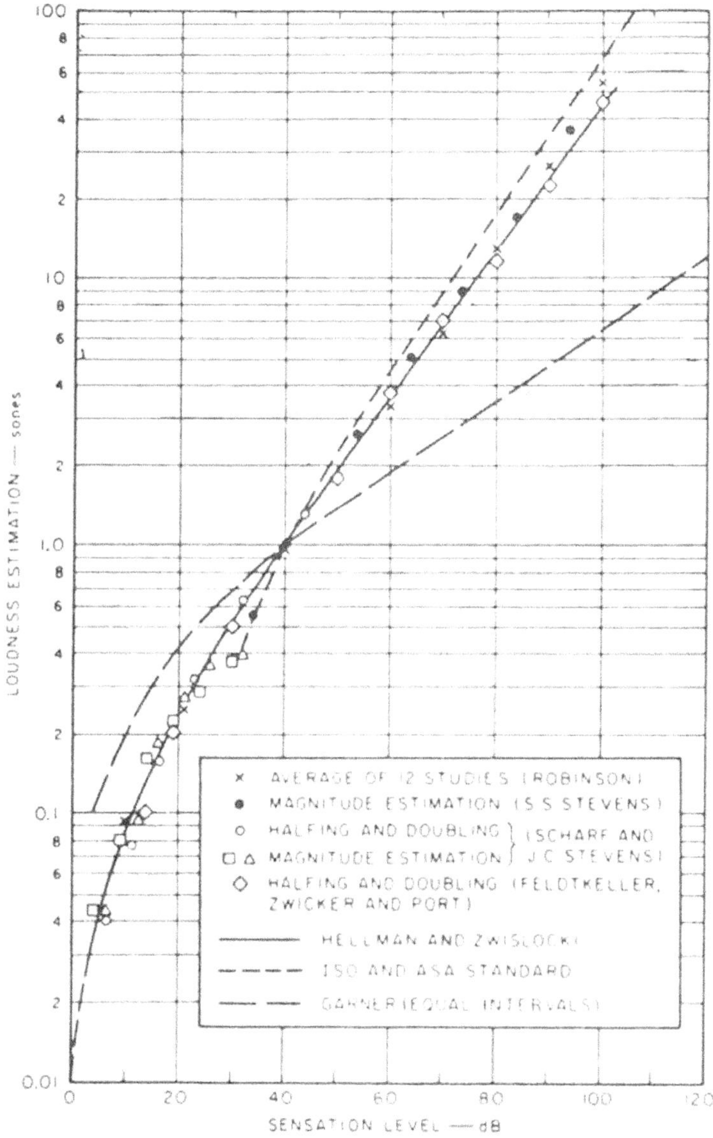

FIGURE 5.4. Comparison of binaural loudness results of several investigators. (From ref. 26.)

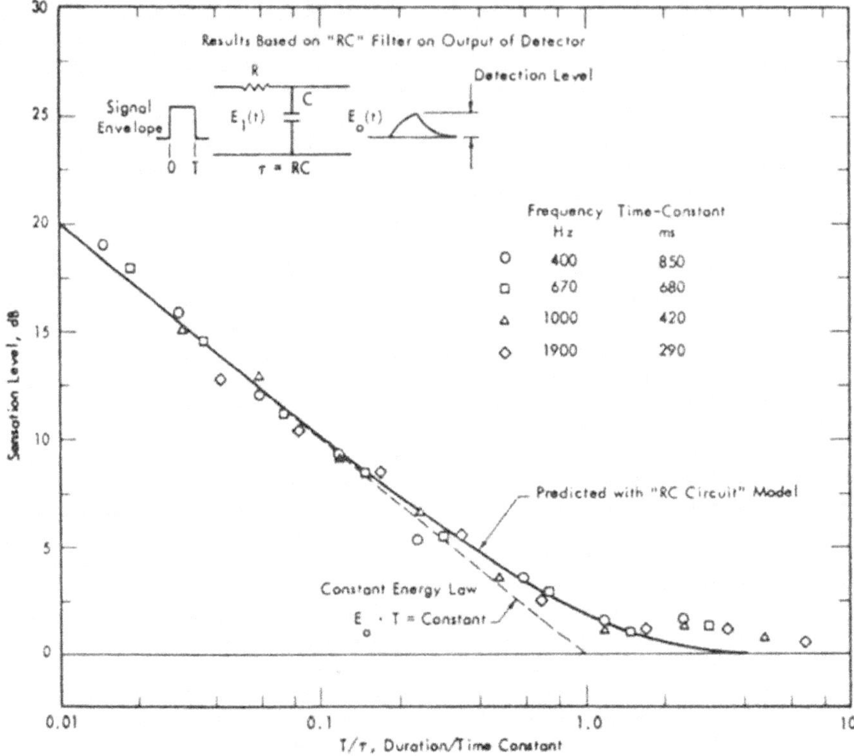

FIGURE 5.5. Measured and predicted (based on resistance-capacitance, RC, filter model) values for change in signal-to-noise ratio of single tone bursts as a function of burst duration (data from ref. 33). (From ref. 34.)

noisiness become rather useless for the assessment of the effects of environmental noise on people.

The concept of both loudness and perceived noisiness exclude from consideration the meaning conveyed by a sound or noise. However, while loudness and perceived noisiness may be similarly related to sound spectra at a moment in time, they are not similarly related to certain important temporal variables in sounds or noises. In brief, as will be shown, loudness is too simple to serve as a general concept for the assessment of noise as unwanted sound.

In real life, people feel annoyed by a noise, independent of its meaning, primarily when the noise interferes with their sleep or auditory communications. This is typical when the noise is a natural and expected part of the living environment, and the people exposed are used to the noise and experience no feelings of fear or the like. Evidence indicates (Borsky, ref. 37) that the feelings of annoyance from noise interference effects stays the same or grows with continued years of living in

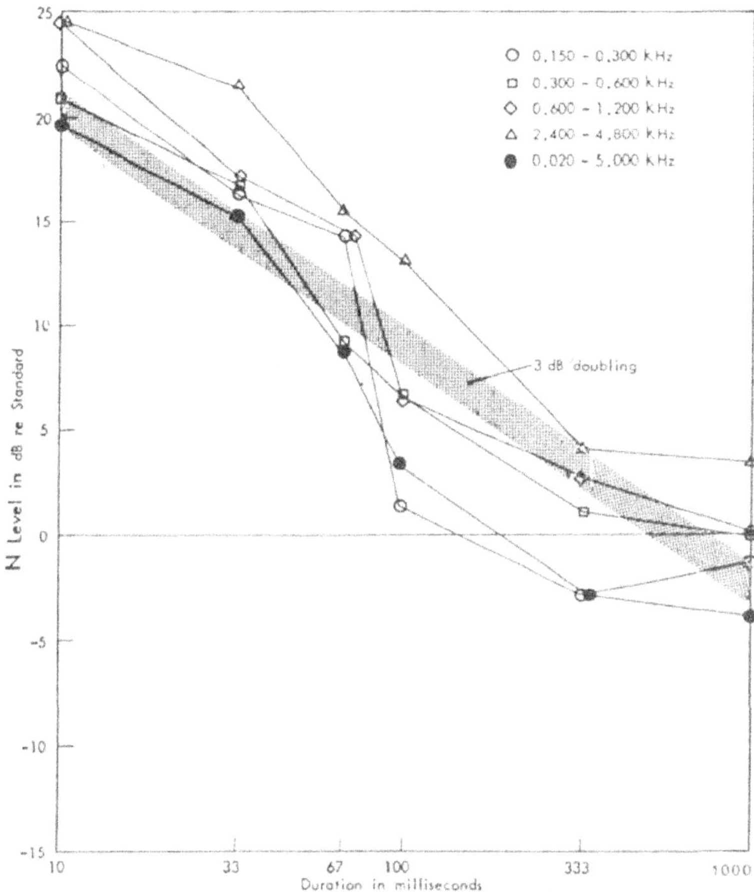

FIGURE 5.6. The sound pressure level of various octave bands at different durations when judged equally noisy (N) as a reference standard octave band of random noise between 0.600 and 1.2 kHz and of 1000-msec duration. (From ref. 35.)

a noisy area even though the people have adapted to the fact that the noise is to be present.

Psychological and sociological factors can usually be reconciled with the general attribute of sound called noisiness. For example, Cederlöf *et al.* (ref. 38) found that propaganda stressing the importance of military aviation and the plans of the government to control and lessen the noise reduced the willingness of citizens near military airports to complain about the aviation noise; the reduction was equivalent to the effect that would have been obtained by lowering the noise levels by 6 dB or so. At the same time, while the willingness to complain can be changed, the number of times the noise masks speech or interferes with sleep

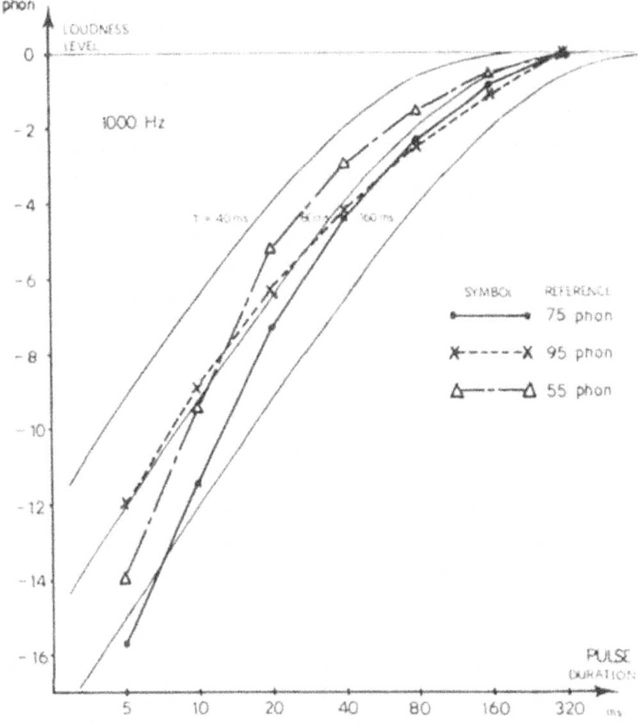

FIGURE 5.7. Loudness level of a 1000-Hz tone pulse as a function of the pulse duration. Shown are results for the three level series and curves conforming to a simple RC (resistance-capacitance) time constant of power τ = 40, 80, and 160 msec. (From ref. 36.)

would not be changed, so that the amount of annoyance should remain substantially unchanged.

The absolute amount of annoyance from noise to be experienced in real life is clearly an empirical question answerable only from physical and psychological tests in communities. The results of such tests are presented for some major environmental noises in chapter 11. However, the basis for the understanding of these noise effects and procedures of noise measurement can presumably be developed from laboratory research and concepts. The remainder of this chapter is concerned primarily with basic concepts and research relating physical measurements of individual noise occurrences with psychological judgments of the noise.

Definition of noisiness

The subjective impression of the unwantedness of a not unexpected sound that does not provoke pain or fear is defined as perceived noisiness, or, more simply, noisiness. As mentioned before, confusion sometimes results from use of the word "noise" as a name for unwanted sound because there are two general classes of "unwantedness."

Unwanted sound in the first category signifies or carries information about the source of the sound that the listener has learned to associate with some unpleasantness not due to the sound per se, but due to some other attribute of the source; the sound of the fingernail on the blackboard perhaps suggests an unpleasant feeling in tissues under the fingernail, a baby's cry causes anguish in a mother, the squeak of a floorboard is frightening as indicating the presence of a prowler, a sonic boom is disturbing because it is an unfamiliar sound. In these cases it is not the sound that is unwanted (although for other reasons it may also be unwanted), but the information it conveys to the listener. This information is strongly influenced by the past experiences of each individual. Because these effects cannot be quantitatively related to the physical characteristics of the sounds, they are rejected from the concept of perceived noisiness.

Unwanted sound in the second category annoys a listener because of the physical content of the sound per se and not because of the meaning, if any, of the noise. Psychological judgment tests have demonstrated that people judge fairly consistently among themselves the unwantedness, unacceptability, annoyance, objectionableness, or noisiness of sounds that vary in their spectral and temporal nature, provided that the sounds do not differ significantly in their emotional meaning and are equally expected. Presumably this consistency is present because people learn through normal experience the relations between the characteristics of sounds and their basic perceptual effects: loudness, masking, arousal from sleep, and, for impulses, startle. Although noise evaluation procedures specific to individual effects, such as speech masking, loudness, and auditory fatigue, are available, a single number rating, including some special correction factors, for the average unacceptability or perceived noisiness of normal environmental noises appears to be an adequate basis, as shown later in the text, for community noise control and management from a physical standpoint. Another hypothesis concerning the concept of perceived noisiness is that even though the absolute levels of noisiness or unacceptability of the noise from a given source may differ somewhat among people, variations in the frequency content, duration, and spectral complexity have the same relative effect on the noisiness perceived by each individual.

Noisiness is obviously synonymous with what is often implied by the word "annoyance." However, annoyance commonly signifies one's reaction to sound based both on its physical nature and its emotional content and novelty (which are excluded from perceived noisiness). The phrase "perceived noisiness," although somewhat redundant, was chosen (Kryter, ref. 5) in an attempt to avoid some of the ambiguity possible from the word annoyance. The word "noisiness" by itself perhaps is an adequate name for this attribute, the word "perceived" merely emphasizing the fact that the quantity of interest is a physiological-psychological entity.

Note that perceived noisiness is a subjective quantity or attribute not to be confused with a particular unit of physical measurement, such as PNdB or dBA, as is discussed further below.

Instructions to subjects

The words used in the instructions to subjects for judgment tests of the acceptability of sounds have some influence on their rating of sounds (Pearsons and Horonjeff, ref. 39), as illustrated in figure 5.8. It is difficult and probably academic to fathom the range of differences shown in figure 5.8, for example, whether the words used really mean different things to different people. In any event, there is no apparent reason why listeners should not be asked to rate directly sounds in terms of their unwantedness, unacceptability, annoyance, or noisiness, as synonyms, rather than to rate their loudness, in the expectation that the latter is only a partial clue to the noisiness or unwantedness of the sounds.

Berglund *et al.* (ref. 40) have also conducted studies on the effect of different instructions (loudness, noisiness, and annoyance) on the judgments of aircraft noise relative to a broadband random (white) noise of 10-sec duration. The aircraft noise durations varied from about 4 sec to 34 sec between the points 10 dB below their maximum level.

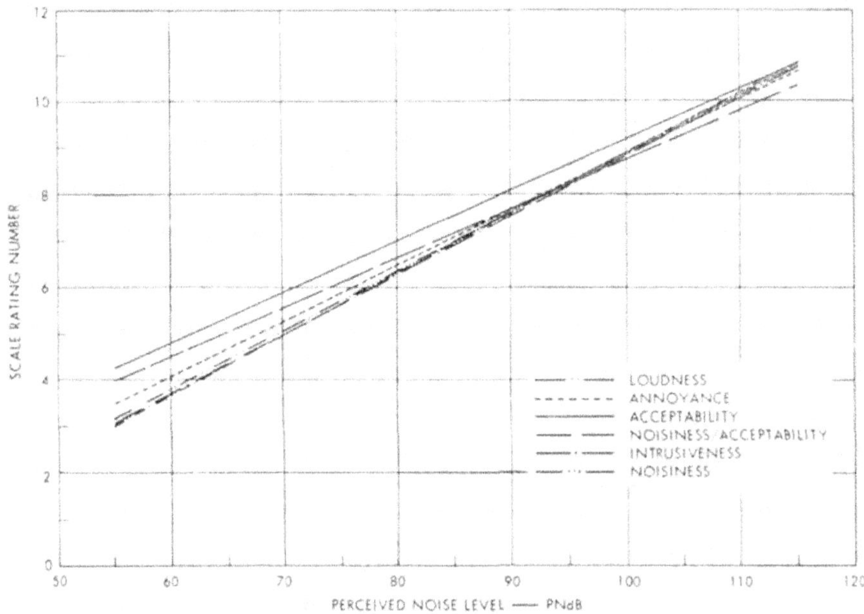

FIGURE 5.8 Mean response ratings of noise stimuli in a laboratory for different rating scales, *i.e.*, different words used to describe acceptability. (From ref. 39.)

These investigators defined

1. Loudness as "the perceptual aspect that is changed by turning the volume knob on a radio set."

2. Noisiness as "the quality of the noise. A jackhammer may be more or less noisy than a motorbike. Music may be loud but not noisy."

3. Annoyance as the nuisance aspect of a noise. "Imagine you hear the noise after a hard day's work while comfortably seated and intending to read your newspaper."

These investigators found large differences between the levels of aircraft noise judged to be equally loud, noisy, or annoying (see fig. 5.9). This is in contrast to the small differences (fig. 5.8) found by Pearsons and Horonjeff (ref. 39).

The findings of Berglund *et al.* are rather startling. For example, figure 5.9 shows that if an aircraft noise at a maximum level of 70 dBA is judged equal in loudness to the 10-sec sample of white noise, the aircraft noise would be judged as annoying as the white noise only when its level has been reduced about 50 dB—presumably a maximum level of 20 dBA. At that level the aircraft noise would be nearly inaudible. Also Berglund *et al.* conclude that $L_{A,\max}$ (or $dBA_{\max}$) is not only sufficient but superior to $L_{A,\text{eq}}$ or $L_{\text{PNdB,eq}}$ (EPNdB) as a means of predicting the judged noisiness of aircraft noise. This differs from the findings of others and also from common experience: that the duration of noise usually has much to do with its acceptability, unwantedness, or noisiness. Some of these differences between the results of Berglund *et al.* and other studies on this subject may arise from their definitions. For example, to paraphrase their instructions to the subjects, a difference in noisiness is a possible difference in the quality of the noise, for exam-

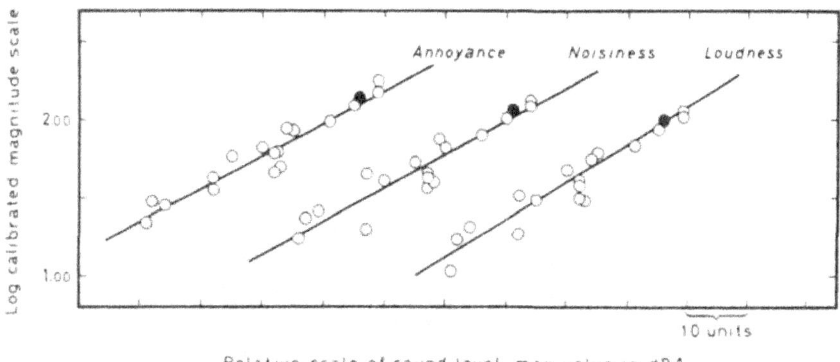

FIGURE 5.9. Psychophysical relations for the three attribute scales. Filled symbols refer to the standard white noise. The regression lines are fitted by the method of least squares. (From ref. 40.)

ple, from a jackhammer and a motorbike, independent of their loudnesses. This concept of quality is a much narrower definition of noisiness than that used in the other studies of perceived noisiness.

Physical Aspects of Sound Directly Related to Noisiness

For practical purposes, the measurable physical aspects of a sound that are most likely to control its noisiness must be determined. To date, six significant features have been identified: (1) frequency spectrum, (2) sound level, (3) spectrum complexity (concentration of energy in pure tones or narrow frequency bands within a broadband spectrum), (4) duration of the total sound, (5) duration of the increase in level prior to the maximum level of nonimpulsive sounds, and (6) the increase in level of impulsive sounds within an interval of 1 sec.

Some physical aspects that might seem important appear to have very secondary effects on people compared with the six physical characteristics mentioned above. Examples of such unimportant factors are Doppler shift (the change in the frequency and sometimes noted pitch of a sound as the sound source moves towards and away from the listener (see Nixon *et al.* (ref. 41) and Ollerhead (ref. 42) and modulation of pure tones (see Pearsons (ref. 43)).

Frequency spectrum

Some experiments were performed in 1943 to pursue the earlier work of Laird and Coye (ref. 44) on the annoyance values of sounds containing different frequencies. The data from these studies, reported by Reese *et al.* (ref. 45) and Kryter (ref. 46), showed that higher frequencies tended to be more annoying than lower frequencies even though they were equally loud.

Kryter (ref. 5) proposed that the subjective unit of noisiness be called the noy parallel to the sone for loudness. A sound of 2 noys was said to be subjectively twice as noisy as a sound of 1 noy; 4 noys, to be four times as noisy as a sound of 1 noy; *etc.* PNdB was coined as the name of a unit of perceived noise level, analogous to the phon for loudness. An increase of 10 PNdB in a sound is equivalent to a doubling of its noy value. The band calculation procedure developed by Stevens for loudness level was adopted (except for the frequency-weighting contours) for the calculation of PNdB. As noted earlier, perceived noisiness and perceived noise level represent the subjective quantity of primary interest. The PNdB is but one unit of physical sound measurement relating to the subjective quantity; dBA, phons, or other units of sound measurement can be, and are, effectively used to predict perceived noisiness.

Kryter and Pearsons (refs. 8 and 47) later obtained further data on equal-noisiness contours (see figs. 5.10 and 5.11) which were proposed for use in place of the contours suggested earlier in 1959. These contours were obtained with

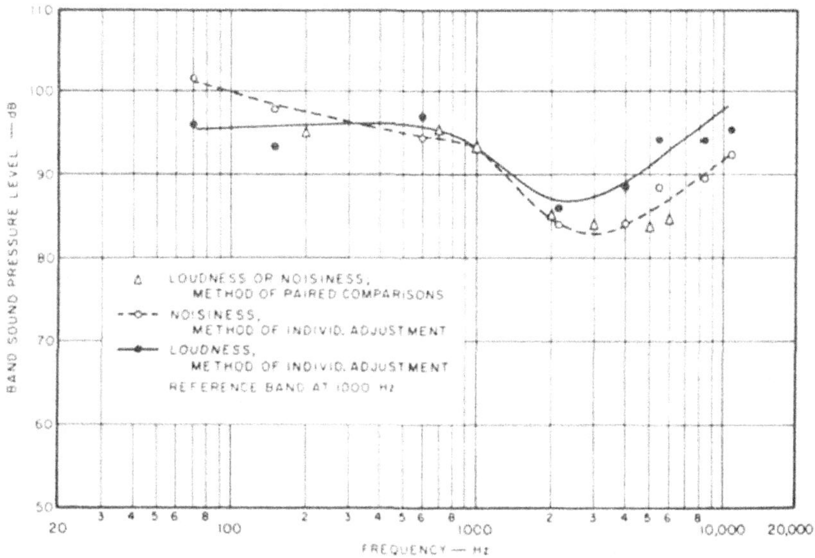

FIGURE 5.10. Equal-loudness and equal-noisiness judgments. (From ref. 8.)

bands of random noise in the middle to higher levels (60 to 100 dB) of intensity; the contours for the very low and highest levels were extrapolated from the lowest and highest experimentally found contours.

Wells (ref. 48) obtained a set of equal-noisiness contours with one-third-octave and full-octave bands of noise. Stevens compiled composite equal-loudness and equal-noisiness contours from various published loudness and noisiness contours, and labeled the contours perceived level, Mark VII (ref. 1). These contours and the contours of Wells (ref. 48) and Ollerhead (ref. 49) are similar in general shape to those obtained by Kryter and Pearsons, as shown in figure 5.11.

Overall frequency weightings

A sound level meter with the overall (OA) frequencies A-weighted can be used with some accuracy as a means of rating the objectionableness of aircraft noises (Kryter (ref. 5), and Young and Peterson (ref. 50). In 1960, Kryter (ref. 51) recommended that the 40-noy contour (first called N, now called D) be used as the basis of a frequency weighting for sound level meters (see D1, table 2.1). Kryter (ref. 52) later suggested that some modifications to the D contour at frequencies below about 250 Hz might be desirable and consistent with the critical bandwidths of the ear at those lower frequencies (see D2, table 2.1).

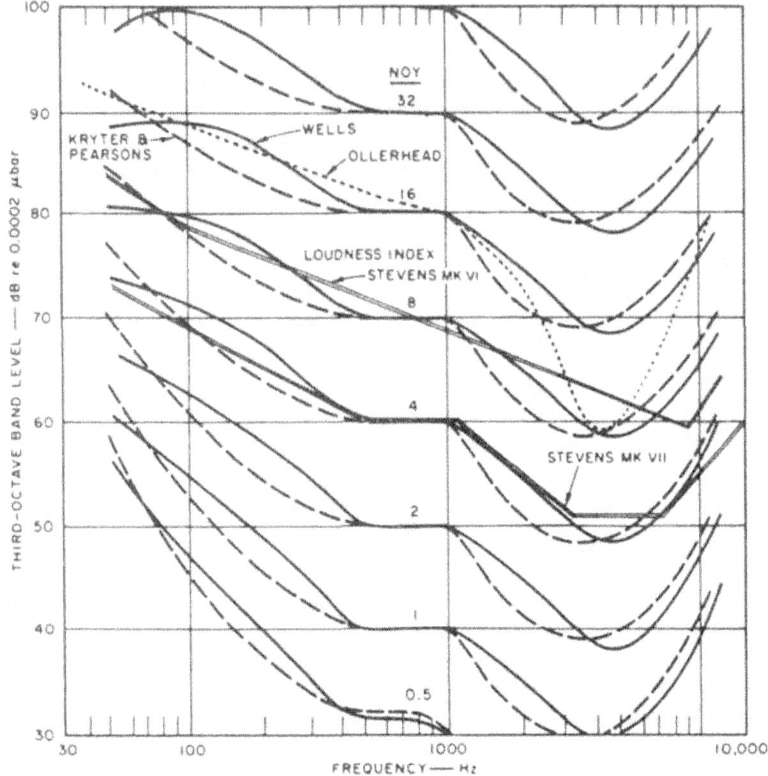

FIGURE 5.11. Equal-noisiness contours as found by Kryter and Pearsons (refs. 8 and 47), Wells (ref. 48), and Ollerhead (ref. 49); and loudness indexes, Mark VI (Stevens, ref. 13), and Mark VII (Stevens, ref. 1).

In many circumstances, A-weighting is justified with broadband noises because of the relative equal weightings, as distinct from absolute weightings, that are earned by these weightings with broadband noises; as seen in figure 2.3, compared with D-weighting, A-weighting overestimates the loudness of mid-frequencies (500 to 2000 Hz) and underestimates the loudness of frequencies above or below that range. Scharf *et al.* (refs. 53 and 54) and Kryter (ref. 55) compared the results of judgments of loudness and noisiness of hundreds of different noise spectra that have been published over the years. Some findings of Scharf *et al.* are shown in table 5.1. For the interpretation of table 5.1, the magnitude of the average differences between the calculated and observed levels can be corrected for, provided that the differences have a small standard deviation. Accordingly the merit of a calculated unit is more a function of the standard deviations than

TABLE 5.1

Calculated Minus Observed Loudness Levels

[From ref. 54; overall means based upon differences for 335 spectra grouped according to spectral type]

By spectral category	A	D1	D2	E	Mark VI	Mark VII	PNL	Zwicker loudness level
Mean of mean differences, dB....	−12.1	−5.3	−5.8	−6.8	−1.2	−8.6	−1.4	3.1
Standard deviation of means, dB...	4.8	4.0	4.3	4.1	3.2	3.2	3.1	3.0
Range, dB	16.0	12.6	13.7	12.6	11.6	11.1	10.8	8.8

the means of the differences shown in table 5.1. Scharf *et al.* reached the following conclusion on the question of frequency weightings:

> One important conclusion, based on a total of over 600 spectra, was that the calculation procedures predicted subjective magnitude with less variability* and with greater validity** than did the frequency weightings.
>
> Among the six frequency weightings studied, the B- and C-weightings were the poorest predictors of subjective magnitude while the D1-, D2-, and E-weightings were the best predictive weighting functions. It was also noted that the A-weighting was less than 0.5 dB more variable than the D1-, D2-, and E-weightings. Among the five calculation procedures studied, Stevens' Mark VI (1961), Mark VII (1972), and Zwicker's (1958) loudness calculation procedures were the least variable, but Perceived Noise Level (Kryter 1959) was almost as reliable. Tone-corrected Perceived Noise Level (following the FAR 36 procedure, 1969) was a somewhat poorer predictor. Mark VI and Perceived Noise Level yielded the calculated values that were closest, on the average, to the observed or judged values, although all of the frequency weightings and computational procedures examined were about equally variable in this respect.

The question as to whether some sound frequencies are inherently noisier than other frequencies of equal loudness remains a matter of some dispute. As shown in figure 5.11, judgments of the relative noisiness vs. loudness of narrow bands of noise or tones indicate that the higher frequencies (above about 1500 Hz to at least 7000 Hz) are considered to be somewhat noisier than loud. The difference is relatively small, however, and in the context of broadband noises (as well as problems encountered in the testing of subjects), the question of possible differences between these two attributes in this regard must be considered academic insofar as general noise assessment procedures are concerned.

---

*The index of variability was the standard deviation of the calculated levels of a group of sounds judged subjectively equal or the standard deviation of differences between calculated and judged levels. These typically ranged from 2 to 4 dB.

**The calculation procedures yielded an absolute calculated level closer to the observed level.

Accordingly, as far as a simple overall frequency weighting for use with a sound level meter is concerned, the conclusion is that D- and E-weightings are slightly better (1 to 2 dB) than the A-weighting for broadband, nonimpulsive noises. For narrow-band noises, D- and E-weightings can be 0 to 10 dB better than the A-weighting in predicting judged loudness or noisiness, depending on their frequency. As we shall see later, the D-weighting is generally superior to the A-weighting in the assessment of impulsive sounds.

Level

Although loudness and noisiness differ in some respects, an assumption of the concept of the noisiness of nonimpulsive noises is that as the intensity of a noise changes, with other factors kept constant, the subjective magnitudes of loudness and noisiness change to a like degree (*e.g.*, a 10-dB increase in the physical intensity causes a doubling, 100-percent increase, of its loudness and its noisiness over a fairly wide range of intensities). There is some experimental proof of this common relation between this subjective scale of noisiness and loudness, but as with loudness, the scale found depends somewhat on the experimental methods used and sounds judged (refs. 42, 56, and 57).

Kryter, in research conducted for the Port Authority of New York and New Jersey on the impact of noise from the Concorde around J. F. Kennedy Airport, found that the "10 dB per doubling" growth of noisiness agrees fairly well with laboratory experiments with aircraft noise up to peak indoor levels of about 80 dBA (fig. 5.12(a)), but that at higher levels (fig. 5.12(b)), the subjective noisiness or unwantedness grows at a somewhat greater rate. This finding is consistent with some attitude survey findings in communities that show a sharp increase in the rate of growth of annoyance as a function of the exposure level of aircraft noise when the level exceeds a certain high level.

Spectral complexity

Noisiness is often judged to be greater for broadband sounds that contain relatively high concentrations of energy in narrow bands (one-third octave or less wide) than for broadband sounds with energy distributed equally over frequency. The judged perceived noisiness of the sounds that have energy concentrated in tones or in narrow bands that exceed adjacent band levels can be estimated, approximately, through corrections applied to the measured sound pressure levels normally used in calculating perceived noisiness. A somewhat simplified version of correction factors developed by Kryter and Pearsons (ref. 58) for this purpose is given in figure 5.13.

Little (ref. 59) suggested another method for correcting calculated PNL sounds that contain pure tones. The magnitude of the correction is a function of the tone-to-noise ratio and frequency of the tone. However, unlike the procedure outlined in the preceding paragraph, only one correction is added to a sound

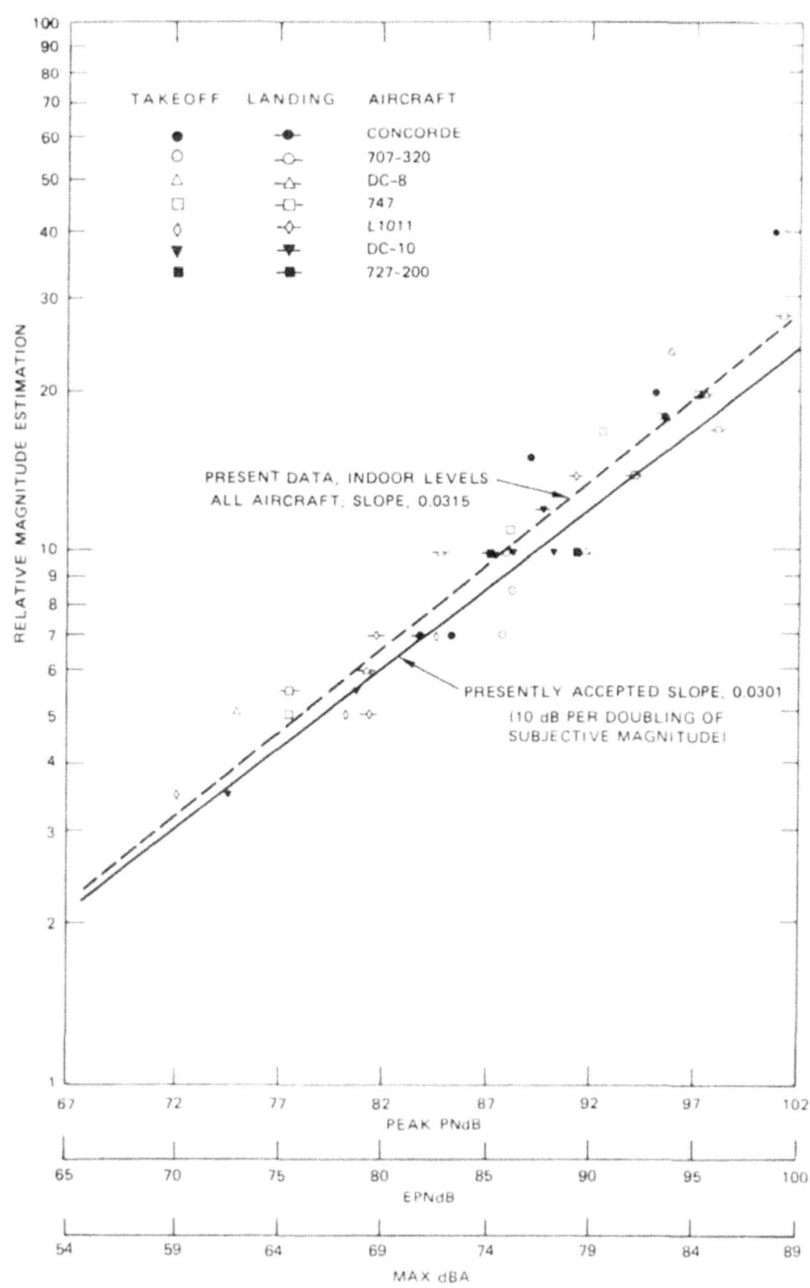

(a) Aircraft noise level of about 60-90 max dBA.

FIGURE 5.12 Relative estimates of subjective noisiness of aircraft noise as a function of level.

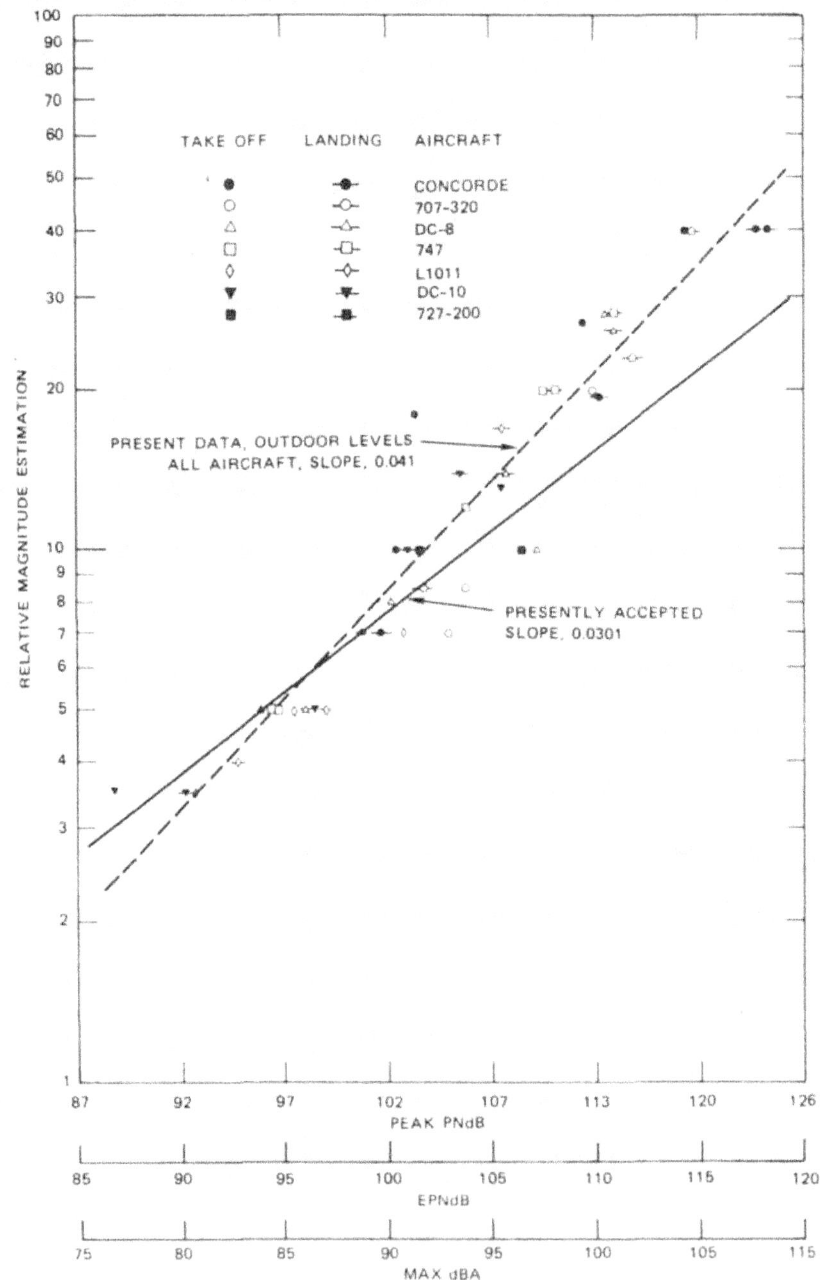

(b) Aircraft noise level of 80–115 max dBA.

FIGURE 5.12. Concluded.

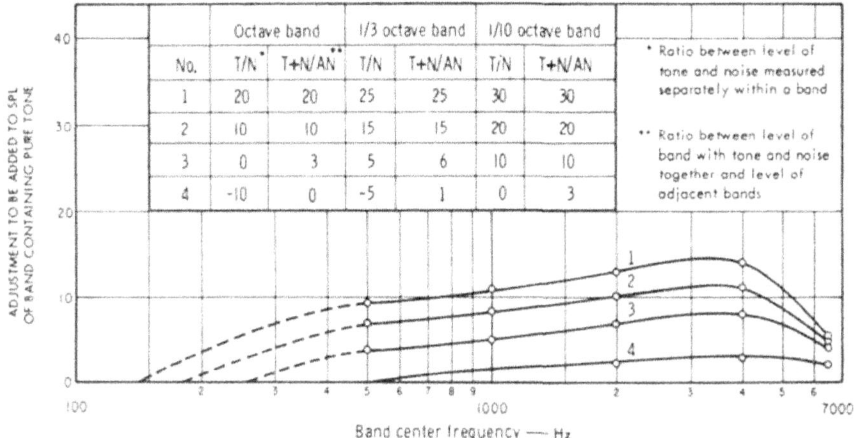

FIGURE 5.13. Correction, in decibels, to be applied to the SPL of a broadband sound containing a pure tone or very narrow band of energy. (From ref. 58.)

even though more than one pure-tone component is present, and the magnitude of the correction is independent of the absolute intensity of the bands of noise. Sperry (ref. 60) quantified the Little method for use by the Federal Aviation Administration (FAA).

In their method for computing the subjective reaction to complex sounds, Wells and Blazier (ref. 61) attempted to account for the effect of pure-tone components on judged noisiness. In the Wells and Blazier approach, the value of one of a family of frequency-weighted contours (given in published graphs) that is tangentially closest to a given sound spectrum is assigned to the actual spectrum of the sound in question. The value is, however, corrected according to the number of one-third-octave bands within 5 dB of the highest contour tangent to the sound spectrum.

Test of tone corrections

Perhaps the most exhaustive tests of the validity of tone corrections for predicting noisiness are those carried out by Pearsons (ref. 43) with single and multiple steady-state and modulated tones. Figure 5.14 illustrates the general spectra of the standard, or reference, sound and the comparison sounds. Some of his findings, shown in figure 5.15, indicate that tone corrections resulted in PNL's for the comparison noises that were comparable to the PNL of the standard sound judged to be equally noisy, whereas PNL's without tone corrections differed considerably. The results of some earlier judgment tests (Kryter *et al.*, ref. 62) with modulated and multiple pure tones are somewhat inconsistent with those findings.

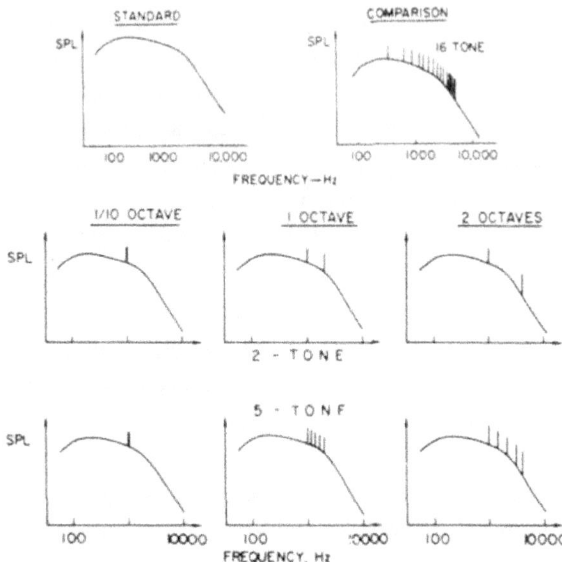

FIGURE 5.14. Spectrum of standard and 16-tone comparison stimuli (upper graphs) and samples of spectra of 2- and 5-tone comparison stimuli (lower graphs). (From ref. 43.)

However, in studies (Kryter, ref. 52) of the noisiness of aircraft sounds, some of which contained pure-tone components, tone corrections did not always appear to be necessary for obtaining the best estimates of judged perceived noisiness. Part of the difficulty appears to be related to the unreliability that is introduced in the detailed band-level measurements required for making the tone corrections.

From their assembly of data from studies of broadband noises that contained some tonal components, Scharf *et al.* (refs. 53 and 54) calculated that the various tone-correction procedures did not improve predictions of the judgments. It must be concluded at this time that only in situations where the temporal and spectral factors are regular or stable can the effects of tones on judged noisiness be predicted from tone-correction procedures.

Duration of a noise

Up to this point, the discussion has been concerned primarily with the calculation of the loudness and noisiness level of sounds of presumably equal durations. The comparisons of the noisiness of sounds of different durations are often of interest and importance.

The human auditory system appears to be able to combine into a perceived entity of loudness or noisiness the distribution of energy present in the sound spectrum in any 1-sec interval of time. Also, humans seem to perceptually inte-

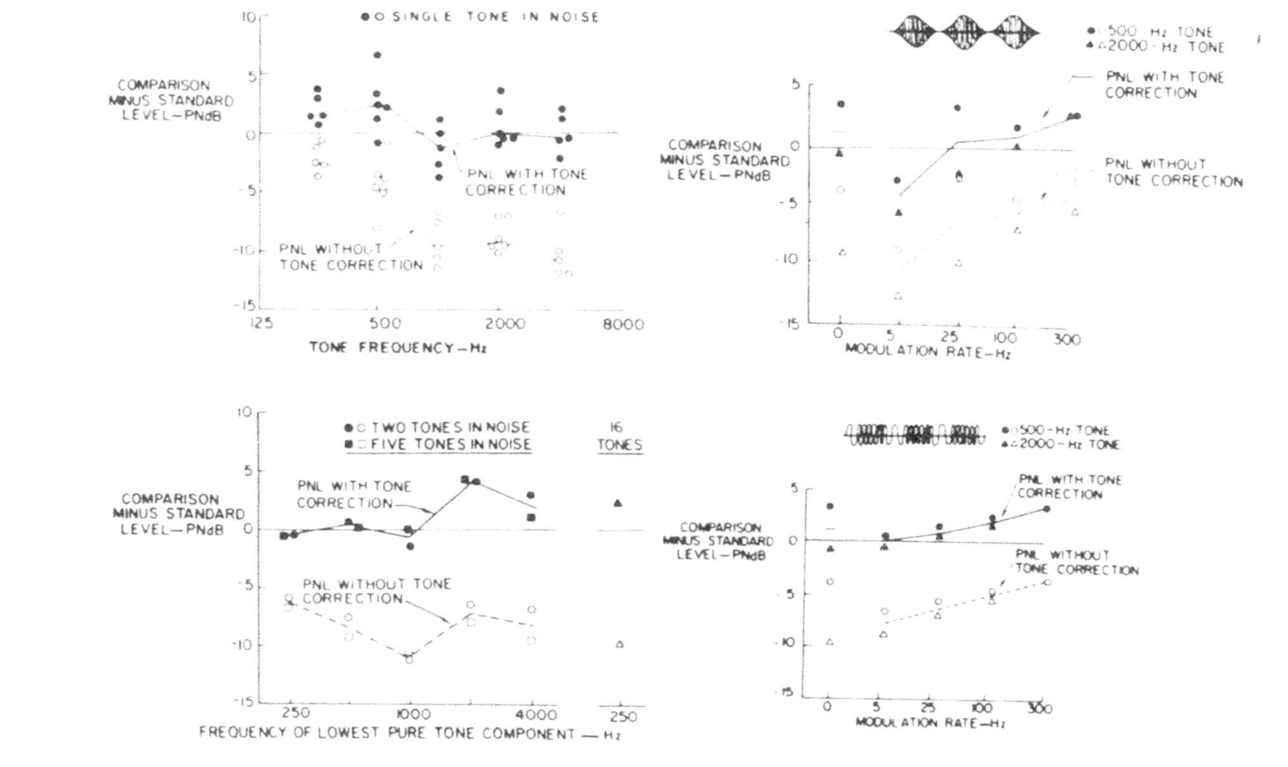

(a) Standard noise without tones, comparison with single and multiple tones.

(b) Amplitude modulated tones (lower graph) and frequency modulated tones (upper graph).

FIGURE 5.15. Difference in PNL when comparison and standard sounds are judged to be equally noisy. Tone corrections according to method of reference 58. (From ref. 43.)

grate, as a continuum, successive intervals of noisiness into an entity of perceived noisiness for the total duration of an identifiable sound. To perform physical integrations of sound as a means of estimating perceived noisiness of the total sound, it is first necessary to decide, among other things, at what threshold level of intensity this integration process should be started.

Threshold of noisiness for speech and sleep interference

As is shown later, in real life the noisiness of sounds, independent of their psychological meaning, is due to the noise interfering with auditory communications, especially speech, and with sleep or rest. House vibrations and rattles induced by the noise serve as an additional major source of annoyance. The latter factor, annoyance from house vibration, is obviously due to more indirect effects of the noise and is discussed separately at the end of this chapter.

The levels of noise that interfere with speech at a conversational level of intensity and with the quality of sleep are recommended as the practical thresholds of noisiness for the assessment of environmental noises. These practical thresholds seem to be approximately the following:

	Steady noise	Peak level of variable (*i.e.*, aircraft, auto) noise
Speech interference	$L_A = 45$ dB	$L_A = 55$ dB
Sleep interference	$L_A = 35$ dB	$L_A = 45$ dB

Figure 5.16 shows the acceptability-unacceptability ratings given by 47 adults to recordings of aircraft flyover noises when heard in a laboratory living room in the presence of speech (a radio program) and when there was no speech present. It is interesting to note that when indoors, and especially when outdoors, and not trying to listen to speech or other weak auditory signals, people typically do not rate aircraft noises as unacceptable until they reach peak levels of 76 to 85 dBA. When listening to speech from a radio at an average level of 64 dBA, the unacceptable peak level of aircraft noise is judged to be about 58 dBA.

Somewhat different results were obtained by Willshire and Powell (ref. 63) in a similar experiment. These investigators asked 48 subjects to rate, among other things the acceptability-unacceptability of recorded aircraft flyover noise while they were seated in a laboratory "living room" engaged in either reverie, conversation, or watching television. The speech levels present were not specified. When watching television, 50 percent of the subjects rated the flyover noise as unacceptable when its peak level was 68 dBA (compared with 58 dBA in fig. 5.16). When engaged in reverie or conversation, 50 percent of the subjects rated the same noise as unacceptable when at a peak level of 78 dBA (compared with 76 dBA for indoor listening when no speech was present, presumably reverie, in fig. 5.16). According to Willshire and Powell, the seemingly higher tolerability of aircraft noise when the subjects were engaged in idle conversation than when

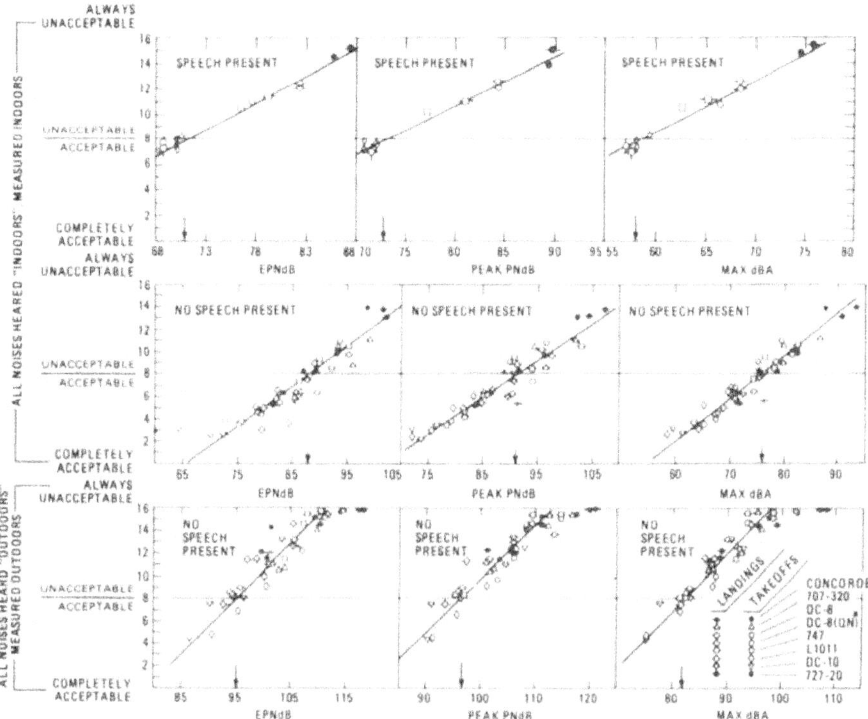

FIGURE 5.16. Mean subjective ratings of acceptability given after each noise occurrence and levels of aircraft flyover noises at listeners' position. The tests were conducted with speech at an average level of 64 dBA, about 10 dB higher than conversational level.

listening to television could have been due to the subjects timing their speech to avoid the aircraft noise and/or raising their vocal effort when the noise was present. For both of the studies just discussed, the noise level specified was present at the position of the listeners.

Lukas (ref. 64) found from an analysis of a number of studies on the effects of noise on sleep that the threshold for noise interference with sleep was about 35 dBA. This area of research, which will be discussed more fully in chapter 11, is complicated by the fact that the threshold of arousal from noise is cyclic, varying from a level of about 35 dBA during a light stage of sleep to 80 dBA or so during a deep stage. Also, the reaction of persons physiologically aroused from sleep depends upon familiarity with, and meaning of, the noises.

For most of the data analyzed by Lukas, the subjects were familiar with (adapted to) the noises involved, none of which had any particular meaning to the subjects. Also, as shown in chapter 11, attitude surveys and complaints of people exposed to noise in real life indicate that the practical threshold for complaints about sleep interference is indeed about 10 dB less than it is for speech interference.

From these, and other considerations discussed in chapter 6, it follows that in the assessment of environmental noise, unwanted sounds are to be measured only when their levels at the ears of the listeners exceed the practical thresholds specified. Only during those periods can a sound be effective as, or perceived as, noise; of course, the sound must also be unwanted for reasons other than its meaning to the listener.

"10 dB down" method

An alternative definition of effective duration is the time a sound is within 10 dB of its maximum or peak level (such as the flyover noise from an aircraft), provided that peak level is greater than 10 dB above the thresholds specified above. This alternative is based on subjective judgment data showing that for sounds which increased to a maximum level at about the same rate as the sound had increased, the energy in the sound below a level of about 10 to 15 dB lower than the maximum level did not appear to contribute significantly to the perceived noisiness of the sound (Kryter and Pearsons, ref. 8).

Often there will be little difference, at least for aircraft type noises, between the magnitudes of duration-integrated noise levels when the threshold of noisiness rule and when the maximum level to 10 dB down rule is used to establish the levels at which the integration process is started. The reason, of course, is that the weaker portions of the sound environment will contribute but a small amount to the integrated noise level in dB.

Equal energy

Given a definition of effective duration, the question remains of how the perceived noisiness increases as a function of its effective duration. Figure 5.6 showed that as the duration of impulsive noises below 1 sec is progressively halved, the intensity must be decreased by about 3 dB in order for the judged noisiness to remain the same. That is, equal energy gave equal judged noisiness.

Figure 5.17 (taken from ref. 65) shows that as the duration of more steady-state noise increases above 1 sec, the equal-energy rule for perceived noisiness is approximately correct, at least up to 100 sec or so. This is in sharp contrast to judged loudness for durations longer than 1 sec. Consistent with the critical summation time of the ear, loudness stays the same or even decreases. Note that in figures 5.6 and 5.17, some of the apparent bend in the data curves may be related to judgments being influenced to some extent by durations of the reference signals, 1 sec and 12 sec, respectively. Incidentally, the impulsive bursts of noise in figure 5.6 (rise time briefer than a few milliseconds) were judged against a standard with an equally short rise time. As shown below, there is an additional contribution to the judged noisiness of sounds with short, or impulsive,

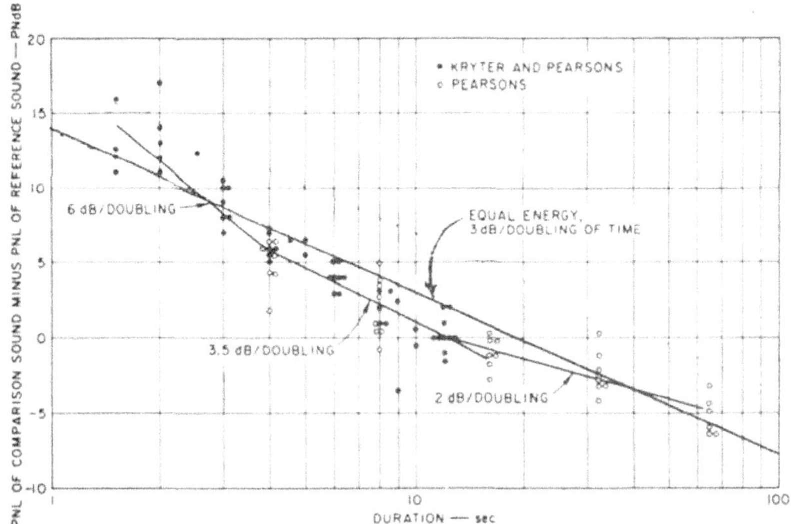

FIGURE 5.17. Relative effect upon PNL of changing duration of a noise from 12 sec. (Data from refs. 8 and 65.)

rise times compared with the noisiness of sounds with slower, or nonimpulsive, rise times. For reasons discussed below, it appears that when a sound increases in level above the threshold level for noisiness (estimated as 45 dBA indoors and 55 dBA outdoors, for brief sounds) at a rate greater than about 10 dB/sec, it is judged impulsive.

Fuller and Robinson (ref. 66) asked subjects to judge the annoyance, unpleasantness, loudness, distractiveness, etc., of recorded traffic noise at essentially a constant level of 85 dBA for durations varying from 5 to 60 minutes. The overall results were consistent with an increase in the subjective unwantedness equivalent to 2 dB for a doubling of duration. This finding, something less than the 3 dB energy rule, is rather like the shape of the data-fit function from about 20 to 60 sec shown in figure 5.17.

Hiramatsu *et al.* (ref. 67) studied the effect on judged annoyance of the duration (30 msec to 90 sec) of white noise at four levels of intensity. Kuwano *et al.* (ref. 68) investigated the judged noisiness of 16 kinds of intermittent and 22 kinds of steady-state noises. The results of the Hiramatsu *et al.* indicated that there was some interaction between the effects of level and duration, but that, taken overall, the slope of the duration effect was 3.4 dB per doubling of duration. Kuwano *et al.* report that energy (3 dB per doubling of duration) was appropriate for expressing the duration effect on the noisiness of steady-state noise, but not entirely so for intermittent noise. For the latter noise its mean energy level plus 10 times the common logarithm of the number of pulses was cited as being more appropriate.

Multiple events

Equal energy, frequency weighted for loudness, seems to be a valid indicator of the relative noisiness, or potential annoyance, due noises of different durations. As seen in chapter 11 from attitude survey data in real life, this equal-energy concept can, within limits, be extended to the summation of annoyance from multiple events.

Duration of onset of nonimpulsive sound

It appeared in figure 5.17 that the longer the duration of a noise, the less wanted it is. More subtle is the apparent fact that the longer the duration in the buildup of the intensity of a noise, the more unacceptable it is, even though the total duration and energy remains the same. Nixon *et al.* (ref. 41) reported that a sound that increases slowly to a given peak level and then decreases rapidly is much more objectionable than one of the same total duration and maximum intensity that increases rapidly and then decreases slowly in intensity (see fig. 5.18). Comparison of the results for the first pair in figure 5.18 (an intensity but no frequency shift) versus those for the second pair (intensity as well as frequency shift) reveals that a shift in frequency, such as would be present with an actual moving sound source (the Doppler shift) does not appear to have a significant effect on the results. Rosinger *et al.* (ref. 69) found that to be judged equally acceptable, the level of signal A of figure 5.18 had to be about 7 dB less than signal B. These investigators suggest that as long as a sound is increasing in intensity, the listeners presume that the source of the sound is approaching and may come dangerously close. Therefore, the onset portion of the sound is judged noisier than the portion that is decreasing in level, even when these two portions are of equal duration and equal energy.

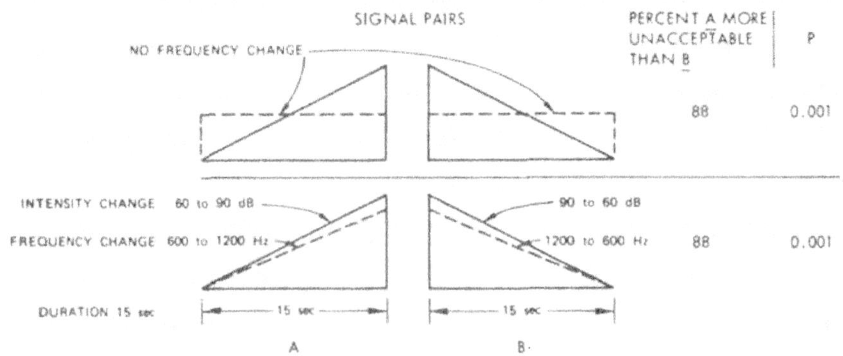

FIGURE 5.18. Temporal intensity and frequency patterns of pure-tone signals used in judgment tests and test results. (From ref. 69.)

The trend in the judgments of these noises is also consistent with the so-called "time-error" of subjective judgments. That term is used to describe the phenomenon that the more recent of two physically equal stimuli is subjectively judged to be the more intense. Accordingly, the sound whose peak level occurred closest (at the end of a sound) might be judged the more intense by an amount equivalent to 1 to 2 dB, other things being equal. However, fear of an oncoming source and, particularly, the longer uncertainty felt by the listener concerning how intense an increasing noise may become seem more reasonable explanations for the phenomenon found by Nixon *et al.* In addition, the effect is generally greater than could usually be explained by the subjective time error.

A series of judgment tests of the relative noisiness of a variety of aircraft noises was conducted at Wallops Island, Virginia (ref. 52). It was found in these tests that the aircraft sounds with relatively long onset portions were judged noisier than aircraft sounds with relatively brief onset portions, even though the two aircraft sounds had the same integrated or effective level. To account for this difference, a correction function was estimated from the data of Nixon *et al.* and then applied to the judgment tests of aircraft noise. This correction procedure improved the prediction of the subjective judgments of some of the aircraft noise. However, because of the meager amount of data available on this phenomenon, the use of such a correction to integrated noisiness does not seem justifiable at this time.

Background noise for judgment tests

The attempt to quantify, in terms of the physical stimulus, the onset duration factor in the calculation of an integrated noisiness level has certain implications for the definition and specification of the temporal-intensity pattern of a standard reference sound and the relation of that sound to a standard reference background noise. Clearly, if the onset duration is a factor in judged perceived noisiness, then the degree to which the standard reference at its maximum level exceeds its initial level will have an influence upon its onset duration and noisiness. For this reason, it is recommended that in laboratory testing with a standard reference sound, a standard background noise be continuously present during the judgment tests of the standard reference sound and any comparison sounds. The level of the background noise should be at least 15 dB below the maximum level at all frequencies of the standard reference and comparison sounds or at least at the practical thresholds of noisiness given previously.

Background noise in real life

There are some data (see chapter 11) and much anecdotal evidence that a sound, such as that heard from an aircraft flying overhead, is not as noticeable in a high background noise as it is in a quiet environment. If one assumes that the average person would adapt to background environmental noise, then increasing the background noise level would make the occasional noises that exceed this level more acceptable.

An alternative interpretation, one more in keeping with surveys of attitudes towards noise in real life, is that the environmental with the high level of background noise is judged unwanted, and the intruding higher level noises are able to add some, but only a relatively small amount of, additional annoyance. In any event, the higher the background noise level, the shorter the onset of an intruding noise, regardless of its final absolute level, and therefore, the lesser the annoyance to be expected from the intruding noise.

Combined noises and long duration test periods

When aircraft noise is heard in the presence of background or competing noises of sufficient intensity, the background noise can be expected to mask to some extent the aircraft noise. Related to this question are some experiments conducted by Namba and Kuwano (ref. 70) in which they independently varied the level of background and intrusive aircraft noise.

Subjects were asked to rate on a 7-point scale (from "not noisy at all" to "intolerably noisy") the noises from moment to moment and also after a 10-minute session of listening. Figure 5.19(a) shows that the judged noisiness increased with $L_{A,eq}$, but that in some conditions, the judged noisiness was somewhat less than that expressed from the same $L_{A,eq}$ of background noise alone. This would seem to suggest that not only did one noise somewhat mask the other but also the overall noisiness was decreased somewhat. However, this finding

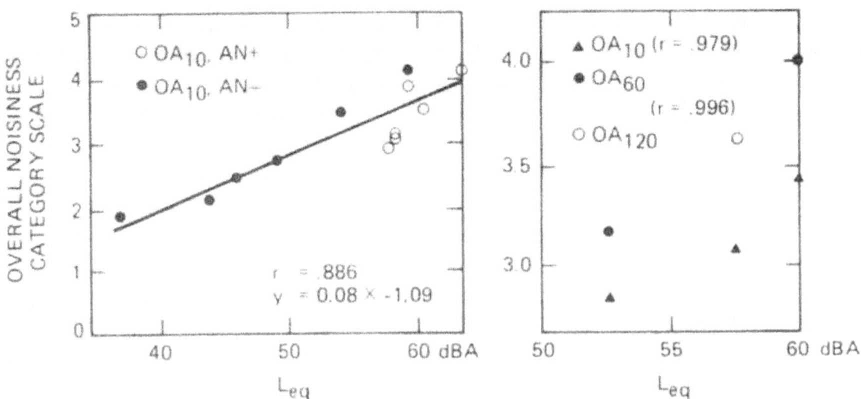

(a) After 10-minute exposures to background noise alone (AN−) and background noise plus aircraft flyover noises (AN+).

(b) After 10-, 60-, and 120-minute exposure to background noise with and without aircraft flyover noises.

FIGURE 5.19. Judgments of overall noisiness. (From ref. 70.)

seems contrary to the steady growth of noisiness as a function of $L_{A,eq}$ of either noise alone. Possible interpretations are that

1. The aircraft noise, because of masking effects, reduced somewhat the perceived noisiness of the background noise, or vice versa.

2. The differences of some of the data points from the solid line on figure 5.19(a) are not statistically significant.

From a practical point of view, it is to be noted that the aircraft noises were varied over rather low levels of intensity (peak levels ranging from 55 to 75 dBA) and that at the higher $L_{A,eq}$, this possible phenomenon seems to disappear. That is, for the three values of $L_{A,eq}$ above 58 dB or so, the combined noise environment is judged to be about equal to or worse than the background noise alone of equal $L_{A,eq}$.

Namba and Kuwano (ref. 70) also conducted experimental sessions of 10-, 60-, and 120-minute duration, of which some contained only steady-state background noise and some contained background plus an aircraft noise every 2 minutes. After each session the subjects rated, among other things, the noisiness overall for each session. Figure 5.19(b) shows that the overall ratings for sessions of a given duration follow the $L_{A,eq}$ hypothesis (that the total energy, regardless of source, controls the judged noisiness) within a session, but that the 60- and 120-minute sessions were judged overall, to be somewhat noisier than the 10-minute session.

It is perhaps important to note that the reliability of how well people judge the noisiness of rather long periods of exposure (up to 120 minutes) appears to be quite high. Namba and Kuwano found correlations between ratings made one month apart of $r = 0.89$ to 0.99 for the different combinations of test conditions.

Noisiness of Impulses

Experience indicates that when the intensity of a sound increases faster than a certain rate, particularly if it reaches certain levels of intensity, this rate of change contributes to an unwanted startle effect, even when the sound is expected. Because rate and duration of onset are clearly not independent, one would perhaps expect this effect to be opposite that found by Nixon *et al.* (ref. 41); that is, shortening the onset duration of a sound would increase subjective noisiness rather than decrease it.

Correction for impulsive level

Beyond a certain rate of change of level, a sound is believed to take on an impulsive characteristic that itself contributes to the perceived noisiness of a sound. This effect may be related to reflex reactions of the autonomic nervous system to sudden, intense sounds. Figure 5.20 depicts a suggested correction

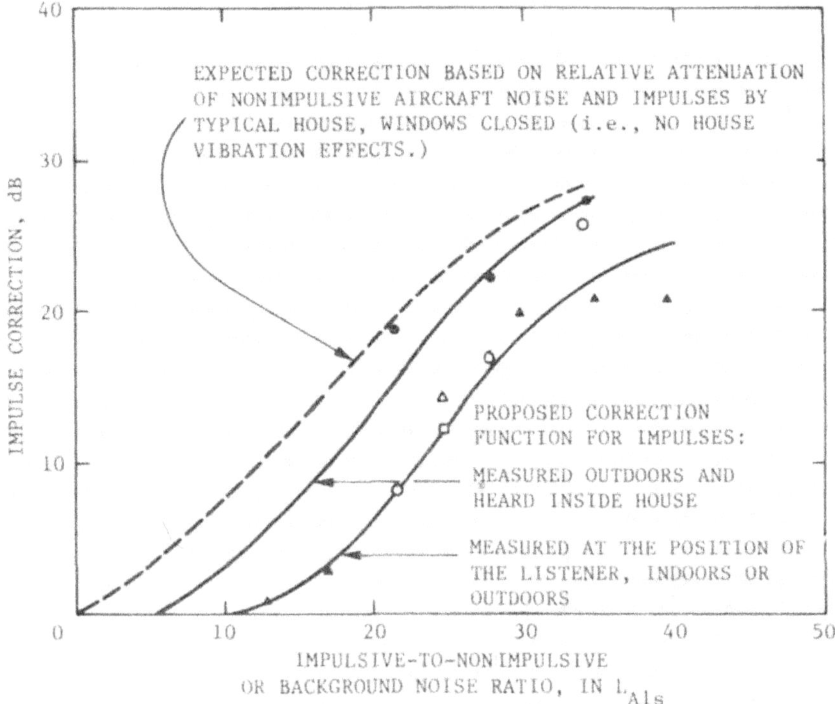

FIGURE 5.20. Judgments of noisiness of impulsive compared with nonimpulsive noise and proposed impulse correction procedure. Data points were obtained with impulses presented in the quiet (≈ 40 dBA).

that can be applied to the level of impulsive segments of sound to take into account this effect. The correction procedure involves determining the ratio of impulsive to background noise for a 1-sec period. An impulse is said to occur when the A-weighted peak sound level within 1 sec, $L_{A,p}$, exceeds by 10 dB or more the average noise level integrated over the preceding or following 1 sec, L_{A1s}, providing that $L_{A1s} > 40$ dB. The lower valued L_{A1s} or 40 dB, whichever is greater, is called the background noise level. If more than one impulse occurs in 1 sec, the average impulse-to-background noise ratio is assigned to that 1-sec period. With this ratio and the graph in figure 5.20, one can find the impulse correction to be added to the measured 1-sec integrated sound level. Note that figure 5.20 is applicable to noise measured in L_D or PNL as well as L_A.

The function shown in figure 5.20 is drawn primarily on the basis of data from an experiment (Kryter *et al.* (ref. 71)) in which subjects judged the relative noisiness of expected sonic booms vs. the sound of a subsonic aircraft (see fig. 5.21 and table 5.2). Shown is a data point from a study by Izumi, as reported by Sutherland and Burke (ref. 34), in which the noisiness of a steady-state pink noise (10-sec duration with a rise and decay time of 2 sec and peak level of 70 dBA) was compared with that of the pink noise interrupted instantaneously. Other data points in figure 5.20 include those from a study by Young (ref. 72) as presented and discussed by Schomer (ref. 73). They are derived from figure 5.22 and the measurements showing that the L_C was 10 dB greater than the L_A of the artillery blasts. Also shown is the average differences between L_A for a reference octave band of random noise (600 to 1200 Hz) and various single impulse sounds as found by Fidell and Pearsons (ref. 74).

Outdoor versus indoor conditions

The impulse correction is about 5 dB greater when the sonic booms and aircraft noise are heard indoors but measured outdoors than when the two noises are measured at the position of the listeners. Schomer (ref. 73) deduced similar conditions for the artillery vs. subsonic aircraft noise study and corrected the levels by "5-10 dB" when plotting figure 5.22. Typically, environmental noise is measured outdoors to predict the resulting annoyance experienced indoors. In the case of impulsive noise it appears in figure 5.20 that the judged acceptability indoors of impulsive noise is 5 dB less than nonimpulsive noise. Although the low-frequency impulses were attenuated 10 dB less than the higher frequency nonimpulsive noise, they were judged indoors to be only about 5 dB less in effective obnoxiousness, compared with the judgments made outdoors.

In the sonic boom studies (ref. 71), subjects were both inside and outside houses exposed to sonic booms; and in the Young study (ref. 72), the test room vibrated from the simulated artillery fire as would a room in a typical house. Because of these and other relevant data, the author believes that the perceptions (proprioceptive, auditory, and visual) of vibrations and rattles contributed 5 dB more unwantedness than expected on the basis of intensity when the booms and artillery fire were heard indoors than when they were heard outdoors.

The impulse startle effect does not appear to cause this additional source of noisiness. The relative startle effect is shown by the shape of the curves in figure 5.20 as a function of impulse-to-background noise ratio. The apparent noise vibration factor is surmised to be a function of the spectra of a noise and to be independent of whether the noise is impulsive or nonimpulsive.

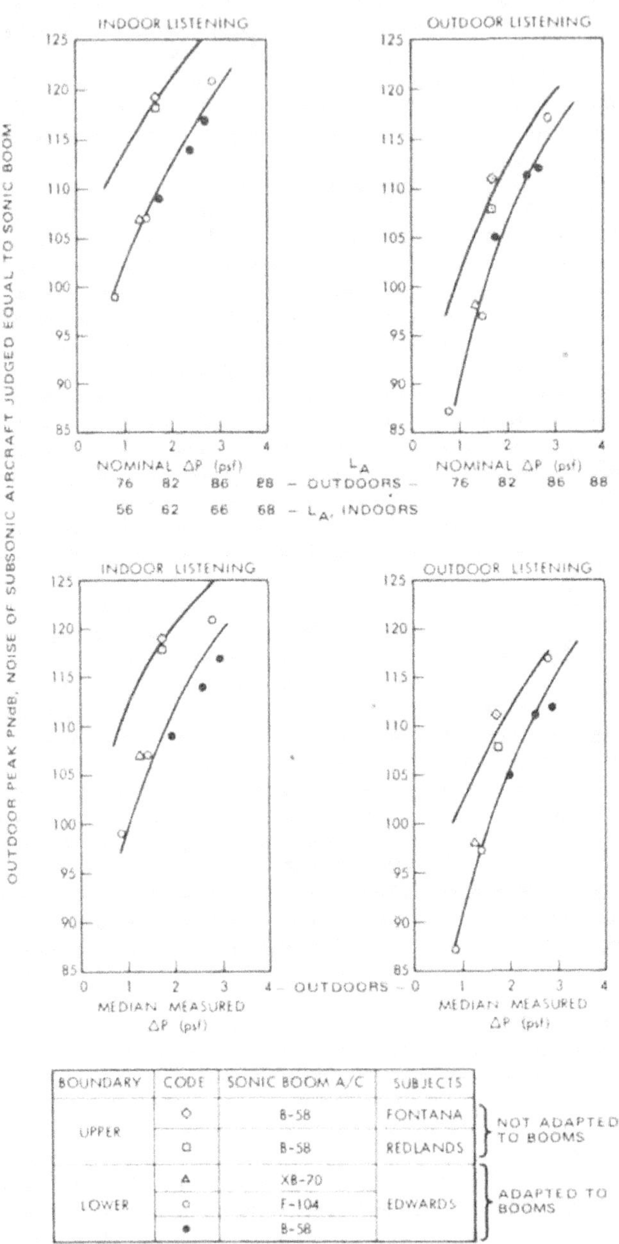

FIGURE 5.21. Level of subsonic aircraft noise judged to be equal in acceptability to sonic booms in the home. Edwards' subjects were adapted to booms from average of 2 year's exposure to 6 booms per day. The abscissa is the level of sonic booms in psf and L_A; on the ordinate, outdoor $L_{A,ex}$ (SEL) ≈ Peak PNL − 7 dB and Indoor $L_{A,ex}$ (SEL) = Outdoor $L_{A,ex}$ (SEL) − 30 dB. (From ref. 71.)

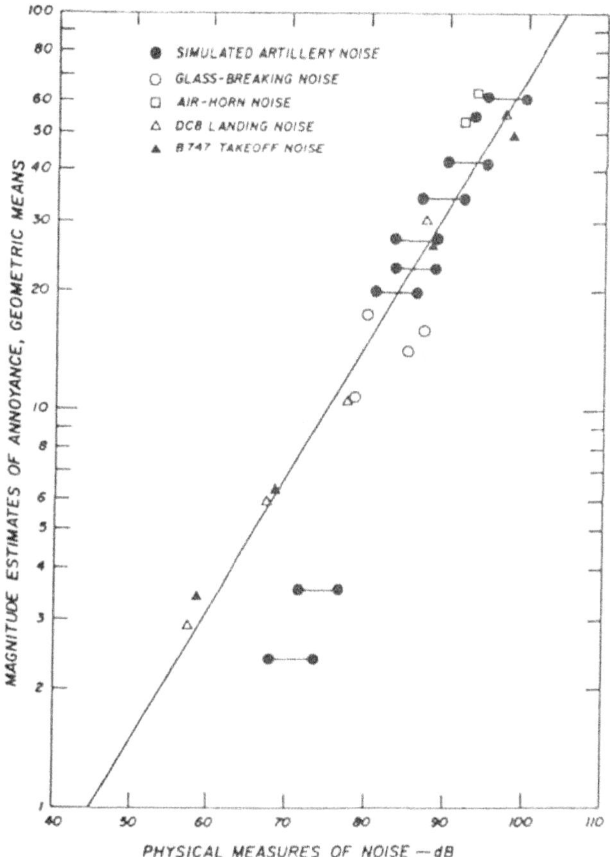

FIGURE 5.22. Relationship between psychological annoyance and physical stimuli. All stimuli are A-weighted L_{ex} (indoors) except the artillery noise which is C-weighted and corrected to account for the reduced building attenuation to low-frequency impulsive noises as compared with the building attenuation to aircraft or air-horn noise. The solid line is a least-square fit to the nonartillery noises. (From ref. 73.)

Research questions

Whether the proposed correction function in figure 5.20 has general validity is open to question. Possible bases for the proposed correction; singly or in combination, are the following: (1) some nonlinear response in the auditory system when the energy flux of a sound is increased above some amount, (2) some response of the autonomic system to the sound, or (3) some purely perceptual psychological response to a sudden, intense sound. In keeping with the previous definitions of perceived noisiness, it is presumed that the impulsive sounds to be evaluated in regard to their perceived noisiness are familiar to the listeners and

TABLE 5.2

Sonic Boom Spectra

[Data from ref. 71]

(a) Nominal outdoor one-third-octave band energy spectra and L_A, L_C, and L_D for XB-70 sonic boom with rise time of 0.005 sec

	ΔP		Nominal outdoor sound energy level, dB																		L		
			Center frequencies, Hz																				
Mission	psi	[a]dB	50	63	80	100	125	160	200	250	315	400	500	630	800	1000	1250	1600	2000	2500	dBA	dBC	dBD_1
1-1	2.91	136.86	102	102	99	96	93	87	73	81	81	71	75	68	67	64	60	58	54	52	85	106	97
2-1	2.55	135.71	101	100	98	95	92	86	72	80	80	70	73	67	66	63	59	57	53	51	84	105	96
10-1	2.41	135.22	101	100	98	95	92	86	72	80	80	69	73	67	65	62	59	57	53	50	83	105	96
15-1	2.18	134.35	100	99	97	94	91	85	71	79	79	69	72	66	64	61	58	56	52	50	83	104	95
14-1	2.10	134.02	100	99	96	94	90	85	71	79	79	68	72	66	64	61	58	55	51	49	82	103	94
9-1	2.09	133.98	100	99	96	94	90	85	71	79	79	68	72	66	64	61	58	55	51	49	82	103	94
6-2	1.78	132.59	98	97	95	92	89	83	69	77	77	67	70	64	62	60	56	54	50	48	81	102	93
5-2	1.19	129.09	95	94	92	89	85	80	66	74	74	63	67	61	59	56	53	50	47	44	77	99	90

(b) Outdoor ΔP, PNL and one-third-octave band energy spectra, and PNL measured in three rooms indoors for XB-70 sonic boom

Location within house[b]	Mission	ΔP, average of 5 outdoor measurements		Outdoor nominal PNL,[c] dB	Recorded indoor sound energy level, dB														Indoor measured PNL,[d] dB
					Center frequencies, Hz														
		psf	[a]dB		50	63	80	100	125	160	200	250	315	400	500	630	800	1000	
B	1-1	2.91	136.86	101	78	74	79	82	77	70	63	60	60	60	54	53	53	54	82
K					73	78	78	73	72	63	63	61	58	57	57	54	54	54	77
D					84	78	75	77	77	71	70	68	61	60	57	57	57	55	82
																			Avg. 80

B	2-1	2.55	135.71	100	70	60	70	69	63	63	56	55	52	53	49	53	52	47	70
K					70	69	75	75	71	63	59	60	57	55	55	55	55	52	76
D					86	76	78	85	76	74	74	69	65	59	59	59	60	56	86
																			Avg. 77
B	10-1	2.41	135.22	100	73	72	80	65	68	65	62	63	61	55	55	56	54	55	78
K					71	70	72	74	72	63	60	66	65	57	63	67	70	72	82
D					76	71	72	73	70	69	69	66	56	57	55	52	52	52	77
																			Avg. 79
B	15-1	2.18	134.35	99	75	74	78	72	68	70	70	61	62	59	54	53	53	52	78
K					71	70	67	70	67	64	63	59	58	56	56	53	53	53	73
D					74	75	73	72	70	70	69	64	61	58	55	54	54	53	77
																			Avg. 76
B	14-1	2.10	134.02	99	69	69	77	68	63	63	65	60	57	56	53	51	50	50	75
K					71	68	65	63	61	59	57	55	54	52	51	51	50	49	67
D					80	71	72	71	68	72	70	64	62	61	57	56	54	52	78
																			Avg. 73
B	9-1	2.09	133.98	99	59	64	67	62	61	59	54	50	49	50	48	46	44	42	65
K					73	69	74	73	71	63	61	58	59	57	53	53	53	51	75
D					76	72	73	68	70	69	67	62	61	59	55	53	52	51	75
																			Avg. 72
B	6-2	1.78	132.59	97	73	69	72	74	75	74	75	60	58	55	53	50	50	48	80
K					77	74	71	72	64	65	63	59	59	56	55	51	51	50	74
D					85	74	69	75	71	69	66	61	61	57	54	53	52	52	80
																			Avg. 78
B	5-2	1.19	129.09	94	68	66	66	65	69	68	59	56	53	52	50	49	49	48	72
K					72	66	61	66	57	59	58	56	52	52	50	50	50	49	68
D					77	69	63	70	67	63	65	60	60	58	56	54	53	54	74
																			Avg. 71

[a] *re 0.0002 μbar.*

[b] *B = Bedroom, K = Kitchen, and D = Dining room.*

[c] *PNL calculated from nominal spectrum of boom from XB-70 with rise time of 0.005 sec and the given peak pressures.*

[d] *PNL calculated from measured spectra.*

are an expected part of their noise environment. This was the case for the judgment tests of sonic booms and the noise from subsonic aircraft represented by the lower curves of figure 5.21 which were used for deriving the data points plotted in figure 5.20.

No systematic data are known to show whether the noisiness penalty is a function purely of impulse-to-background noise ratio, as shown by the abscissa of figure 5.20, or that it is also influenced by the absolute levels involved. In the sonic boom and the artillery fire studies, the background noise was low, 40 to 50 dBA. Whether an impulse-to-background noise ratio of 20 with a background level of 55 dBA (earning an impulse correction penalty of about +8 dB according to fig. 5.20) would be the same with the same ratio but a background noise level of, say, 80 dBA is unknown, in the author's opinion. The effect should be about the same, of course, if the masking effects of the background noise on the impulsive noisiness is about the same with constant ratios. That would be a reasonable expectation within limits.

Method of C-weighting

Table 5.2 shows that using C-weighting instead of A-weighting for measuring the energy in the sonic boom impulses increases the relative level by some 21 dB. This is similar to the 19-dB difference found by Young (ref. 72) between L_A and L_C for artillery fire impulses. Averaged over the range of impulse levels involved, the numerical values of the C-weighted energy level of the impulsive noise is thus about the same as the A-weighted energy level of the nonimpulsive noise judged to be equally acceptable. Accordingly, measuring impulsive sounds with C-weighting and nonimpulsive sounds with A-weighting would provide levels that would be about numerically equal when judged to be equally acceptable. Schomer (ref. 73) has proposed that L_C be used in this way. However, this finding is possibly unique to the particular impulses that were judged, and impulses with different spectra could give different results.

It is more logical (but with present sound level instruments less practical) to use A-weighting energy or some similar loudness-noisiness frequency weighting, for all noises, impulsive or nonimpulsive, and to add to the impulse energy level in 1-sec intervals a correction value in accordance with the average contribution to noisiness of expected impulsive sounds. In brief, to use the proposed corrections shown in figure 5.20. The possible applicability of this correction procedure to the assessment of the acceptability of multiple impulses is examined below.

Repeated impulses

The recommendation that sound energy be integrated within a 1-sec period of time, as the minimum, would seem to be inappropriate for closely occurring impulses that are so short that more than one can occur within a second. However, this measurement time interval appears to be of practical accuracy even here since the loudness and noisiness of pulses shorter than 1 sec depend upon their energy summed over 1 sec.

As shown in figure 5.23 from Izumi (ref. 75), each doubling of a number of impulses of a constant duration required that the burst energy level be decreased by about 3 dB to maintain equal subjective judgments. This is in agreement with earlier data on loudness of 1-msec triangular transients of varying rise times and repetition rates (Carter, ref. 76).

Impulsive sound coming from helicopters (the so-called "blade slap") has been studied in the context of practical procedures for assessing aircraft noise. It has been difficult, however, to generalize the results from the various studies. A detailed review of helicopter noise studies through about 1978 is given in Sutherland and Burke (ref. 34). It appears from the results of these studies that a correction to the $L_{A,eq}$ of helicopter noise with blade slap relative to helicopter noise without blade slap ranges from about −8 to −2 dB. This underestimation by $L_{A,eq}$ of the perceived noisiness is perhaps to be expected, not because of impulsiveness *per se*, but because the A-weighting underweights the relative contribution to loudness and noisiness of the low frequencies, frequencies augmented in level by the blade slap noise (see difference in D versus A, table 2.1). Indeed, when sound level is expressed as $L_{PNdB,eq}$ or $L_{D,eq}$, which more appropriately weights the frequency spectrum, the required correction becomes only −2 to +4 dB rather than −8 to −2 dB (Powell, ref. 77). Also, Galloway (ref. 78) found that the rate of blade slap (from 10 Hz to 30 Hz) contributed to judged noisiness

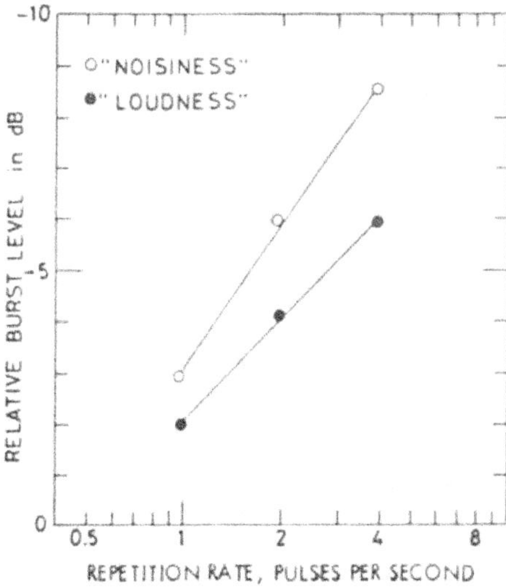

FIGURE 5.23. Comparison of loudness and noisiness versus repetition rate for a burst time fraction (on time divided by on time plus off time) of 0.063. (Data from ref. 75.)

over and above that predicted from L_{eq}. It appeared that increasingly the rate of slap, with total A-weighted energy constant, increased the perceived noisiness by the equivalent of 4 dB or so. Powell (ref. 79) conducted a study of the judged noisiness of the sound from helicopters and a fixed wing aircraft. The rate of blade slap of the helicopter noise was systematically varied. Powell found that methods presently proposed by the ISO for predicting from A-weighted measures the noisiness of helicopters do not improve the prediction obtained with $L_{PNdB,eq}$ (EPNdB). Some further references to research on helicopter noise are presented in chapter 11.

Rate and on-off times

Izumi (ref. 80) conducted a study of the effects of the rate and duration of interruption on the relative perceived noisiness of a 20-sec period of pink noise (onset and offset times of 30 and 200 msec, respectively, *i.e.*, impulsive sounds). In these tests, 20-sec periods of uninterrupted pink noises were judged against 20-sec periods of the noise with all combinations of on times from 0.125 to 1 sec and off times from 0.125 to 1 sec. Peak levels of the uninterrupted reference noise were 60, 70, and 80 dBA.

Izumi developed graphic and mathematical formulas to show the contribution to perceived noisiness of the parameters of BTF (burst time fractions, equal to on time divided by on time plus off time) and RR (repetition rate), as found in his studies. The following formula represents results he obtained in a later study (Izumi, ref. 81) with interrupted vs. uninterrupted 10-sec periods of pink noise:

$$L_{RB} = 6 \log_{10}\mathrm{BTF} + (10 \log_{10}\mathrm{RR} + 10)(1 - e^{-15T_{\mathrm{off}}})$$

where—L_{RB} = Relative A-weighted noise level of burst, dB
BTF = Burst time fraction
RR = Repetition rate, sec^{-1}
T_{off} = Off time, sec

Also, for multiple, brief (less than 1 sec) impulses such as those studied by Izumi, the aural reflex could possibly affect judged loudness and noisiness in somewhat confusing ways. That is, the aural reflex, as discussed in chapter 3, could be activated to different degrees for given burst-time fractions depending upon the actual durations of the off time between impulse bursts of differing durations.

The new procedure proposed (and apparently Izumi's methods) requires a time history of A-weighted, peak, and 1-sec integrated sound pressure levels, values not directly available from standard sound level meters. The use of an overall A-weighting filter (not a sound level meter) in conjunction with an oscilloscope provides an appropriate means of finding the peak level, and the 1-sec energy levels can, of course, be calculated or estimated from time history records of the impulses. Future sound level meters using digital forms of analysis no doubt will be able to measure sound energy over 1-sec periods of time (see Kundert, ref. 82), as can present computer-aided sound measurement systems.

Summary of impulse variables

To recapitulate, there appear to be two major subjective factors to be considered in predicting from physical measures the acceptability of impulsive sounds that are an expected part of the environment, independent of any meaning the sounds may have.

1. Noisiness from loudness is found to be predictable from its frequency-weighted (according to some loudness function such as dBA, dBD, PNdB, or phon) energy integrated over 1 sec. As with nonimpulsive sound, impulses continuing to be repeated for periods longer than 1 sec do not increase in loudness, although their unacceptability or noisiness continues to grow on an integrated energy basis.

2. Noisiness from impulsiveness appears to be an objectionable, startle-related reaction due to expected large and rapid changes during 1 sec in noise level.

Only when one wishes to consider the noisiness of impulsive sounds along with, or in comparison to, that of nonimpulsive sounds is the impulse correction given in figure 5.20 required. To achieve a general physical measure of noise dosage that is consistently relatable to the perceived noisiness of the environment, it is important to add a special "psychological" weight to the impulse segments.

The approximate equivalence in annoyance between sonic booms (measured in pounds per square foot) from aircraft flying at supersonic speeds and the flyover noise (measured in SEL, $L_{A,ex}$, or EPNdB) shown in figure 5.21 are used later in chapters 11 and 12 to estimate the impact of sonic booms upon people in residential communities.

Round Robin Study of Loudness of Impulses

A special study was conducted under the auspices of the International Standardization Organization, ISO (Pedersen *et al.*, ref. 36) in which a variety of recorded impulsive sounds were administered to subjects in 21 laboratories in 11 different countries. The subjects were asked to judge the "loudness" of the impulsive sounds related to that of the reference sounds, as shown in table 5.3.

Level comparisons with reference signals

The average decibel attenuation required in the level of a comparison impulse so that a given panel of subjects considered the impulse equal in loudness

TABLE 5.3

List of Noise and Reference Signals in Round Robin Study

[From ref. 36]

Group	Noise no.	Noise		Reference	
		Source	Duration, msec	Signal	Duration, msec
i	1	Puncher	1000	1/3 octave noise	1000
	2	Cement mill	↓	↓	↓
	3	Teletype			
	4	Pneumatic hammer (silenced)			
	5	Pneumatic drill			
	6	Outboard motor			
	7	Hammer and anvil			
	8	Ram			
	9	Puncher			
ii	10	Typewriter		1-kHz tone	80
	11	Hammer and anvil			320
	12	Ram			160
	13	Gun			10
	14	Sonic boom			320
iii	15	1-kHz tone	160	1-kHz tone	320
	16	↓	80	↓	160
	17		40		80
	18		20		40
	19		10		20
	20		5		10
	21	1-kHz tone	1000	1/3 octave noise	1000

to the reference signal was called the ELA (equal loudness attenuation). The comparative results among the various panels of subjects are shown in figure 5.24. Except for the group iii impulses, the data show a large (15 to 20 dB) range in ELA for a given impulse among the 21 different panels of subjects a range, interestingly, about equal to the impulse correction developed in figure 5.20.

The exceptions, group iii impulses, consisted of a burst of the same tone at twice the duration, and not of impulses of complex frequency spectra with somewhat variable durations (group i and ii) and repetitions (group i). It is hypothesized that since the group i and ii impulses were more complex than those in group iii, their relative differences from the reference signals with respect to potential startle or noisiness were more clearly perceived by some of the subjects.

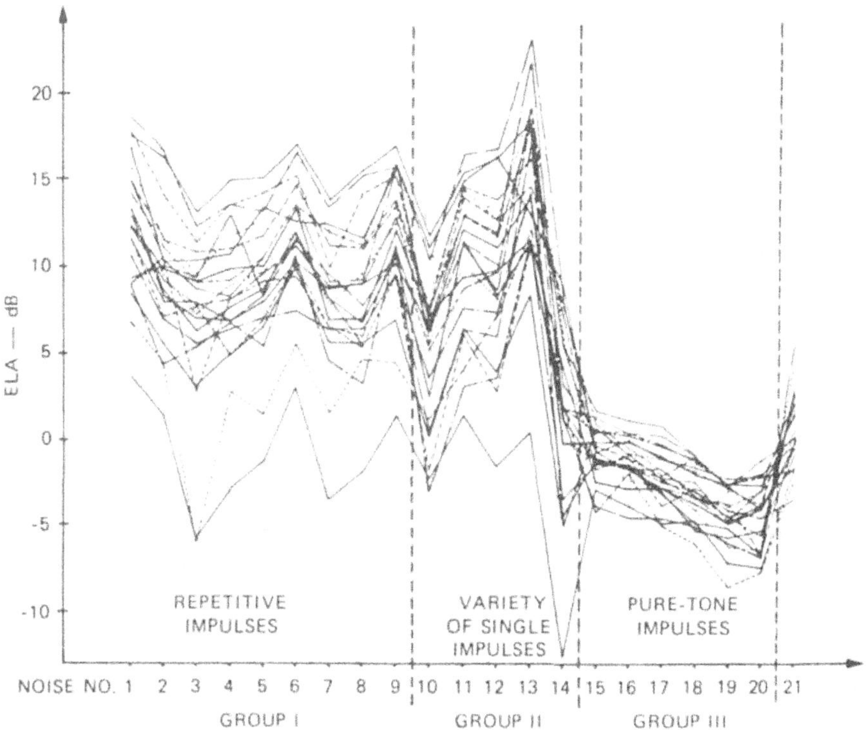

FIGURE 5.24. Mean values of equal loudness attenuations (ELA) from subjective measurements (level of 75 dBA) as reported in Round Robin Study.

The instructions to the subjects were unfortunate in the context of this study because the word "loudness," and no doubt its counterparts in other languages, can be interpreted ambiguously. Some subjects may have taken the instruction to mean that they should subjectively judge the peak physical intensity in the noise in relation to the reference noise; others may have judged the summed or averaged subjective intensity of the comparison versus the reference noises; other subjects might well have interpreted that the intent of the study was to have them judge how acceptable they felt the impulses were relative to the reference stimulus. Extracting from the subjects quantitative data on only the loudness of an impulse is made difficult because the perceived obnoxiousness of impulses could at times lead to confusing results, depending on whether or not subjects, when instructed to judge "loudness," interpreted "loudness" to include obnoxiousness. Note that Izumi found a consistent difference of several decibels between loudness and noisiness judgments of impulses (the noisiness was greater than the loudness, see fig. 5.23).

Relative contributions of spectra and duration

However, although in the Round Robin Study the subjective aspect of impulsiveness *per se* was seemingly a floating variable among the different panels of subjects, the relative effects of spectra and duration upon the judgments were fairly consistently revealed. That is, the relative differences in ELA between different impulses were somewhat similar for all the panels of subjects (see fig. 5.24). The investigators had slightly different methods for calculating the variability of the ELA data; one was called a variance shape deviation criterion (log S^2) and the other a mean square (log MS) deviation criterion. The results plotted in figure 5.25 show the general magnitude of the errors or deviations of the physical measures to predict the judgments of loudness according to the two criteria measures.

These results demonstrate several things that in general agree with previous studies on these matters:

1. The subjective magnitude of loudness (or noisiness if the equivalent was judged) is more closely a function of stimulus energy, frequency weighted, for stimuli of up to 1 sec in duration (the maximum duration judged) than it is of the other meter measurements. This is shown by the generally smaller criteria errors found for the integration (energy) measures. (Note the use of the phrase "short time integration" in figures 5.25(b) and (d), those concerned with the single impulses of noise. Short time integration means that the integration time was as long as the duration of either the impulse or the reference signal, whichever was the longer. This being the case, the same results would, of course, have been obtained had a 1 sec integration time been used throughout.)

2. Stevens and Zwicker phons are generally more accurate in predicting subjective judgments than are overall frequency weightings, such as linear, A, B, C, or D, the apparent reason being that the phon is calculated from band spectral measures and thereby reflects the contribution to the subjective judgments of relative differences in the bandwidths of the different noises.

3. The D-weighting is generally more accurate in predicting the loudness judgments according to the mean square deviation criterion than the A-weighting or any of the other overall frequency weightings for the quasi-stationary noises (fig. 5.25(c)) or for the single pulses (fig. 5.25(d)). However, with variability scores based on the variance shape criterion, the A-weighting predicts the judgments of quasi-stationary noises with somewhat more accuracy than do the other frequency weightings, except when using impulse and peak meter functions; however, according to the variance shape criterion scores, all the frequency weightings, except for linear, predict the judgments of single impulses about equally well. The superior performance of the D-weighting over the other weightings, when it occurs, is probably to be expected because the other weightings, including A, do not reflect the approximate 10 dB "hump" in the equal-loudness and equal-noisiness contours in the 1 to 4 kHz region. The phon units also reflect

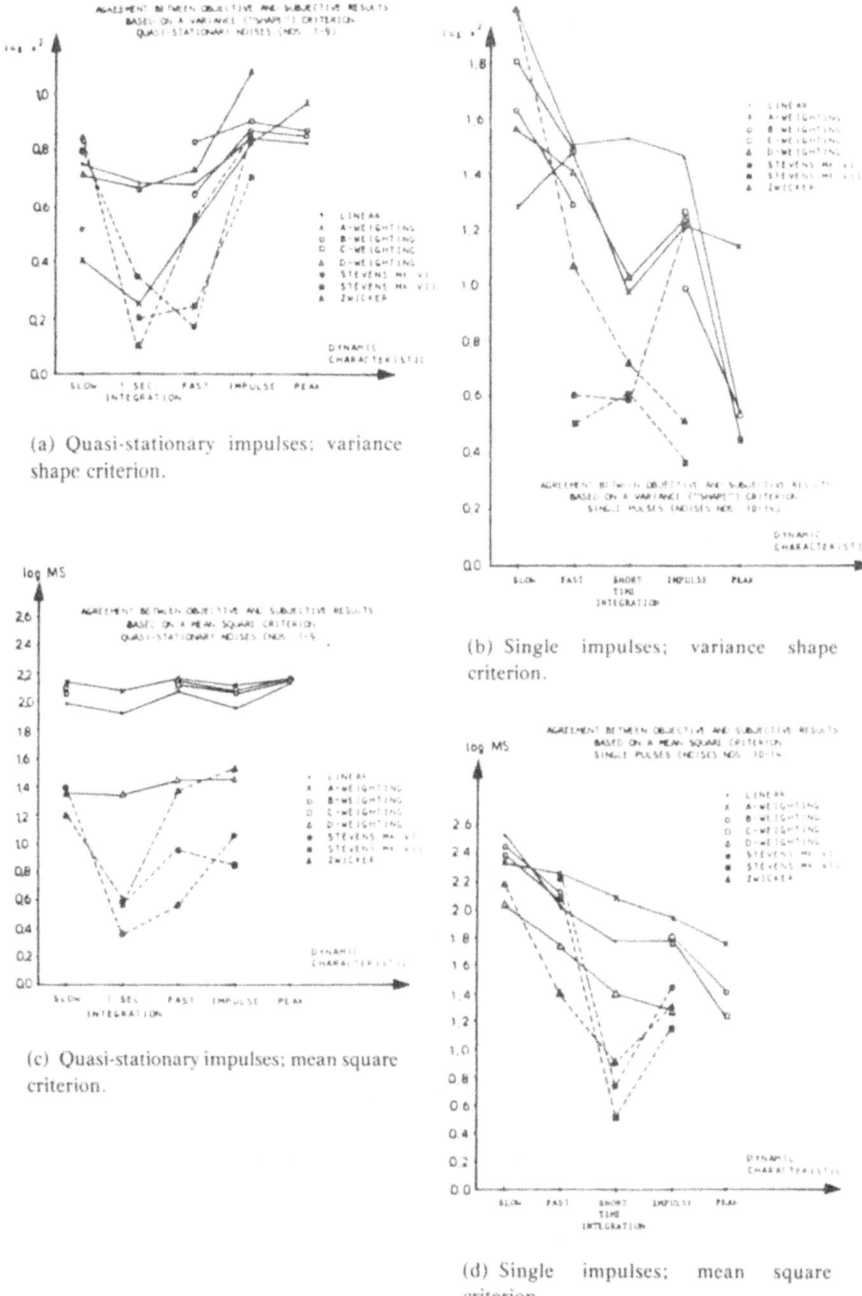

(a) Quasi-stationary impulses; variance shape criterion.

(b) Single impulses; variance shape criterion.

(c) Quasi-stationary impulses; mean square criterion.

(d) Single impulses; mean square criterion.

FIGURE 5.25 Relation between two error criteria and dynamic characteristics of sound level measurements, including integrated energy. Parameter is frequency weighting procedure. (From ref. 36.)

this 1 to 4 kHz hump in equal-loudness contours, a fact probably also contributing to the better performance of those units of noise measurement in predicting the judgments of the impulses.

In spite of these results, in the report of the Round Robin Study the A-weighting is recommended for the assessment of environmental noise in general for the following reasons: (1) a sound level meter with an overall frequency weighting is much more practical for everyday use than the instrumentation required to obtain the spectral data needed to calculate phons, and (2) the A-weighting is generally used as a standard method for the assessment of steady-state noises. However, the D-weighting, although not incorporated in most sound level meters, also generally better assesses the perceived loudness and noisiness of more steady-state noises than does the A-weighting, as was discussed earlier.

Sound Level Meter Problems—Energy and A-Weighting

Theoretically, measurements of single or multiple impulses with a frequency-weighted sound level meter with a 1-sec integration time should be an appropriate basis for estimating the perceived noise level, loudness level, or damage risk level to hearing (to be discussed in chapter 9). For one thing, the spectra of the impulses are generally broadband, and secondly, the frequency weighting could be made to appropriately weight all frequency components contributing to the auditory functions mentioned.

However, standard sound level meters do not provide a true integration of the energy to be found in impulses or any short duration sound. We found, for example, that calculated (computer-aided Fourier transforms) A-weighted energy, integrated over 1 sec, of sonic boom impulses (durations of 100 to 200 msec, with rise time portions of 10 msec or so) was 30 to 50 dB greater than that readable from a standard sound level meter set on A-weighting and fast or slow meter action (Kryter *et al.* ref. 71).

Unfortunately, sound level meters integrate and average the sound not on an energy basis (as apparently does the ear), but in an exponential manner. For example, a brief (few milliseconds) impulse that occurs at the beginning of a 1-sec period is weighted on the output of the fast meter only 0.00033 times as much as the same sound recurring at the end of the 1-sec period (Young and Cowen, ref. 83). Why this is done is not explained, but it can result in meter readings that are 10 to 50 dB below the true integrated 1-sec average sound energy for impulses, depending on their spectra.

Perhaps for this reason, some sound level meters have incorporated briefer and briefer time constants to reduce the role of the exponential factor. However, the use of such peak reading meters introduces other errors related to frequency weighting and frequency band limitations. Also, this solution of getting higher sound pressure levels for brief impulses does not address the fundamental problem of measuring proportionally over 1 sec, as apparently does the ear, the frequency-weighted energy in sound whether impulsive or nonimpulsive.

An important contribution of the Round Robin Study was the analysis made by Pedersen *et al.* (ref. 36) of the role of the dynamic characteristics of sound level meters as a factor in the prediction from level measurements of the judged loudness (or possibly noisiness) of the impulses. Although it can be deduced from data such as that presented in figures 5.5, 5.6, and 5.7, figure 5.25 explicitly shows that the so-called "slow," "fast," "impulse," and "peak" values from present day sound level meters are not, by and large, appropriate for predicting judgments of impulses, whereas in conformity with present and previous research, energy integrated over 1 sec is.

Howes (ref. 84), observing the results of psychological and physiological experiments, described mathematically the transforms between the effective energy and judged loudness of a steady sound of a given duration. In essence, he described the critical band for frequency, the critical summation time, and the frequency bandwidth and shaping imposed by the auditory system as being, up to a point, a linear, energy transform system (except for the growth of loudness with level). Howes' formulations are probably sufficiently general to cover impulsive sounds (although he does not claim to do so); at least they express the importance of measuring the energy in a sound as a correlate of loudness.

As alluded to earlier, the A-weighting in sound level meters can be a problem in the assessment of the loudness or noisiness of sounds of certain spectral shapes. For example, noises having most of their energy in the frequency regions either around 100 Hz and below or around 3000 Hz and above are judged about equally loud or noisy as a third noise having its maximum energy in the frequency region around 1000 Hz and an L_A level that is 8 dB or so less than the level of the 100 Hz noise and 8 dB or so greater than the level of the 3000 Hz noise. This deficiency of the A-weighting function is generally not manifested, to its full extent because many, if not most, everyday noises have most of their energy in the frequency region between 100 and 3000 Hz. There are, of course, exceptions. This deficiency in sound level meters could be largely rectified by the substitution of the D- or E-weighting for the A-weighting that is presently used.

House Vibration and Noisiness

Another factor that contributes to annoyance experienced from intense noise in everyday life is house vibrations and rattles. This annoyance is obviously more indirectly engendered in a person and involves more senses than that from direct auditory system stimulation by the noise.

However, house vibration is found to be a rather universal source of annoyance, and its magnitude is to some extent predictable from spectral and duration variables of sounds in much the same way as is directly perceived auditory noisiness. Its magnitude is inversely related to the spectrum shape because most house structures show increased vibration as sound frequency decreases, especially below several hundred hertz, whereas the loudness and noisiness from the sound coming directly from a source becomes progressively less at the lower frequencies.

Structure displacement

Studies were conducted by the National Aeronautics and Space Administration (ref. 85) in which a mechanical piston (shaker) was attached to a point on a house wall (see fig. 5.26). This shaker was capable of applying a constant vibrating force to the wall at different frequencies. Accelerometers attached to the wall provided measurements of vibration responsiveness of the wall to vibration imposed at different frequencies. As seen in figure 5.26, a given displacement (in

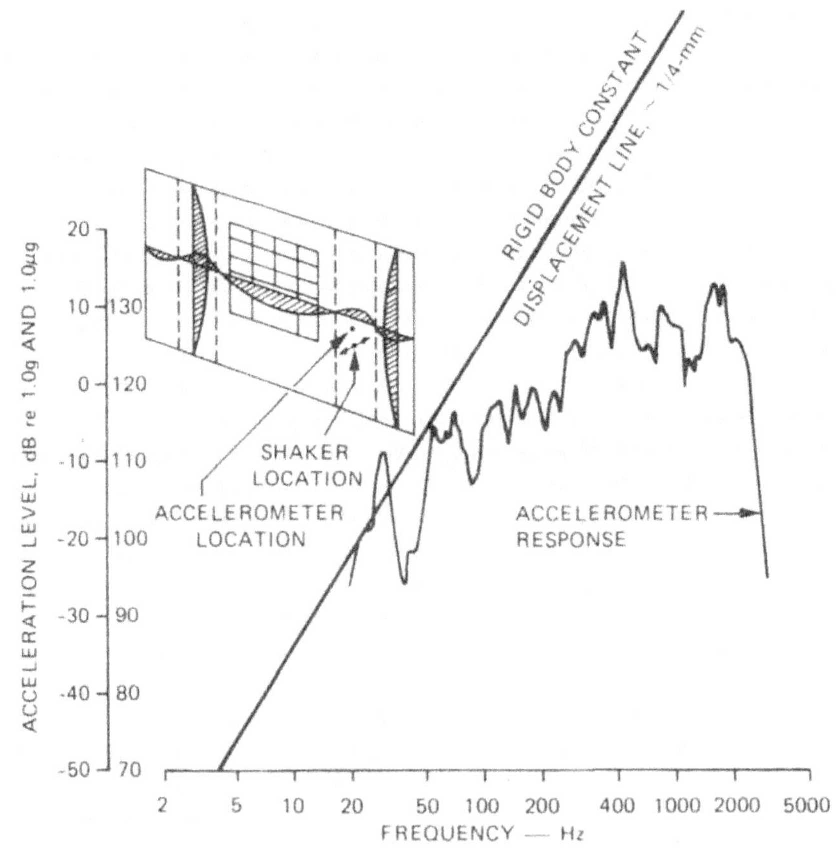

FIGURE 5.26. Vibration responsiveness of a house wall to vibration imposed at different frequencies. (NASA data, ref. 85.)

millimeters) of a body is effected with less acceleration, 12 dB less per octave, as the frequency of vibration is lowered. However, walls and most other elements of house structures are displaced less at higher frequencies than at lower. Figure 5.26 shows that maximum displacement for a given acceleration force (kept constant at the different shaker frequencies) occurs only for frequencies causing acceleration below 100 Hz or so; above that frequency of imposed vibration, the acceleration level of the wall progressively falls below that required to achieve the same amount of wall displacement, 0.25 mm.

This is not to say that all parts and objects in a house vibrate most readily at frequencies below 125 Hz or so; but it is to be expected that the creaks of a house and the shaking of objects on walls, shelves, and floors (including people) are more likely to occur from a noise whose spectrum contains considerable energy in this lower frequency region. This being the case, it follows that while the D- or A-weighted energy may be a reasonable index to loudness and noisiness therefrom, its magnitude does not directly relate, perhaps even conversely, to the annoyance or noisiness to be experienced because of house shaking and rattles.

The low-frequency content of some sounds (such as sonic booms, distant cannon firings, and factory machinery rumblings) can be expected to cause as much (if not more) annoyance from possible house vibration effects as from the noisiness due to loudness or speech or sleep interference effects from the higher frequency content. However, more common noises (such as aircraft and street traffic) containing sufficient low-frequency energy to cause annoyances from house vibrations have energy at the higher frequencies sufficient to cause somewhat more annoyance because of loudness, noisiness, and speech and sleep interference.

A-weighting

Because A-weighting gives such large negative weightings to low frequency (*e.g.*, −20 dB at 100 Hz and −30 dB at 50 Hz), large differences in the A-weighted energy of noise with different spectra are possible even when they have about equal amounts of low-frequency energy. For example, it is seen in figure 5.27 that if the sound pressure levels of one-third-octave bands below about 100 Hz were set to be about equal for the three noises (by increasing the level of the noise of the 747 aircraft by 10 dB), the L_A-level of the boom would be 20 to 30 dB less than that of the noise from the subsonic aircraft (108 dB for the subsonic Concorde, 107 dB for the 747, and 82 dB for the sonic boom).

The house vibration factor would seemingly complicate attempts to achieve a single frequency-weighted measure of sound energy for purposes of predicting the acceptability of noises of all spectra. However, simple vibration penalty procedures apparently can be used with A-weighted or, with proper adjustments, with D- or E-weighted, energy measures to take this factor into account for some common impulsive and nonimpulsive noises.

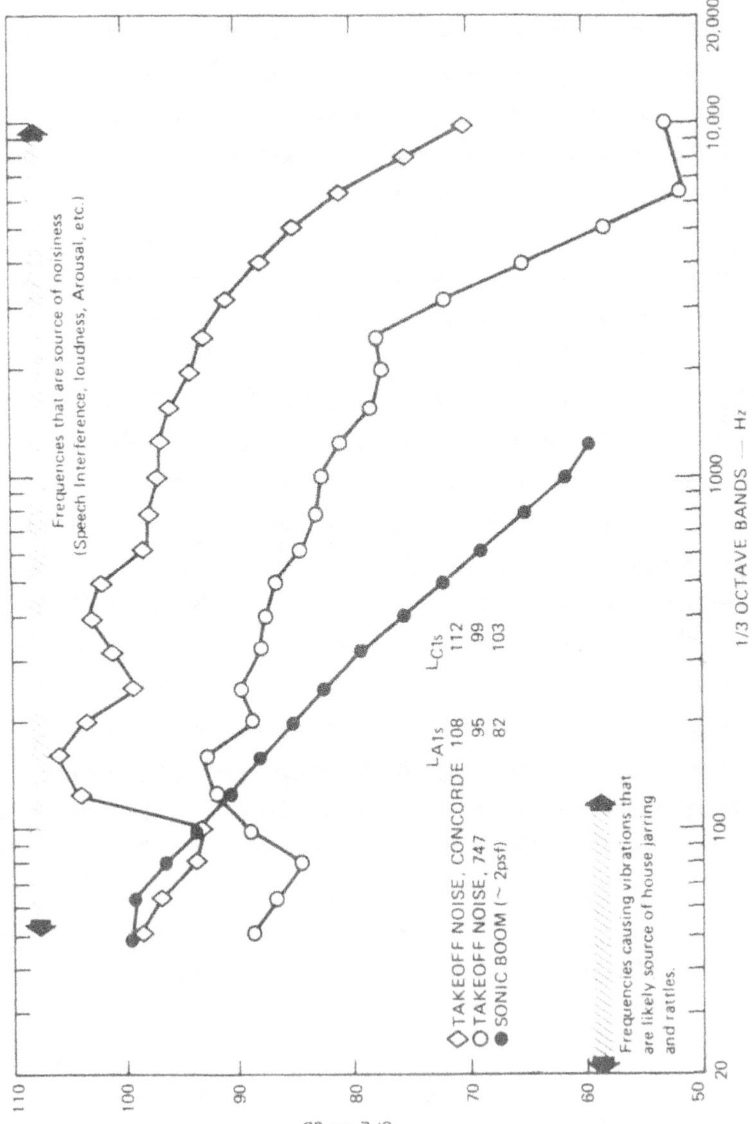

FIGURE 5.27. Relations between typical peak spectrum levels of aircraft noises and a sonic boom.

Impulses and house vibration

The data in figure 5.20 include both the startle and vibration contributions to the noisiness of expected impulses. The house vibration is not due, of course, to impulsiveness *per se*, but to the generally intense low-frequency content of impulse (see for example, that for a sonic boom in fig. 5.27). As discussed earlier, and as indicated by the difference between the dashed and middle curve in figure 5.20, the vibration effects contribute additional annoyance from the sonic boom impulses, an amount equivalent to about a 5-dB increase in sound pressure level.

Nonimpulsive noises

Most, but not all, nonimpulsive noise (*e.g.*, from aircraft and automobile) contains effectively more energy in the higher frequency regions affecting loud-

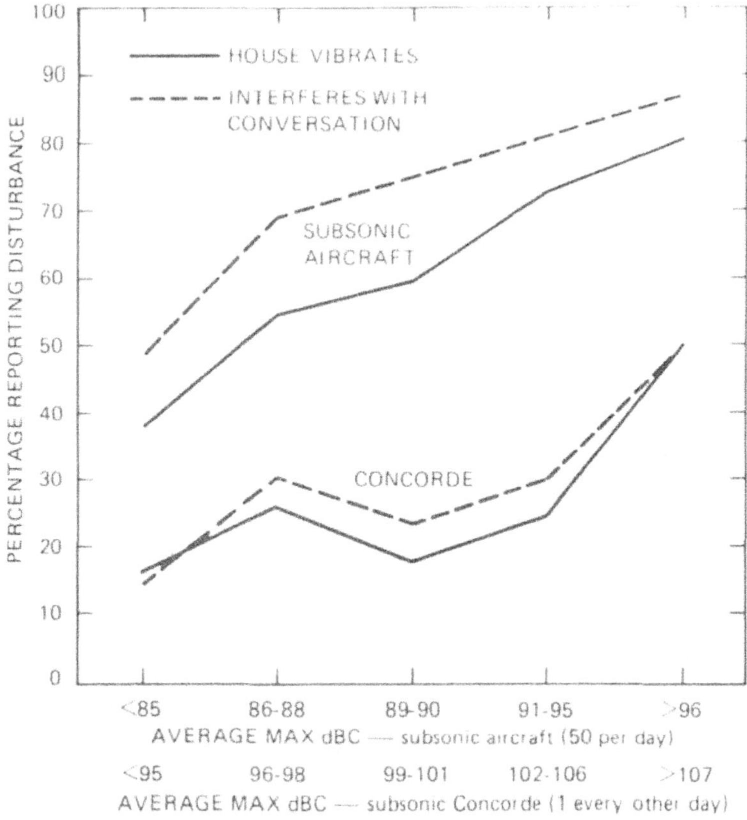

FIGURE 5.28. Reported disturbances to conversation and house vibration near Heathrow Airport, London. (Data from ref. 86.)

ness, speech, and sleep than in the low frequencies causing house vibrations. As a result, the perceived noisiness is generally due more to the direct auditory effects than to house vibrations. However, as seen in figure 5.28, as aircraft noise increases in sound level the disturbance from house vibrations progressively increases relative to that from noise interference with conversation. Accordingly, it would appear from these somewhat meager data that a house vibration annoyance penalty to take into account this increasing source of noise-induced annoyance is appropriate at these higher levels of typical aircraft noise.

This is not to suggest that this level might be considered a threshold level for causing noticeable vibration effects in typical houses from, for example, aircraft noise. Indeed, as seen in figure 5.28 from McKennell (ref. 86), significant disturbance in this regard is present at levels less than 80 dBA (85 dBC), as is also shown in attitude survey data about aircraft noise presented in chapter 6. However, above a level of about 85 dBA (90 dBC), the house vibrations reach in some typical houses a stage of progressively higher noticeability. At a level of about 110 dBC the house vibration effects are equally disturbing as speech interference. This is equivalent to a shift to the left of about 10 dB for vibration effects relative to speech interference effects, in comparison to the difference between the relative disturbing effect of vibration and speech interference when the aircraft noise is at a level of about 85 dBA.

Vibration penalty for nonimpulsive noise

The data in figures 5.20 and 5.28 are used as a basis for the correction procedure given in figure 5.29. Figure 5.29 provides the penalty in decibels to be added to measured L_A values of nonimpulsive broadband noises when they are up

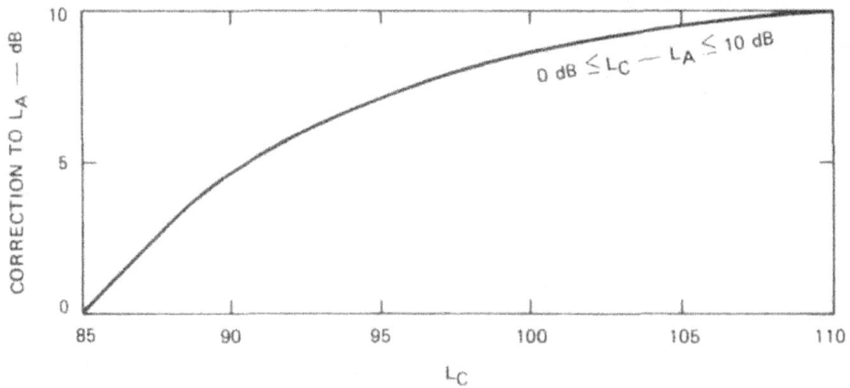

FIGURE 5.29. Correction to L_A to be applied to nonimpulsive broadband noises when L_A is up to 10 dB less than L_C, and $L_C \geq 85$ dB. The result is presumably predictive of annoyance due to both direct auditory effects and house vibration.

to 10 dB less than their L_C values and have an L_C level of 85 dB or greater. This suggested range comes from the general fact that noises falling in that range almost surely have spectra that sufficiently peak in the lower frequency region to potentially cause undue annoyance from house vibrations. A difference between L_C and L_A larger than 10 dB may well represent narrowband spectra, the vibration effects of which are not predictable from such a general procedure.

Schomer (ref. 73) also suggests calculating C-weighted L_{dn} to be used as a means of rating the house vibration and other effects of impulsive noise, as mentioned before. The presently proposed procedure (fig. 5.29) attempts to keep the same frequency-weighted unit (A or similar) as a basic part of a completely general measure for environmental noise, impulsive and nonimpulsive.

Summary of Concept of Perceived Noisiness

1. The primary goal of most noise level measurements is to assess, or to predict the unwantedness, disturbance, objectionableness, undesirability, unacceptability, perceived noisiness, or simply noisiness of the sound environment in real life. And, as stressed before, for practical reasons this is to be done independently of any meaning the different sounds have to different people. A secondary goal of research in this area has been to develop a simple, unitary physical measure of noises that is highly correlated with basic society-wide factors (such as startle, interference with speech communications and sleep, duration, and general loudness) that cause a sound to be judged as unwanted.

2. The exclusion of the effects of the meaning of sounds upon their judged unwantedness is a fundamental requirement of the definition and use of the phrase "perceived noisiness" as a judgeable attribute of sound. The attribute of noisiness includes, in addition to loudness, some perceived effects of impulsiveness and duration of sounds upon their unwantedness that are not generally perceived as a part of the attribute of loudness. Also a premise of the concept of noisiness is that it is applicable to the assessment of only noises, including impulsive noises, that are an expected normal part of an environment.

3. Whether there are wanted speech signals present along with noise and whether the subjects' environment is indoors or outdoors significantly (10 to 15 dB) affect the absolute, but not the relative, judgments of acceptability-unacceptability of different noises.

4. The specification of absolute levels of acceptability of noise exposures under different listening conditions (*e.g.*, indoors vs. outdoors) is more a matter for attitude survey research in real-life situations (see chapter 11) than for the rather basic laboratory and field research described in this chapter. However, the results of the laboratory research are generally consistent with the findings collected from studies of the reactions of people to noise in real life.

5. As a factor determining annoyance, the meaning of a sound may be argued to be more important and influential than its noisiness, as defined. This does not diminish the need for assessing noise with respect to more universal, acoustic factors. Also, the degree that the meaning of sounds may make noisiness at times a limited definition of the acceptability of a sound, makes loudness an even more limited description of the acceptability of sounds in real life because loudness does not include some of the temporal factors important to the relative acceptability of noises.

Summary of Noise Measurement Procedures

1. For broadband sounds, perceived loudness and noisiness are judged to be about the same as a function of solely spectral frequency. The D or E overall-frequency weightings are somewhat more accurate than A; and phons and PNdB are yet more accurate predictors of perceived loudness or noisiness of broad-band and, especially, narrow-band noises.

2. The critical summation time (sometimes called the time constant) of the ear appears to be about 0.3 sec for suprathreshold noises and 1 sec for noises near masked threshold for the perception of loudness or noisiness of sounds. For reasons of practicality and simplicity, it is proposed that all measurements of noise level, impulsive or nonimpulsive, be made with a basic integration time of 1 sec.

3. When either an impulsive or a nonimpulsive burst of noise continues for a longer period than about 1 sec, its loudness remains more or less constant, whereas its perceived noisiness continues to grow with time.

4. When expected impulses exceed a certain level and a certain background noise level, they are judged to be significantly noisier or more unwanted than a nonimpulsive noise having the same total integrated energy. A correction procedure for equating frequency-weighted impulse energy (integrated over 1 sec) to nonimpulsive energy is proposed.

5. Displacement of elements in typical houses, causing vibrations indirectly and directly sensed by the occupants, occurs primarily from sound energy below about 125 Hz and contributes to the perceived noisiness, or at least unwantedness, of a noise exposure over and above that perceived because of the higher frequency energy in the noise. A correction procedure is proposed for assessing this typical house vibration factor in predicting from physical measures the overall noisiness of intense sounds.

6. Laboratory and field research studies indicate that the perception of the noisiness or unwantedness of impulsive or nonimpulsive noises in real life should be predictable, in a statistical sense, from A-weighted (preferably D- or E-weighted) sound energy measured in 1-sec intervals. In addition quantitative cor-

rections for the extra noisiness due to possible impulsiveness and house vibration from the noise are needed for this purpose. These corrections are also applicable to noise measured in PNdB.

References

1. Stevens, S. S.: Perceived Level of Noise by Mark VII and Decibels (E). J. Acoust. Soc. America, vol. 51, no. 2, pt. 2, Feb. 1972, pp. 575-601.
2. Fletcher, H.; and Steinberg, J. C.: The Dependence of the Loudness of a Complex Sound Upon the Energy in the Various Frequency Regions of the Sound. Phys. Rev., second ser., vol. 24, no. 3, Sept. 1924, pp. 306-317.
3. Fletcher, Harvey; and Munson, W. A.: Loudness, Its Definition, Measurement and Calculation. J. Acoust. Soc. America, vol. V, no. 2, Oct. 1933, pp. 82-108.
4. Stevens, S. S.: The Measurement of Loudness. J. Acoust. Soc. America, vol. 27, no. 5, Sept. 1955, pp. 815-829.
5. Kryter, Karl D.: Scaling Human Reactions to the Sound From Aircraft. J. Acoust. Soc. America, vol. 31, no. 11, Nov. 1959, pp. 1415-1429.
6. Robinson, D. W.; and Whittle, L. S.: The Loudness of Octave-Bands of Noise. Acustica, vol. 14, no. 1, 1964, pp. 24-35.
7. Zwicker, E.; and Feldtkeller, R.: Das Ohr als Nachrichtenempfanger. Auflage, 2, S. Hirzel, Stuttgart, 1967.
8. Kryter, Karl D.; and Pearsons, Karl S.: Some Effects of Spectral Content and Duration on Perceived Noise Level. J. Acoust. Soc. America, vol. 35, no. 6, June 1963, pp. 866-883.
9. Churcher, B. G.; and King, A. J.: Performance of Noise Meters in Terms of the Primary Standard. J. Inst. Elec. Eng. (London), vol. 81, July 1937, pp. 57-90.
10. Beranek, L. L.; Marshall, J. L.; Cudworth, A. L.; and Peterson, A. P. G.: Calculation and Measurement of the Loudness of Sounds. J. Acoust. Soc. America, vol. 23, no. 3, May 1951, pp. 261-269.
11. Stevens, S. S.: Calculation of the Loudness of Complex Noise. J. Acoust. Soc. America, vol. 28, no. 5, Sept. 1956, pp. 807-832.
12. Stevens, S. S.: Concerning the Form of the Loudness Function. J. Acoust. Soc. America, vol. 29, no. 5, May 1957, pp. 603-606.
13. Stevens, S. S.: Procedure for Calculating Loudness: Mark VI. J. Acoust. Soc. America, vol. 33, no. 11, Nov. 1961, pp. 1577-1585.
14. American National Standard Procedure for the Computation of Loudness of Noise. ANSI S3.4-1972 (R-1968), American Nat. Stand. Inst., Inc., Mar. 26, 1968.
15. Method for Calculating Loudness Level. ISO R532-1967, American Nat. Standards Inst., Inc.

16. Zwicker, Von E.: Ein Verfahren zur Berechnung der Lautstarke (A Means for Calculating Loudness). Acustica, vol. 10, no. 1, 1960, pp. 304-308.
17. Zwicker, Von E.: Uber Psychologische und Methodische Grundlagen der Lautheit (Psychological and Methodical Foundations of Loudness). Acustica, vol. 8, 1958, pp. 237-258.
18. Zwicker, E.; Flottorp, G.; and Stevens, S. S.: Critical Band Width in Loudness Summation. J. Acoust. Soc. America, vol. 29, no. 5, May 1957, pp. 548-557.
19. Robinson, D. W.: Statistical Aspects of the Relation Between Binaural and Monaural Thresholds. Acustica, vol. 11, no. 4, 1961, pp. 185-190.
20. Egan, James P.; and Meyer, Donald R.: Changes in Pitch of Tones of Low Frequency as a Function of the Pattern of Excitation Produced by a Band of Noise. J. Acoust. Soc. America, vol. 22, no. 6, Nov. 1950, pp. 827-833.
21. Ehmer, Richard H.: Masking by Tones vs. Noise Bands. J. Acoust. Soc. America, vol. 31, no. 9, Sept. 1959, pp. 1253-1256.
22. Carter, N. L.; and Kryter, K. D.: Masking of Pure Tones and Speech. J. Aud. Res., vol. 2, 1962, pp. 68-98.
23. Stevens, S. S.: On the Validity of the Loudness Scale. J. Acoust. Soc. America, vol. 31, no. 7, July 1959, pp. 995-1003.
24. Gzhesik, Ya; Lempkovski, A.; Turchinski, B.; Fazonovich, Ya.; and Shimchik, K.: Comparison of Methods for Evaluating Loudness From Data Published During the Period 1930-1957. Soviet Phys.—Acoust., vol. 6, no. 4, Apr.-June 1961, pp. 421-441.
25. Reynolds, George S.; and Stevens, S. S.: Binaural Summation of Loudness. J. Acoust. Soc. America, vol. 32, no. 10, Oct. 1960, pp. 1337-1344.
26. Hellman, Rhona; and Zwislocki, J.: Monaural Loudness Function at 1000 cps and Interaural Summation. J. Acoust. Soc. America, vol. 35, no. 6, June 1963, pp. 856-865.
27. Kryter, Karl D.: Review of Research and Methods for Measuring the Loudness and Noisiness of Complex Sounds. NASA CR-422, 1966.
28. Garner, W. R.: Half-Loudness Judgments Without Prior Stimulus Context. J. Exp. Psychol., vol. 55, 1958, pp. 482-485.
29. Garner, W. R.: The Development of Context Effects in Half-Loudness Judgments. J. Exp. Psychol., vol. 58, 1959, pp. 212-219.
30. Wolsk, David: Discrimination Limen for Loudness Under Varying Rates of Intensity Change. J. Acoust. Soc. America, vol. 36, no. 7, July 1964, pp. 1277-1282.
31. Kwiek, M.: Investigation of the Relation Between the Hearing Sensitivity and the Intensity of Sinusoidal Tones by Differential Methods. The Friends of Sciences Soc., Poznan, ser. A, vol. 6, 1953, p. 329.
32. Garner, W. R.: A Technique and a Scale for Loudness Measurement. J. Acoust. Soc. America, vol. 26, no. 1, Jan. 1954, pp. 73-88.
33. Garner, W. R.; and Miller, G. A.: The Masked Threshold of Pure Tones as a Function of Duration. J. Exp. Psychol., vol. 37, 1947, pp. 293-303.

34. Sutherland, L. C.; and Burke, R. E.: Annoyance, Loudness, and Measurement of Repetitive Type Impulsive Noise Sources. Rept. EPA-550/9-79-103, U.S. Environ. Prot. Agency, Nov. 1979. (Available from NTIS as PB 82-138 785.)
35. Fidell, Sanford; Pearsons, Karl S.; Grignetti, Mario; and Green, David M.: The Noisiness of Impulsive Sounds. J. Acoust. Soc. America, vol. 48, no. 6, pt. 1, Dec. 1970, pp. 1304-1310.
36. Pedersen, O. Juhl; Lyregaard, P. E.; and Poulsen, T.: The Round Robin Test on Evaluation of Loudness Level of Impulsive Noise. ISO/TC 43/SC 1/SG 'B' (Secretariat-15) 23, Acoustics Lab., Tech. Univ. of Denmark, Sept. 1977.
37. Borsky, Paul N.: Community Reactions to Air Force Noise. Parts I and II. WADD Tech. Rept. 60-689 (I) and (II), U.S. Air Force, Mar. 1961.
38. Cederlöf, R.; Jonsson, E.; and Sörensen, S.: On the Influence of Attitudes to the Source on Annoyance Reactions to Noise. Nord Hyg. Tidsk., vol. 48, 1967, pp. 16-59.
39. Pearsons, Karl S.; and Horonjeff, Richard D.: Category Scaling Judgment Tests on Motor Vehicle and Aircraft Noise. FAA-DS-67-8, July 1967.
40. Berglund, Brigitta; Berglund, Ulf; and Lindvall, Thomas: Scaling Loudness, Noisiness and Annoyance of Aircraft Noise. J. Acoust. Soc. America, vol. 57, no. 4, Apr. 1975, pp. 930-934.
41. Nixon, Charles W.; Von Gierke, H. E.; and Rosinger, George: Comparative Annoyances of "Approaching" Versus "Receding" Sound Sources. J. Acoust. Soc. America, vol. 45, no. 1, Jan. 1969, p. 330.
42. Ollerhead, J. B.: Subjective Evaluation of General Aviation Aircraft Noise. Tech. Rep. NO-68-35, FAA, Apr. 1968.
43. Pearsons, Karl S.: Assessment of the Validity of Pure Tone Corrections to Perceived Noise Level. Progress of NASA Research Relating to Noise Alleviation of Large Subsonic Jet Aircraft, NASA SP-189, 1968, pp. 573-586.
44. Laird, Donald A.; and Coye, Kenneth: Psychological Measurements of Annoyance as Related to Pitch and Loudness. J. Acoust. Soc. America, vol. 1, no. 1, Oct. 1929, pp. 158-163.
45. Reese, T. W.; Kryter, K. D.; and Stevens, S. S.: The Relative Annoyance Produced by Various Bands of Noise. IC-65, Psycho-Acoustic Lab., Harvard Univ., Mar. 1944.
46. Kryter, K. D.: Loudness and Annoyance-Value of Bands of Noise. Oralism and Auralism: Transactions of the 30th Annual Meeting of the National Forum on Deafness and Speech Pathology, 1948, pp. 26-28.
47. Kryter, K. D.; and Pearsons, K. S.: Modification of Noy Tables. J. Acoust. Soc. America, vol. 36, no. 2, Feb. 1964, pp. 394-397.
48. Wells, R. J.: Recent Research Relative to Perceived Noise Level. J. Acoust. Soc. America, vol. 42, no. 5, Nov. 1967, p. 1151.
49. Ollerhead, J. B.: The Noisiness of Diffuse Sound Fields at High Intensities. FAA-NO-70-3, Aug. 1969. (Available from DTIC as AD 708 816.)

50. Young, Robert W.; and Peterson, Arnold: On Estimating Noisiness of Aircraft Sounds. J. Acoust. Soc. America, vol. 45, no. 4, Apr. 1969, pp. 834-838.
51. Kryter, K. D.: Concepts of Perceived Noisiness. Their Implementation and Application. J. Acoust. Soc. America, vol. 43, no. 2, Feb. 1968, pp. 344-361.
52. Kryter, K. D.: Possible Modifications to the Calculation of Perceived Noisiness. NASA CR-1636, 1970.
53. Scharf, B.; and Hellman, R.: Comparison of Various Methods for Predicting the Loudness and Acceptability of Noise. Part 2: Effects of Spectral Pattern and Tonal Components. Rept. EPA-550/9-79-102, U.S. Environ. Prot. Agency, Nov. 1979. (Available from NTIS as PB 82-138 702.)
54. Scharf, B.; Hellman, R.; and Bauer, J.: Comparison of Various Methods for Predicting the Loudness and Acceptability of Noise. Rept. EPA-550/9-77-101, U.S. Environ. Prot. Agency, 1977.
55. Kryter, Karl D.: The Effects of Noise on Man. Academic Press, Inc., 1970.
56. Broadbent, D. E.; and Robinson, D. W.: Subjective Measurements of the Relative Annoyance of Simulated Sonic Bangs and Aircraft Noise. J. Sound & Vib., vol. 1, no. 2, Apr. 1964, pp. 162-174.
57. Parnell, John E.; Nagel, David C.; and Parry, Hugh J.: Growth of Noisiness for Tones and Bands of Noise at Different Frequencies. FAA-DS-67-21, Dec. 1967.
58. Kryter, K. D.; and Pearsons, K. S.: Judged Noisiness of a Band of Random Noise Containing an Audible Pure Tone. J. Acoust. Soc. America, vol. 38, no. 1, July 1965, pp. 106-112.
59. Little, John W.: Human Response to Jet Engine Noises. Noise Contr. Shock Vib., vol. 7, no. 3, May-June 1961, pp. 11-13.
60. Sperry, William C.: Aircraft Noise Evaluation. Rep. FAA-NO-68-34, Sept. 1968. (Available from DTIC as AD 676 230.)
61. Wells, R. J.; and Blazier, W. E., Jr.: A Procedure for Computing the Subjective Reaction to Complex Noise. Congress Report I of the Fourth International Congress on Acoustics—Copenhagen, 1962, Organization Committee of the Fourth International Congress on Acoustics, 1962, Paper L24.
62. Kryter, K. D.; Pearsons, K. S.; and Woods, B.: Preliminary Study on the Effect of Multiple and Modulated Tones on Perceived Noise. NASA CR-69606, 1965.
63. Willshire, Kelli F.; and Powell, Clemans A.: Effects of Activity Interference on Annoyance Due to Aircraft Noise. NASA TP-1938, 1981.
64. Lukas, J. S.: Noise and Sleep, and Some Effects of Noise on Children. Paper presented at the Second International Meeting on Noise and the Community. (Buenes Aries, Argentina). 1977.
65. Pearsons, Karl S.: The Effects of Duration and Background Noise Level on Perceived Noisiness. ADS-78, FAA, Apr. 1966.

66. Fuller, H. C.; and Robinson, D. W.: Temporal Variables in the Assessment of an Experimental Noise Environment. NPL Acoustics Rept. Ac-72, British A.R.C., Feb. 1975.
67. Hiramatsu, K.; Takagi, K.; and Yamamoto, Y.: The Effect of Sound Duration on Annoyance. J. Sound & Vib., vol. 59, no. 4, Aug. 22, 1978, pp. 511-520.
68. Kuwano, S.; Namba, S.; and Nakajima, Y.: On the Noisiness of Steady State and Intermittent Noises. J. Sound & Vib., vol. 72, no. 8, Sept. 8, 1980, pp. 87-96.
69. Rosinger, George; Nixon, Charles W.; and Von Gierke, Henning E.: Quantification of the Noisiness of "Approaching" and "Receding" Sounds. J. Acoust. Soc. America, vol. 48, no. 4, Oct. 1970, pp. 843-853.
70. Namba, Seiichiro; and Kuwano, Sonoko: An Experimental Study on the Relation Between Long-Term Annoyance and Instantaneous Judgment of Level-Fluctuating Sounds. Proceedings Inter-Noise 79—Volume II, Stefan Czarnecki, ed., Inst. Fundam. Technol. Res. Polish Acad. Sci., c.1979, pp. 837-842.
71. Kryter, K. D.; Johnson, P. J.; and Young, J. R.: Psychological Experiments on Sonic Booms. Annex B of Sonic Boom Experiments at Edwards Air Force Base, NSBEO-1-67 (Contract AF 49(638)-1758), CFSTI, U.S. Dep. Com., July 28, 1967.
72. Young, J. R.: Measurement of the Psychological Annoyance of Simulated Explosion Sequences (Second Year). Contract DACA 23-74-C-0008, Stanford Res. Inst., Feb. 1976.
73. Schomer, Paul D.: Evaluation of C-Weighted L_{dn} for Assessment of Impulse Noise. J. Acoust. Soc. America, vol. 62, no. 2, Aug. 1977, pp. 396-399.
74. Fidell, Sanford; and Pearsons, Karl S.: Study of the Audibility of Impulsive Sounds. NASA CR-1598, 1970.
75. Izumi, Kiyoto: Two Experiments on the Perceived Noisiness of Periodically Intermittent Sounds. Noise Contr. Eng., vol. 9, no. 1, July-Aug. 1977, pp. 16-23.
76. Carter, N. L.: Effect of Rise Time and Repetition Rate on the Loudness of Acoustic Transients. J. Sound & Vib., vol. 21, no. 2, Mar. 22, 1972, pp. 227-239.
77. Powell, Clemans A.: A Subjective Field Study of Helicopter Blade-Slap Noise. NASA TM-78758, 1978.
78. Galloway, William J.: Subjective Evaluation of Helicopter Blade Slap Noise. Helicopter Acoustics, NASA CP-2052, Part II, 1978, pp. 403-418.
79. Powell, Clemans A.: Subjective Field Study of Response to Impulsive Helicopter Noise. NASA TP-1833, 1981.
80. Izumi, Kiyoto: Two Aspects of the Perceived Noisiness of Intermittent Sounds. Inter-Noise 75 Proceedings, Ken'iti Kido, ed., Tohoku Univ. (Sendai 980, Japan), 1975, pp. 453-456.

81. Izumi, Kiyoto: The Startle Effect and the Perceived Noisiness of Periodically Intermittent Sounds. Inter-Noise 77—Noise Control: The Engineer's Responsibility, Eric J. Rathe, ed., Swiss Federal Inst. of Technol. (Zurich, Switzerland), 1977, pp. B 363-B 368.
82. Kundert, Warren R.: How To Measure Impulse Noise. Sound & Vib., vol. 15, no. 3, p. 13, 1981.
83. Young, R. W.; and Cowen, S. J.: Responses of Sound Level Meters to Impulsive Sounds. J. Acoust. Soc. America, vol. 59, suppl. 1, Spring 1976, paper KK11.
84. Howes, Walton L.: Overall Loudness of Steady Sounds According to Theory and Experiment. NASA RP-1001, 1979.
85. Carden, Huey D.; and Mayes, William H.: Measured Vibration Response Characteristics of Four Residential Structures Excited by Mechanical and Acoustical Loadings. NASA TN D-5776, 1970.
86. McKennell, A. C.: Community Response to Concorde Flights Round London (Heathrow) Airport. Social & Community Planning Research, Mar. 1977.

Chapter 6
Presbycusis, Sociocusis, and Nosocusis

Introduction

Sound pressure levels, expressed in terms of hearing level (HL), of pure tones barely detectable in the quiet reveal the following effects on the threshold sensitivity of the auditory system: aging (presbycusis); exposure to the sounds and noises of everyday living (sociocusis); and various common pathological conditions of the ear from other causes (nosocusis). Historically, sociocusis refers only to nonindustrial noise-induced threshold shift. Shifts caused by noise at the work place are identified as industrial noise-induced threshold shifts. (See ref. 1.)

A significant number of studies and analyses of hearing levels found in general populations have been published over the past ten years or so. However, some uncertainty remains as to how to quantitatively interpret the contribution of several important variables to the hearing sensitivity of various segments of the general population. It is the purpose of this chapter to review published pure-tone hearing-level surveys and to derive, as warranted, new generalizations with respect to the variables of age, sociocusis, sex, race, and, to a limited extent, some common otological disorders.

Quantitative knowledge of the effects of these variables upon hearing levels has application to the resolution of two important practical matters:

1. The establishment of a baseline of normal hearing. This is needed to identify limits of exposure to noise that will guard populations from noise-induced hearing loss (shifts in HL).

2. The determination of the relative detrimental effects of sociocusis, presbycusis, and noise at the workplace on hearing in individuals. This is, of course, critically related to the determination of workmen's compensation for possible damage to hearing from exposure to noise.

Methodological Problems of Hearing Surveys

The presentation of some information about the methods and procedures involved in the selection of subjects and in the measurement of HL is in order before presenting the data pertaining to sociocusis and presbycusis. This information should be helpful in understanding some of the apparent complexities in the data that follow. The reference zero sound pressure levels of all the HL values given in this report are relative to ISO-389 (ref. 2), unless otherwise indicated. (See refs. 3 and 4.)

Nosocusis

Usually, but not always, persons collecting HL data for the assessment of presbycusis and sociocusis exclude the data for certain individuals when there is a history of exposure to intense noise or gunfire; when the eardrum is scarred or other evidence of middle-ear disease is present; when the ears are impacted with wax; or when there are indications of otological trauma from blows to the head, use of ototoxic drugs, and other sensorineural disorders.

Robinson *et al.* (ref. 5) investigated the general magnitude of the effects on HL of some of the otological conditions that are sometimes used as criteria for the rejection of data in studies of presbycusis and sociocusis. As shown in table 6.1, the effect of wax in the ears and upper-respiratory infection is rather small—of the order of 1 to 3 dB.

Children (ages 6 to 11) with apparent otological problems, at least upon otoscopic examination, were found in a large population survey (ref. 6) to have HL's about 5 dB greater than persons classified as otologically normal. (See fig. 6.1.) However, in a random selection of people in the United States, somewhat less than 25 percent of the population of both children and adults had otoscopically abnormal ears. (See refs. 6 and 7.) Presumably, eliminating the otologically abnormal subjects would reduce the average HL of the entire population by about 1 to 2 dB (about 25 percent of the 5 dB indicated in fig. 6.1).

Somewhat larger differences were found by Burns *et al.* (ref. 8) in a clinical study of about 700 60- to 64-year-old factory workers. Table 6.2 shows that various pathological conditions of the middle or inner ear identified during otological screening cause about a 10- to 15-dB shift in mean HL relative to the mean HL of normal ears that have not been exposed to any military gunfire. The difference between all subjects and normal subjects with no exposure to gunfire is of the order of 6 dB over all frequencies.

Thus, it appears from reference 8 that a difference of about 6 dB could be expected between the mean HL of an otologically unscreened population and

TABLE 6.1

Effect of Wax and Upper-Respiratory Infection Upon Hearing Levels

[From ref. 5]

Difference in hearing loss, dB	Number of subjects (N)	Frequency, kHz						
		0.5	1	2	3	4	6	Average of 0.5 to 6
With wax minus after wax removed	4	5.1	2.4	2.8	3.0	−1.4	−1.7	1.7
Group with wax minus group free from wax	108	0.5	1.2	2.1	0.5	−1.9	1.6	0.7
Before and after upper-respiratory infection	4	6.5	5.6	−0.1	3.5	0.4	0.3	2.7

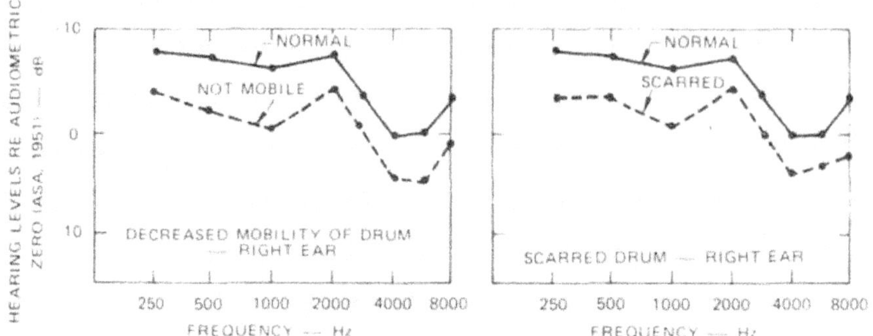

FIGURE 6.1. Average hearing levels in the right ear for U.S. children with normal and significantly abnormal otoscopic findings. (From ref. 6.)

that of a group of otologically and gunfire-exposure screened persons from the same population. However, the prevalence of otoscopic abnormalities is near 60 percent in the industrial-worker population, as compared with about 25 percent in the general (U.S.) population; the difference in HL between the total population and the abnormal group is presumably proportionately different (6 dB compared with 1 to 2 dB).

Practice effects of audiometry

Robinson, Shipton, and Whittle (ref. 9) studied the effects on HL of the learning or practice that takes place with automatic audiometry. Robinson first presented to subjects a series of 15 test items (tones) to achieve a total audiogram; this series was then repeated. One week later, the two audiograms were repeated.

Figure 6.2 shows that the threshold level of the first item of the first series was, on the average, about 7 dB higher than it was when the series was repeated on the same day. A week later, the first item of a series was only about 3.5 dB above its repeat level. The practice effect for items after the first few items in the series was reduced to about 1 to 3 dB.

With manual audiometry, the practice effect appears to be smaller. In reference 10, for example, it was found that more than 95 percent of the repeat thresholds of the first test item (1000 Hz) were identical or no more than ±5 dB from repeat of the same item six items later. The average difference was near zero (private communication with J. Roberts, U.S. Dept. of Health, Education, and Welfare, National Center for Health Statistics, Rockville, Maryland).

Reliability of repeat audiograms

A second, perhaps more important, question concerns the effect, if any, of practice upon audiograms repeated on different days within a few weeks. Robin-

TABLE 6.2

Relative Hearing Levels of Otologically Normal Groups, Pathological Groups, and Total Group

[Values are differences of mean hearing levels, left and right ears averaged, between otological normals, those with no experience with gunfire, and other groups; from ref 8]

Groups compared	Number in group	Frequency, kHz					
		0.5	1	2	3	4	6
Normals[a]	291	0	0	0	0	0	0
Normals[a] — normals[b]	93	1.3	1.5	1.8	1.9	2.5	4.5
Conductive[c] — normals[a]	133	13.5	13.3	15.7	12.8	13.4	14.3
Sensorineural[d] — normals[a]	36	8.4	9.1	13.1	16.6	16.7	16.2
Doubtful etiology[e] — normals[a]	170	8.3	10.1	14.0	13.1	11.1	12.2
All pathology — normals[a]	339	10.4	11.3	14.6	13.3	12.6	13.4
All subjects — normals[a]	723	5.0	5.5	7.2	6.5	6.2	6.9

[a]*Normal otologically; no experience with gunfire.*
[b]*Normal otologically; some experience with gunfire.*
[c]*Abnormal eardrums.*
[d]*Tinnitus, trauma, ototoxic drugs.*
[e]*"Hearing trouble," opaque drums, dizziness.*

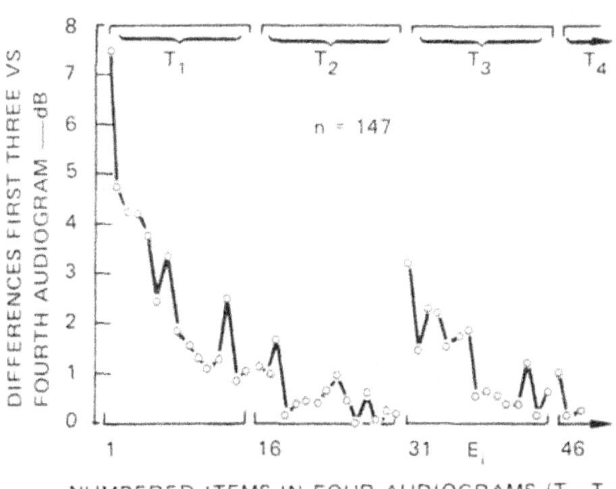

FIGURE 6.2. Average learning curve for whole group for first three audiograms and part of fourth. Values are pegged to corresponding frequency and ear in last 12 elements of fourth audiogram. T_1 and T_2 are day 1; T_3 and T_4 are day 2, one week later. (From ref. 9.)

son *et al.* (ref. 9) found that test/retest variations (over four test series) were generally quite small when using automatic audiometers for all but a few subjects. (See fig. 6.3.) Relatively practiced subjects are able to repeat manually obtained audiograms from one week to the next with an average difference of about 1 dB. (See ref. 11.)

Large Population Surveys of Hearing Level

Surveys in United States

The surveys of the hearing of U.S. citizens made by the U.S. Public Health Service (USPHS) in 1960-62, 1963-65, and 1971-75 provide a comprehensive bank of data relevant to the questions at hand. (See refs. 6, 7, 10, and 12 to 16.) Several features of these USPHS surveys are as follows:

1. The numbers of subjects were large (about 6700 adults in each of the 1960-62 and 1971-75 surveys and 7119 children (ages 6-11) in the 1963-65 survey).

2. The subjects were selected at random with respect to geographic and demographic variables.

3. HL data were reported for the subjects regardless of their occupation, history of noise exposure, or otological disease.

In short, the data represent a cross-sectional baseline for describing pure-tone hearing-level characteristics of the general population of the United States. The 1971-75 survey (ref. 16) also included some speech intelligibility tests.

Differences in hearing level between 1960-62 and 1971-75 U.S. surveys

For pure-tone audiometric test frequencies of 500, 1000, 2000, and 4000 Hz, figure 6.4 shows the cumulative percentage of the population tested having hearing losses shown on the abscissa, with age as the parameter. Figure 6.4 and table 6.3 show that, for the median (50th percentile), there is fairly close agreement between the HL's as found for the different decades of ages between the 1960 and 1970 periods, except, perhaps, for the 500-Hz and 4000-Hz test frequencies (especially for the 70-year-old group). The decades of an age group, incidentally, represent the average for a span of ±5 years. That is, the 70-year-old group covers the range of 65 to 74 year olds.

The reduced hearing at 4000 Hz in the 1960-62 survey relative to the 1971-75 period may be because about 30 percent of the 70-year-old men of the 1960-62 survey were in the military services during World War I, but the 70-year-old men of the 1971-75 testing period were not involved in the war. The 70 year olds in the 1971-75 survey were for the most part too young, less than 18 years of age, to have been in the military services in World War I (1916-18). The 70 year olds

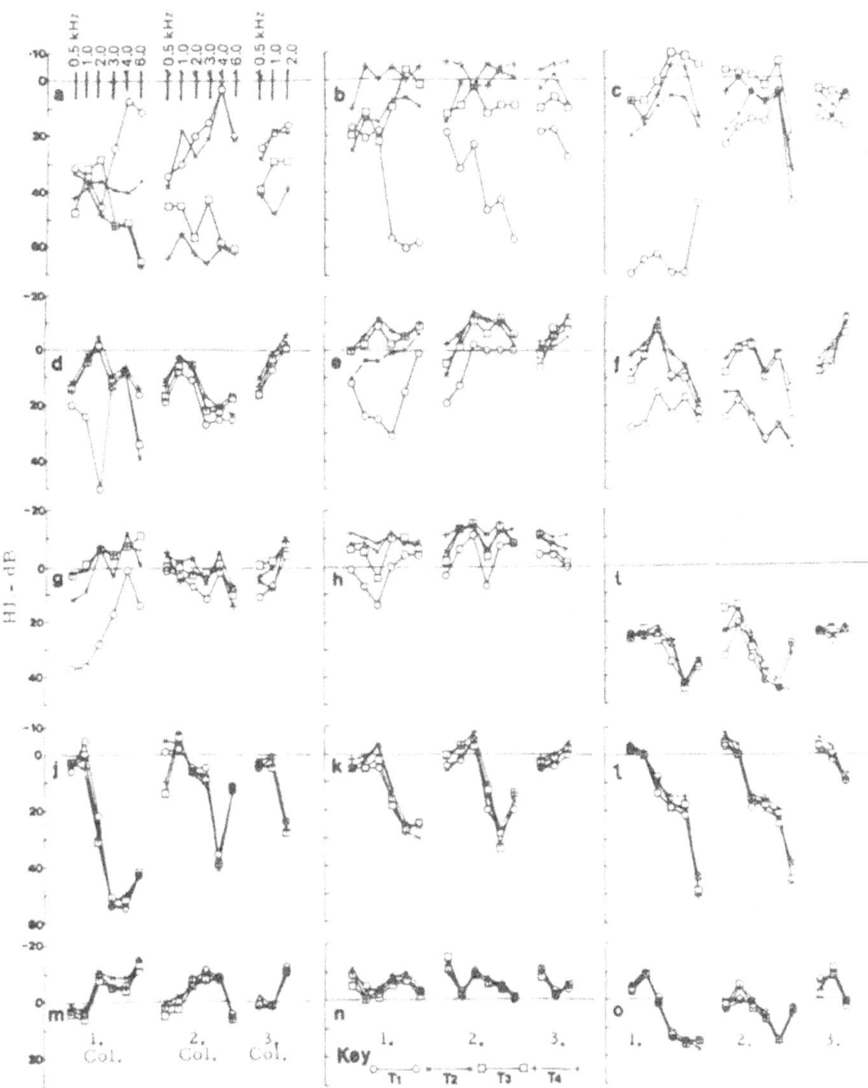

FIGURE 6.3. Repeat audiograms of worst (a to h) and best (i to o) subjects, in ascending frequency order: column 1 is data for 0.5 to 6 kHz, left ear; column 2 is for 0.5 to 6 kHz, right ear; and column 3 is for 0.5 to 2 kHz, left ear repeat. T_1 and T_2 are day 1; T_3 and T_4 are day 2, one week later. (From ref. 9.)

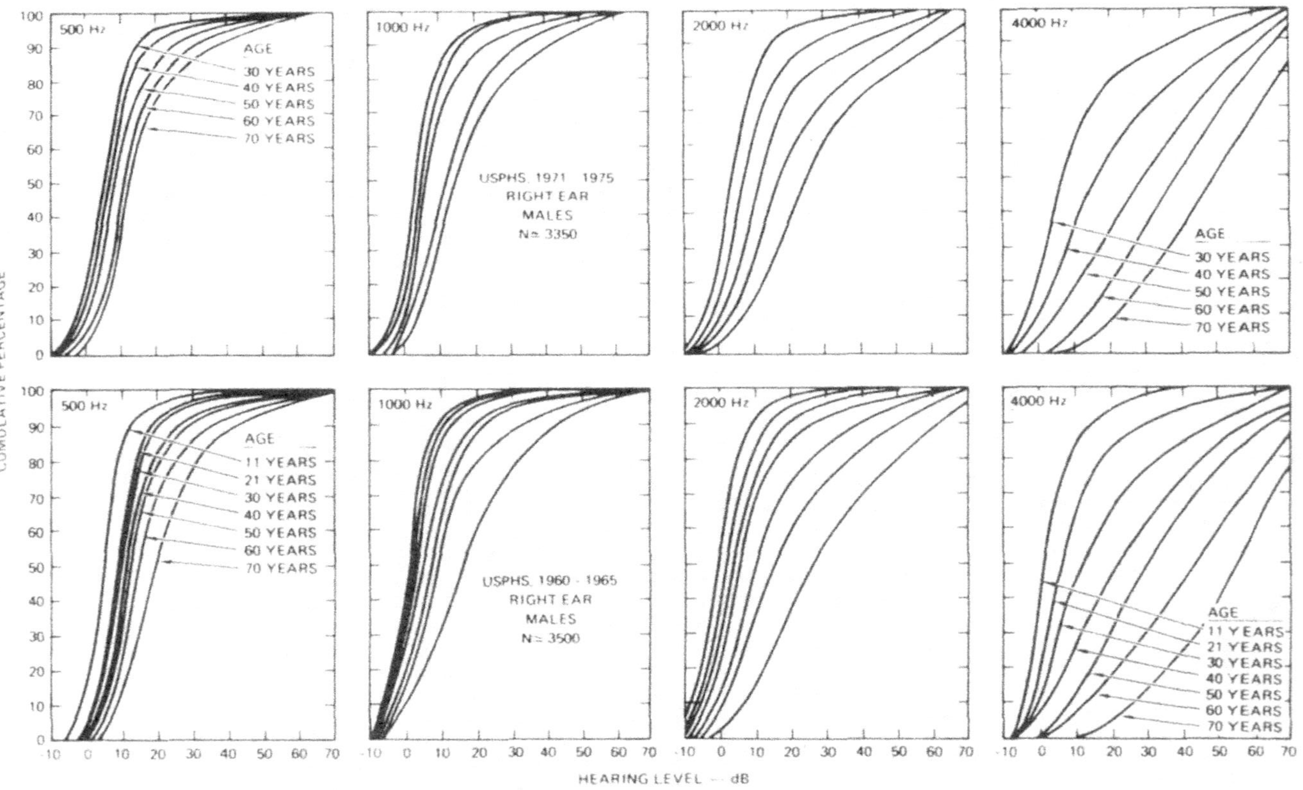

FIGURE 6.4. Cumulative percentages of population below specified hearing levels. Based on USPHS surveys with age as the parameter. Top panels for Rowland (1971–75) survey (ref. 16), bottom panels for Glorig and Roberts (1960–62) survey (ref. 10), and, age 11 yrs, Roberts and Huber survey (ref. 13).

TABLE 6.3

HL at the 50th Percentile as a Function of Age for USPHS Surveys of 1960-62 and 1971-75

		1960-62 HL's minus 1971-75 HL's, dB, for age				
Frequency, Hz	Year	30	40	50	60	70
500	1960	10	13	14	15	20
	1971	5	6	7	11	12
Difference		5	7	7	4	8
1000	1960	4	5	8	10	17
	1971	4	5	8	12	15
Difference		0	0	0	−2	2
2000	1960	3	5	10	15	26
	1971	4	7	11	17	24
Difference		−1	−2	−1	−2	2
4000	1960	12	22	31	43	56
	1971	7	16	29	38	49
Difference		5	6	2	5	7

in the 1960-62 survey would have been of prime military age. Accordingly, it is probable that a significant portion of the 70-year-olds in the 1971-75 survey suffered less noise-induced threshold shift because of exposure to gun noise in basic training and combat. For this reason, in further analysis of these USPHS data in this report, the HL for the 70 year olds are taken only from the 1971-75 survey.

It is also possible that some, perhaps all, of the differences at 500 Hz and 4000 Hz between the 1960 and 1970 surveys are due to uncertainties involved in converting the HL data for the 1960 surveys from the old 1951 standard for audiometric zero (ref. 3) to those of the ISO standard (ref. 2) used in the 1970 surveys. The conversion factors used are indicated in the discussion of table 6.4. The uncertainties arise from the fact that the audiometer earphones used in the 1960 surveys were so-called "TDH-39," whereas the conversion factors are based on so-called "W.E. 705-A" earphones.

Age of onset of presbycusis and sociocusis

It is clear from the 1960-62 and 1971-75 surveys that there is a gradual reduction in hearing sensitivity with increasing age, especially after the age of about 20. There is some indication (in the difference between the 11- and 21-year-old group) that this trend may start at an even earlier age.

Several thousand of the same subjects were tested at the ages of 7, 11, and 16 years. (See ref. 17.) The results indicate an improvement (fig. 6.5(a)) in hearing sensitivity from the ages of 7 to 11 years, followed by a decline from the ages of 11 to 16 years. (Also see refs. 18 and 19.)

Eagles *et al.* (ref. 20) tested about 4000 children ranging in age from 5 to 14 years. A pattern of hearing sensitivities similar to that found in reference 17 is shown in figure 6.5(b). That is, there was an improvement in HL of several deci-

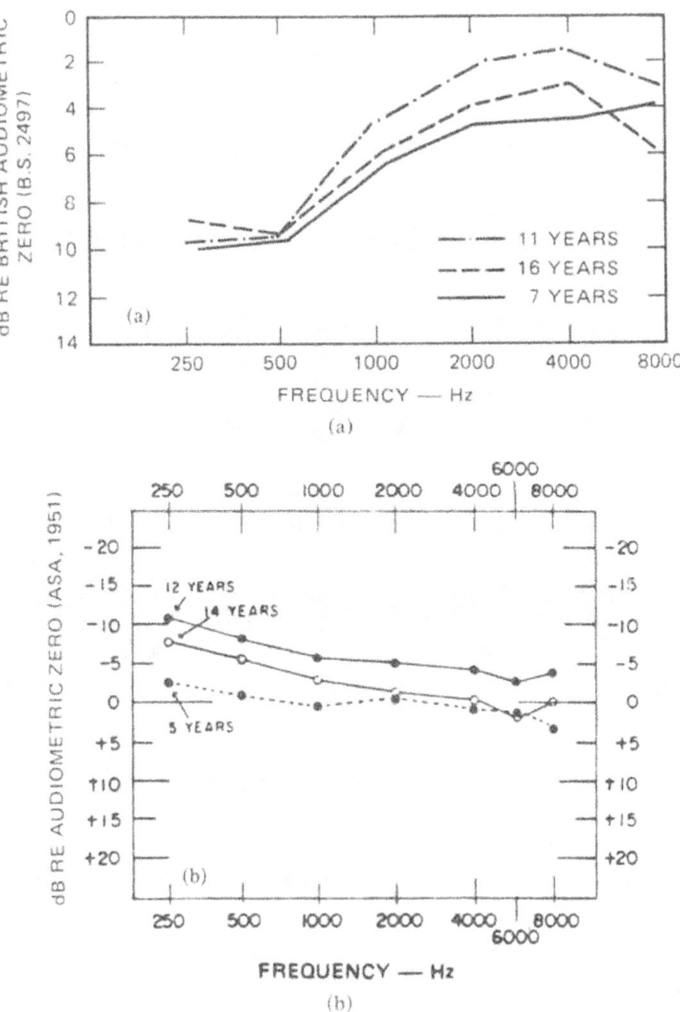

FIGURE 6.5. Hearing sensitivity tests in children. Graph (a) shows median thresholds (data from ref. 17), graph (b) shows mean thresholds (data from ref. 20).

bels for children between the ages of 5 to 12 years, followed by a decline of about 2 dB at the age of 14 years. As shown in figure 6.6, an improvement of several decibels in HL sensitivity from 7 to 11 years of age was also found in the USPHS survey of 1963–65. The differences in the general shape of the audiograms as a function of frequency are largely due to differences between the British and U.S. audiometer reference calibrations and test environments.

On the other hand, it was determined in reference 21 that the average audiometric threshold for both boys and girls from 12 to 17 years of age was about 1 dB lower than from 6 to 11 years of age. Reference 21 involved only 224 subjects, however, and the age-group averaging may have obscured the trends reported in the aforementioned studies. Figures 6.7(a), (b), and (c) suggest that, although both boys and girls are exposed to "social noises," boys are exposed to more noise than are girls; and some relation, though statistically insignificant, may exist in young people between the amount of exposure to everyday noises or sounds and hearing thresholds.

In subsequent studies, these investigators (refs. 22 and 23) measured the A-weighted energy in the noise (actually all the sound energy) to which 127 people (ages 7 to 20 years) were exposed during typical 24-hour days. The sound energy was measured by means of small microphone recording dosimeter devices

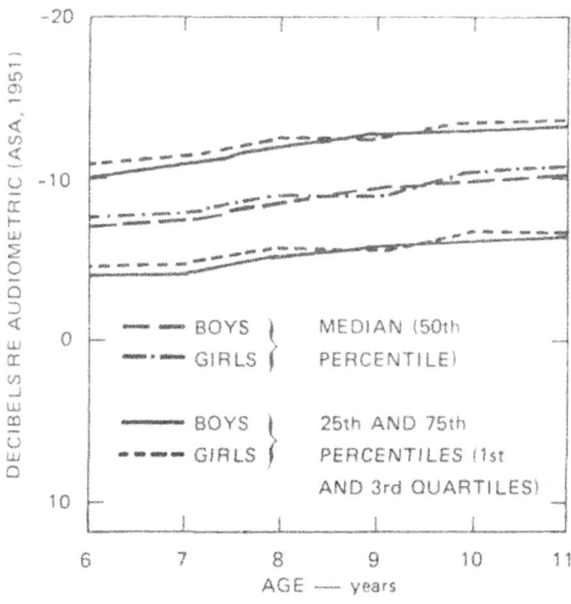

FIGURE 6.6. Medians and quartiles from distribution of children from 6 to 11 years of age by hearing threshold levels for speech (average of pure-tone levels at 500, 1000, and 2000 Hz) in the better ear. (From ref. 13.)

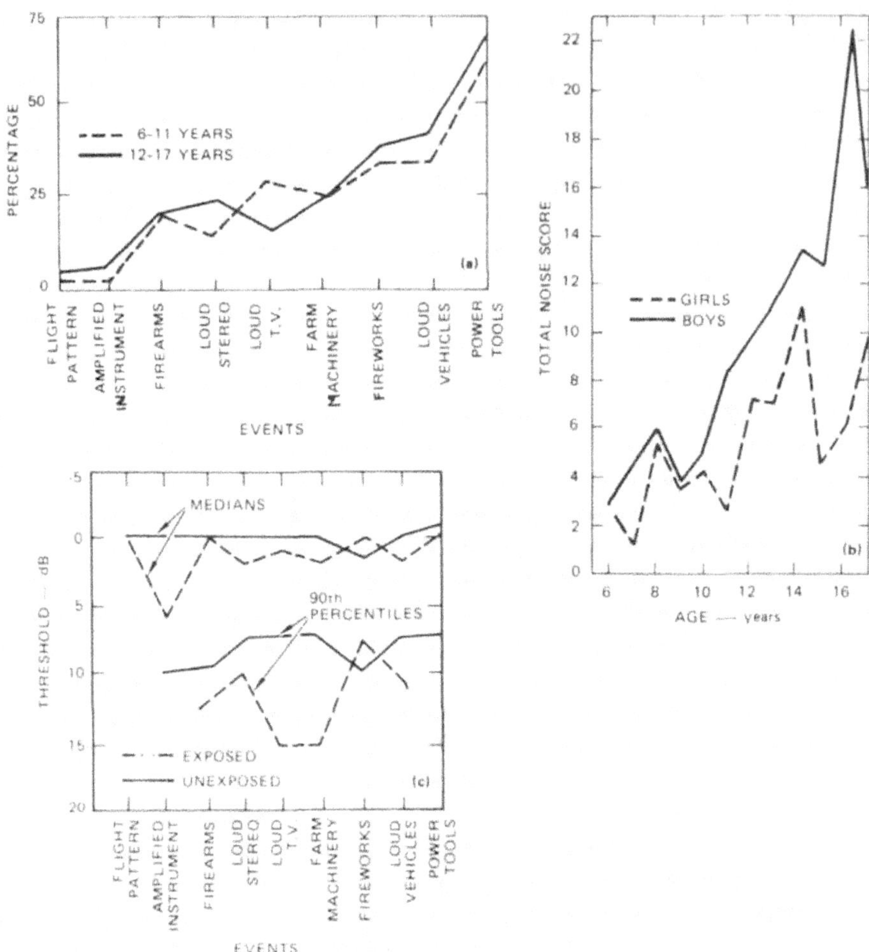

FIGURE 6.7. Hearing and noise exposure data for subjects 6 to 11 and 12 to 17 years of age. (a) Percentage of subjects 6 to 11 years old and 12 to 17 years old reporting exposure to specific noise events. (b) Median total noise 6-month-interval scores for boys and girls from the noise exposure histories obtained from questionnaires (higher scores indicate greater exposure to noise). (c) Left ear, auditory threshold level medians, upper curves, and 90th percentiles, lower curves, at 4000 Hz in 12 to 17 year olds exposed or not exposed to specific noise events. (From ref. 21.)

worn by each person. The dosimeter measurements bore out the previous estimates that males were generally exposed to more sound energy than were females, about 78 L_{eq} per 24 hours for females, opposed to 80 for males. There were no significant differences between age and exposure level. There was a small positive correlation between HL at 4000 Hz and exposure level for females, but not for males.

Cozad *et al.* (ref. 24) found, from audiograms given to 18 600 students, a steady increase in the percentage of students having sensorineural hearing loss with age (6 to 18 years). The incidence of this loss was about four times greater in boys than in girls. These investigators concluded that these losses were primarily caused by the noise from things like firearms and farm machinery (most of the students were from rural areas).

On the basis of data showing a decline in recent years in the HL's of students entering some particular colleges, reference 25 indicates that an increase in sensorineural hearing loss is taking place among young adolescents. The presumed reason is increased exposure to such things as noise and sound from transportation vehicles and hi-fi music.

Although the evidence is not entirely consistent, the following appear to be true:

1. Hearing sensitivity increases with maturation up to the age of about 16 years.

2. Sociocusis starts in the early teens, with boys being more exposed to intense noises and sounds than are girls.

Survey in Scotland

Hinchcliffe (ref. 26) measured hearing thresholds of about 645 otologically normal persons selected at random from a 136-square-mile area in Scotland. Figure 6.8 shows some of the data reported for the 25th percentile, 50th percentile, and 75th percentile and smoothed, extrapolated curves drawn in this paper to the data points. Unlike the USPHS survey subjects, the subjects in this survey were clinically screened. However, as can be seen by comparing figure 6.8 with figure 6.9, the general relations between hearing levels of males and females appear to be quite similar, in comparable age groups, for the surveys conducted in Scotland and the United States.

Composite analysis of selected surveys

Robinson and Sutton (ref. 27) in an attempt to obtain a quantitative description of presbycusis uncontaminated by otological diseases and noise exposure, analyzed the findings of a large number of hearing-level surveys in industrialized countries. They chose 11 of these surveys (included was part of the Hinchcliffe study, but not the USPHS surveys) as best-suited for revealing presbycusis on the basis of the subjects having been screened for otological abnormalities and, to varying extents, for histories of exposure to intense noise. Their analyses revealed quantitative rates of change with aging in HL as functions of audiometric frequency and sex of subjects. Mathematical formulae derived from these relations were used to calculate extensive tables relating pure-tone audio-

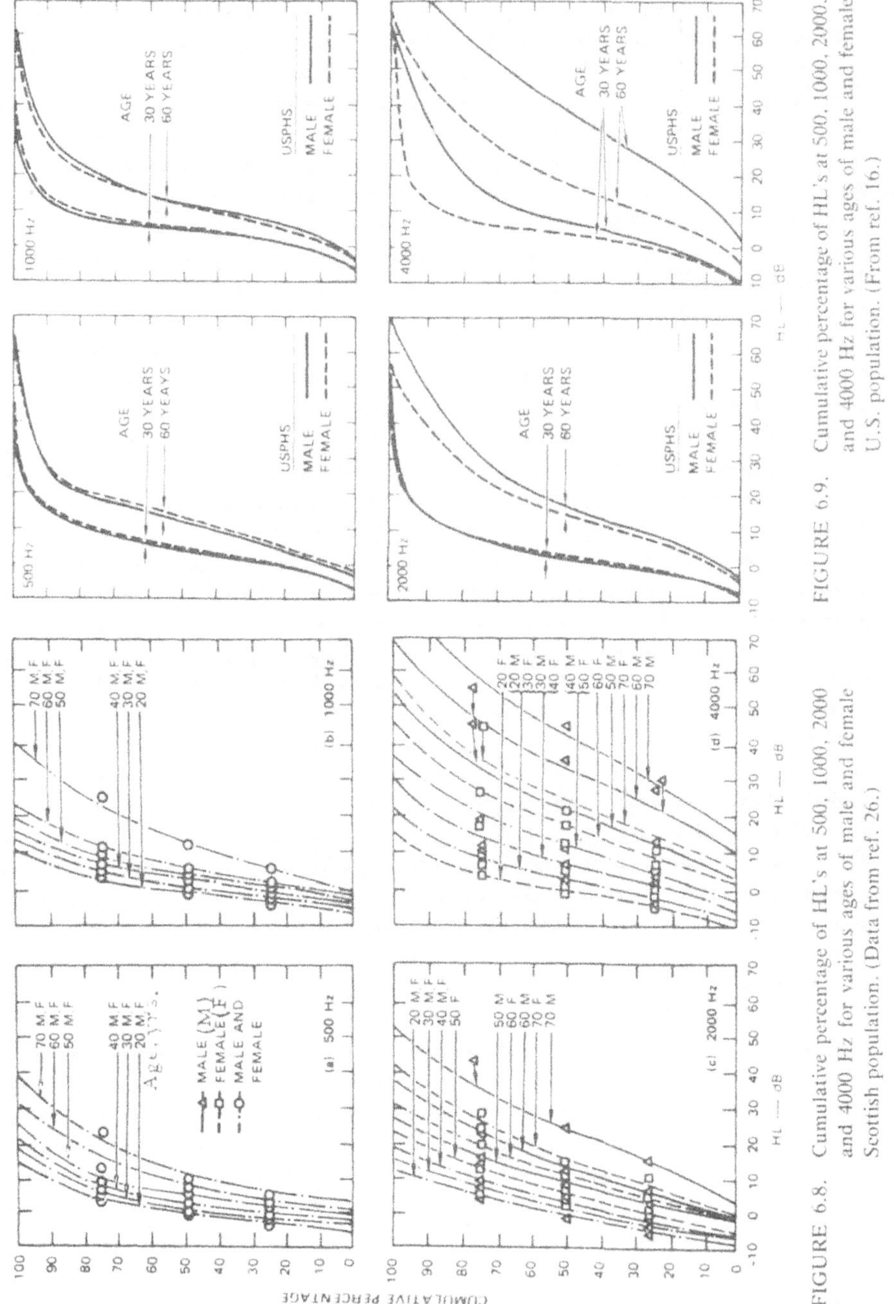

FIGURE 6.8. Cumulative percentage of HL's at 500, 1000, 2000 and 4000 Hz for various ages of male and female Scottish population. (Data from ref. 26.)

FIGURE 6.9. Cumulative percentage of HL's at 500, 1000, 2000, and 4000 Hz for various ages of male and female U.S. population. (From ref. 16.)

metric threshold for males and females to age. (See ref. 28.) The formulae and necessary factors to make these calculations are as follows (ref. 28):

Median values of the age correction ΔH_{md} are generated by the formula

$$\Delta H_{md} = H_{md,N} - H_{md,18} = a(N - 18)^2$$

where $H_{md,N}$ is the median value of hearing threshold level at age N years. The coefficient a depends on frequency and sex as given in table 6.4.

Values of the age correction ΔH_p for other percentile p, are generated from the median value ΔH_{md} using the formulae below:

$$\Delta H_p = \Delta H_{md} + k(b_u + 0.445\ \Delta H_{md}) \qquad \text{for } 1\% \leq p \leq 50\%$$

$$\Delta H_p = \Delta H_{md} - k(b_l + 0.356\ \Delta H_{md}) \qquad \text{for } 50\% \leq p \leq 99\%$$

where k is a factor determined by the percentile as given in table 6.5, and b_u, b_l are constants depending on frequency and sex as given in table 6.4. The symbol p is defined so that low values correspond to high values of hearing threshold level (*i.e.*, percentile of population exceeding a given hearing threshold level).

Figure 6.10 shows data taken from references 27 and 28 for males and females, ages 30 and 65 years. These data are chosen to permit comparison with figures 6.8 and 6.9, which include comparable data from the surveys conducted in Scotland and the United States. While there are some differences, many of the comparable functions are similarly placed.

TABLE 6.4

Values of the Parameters a, b_u, and b_l

[From ref. 28.]

Frequency, kHz	Males			Females		
	a	b_u	b_l	a	b_u	b_l
0.125	0.0030	7.23	5.78	0.0030	6.67	5.34
0.25	0.0030	6.67	5.34	0.0030	6.12	4.89
0.5	0.0035	6.12	4.89	0.0035	6.12	4.89
1	0.0040	6.12	4.89	0.0040	6.12	4.89
1.5	0.0055	6.67	5.34	0.0050	6.67	5.34
2	0.0070	7.23	5.78	0.0060	6.67	5.34
3	0.0115	7.78	6.23	0.0075	7.23	5.78
4	0.0160	8.34	6.67	0.0090	7.78	6.23
6	0.0180	9.45	7.56	0.0120	8.90	7.12
8	0.0220	10.56	8.45	0.0150	10.56	8.45

TABLE 6.5

Values of Percentile Parameter k

[From ref. 28.]

p	k	p	k
1,99	2.326	26,74	0.643
2,98	2.054	27,73	0.613
3,97	1.881	28,72	0.583
4,96	1.751	29,71	0.553
5,95	1.645	30,70	0.524
6,94	1.555	31,69	0.496
7,93	1.476	32,68	0.468
8,92	1.405	33,67	0.440
9,91	1.341	34,66	0.413
10,90	1.282	35,65	0.385
11,89	1.227	36,64	0.359
12,88	1.175	37,63	0.332
13,87	1.126	38,62	0.306
14,86	1.080	39,61	0.279
15,85	1.036	40,60	0.253
16,84	0.995	41,59	0.228
17,83	0.954	42,58	0.202
18,82	0.915	43,57	0.176
19,81	0.878	44,56	0.151
20,80	0.842	45,55	0.126
21,79	0.806	46,54	0.100
22,78	0.772	47,53	0.075
23,77	0.739	48,52	0.050
24,76	0.706	49,51	0.025
25,75	0.675	50	0.000

Survey in Sudan (Mabaan tribe)

Rosen *et al.* (ref. 29) conducted a survey of the threshold of hearing of 541 persons selected at random from the Mabaan tribe living in Sudan. The tribe is black, primitive in culture, and living in a rural environment with no guns or other sources of noise. The results of this study are of particular interest and import because the subjects live in an apparently noise-free environment and, therefore, should be free of any sociocusis. Also, only 12 of the 541 subjects had any otological abnormalities or history of conditions causing deafness. The median HL's were reported in reference 29, and some data for the 10th percentile and 90th percentile from the same bank of data were reported in reference 30. Figures 6.11 and 6.12 are based on the data from these two reports. In comparison with the HL surveys conducted in industrialized countries, the effect of aging

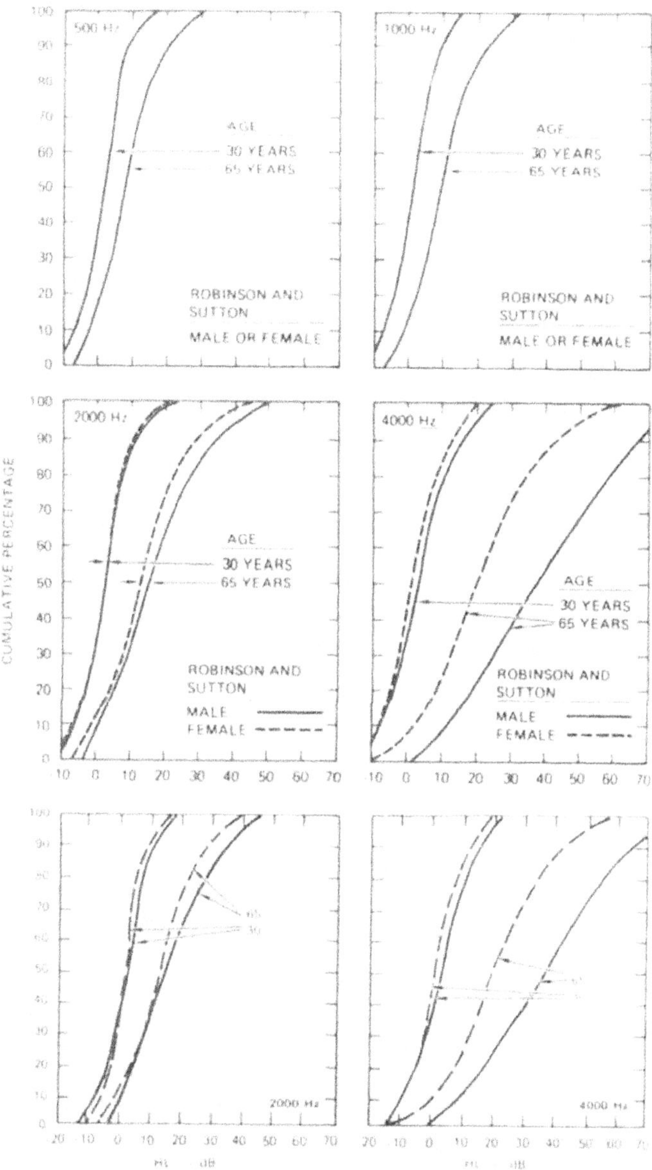

FIGURE 6.10. Cumulative percentage of HL's at 500, 1000, 2000, and 4000 Hz for various ages of male and female population. (Robinson and Sutton analysis (ref. 27) and Shipton (ref. 28).)

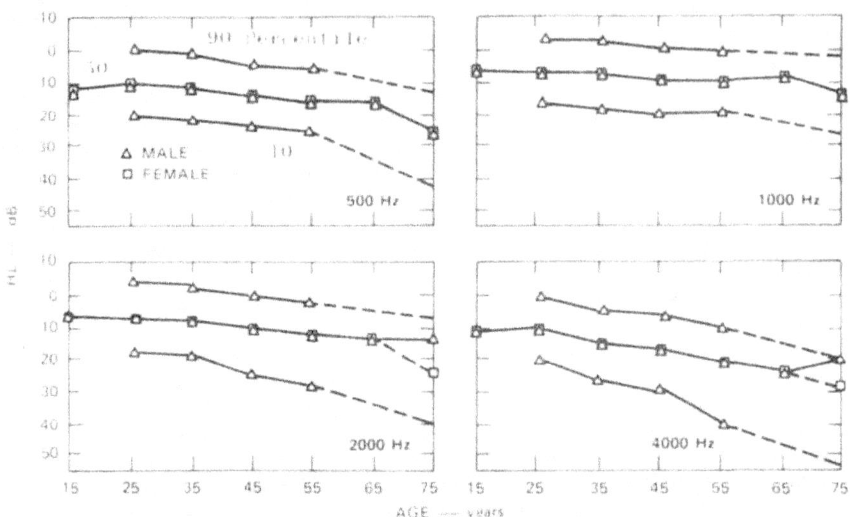

FIGURE 6.11. Hearing losses for males and females (Mabaan tribe) as a function of age; 50th percentile data are from reference 29; 10th percentile and 90th percentile data are based on reference 30. Parameters are frequency and percentile.

upon HL is practically the same for men as for women, and the effect progresses more slowly. As shown in figures 6.11 and 6.12 (left-hand graphs) the 50th percentile HL at all the test frequencies for the young Mabaans (15 to 35 yrs) are at about +10. This indicates that, even though in the older age groups the hearing sensitivity of Mabaans compares favorably with that of females from industrialized societies, the hearing sensitivity of young Mabaans before the start of presbycusis is about 10 dB less than it is for young, predominantly white, people from industrialized societies. Bergman (ref. 30) stresses this point in a paper on the Mabaan investigation.

There are several possible explanations for this apparent anomaly:

1. There are some inherent physiological differences in the characteristics of the auditory systems of the black and white races. This is probably not a valid explanation, in that, as is discussed subsequently, no consistent differences are found in the hearing sensitivity of young or old blacks and whites in the United States.

2. There were present some subject-selection or test-condition artifacts in the Mabaan study. There is nothing, however, in the protocols of the study to suggest any problems of this sort, except perhaps that the testing was done in an environment that was not as quiet as the investigators believed it to be. The fact that the apparent anomaly occurred about equally at all test frequencies indicates that this latter possibility is not a good explanation.

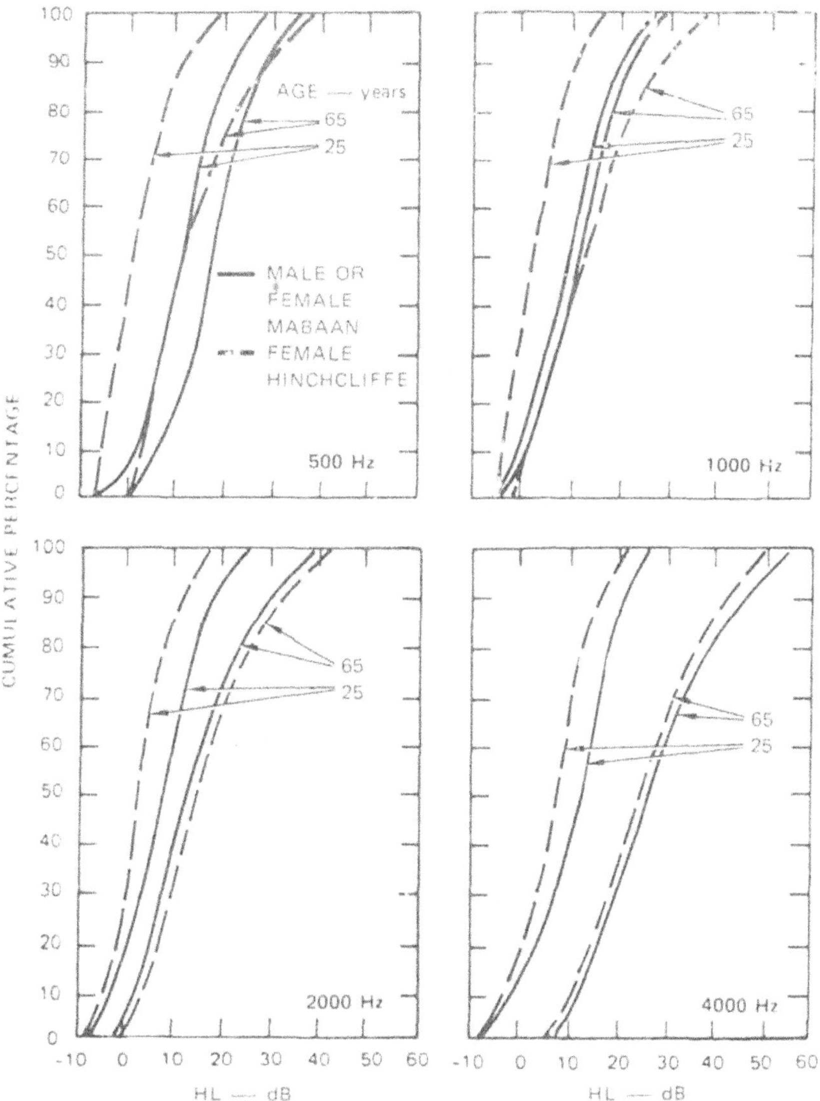

FIGURE 6.12. Cumulative percentages of HL's at 500, 1000, 2000, and 4000 Hz (based on refs. 26, 29, and 30) and at 14 000 Hz (from ref. 31) for various age and cultural groups.

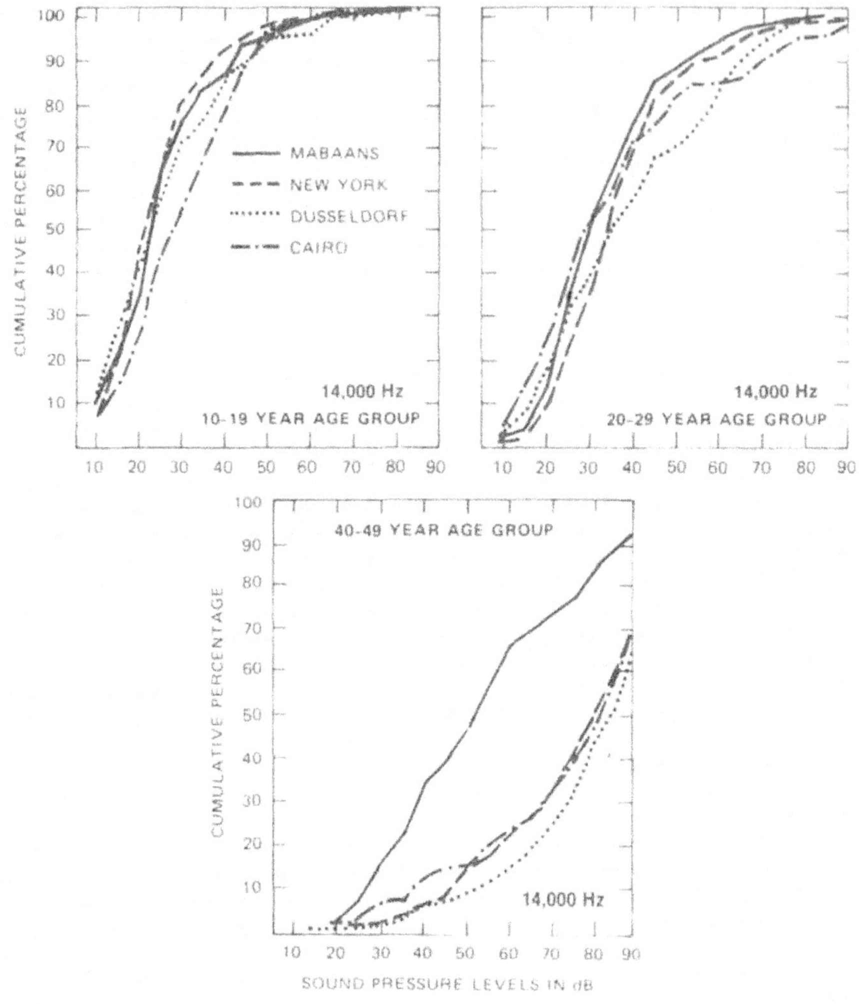

FIGURE 6.12 Concluded.

3. There was a constant 10-dB calibration error in the audiometric data. This possibility is suggested by the following sentence from reference 31: "The hearing loss dial, instead of ending with a reading of −10 by the American standard, was designed for a maximum low of −20 dB referenced to American Standard Zero."

On this last point, the audiometers were especially designed to have 10 dB more attenuation of the test tones than normal, and in accordance with earlier calibrations made at the factory, the annotated audiometer dial settings made by the audiometrist at the testing site were later converted to sound pressure levels.

The possibility of the error would come from inadvertently changing the numbers on the audiometer dial, so that a reading of −10 at 1000 Hz, for example, gave an SPL not of −3.5 dB, as normal, but −13.5 dB, or 20 dB below standard reference zero for U.S. audiometers. (See table 6.6.)

It follows from a study of HL's at 14 000 Hz of Mabaans and of people from industrialized societies (ref. 31) that the differences between the HL's of young whites and Mabaans at the lower test frequencies can possibly be attributed to audiometer calibration problems. (Compare the left- with the right-hand graphs of fig. 6.12.) In the study of HL's at 14 000 Hz, all the subjects were tested on the same audiometer, and the right-hand graphs of figure 6.12 show that the young, but not older, groups had about the same hearing losses, whether from the Mabaan tribe or from a modern city.

If, by any stretch of the imagination, this possible calibration error were the case, the results of the Mabaan study could be expressed as follows:

1. The hearing sensitivity at all sound frequencies of young, black Mabaans is essentially the same as that of young blacks or whites in industrialized societies.

2. There are no differences in the hearing thresholds of males and females in a noise-free society.

TABLE 6.6

Recommended Reference Equivalent Threshold Sound Pressure Levels in Decibels

[See ref. 2 for types of earphones and artificial ear or coupler used for calibration purposes.]

Frequency, Hz	Reference equivalent threshold sound pressure levels in dB relative to 2×10^{-5} N/m² (2×10^{-4} dyn/cm²)				
	France	Germany	U.K.[a]	U.S.A.[b]	USSR
125	44.5	47.5	47	45.5	55
250	27.5	28.5	28	24.5	33
500	11.5	14.5	11.5	11	14.5
1000	5.5	8	5.5	6.5	8.5
1500	4.5	7.5	6.5	6.5	8.5
2000	4.5	8	9	8.5	9
3000	6	6	8	7.5	10.5
4000	8	5.5	9.5	9	11.5
6000	17	8	8	8	18.5
8000	14.5	14.5	10	9.5	9.5

[a] *To convert pre-1969 levels to ISO, add to data: 0, 5, 3.5, 3, 0, 0, 0, 0.5, 0, 2 (ref. 3). U.K. does not specify levels for 125, 1500, 3000, or 6000 Hz.*

[b] *To convert pre-1969 levels to ISO, add to data: 0, 14.5, 14, 10, 0, 8.5, 0, 6, 0, 12.5 (ref. 4). U.S.A. does not specify levels for 125, 1500, 3000, or 6000 Hz.*

3. Presbycusis progresses at a slower rate in a noise-free society than in an industrialized, relatively noisy society.

When there is a question of audiometry calibration, it is customary to take the median of the HL's of the otologically normal 18 to 21 year olds in a population as representing zero HL, or the reference sound pressure level for normal hearing. Figure 6.13 illustrates these findings. The data reported for the Ma-

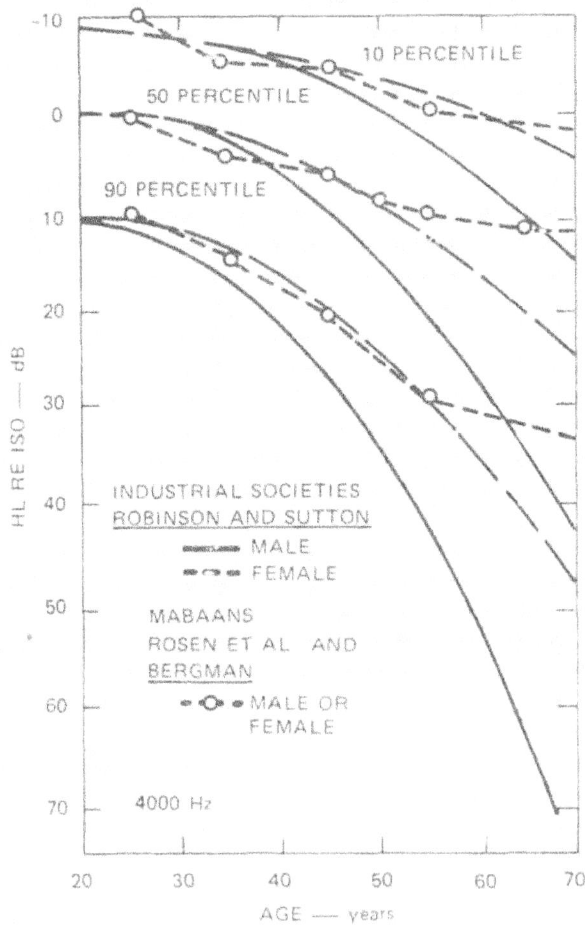

FIGURE 6.13. Similarity of HL's for young people from the studies of Mabaans and industrialized societies, and the differences in HL's for the older age groups. Test frequency is 4000 Hz. Raw data points for Mabaans are plotted 10 dB higher (re 15 to 25 year olds) than reported, on assumption of audiometer calibration error. (Data from analysis of refs. 27, 29, and 30.)

baans have been adjusted for the suggested audiometer calibration problem. The data for persons from industrialized societies are taken from reference 28.

Further support for the belief that the hearing thresholds for young persons from nonindustrial, noise-free societies are equal to or perhaps even better than those of young persons from industrialized societies, is found in reference 32. This investigator (ref. 32) reported that median HL's for the 18 to 24 year-old natives of certain tribes in Zaire was about −5 over the frequency range of 500 to 4000 Hz.

Some so-called nonindustrialized societies can be noisy because of firearms (see ref. 33), and some nonindustrialized societies may suffer from disease and malnutrition that can cause elevated hearing thresholds. Problems involved in finding otological normal and non-noise-exposed populations are discussed in references 34 and 35.

Comparison of hearing sensitivity in males and females

It has been argued that a primary reason that the HL's in females are generally lower than those in males in industrialized societies is that the males, whether at work or at play (especially hunting with firearms), are exposed to higher intensities of sound and noise than are the females, and therefore suffer from more noise-induced hearing losses. This notion, which is discussed in more detail subsequently in this paper, is supported by the aforementioned lack of difference in the HL's of males and females of the Mabaan tribe. These male-female differences are greatest at the audiometric test frequency of 4000 Hz, the frequency region where noise-induced deafness is most prevalent. This latter point is a necessary, but not conclusive, argument, because aging effects also first affect the HL's in the frequency region of 4000 Hz (see figs. 6.4, 6.8, and 6.9), or possibly higher but generally not tested frequencies.

Figure 6.14 shows the 50th percentile HL's for males and females, age 65, as a function of frequency, as found in the USPHS data and as determined by Robinson and Sutton (ref. 27). Previous presbycusis curves developed in references 1 and 36 for 50th-percentile data are also shown. The male-female differences in HL are similar among the various studies shown except for the ISO curve. The ISO curve does not distinguish between male and female HL's and is grossly out of line (shows poorer hearing) with the other curves in terms of absolute values of HL to be expected at age 65.

Figure 6.15 shows the differences, for several of the major studies mentioned previously in the HL's for males and females as a function of frequency with age as a parameter. Figure 6.16 summarizes the differences in HL's for males and females as a function of age and percentile for several studies. Because the differences in male and female HL's for test frequencies below 2000 Hz are negligible in figures 6.14 and 6.15, only the results at 2000 Hz and 4000 Hz are given in figure 6.16. In figures 6.11 and 6.15, the difference between HL's for

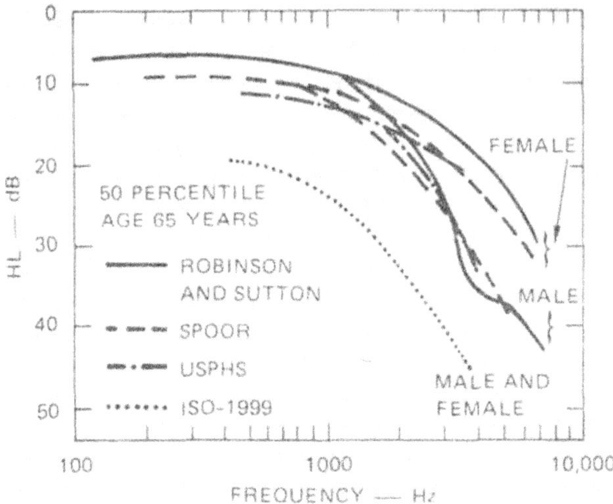

FIGURE 6.14. Fiftieth percentile HL's of males and females at age 65 years as a function of frequency. (From analyses of refs. 1, 16, 27, 28, and 36.)

males and females of the Mabaan population is essentially zero. However, it is shown in figure 6.16 that, in surveys of populations from primarily the United States and European countries, the differences are 15 to 25 dB at 4000 Hz by the age of 65 years (50th percentile).

The formulations of Robinson and Sutton (ref. 27) of the roles of sex and age in hearing levels of nonindustrial-noise-exposed people in industrialized societies are recommended because of their generality and mathematical expression. However, whether these formulations represent presbycusis as such is questionable, as is discussed subsequently.

Racial differences in hearing sensitivity

Figure 6.17 shows the cumulative percentages of whites and blacks, ages 30 and 60 years, with HL's equal to or greater than the values along the abscissa, as found in the USPHS survey of 1971-75, at 500, 1000, 2000, and 4000 Hz. Although there is some suggestion that blacks have somewhat poorer hearing than whites at 500 Hz, and possibly at 1000 Hz, in the higher percentile (least sensitive hearing) groups, there are no clear differences in the rest of the data.

As shown in figure 6.18(a), no differences between black children and white children (6 to 11 years) are apparent at test frequencies from 250 to 8000 Hz as tested in the USPHS survey. Also, there were no consistent differences in hearing sensitivity of white and black, or male and female children in the age range of 5 to 14 years. (See ref. 37.) Their findings are shown in figure 6.18(b).

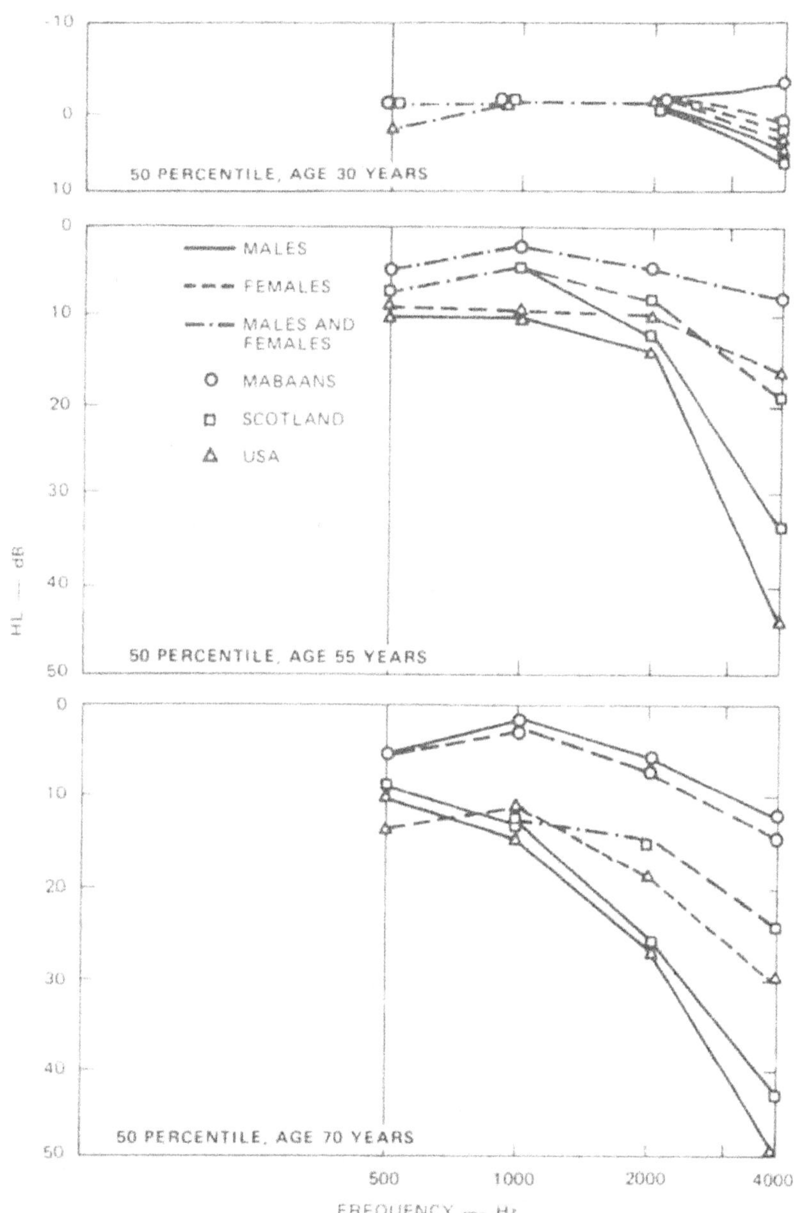

FIGURE 6.15. Fiftieth percentile HL's as a function of frequency and age referenced to young ears of random selection of males and females from Sudan (refs. 29 and 30), Scotland (ref. 26), and the United States (ref. 16). Mabaan data are HL's referenced to 15 to 25 year olds.

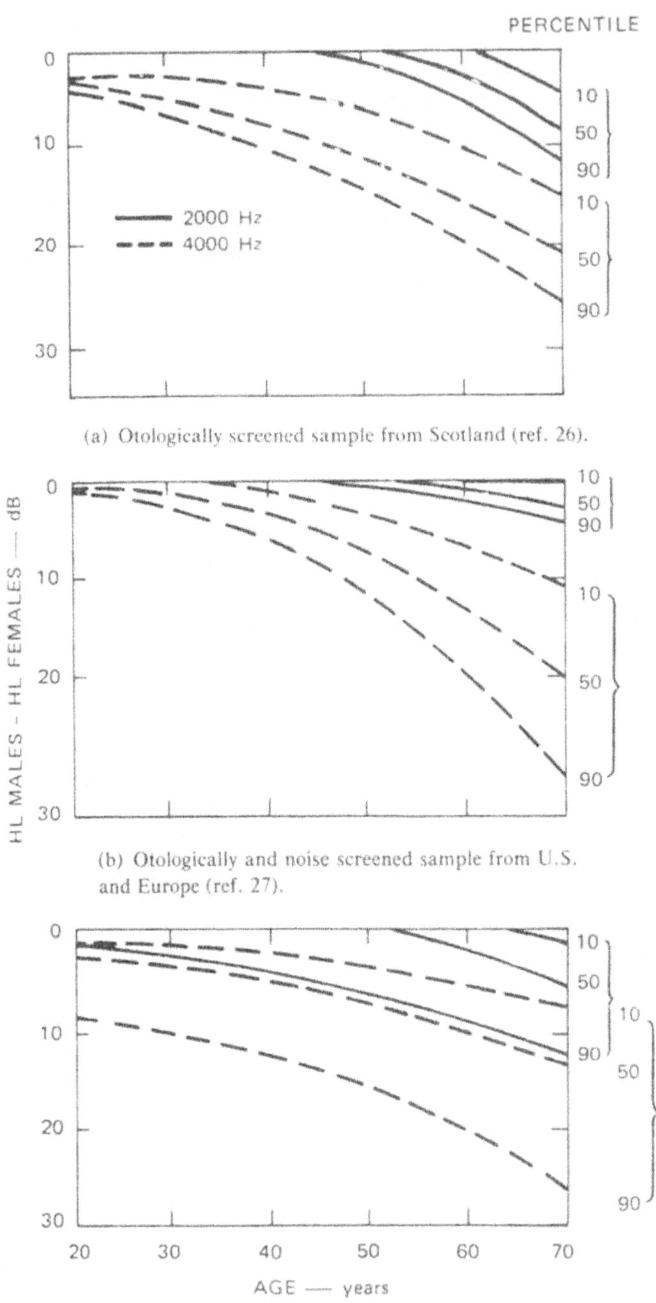

(a) Otologically screened sample from Scotland (ref. 26).

(b) Otologically and noise screened sample from U.S. and Europe (ref. 27).

(c) Unscreened sample of U.S. population (USPHS).

FIGURE 6.16. Difference between HL's of males and females as a function of age.

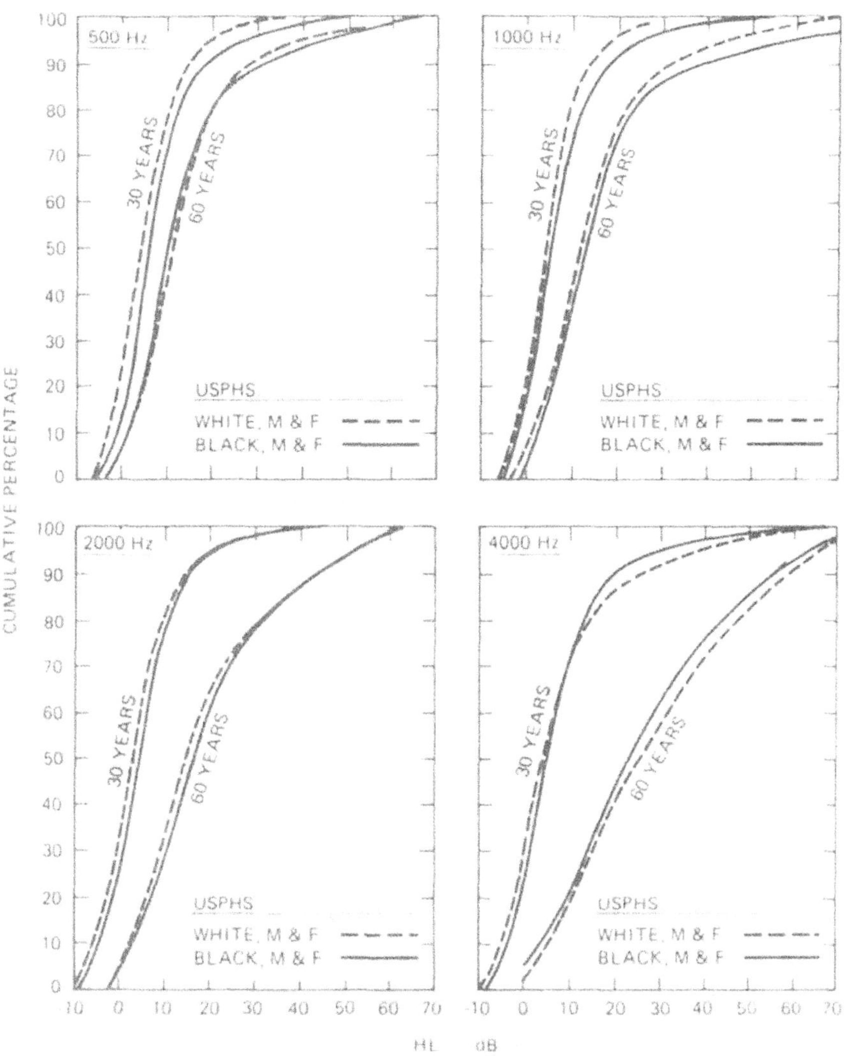

FIGURE 6.17. Cumulative percentages for white and black population, average of males and females ages 30 and 60 years at 500, 1000, 2000 and 4000 Hz. (*Data from ref. 16.*)

Data from reference 38 on the hearing sensitivity of white males, and data from reference 39 on black males from the state of North Carolina are shown in figure 6.19. These men were all selected because they were free of prior or current significant exposure to intense industrial noise. Interestingly, the HL's of the black males in reference 39 are similar to those for white males. (See ref. 16, 26, and 27.) The reason, or reasons, that the hearing sensitivity of the nonindus-

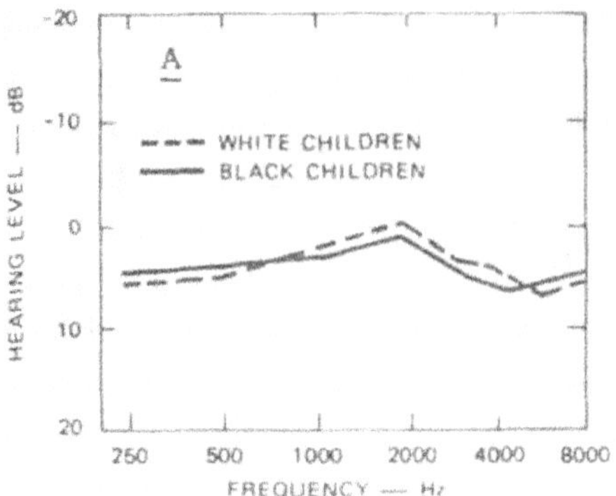

(a) Ages 6 to 11 years—6100 white, 987 black. (From ref. 15.)

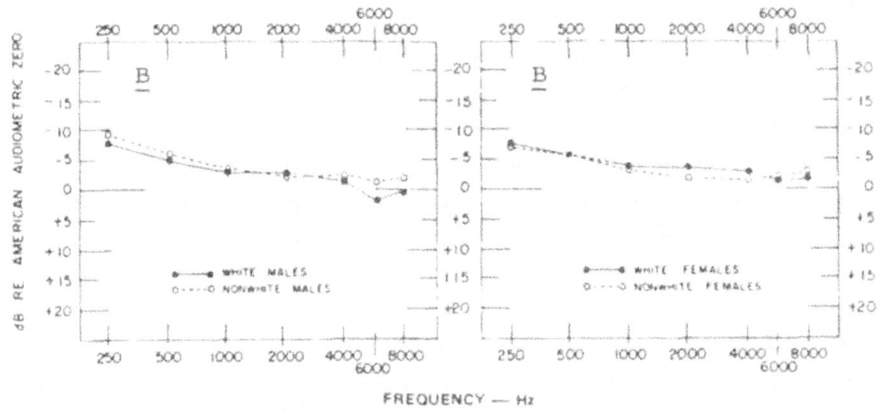

(b) Ages 5 to 14 years. (From ref. 37.)

FIGURE 6.18. Mean hearing levels of children.

trial-noise-exposed white males in reference 39 is so inferior to that of black males from the same geographic region—and from that of white males as found in other surveys—can only be surmised. Royster *et al.* did not exclude subjects who had any other otological diseases or disorders, or who were exposed to gunfire from hunting or the military services, or to machinery noises in farming or shopwork. It is a possibility that the white subjects in the Royster and Thomas study are not representative of the general white population of North Carolina. Many of the white subjects came from volunteers found at a large shopping mall. Whether such volunteers for tests of hearing level are representative of the general population is a matter of uncertainty. Indeed, this finding of Royster and Thomas is reminiscent of the surveys of hearing conducted of volunteers from

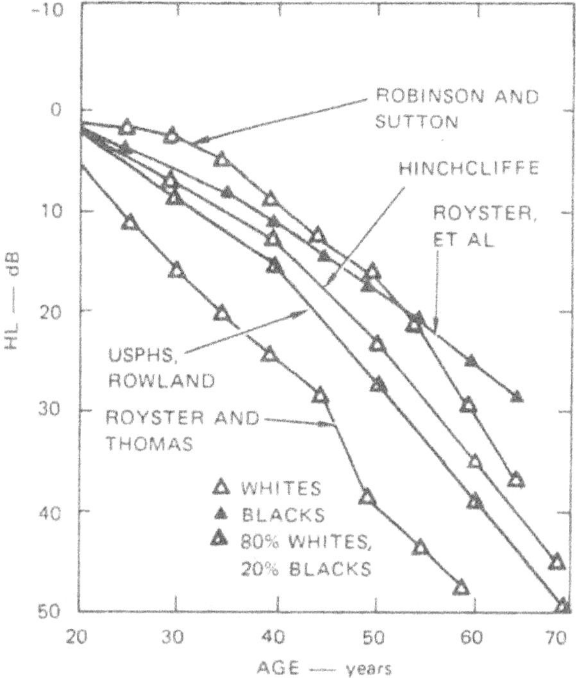

FIGURE 6.19. Black males (ref. 39) have HL's similar to white males of general surveys, but white males (ref. 38) have generally poorer hearing than black males or white males in other surveys. (See refs. 16, 26, and 27.) All the data are for essentially nonindustrial-noise-exposed males.

attendees at State and World Fairs, where it was found that the hearing of the volunteers turned out to be about 10 dB less sensitive than was later found in persons selected more randomly from the general population.

Karsai *et al.* (ref. 40) found that in New York City at test frequencies above 2000 Hz, black longshoremen had somewhat better hearing than did white longshoremen of various ethnic backgrounds. Again, possible differences in sociocusis and nosocusis between the white and black groups could be factors contributing to these findings. In any event, there are not sufficient data available to show that there exists any significant inherent difference in the sensitivity of hearing between the black and white races.

Presbycusis in Industrial and Nonindustrial Societies

A number of functions showing presbycusis have been proposed as a baseline for assessing the amount of hearing loss, if any, induced by the noise of the workplace. Samples of these so-called non-noise presbycusis functions are shown in figure 6.20. Curve A of figure 6.20 is for otologically screened, nonindustrial-noise-exposed males and females (refs. 41 and 42). Curve B is from an analysis of 11 surveys of nonindustrial-noise-exposed, otologically screened males (refs. 27 and 28). Curve C is for an unscreened random male sample of the U.S. population, right ear (ref. 16). Curve D is for males in industry who were otologically screened and exposed to nonindustrial noise and industrial noise less than 80 dBA (ref. 43). Curve E is for an unscreened, large sample (N = 120 000) of military and civilian males and females at U.S. military bases (ref. 44). Curve F is for unscreened, nonindustrial-noise-exposed, white males (ref. 38). Curve G is for males in industry who were otologically screened and exposed to nonindustrial noise and industrial noise less than 75 dBA (ref. 45). Curve H is for otologically screened, nonindustrial-noise-exposed males in industry (ref. 46). Curve I is for otologically screened, nonindustrial-noise-exposed workers in industry (ref. 47).

Variations in otological screening

The following are possible explanations for some of the variations (a range of 30 dB in the mid-age range) among the functions shown in figure 6.20. The high location of the curve proposed in reference 41 is, perhaps, related to the relatively small number (97) of subjects and to the strict otological and non-noise exposure (in and outside the workplace) criteria involved in the selection of subjects. Also, the application of certain mathematical smoothing and modeling procedures applied to the raw data may have contributed to the position of the curve proposed by Burns and Robinson (ref. 41).

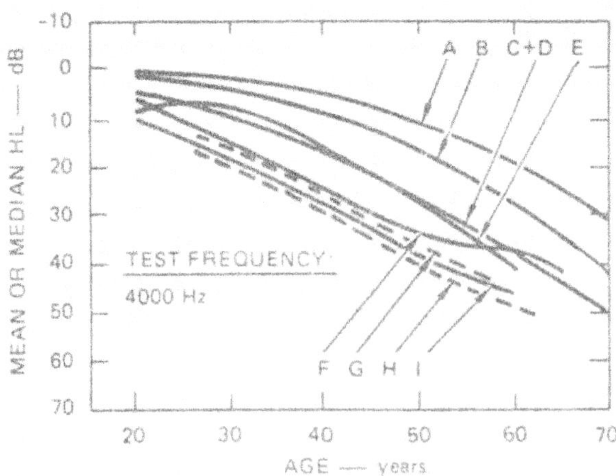

FIGURE 6.20 Variety of presbycusis functions from industrial and nonindustrial settings with a test frequency of 4000 Hz.

Also not to be overlooked are the aforementioned findings of Burns *et al.* (ref. 8) (table 6.2) that indicate that the HL's of 60 to 64-year-old industrial workers are elevated about 10 to 15 dB because of otological problems, presumably not attributable to the noise, and that the prevalence of these disorders is greater than that found in the general population. This would suggest that differences in the degree of otological screening of these non-noise-exposed populations could cause significant differences in the HL averages found for the different populations.

Non-noise-exposed workers as a misnomer

A somewhat insidious procedure has occurred in surveys in industrial settings where male office workers were classified for presbycusis analysis as being non-noise-exposed. It was later discovered that many of these men had previously worked in nonoffice jobs where the noise was intense. (See discussion by Bergman (ref. 30) of HL surveys of Glorig and Nixon (ref. 48).) The Glorig and Nixon analysis of presbycusis in industry, probably the basis for the recommendation in reference 1 (see fig. 6.14), appears to have included some misnomered data of this sort.

On- and off-the-job noise exposures of factory workers

The divergencies among all the various functions in figure 6.20 are, no doubt, due to some extent to differences among the studies in criteria of subject selection and what constitutes "nonnoise" environments. Indeed, in some of the surveys shown in figure 6.20, the definitions of "non-noise-exposed" are rather farfetched in light of what is now known about noise-induced hearing loss. In reference 43, for example, industrial noise conditions of up to 80 dBA, (70 dBA in ref. 45), were classified as non-noise conditions in the workplace.

Another important factor may be present in these industrial surveys that make the non-noise subjects unrepresentative of the general population. The notion is that the workers in noisy and not-so-noisy jobs in industry are exposed to more nonindustrial noise, particularly in the age range of about 18 to 30 years, than is the general population. The contributing reason could be differences in life-style, including more involvement in the off-the-job use of machinery, power tools, manual training, or guns in hunting and military service, on the part of factory workers than on the part of nonfactory workers. In addition, it is conceivable that the general ambient noise environment in some industries contributes to sociocusis for some industry personnel, even though the noise at their specific work location in industry is not intense.

In any event, it is clear from figure 6.20 that the presbycusis function for the general population cannot be found from surveys conducted of non-noise-exposed people in factories or industries; nor, for that matter, can it be found from a HL survey of subjects not selected on a random basis, as in reference 38.

Idealized Functions for Presbycusis and Sociocusis

Assumptions for presbycusis and sociocusis-nosocusis analysis

The following assumptions are presented as a framework for the analysis and discussion to follow.

1. There are no significant inherent racial differences in the sensitivity of hearing.

2. There are no significant inherent differences, as measured by pure-tone audiometry between males and females, in sensitivity of hearing or in the effects of aging thereon, or in the prevalence of nosocusis.

3. Because of the prevalence and intensity of sounds and noises in everyday, nonworkplace environments in industrialized societies, the HL's measured in surveys of hearing sensitivity in these societies reflect the joint effects of presbycusis and sociocusis, even when the subjects are screened to exclude those exposed to intense noise at their place of work.

4. In industrialized societies, the males in nonindustrial (white collar) occupations generally receive less noise exposure that is not workplace related (*e.g.*, farming, shopwork, firing of guns in hunting and the military services) than do males who work in factories.

5. In industrialized societies, males are usually exposed to more intense noise, nonworkplace and workplace, than are females.

6. The effects of presbycusis and noise-induced threshold shift on HL are additive. (See ref. 41.) The degree of "additivity" of nosocusis effects to the effects of aging and noise exposure is not known. Insofar as nosocusis consists of middle ear diseases, nosocusis could afford some protection against, and possibly augment in some conditions, the insults to the inner ear from exposure to noise.

Audiometric test frequencies

As shown by the surveys discussed previously, a shift in hearing threshold at 4000 Hz is the most sensitive, of the usual 500- to 4000-Hz test frequencies, indicator of the possible presence of presbycusis and noise-induced hearing losses. However, there is a practical interest, regarding the assessment of noise-induced hearing loss for workmen's compensation purposes, in knowing the average magnitude of these possible effects at 500, 1000, and 2000 Hz. Accordingly, the subsequent analysis attempts to show presbycusis and sociocusis at 2000 and 4000 Hz, and the average for 500, 1000, and 2000 Hz.

Survey data used

To test the hypotheses and assumptions given previously, HL data have been assembled in subsequent figures from the studies of USPHS (United States); Hinchcliffe (Scotland); Robinson and Sutton (mostly European and United States), and Rosen *et al.* (Mabaans of Sudan). Comparison of the Hinchcliffe and USPHS surveys provides some indication of the importance of otological screening of the results of such surveys. However, as shown previously, the generally small magnitude of effects and incidence of otological disorders in the USPHS studies suggests that the effect of otological screening would be difficult to measure with averaged data for these two studies.

Exposure to industrial-type noises—from tractors, trucks, and power tools, not to mention hunting—is prevalent in rural areas of industrialized countries. In the USPHS survey, no essential differences were found between the HL results for urban and rural areas (fig. 6.21). Therefore, no large difference is necessarily expected between the Hinchcliffe study (rural area) and the USPHS surveys (urban and rural areas).

As shown in figure 6.22, except for the 2000-Hz and 4000-Hz, 90th-percentile curves for males, there are only small differences among the smoothed, idealized curves for the USPHS, Hinchcliffe, and Robinson and Sutton studies. It does appear, however, for the 90th percentile (those persons with the largest hearing losses), that the amount of loss is 5 to 10 dB greater at 2000 Hz and 4000 Hz for the USPHS surveys than for those found by Robinson and Sutton in their analysis of surveys of otologically screened persons and largely nonindustrial-noise-exposed persons.

That the differences between the USPHS studies, the Hinchcliffe studies, and the Robinson and Sutton studies are not larger is surprising, inasmuch as there was no screening of any sort in the USPHS surveys. Some of the subjects in

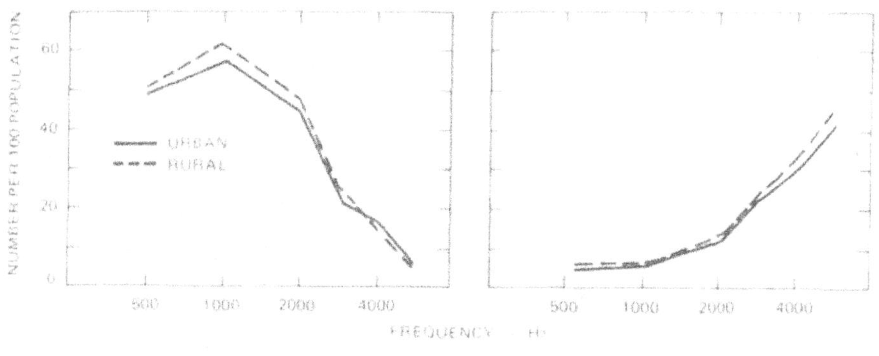

(a) With some hearing handicap. (b) In better ear.

FIGURE 6.21. Rates for adults having HL's better than normal. (From ref. 12.)

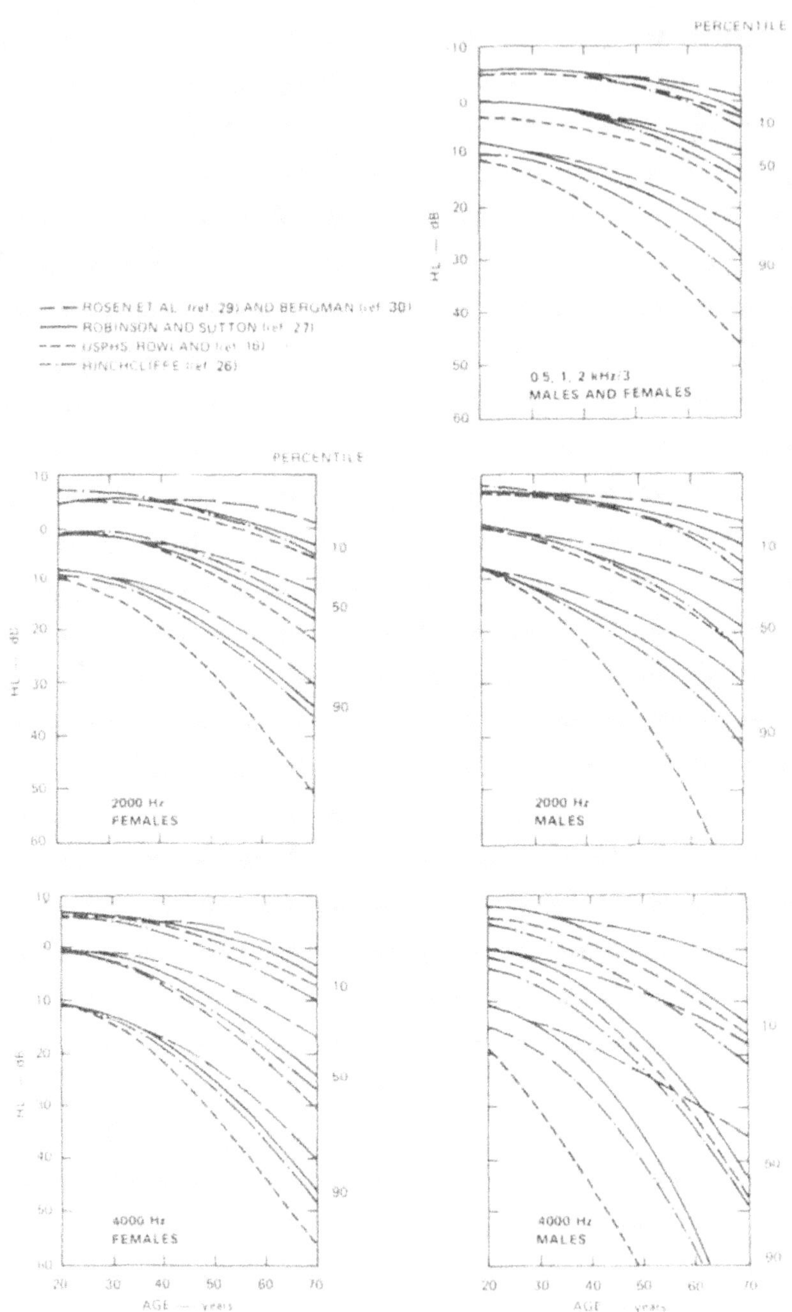

FIGURE 6.22. Smoothed curves showing HL at various test frequencies as function of age as found in several major surveys. Parameter is percentile.

the USPHS surveys no doubt were persons exposed to intense industrial noise. Some, as discussed previously, suffered from some otological disease or injury. However, the incidence of otological disease and industrial-noise-induced hearing losses are, apparently, sufficiently few to have little impact upon the distribution of HL's in the general population compared with that of screened populations, except at the higher frequencies and percentiles.

As noted by Robinson and Sutton (ref. 27), there are, due to force of circumstances, some uncertainties and ambiguities in the data used in their analyses. For example, in 4 of the 11 studies, audiometric reference zero was not translatable to the ISO 389 reference used for the other seven studies. Four of the studies provided no information about dispersion of the data around the given means or medians; four provided standard deviations; two gave 25th, 50th, and 75th percentiles; and one gave the total range of HL's. The number of subjects in these studies ranged from 120 to 539. These numbers are relatively small considering that they were generally divided into males and females and four to six decades of age brackets. Also, the possible presence of intense work-noise or everyday-noise exposures for the subjects in some of the studies is unknown.

Presbycusis plus sociocusis-nosocusis

Similarity of the functions synthesized by Robinson and Sutton (included were some of Hinchcliffe's raw data) to the more strictly empirical ones present in the USPHS and Hinchcliffe surveys presented herein indicates that something approaching the true, typical presbycusis-plus-sociocusis functions for males and females of the general population of industrialized societies has been realized in their formulations. The 90th percentile part of the distribution of HL's for 2000 Hz and 4000 Hz are clearly more affected by nonpresbycusis factors, as would be expected because of the greater sensitivity of the ear to hearing loss at those frequencies.

In the analyses to follow, the Robinson and Sutton formulae relating pure-tone audiometric thresholds to age, as in Shipton's tables, are used exclusively to represent those relations for the general population in industrialized societies. The functional relations proposed by Robinson and Sutton, original or as modified, do not depict what, from a purely physiological point of view, should be called presbycusis. They reflect the joint effects of aging and exposure to the noise in modern societies (sociocusis) as well as some nosocusis.

Finally, the hearing sensitivity for all the test frequencies at all the percentiles shown for people from the Mabaan tribe declines less with aging than it does for people from industrialized societies. (See fig. 6.22.)

Typical presbycusis plus sociocusis-nosocusis

Figure 6.23 shows the HL's for typical presbycusis plus sociocusis for males and females as a function of frequency. Age and percentile are the parameters.

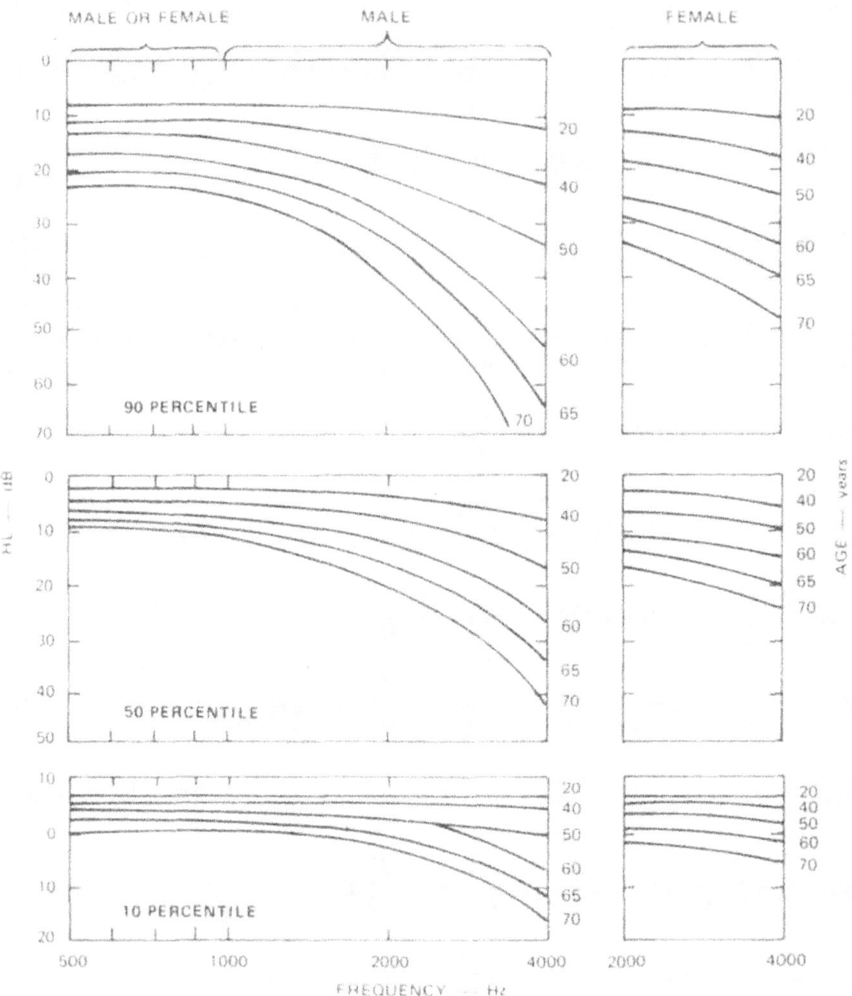

FIGURE 6.23. Idealized typical presbycusis plus sociocusis-nosocusis as function of frequency. (See ref. 28.) Parameters are age and percentile.

Separate functions are needed to represent male and female HL's for test frequencies above about 1500 Hz, but at the lower frequencies the HL's for the two sexes are essentially the same. The magnitude in dB of typical (that from industrialized societies) presbycusis plus sociocusis-nosocusis is shown as a function of age for males in figure 6.24(a) and for females in figure 6.24(b). The parameters of these figures are percentile (10, 50, 90) and frequency (4000 Hz, and the average for 0.5, 1, and 2 kHz).

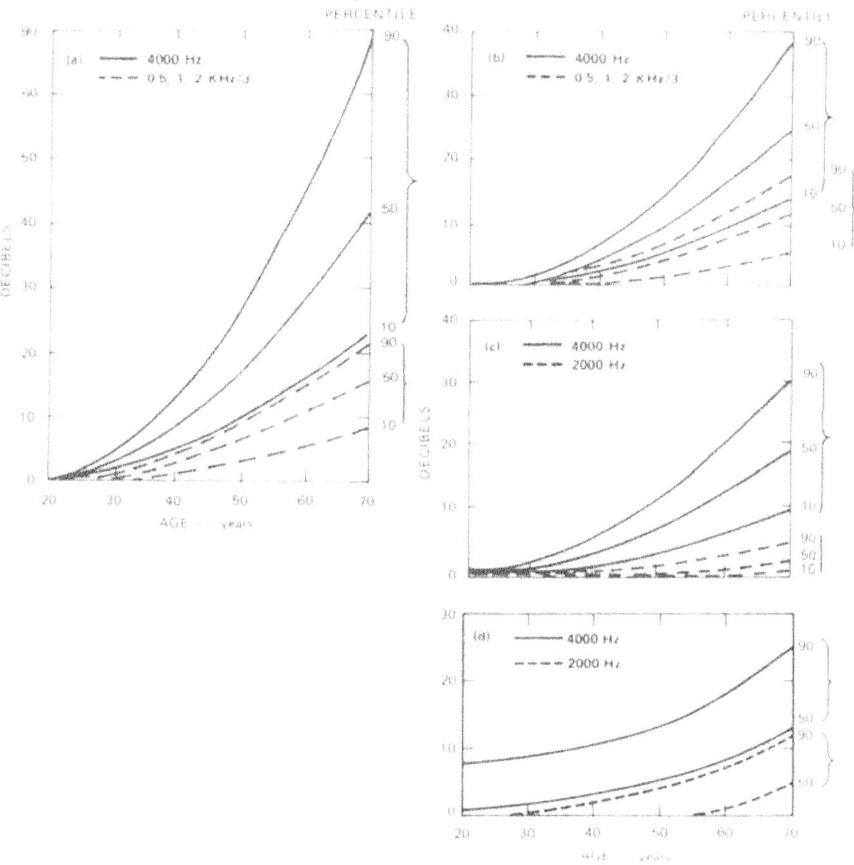

FIGURE 6.24. Magnitude of typical presbycusis plus sociocusis-nosocusis for males (a) and females (b) in industrialized society, and magnitude of typical sociocusis-nosocusis in industrialized society ((c) and (d)). Parameter is percentile. (Data for (a), (b), and (c) are from ref. 27, and data for (d) are from ref. 6.)

Typical presbycusis

It is generally assumed that females are exposed to less noise than men and that the inherent differences between male and female in hearing threshold levels are negligible. If this is the case, the magnitude of typical presbycusis is represented by the difference between the HL's for 20-year-old females and the HL's for older females. (See fig. 6.24(b).)

Typical sociocusis-nosocusis

The differences in pure-tone hearing levels between males and females, in comparable age groups, is a measure of the everyday-noise-induced hearing loss in men. These differences are shown in figure 6.24(c). This function is the general magnitude, at the test frequencies and percentiles shown, of the losses in pure-tone hearing sensitivity that occur because of everyday noises in the industrialized societies—"typical sociocusis."

One of the differences between the Robinson and Sutton and USPHS functions of HL is the differences at 2000 and 4000 Hz between males and females. (See fig. 6.22.) Inasmuch as the USPHS data represent large randomly selected samples of the general, unscreened population, the difference between HL's for males and females in comparable age groups is perhaps as good a measure of typical sociocusis-nosocusis as that derived from the Robinson and Sutton functions. Typical sociocusis is somewhat less at 4000 Hz, but somewhat greater at 2000 Hz, for the USPHS (fig. 6.24(d)) than for the Robinson and Sutton analysis (fig. 6.24(c)). At other test frequencies the differences are negligible.

Pure presbycusis

Figure 6.25 shows the HL's to be expected (with age as the parameter) of persons free of sociocusis or nosocusis at several percentiles for males and fe-

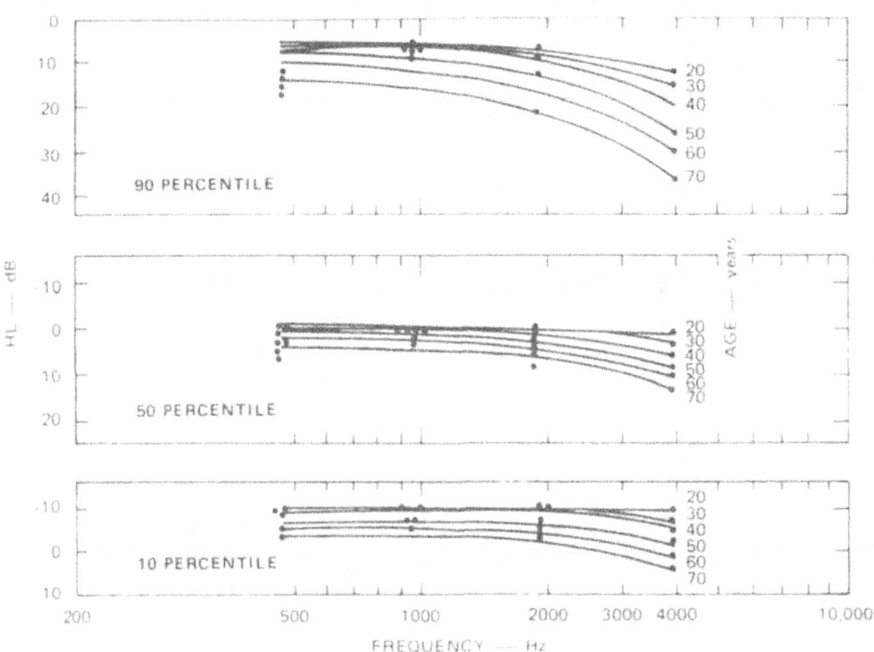

FIGURE 6.25. Pure presbycusis at 10th percentile, 50th percentile, and 90th percentile as a function of frequency. Smoothed averages of male and female HL's of Mabaans relative to the HL's of 15- to 25-year-old Mabaans. (See refs. 29 and 30.) Parameter is age.

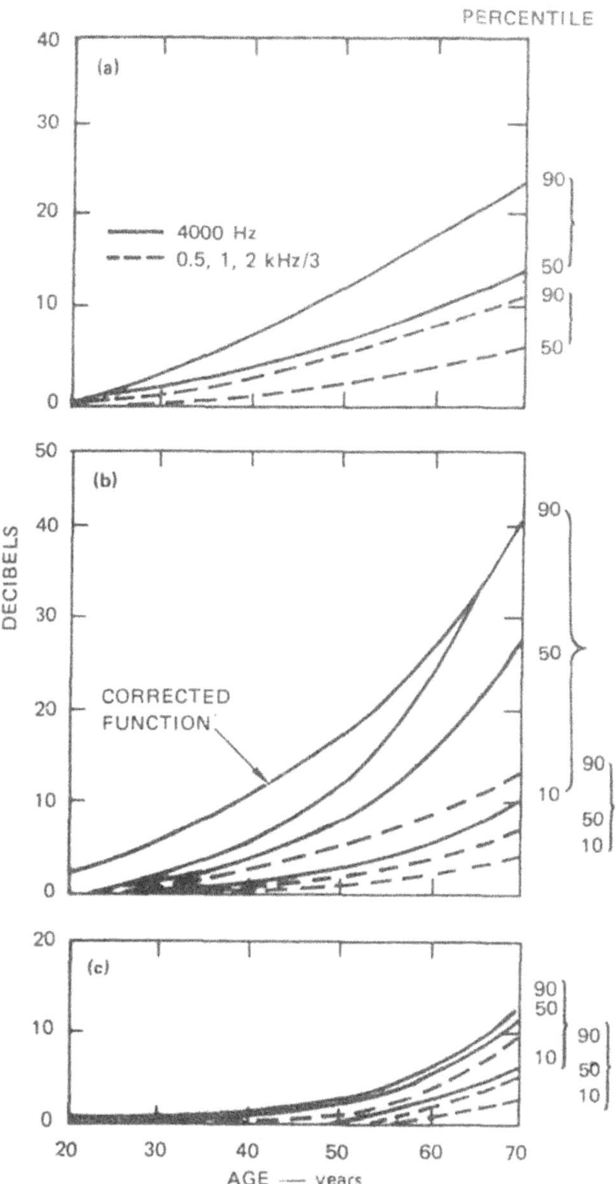

FIGURE 6.26. Magnitude of pure presbycusis for males or females in nonindustrial society (a) and pure sociocusis-nosocusis in industrialized societies of males (b) and females (c). Parameter is percentile. (From refs. 27, 29, 30, and 31.)

males as a function of frequency. As previously mentioned, the small amount of otological disorders and the lack of exposure to any intense sounds or noises in the Mabaan society is presumed to justify the conclusion that the functions in figure 6.25 reflect the effects of aging *per se* on hearing. The magnitude of this process in dB is shown in figure 6.26(a).

Pure sociocusis-nosocusis

The differences for comparable age groups, frequencies, and percentiles between HL's of Mabaan society and industrialized societies presumably show the magnitudes, in dB, of what is herein called "pure sociocusis-nosocusis." The results are shown in figure 6.26(b) for males and in figure 6.26(c) for females.

As expected, pure sociocusis-nosocusis is greater than typical sociocusis-nosocusis (compare fig. 6.24(c) with fig. 6.26(b)) and is much smaller for females than for males. If the data and assumptions involved are valid, females in industrialized societies are indeed subjected to less intense sounds and noises than are males and perhaps also suffer less nosocusis.

Summary

1. Race and sex do not appear to be significant factors in determining the threshold of hearing sensitivity.

2. Surveys conducted in industrialized societies on the thresholds of hearing of workers selected other than randomly from industry who do not work in intense noise, or of persons not randomly selected from relatively small geographic areas, provide functions for presbycusis plus sociocusis-nosocusis that are generally about 10–20 dB below (less sensitive hearing) that function for the general, randomly selected populations of a country. These differences are probably due to differences in the prevalence of nosocusis and sociocusis between factory workers and the general population.

3. General mathematical formulae derived by Robinson and Sutton to be predictive of presbycusis appear to be predictive of presbycusis plus sociocusis-nosocusis as found in industrialized societies.

4. Surveys of pure-tone thresholds of hearing for relatively large samples of males and females from industrialized societies and from a nonindustrial, noise-free society provide a base, along with certain assumptions, for deducing the quantitative effects on hearing thresholds of presbycusis and sociocusis-nosocusis and the joint effects of presbycusis and sociocusis-nosocusis.

5. Females in industrialized societies are apparently exposed to fewer noises or sounds that are not work-related than are the males, and they experience appreciably less sociocusis or nosocusis, or both, than do the males.

6. There is some evidence that the prevalence of nosocusis identified by otoscopic examination in industrialized societies increases the median HL's at all audiometric test frequencies by 1 to 2 dB for the general population (25 percent affected) and by about 6 dB for factory workers (60 percent affected).

References

1. Acoustics—Assessment of Occupational Noise Exposure for Hearing Conservation Purposes. ISO 1999-1975 (E). Int. Organ. Stand.
2. Acoustics—Standard Reference Zero for the Calibration of Pure-Tone Audiometers. ISO-389-1975, Int. Org. Stand.
3. American Standard for Audiometer for General Diagnostic Purposes. ASA Z24.5-1951, American Nat. Stand. Inst., Inc.
4. British Standards Institute: The Normal Threshold of Hearing for Pure Tones by Earphone Listening. B. S. 2497, 1954.
5. Robinson, D. W.; Shipton, M. S.; and Hinchcliffe, R.: Normal Hearing Threshold and Its Dependence on Clinical Rejection Criteria. NPL Acoustics Rep. Ac 89, British A.R.C., Feb. 1979.
6. Roberts, Jean; and Federico, John V.: Hearing Sensitivity and Related Medical Findings Among Children. DHEW Publ. No. (HSM) 72-1046, Series 11, No. 114, U.S. Dep. Health, Educ., & Welfare, Mar. 1972.
7. Roberts, Jean: Hearing Status and Ear Examination—Findings Among Adults. PHS Publ. No. 1000, Series 11, No. 32, U.S. Dep. Health, Educ., & Welfare, Nov. 1968.
8. Burns, W.; Robinson, D. W.; Shipton, M. S.; and Sinclair, A.: Hearing Hazard From Occupational Noise: Observations on a Population From Heavy Industry. NPL Acoustics Rep. Ac 80, British A.R.C., Jan. 1977.
9. Robinson, D. W.; Shipton, M. S.; and Whittle, L. S.: Audiometry in Industrial Hearing Conservation—II. NPL Acoustics Report Ac 71, British A.R.C., Jan. 1975.
10. Glorig, Aram; and Roberts, Jean: Hearing Levels of Adults by Age and Sex: United States, 1960-1962. Report PHS-PUB-1000-SER-11-11, Nat. Cent. Health Serv. Res. and Dev., Oct. 1965. (Available from NTIS as PB 267 177/4.)
11. Hartley, B. P.; Howell, R. W.; Sinclair, A.; and Slattery, D. A.: Subject Variability in Short-Term Audiometric Recording. British J. Ind. Med., vol. 30, no. 3, July 1973, pp. 271-275.
12. Roberts, Jean; and Bayliss, David: Hearing Levels of Adults by Race, Region, and Area of Residence. PHS Publ. No. 1000, Series 11, No. 26, U.S. Dep. Health, Educ., & Welfare, Sept. 1967.
13. Roberts, Jean; and Huber, Paul: Hearing Levels of Children by Age and Sex. PHS Publ. No. 1000, Series 11, No. 102, U.S. Dep. Health, Educ., & Welfare, Feb. 1970.

14. Roberts, Jean: Hearing Levels of Children by Demographic and Socioeconomic Characteristics. DHEW Publ. No. (HSM) 72-1025, Series 11, No. 111, U.S. Dep. Health, Educ., & Welfare, Feb. 1972.
15. Roberts, Jean: Hearing and Related Medical Findings Among Children: Race, Area, and Socioeconomic Differentials. DHEW Publ. No. (HSM) 73-1604, Series 11, No. 122, U.S. Dep. Health, Educ., & Welfare, Oct. 1972.
16. Rowland, Michael: Basic Data on Hearing Levels of Adults 25-74 Years. DHEW Publ. No. (PHS) 80-1663, Series 11, No. 215, U.S. Dep. Health, Educ., & Welfare, Jan. 1980.
17. Richardson, K.; Hutchison, D.; Peckham, C. S.; and Tibbenham, A.: Audiometric Thresholds of a National Sample of British 16-Year-Olds: A Longitudinal Study. Dev. Med. & Child Neurol., vol. 19, 1977, pp. 797-802.
18. Penniceard, R.; Richardson, K.; Peckham, C. S.; Hutchison, D.; and Tibbenhaum, A.: Audiometric Thresholds of British 16-Year-Olds. Dev. Med. Child Neurol., vol. 20, no. 1, Feb. 1978, pp. 115-116.
19. Barr, B.; Anderson, C.; and Wedenberg, E.: Epidemiology of Hearing Loss in Childhood. Audiology, vol. 12, Sept.-Dec. 1973, pp. 426-437.
20. Eagles, E. L.; Wishik, S. M.; Doerfler, L. G.; Melnick, W.; and Levine, H. S.: Hearing Sensitivity and Related Factors in Children. Laryngoscope Suppl., 1963, pp. 1-220.
21. Roche, Alexander F.; Siervogel, R. M.; Himes, John H.; and Johnson, Daniel L.: Longitudinal Study of Human Hearings: Its Relationship to Noise and Other Factors. I. Design of Five Year Study; Data From First Year. Rep. AMRL-TR-76-110, U.S. Air Force, Mar. 1977. (Available from DTIC as AD A040 168.)
22. Roche, A. F.; Siervogel, R. M.; and Himes, J. H.: Longitudinal Study of Hearing in Children: Baseline Data Concerning Auditory Thresholds, Noise Exposure, and Biological Factors. J. Acoust. Soc. America, vol. 64, no. 6, Dec. 1978, pp. 1593-1601.
23. Siervogel, R. M.; Roche, A. F.; Johnson, D. L.; and Fairman, T.: Longitudinal Study of Hearing in Children II: Cross-Sectional Studies of Noise Exposure as Measured by Dosimetry. J. Acoust. Soc. America, vol. 71, no. 2, Feb. 1982, pp. 372-377.
24. Cozad, Robert L.; Marston, Larry; and Joseph, Donald: Some Implications Regarding High Frequency Hearing Loss in School-Age Children. J. Sch. Health, vol. 44, no. 2, Feb. 1974, pp. 92-96.
25. Lipscomb, David M.: Environmental Noise is Growing—Is It Damaging Our Hearing? Clin. Pediatr., vol. 11, no. 7, July 1972, pp. 374-375.
26. Hinchcliffe, R.: The Threshold of Hearing as a Function of Age. Acustica, vol. 9, no. 4, 1959, pp. 303-308.

27. Robinson, D. W.; and Sutton, G. J.: A Comparative Analysis of Data on the Relation of Pure-Tone Audiometric Thresholds to Age. NPL Acoustics Rep. Ac 84, British A.R.C., Apr. 1978.
28. Shipton, M. S.: Tables Relating Pure-Tone Audiometric Threshold to Age. NPL Acouties Rep. Ac 94, British A.R.C., Nov. 1979.
29. Rosen, S.; Bergman, M.; Plester, D.; El-Mofty, A.; and Satti, M.: Presbycusis Study of a Relatively Noise-Free Population in the Sudan. Ann. Otol., Rhinol., & Laryngol., vol. 71, 1962, pp. 727-743.
30. Bergman, M.: Hearing in the Mabaans—A Critical Review of Related Literature. Arch. Otolaryngol., vol. 84, no. 4, Oct. 1966, pp. 411-415.
31. Rosen, Samuel; Plester, Dietrich; El-Mofty, Aly; and Rosen, Helen V.: High Frequency Audiometry in Presbycusis. Arch. Otolaryngol., vol. 79, 1964, pp. 18-32.
32. Muyunga, C. Y. K.: Prospects for Hearing Surveys Among Isolated Populations in Zaire. Hearing Threshold Levels of "Isolated" Human Populations: Report of a Special Conference. Publ. No. NRCC/CNRC 16738, Natl. Res. Counc. of Canada, Apr. 1976, pp. 59-61.
33. Baxter, J. D.; and Ling, Daniel: Ear Disease and Hearing Loss Among the Eskimo Population of the Baffin Zone. Canadian J. Otolaryngol., vol. 3, no. 2, 1974, pp. 110-112.
34. Hinchcliffe, R.: Presbycusis in the Presence of Noise-Induced Hearing Loss. Occupational Hearing Loss, D. W. Robinson, ed., Academic Press, Inc., 1971, pp. 173-178.
35. Hearing Threshold Levels of "Isolated" Human Populations. Report of a Special Conference, NRCC 16738, Environmental Secretariat, National Research Council of Canada, 1978.
36. Spoor, A.: Presbycusis Values in Relation to Noise-Induced Hearing Loss. Int. Audiol., vol. 6, no. 1, 1967, pp. 48-57.
37. Eagles, E. L.; Wishik, S. M.; and Doerfler, L. G.: Hearing Sensitivity and Ear Disease in Children: A Prospective Study. Laryngoscope Suppl., 1967, pp. 1-274.
38. Royster, L. H.; and Thomas, W. G.: Age Effect Hearing Levels for a White Nonindustrial Noise Exposed Population (NINEP) and Their Use in Evaluating Industrial Hearing Conservation Programs. J. American Ind. Hyg. Assoc., vol. 40, no. 6, June 1979, pp. 504-511.
39. Royster, L. H.; Driscoll, D. P.; Thomas, W. G.; and Royster, J. D.: Age Effect Hearing Levels for a Black Nonindustrial Noise Exposed Population (NINEP). American Ind. Hyg. Assoc. J., vol. 41, Feb. 1980, pp. 113-119.
40. Karsai, Lilly Klein; Bergman, Moe; and Choo, Young Bin: Hearing in Ethnically Different Longshoremen. Arch. Otolaryngol., vol. 96, Dec. 1972, pp. 499-504.

41. Burns, W.; and Robinson, D. W.: Hearing and Noise in Industry. Her Majesty's Stationery Office, 1970.
42. Robinson, D. W.; and Shipton, M. S.: Tables for the Estimation of Noise-Induced Hearing Loss. NPL Acoustics Rep. Ac 61 (Second ed.), British A.R.C., June 1977.
43. Lempert, Barry L.; and Henderson, T. L.: Occupational Noise and Hearing 1968-1972. NIOSH-TR-201-74, Nat. Inst. Occupat. Saf. and Health, 1973. (Available from NTIS as PB 232 284.)
44. Sutherland, H. C.; and Gasaway, D. C.: Current Hearing Threshold Levels for USAF Noise Exposed Personnel—One Year's Reportings. Paper presented at American Speech and Hearing Association Meeting (Houston, Texas), Nov. 22, 1976.
45. Yerg, R. A.; Sataloff, J.; Glorig, A.; and Menduke, H.: Inter Industry Noise Study—The Effects Upon Hearing of Steady State Noise Between 82 and 92 dBA. J. Occup. Med., vol. 20, no. 5, May 1978, pp. 351-358.
46. Riley, E. C.; Sterner, J. H.; Fassett, D. W.; and Sutton, W. L.: Ten Years' Experience With Industrial Audiometry. J. American Ind. Hyg. Assoc., vol. 22, 1961, pp. 151-159.
47. Cohen, Alexander; Anticaglia, Joseph R.; and Jones, Herbert H.: Noise Induced Hearing Loss—Exposures to Steady-State Noise. Sixth Congress on Environmental Health, American Med. Assoc., Apr. 1969.
48. Glorig, Aram; and Nixon, James: Distribution of Hearing Loss in Various Populations. Ann. Otol., Rhinol., & Laryngol., vol. 69, no. 2, June 1960, pp. 497-516.

Chapter 7
Noise-Induced Hearing Loss and Its Prediction

Introduction

Since time immemorial, people working at noisy occupations have suffered deafness. However, it was not until electronic instruments became available for measuring hearing sensitivity and level of noise exposure that research could be undertaken to quantify the relations between exposure to noise and damage to hearing. That research continues although much is now known on this subject.

There are two somewhat different questions to be answered from data showing a permanent shift in the threshold of hearing (*i.e.*, hearing loss) because of exposure to noise. The most fundamental question is what is the maximum level of daily noise exposure over a period of many years that will not cause a measurable hearing loss (beyond that due to aging) in the undiseased ear typical of the general population? From such a determination, recommendations can be made as to tolerable noise exposure limits for people with normal ears.

The second question, and one of practical interest to industry, is what level of industrial noise exposure will cause measurable noise-induced hearing loss in factory workers? Such a determination would help establish possible liability for noise-induced hearing loss in industry. As seen in the next two chapters, the answers to these two questions from an analysis of presently available data are somewhat different.

First to be discussed in the present chapter are selected data on hearing level (HL) and present state-of-the-art solutions related to the measurement and prediction of noise-induced permanent threshold shift (NIPTS) and temporary threshold shift (TTS). Following that, a brief summary of some articles and reports published since 1963 on procedures for predicting NIPTS from exposure to noise is given.

Sociocusis and Industrial HL Data

As discussed in some detail later, it is expected that the "energy" in the sounds and noises of everyday living and of the workplace are additive in terms of their effect on the threshold of hearing sensitivity. If so, the contribution of the

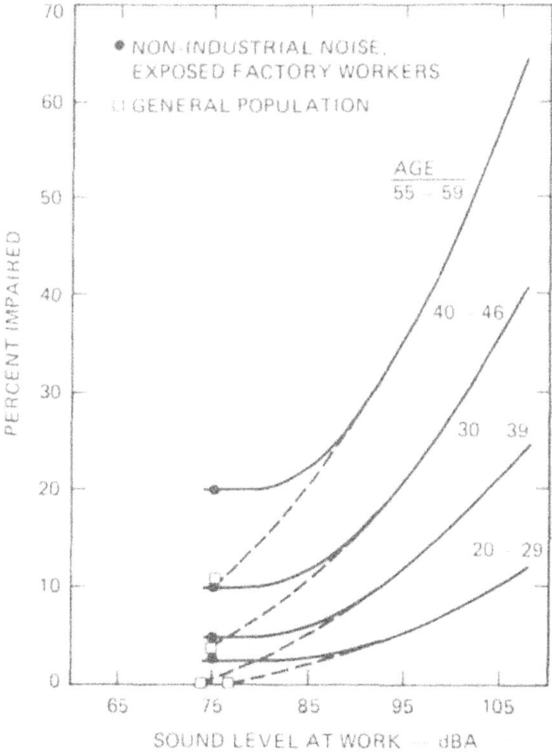

FIGURE 7.1. Percent of workers who work in different levels of noise (after ref. 1) and of general U.S. population (after ref. 2) of different ages (and years of exposure) with average HL >25 dB at 0.5, 1, and 2 kHz.

nonindustrial, everyday noise to a noise-induced threshold shift would be insignificant for those workers exposed to industrial noise capable of causing large industrial noise-induced permanent threshold shift (INIPTS) but not to those workers exposed to noise causing small amounts (up to 10 dB or so) of INIPTS. It is the noise exposures that are additive, not the threshold shifts in HL.

This additivity of exposure to industrial noise with the apparently high (compared with that for the general population) amount of exposure to everyday noise found in factory workers would explain the unusual shape of the function relating intensity level of industrial noise and amount of NIPTS. As shown in figure 7.1 (from refs. 1 and 2), this function does not trail off below industrial noise levels of about 85 dBA, as would be expected if the threshold shifts of the workers were a function only of the industrial noise and their hearing thresholds before INIPTS were the same as those of the general population. It is suggested that 5 to 10 dB of excess sociocusis is present in the average industrial workers. This conclusion is also based on the data presented in figure 6.20 and on data discussed later in this chapter.

Nosocusis and Industrial HL Data

A differential in nosocusis between the general population and that of factory workers is a possible factor that could lead to an overestimation of NIPTS *per se*. This is the result of using the HL's of the general population as a reference base for comparison with the HL's reported for industrial audiometry. In this regard, otological problems were found in less than 25 percent of the general U.S. population as a whole, and the average magnitude of their effects upon HL's was about 5 dB of loss compared with normal ears. This would presumably result in the average HL's of the abnormal ears to be but 1 to 2 dB greater than the average HL's of the normal and abnormal ears combined. (See chapter 6.)

On the other hand, Burns *et al.* (ref. 3) found that the average HL's of the otologically normal subjects never exposed to gunfire in a group of workers in a noisy industry were about 10 to 14 dB higher than the average HL for all otologically normal people not exposed to gunfire. Otologically normal subjects not exposed to gunfire had HL's about 6 dB higher than the average for the entire group of workers. (See table 7.1.) About 60 percent of the total group of workers had pathological ears, as compared with about 25 percent of the general U.S. population. Of course it is reasonable to presume that the deviations from expected HL's (as shown in table 7.1) in persons with certain pathological conditions are determined (for particular cases) primarily by the pathological conditions and not from an interactive effect with the noise. As noted by Burns *et al.*, there are not sufficient data available to provide any definitive answers to this question of possible interaction between ear disease and noise on HL. Indeed, there are some pathological conditions that would reduce conduction of the noise to the inner ear and thereby reduce NIPTS *per se*.

TABLE 7.1

Hearing Levels of Workers in a Noisy Industrial Environment

[Data from ref. 3]

(a) Categorization of subjects

Group Description	Subgroups	Clinical description	No. of subjects*		
			a	b	a + b
	1	Normal otological exam and history	226	63	289
	2	As 1 but ears syringed shortly prior to audiometry	65	30	95
Normal (N)	1 + 2		291	93	384
	7	White patch on drum(s)	5	4	9
	8	Scarred drums and macerations	24	6	30
	9	History of otorrhoea	74	20	94
Conductive (C)	7 + 8 + 9		103	30	133
	11	Tinnitus	16	10	26
	12	Trauma-ototoxic drugs	5	5	10
Sensorineural (S)	11 + 12		21	15	36
	3	Bilateral "hearing trouble"**	24	6	30
	4	Drums not translucent	33	8	41
	5	Drums opaque	23	6	29
	6	Combination of 3 and 5	9	4	13
	10	Dizziness	18	5	23
	13	Uncertain; possible wax, catarrh, unilateral "hearing trouble"	29	5	34
Doubtful etiology (D)	3 + 4 + 5 + 6 + 10 + 13		136	34	170
All pathological (P) cases (C + S + D)	3-13		260	79	339
All subjects (N + P)			551	172	723

**Subjects with: a—no recorded gunfire history; b—known military gunfire history.*
***Self-reported.*

TABLE 7.1 Concluded

(b) Relative hearing levels

Groups compared	Differences between HL's, dB, for frequencies, kHz, of					
	0.5	1	2	3	4	6
C(a + b) − N(a)	13.5	13.3	15.7	12.8	13.4	14.3
S(a + b) − N(a)	8.4	9.1	13.1	16.6	16.7	16.2
D(a + b) − N(a)	8.3	10.1	14.0	13.1	11.1	12.2
P(a + b) − N(a)	10.4	11.3	14.6	13.3	12.6	13.4
(N + P)(a + b) − N(a)	5.0	5.5	7.2	6.5	6.2	6.9

Additional indication of the seemingly large amount of nosocusis in factory workers, at least those in noisy industries, was found by Sulkowski (ref. 4) as shown in table 7.2. It is shown that some 33 percent of the total group of workers having significant hearing losses (HL's greater than 30 dB average at 1000, 2000, and 4000 Hz) were deemed to have losses ascribable to causes other than noise. Unfortunately, more data on the prevalence of otological problems among workers in "non-noisy" occupations and in the so-called "control groups" are not available.

In any event, it could be argued that the prevalence and degree of otological problems in factory workers unrelated to noise are such as to result in 5-dB-higher HL's, on the average, than in the general population.

A study relevant to the question of the additivity of nosocusis and NIPTS (and perhaps also to the question of additivity of presbycusis and NIPTS) was conducted by Howell (ref. 5). He found that deterioration in the HL's of steelworkers exposed to noise levels was nearly the same, after age standardization, over a 7-year exposure period regardless of the initial HL of the worker. For example, for men with initial-age-corrected HL's of less than 2 dB, 12 to 26 dB, and greater than 26 dB, the average shifts over the frequency range 0.5 to 6 kHz for the 7-year period were, respectively, 7.5, 8.7, and 7.1 dB (in exposure levels of greater than 100 dBA); 7.8, 6.8, and 7.3 dB (in exposure levels of 90 to 99 dBA); and 6.2, 5.0, and 5.2 dB (in exposure levels of less than 90 dBA). If the differences in initial HL were attributable, at least in part, to nosocusis (the 449 subjects were not screened for otological abnormalities), these results would indicate some additivity of the effects of noise and nosocusis on HL.

Some results consistent with the hypothesis of the additivity of nosocusis and NIPTS can be seen in a number of HL surveys conducted in factory settings. For example, a difference of 8 dB between the HL (at 4000 Hz) for 20-year-old workers in so-called non-noise jobs in factories and the HL's of the general U.S. population (age 20 years) is typical, and this difference gets to be about 10 to 15 dB for

TABLE 7.2

Motives for Elimination of Occupational Etiology

[Data from ref. 4]

Motives other than exposure to noise	Number of cases
Nonorganic hearing loss	55
Otosclerosis	12
Tympanosclerosis	14
Undergone conservative and radical operations of the middle ear	21
Chronic purulent otitis media	39
Congenital deafness	2
Meniere disease	15
Lack of essential noise exposure	8
Total*	166

* *Total sample of compensable cases was 496; 330 were diagnosed as being the result of noise.*

the older age groups. (See chapter 6, fig. 6.20.) Perhaps some of this difference can be attributed to excess nosocusis in industrial workers relative to that in the general U.S. population.

Two Major Studies on NIPTS From Industrial Noise

When defined as the probable difference between the HL of a person before exposure to a noise environment and their stabilized HL after, the best delineation of the relation between exposure to noise and hearing loss appears to be found in data and analyses in the papers of Baughn (refs. 6 and 7) and of Burns and Robinson (ref. 8). However, two somewhat different philosophies regarding the assessment of industrial noise-induced shifts in hearing level were followed by Baughn and by Burns and Robinson and, as we shall see, there are some problems and difficulties involved in the interpretation of the basic data in both studies.

I should first mention a large retrospective study of the auditory thresholds of workers in noisy industries in Austria (Raber, ref. 9). Raber reported that even after 25 years of exposure in steady noise of 110 dBA, the incidence of workers with an average HL greater than 25 dB at 0.5, 1, and 2 kHz was only about 12 percent as compared with 10 percent for people not exposed (*i.e.*, less than 80 dBA). According to other studies of NIPTS in industry, some 85 percent (compared with Raber's 12 percent) of the population exposed to 110 dBA noise for 25 years should have HL's that exceed an average of 25 dB at 0.5, 1, and 2 kHz. A full explanation of why the Austrian data are so disparate does not seem possible. Perhaps one major reason is that all the data for persons with a bone-conduction hearing loss that was, at any frequency, 15 dB less than the air-conduction hearing loss were removed from the data base regardless of the absolute air- and bone-conduction hearing losses at other test frequencies. In brief, it appears probable that many of the subjects with NIPTS were inadvertently eliminated from the data base.

Royster and Thomas (ref. 10) suggest that the HL's at 3000 to 6000 Hz may be relatively improved (less INIPTS) by long-term "learning" to take annual audiometer tests administered in hearing conservation programs. However, if long-term practice effects are a significant factor in industrial audiometry, it would seem that it should affect all test frequencies, not just those above 3000 Hz. In the studies presented below, all or nearly all the HL data bases represent initial or second repeat audiograms. Possible long-term audiometer-test "learning" effects, if such exist, are not involved.

Baughn study

Baughn (refs. 6 and 7), after certain averagings and normalizations of the raw data, reported idealized HL data for workers in a large industrial plant who were exposed to certain known levels of more or less steady-state noise. No screening of the workers for otological problems or off-the-job noise exposures were done. The work force involved was very stable in job location and content. As a result, data were available for a large number of subjects (6835) for exposure durations up to about 45 years in 3 average levels of noise exposure (78, 86, and 92 dBA) for 8 hours per workday.

As discussed previously, the average worker appears to have 10 dB more sociocusis at 3000 and 4000 Hz than the general population. Thus, in a retrospective study such as Baughn's, use of the average HL of the general population to represent the average HL of industrial workers would lead to deducing more industrial NIPTS than is justified. That is, the worker with an HL at 3000 to 4000 Hz of up to about 10 dB can be expected to have that HL because of excess sociocusis–nosocusis.

Taking the above factors into consideration, it seems appropriate to adjust somewhat the low HL's of Baughn's data base. Figure 7.2 shows a function applied to Baughn's data in the analysis to follow to account for the excess sociocusis at higher test frequencies and audiometry "errors" at lower test frequencies. Some low-frequency masking noise was present in the test booth (ref. 7), and excess sociocusis is suspected to have caused some loss at high frequencies.

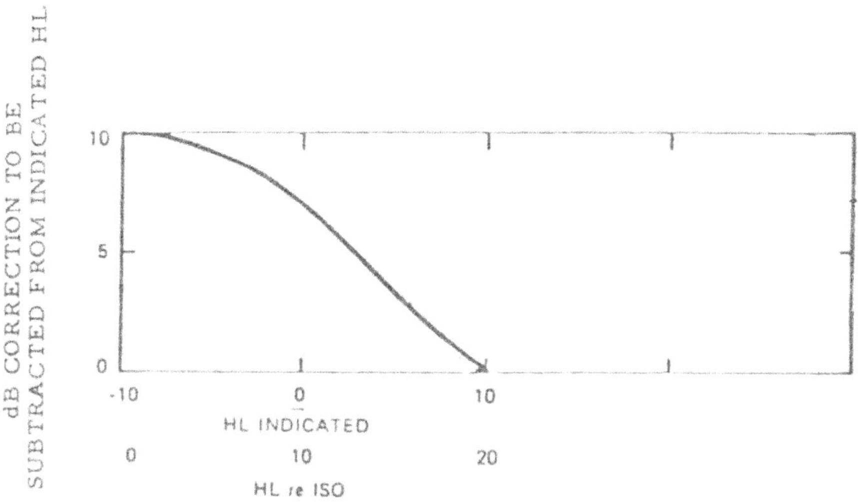

FIGURE 7.2. Estimated correction to be used with Baughn's HL data.

Burns and Robinson study

Burns and Robinson (ref. 8) studied 759 subjects (422 males and 337 females) exposed to 4 average classes of industrial noise: (1) less than 87 dBA; (2) 87 to 90 dBA; (3) 91 to 97 dBA; and (4) greater than 97 dBA. The maximum exposure was about 49 years. In addition, 97 subjects not exposed to noise were studied. The age range was about 20 to 60 years. Unlike Baughn's subjects, Burns and Robinson's subjects, including the 97 non-noise-exposed subjects, were selected to have no otological problems, to have no histories of such problems, to be taking no ototoxic drugs, and to have had no exposure to gun noise.

Which of these two approaches (total population of workers versus highly screened population of workers) is best for the problem at hand depends to some extent upon the use or uses to which the results are to be put. However, both approaches are faced with fundamental questions of what are the interactions, with respect to shifts in hearing sensitivity, between exposure to noise and nosocusis or presbycusis. As discussed above, Howell's data suggest that nosocusis and NIPTS can be treated as being additive, statistically speaking.

A major problem in the interpretation of retrospective studies such as Baughn's and Burns and Robinson's is the lack of knowledge about the distribution HL's prior to exposure of the population of workers involved in the studies as a function of their age. Burns and Robinson would seem to have less uncertainty than Baughn in this regard because their 97 non-noise-exposed control subjects were free of otological pathologies. However, a recent analysis by Robinson and Sutton (ref. 11) of the distribution of HL as a function of age in populations screened for otological problems and exposure to "excess" noise is considerably different at 4000 Hz in the older group from that depicted in the Burns and Robinson study, as shown in figure 7.3 (refs. 8 and 11 to 15). Note that except for the 50th percentile, the Burns and Robinson data have been treated according to mathematical simplifications not involved in the other functions shown.

Burns and Robinson concluded that the daily noise energy in decibels, summed on an energy basis over the years of exposure, could be used to predict noise-induced hearing loss. This noise exposure variable was labelled E_A. Burns and Robinson identified the relation, for separate audiometric test frequencies, between E_A and the HL's of males and females over the percent distribution of ear sensitivities. The basic formula is as follows (from ref. 11):

$$H' = 27.5\ \{1 + \tanh[(E_A + k - \lambda + u)/15]\} + u + F$$

where H' is the HL exceeded only by a percentile of the population, F is given in a table of age correction for presbycusis, E_A is the energy emission level of the noise over years of exposure, k and λ are factors pertaining to audiometric test frequency and sex of the subject, and u is a normal percentile distribution of hearing levels for young normal ears around the median and having a standard deviation of 6 dB. Robinson and Shipton calculated what they called expected hearing loss H after omitting F from the above formula:

$$H = 27.5\ \{1 + \tanh[(E_A + k - \lambda + u)/15]\} + u$$

Some of their tables of H and related information are given in the appendix to this chapter.

Robinson and Shipton emphasize that except for the median (or 50th percentile), the H values from their model are properly predictive of the expected distribution of HL's of otological- and noise-screened populations exposed to a given E_A only when used in conjunction with the age (presbycusis) function F and sex factors derived for their subjects. One reason is that the distribution statistics around the 50th percentile of the non-noise exposed, 20-year-old subjects were taken as a normal distribution with a standard deviation of 6 dB (u of the above formulas). Further, this same distribution statistic became incorporated with the measured HL data, regardless of the age of the population involved.

It is well known that the distribution of HL's becomes progressively skewed above the 50th percentile at older ages because of presbycusis. In the Burns and Robinson model, these deviations from the distribution of hearing threshold sensitivities as a function of age are part, along with any noise-induced hearing loss, of the deduced and tabulated expected hearing loss H. At the higher percentiles of hearing sensitivities this "excess" presbycusis could amount to a significant amount of the predicted hearing loss compared with the part due to the noise exposure (NIPTS). However, this presbycusis skewness factor presumably affected to a somewhat similar degree the actual distributions of the HL data for

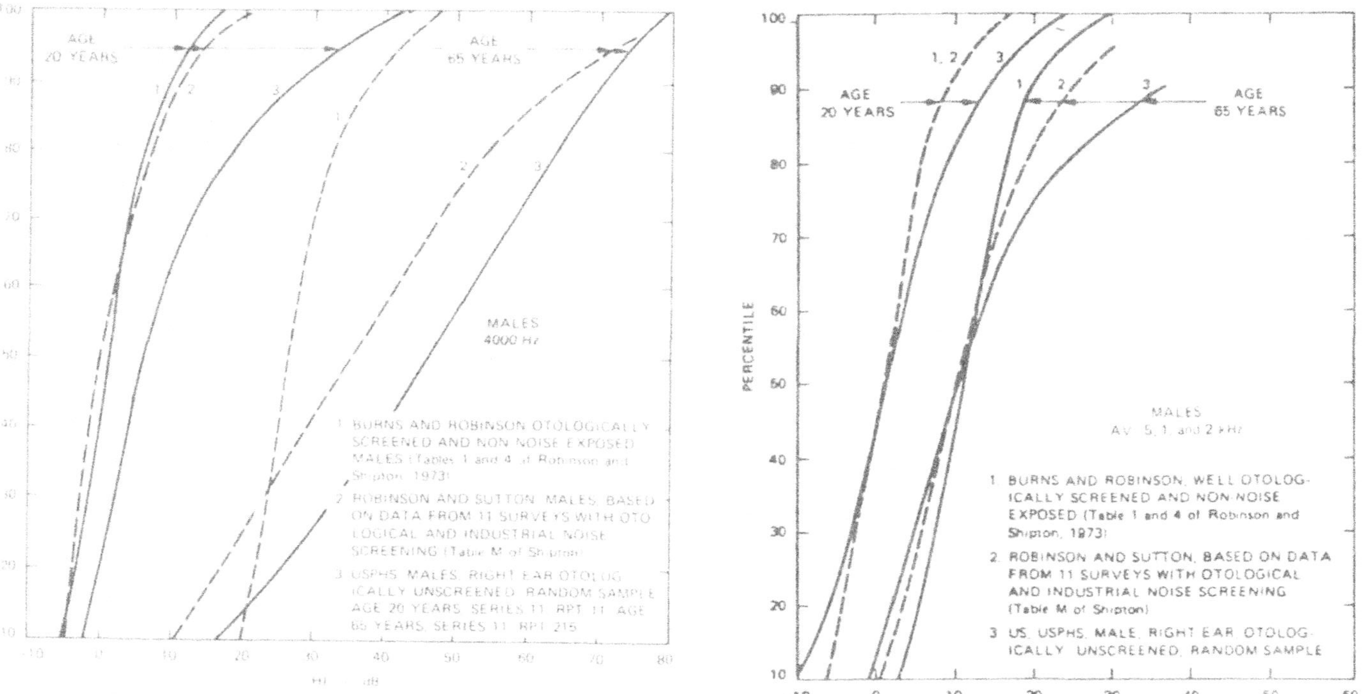

FIGURE 7.3. Presbycusis effect on HL for various subject types. (Data from Burns and Robinson (ref. 8), Robinson and Shipton (ref. 12), Shipton (ref. 13), Robinson and Sutton (ref. 11), USPHS report 11 (ref. 14), and USPHS report 215 (ref. 15).)

non-noise- and noise-exposed groups at a given age. Accordingly, for a given age group and percentile, the differences in measured hearing loss from one step in E_A to the next should be indicative of solely NIPTS due to the decibel step increase in E_A.

These possible limitations and problems of specifying absolute values of HL for workers exposed to noise do not necessarily invalidate the general findings of the Baughn nor the Burns and Robinson studies with respect to threshold shifts *per se* or to NIPTS. As shown by Rop and Raber (ref. 16) and Macrae (ref. 17), the effects of presbycusis on HL are additive with NIPTS. Thus, a shift in HL of 10 dB because of presbycusis plus 10 dB NIPTS will cause a change of 20 dB in HL. Or, in another way, a 10-year total exposure to a given noise will apparently cause the same amount of NIPTS regardless of the age of persons equally exposed to noise (within a 40 dB NIPTS limit to be prescribed and some absolute limits on HL due to aging).

This apparent independence of presbycusis and NIPTS may be the result of possibly separate physiological mechanisms being involved. These mechanisms are a metabolic blood-supply change in the auditory system in presbycusis and some mechanical chemical actions on the hair cells and other tissues of the cochlea imposed by intense sound. Both types of insult presumably lead to injury or disfunction in the structures involved, including attached nerve cells. Also, in support of this conclusion is the finding that the amount of TTS from exposure to noise was essentially the same for older people with presbycusis as for young adults without presbycusis (ref. 18).

Analysis of Combined Studies

The Baughn data and analysis and the Burns and Robinson data and analysis taken together perhaps provide a more valid, realistic base upon which to derive NIPTS functions as well as predictions of actual HL than is likely from either data base alone. The arguments for averaging NIPTS functions derived from the Burns and Robinson and those derived from the Baughn and USPHS studies are: (1) although in the Baughn and USPHS comparisons there is some uncertainty that the populations of non-noise-exposed versus noise-exposed persons were strictly comparable, the thousands of subjects and the stable noise and work conditions involved supports the reliability and validity of the basic data; and (2) in the Burns and Robinson study, the number of subjects was sufficiently small, considering the large range of ages and noise exposure conditions involved, so as to possibly limit the reliability of the basic data and the generality of the model derived.

The necessity of going through the sometimes questionable rationalizations and steps required to utilize the data of retrospective studies of NIPTS is, of course, unfortunate. However, the Baughn and USPHS data and the Burns and Robinson data and related analyses appear to be the best that are available on the question at hand. Furthermore, imposition of noise control and hearing conservation programs in many industries in many countries over the past 10 years or so make somewhat remote the possibility of performing a meaningful retrospective study of the effects in industry of noise on the unprotected ear. An analysis of the combined findings of these two studies follows.

Results at 4000 Hz and average at 0.5, 1, and 2 kHz

Figure 7.4 shows the cumulative percentage of 65-year-old males for 4000 Hz and for the average of 0.5, 1, and 2 kHz having HL's equal to or less than the HL indicated on the abscissa after an 8-hour workday, 45-year career in no noise and in various levels of noise. Although there are some differences between the exposure distributions of HL for the Burns and Robinson and for the Baughn and USPHS data, the effect of the noise is to shift upwards (cause a loss in hearing sensitivity) the HL's over the entire distribution of hearing sensitivity.

Smoothed data for groups exposed to different levels of industrial noise and to no-noise conditions according to the studies of Burns and Robinson and of Baughn and USPHS are shown as a function of years of exposure (age) in the upper panels of figure 7.5(a) for 4000 Hz and in figure 7.5(b) for the average of 0.5, 1, and 2 kHz. NIPTS according to the two studies is plotted in the bottom panels of the figures. For the Burns and Robinson study, NIPTS is taken as the difference between the HL of the non-noise exposed versus the HL related to the E_A calculated for specified noise exposure conditions. NIPTS for the Baughn data is taken as the difference between the HL's of the USPHS presbycusis curves versus the HL's for the noise exposure conditions for the Baughn study.

In general, somewhat greater NIPTS is found for the HL's of Baughn's noise-exposed people versus the HL's of the general public than is found for comparable ages and exposure conditions in the Burns and Robinson analysis. An exception is 4000-Hz data at the 90th percentile data. Also, the 90th percentile NIPTS data from the Baughn study appear to be constrained by limitations imposed on the measured HL's by the maximum range of hearing.

Generalized NIPTS

Figure 7.6 shows NIPTS at 4000 Hz and average 0.5, 1, and 2 kHz as a function of A-weighted noise level in decibels for steady-state noise lasting 8 hours per workday for 45 years exposure as found in the data for Burns and Robinson and for Baughn versus USPHS. Also shown on these graphs are the K values (the maximum DL's that give 0 dB NIPTS) and two dashed lines that seem to fit the trends of the curves based on the data.

Both the Baughn and the Burns and Robinson analysis show a rather slow rise as a function of increasing noise level in NIPTS up to about 10 dB, at which point the rise becomes more rapid. These NIPTS straight-line functions are shown up to 40 dB.

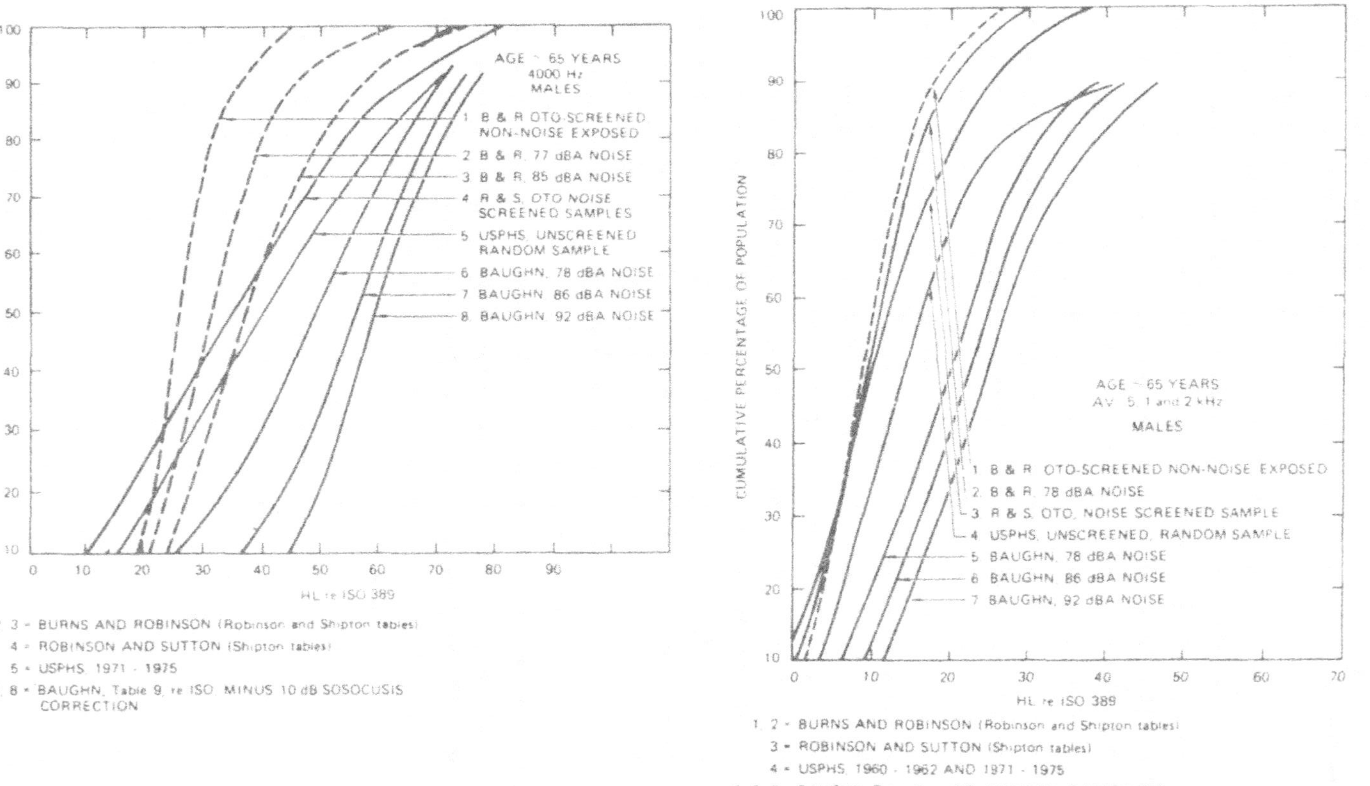

FIGURE 7.4. Comparison of data from Burns and Robinson (ref. 8), Robinson and Sutton (ref. 11); USPHS reports (refs. 14 and 15), and Baughn (refs. 6 and 7) for HL as a function of sensitivity to sound.

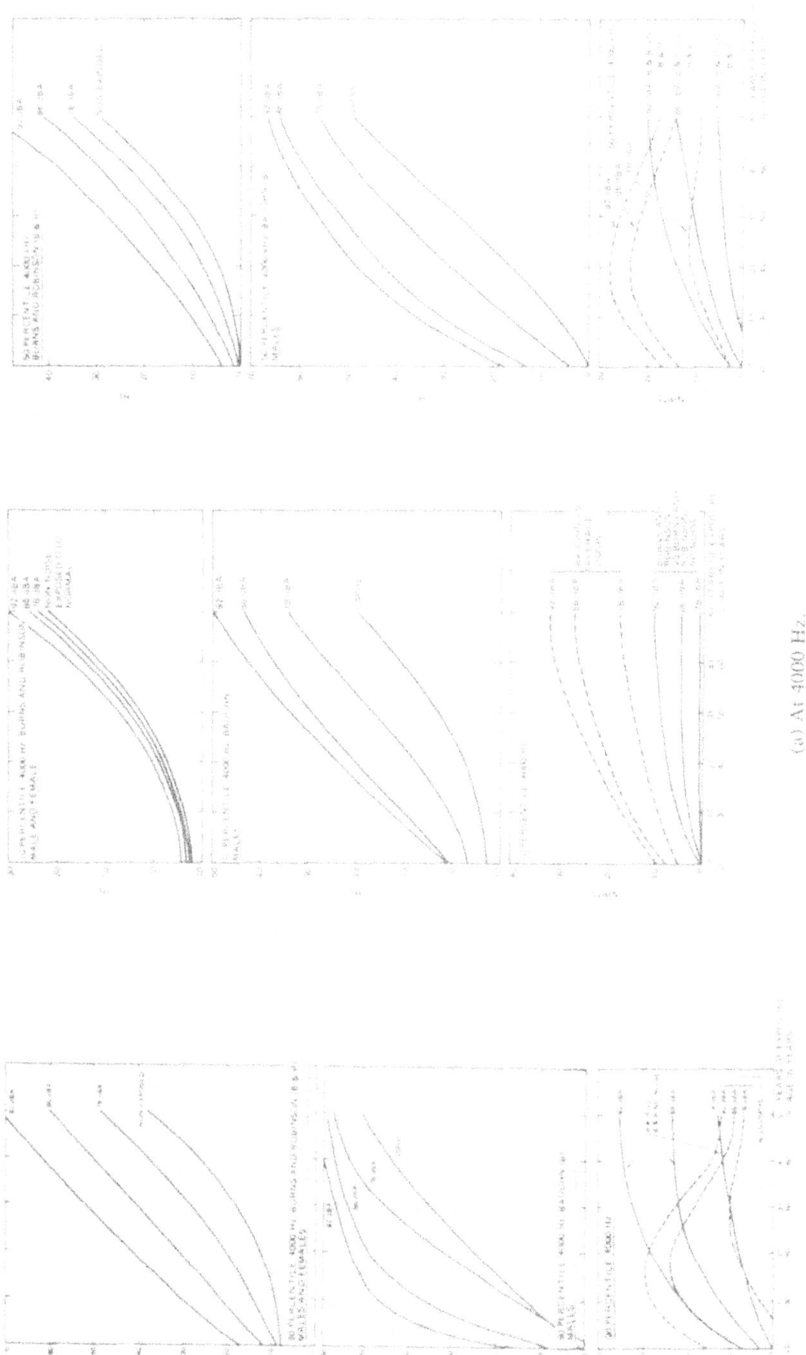

(a) At 4000 Hz.

FIGURE 7.5. HL and NIPTS as functions of age or exposure for data from Burns and Robinson (ref. 8), Baughn (refs. 6 and 7), USPHS (refs. 14 and 15), and Robinson and Sutton (ref. 11).

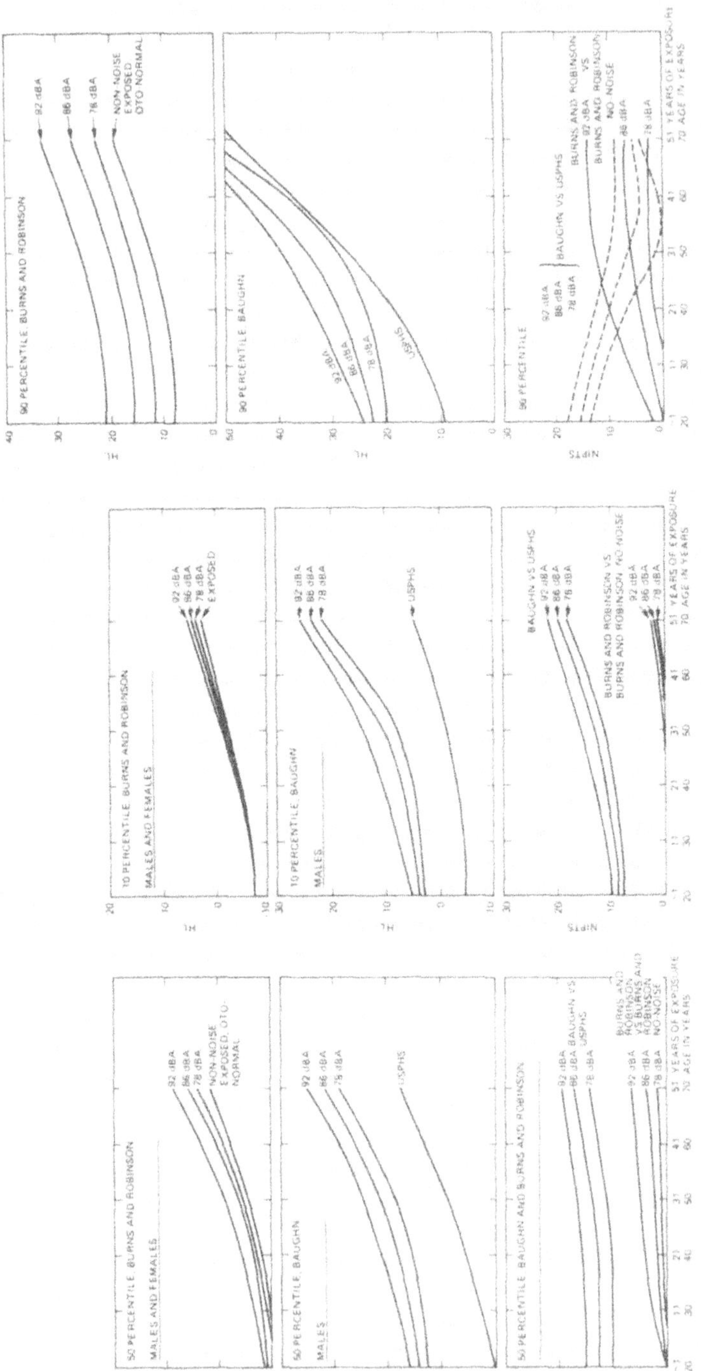

(b) Average at 0.5, 1, and 2 kHz.

FIGURE 7.5. Concluded.

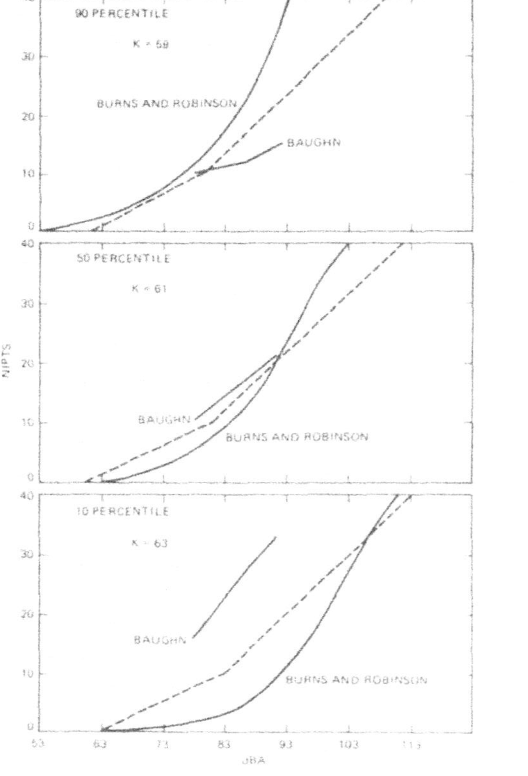

(a) At 4000 Hz.

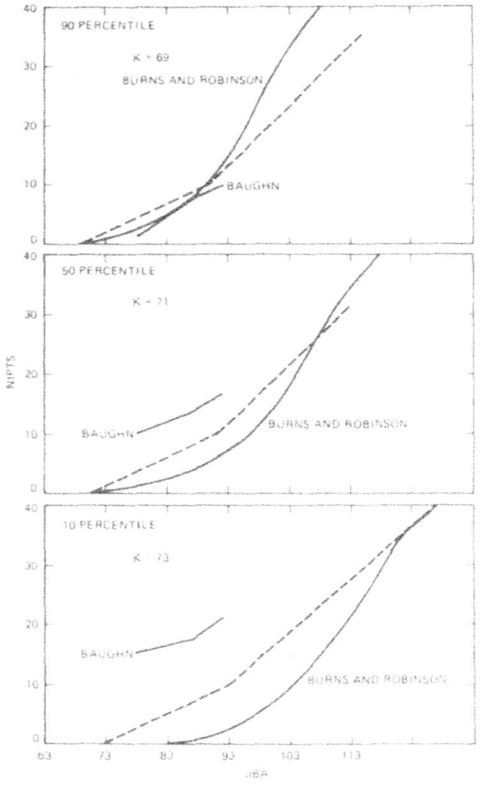

(b) Average at 0.5, 1, and 2 kHz.

FIGURE 7.6. NIPTS (in decibels) for males with 45 years exposure as a function of A-weighted noise level based on 8-hour-per-workday exposure for data from Burns and Robinson (ref. 8) and Baughn (refs. 6 and 7). The upper portion of the dashed curves represents a slope of 20 log SP + k (or 10 log $SP^2 + k$) and the lower portion represents a slope of 10 log SP + k (or 5 log $SP^2 + k$).

As shown later, for exposures that can cause TTS or NIPTS greater than about 40 dB, the recovery and hearing-loss processes in the ear appear to behave differently than they do for exposures to less intense noise. Further, there is a natural limitation to the range of hearing sensitivity that makes the course of growth of NIPTS above that value somewhat academic.

The use of straight lines rather than some possibly more precise curvilinear fit to the basic data is perhaps justified on the grounds of simplicity in the derivation of a procedure for predicting NIPTS and HL to be outlined below. Also, the basic data from both the Baughn and the Burns and Robinson investigation have been processed and idealized to such an extent it is perhaps questionable that more precise modelling is likely to be meaningful.

As shown in figure 7.6, a change of 1 dB in SPL is required to cause about a 0.5-dB change in NIPTS when NIPTS is between 0 and 10 dB; between about 10 and 40 dB NIPTS, a 1-dB change in SPL effects about a 1-dB change in NIPTS. The slope of the 0- to 10-dB curve (above some given intercept k) can be represented by 10 log SP + k (where SP is sound pressure), and the slope of the 10- to 40-dB curve is 20 log SP + k. This means a 20-dB change in SPL is required to go from 0 to 10 dB NIPTS, and an additional 30-dB change in SPL is needed to go from 10 to 40 dB NIPTS.

Two major variables causing NIPTS are the intensity and the duration of exposure to a noise, the product of these being sound energy. At first glance, equal noise energy appears to be, at least for exposures causing no more than about 40 dB NIPTS, a valid measure for predicting NIPTS from near-daily exposures to a given noise environment. That is, a fractional change in the number of days of exposure of 8 to 10 hours per day will have the same general effect on the magnitude of NIPTS as will a similar fractional change in sound pressure level. However, the magnitude of the change will depend to some extent upon the amount of NIPTS involved.

Figure 7.7 shows that when exposure duration in years Y is expressed in decibels (10 log Y), there is about a 1-dB increase in NIPTS per 1-dB increase in exposure duration in years when NIPTS is above 10 dB, but only about a 0.5-dB increase (5 log Y) in NIPTS per 1-dB increase in exposure duration when NIPTS is between 1 and 10 dB. This is consistent with the growth of NIPTS as a function of sound pressure level and, apparently, with the equal-energy concept. However, note that the daily industrial noise exposures are for an 8- to 10-hour-workday exposure to more or less steady-state noise. As we shall see, the temporal pattern of the exposure to noise within the 8- to 10-hour day can have an effect on TTS (and presumably NIPTS) that does not follow the equal-energy rule. Accordingly, the "equal-energy" results described above for NIPTS from exposure to industrial noise agree only fortuitously with the equal-energy concept, and other temporal patterns of daily noise exposures need not follow the equal-energy rule.

A prediction of the maximum noise level for an 8-hour daily exposure that will not produce measurable (*e.g.*, less than 5 dB) NIPTS after, say, 45 years of

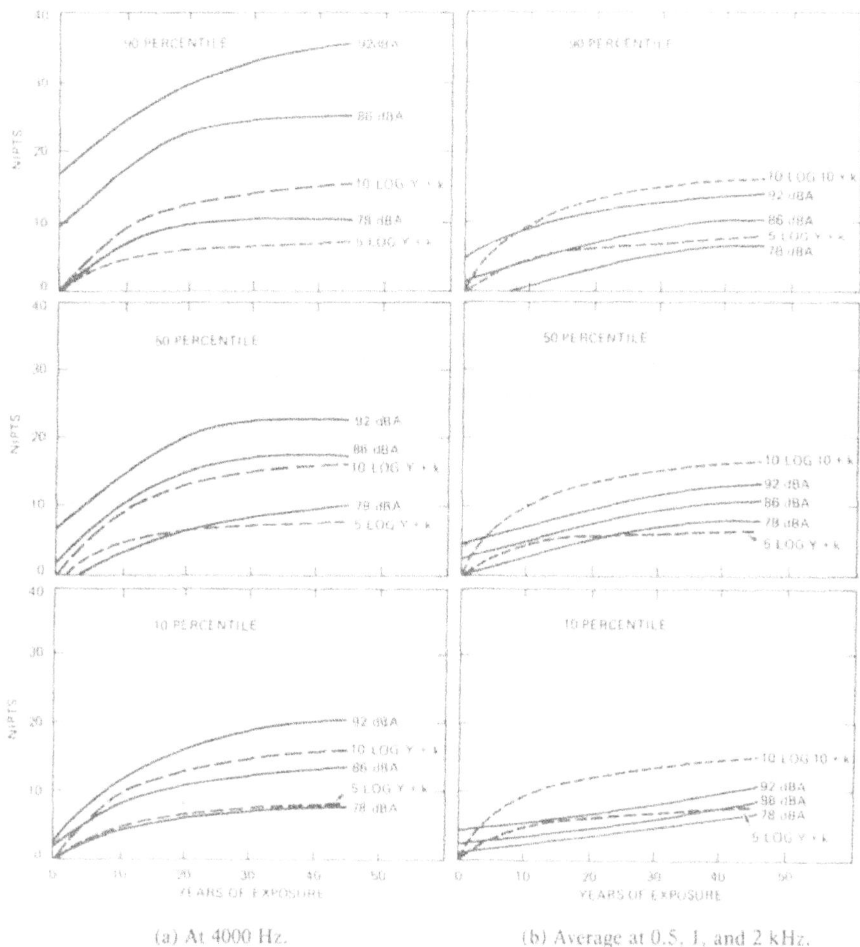

(a) At 4000 Hz.

(b) Average at 0.5, 1, and 2 kHz.

FIGURE 7.7. NIPTS in decibels at 90, 50, and 10 percentiles as a function of years of exposure for average of Burns and Robinson (ref. 8) and Baughn (refs. 6 and 7) data. Parameter is noise level.

almost daily exposure can be extrapolated from the Baughn and the Burns and Robinson data. As shown in figures 7.6 and 7.7, this level depends on the percentile of interest in the distribution of hearing sensitivities. For the 10th percentile (more sensitive ears), the level is quite low, from about 63 to 75 dBA depending on audiometric test frequency. These levels are reasonably close to those that would be predicted from consideration of the TTS-PTS (permanent threshold shift) relations (discussed later) that have been observed in or extrapolated from research on TTS and on "effective quiet" (the maximum level of noise between bursts of more intense noise that will not retard recovery from the intense noise).

Predicting NIPTS From TTS Data

The most extensive data on NIPTS (such as those of Baughn and of Burns and Robinson) are related to exposure to more or less steady-state, broadband noise measured over all frequencies with an A-weighted sound level meter for an 8- to 10-hour workday. Lunch and normal rest breaks are considered as constant factors. That these factors are probably significant (although presumably constants in the industrial context) is shown in figure 7.8 from Hetu and Parrot (ref. 19). This figure shows considerable recovery effects on HL, especially after the lunch break.

Studies of the effects of industrial noise and NIPTS do not lend themselves to the derivation of detailed models of the effects of widely different frequency spectra or of restricted bandwidths. Neither are they useful for deriving models of the effects of exposure to noises of varying intensity (durations of daily expo-

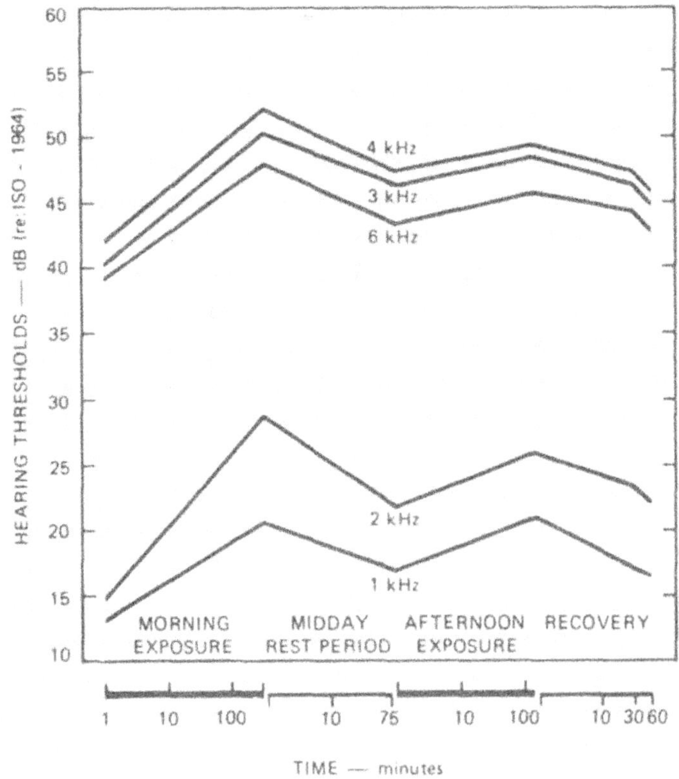

FIGURE 7.8. Evolution of mean hearing thresholds throughout the workday. Parameter is audiometric test frequency. (Data from ref. 19.)

sures shorter than about 8 hours) of exposures to intermittent noise, or, as a somewhat special case, of exposure to so-called impulsive sounds. For quantifying the approximate effects of these variables, it is necessary to turn to studies done in the laboratory on temporary threshold shifts (TTS) from exposures to noises under well-controlled conditions. The justification for doing so rests on the assumption that TTS is a precursor of NIPTS and on the finding that such models could be used to predict NIPTS data in certain specified industrial noise studies. Kryter (ref. 20) noted that the average TTS_2 (TTS measured 2 minutes after exposure) from an 8-hour exposure to noise in young, normal ears was similar in magnitude to the average NIPTS found in workers after 10 to 20 years exposure to about the same level of noise in industry. Since then, other verifications have been published (see fig. 7.9), and some comparable effects on PTS and TTS for impulses and interrupted noise will be shown below. There seems to be no data available to seriously challenge this basic assumption, at least within the restrictions of noise conditions that consist of daily noise-exposure periods of 8 to 10 hours or so and that cause of the order of no more than about 40 dB TTS_2 in the normal ear. Data supposedly contrary to this assumption are discussed later are relation to data on TTS and NIPTS from impulse noise. Kraak *et al.* (ref. 21) proposed that rather than TTS *per se* it is "the time integral over TTS during and after exposure that gives a true measure of the stress on hearing."

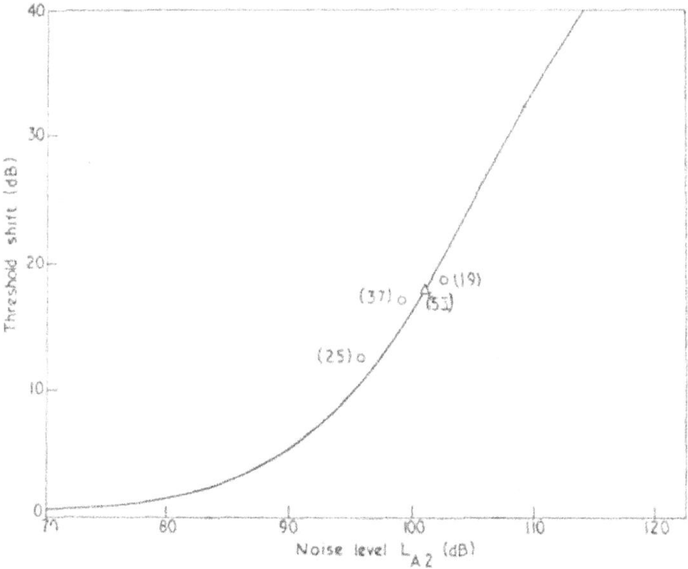

FIGURE 7.9. Relation between TTS at 2 min after exposure (data symbols) and presumed noise-induced hearing loss after 10 years of exposure to the same noise (curve). Frequency is 4 kHz. Values in parentheses indicate number of subjects. (From ref. 8.)

Spectral differences in steady-state noises

Figure 7.10 (ref. 22) shows the general differences in effectiveness of different parts (octave bands) of the sound spectrum to cause TTS and, presumably, PTS. As more bands in a broader band noise reach the contours shown, the hearing loss will extend over a wider and wider frequency range to which the ear is sensitive. However, hearing loss in any one frequency region should not be significantly greater than that expected from a band of noise located about an octave below that particular frequency region.

As shown in figure 7.11(a), for the test frequencies of 250 to 4000 Hz, the differences in TTS from the widely varying spectra (shown in fig. 7.11(b)) range up to a maximum of about 6 dB when averaged over the two tests (ref. 23). Although for some purposes this could be considered a significant difference, it is surprisingly small considering the wide differences in the distribution of sound energy over the frequency spectra involved (−6 dB per octave, 0 dB, and 6 dB per octave). Burns and Robinson (ref. 8) reported somewhat similar findings (see fig. 7.12) for NIPTS from noises having a range of spectral slopes from about −8 to 9 dB per octave.

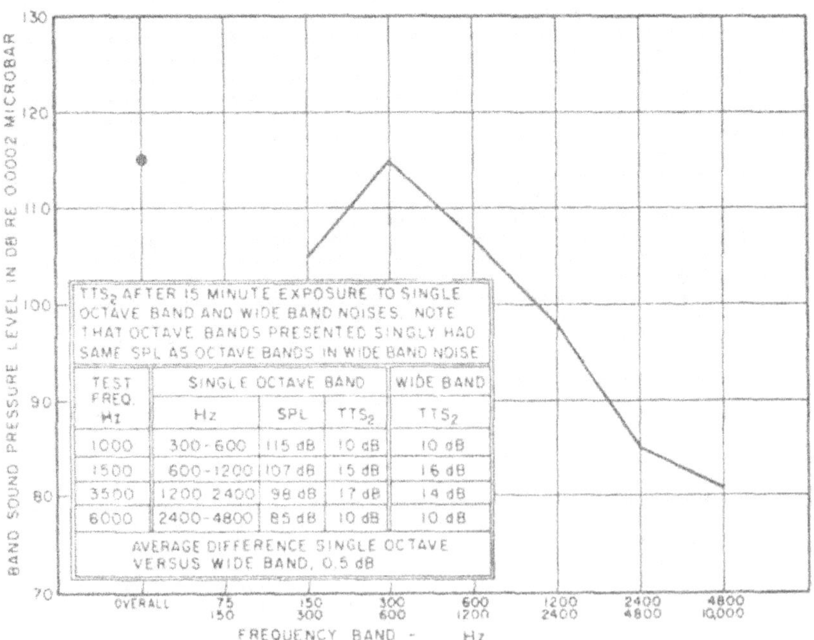

TTS$_2$ AFTER 15 MINUTE EXPOSURE TO SINGLE OCTAVE BAND AND WIDE BAND NOISES. NOTE THAT OCTAVE BANDS PRESENTED SINGLY HAD SAME SPL AS OCTAVE BANDS IN WIDE BAND NOISE

TEST FREQ.	SINGLE OCTAVE BAND			WIDE BAND
Hz	Hz	SPL	TTS$_2$	TTS$_2$
1000	300-600	115 dB	10 dB	10 dB
1500	600-1200	107 dB	15 dB	16 dB
3500	1200 2400	98 dB	17 dB	14 dB
6000	2400-4800	85 dB	10 dB	10 dB

AVERAGE DIFFERENCE SINGLE OCTAVE VERSUS WIDE BAND, 0.5 dB

FIGURE 7.10. Octave-band spectrum of wideband noise and comparison of TTS$_2$ from wideband and octave-band noise. (From ref. 22.)

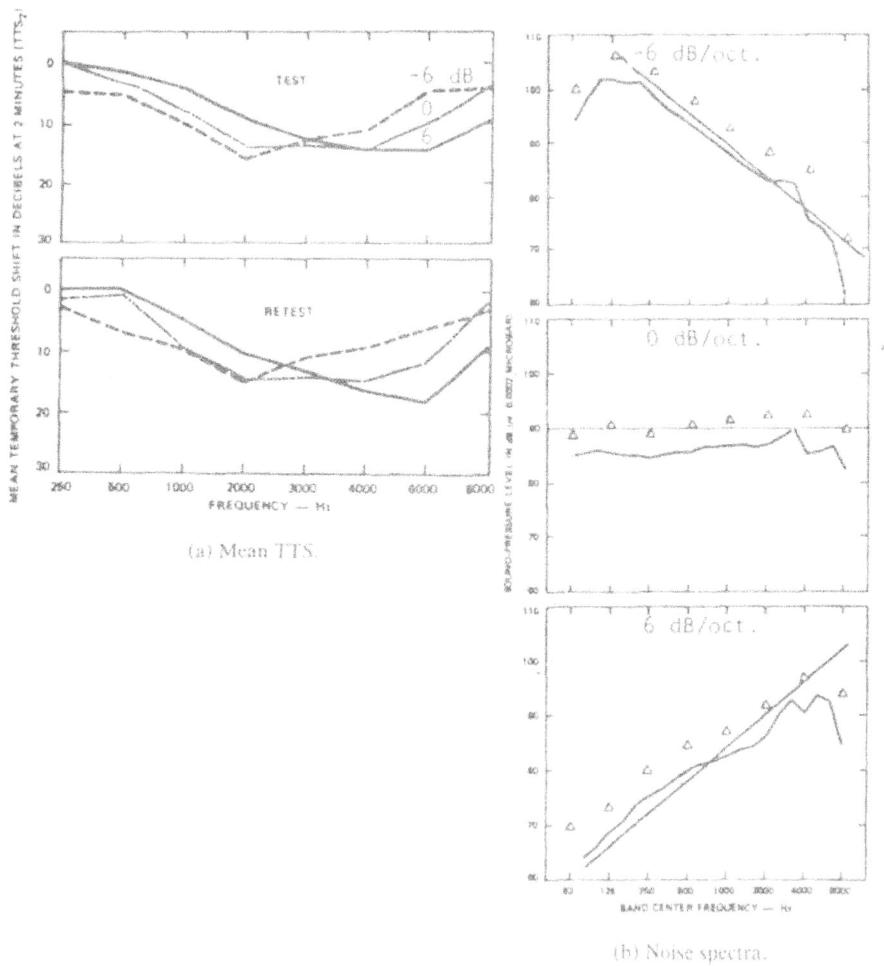

(a) Mean TTS.

(b) Noise spectra.

FIGURE 7.11. Mean temporary threshold shift in hearing level for different tonal frequencies (corrected to 2 min) following exposure to three noise spectra. (From ref. 23.)

Robinson (ref. 24) and Martin (ref. 25) are impressed with the 4000 Hz dip in the pattern of audiometric hearing levels for men exposed to a seemingly wide variety of spectra of industrial noises equated in A-weighted decibels. The suggestion of Robinson is that TTS data, which show a maximum TTS at an audiometric test frequency one-half to one octave or so above the frequency of an octave-band noise stimulus, is not consistent with the rather universal 4000-Hz dip found in NIPTS. There is perhaps no such inconsistency when it is recognized that these industrial noises are of broadband spectra and that the energy in the octave bands centered around 1000 to 2000 Hz probably did not differ very much for a given level in A-weighted decibels of most industrial noises. For example, the difference in the energy for the octave band 1000 to 2000 Hz is but 1 to 3 dB for noises having 6- to −6-dB slopes per octave, the A-weighted levels are equal. (See Fig. 7.11(b).) According to the general finding that a threshold shift occurs most readily at 4000 Hz and that it is most readily caused by noise energy in the frequency region about one octave lower (1000 to 2000 Hz or so), it is likely that the somewhat similar pattern in audiometric threshold shifts for the industrial noises is because when they are about equal in terms of A-weighted level, they are also equal in energy in the frequency region of 1000 to 2000 Hz.

Also, as shown in figures 7.11(a) and 7.12, the patterns of TTS and NIPTS at different audiometric test frequencies are somewhat similar, the rising-spectra noises causing somewhat more TTS and PTS at the higher frequencies than do the falling-spectra noises when the noises are equated A-weighted decibel level. There is no question, of course, that the ear is more susceptible to both TTS and PTS in the frequency region around 4000 Hz.

Spectral differences in impulse noises

To relate the physical characteristics of so-called impulse noise to TTS and PTS has been difficult. Major difficulties are a lack of measurements of the rele-

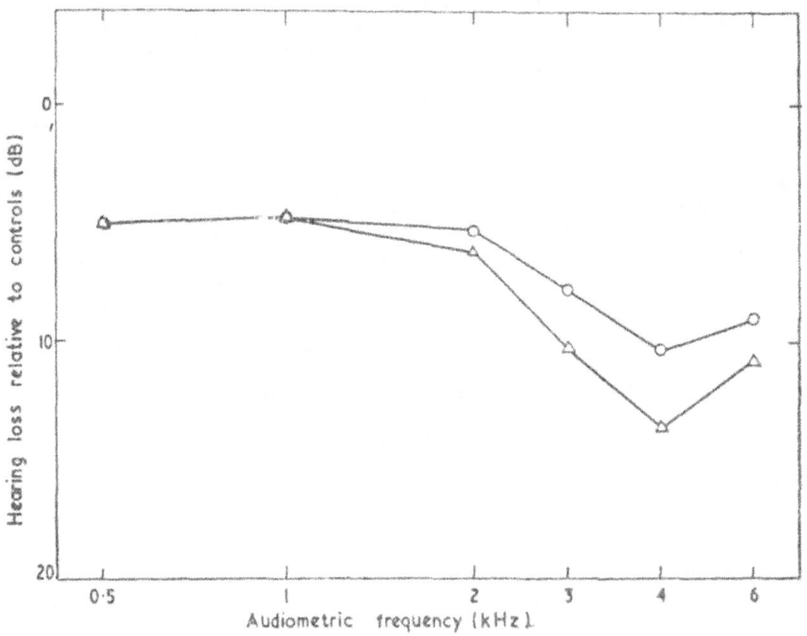

FIGURE 7.12. Shape of average audiogram for large groups. Upper curve is for rising spectra (489 subjects) and lower curve is for falling spectra (219 subjects). (From ref. 8.)

vant physical parameters of impulses and the relatively few (compared with steady-state noise) systematic studies of TTS and PTS from impulses. In addition, high-intensity impulses with proper rise times are difficult to generate under controlled laboratory conditions suitable for listening tests with humans.

Kryter (ref. 26) proposed that distinctions in the effects upon the cochlea commonly surmised to exist between impulsive and more steady-state sounds largely disappear under certain conditions as follows:

1. The spectra of impulses and steady noises are known and weighted in the same way, either on an octave-band or similar band basis or by an overall frequency weighting, such as A-weighting.

2. The sound energy is integrated over time periods similar to the apparently optimum processing time of the ear (about 1 sec) with respect to loudness and TTS.

Graphic means coupled with oscillograph traces or other means of estimating peak SPL, rise time, and duration can be used to find the spectra of impulses in spectrum level. (See figs. 7.13 and 7.14.) These spectra can, of course, be converted to overall A-weighted SPL in decibels, octave band, or other levels. Figures 7.13 and 7.14 are used as follows:

1. Use figure 7.13 with the rise time T_R and duration D of an impulse to find the frequency f_b (from T_R) of the point at which the slope of the spectrum "breaks" from -6 to -12 dB per octave or the frequency f_p (from D) at which the spectrum reaches its peak intensity.

2. Use these two frequencies in figure 7.14 according to the type of impulse involved in order to find the spectrum of the impulse.

A more practical procedure would seem to be to use a standard sound level meter (SLM) A-weighted with slow-meter action. However, an SLM does not provide an energy integration process that approaches that for the ear, as revealed by tests of loudness and TTS for impulsive noise. The A-weighting, as noted earlier, approaches the frequency response characteristics of the ear with respect to loudness judgements and TTS, but unfortunately the filtering and dynamics of a standard SLM integrate sound energy on an exponential and not on a linear basis as the ear apparently does. (See chapter 5.)

Atherley and Martin (ref. 27) proposed a procedure for converting oscilloscopic traces, or peak levels of impulsive sounds, to equivalent A-weighted levels. Their formula for this procedure is as follows:

$$L_{A,\mathrm{eq}} \cong 85.4 + 20\log(P_h) + 10\log(N_a) - 10\log(k) + 10\log(1 - e^{-2k/N})$$

where P_h is peak sound pressure in pascals, N_a is average repetition rate per working day, k is the decay constant in reciprocal seconds (equal to $1/t_e$, where

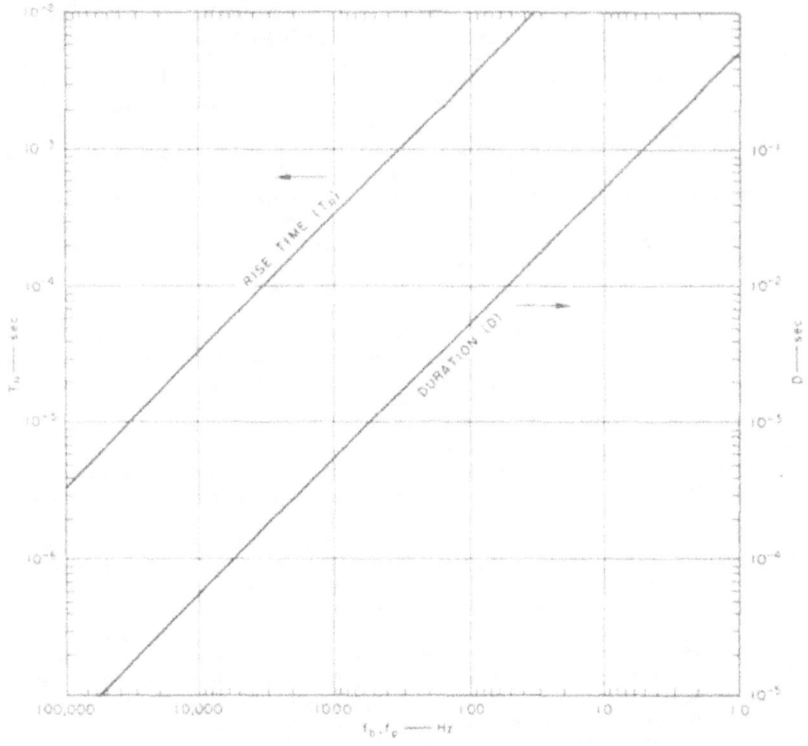

FIGURE 7.13. Frequencies as functions of duration and rise time of an impulse. (From ref. 26.)

t_e is the time in seconds taken for the pressure envelope to decay to $0.37\,P_h$), and N is number of impulses per second.

This can also be written as

$$L_{A,\mathrm{eq}} \cong 85.4 + 20\log(P_h) + 10\log(d_t) + \log(1 - e^{-2/d_t})$$

where d_t represents the fraction of 1 sec that the peak sound pressure is between peak and 8.7-dB downpoint.

Rise time of the impulse is not considered in this formulation so that spectral differences among impulses due to differences in rise time are not reflected in the calculated level. Martin (ref. 25) makes the assumption that his procedure will be applied to impulses having a broad frequency spectrum like those from punch presses. This is probably a reasonable assumption for most industrial and perhaps military (gunfire) noise environments but probably not appropriate for spark gap, pulsed tones, and the like.

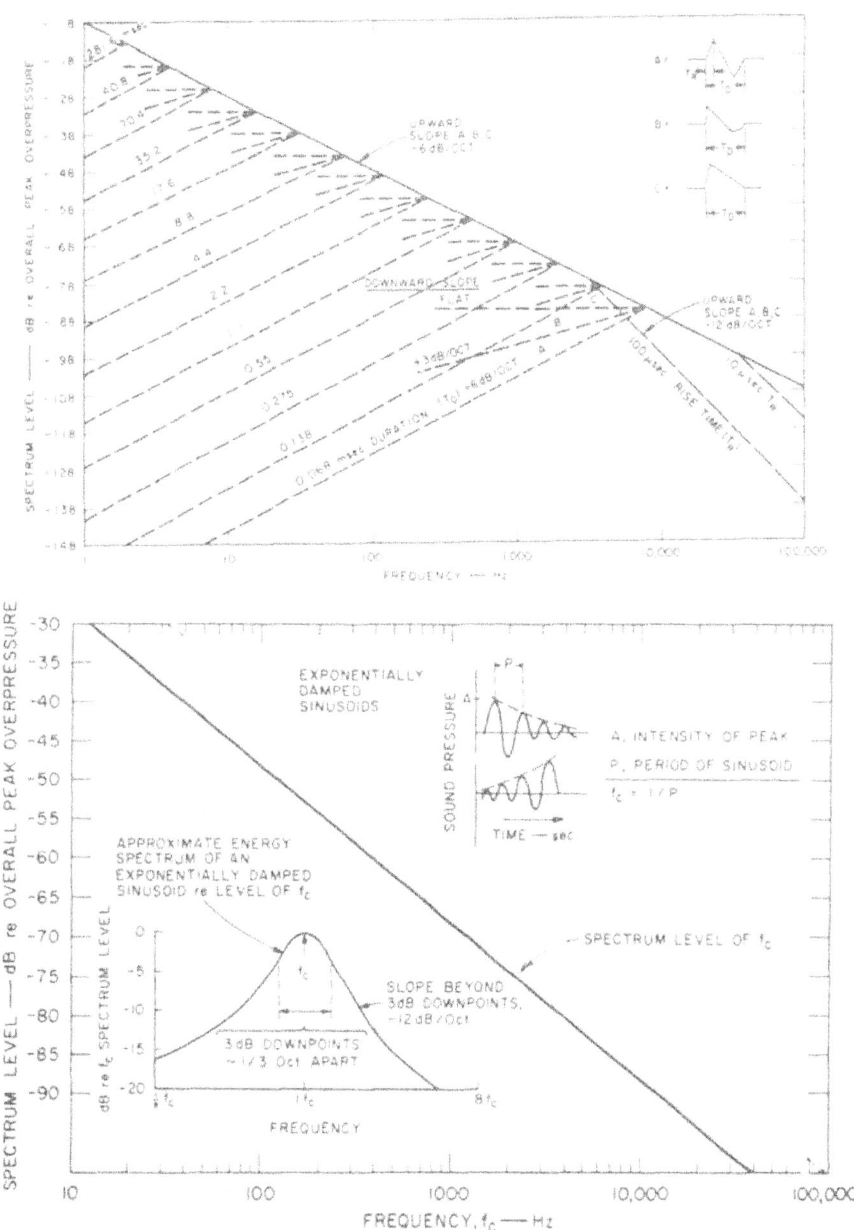

FIGURE 7.14. General spectrum level envelope of impulses having various waveforms and approximate spectrum level of impulses having waveform of an exponentially damped sinusoid. (From ref. 26.)

TTS from impulses of differing spectra

Carter and Kryter (ref. 28) and Hecker and Kryter (ref. 29) reported the acoustic waveform and the spectra of impulses (presented by means of a loudspeaker and an earphone) used in two different laboratory studies of TTS. (See fig. 7.15.) The measured spectra on figure 7.15 were determined by rather complex means whereby the waveform, as photographed from an oscilloscope, was periodically scanned electronically and the resulting modulated wave then passed through an envelope detection device and a band-pass filter. The procedures in figures 7.13 and 7.14 were used for obtaining an estimation of the spectra of the impulses, as is also shown in figure 7.15.

Kryter and Garinther (ref. 30) also reported the spectra and waveforms of gunfire noises. (See fig. 7.16.) These measured spectra were determined with the same means and equipment used for determining the impulse spectra of figure 7.15, and the estimated spectra for these impulses were determined by the use of figures 7.13 and 7.14. There appears to be reasonable agreement for these studies between the spectra obtained by actual measurement and by estimation from the graphs in figures 7.13 and 7.14.

Interestingly, the frequencies at which maximum threshold shift from gunfire noises occurred (4000 to 6000 Hz; see fig. 7.17) are consistent with the observation that the maximum threshold shift for steady-state sounds occurs about one octave above the frequency band containing the greatest energy (around 2000 Hz; see fig. 7.16). Also, since the impulses in the Kryter and Garinther study are probably representative of all types of gunfire with respect to rise time and duration, the general shape of the sound spectra and the threshold shifts for a given peak SPL and number of firings should be similar for most guns.

The importance of the spectral distribution of energy in impulses to threshold shift is well illustrated by a study performed by Fletcher and Loeb (ref. 31). The impulses, generated by an electrical spark gap, had the waveforms shown in figure 7.18. The spectra of these impulses were estimated with the graphic procedure of figures 7.13 and 7.14. Figure 7.18 shows that the spectrum of the longer duration impulse peaks at about 5000 Hz and at about 13 000 Hz for the impulse for shorter duration. The pattern of TTS (shown in fig. 7.19) is what one would predict, namely, the maximum threshold shift occurs at a higher frequency than the peak of the spectrum, and the greater the band levels the greater the amount of TTS.

SPL and growth of TTS

The growth of TTS as a function of SPL follows a pattern very similar to that followed by NIPTS in the Baughn and the Burns and Robinson data. (See fig. 7.6.) That is, about 0.5 dB TTS or NIPTS occurs per 1 dB SPL for exposures that cause up to 10 dB TTS or NIPTS, and about 1 dB TTS or NIPTS occurs per

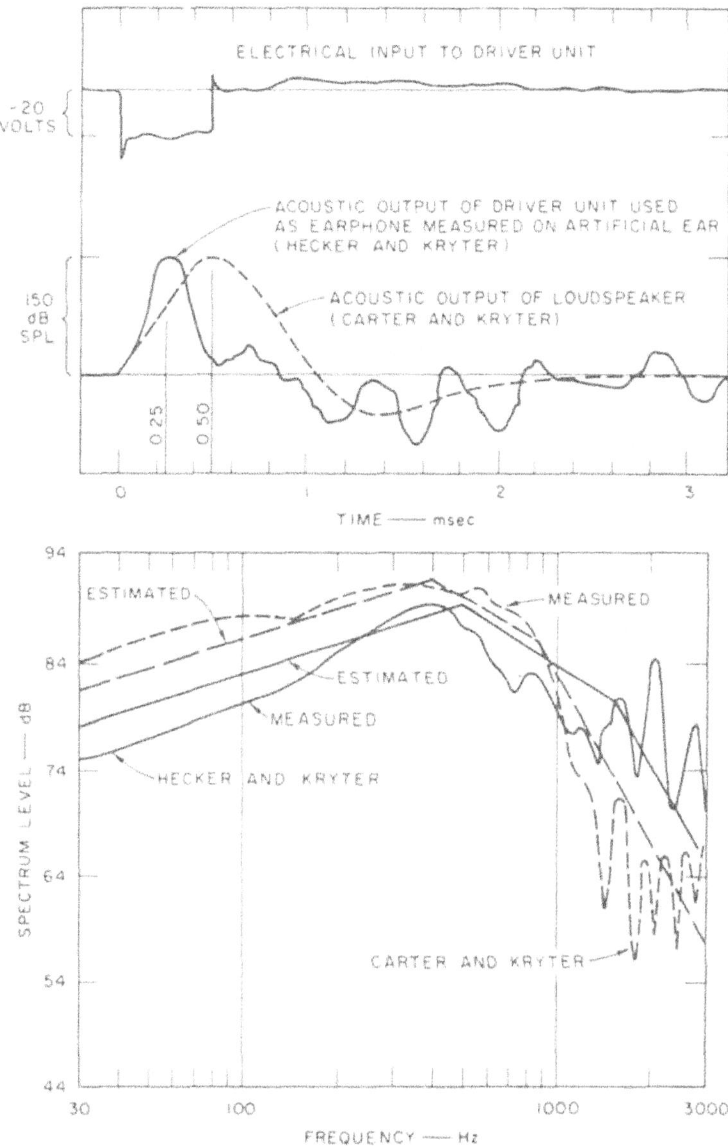

FIGURE 7.15. Waveforms and measured and estimated spectra of impulses used in some studies of TTS from impulses. (Data from Carter and Kryter (ref. 28) and Hecker and Kryter (ref. 29).)

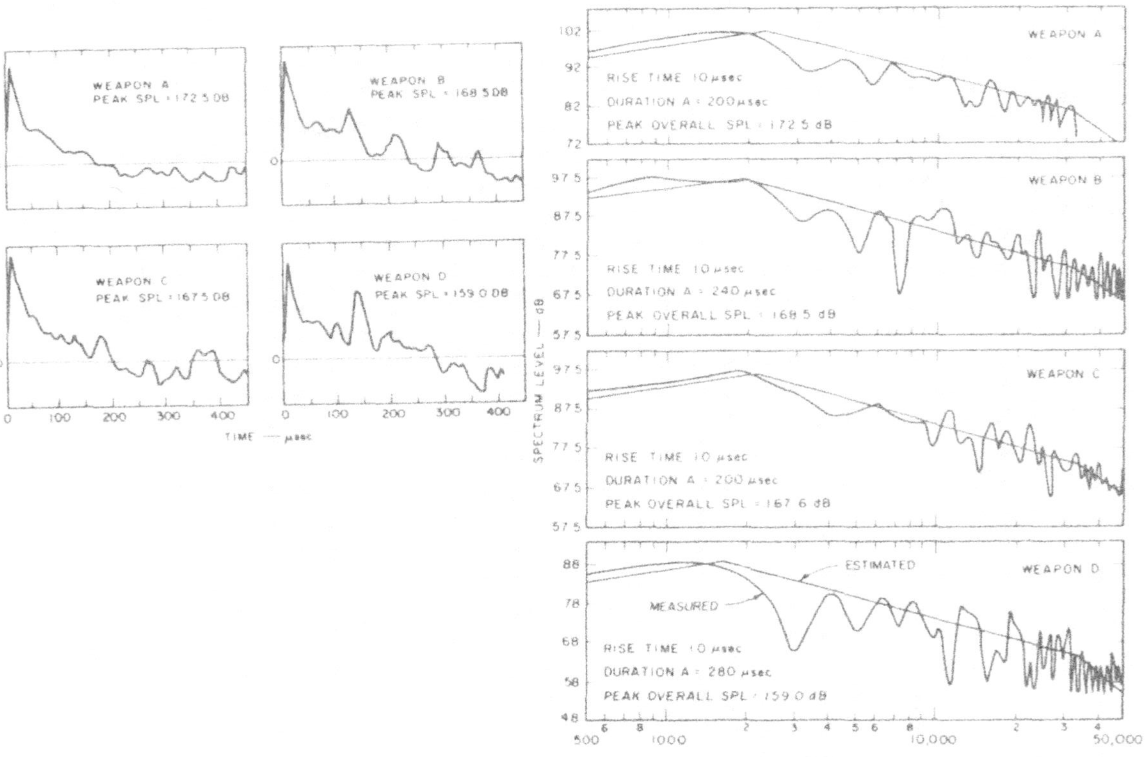

FIGURE 7.16. Measured and estimated spectra of gunfire impulses in reference 30. Pressure-time waveforms are shown in left hand panels. Peak levels of the waveforms have been adjusted to have approximately equal peak amplitude. (From ref. 30.)

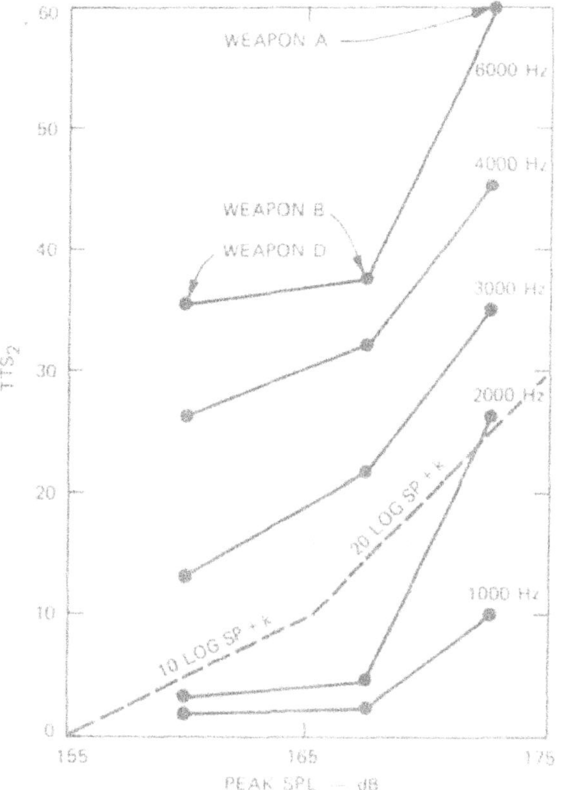

FIGURE 7.17. TTS_2 for the 75th percentile as a function of peak SPL for 100 trigger pulls at the rate of one every 5 sec. (Data from ref. 30.)

1 dB SPL for exposures causing 10 to 40 dB TTS or NIPTS. This is shown in figure 7.20 (refs. 32 and 33).

The data in the lower graph of figure 7.20 are not consistent in magnitude with data obtained with somewhat similar noise exposures by Schori (ref. 34). The difference is probably attributable to the use of earphones by Schori and loudspeakers by Brownsey (ref. 33) with the attendant problems of calibrating noise levels in terms of real-ear, open-field listening. Even so, Brownsey's data appear to be somewhat higher than would be expected from the noise levels indicated. However, the relative changes in TTS because of level and duration changes should be realistic.

Impulsive noises often have a characteristic other than briefness that sets them apart from many common bursts of more steady-state industrial noises. That is, they reach peak levels that are considerably more intense than the peak levels of steady-state noise. Dieroff (ref. 35) estimates from his studies of indus-

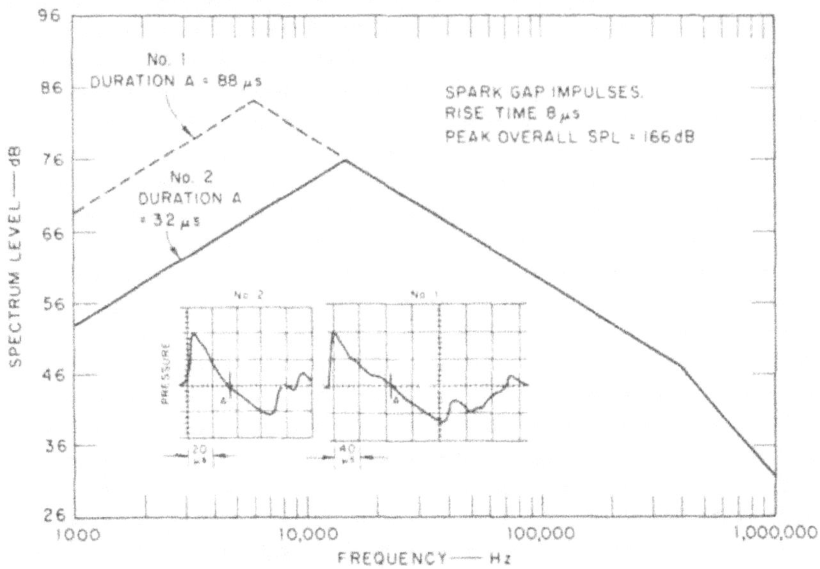

FIGURE 7.18. Waveforms and estimated spectra of spark gap used in reference 31. (From ref. 26.)

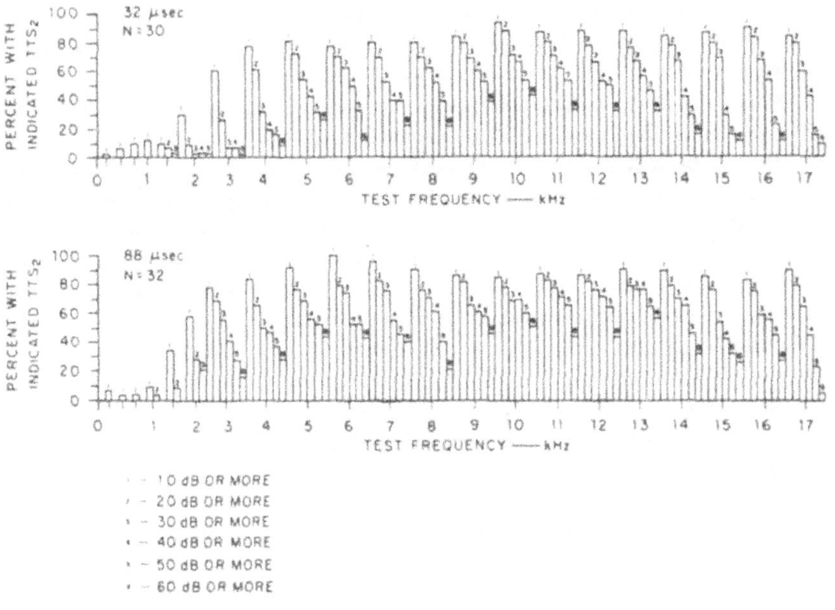

FIGURE 7.19 TTS from spark gap impulses. (From ref. 31.)

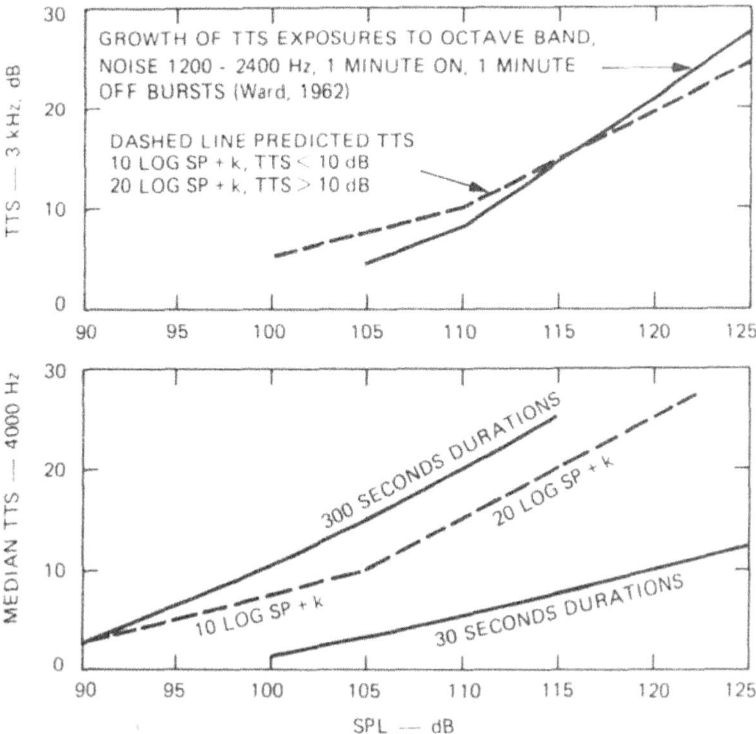

FIGURE 7.20 Growth of TTS as a function of SPL. Parameter is duration of noise. (Data of upper graph from ref. 32; data of lower graph from ref. 33.)

trial workers that there may be a critical peak level of about 150 dB above which damage to the organ of Corti progresses at a more rapid rate than would be expected from a strict energy equation.

Clearly some nonlinear acceleration of damage to the ear does occur at high pressure levels, as was shown in figure 7.17. Here the rate of growth of TTS was shown to be indeed quite large as a function of instantaneous peak pressures for the gunfire impulses involved (which had unusually fast rise times and were in free field at a level of 160 to 170 dB).

At these high sound pressures, especially for pulses with slow rise times, the eardrum may be ruptured, to some extent affecting measured HL's. However, such rupturing appears to provide some protection to the inner ear so that after the drum heals the HL often shows less shift than occurs in subjects with unruptured eardrums.

What stress reactions and physiological mechanisms involved in the inner ear are reflected in TTS from exposure to noise can only be definitely answered with further research. (See Henderson *et al.* (ref. 36) and Hamernik *et al.* (ref.

TABLE 7.3

TTS_2 Average at 1, 2, and 3 kHz as Found in or Estimated from Various Studies of Threshold Shift from Gunfire Noise

[1 to 10 sec between impulses; data from ref. 26]

Study	Peak SPL grazing to ear, dB	Listening condition	No. of rounds	TTS_2 equaled or exceeded by 25 percent of people, dB		No. of subjects
				Measured	Corrected[a]	
Coles and Rice (ref. 38)	160	Open field	10–50	7	19	20
Coles and Rice	160	Reverberant	10–50	10	22	20
Coles and Rice	159	Open field	20–50	7[b]	19	20
Coles and Rice	159	Reverberant	20–50	19[b]	31	20
Elwood *et al.* (ref. 39)	161	Open field	20	7[c]	20	12
Elwood *et al.*	173	Open field	1	7[c]	47	12
Acton *et al.* (ref. 40)	138	Open field	100	0	0	19
Acton *et al.*	138	Reverberant	100	5[d]	5	19
Murray and Reid (ref. 41)	159	Open field	100		20[e]	[f]
Murray and Reid	176	Open field	10	5	27[g]	1
Murray and Reid	181	Open field		17	48	1
Goldstone and Smith (ref. 42)	158	Open field	25	5	17	30
Kryter and Garinther, weapon D (ref. 30)	159	Open field	100	10	10	30
Kryter and Garinther, weapon B	168	Open field	100	22	22	36
Kryter and Garinther, weapon A	173	Open field	100	55	55	8

[a] *Corrected to 100 rounds by adding 20 times the common logarithm of the number of rounds.*

[b] *TTS data were not given at 1 kHz: the average for 1, 2, and 3 kHz was taken to be the TTS at 2 kHz.*

[c] *Data were given as TTS average at 2 to 6 kHz; this average multiplied by 0.3 was taken as TTS average at 1, 2, and 3 kHz.*

[d] *Authors stated that "noise approached an auditory hazard for about 5 percent of people."*

[e] *Estimated from authors' statement that 159 peak SPL's "commonly" (taken to mean in 50 percent of people) caused 40 dB Peak TTS after 100 rounds or more.*

[f] *Not known.*

[g] *Average 512–8192 Hz TTS_{15} corrected to TTS_2 by adding 10 dB and then to average TTS_2 at 1, 2, and 3 kHz by multiplying result by 0.5.*

37).) By and large, except for unusual circumstances in industry and for the firing of some guns, impulsive and steady-state noises are such that TTS_2 (and eventual NIPTS) is typically 40 dB or less.

Table 7.3 presents a summary of much of the data (refs. 30 and 38 to 42) that can be used to show (or estimate) the average amount of threshold shift at 1000, 2000, and 3000 Hz for 100 gunfire noise impulses with 1 to 10 sec between impulses. In order to interpret the data from the various studies, conversion of the threshold shift data to a common set of exposure conditions is necessary. The rules for doing so are given in the table. Although the validity of such conversions may be questioned, their use presumably permits a more meaningful and effective use of the data than is otherwise possible. Most of the data given in table 7.3 were also used in a comparative analysis by Smoorenburg (ref. 43) of TTS from impulse noise. Smoorenburg's findings and conclusions in his analysis of these data and concepts related to TTS and PTS from exposure to noise are consistent with those reached here.

Plotted on figure 7.21 are data from a number of TTS studies in which SPL was a variable. The following table gives the references and pertinent information for the curves in figure 7.21.

Curve in fig. 7.21	Reference	Frequency at which TTS measured, kHz	Frequency of noise, kHz	Duration of noise, minutes
1	44	4	2-4	55
2	45	4	1.2-2.4	480
3	46	4	1.2-2.4	47
4	47	4	(a)	20
5	48	4	(b)	—
6	49	4.6	(a)	12
7	50	2-4	2	2
8	51	4	(a)	3
9	50	0.5, 1, and 2	0.5	24
10	50	4-8	4	2
11	52	3.4	1.4-2.8[b]	1
12	53	4	(c)	75
13	30	1, 2, 3, and 4	(d)	—
14	29	2 and 4	—	(e)
15	41	0.5-8	(f)	—

[a] *White noise.*
[b] *Pulsed tone.*
[c] *Clicks.*
[d] *Gunfire, 100 rounds.*
[e] *100 impulses.*
[f] *Gunfire, 10 rounds.*

Although there are few data points for TTS less than 10 dB, it is proposed that the relations 10 log SP + *k* for these lower values and 20 log SP + *k* for higher TTS values are in keeping with the probable course of TTS values as a function of SPL. Kryter (ref. 26) previously suggested that 20 log SP + *k* was appropriate for all TTS values of the data shown in figure 7.21. The more recent data shown in figures 7.17 and 7.20 also indicate that the two segment relation (10 log SP + *k* and 20 log SP + *k* depending on TTS level) is more appropriate. Furthermore, as mentioned previously, these general relations between TTS and sound pressure level are the same as those shown earlier in the text for NIPTS and sound pressure level. (See fig. 7.6.)

Threshold level for TTS

As mentioned previously, it appears that the noise that causes 5 dB TTS_2 in the young, normal ear after 8 hours exposure will cause the same amount of NIPTS after about 10 years of near daily exposure. This would extrapolate to about 12 dB according to the NIPTS data of Baughn (refs. 6 and 7) and of Burns and Robinson (ref. 8), with 45 years rather than just 10 years of exposure. Also, it is to be expected that the ears more susceptible to threshold shift (*e.g.*, the 90th percentile) would have about 7 dB more NIPTS than the median or mean of the distribution of people. Accordingly, to achieve a 5-dB NIPTS for the 90th-percentile, 45-year exposure, the noise would have to be about 10 dB less than the noise which gives a TTS_2 of 5 dB for the average (50th percentile) of young, normal ears. Robinson and Sutton found (see table 8 of ref. 11) that median HL's (50th percentile) are generally about 1 or 2 dB lower than mean HL's for the same distribution. Be-

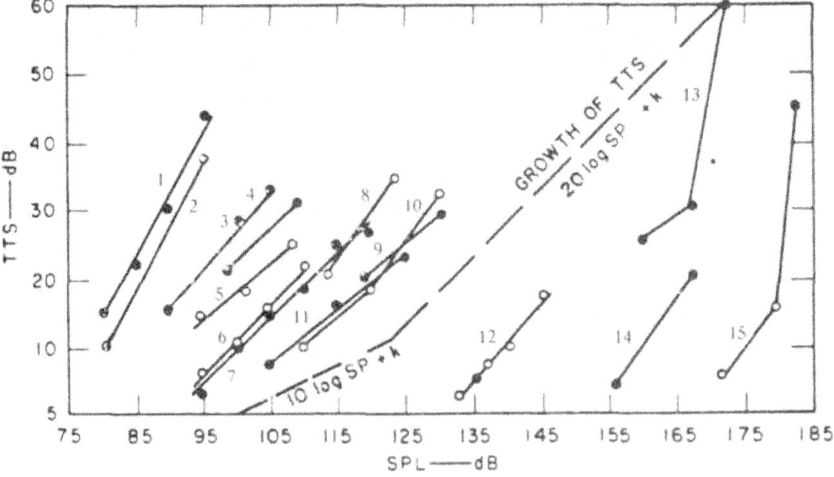

FIGURE 7.21. Growth of TTS as a function of the sound intensity. (Data from ref. 26.)

cause of the variability of this relationship as a function of test frequency and the generally small magnitude of the difference, means and median data will be taken as equal for most purposes in this paper. The difference between the median and the 90th percentile for normal ears is generally about 7 dB for TTS (Kryter *et al.*, ref. 54) and about 10 dB for NIPTS. (See fig. 7.3.)

With these extrapolations, it can be deduced from a study by Ward *et al.* (ref. 55) that a broadband noise at about 63 dBA will cause persons in the 90th percentile about 5 dB NIPTS at 4000 Hz after 45 years of exposure for 8 hours per workday. Ward *et al.* found that the average TTS_2 for an 82-dBA "magenta" noise (noise with a spectrum slope of −5 dB per octave) was about 4 dB at 4000 Hz. (See fig. 7.22.) Allowing 1 dB additional TTS for extending the duration of the experimental exposure from 5 to 8 hours, we find that the maximum level for people in the 90th percentile to have no more than 5 dB NIPTS would be about 63 dBA (a maximum octave-band level, in the broadband noises, of about 58 dB). That is, a 7-dB reduction from the 82-dBA level to a level of 75 dBA would be required in order to prevent more than 5 dB TTS in the 90th percentile ear after an 8-hour exposure, or more than 5 dB PTS after 10 years of near daily exposure. An additional 12-dB reduction in level would be required to produce no more than 5 dB PTS after 45 years of near daily exposure, further reducing the required level from 75 dBA to 63 dBA.

These findings are quite consistent with earlier data and extrapolations made of TTS octave-band exposure (ref. 56) and NIPTS (refs. 2 and 57) if we take

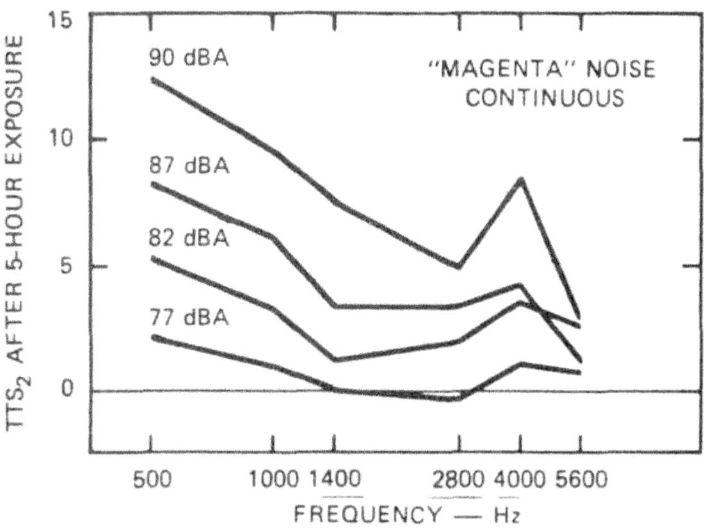

FIGURE 7.22. Average TTS_2 in decibels produced at several test frequencies by 5 hours continuous exposure to "magenta" noise. Parameter is overall A-weighted sound level. (From ref. 55.)

into account information regarding the range of hearing sensitivities among normal ears and the relatively slow growth of TTS from 0 to 10 dB. Rossi *et al.* (ref. 58) found that a 2-hour exposure to noise (recorded road traffic) at a level of about 67 dBA caused some measurable TTS at 2000 Hz. All these findings are consistent with studies of effective quite.

Effective Quiet

Various investigators have studied the effect on TTS of exposures to rather intense noise that was interrupted with periods of less intense noise, including quiet. The presumption was that when the less intense noise failed to cause a decrease in TTS from the more intense noise its level was at effective quite, a level that presumably could be continued indefinitely without effecting any TTS or PTS.

However, such an assumption must be qualified to some extent, as shown in figure 7.23. Here effective quiet is taken as the maximum level of noise present

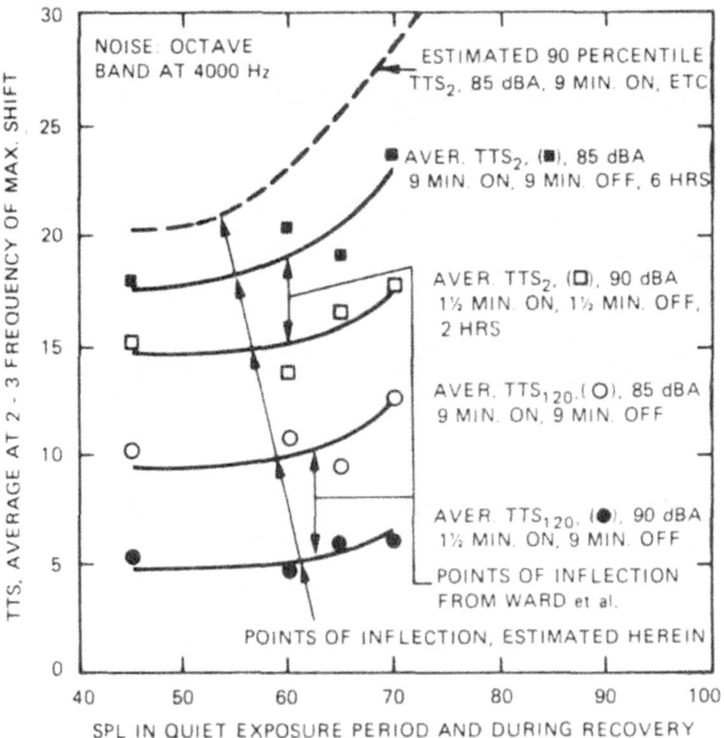

FIGURE 7.23. TTS as a function of SPL of lower level of noise alternated with higher levels of noise. (Data points from ref. 55.)

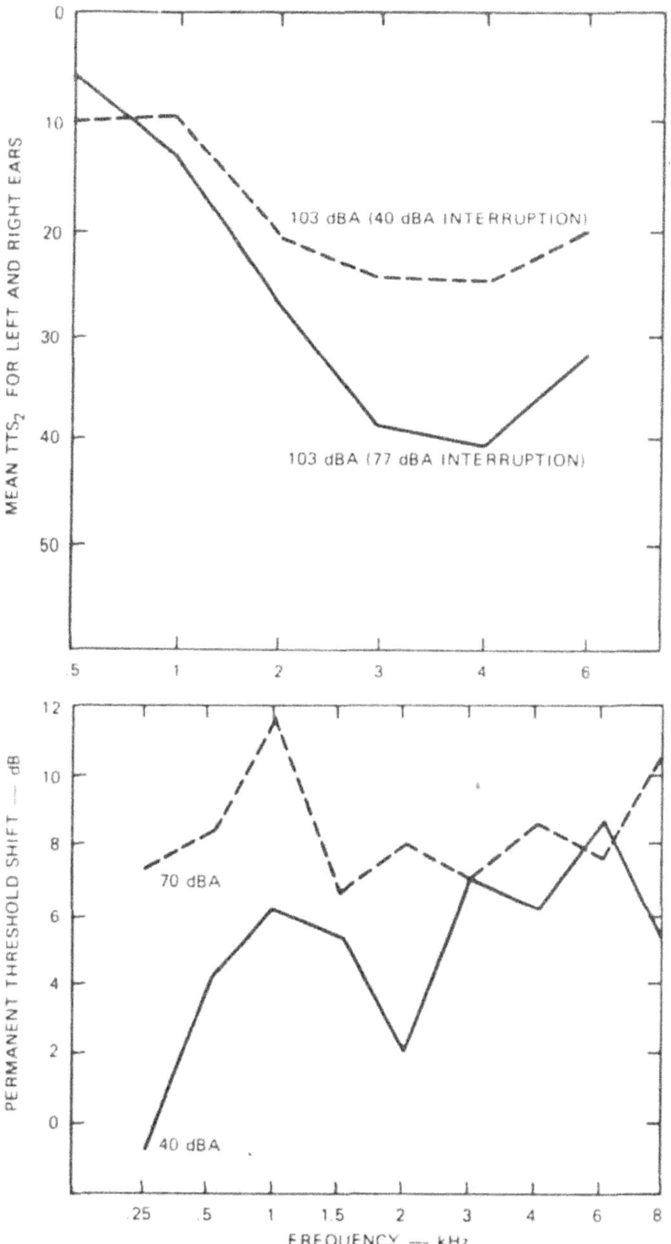

FIGURE 7.24. TTS in humans (upper graph, from ref. 60) and PTS in chinchillas (lower graph, from ref. 61). Humans were exposed for 2 hours to 15-minute bursts of noise at 103 dBA with 2-minute interruptions at 40 or 77 dBA. Chinchillas were exposed for 8 hours to 110 dBA noise followed by 7 days recovery in 70 dBA and 40 dBA.

TABLE 7.4

TTS From Exposure to Steady-State and Impulse Noises

[15 male subjects; maximum energy, 500 to 1000 Hz; data from ref. 62]

Exposure	Median TTS, dB, for test frequency, Hz, of —					
	1000	1500	2000	3000	4000	Average
[a]*I*, 124–127 dB	8.0	8.0	6.0	5.0	3.0	6.0
I + [b]*S*, 110 dB	4.0	9.0	9.0	5.0	7.0	6.8
I + *S*, 100 dB	2.5	4.0	5.5	2.5	1.5	3.2
I + *S*, 90 dB	1.5	4.5	3.5	2.0	1.5	2.6
S, 110 dB	5.0	11.0	9.0	5.0	3.0	6.6
S, 100 dB	3.0	7.0	7.0	3.0	1.0	4.2
S, 90 dB	1.0	3.0	3.0	1.0	0	1.6

[a] *Impulse.*
[b] *Steady state.*

during the "quiet" period of exposure and during recovery without causing any increase in TTS from the 85-dBA and the 90-dBA noise. It is shown that the level at which the noise becomes as effective as quiet appears to be somewhat lower as the amount of TTS generated by the higher level noise becomes greater. I have added an estimated function for the 90th percentile of the population of subjects as well as the curves to the data points of figure 7.23. As shown, Ward *et al.* (ref. 55) estimated somewhat higher levels for effective quiet. The differences are not great, and allowing 5 to 10 dB for differences in ear sensitivity from the average TTS data points shown, it appears that effective quiet is somewhere between 50 and 60 dB for the 90th percentile of the population for this octave band. (This correlates with a broadband overall noise level of about 55 to 65 dBA.)

Hetu *et al.* (ref. 59) reported that an ambient level of 50 dBA was effective quiet for a study in which 13 dB TTS_2 was produced and that 70 dBA was effective quiet for a study in which 26 dB TTS_2 was produced. This trend appears to be somewhat at odds with the trend noted above for the data of Ward *et al.* in which the level required for effective quiet decreased as TTS increased. Schmidek *et al.* (ref. 60) (using human subjects) and Lipscomb (ref. 61) (using chinchillas) found that effective quiet was somewhere between 40 and 70 dBA. (See fig. 7.24.)

That further research on effective quiet is required is indicated by the findings of Cohen *et al.* (ref. 62) as shown in table 7.4. The 6-dB TTS from 15 minutes of exposure to impulses was reduced to about 3 dB when a 500- to 1000-Hz octave band of noise was also present at levels of 100 and 90 dB. The steady-state noise at a level of 110 dB resulted in a TTS of about 7 dB whether the impulse noise was present or not. The decrease in TTS with decrease in level of the steady-state noise is, of course, to be expected. However, that the TTS from the impulse noise presented alone should be greater in the quiet than in the 90 and 100 dB steady-state noise is seemingly anomalous; it is probably related to ubiquitous operation of the aural reflex.

Effects of Exposure and Recovery Times on TTS and NIPTS

Effective level of bursts of noise

Effective "quiet" and effective "level" are descriptors for the reaction of the auditory system to interrupted noise exposures within the typical 8-hour exposure day. The contentions of Burns and Robinson (ref. 8), Robinson (ref. 24), Atherley and Martin (ref. 27), and Martin (ref. 25) have been that these effects are minimal and can, for practical purposes, be ignored in estimating eventual NIPTS from daily dosages of noise. Their position is that the total amount of A-weighted noise energy present during the 8-hour day is adequate for predicting NIPTS regardless of whether it is continuous or interrupted, that is, the so-called equal-energy concept.

Martin concludes that data showing less TTS from interrupted impulse noises than to be expected on an equal-energy basis must mean that there is no predictive relation between TTS and NIPTS. However, as shown later, probably no such inconsistencies exist when you consider certain characteristics of the ear in response to brief bursts of noise.

Ward (ref. 63) presented data showing the relative effectiveness of intermittent noise of equal energy as a cause of TTS_2 after a total exposure period of 8 hours. Some of his data are plotted in figure 7.25. According to the all-pervasive equal-energy hypothesis, the TTS should have been the same for all the data points in figure 7.25. The vertical ordinate shows how many decibels the total exposure energy would have to be reduced to make the various on-off fractions and cycle durations equally effective in causing TTS_2 after 8 hours of exposure to the different conditions.

For figure 7.25, the results found by Ward for the 1000-Hz and 4000-Hz octave bands have been averaged, even though the 1000-Hz band was somewhat less effective in causing TTS_2 on an equal-energy basis than the 4000-Hz band. However, as shown, the 250-Hz band was much less effective than the higher frequency bands and about equally so for all exposure-cycle periods tested. The reason for this lesser effectiveness of the lower frequency bands is perhaps related to the greater response of the aural reflex to lower frequency bands of noise.

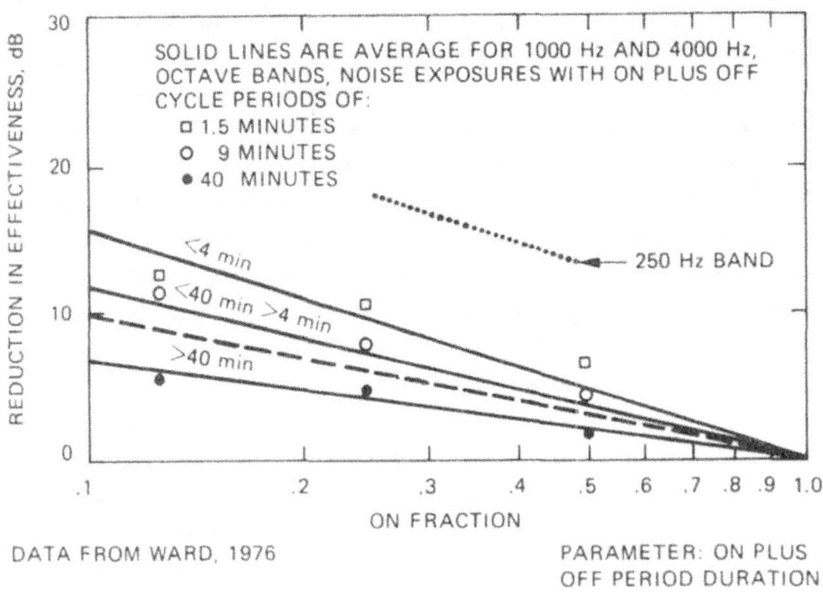

FIGURE 7.25. Effectiveness of octave bands of noise in causing TTS at test frequencies most affected by a particular band as a function of on-fraction (ratio of on time to total time). Total noise exposure duration of 8 hours. (Data from ref. 63.)

This factor indicates the appropriateness of giving less weight, as does A-weighting, to lower frequency sound than to higher frequency sound for the assessment of potential noise-induced threshold shifts. (The A-weighting gives about a 10-dB negative weight at 250 Hz compared with that given at 1000 Hz.) It would seem fairly realistic to use the solid-line dashed-line functions shown in figure 7.25 as a means of correcting equal-energy noise-dosage levels of intermittent noises based on A-weighted SPL's in the prediction of threshold shifts from exposure to noise.

These data taken in conjunction with some PTS data and some data on recovery from TTS provide factual support for a concept called equivalent recovery from daily noise exposures. A simple procedure for equating continuous 8-hour noise with intermittent or short-duration noises is presented later. The dashed line in figure 7.25 is used in that procedure.

NIPTS from irregular noise

Johansson *et al.* (ref. 64) studied NIPTS and TTS from exposures to industrial noises having continuous and intermittent characteristics. (See fig. 7.26.) They found that for both NIPTS and TTS there was less shift from exposure to the irregular noise than from the continuous noise (both had the same spectrum) when the two were equated for equal energy during daily exposure. The difference was equivalent to about 5 dB in effectiveness as a cause of NIPTS.

Figure 7.26 shows that the irregular noise varied between levels of about 105 and 75 dBA with periods of the more intense level being about half as long as the periods at the lower level. This is in general agreement with the data in figure 7.25 showing that a noise with an "on-plus-off" duration of more than 10 sec and an on-off fraction of 0.4 is about 6 dB less effective as a cause of TTS than would be expected purely on the basis of the amount of noise energy involved. Figure 7.27 illustrates some other types of industrial noises involving various durations of intermittent or irregular noises (refs. 4 and 65).

Effective level of impulses

The temporal spacing between impulses also influences the threshold shifts experienced. The tolerable exposures proposed for gun impulses usually presume, or are at least consistent with, intervals of 2 to 10 sec between rounds. When the interval between impulses or bursts of noise become less than about 5 sec, the TTS does not increase and sometimes decreases.

Also, when the interval between impulses increases beyond a certain point, the recovery from auditory fatigue between impulses permits a net decrease in TTS for a given total number of impulses or energy. There is, however, conflicting data as to what is the exact interval at which recovery from threshold shift first starts. For example, Ward *et al.* (ref. 53) found in one study that impulses separated by slightly more than 2 sec caused less TTS than when the interval was less

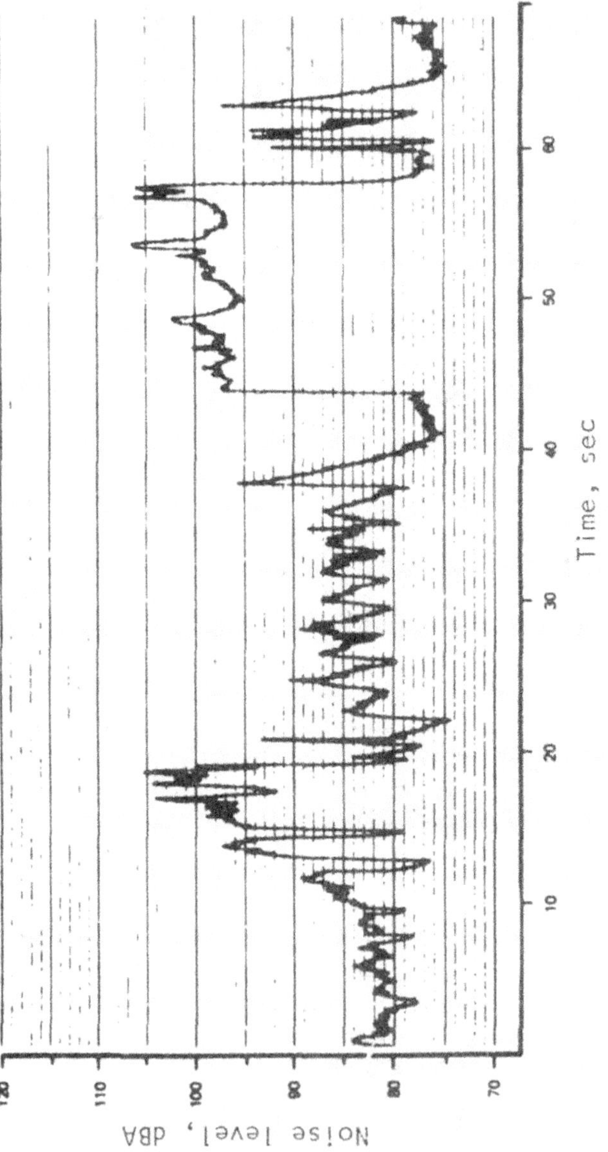

FIGURE 7.26. Example of irregular noise present in a steel welding plant and used in laboratory tests for TTS. (From ref. 64.)

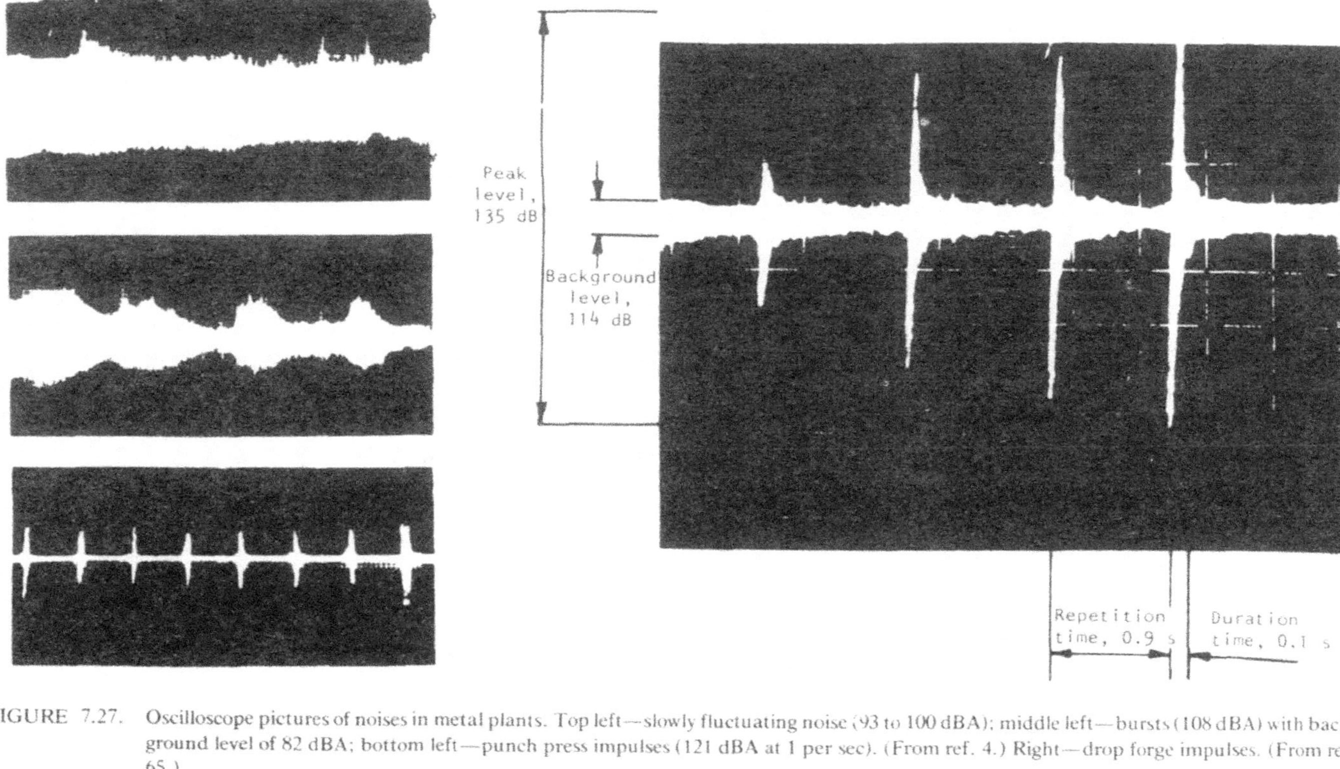

FIGURE 7.27. Oscilloscope pictures of noises in metal plants. Top left—slowly fluctuating noise (93 to 100 dBA); middle left—bursts (108 dBA) with background level of 82 dBA; bottom left—punch press impulses (121 dBA at 1 per sec). (From ref. 4.) Right—drop forge impulses. (From ref. 65.)

than 1 sec but in a second study with the same impulses, Ward (ref. 32) concludes that intervals between impulses as long as 9 sec do not influence TTS.

Most of the evidence (fig. 7.28) suggests that perhaps only after about 10 sec of relief from intense auditory stimulation does the ear start recovering from auditory fatigue. That is to say, the ear is physiologically refractory for 5 to 10 sec following a sufficiently intense stimulation.

The decrease in TTS shown in figure 7.28 for a given number of impulses or noise bursts when the interval between them is less than about 1 sec is possibly related to sustained action of the aural reflex. But also the aforementioned approximate 1-sec integration time of the ear means that the effective energy in very brief impulses or bursts of impulses is reduced relative to the instantaneous peak level of the impulses. Indeed, the data in figure 7.28 for interruptions of less than 1 sec indicate that for TTS this "time constant" of the ear is about 0.5 seconds rather than 1 sec. However, for the sake of simplicity and consistency with present-day noise measurement techniques, a 1-sec integration time period is presumed for the assessment of various effects of noise on people, including damage to hearing.

PTS in laboratory studies

Henderson and Hamernik (ref. 66) conducted a series of studies with chinchillas in which behavioral, electroneural, and anatomical measures of the effects of noise on hearing and on the ear were made. These studies provide data on effective quiet, effect of different durations of intervals between impulses, and applicability of the equal-energy concept to the assessment of a 1-day exposure to short-duration, non-steady-state noise.

Figure 7.29 shows that with 60 sec between impulses little PTS is caused from 50 impulses, but that with 10 seconds between impulses large shifts are found. This difference can to some extent be attributed to the greater time available for some recovery from the auditory insult caused by the impulses to take place. The results are somewhat like those found for the TTS with humans and underscore the inadequacy of "equal energy" of interrupted exposures to noise as a cause of TTS and NIPTS.

Impulses plus continuous noise

An apparently similar result was also demonstrated in another experiment by Henderson and Hamernik (fig. 7.29). Here we see that a continuous noise and an impulsive noise (50 impulses at 1 per minute) caused about 0-dB PTS when either was presented alone but caused 10 to 40 dB when both were presented together. Henderson and Hamernik attribute this result to a synergistic effect of the two noises on PTS. An alternative explanation for at least some of this effect which is consistent with aforementioned results of studies of effective quiet and equivalent recovery is that the recovery from auditory fatigue or damage to be ex-

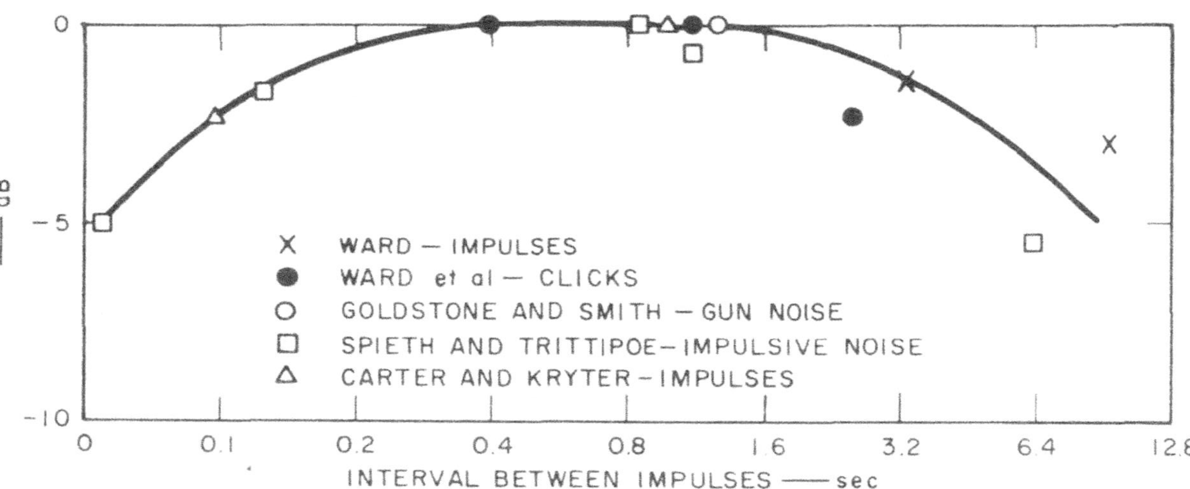

FIGURE 7.28. TTS for a given number of impulses with intervals shorter or longer than 1 sec between impulses minus TTS for same number of impulses with 1-sec intervals. (From ref. 26; data taken from Ward (ref. 32), Ward *et al.* (ref. 53), Goldstone and Smith (ref. 42), Spieth and Trittipoe (ref. 47), and Carter and Kryter (ref. 28).)

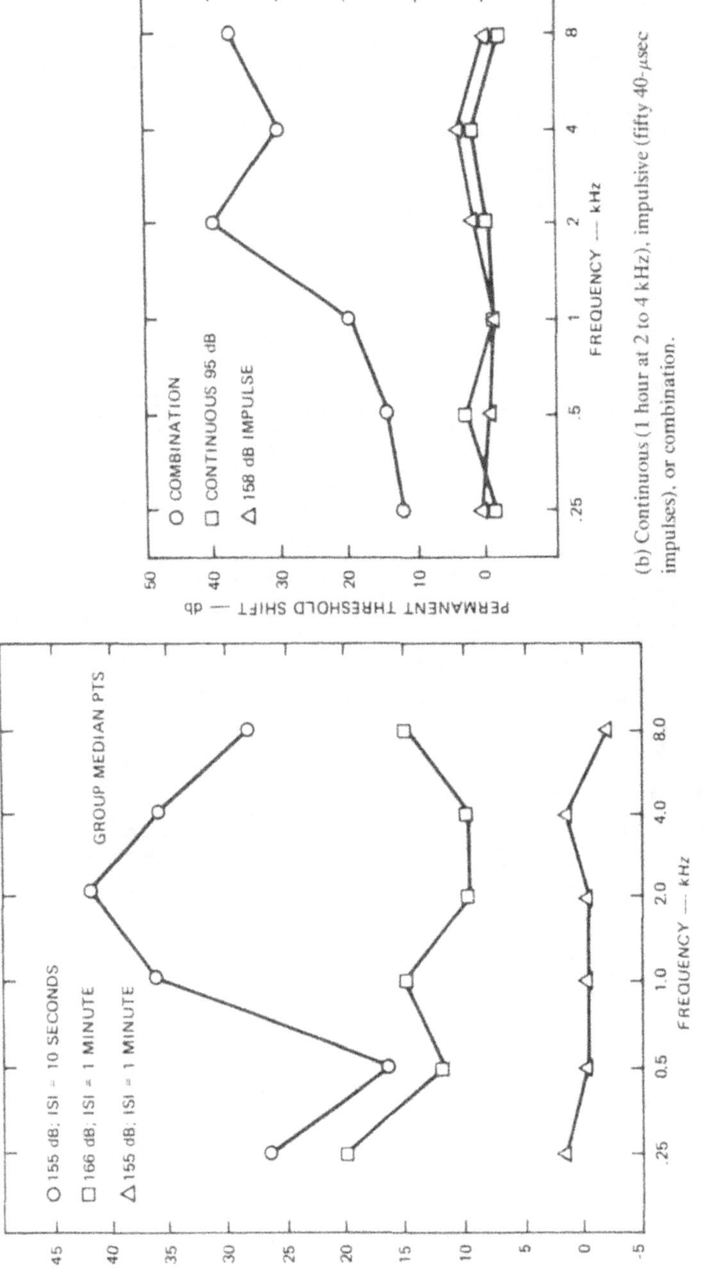

(a) Fifty 1-msec impulses at varying interstimulus intervals (ISI's).

(b) Continuous (1 hour at 2 to 4 kHz), impulsive (fifty 40-μsec impulses), or combination.

FIGURE 7.29. Median ($N = 5$ chinchillas) PTS after exposure to impulsive and continuous noise. (From ref. 66.)

pected in the minute between each impulse is prevented by the presence of continuous noise at a fairly high level of intensity. Note that the absolute levels in decibels of these impulses and this continuous noise are not directly indicative of their relative energy levels because of differences in their spectra and durations.

Impulses in industry

Sulkowski *et al.* (ref. 67) found that the Burns and Robinson (ref. 8) and the Atherley and Martin procedure (ref. 27) tended to give overestimates of the PTS from impulsive noise for frequencies above 2000 Hz for less than about 15 years of exposure and to give underestimates of that found for greater than about 15 years of exposure. At test frequencies below 2000 Hz PTS was generally underestimated. Sulkowski *et al.* note that the impulses had broadband spectra. Possibly, the ubiquitous aural reflex, with its greater protective action for low frequencies than for frequencies above 2000 Hz, is involved in these interesting and somewhat complex findings.

Martin (ref. 25) calculated the A-weighted energy present in impulsive noises in drop-forge factories studied by Guberan *et al.* (ref. 68), by Atherley and Martin (ref. 27), and by Ceypek *et al.* (ref. 65). (Fig. 7.27, from Ceypek *et al.*, fairly represents the temporal nature of the impulses involved in these studies.) Martin found that the NIPTS reported for otologically normal workers not exposed to loud noise for these three studies was in close agreement with NIPTS predicted by the Burns and Robinson equal-energy calculation procedure developed for steady-state noise and discussed earlier. On the basis of this finding, Martin (ref. 25) and Robinson (ref. 24) conjecture that the results of studies showing the efficacy of the interruptions in exposures to bursts of noise and impulses on recovery from TTS were not indicative of what the effects would be with respect to permanent threshold shifts. Indeed, these authors propose that the equal-energy concept is sufficiently well demonstrated that standardization for the prediction of NIPTS from all types of industrial noises is justified. However, for these impulse noise conditions in which the impulses occurred at a rate of about 1 per second to 1 per 5 sec, one would predict, from TTS data (fig. 7.28), that no recovery would occur from impulses with such a brief recovery period between them.

In short, the PTS found in the drop-forge factories is not at odds with TTS findings, as perceived by Martin and by Robinson. Further, Martin and Robinson do not consider the results of PTS studies (Johansson *et al.* and Henderson and Hamernik) given above which show that when the recovery periods were longer between bursts of noise (about 20 sec to 1 minute) the equal-energy method of noise measurement would significantly overestimate the PTS found.

In conclusion it appears that a general and valid procedure for measuring noise during a daily 8-hour exposure period or equivalent for purposes of predicting NIPTS cannot consist of equal energy *per se*. Some corrections or adjustments should be given to temporal factors such as "off" periods and their dura-

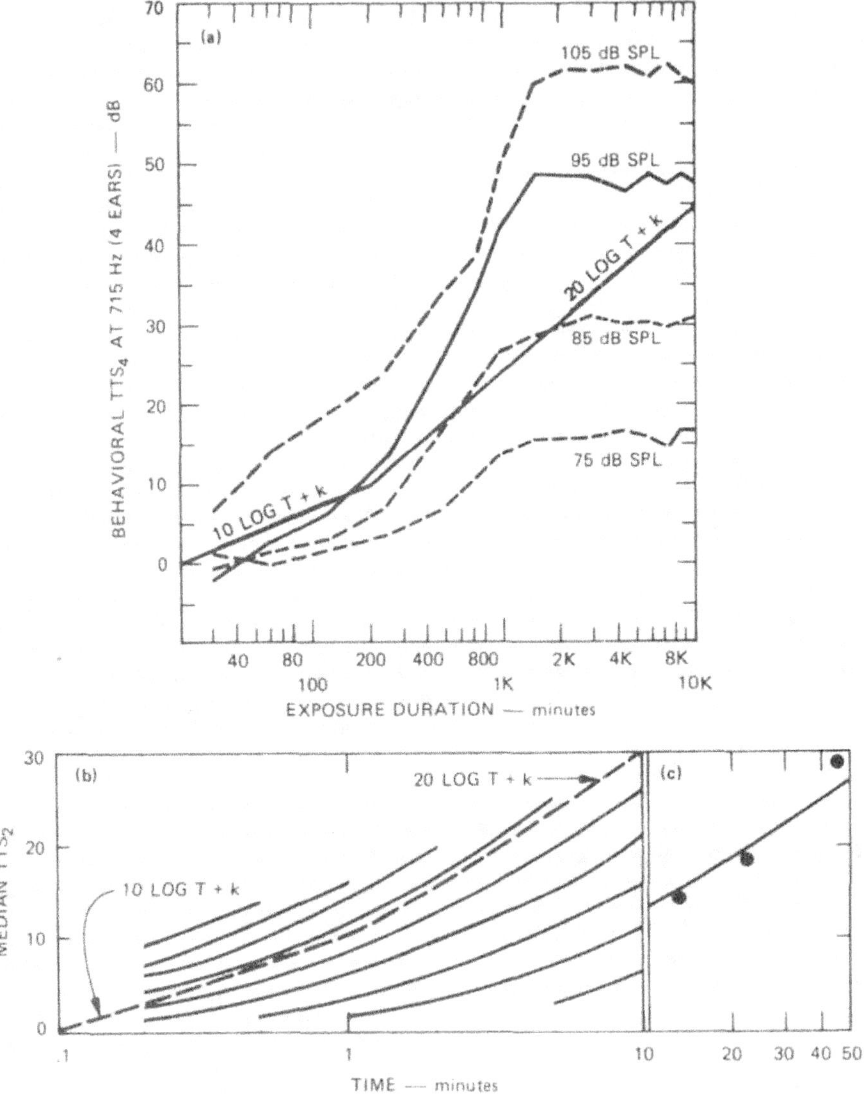

FIGURE 7.30. General growth of average TTS for the following: (a) chinchillas during exposure to an octave band of noise (Center frequency = 500 Hz) (data from ref. 70); (b) humans as a function of duration of exposure to steady-state noise (data from ref. 33); and (c) humans as a function of duration of exposure to a noise (data from ref. 73).

tions. Information from TTS and NIPTS studies, when interpreted in certain ways, provide some consistent basis for these corrections. The equal-energy concept is here hypothesized to be valid only when the interruptions between bursts of a given noise are about 10 sec or less in duration (*i.e.*, when the ear is "refractory") and when the daily exposure period is not significantly shorter than about 8 hours.

TTS as a function of exposure duration

As a function of exposure duration, TTS grows within a 10-hour period to an asymptotic level, depending on the intensity of the noise (refs. 69 to 72). The general nature of this growth of TTS is shown in figure 7.30 (refs. 33, 70, and 73). Figure 7.30 shows about 1 dB increase in TTS per 1 dB exposure over time T ($10 \log T + k$) up to about 10 dB TTS and about 2 dB TTS per 1 dB exposure over time ($20 \log T + k$) for 10 to 40 dB TTS (or to its asymptote, if that is less than 40 dB).

Plotted on figure 7.31 are TTS data from a number of studies in which exposure duration was a variable. The following table gives the references and pertinent information for the curves in figure 7.31.

Curve in fig. 7.31	Reference	Frequency at which TTS measured, kHz	Frequency of noise, kHz	SPL of noise, dB
1	30	4	(a)	—
2	50	2, 4, and 8	2	120
3	47	4	(b)	108
4	55	4	(c)	—
5	44	4	2-9	90
6	46	4	1.2-2.4	100
7	45	4	(b)	100
8	50	0.5-2	0.5	120
9	48	4	(d)	—
10	57	3, 4, and 6	(b)	90
11	56	1	(b)	106
12	41	0.5-8	(e)	—
13	30	4	(a)	—

[a] *Gunfire, 1 per 5 sec.*
[b] *White noise.*
[c] *Clicks, 25 per minute.*
[d] *Pulsed tones.*
[e] *Gunfire, 1 per 15 minutes.*

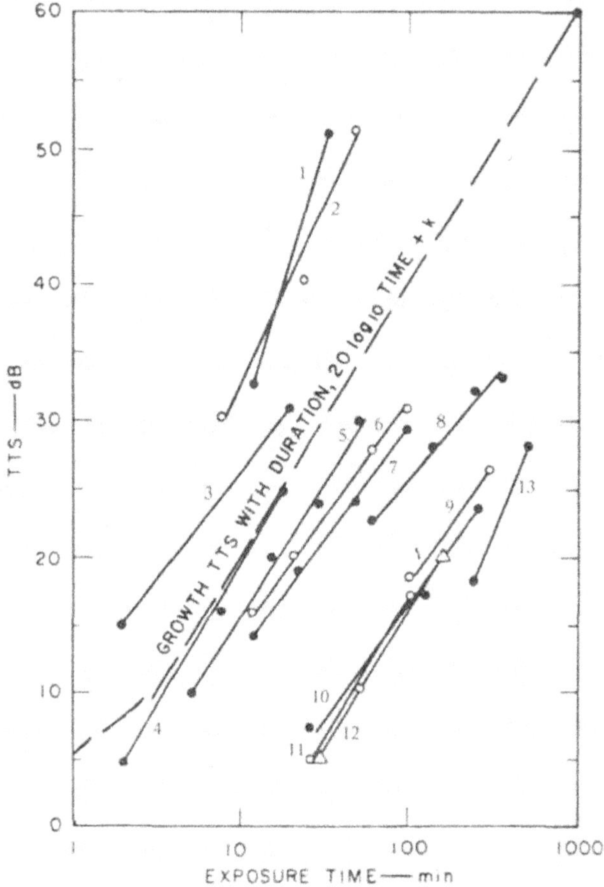

FIGURE 7.31. TTS as a function of the duration of the noise. (From ref. 26.) The slope of the lower portion of the dashed curve represents $10 \log T + k$.

A line representing $20 \log T + k$ fits the TTS data from 10 to 40 dB reasonably well. The data points below 10 dB are too few to draw a best fit curve to 0 dB TTS, as was done earlier in reference 26. On the basis of the findings shown in figure 7.30, a line representing $10 \log T + k$ is now suggested as being appropriate for TTS values from 0 to 10 dB.

Figures 7.30 and 7.31 both indicate that, in terms of energy, the rate of growth of TTS as a function of exposure duration is twice that found for sound pressure level, as shown in figures 7.17, 7.20, 7.21, and, for PTS, figure 7.6. Although firm data for PTS from daily exposure to workday noise of different durations are lacking, we can conclude that the equal-energy concept (duration and sound pressure level are interchangeable as causes of threshold shift) is fundamentally incorrect.

Recovery from TTS

The auditory fatigue or incipient trauma that is carried over from the end of one workday period to the next day is the important aspect of the growth of NIPTS. Accordingly, if the period away from the noise is extended, then additional time for recovery of the auditory system from the effects of the noise is available within the 24-hour day.

Recovery from TTS_2 values up to about 40 dB follows a uniform course, as shown in figures 7.32 and 7.33. The relation between recovery time T_{rec} and TTS_2 is approximately a 1-dB drop in TTS per 1 dB of recovery over time, that is, 10 log T_{rec} + k. This, as shown in figures 7.30 and 7.31, is about half the rate of the growth of TTS with time (*e.g.*, a 2-dB growth per 1 dB of exposure over time when the TTS exceeds about 10 dB). In order to acquire any TTS, the rate of recovery from auditory fatigue must proceed at a slower rate than its growth.

Some TTS data (see ref. 76) show a delayed recovery for ears having initially high TTS_2 values. This is not inconsistent with the belief that noise exposures capable of causing a TTS_2 greater than about 40 dB represent an overload to the ear that has some accelerated "nonlinear" adverse effect on the auditory system compared with the more linear effect from lower dosage levels.

Accumulation of trauma

This 40-dB TTS ceiling for linear growth of PTS is a TTS level for which there is apparently complete recovery within 16 hours, the period typically available for recovery from the 8-hour workday noise exposure. Why do daily exposures to noise causing TTS from which there is, or presumably should be, complete daily recovery eventuate in PTS of comparable magnitude? One probable reason is that the measures of audiometric thresholds are too simple and do not

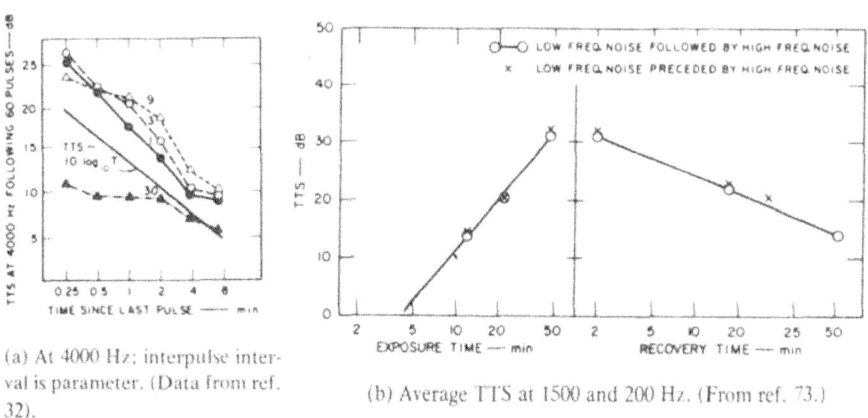

(a) At 4000 Hz; interpulse interval is parameter. (Data from ref. 32).

(b) Average TTS at 1500 and 200 Hz. (From ref. 73.)

FIGURE 7.32. Recovery time of TTS.

reflect the adverse effects of the noise on the auditory system that are taking place in an accumulative manner. A second possible reason is that the normal sounds and noises to which people are exposed away from work are sufficient to prevent the full recovery from the daily TTS incurred at the workplace. As discussed before, effective quiet is indeed a rather low level of sound or noise, perhaps on a continuous basis of around 60 dBA.

It appears that the ear can operate indefinitely without undue auditory fatigue in an acoustic environment having a range of about 60 dB from its threshold of sensitivity. It appears further that it can operate over a range in sound energy of at least another 60 dB (up to about 120 dB above absolute threshold level in the quiet), but with some progressive deterioration due to inadequate recovery processes. It also seems a sound at a level about 155 dB above normal threshold level can cause some immediate damage to the ear and permanent shift in its threshold of sensitivity.

Corrections for interruptions in and durations of daily noises

If we take TTS_2 up to about 40 dB after an 8-hour daily exposure to continuous noise as predictive of NIPTS, other shorter daily exposure periods and interrupted noises must be equated for their relative effectiveness in causing NIPTS. For more or less uninterrupted noise, this equation is reasonably well accomplished by use of the recovery function (dashed line) in figure 7.33. The following table gives the references and pertinent information for the curves in figure 7.33.

Curve in fig. 7.33	Reference	Brief description of noise presented
1	41	Gunfire
2	49	2 minutes white noise; TTS measured at 4 kHz
3	47	20 minutes white noise at 108 dB
4	41	Gunfire
5	49	12 minutes white noise; TTS measured at 3 kHz
6	52	1.4-2.8 kHz noise
7	74	White noise
8	56	106 dB noise; TTS measured at 1 kHz
9	75	Continuous tone; TTS measured at 4 kHz
10	75	Impulsive tone; TTS measured at 4 kHz
11	42	Gunfire; TTS measured at 4 kHz

In brief, effective exposure (*i.e.*, 8 hours) is proportional to 20 log T_{ex} (where T_{ex} is exposure time) (see fig. 7.31) minus 10 log T_{rec} (see fig. 7.33). Thus, for halving of duration the measured noise energy would have to be 6 dB less to be equivalent in effect to the continuous 8-hour exposure, 3 dB less because of the reduction of en-

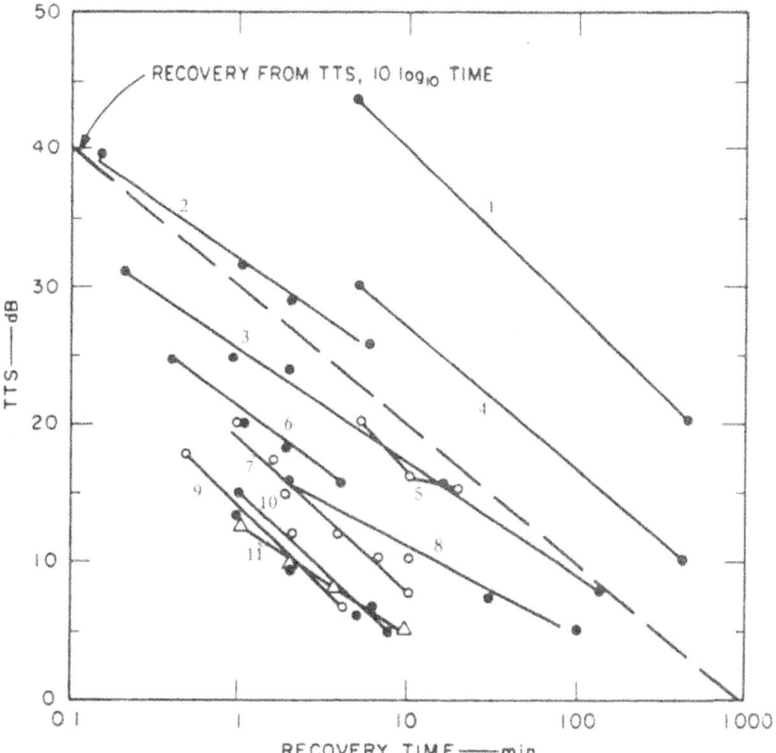

FIGURE 7.33. TTS as a function of postexposure time in minutes. (From ref. 26.)

ergy with time and 3 dB less because of the additional recovery (to the end of 8 hours) afforded by the shorter 4-hour exposure.

With interrupted noise, the equation would also involve the measured noise energy over the 8-hour day corrected by the equivalent recovery rule shown by the dashed line in figure 7.25. There is the proviso that interruptions between repeated bursts of a given noise must exceed 10 sec; if the interruption is less than 10 sec the noise is considered as being "on" during the "off" period.

Better prediction could presumably be accomplished by using the solid-line functions of figure 7.25 for the different on-off ratios. These on-off ratios are proportional, in real time, to total time T_{tot} minus off time T_{off} divided by total time, or simply real off time for a given total time (*e.g.*, 8 hours). The dashed line is drawn to show that the effective recovery in decibels is $-10 \log (T_{tot} - T_{on})$, or $-10 \log T_{off}$. This is equivalent to a simple treament of the energy involved and is easy to implement in the measurement of noise environments. Indeed, the effective exposure for predicting NIPTS from daily exposures to either continuous or to intermittent noise can be related to SPL + 20 log T_{on} − 20 log T_{off}. The factor of −20 log T_{off} consists of −10 log T_{off} because of the reduced energy in the off periods plus −10 log T_{off} because of the excess recovery during the off periods.

From the proposed formulation as well as from the empirical data, it does not matter whether the off time is continuous or interrupted during the 8-hour day. In either case, the recovery process continues and is equally effective. For example, the level of a noise of 8 hours duration per workday could be increased by 6 dB and cause no additional PTS provided its duration is decreased to 4 hours, either by reducing the total work period by 4 hours or by introducing "off" periods (longer than 10 sec each) which total 4 hours. This, of course, is in reasonably close agreement with the "5 dB exchange" that would be allowed in some noise assessment procedures, such as the U.S. Department of Labor Occupational Safety and Health Administration (OSHA) regulations.

Definitions for noise and off time

Again note that a noise, according to the procedure being proposed, is defined as a given SPL (±2.5 dB) that exceeds effective quiet with a broad frequency spectrum that is A-weighted and with a 1-sec reference duration (*i.e.*, energy in 1-sec periods). Accordingly, noises differ from each other only when their levels differ by 5 dB; no other distinction is made. A noise is "off" when replaced by another SPL for a period of more than 10 sec.

A descriptive concept for these definitions is that the recovery processes in the ear 10 sec after an exposure to any given level of noise continue whether followed by less intense or by more intense noise. The state of "auditory fatigue" at the end of 8 hours is proportional to the energy sum of the noises (levels) above effective quiet during a day, each noise (level) being corrected for true equivalent effectiveness (duration and recovery) to effectiveness of a steady-state noise continuously present for 8 hours.

It should be emphasized that the effects of pure tones or of narrowband spectra upon PTS at different frequencies are not necessarily predicted by the use of the A-weighted decibel or by the procedures proposed herein. Tentative procedures for predicting such specific effects are described elsewhere (refs. 2 and 26).

Interruptions in days or years of exposure

The effects on NIPTS of long interruptions (weeks, months, or years) in daily exposures to noise are surmised to be predictable in accordance with the effective equal-energy concept proposed herein, exposures within the same day being equated for effective recovery from auditory fatigue. As discussed earlier, the equal-energy procedure appears to be adequate for predicting NIPTS from long periods of daily exposures to a given noise environment. The concept involved is that the auditory trauma experienced each day is carried over and accumulates over a lifetime. A corollary to this concept is that all the recovery that is to take place from a daily exposure to noise occurs during the 16 hours of "rest" following the 8-hour exposure period. As discussed above, this is not inconsistent with TTS data, provided the threshold shift is not in excess of about 40 dB from an equivalent 8-hour exposure. This is also not to suggest that the acuity of hearing at the end of the 16 hours of rest after exposure will not show some further reduction with time. However, some of this reduction, when it does occur, could conceivably be related to some elevation of HL due to sociocusis and nosocusis.

In keeping with this concept, long interruptions in exposure to the noise would reduce the NIPTS at the end of the work career by an amount proportional to the total noise energy present over the total career. In any event, "equal energy" of exposure to noise when measured over days or years appears to be the most reasonable and workable assumption that can be deduced from available data, such as that from the studies of Baughn (refs. 6 and 7) and of Burns and Robinson (ref. 8). For our purposes, the energy in careers of exposure to truly equivalent noise levels is expressed as an amount equivalent to an exposure career of 50 years.

Anchoring equivalent noise exposures and TTS

The amount of NIPTS measured 16 hours after 10 years of exposure to a given steady-state noise (for 8 hours a day, 5 days a week) is about numerically equal to the TTS_2 measured in young normal ears after exposure to the given 8-hour noise environment. Further, a change in NIPTS proportional to energy will occur with fewer or more years of exposure. This empirical finding with respect to NIPTS does not mean or imply that the changes in HL due to TTS follow a uniform increase up to the end of the work period; it likewise does not mean that similarities in the absolute numerical magnitudes of TTS and NIPTS are important. Indeed, as shown by Hetu and Parrot (ref. 19) (see fig. 7.8), this uniform increase is not the case because of interruptions such as the lunch break during the workday. This finding casts some doubt on the appropriateness of the design of some TTS experiments (*i.e.*, 8-hour exposure without breaks from the noise would tend to overestimate potential NIPTS from 8 hours of industrial noise exposure with a break for lunch). However, the variety of intensity-temporal patterns of workday noise exposure in real life makes some uncertainty in this regard impossible to avoid.

Noise and vibration

Besides the normal air path, a possible path for sound to reach the inner ear is *via* bone or tissue conduction from vibrating the body or portions thereof. In general, this is a very inefficient way to stimulate the ear but, nevertheless, in some situations it could be a factor in causing TTS and NIPTS.

Guignard and Coles (ref. 77) found no significant effects of whole body vibration on TTS either when given alone or in the presence of intense noise that caused some TTS. However, Okada *et al.* (ref. 78) (see fig. 7.34(a)), Yokoyama *et al.* (ref. 79), and Kile and Wurzbach (ref. 80), measuring TTS in human subjects, found 5 dB TTS or so from vibrating either the whole body or just the arm. Hamernik *et al.* (ref. 81) (see fig. 7.34(b)) measured 5 to 10 dB PTS in chinchillas from vibrating the whole body.

Vibration and noise together caused more TTS or PTS than did the noise alone or the vibration alone, indicating an additive or, perhaps in the Hamernik *et al.* study, something of a "synergistic" effect. The latter finding could conceivably be due to a kind of acoustic effect upon the ear similar to the presence of a continuous "noise" from the vibrations that inhibited recovery from the airborne impulses.

It might be noted that the acceleration rates and frequency of the vibrations are important variables affecting threshold shifts. In humans, the greatest effects on TTS (which covered a wide range of audiometric test frequencies much like that from exposure to an airborne noise with a broad frequency band) came from a vibration frequency of about 5 Hz and required an acceleration of greater than 100 cm/sec^2, according to Okada *et al.* Higher frequencies of vibration (*e.g.*, 20 Hz) caused less TTS and higher accelerations caused more TTS.

Perhaps a straightforward explanation of these findings can be made in terms of bone conduction and the acoustic spectrum of vibration impulses. At the rapid rates of acceleration involved in most of the studies, the vibrations would give, *via* bone conduction, an effective acoustic impulse in the cochlea with rise times of 1 msec or so. This would represent a broad-spectrum impulsive noise as determined from figures 7.13 and 7.14. The decline in effectiveness in causing TTS with more vibrations per second above 5 Hz could be primarily a function of the impedance of the body structures to vibration as a function of frequency.

In any event, the state of knowledge of bodily vibrations upon NIPTS is too meager to attempt its inclusion in NIPTS prediction models or procedures. Also, fortunately, it does not appear to be a very frequently occurring factor in NIPTS because the high rates of acceleration and amounts of displacement required to be effective in this regard are not common in even noisy industries.

Proposed New Procedures for Predicting NIPTS

In order to formulate a general model for predicting NIPTS, procedures for quantitatively expressing the equivalent 8-hour exposures for steady-state and for intermittent noises for different numbers of years of exposure must be provided. Based on the experimental findings presented herein, procedures are now described for calculating the equivalent exposure values of a wide variety of types of noises and exposure conditions. Because of the complex noise measurements sometimes required for exact descriptions of interrupted noises, certain simplified procedures are also suggested.

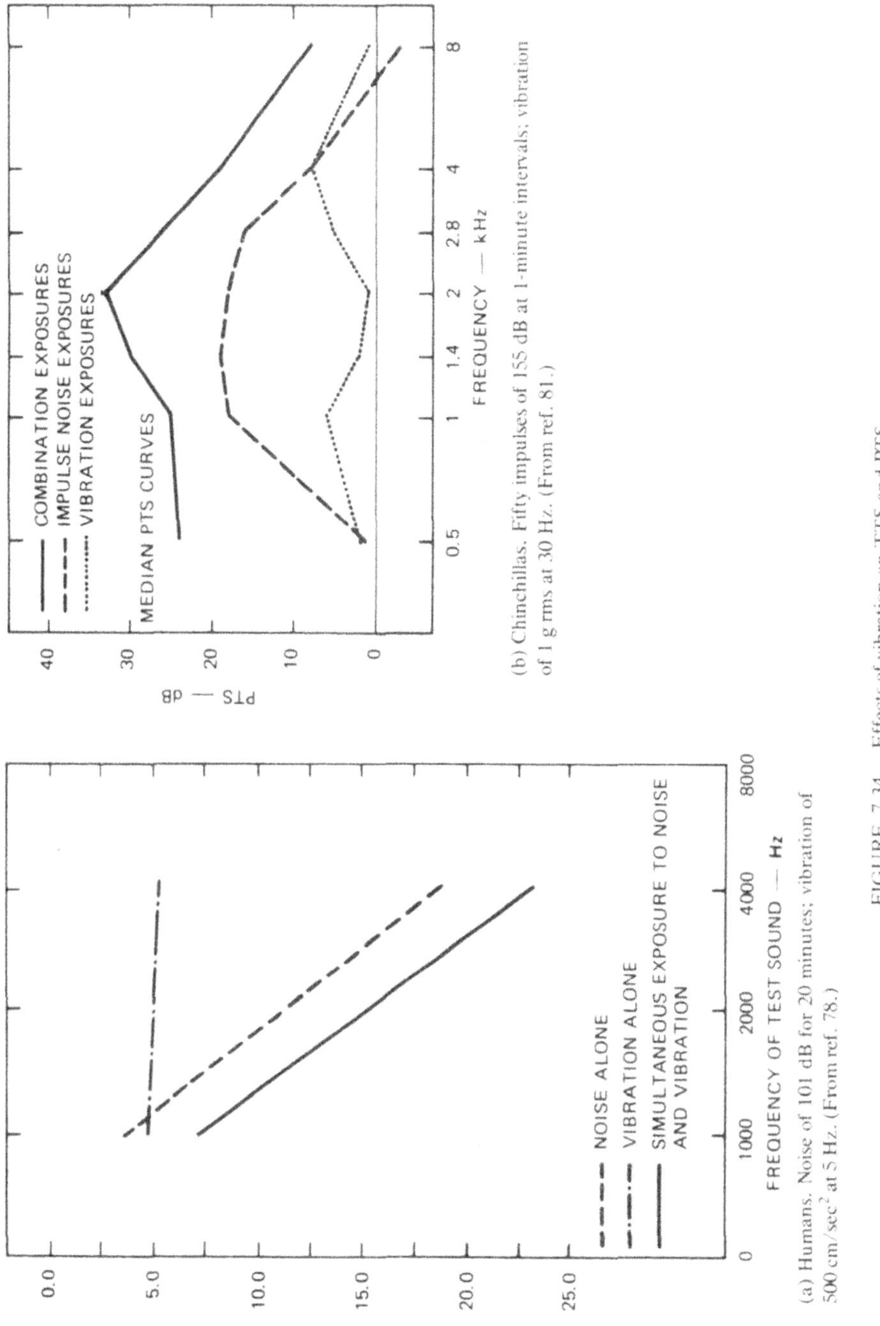

(a) Humans. Noise of 101 dB for 20 minutes; vibration of 500 cm/sec^2 at 5 Hz. (From ref. 78.)

(b) Chinchillas. Fifty impulses of 155 dB at 1-minute intervals; vibration of 1 g rms at 30 Hz. (From ref. 81.)

FIGURE 7.34. Effects of vibration on TTS and PTS.

Definitions, procedures, and assumptions

The definitions, procedures, and assumptions used for predicting NIPTS are given below.

1. For our purposes, a noise is defined as the 1-sec average A-weighted SPL ± 2.5 dB which exceeds a level capable of causing no more than 5 dB NIPTS in no more than 10 percent of the people after 8 hours per day, 5 days a week, for 50 years of exposure. No distinction is made between so-called impulsive and non-impulsive periods of noise, except that the 1-sec A-weighted SPL's can be measured by means of a standard sound level meter for most nonimpulsive periods of noise but not for impulsive periods.

2. The growth of NIPTS is taken to occur at the rate of 0.5 dB per 1 dB of SPL up to 10 dB NIPTS and at the rate of 1 dB NIPTS per 1 dB of SPL from 10 to 40 dB NIPTS. The relation between exposures and NIPTS for exposures that cause more than about 40 dB TTS_2 at any frequency in 50 percent of young, normal ears after a typical day's exposure is not to be predicted with this procedure. (The proposed procedure could be expected to underestimate the amount of NIPTS to be expected from exposures that cause more than 40 dB TTS_2.)

3. For estimating the true equivalent hearing-damage risk exposure to noises that are shorter than 8 hours and/or interrupted by another noise or by quiet, corrections to their total sound energy are made to take into account recovery from auditory fatigue that occurs during the times each noise is off. If the interruptions are shorter than 10 sec, the durations are added to the durations of the given noise. (No recovery is presumed to occur during interruptions of less than 10 sec.)

4. The true equivalent hearing-damage-risk exposures to each noise during the 8-hour day are summed on an energy basis to arrive at a total daily noise exposure for predicting NIPTS.

5. NIPTS is presumed to grow at the rate of 10 log *Y*, where *Y* is the number of years of exposure regardless of any interruptions that may occur between years of exposure.

6. No significant differences in susceptibility of NIPTS for true equivalent hearing-damage-risk exposures to noise are presumed for males and females. However, a greater amount of sociocusis in males than in females is presumed to be present in modern societies and to effect so-called presbycusis functions for males and females accordingly.

7. There is no difference presumed in susceptibility to TTS or PTS as a function of age of adults.

8. Presbycusis and NIPTS are presumed to be additive. Typical presbycusis as derived by Robinson and Sutton (ref. 11) and in U.S. Public Health surveys (refs. 14 and 15) depict the HL's of the general population with no otological disorders and no exposure to loud noise.

9. Effective noise exposure level for damage to hearing (DL) is the workday A-weighted sound energy modified by equivalent recovery for less than 8 hours of exposure per workday over a work career of a given number of years. This DL can be used to predict the NIPTS and HL values to be expected from a given noise environment for various audiometric test frequencies and percentages of the population. A given calculated DL will represent a damaging effect on the hearing of people equivalent to that expected from a steady-state noise for 8 hours per workday for 50 years at an A-weighted SPL of the same numerical value as the given DL.

10. The distribution of HL's in a given population of a specified age is predictable from the distributions of HL for otologically normal male and female ears from that population corrected for NIPTS, including that due to excess sociocusis plus any excess nosocusis.

Formulas and graphs

The equivalent DL of a noise or noises is the daily effective noise energy corrected for excess recovery and for up to 50 years of almost daily exposure. The formulas for DL are as follows:

1. Equivalent DL of a noise for 50 years of almost daily exposure is

$$\mathrm{DL}_{N_1} = L_{A,N_1} - 20\log(T - t) - 10\log(50/Y)$$

where

L_A	A-weighted SPL energy over 1 sec ±2.5 dB, dBA
N_1	L_A above 60 dBA (conservative estimate of threshold of damage risk to hearing)
t	duration of L_{A,N_1} during typical daily period T
T	daily period, nominally 28 800 sec (8 hours) but can be extended to 43 200 sec (12 hours) with some increase in uncertainty as to validity of predictions of NIPTS values
Y	years of exposure, total number of typical weeks (including up to 4 weeks "vacation" per year) per career of exposure to a level of noise divided by 52

2. Equivalent DL of noises for 50 years of almost daily exposure is

$$\mathrm{DL} = \sum_{N_1}^{N_n} 10 \log \mathrm{DL}_{N_1 - N_n}$$

where N_1 is the lowest and N_n is the highest value of L_A present during a typical day of exposure to noise

3. NIPTS at given test frequencies and population percentiles is

$$\text{Percentile NIPTS}_{\text{freq}} = \mathrm{DL} - K/2 + [\mathrm{DL} - (K + 20)]$$

where K is the maximum DL that gives 0 dB NIPTS (8-hour workday for 50 years of exposure) for a specified audiometric test frequency and percentile of the exposed population. The expression $K/2$ is limited to 0 to 10 and $K + 20$ is limited to positive values. These limits reflect the fact that NIPTS grows, as a function of exposure level, at about one-half the rate from 0 to 10 dB NIPTS as from 10 to 40 dB NIPTS, as discussed earlier.

4. HL at given test frequencies and population percentiles is

$$\text{Percentile HL}_{\text{freq}} = \text{Percentile NIPTS}_{\text{freq}} + \text{Percentile HL}$$

for a given population and sex.

Use of DL to predict NIPTS and HL

Figure 7.35 provides the means for finding, from a DL, the magnitude of likely hearing damage in terms of NIPTS for different audiometric test frequencies and the percentages of people to be affected thusly. For example, if a group of men are exposed to a DL of 95, 10 percent of the group could be expected to have a NIPTS of 15 dB or greater at 1000 Hz (or for the average NIPTS at 0.5, 1 and 2 kHz). This DL of 95 dBA is found by dropping a perpendicular from the 90th percentile intersection with the curve for 1000 Hz to a DL of 95 on the abscissa; the right-hand ordinate indicates a NIPTS of 15 dB for that point. A NIPTS of about 26 dB or greater for 10 percent of the population is found at 4000 Hz for the same DL.

The locations of the functions for 4000 Hz and the average of 0.5, 1, and 2 kHz are based on the generalized average NIPTS's for those frequencies derived from the Baughn and from the Burns and Robinson data (refs. 6 to 8). (See fig. 7.6.) The locations of the functions for the frequencies of 500, 1000, 2000, and 3000 Hz are estimates, taking into account the separation of the empirically found functions for 4000 Hz and the average for 0.5, 1, and 2 kHz, as well as the

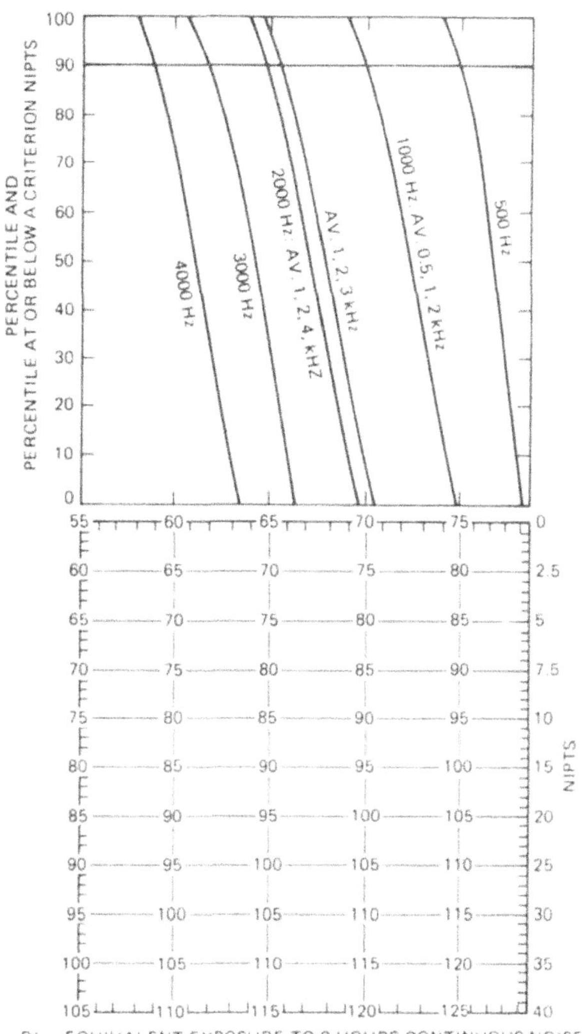

TABLE OF K's								
	kHz							
%ile	.5	1	2	3	4	AV. .5, 1, 2	AV. 1, 2, 3	AV. 1, 2, 4
90ile	75	70	65	62	59	70	66	65
50ile	77	72	67	64	61	72	68	67
10ile	79	74	69	66	63	74	70	69

CRITERION. 0 dB
NOISE INDUCED PERMANENT THRESHOLD SHIFT (NIPTS) FOR GIVEN PERCENTILES

FIGURE 7.35. Graph for estimating percentages of exposed population at or below a specified criterion of NIPTS at various audiometric test frequencies for a given DL.

similar nature of the data for the functions for the individual test frequencies as deduced by Burns and Robinson. A table of *K* values (0 dB NIPTS for the 90th, 50th, and 10th percentiles) is given in figure 7.35.

As mentioned earlier, and discussed in detail later, non-workplace-noise exposures (beyond those to be ascribed to typical sociocusis) that can be measured or estimated should be added to workplace noise in the calculation of DL and the prediction of NIPTS therefrom. HL values to be expected in populations exposed to a given DL are found by adding the expected NIPTS, in decibels, to the presbycusis functions found by Robinson and Sutton for non-noise-exposed, otologically normal ears, as shown in figure 7.36 and in tabular form in the appendix to this chapter.

Some further adjustment or correction to the resulting HL's could be made for excess nosocusis that appears to be present in certain populations. However, this correction is questionable in view of lack of information available regarding

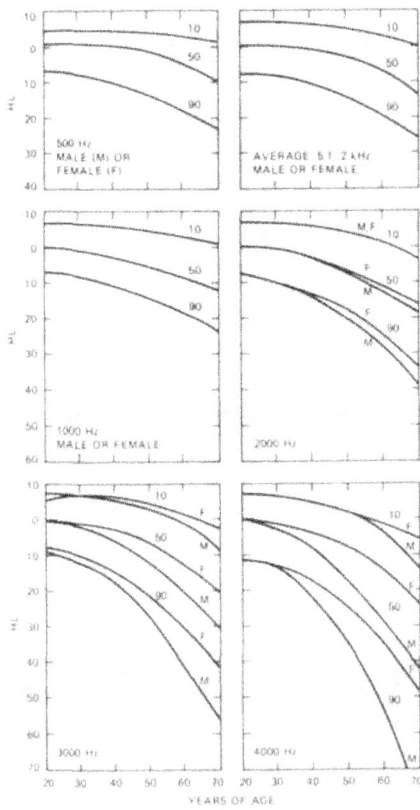

FIGURE 7.36. HL as a function of age for males and females at various test frequencies for 10th, 50th, and 90th percentile. (Data from ref. 11.)

the possible additivity-subtractivity in threshold shifts due to various otological diseases and NIPTS. All things considered, it can be recommended that to estimate the distribution of HL's to be expected in a population with typical amounts and kinds of nosocusis effects (*i.e.*, otologically unscreened groups), the NIPTS expected from a given DL should be added to the presbycusis functions found in the USPHS surveys (refs. 14 and 15) for random samples of males and females unscreened for otological disease or for exposure to noise.

Predicted and measured "everyday noise" exposure

The approximate amount of permanent HL shift (PTS) beyond that due to presbycusis at age 70 years in males presumably not exposed to industrial noise is probably due to everyday noise (typical sociocusis) and typical nosocusis. Typical sociocusis-nosocusis is taken as the difference between HL's for males versus females of the general, industrialized population. (See fig. 6.24.) For example, the difference between the 4000-Hz HL's of the 50th percentile of males versus females at age 70 years is about 18 dB.

Studies have been conducted in which individuals wore a small device, called a dosimeter, that measured the sound energy $L_{A,eq}$ present near the wearer's ear during a typical 24-hour day. For example, 76 dB was the average daily 24-hour $L_{A,eq}$ for 31 days for a young adult male office worker (ref. 82); 77 dB was the average for 16 adult males at an Air Force base (ref. 83); and 73.3 dB was the average for 54 subjects age 5 to 54 years (including housewives, students, office workers, and factory-commercial workers) (ref. 84). Standard deviations, when obtained, were of the order of 6 dB. It is to be noted that different dosimeters worn by the subjects in these studies did not measure the energy in certain kinds of sounds in the same way; systematic differences of 6 dB or so were observed between instruments when measuring the same sound (ref. 83).

An extensive dosimeter study was conducted by Nimura and Kono (ref. 85). Their findings, some of which are shown in figure 7.37, are consistent with the other cited dosimeter studies of everyday sound and noise. The average overall $L_{A,eq}$ of these dosimeter studies is estimated to be about 75 dBA for adult males, the same $L_{A,eq}$ deducted from the DL in figure 7.35 as the basis for the 9.5 dB of sociocusis (for 50th percentile males). However, several points need to be made about this close agreement. First, it is doubtful that the dosimeter $L_{A,eq}$ for 24 hours is much different than the $L_{A,eq}$ for 8 hours of daylight for some typical 8-hour period during the day. In particular it is not likely that the sounds or noises during the nighttime and other rest periods during the typical 24-hour day contribute, on an energy basis, to the measured $L_{A,eq}$. Second, everyday sounds and noises are generally interrupted or intermittent in nature. According to the research studies of temporary threshold shifts, exposure to intermittent sounds or noises does not cause as much threshold shift as do the steady-state noises found in factories. It would seem probable that the 24-hour $L_{A,eq}$ dosimeter measures of 80 dBA found for everyday intermittent activities would be the equivalent, as far

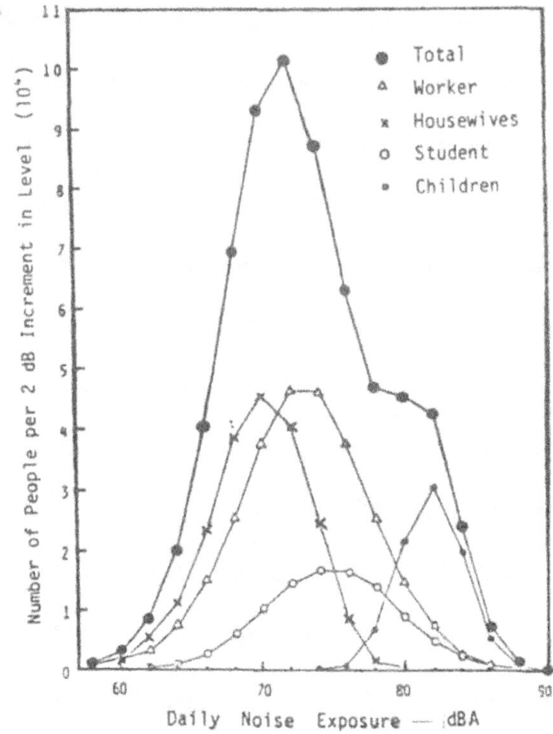

FIGURE 7.37. Estimated distributions of people in a Japanese city (Sendai) experiencing different levels of 24-hour per day exposures to everyday noise as measured by body-worn dosimeters. (From ref. 85.)

as damage to hearing is concerned, to steady-state workplace noise of 75 dBA for 8 hours.

Of special interest in the Nimura and Kono data (fig. 7.37) are the high levels, exceeding 80 dBA, found for the children and the relatively low levels, around 70 dBA, found for housewives. The latter is, of course, in agreement with the notion that women suffer less sociocusis than men. As for the high levels for children, one must wonder how much the measured $L_{A,eq}$ is due to voice sounds from both the person wearing the dosimeter and from other talkers. As seen in table 4.1, for "loud" and "shout" voice levels the speech levels ranged from 73 to 85 dBA even at a distance of 1 m from the talker. Speech sounds are, however, irregular and interrupted in level and, presumably, do not cause appreciable auditory fatigue because of low energy and recovery periods.

On the assumption that the effects of sociocusis, nosocusis, and presbycusis are additive, it follows that the 50th percentile, 4000-Hz HL of 41 dB for 70-year-old men in industrialized societies (see fig. 6.24(a)) comes from 13 dB pure presbycusis (see fig. 6.26(a)), plus 9.5 dB pure sociocusis (DL of 80 dBA), leaving 18.5 dB apparently due to nosocusis. (Fig. 6.26(b) shows 28 dB for sociocusis plus nosocusis in males.) For 70-year-old-females, the typical 50th percentile, 4000-Hz HL of 25 dB (see fig. 6.24(b)) is presumably due to 13 dB pure presbycusis (see fig. 6.26(a)), plus 7 dB pure sociocusis (DL of 75 dBA), leaving 5 dB for nosocusis. (Fig. 6.26(c) shows 12 dB for typical sociocusis plus nosocusis in females.)

Sociocusis is apparently no greater a hazard to males than females (according to the graphs in figure 7.36) for hearing thresholds at frequencies below about 2000 Hz, although females in industrialized societies show a small amount of apparent sociocusis when their hearing thresholds are compared with those of males or females from a noise-free society. This apparent fact of significant amounts of sociocusis-nosocusis in industrial societies does not affect the calculation of NIPTS to be expected from exposure to an industrial noise environment in that

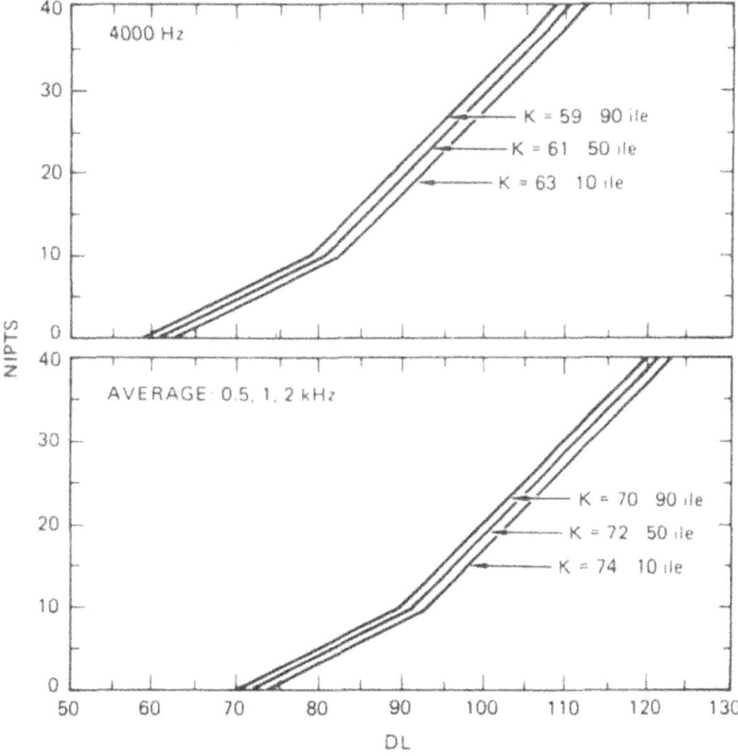

FIGURE 7.38. NIPTS as idealized function of DL for steady-state noise of 8 hours per workday for 50 years of exposure. Based on average of data from refs. 6 to 8.

this sociocusis-nosocusis factor is included in typical presbycusis functions used to derive the NIPTS values.

NIPTS criterion

Figure 7.38 illustrates the amount of NIPTS to be expected as a function of DL at 4000 Hz (top graph) and at the average of 0.5, 1, and 2 kHz (lower graph) at the 10th, 50th, and 90th percentiles of an exposed population. Also indicated on the figure is a suggested criterion for a measurable (5 dB) amount of NIPTS. The DL exposures that would meet this or other possible criteria can be found by dropping perpendiculars to the abscissa.

HL criteria

The left-hand and middle panels of figure 7.39 show the HL at 4000 Hz and the average of 0.5, 1, and 2 kHz to be expected at given percentiles of populations, at 70 years of age, exposed to various levels of DL. The left-hand panel is for populations free of ear disease (other than NIPTS) as found by the Robinson and Sutton analysis of a number of HL surveys; the middle panel is for the general U.S. population unscreened for any otological disorders or exposure to noise. The right-hand panel shows the HL to be expected from exposure to several levels of E_A as prescribed by Burns and Robinson (ref. 8).

In contrast to DL, the E_A procedure shows a greater shift at the higher percentiles than at the lower percentiles for a given level of exposure to noise. This difference is due to the combination of the effects of aging and noise exposure in the Burns and Robinson modeling of HL data and to the assumption in the derivation of DL that the susceptibility of the ear to TTS and NIPTS is largely independent of age and that the skewness in HL distribution at the higher percentiles is due to the aging process.

Criteria of "acceptable" hearing damage

For each decibel increase in DL (see fig. 7.40), there is an increase of about 2 percent in the number of exposed people who will exceed some specified "fence" of HL at an average of 0.5, 1, and 2 kHz. If the average of the thresholds at 0.5, 1, and 2 kHz is taken as a measure of relative hearing ability and an acceptable percentage of the population to be affected is set at 10 percent (those above the 90th percentile), it is shown in figure 7.38 that for a criterion NIPTS (after 50 years of exposure) of 5 dB or less, the DL is about 80 dB; with the same criterion but the threshold at 4000 Hz as the measure of hearing ability, the DL level is 69 dB. For the median (50th percentile) of the population, the DL noise exposures could be increased by 2 dB for both thresholds with the same criterion NIPTS (5 dB or less).

If the average of the HL's at 0.5, 1, and 2 kHz of 15 dB is taken as being indicative of impaired hearing, it is shown in figure 7.39 that by the age of 70 years,

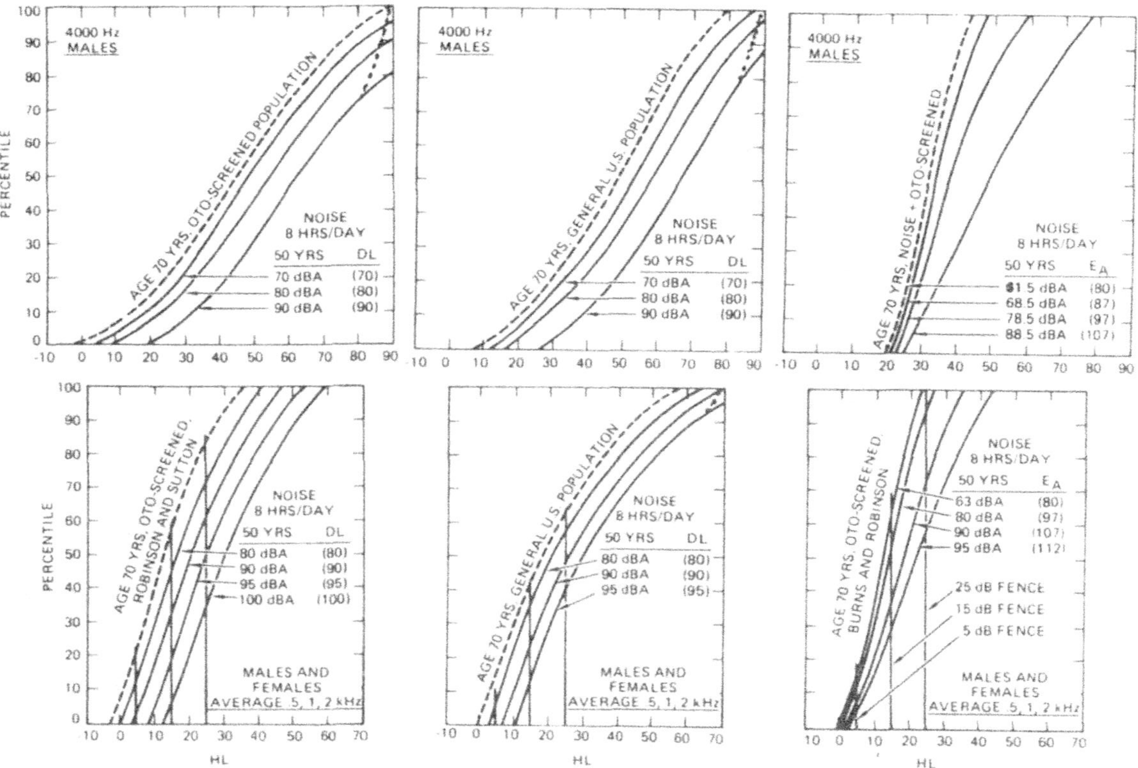

FIGURE 7.39 HL's at or below percentiles of populations as a function of DL for otologically normal populations (data from ref. 11) and for general U.S. population (data from ref. 15). Also presented are HL's as predicted by method of Burns and Robinson. (Data from ref. 8.) Dotted lines represent upper limits for HL.

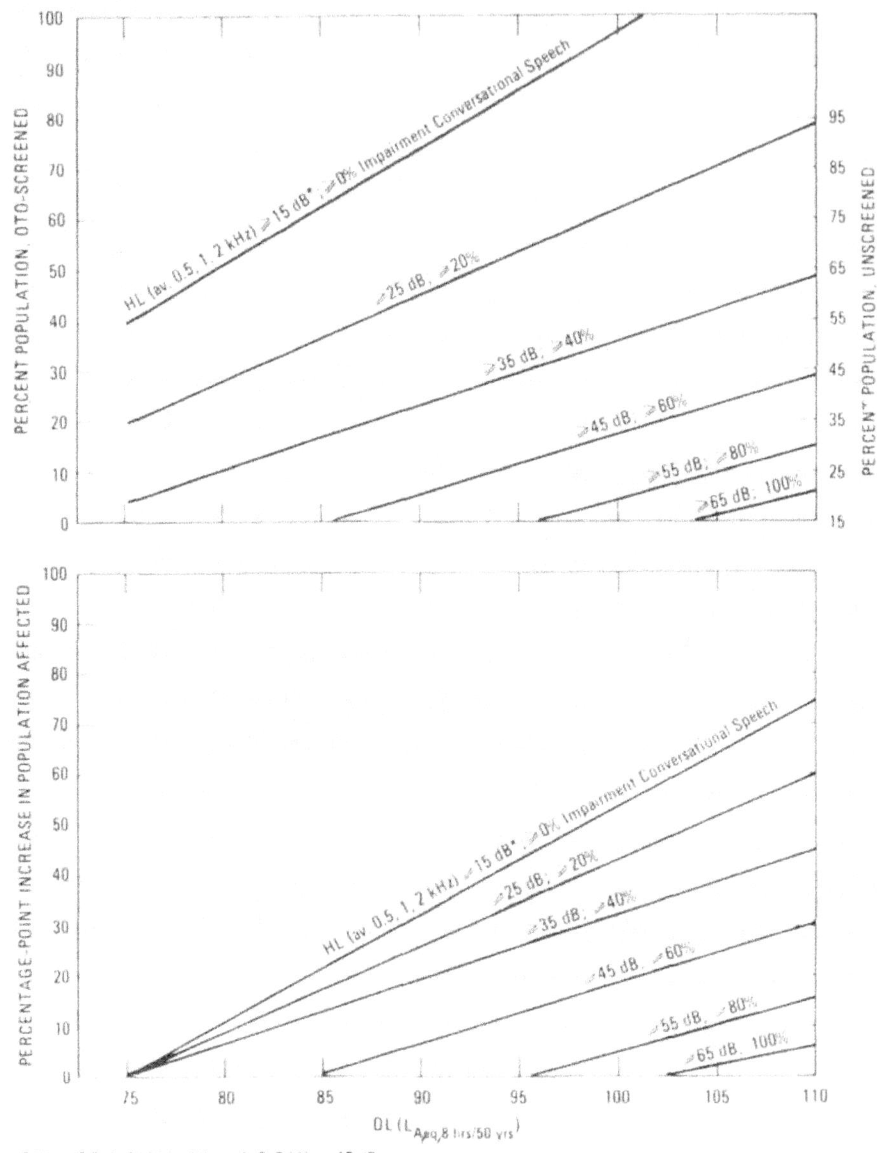

FIGURE 7.40. Percent of otologically screened and general populations exceeding fence levels and percent impairment for conversational speech as a function of noise exposure in DL.

55 percent of the general population (40 percent of the otologically screened population) will have average HL's that exceed 15 dB at those frequencies. However, 65 percent of the general population (55 percent of the otologically screened population) exposed to DL of 80 dB will exceed that 15-dB criterion fence of impairment, an increase of 10 percentage points of the population at that degree of risk.

Figure 7.40 shows, as a function of noise exposure level, the percent of the otologically screened and the general populations exceeding (and the percentage point increase in percent exceeding) various fence levels, the average at 0.5, 1, and 2 kHz, and the associated percent impairment for conversational speech in the quiet. (The latter data are developed in chapter 8.) From figures 7.36, 7.37, 7.39, and 7.40, the DL exposures required to meet a variety of quantitative descriptions or criteria of hearing damage and percent populations at risk from exposure to noise can be deduced.

Changing the fence criterion does not appreciably change the increase in the percentage points of the population at risk. Of course, the absolute numbers of people exceeding a fence decrease as the fence is increased, the decrease being proportional to the ratio of the higher percentile at risk at the higher fence to the lower percentile at risk, at the lower fence. From figures 7.36, 7.37, 7.39, 7.40, and 7.41, other DL levels required to meet a variety of quantitative descriptions or criteria of hearing damage and percent population at risk from exposure to noise can be deduced.

The concept of "safe" noise limits for industry is confusing. For example, the notion that an exposure at a noise level of, say, 85 dBA for 8 hours per workday will protect all but 10 percent of the most tender eared of the workers from more than 10 dB NIPTS at 4000 Hz is misleading. For practical purposes (see fig. 7.39), the entire population of persons exposed to a given noise level suffers, after a work career of 50 years or so, nearly the same amount of NIPTS (the difference in fig. 7.39 between the non-noise-exposed, general population and the noise exposed). This "taking" of hearing capacity by the noise from workers with good to excellent hearing (better than some NIPTS or "fence" defined as representing adequate hearing) should perhaps not be overlooked in the assessment of the impact of noise on hearing.

Summary and implications of shifts in threshold of hearing

For illustrative purposes, this summary is concerned with hearing sensitivity, measured as HL, at 4000 Hz for the 50th percentile of the 70-year-old population screened to have otologically normal ears. The following conclusions can be drawn:

1. Pure presbycusis causes about a 13-dB elevation in HL (fig. 6.26(a)).

2. The combined effects of sociocusis and nosocusis in industrialized societies further elevate HL by about 28 dB for men (fig. 6.26(b)) and 12 dB for women (fig. 6.26(c)).

3. Dosimeter measures of everyday noise indicate that men are exposed to $L_{A,eq}$ for 24 hours of about 75 dB (estimated as equivalent to $L_{A,eq}$ for 8 hours of about 80 dB steady-state noise), and women are exposed to a level 5 dB lower. (See fig. 7.37.)

4. According to the DL prediction procedure, these everyday noise exposures would cause a NIPTS of about 9.5 dB in men and about 7 dB in women and, according to the E_A prediction, a NIPTS of about 7.5 dB in men and 2.5 dB in women. (See appendix tables 7.A1 to 7.A4.)

5. Assuming that sociocusis, nosocusis, and presbycusis are, to a first approximation, additive in their effects on hearing thresholds, it is deduced that by the age of 70 years, typical nosocusis in industrialized societies amounts to about 18 dB for males and 5 dB for females.

However, this quantitative model of factors affecting the threshold of hearing is particularly weak with respect to deducing the relative effects of nosocusis and sociocusis. As discussed earlier, nosocusis is an ill-defined complex of disease conditions, some of which (such as conductive loss from otosclerosis) should provide some protection from NIPTS rather than being additive to it. Indeed, it was suggested by Rosen *et al.* (see chapter 6) that what is here referred to as nosocusis is not due to infections or otosclerosis, but is a general, age-related, systemic-cardiovascular disease condition that apparently affects, among other things, the threshold of hearing in industrialized societies compared with that of a particular nonindustrial society (*i.e.*, the Mabaans). The Mabaans were not only free from exposure to noise, they were also remarkably free of cardiovascular disease. Also, with respect to sociocusis, it is possible that men are exposed to higher noise levels from shooting guns, sports, and hobby activities than are presented in the dosimeter measurements reported for everyday noise; if so, the relative amount of sociocusis would be somewhat overestimated. Whatever the mixture of sociocusis-nosocusis factors might be in the data at hand, it appears that men experience about twice the shift in threshold of hearing because of sociocusis and two to three times the shift because of nosocusis than do women.

In view of these amounts of sociocusis, the question arises as to what limits might be placed on noise at the workplace. The effects of the two noises, non-workplace and workplace noise, are presumably additive. Specifically, for example, the typical (50th percentile) male after a 45- to 50-year career of exposure to 80 dBA steady-state workplace noise will experience, according to the DL prediction procedure, about 9.5 dB NIPTS because of that noise (7.5 dB according to E_A). If the worker is, in addition, exposed to the equivalent of 8 daily hours of 80-dBA steady-state noise because of activities outside the workplace, he can be expected to suffer an additional 3 dB of NIPTS according to the DL procedure (3.0 dB according to E_A). The expected HL rises from 41 to 44 dB. If the workplace noise was at a level of 90 dBA, the expected NIPTS from that source would be about 19 dB according to the DL procedure (20.0 dB according to E_A), giving a HL of 51 dB. With this level of workplace noise exposure, the estimated typical sociocusis noise (the equivalent $L_{A,eq}$ for 8 hours at 80 dB) becomes insignificant as a cause of NIPTS. It appears steady-state workplace noise could not exceed about 70 dBA for it to cause less than a 1-dB shift in the threshold of hearing at 4000 Hz for 50 years of exposure to the 50th percentile of the population. This follows, of course, from the fact that the total energy of the 80 dB of sociocusis noise plus the 70 dB of workplace noise is only 80.4 dB. (See fig. 2.5.)

It is usually difficult to specify the magnitude of the separate contributions made by presbycusis, nosocusis, sociocusis, and workplace (or some other specified noise exposure condition) to the measured HL of individual persons. Also, as mentioned previously, the significance of shifts in the threshold of hearing to human behavior and auditory functions is yet another important matter. These matters are discussed in chapter 8.

Examples of Predicting HL and NIPTS

The data from several studies of hearing threshold following exposure to industrial noise are used to examine the relative accuracy of the DL and E_A procedures in predicting HL and NIPTS.

Data of Burns *et al.*

Burns *et al.* (ref. 3) found that the mean HL's of 723 workers from a heavy industry (steel mills) was about 6 dB higher at all audiometric test frequencies from 0.5 to 6 kHz than the mean HL's of 291 of the same workers screened to be otologically normal and not having been exposed to gunfire. (See table 6.2.) Table 7.5 shows the measured HL data for unscreened and screened non-noise-exposed populations. Also shown are the HL's for the various groups as predicted from E_A and DL and the differences between predicted and measured HL's. It is shown that DL predicts the HL of both the screened and unscreened workers to be about 1 to 2 dB higher than does E_A when combined with any of the three non-noise-exposed populations—the unscreened U.S. population, the otologically screened populations of Robinson and Sutton (ref. 11), or the screened population of Burns and Robinson (ref. 8). Both the E_A and DL procedures predict about equally well (within 1 dB) the HL's expected on the basis of the Burns and Robinson non-noise-exposed people. However, this finding is somewhat gratuitous in that the 86.9-dBA workplace noise was estimated as that probably present in the factories on the basis of the observed HL of the screened factory workers and the Burns and Robinson prediction procedure. No noise measurements were actually made in the factories.

TABLE 7.5

Measured Mean and Predicted Median HL of Workers in Heavy Industry

[Measured data for males, avg. age of 62.5 years, with 45 years exposure to 86.9 dBA; data from ref. 3]

Parameter	Value, dB, at frequency, kHz, of—					Average difference
	0.5	1	2	3	4	
Unscreened workers versus unscreened U.S. population						
1. Measured mean HL of unscreened workers (723)	15.7	16.6	24.3	38.9	45.9	
2. Measured mean HL of unscreened U.S. population[a]	11	12	19	28	38	
3. Median HL predicted with E_A procedure[b]	1.9	3.0	6.5	12.0	14.7	
4. Median HL predicted with DL procedure[c]	4.7	7.2	9.7	12.5	15.5	
5. Row 1 minus (row 3 plus row 2)	2.8	1.6	−1.2	−1.1	−6.8	−0.94
6. Row 1 minus (row 4 plus row 2)	0	−2.6	−4.4	−1.6	−7.6	−3.24
Screened workers exposed to gunfire versus screened, non-exposed U.S. population						
7. Measured mean HL of screened workers (291)	10.7	11.1	17.1	32.4	39.7	
8. Age factor for HL of screened, non-exposed population[d]	6.9	7.9	13.9	22.8	31.7	
9. Age factor for HL of screened, non-exposed population[e]	8.8	9.3	12.4	16.0	23.2	
10. Row 7 minus (row 3 plus row 8)	1.9	.2	−3.3	−2.4	−6.7	−2.06
11. Row 7 minus (row 4 plus row 8)	−0.9	−4.0	−6.5	−2.9	−7.5	−4.36
12. Row 7 minus (row 3 plus row 9)	0	−1.2	−1.8	4.4	1.8	.64
13. Row 7 minus (row 4 plus row 9)	−2.8	−5.4	−5.0	3.9	1.0	−1.66

[a] *Based on ref. 15.*
[b] E_A = *104.9 dBA, calculated from 86.9 + 16.5 (from table 7.A1) + 1.5 (for males); from this value, HL is calculated from table 7.A2.*
[c] *DL = 86.4 dBA, calculated from 86.9 − 0.5 (from table 7.A1); from this value, HL is calculated from table 7.A4.*
[d] *From ref. 11 as given in ref. 13.*
[e] *From ref. 8 as given in ref. 12. (See table 7.A3.)*

TABLE 7.6

Measured and Predicted NIPTS for Foundry Workers

[Mean NIPTS at 4000 Hz as measured in ref. 86; median NIPTS predicted with procedure of Burns and Robinson (ref. 8) and with DL (proposed) procedure]

(a) Basic data

Parameter	NIPTS, dB, for age (years of exposure) of—			
	18-29(5)	30-39(15)	40-49(25)	50-65(38)
Cold mill, 86.8 dBA level				
Observed mean	5.0	7.0	17.0	18.0
E_A	95.3	100.1	102.3	104.1
Median predicted with Burns and Robinson procedure	5.1	8.8	11.3	13.5
Mean predicted with Burns and Robinson procedure	6.2	10.2	12.7	14.9
DL	76.8	81.6	83.8	85.6
Median predicted with DL (proposed) procedure	7.9	10.6	12.8	14.6
Slinger, 86.0 dBA level				
Observed mean	10.0	15.0	15.0	22.0
E_A	94.5	99.3	101.5	103.3
Median predicted with Burns and Robinson procedure	4.6	8.1	10.4	12.5
Mean predicted with Burns and Robinson procedure	5.7	9.5	11.8	13.9
DL	76.0	80.8	83.0	84.8
Median predicted with DL (proposed) procedure	7.5	9.9	12.0	13.8
Furnace, 89.0 dBA				
Observed mean	2.0	4.0	11.0	27.0
E_A	97.5	102.3	104.5	106.3
Median predicted with Burns and Robinson procedure	6.6	11.3	14.1	16.8
Mean predicted with Burns and Robinson procedure	7.9	12.7	15.5	17.9
DL	79.0	83.8	86.0	87.8
Median predicted with DL (proposed) procedure	9.0	12.8	15.0	16.8

(b) Comparison of data

Noise type	Difference (predicted minus observed) between NIPTS for age (years of exposure) of—				Average difference
	18-29(5)	30-39(15)	40-49(25)	50-65(38)	
Burns and Robinson (E_A) procedure					
Cold mill: median	0.1	1.8	−5.7	−4.5	[a]−4.34
mean	1.2	3.2	−4.3	−3.1	[a]−3.01
Slinger: median	−5.4	−6.9	−4.6	−9.5	[a]−4.34
mean	−4.3	−5.5	−3.2	−8.1	[a]−3.01
Furnace: median	4.6	7.3	3.1	−10.2	[b]−2.49
mean	5.9	8.7	4.5	−9.1	[b]−1.18
DL (proposed) procedure					
Cold mill, median	2.9	3.6	−4.2	−3.4	[a]−2.49
Slinger, median	−2.5	−5.1	−3.0	−8.2	[a]−2.49
Furnace, median	7.0	8.8	4.0	−10.2	[b]−0.86

[a] *Cold mill and slinger averaged together.*
[b] *Average of all three.*

Data of Martin *et al.*

Another comparison between the relative accuracy of these two prediction procedures is shown in table 7.6 with NIPTS data from Martin *et al.* (ref. 86). (See fig. 7.41.) As indicated in table 7.6 the Burns and Robinson procedure underpredicts the NIPTS at 4000 Hz from the noise in the three work areas by 2.49 dB for the median and 1.18 dB for the mean. The proposed procedure underpredicts the median by 0.86 dB.

Both procedures overpredict NIPTS from the furnace noise, except for the 50- to 65-year-old group. Except for the oldest group, this could to some extent be due to the fact the furnace noise environment was rather irregular in level, a major mode being around 82 dBA and others in the region of 92 to 98 dBA. (See fig. 7.42.) This would indicate that more effective recovery occurred with the irregular levels than with the more steady-state noises in the cold mill and slinger environments.

The noise measurements presented do not easily permit calculation of daily effective energy and recovery factors involved in the calculation of DL, but it was estimated with the aid of figure 7.42 that corrections would reduce the DL by about 2 dB or so for the furnace noise. This would reduce somewhat the magni-

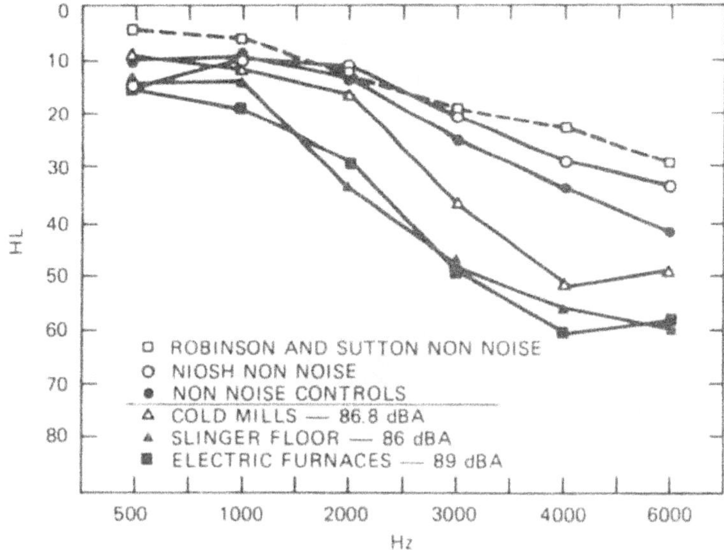

FIGURE 7.41. Average HL for various noise-exposed workers in a foundry at ages 50 to 65 years. NIPTS was taken as the difference between the "non-noise" and the "noise-exposed" groups. (Data from ref. 86.)

tude of the predicted NIPTS; in any event neither of the predictive schemes are very accurate in predicting NIPTS from the furnace noise.

Interindustry noise study

A carefully controlled study was conducted for a group of industries in the United States on the effects on hearing of steady-state noise (ref. 87). From some 78 000 workers, 155 males and 193 females were selected who were exposed to noise from 82 to 92 dBA, plus 96 males and 132 females who were not exposed to on-the-job noise greater than 75 dBA. The workers were screened otologically and for previous noise exposures. They had worked for at least 3 years in steady-state noise with a median of 15 years of noise exposure. Table 7.7 gives the HL's for males, right ear, at 4000 Hz. The years of exposure are estimated for each age grouping. This table shows that the Burns and Robinson procedure underpredicts median NIPTS by 0.80 dB, whereas the proposed method overpredicts median NIPTS by 1.18 dB.

One of the conclusions reached in reference 87 was that there was no apparent difference in NIPTS for men exposed to 82 dBA compared with those exposed to 92 dBA noise. However, Schori and Johnson (ref. 88) reanalyzed the data and concluded that a 90-dBA level would cause on the average about 6 dB more NIPTS after long-term exposure than would a level of 85 dBA.

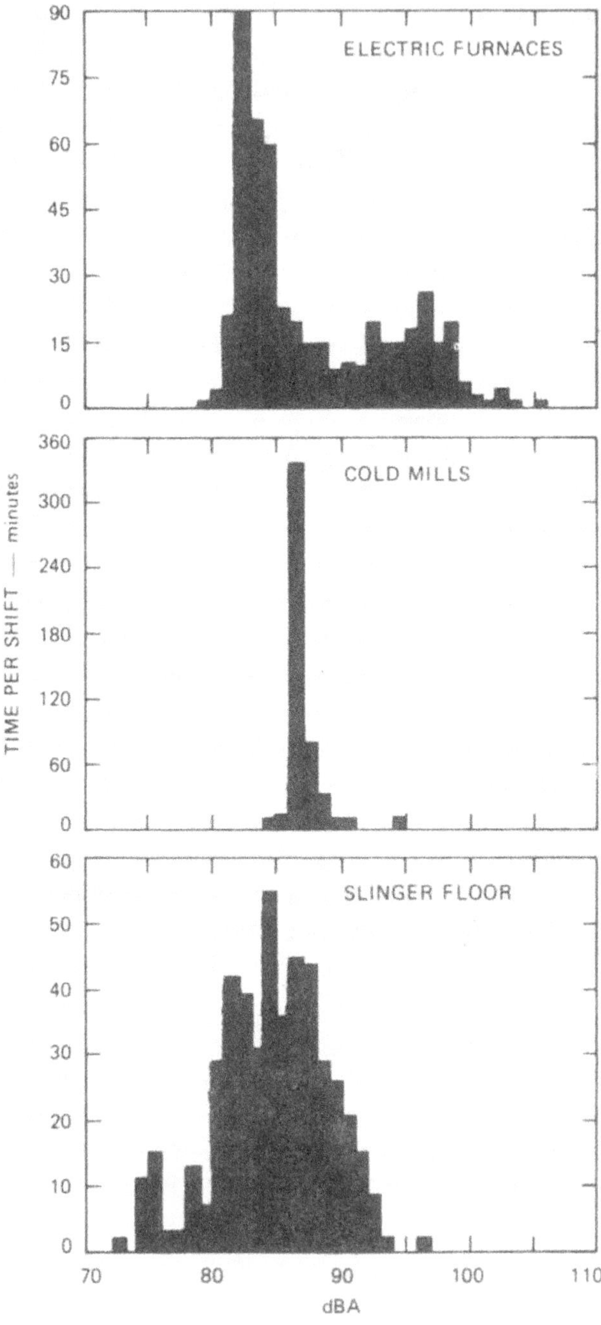

FIGURE 7.42. Daily exposure time as a function of noise level. (From ref. 86.)

TABLE 7.7

Mean HL and Measured and Predicted NIPTS for Males

[Steady-state noise of 82 to 92 dBA; 96 non-noise-exposed and 155 noise-exposed subjects; measured data from ref. 87]

Age, years	Estimated exposure, years	Measured mean HL, dBA, for—		Measured NIPTS, dBA, at 4000 Hz	Mean NIPTS, dBA, predicted with—		Median NIPTS, dBA, predicted with Burns and Robinson procedure	E_A, dB	DL, dBA	Difference[a] in NIPTS, dB, for—		
		Non-noise-exposed subjects	Noise-exposed subjects		DL (proposed) procedure	Burns and Robinson procedure				DL (proposed) procedure, mean	Burns and Robinson procedure, mean	Burns and Robinson procedure, median
≤30 (av. 25)	5	13.0	14.0	1.0	8.0	6.3	5.2	94	77	7.0	5.3	4.2
31-40 (av. 35)	10	15.8	23.4	7.6	9.5	8.7	7.4	97	80	1.9	1.1	−.2
41-50 (av. 45)	20	24.0	39.0	15.0	12.0	11.7	10.3	100	83	−3.0	−3.3	−4.7
≥51 (av. 58)	30	35.6	50.6	15.0	13.8	13.9	12.5	102	84.8	−1.2	−1.1	−2.5
Average difference										1.18	0.50	−0.80

[a] *Positive values indicate an overprediction and negative values indicate an underprediction.*

Summary of comparison of proposed procedure and Burns and Robinson procedure

As shown in table 7.8, the Burns and Robinson procedure and the procedure proposed herein have varying degrees of accuracy in predicting the means or medians of the several examples of measured NIPTS from exposures to steady-state noise for 8 hours per workday. The proposed (DL) procedure in general overestimates, compared with E_A, the estimated measured HL by about 3 dB in the Burns *et al.* study of hearing in a heavy industry. On the other hand, in the other two studies examined, the E_A procedure underestimates by about 2 dB the measured NIPTS, and DL is fairly accurate.

Although the E_A and DL procedures could be expected to predict somewhat similarly the NIPTS from steady-state noises, greater differences could be expected for daily exposures to interrupted noises. In those cases, the E_A procedure would presumably tend to overpredict NIPTS, because that procedure does not make any allowance, as does the proposed method, for the apparent reduction in the impact on the auditory system of interrupted (or intermittent) noises during the typical day in comparison with basically steady-state noise of equal energy.

The uncertainties involved in selecting a valid (or at least representative) control (non-noise exposed) reference group, variations in nosocusis among both the control and noise-exposed groups, and uncertainties regarding noise dosages make the validation of such predictive procedures difficult.

TABLE 7.8

Summary of Accuracy of Procedures for Predicting HL and NIPTS

[Positive values indicate overprediction and negative values indicate underprediction]

	Average difference between measured and predicted median (mean) values, dB, at 4 kHz from—		
Procedure	HL data of ref. 3 (table 7.5)	NIPTS data of ref. 86 (table 7.6)	NIPTS data of ref. 87 (table 7.7)
E_A (Burns and Robinson, ref. 8):			
Unscreened workers vs. population	−6.8		
Screened workers vs. screened population	−6.7		
Screened workers vs. control population	1.8	−2.49 (−1.18)	−0.80 (0.50)
DL (proposed):			
Unscreened workers vs. population	−7.6		
Screened workers vs. screened population	−7.5		
Screened workers vs. control population	1.0	−.86	1.18

Other studies of NIPTS

As mentioned previously, Johansson *et al.* (ref. 64) found that an irregular noise (see fig. 7.26) caused about 5 dB less NIPTS than is to be expected from steady-state noise of equal energy. The application of the equivalent recovery contribution to DL would provide about a 4 dB lower value than from equivalent energy alone or, presumably, would provide a prediction within about 1 dB of measured NIPTS.

Passchier-Vermeer (ref. 89) also found, in an analysis of data from a number of industrial studies on NIPTS, that the Burns and Robinson prediction procedures generally seemed to indicate underprediction of NIPTS from steady-state noise (as was the case in the comparison of the Baughn study with the Burns and Robinson study made earlier in the paper). For this reason, their equal-energy procedure might be expected to predict somewhat better the effects of intermittent noise because of the presence of greater effective recovery for energies equal

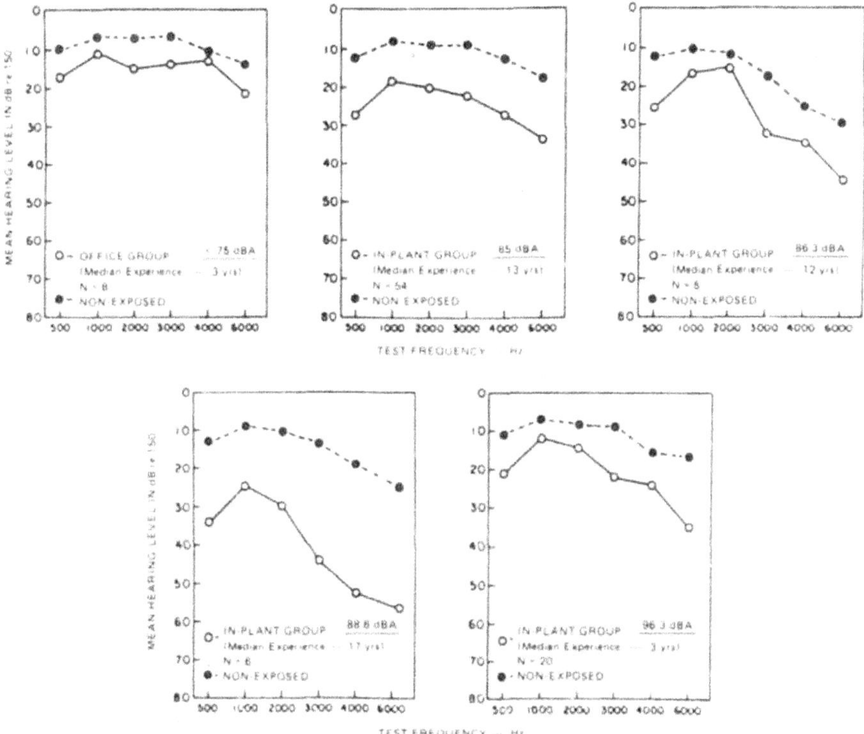

FIGURE 7.43. Mean hearing levels for workers in different job locations compared with non-noise-exposed groups equated in number, age, and sex composition. (From ref. 90.)

to steady state. Passchier-Vermeer reported that this seemed to be true for higher but not for lower levels of intermittent noise.

Figure 7.43, from Cohen *et al.* (ref. 90) illustrates the effects of exposure to noise as found in a number of studies conducted in industry. As shown in the figure, the amounts of PTS for the average person found in these examples are not inconsistent with the deductions reached in the present analysis of the Baughn and the Burns and Robinson data and of the prediction procedure developed herein. That is, noise levels in the range of 85 to 90 dBA will cause on the average more than 20 dB NIPTS at 4000 Hz with exposures of only 10 to 20 years.

Nonindustrial Noise

Identifying and quantifying the exposures of individuals to intense sounds and noises outside the workplace is generally difficult. For the most part these exposures may be relatively brief and infrequent so that their effect is perceived as being temporary, if noticed at all.

Noise in vehicles and recreational noise

Figure 7.44 shows examples of the levels of noise present in or near various modes of public and recreational transportation (refs. 91 to 94). Clearly the laboratory and field data on NIPTS (discussed earlier) would indicate that there is potential damage to hearing from these and other recreational noises, including small-arms fire (ref. 95) and firecrackers (refs. 96 and 97), and from exposure to vehicular noise of people outside the vehicles (refs. 58 and 98 to 100). Indeed, it is not surprising that sociocusis from these non-workplace noise environments appears to amount to about a 25-dB NIPTS at 4000 Hz by the age of 70 years or so. (See also the non-workplace, "everyday" noise dosage studies given in refs. 82 to 85.)

Music

A sound (and possible noise) that is of considerable interest is music, especially "rock and roll." This type of music is electronically amplified and presented *via* loudspeakers at levels considerably higher than those found with symphonic music, as shown in figure 7.45 (refs. 101 to 103).

Figure 7.45(a) gives the preexposure resting HL's of some rock and roll musicians and of a nonperforming control group of listeners. These musicians appear to have a permanent shift in their hearing thresholds, presumably because of their playing rock and roll music. Further, the musicians experienced some TTS_2 after an 85-minute performance.

Numerous other studies (refs. 104 to 134) of the effects of rock and roll music have also shown TTS of magnitudes comparable to those shown in figure 7.45

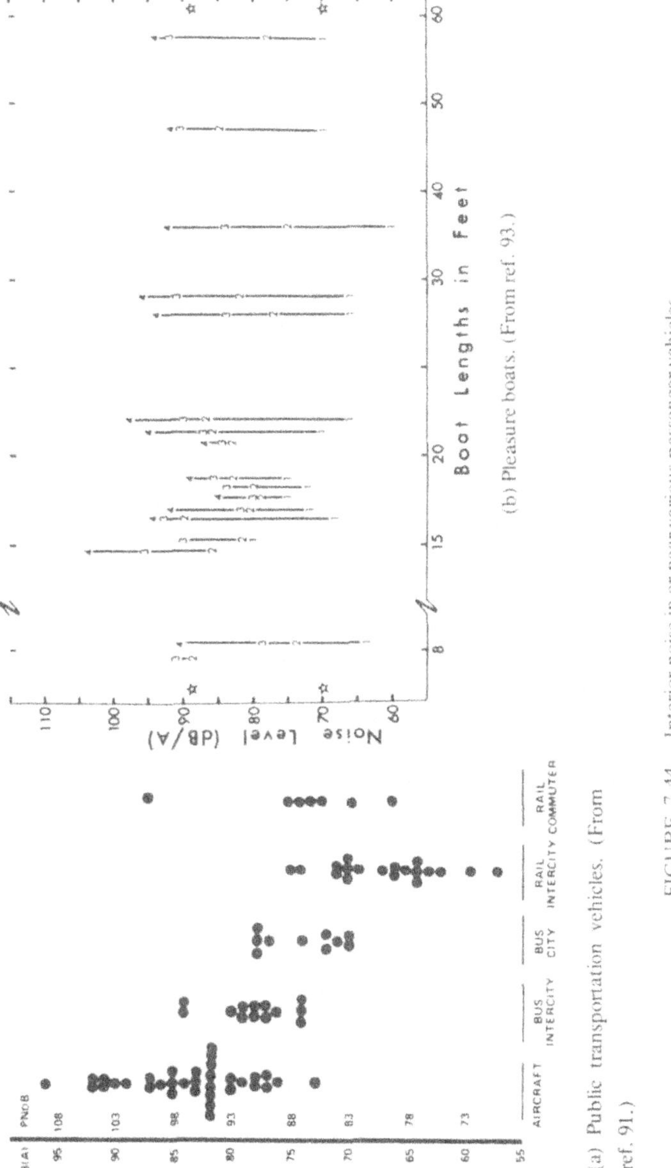

(a) Public transportation vehicles. (From ref. 91.)

(b) Pleasure boats. (From ref. 93.)

FIGURE 7.44. Interior noise in or near various passenger vehicles.

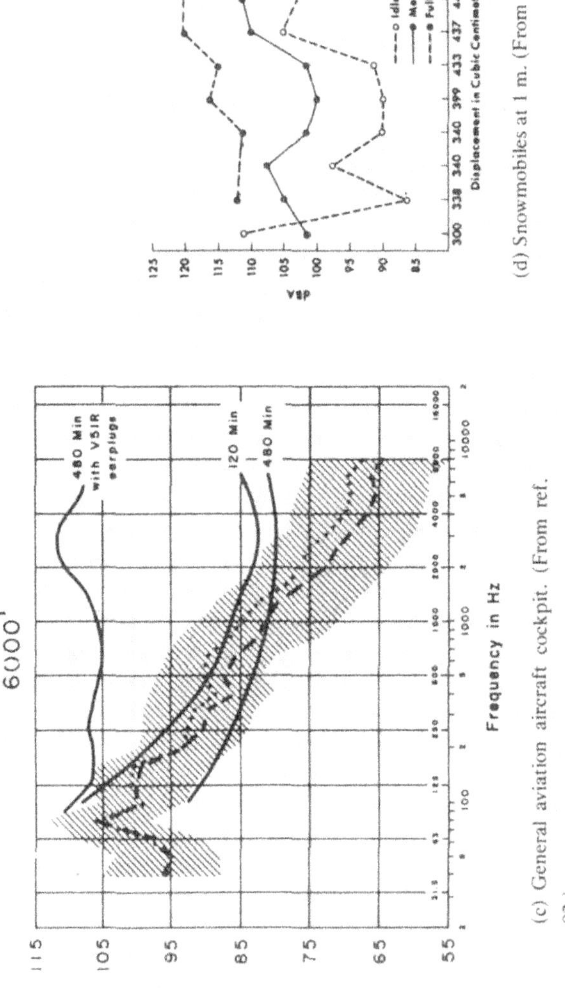

(c) General aviation aircraft cockpit. (From ref. 92.)

(d) Snowmobiles at 1 m. (From ref. 94.)

FIGURE 7.44. Concluded.

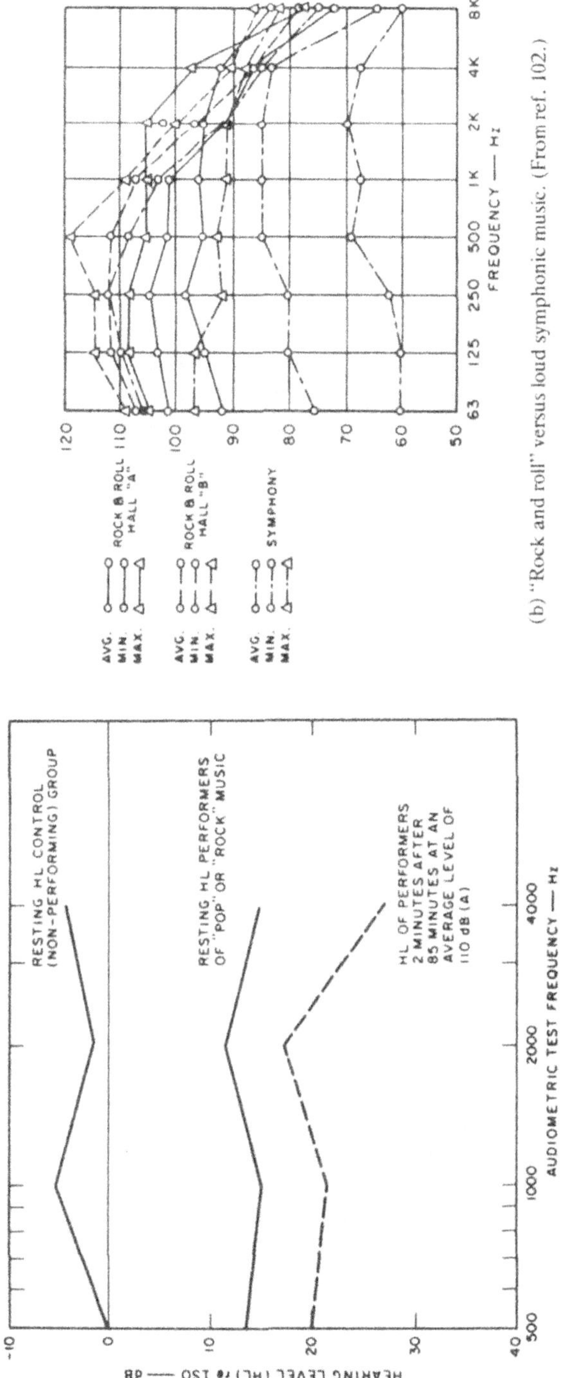

(a) Resting and post-performance HL. (Data from ref. 101.)

(b) "Rock and roll" versus loud symphonic music. (From ref. 102.)

FIGURE 7.45. Hearing levels and SPL's for various types of music.

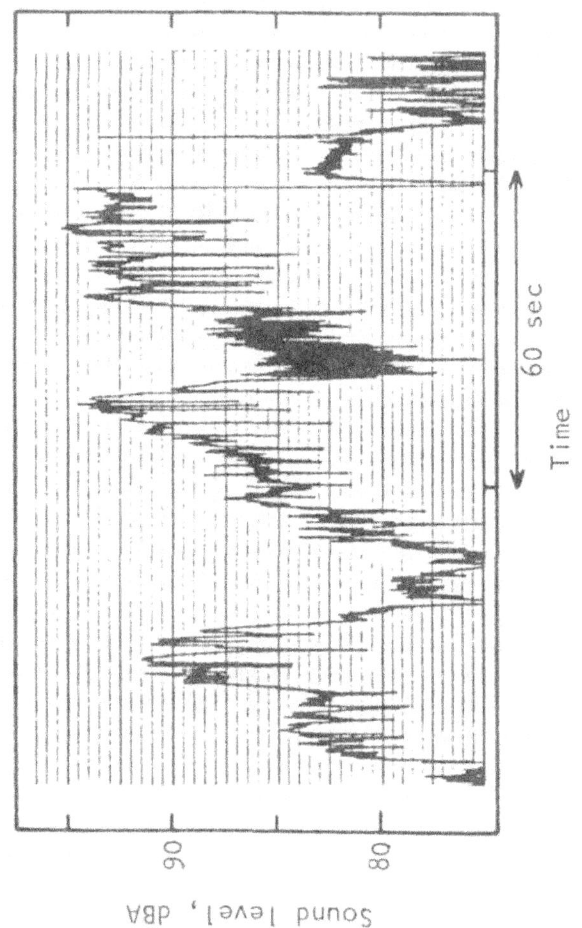

(c) "Heavy" symphonic music. (From ref. 103.)

FIGURE 7.45. Concluded.

when the intensity levels, the durations of exposure, and the times of measuring TTS are taken into account. However, average permanently elevated HL's of the magnitude shown in figure 7.45 have not always been found among groups of rock and roll musicians. The reasons for inconsistent findings with respect to permanent threshold shift from exposure to rock and roll music are as follows: first, the exposure periods are probably highly variable in duration and frequency among performers; second, the people involved tend to be young and have but few years of exposure to the music; and third, the music itself is somewhat variable or intermittent during a playing session.

There is reason to believe from the sound levels, probable durations, and frequencies of exposure involved that the audiences of rock and roll concerts are not likely to incur PTS therefrom. However, some musicians are likely to exhibit some music-induced PTS if performances are nearly daily for even a few years.

In 1974, Whittle and Robinson (ref. 135) reviewed studies of effects of "pop" music on TTS and PTS and concluded that such music need not be treated any differently than industrial noise in assessing its damage risk to hearing. However, calculating the typical long-term dosage for exposure to such intense music usually will be difficult.

The average levels of loud symphonic music recently measured by Jansson and Karlsson (ref. 103) are similar to those shown for symphonic music in figure 7.45(b). In addition, Jansson and Karlsson recorded the short-time variations in level for what they called "heavy" symphonic music (see fig. 7.45(c)) and determined its L_{eq}. The L_{eq} of the music indicated that, after 10 hours per week in front of the trumpets and 25 hours per week in a typical orchestral position, the exposure to the sound would be the equivalent of the maximum damage-risk exposure levels allowable according to Swedish standards for exposure to industrial noise. (The standards do not allow for recovery due to intermittency.)

However, these investigators found that the musicians had normal hearing, even though they had generally been exposed to the music for longer periods per week than those specified for damage risk. These results could be because of the intermittent nature of the music. As discussed above, less auditory fatigue and threshold shift would be expected from intermittent industrial noise or other sound than from more steady-state noise or sound of equal L_{eq}.

Summary of Procedures for Predicting NIPTS and HL

Over the past 20 years or so, several attempts have been made to organize data and concepts of noise-induced hearing loss, temporary and permanent, into graphs and mathematical expressions for predicting what the magnitude of these effects will be for a wide variety of possible noise-exposure conditions. As a matter of possible historical interest, these procedures are briefly summarized below in chronological order of their publication.

Kryter (1963 to 1965 and 1966)

A damage-risk criterion for exposures to steady-state noise was proposed by Kryter (refs. 20 and 22) as follows: average HL's of no more than 10 dB at 1000 Hz or below 15 dB at 2000 Hz, or 20 dB at 3000 Hz or above shall be considered as the maximum HL's that indicate an ability to correctly perceive sentences heard under optimum listening conditions. On the basis of laboratory experiments on temporary threshold shifts (TTS's) and some data on permanent threshold shifts (PTS's) in workers exposed to industrial noise, graphs (sometimes called "damage-risk curves") were prepared to show the general spectra, the durations of exposure to steady-state noise (up to 8 hours per day), and the years of exposure required to achieve the criterion mentioned in the average normal ear or, more properly, in no more than 50 percent of normal ears. On behalf of the Committee on Hearing, Bioacoustics, and Biomechanics (CHABA) of the National Research Council-National Academy of Sciences, these formulations were later combined with a calculation procedure developed by Ward *et al.* (ref. 136) that presumably allowed corrections to be made for the beneficial effects, relative to steady-state noise, of temporal interruptions in the noise. (See Kryter *et al.* (ref. 54).)

Baughn (1966 and 1973)

The strengths and weaknesses of Baughn's data (refs. 6 and 7) and analyses thereof have been discussed previously. Baughn did not develop tables of NIPTS but only of the distributions of HL's to be expected from exposure to 8 hours per day of steady-state noise from 78 dBA to 92 dBA for various numbers of years of exposure. Baughn recommended that the HL's of his subjects exposed to the 78-dBA noise be considered as representative of those for typical non-noise-exposed males 18 years old or so. The possibility that factory workers suffer more sociocusis and nosocusis than the general population is the major aspect of Baughn's data and analyses requiring consideration in their application to standards.

Robinson (1968) and Burns and Robinson (1970)

This program of research (ref. 8) on noise-induced threshold shifts and the mathematical generalizations drawn from references 8, 12, and 137 set a benchmark for the development of assessment models of noise-induced hearing loss.

The noise emission E_A assessment procedure derived by Burns and Robinson ostensibly provides estimated losses in hearing sensitivity at different test frequencies because of exposure to noise confounded with losses because of presbycusis. For this reason, Robinson and Shipton emphasize that the hearing losses as calculated by their formulas and tabulated should be used only in conjunction with the age and sex corrections provided in their report in order to predict the distribution of a noise-exposed population of otologically normal ears such as modeled in their study.

However, when corrected for the generalized percentile distribution of hearing sensitivities derived by Burns and Robinson, the increments in hearing loss as a function of E_A because of noise for a given number of years of exposure and at a given percentile represent increments in NIPTS. As such they can presumably be applied, within limits, to the distributions of HL's of populations of known otological characteristics in order to estimate the distribution of HL's to be expected in portions of the populations exposed to more or less steady-state noise.

Passchier-Vermeer (1968 to 1974)

Passchier-Vermeer (refs. 138 and 139) plotted on common sets of coordinates the HL data reported in 19 studies of the hearing of workers exposed to steady-state noise. The general magnitudes of hearing losses for the 25th, the 50th, and the 75th percentile of the noise-exposed groups as functions of noise level and years of exposure were derived from these graphs for different audiometric test frequencies.

The basic data (which did not include the data of Baughn and of Burns and Robinson) were quite variable among the different studies. This is not surprising in view of the small number of subjects involved in most of the studies (12 out of 19 had 50 or fewer subjects), and the audiometer zero used in some of the larger studies (*e.g.*, Glorig and Nixon (ref. 140)) was probably different than that for others. Figure 7.46 and tables 7.9 and 7.10 are those developed by Passchier-Vermeer for estimating NIPTS from exposure to noise.

For some specified conditions and test frequencies, there is reasonable agreement among NIPT's, as estimated with the models of Passchier-Vermeer and of Robinson and with that developed earlier in this paper from a combination of Baughn and Burns and Robinson data. However, for other conditions, large differences are found.

Coles *et al.* and CHABA (1968)

Coles *et al.* (ref. 141) recommend a procedure for the assessment of damage risk to hearing from only impulse noise. They recommended that impulse noises be divided into two types (see fig. 7.47) and that tolerable peak sound pressure levels for these two types would be those shown in figure 7.47, left-hand graph. By "tolerable," Coles *et al.* meant that no more than 25 percent of the people would have more than 10 dB NIPTS at 1000 Hz, 15 dB at 2000 Hz, and 20 dB at 3000 Hz if exposed to 50 to 200 impulses per day.

The use of A and B types is based on the observation by Coles *et al.* that guns fired in an enclosure or under some reverberant conditions cause more TTS or NIPTS than in the free field. This is undoubtedly a valid observation, but it is suggested that the basic physical parameters controlling these auditory fatigue effects are best represented by means of spectrum classifications and not by the

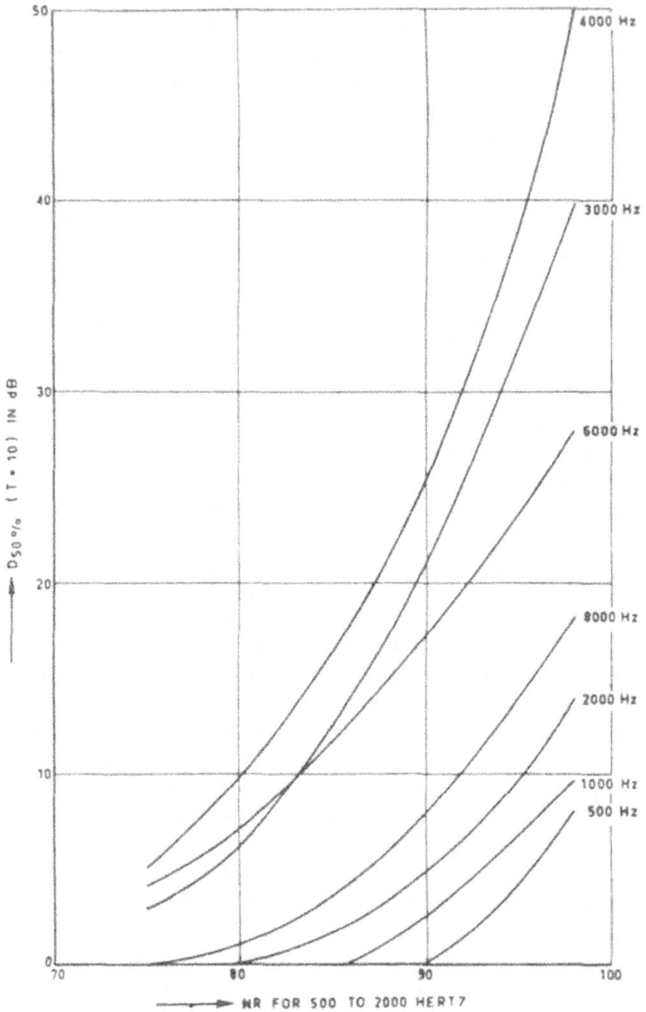

FIGURE 7.46. Median hearing loss ($D_{50\%}$) caused by exposure of 10 years ($T = 10$). Noise rating (NR) plus 5 dB gives approximate level of most broadband noises. (From ref. 139.)

A and B types. Keeping the rise time and the peak overpressure constant but making the impulse a form of damped sinusoid by means of reverberation would have the effect of modifying the spectrum of the sound in various significant ways and of increasing the amount of energy present at the ear at certain frequencies.

Coles *et al.* also suggest that the ear is about 5 dB less tolerant to an impulse approaching the ear canal directly (at normal incidence) than at grazing inci-

TABLE 7.9

Yearly Increase After 10 Years of Exposure of Median Hearing Losses Caused by Exposures to Noise

[Data from ref. 139]

Frequency, Hz	Increase of median hearing loss for exposure greater than 10 years, percent per year
500	2.0
1000	2.5
2000	10.0
3000	1.0
4000	0
6000	0[a]
	0.28[b]
8000	0[a]
	0.37[b]

[a] *NR* < *92 dB.*
[b] *NR* > *92 dB.*

dence, the latter being the more typical case for a person firing a gun. This recommendation, according to Coles *et al.*, is made on the basis of the following: (1) measurements by Golden and Clare (ref. 142) showing that the pressure from a gunshot at the position of the eardrum in an artificial auditory canal is about 6 dB greater with normal incidence than with the grazing incidence of the wave at the opening of the canal, and (2) some TTS data obtained by Hodge *et al.* (ref. 143) that indicate that about 6 dB more TTS occurs from the normal than from the grazing incident impulse from a gunshot.

Subsequent to the review of Coles *et al.*, a CHABA document (Ward, ref. 144) also proposed damage-risk exposures to impulses (gunfire). The tolerable limits are very similar to those proposed in figure 7.47 except the levels are lowered by 10 dB to protect the top 5 percent of the population. In addition, the following is proposed: (1) an equal-energy adjustment should be made for the number of impulses (*i.e.*, −3 dB, for each doubling of the number of impulses above 100 per day and 3 dB for each halving of the number of impulses below 100); and (2) set an upper limit of 179 dB for any type of impulse or condition of listening.

Kryter (1970 and 1973)

Kryter (ref. 26) proposed that the relation of peak SPL to TTS for gun noise can perhaps be approximated by the continuous functions shown in figure 7.48.

TABLE 7.10

Corrections for Hearing Loss at Differing Percentiles of Population

[Data from ref. 138]

Noise rating for 500 to 2000 Hz	Number of decibels to be added to median hearing loss in order to calculate 75th percentile loss for frequency, Hz, of—						
	500	1000	2000	3000	4000	6000	8000
75	0	0	0	0	4	0	0
80	0	0	1	0	3.5	1	1
85	0	0	2	2.5	3	2.5	2
90	0	0	3	4.5	2	3.5	3
94	0	0	4.5	4.5	.5	4	3
98	0	0.5	7	4.5	0	5	3

Noise rating for 500 to 2000 Hz	Number of decibels to be subtracted from median hearing loss in order to calculate 25th percentile loss for frequency, Hz, of—						
	500	1000	2000	3000	4000	6000	8000
75	0	0	0	1	5	1	0
80	0	0	0	1	5	3.5	0
85	0	0	.5	2.5	5	6	0
90	0	0	3	3.5	4	7	0
94	.5	0.5	4	3.5	2	7.5	0
98	1.5	1.5	5	3.5	1	8	0

* *NR + 5 ≅ 5 dBA.*

Except for a few data points, the long curve drawn in figure 7.48 seems to be a reasonably good estimate of the temporary threshold shift effects (given in table 7.3) of gun noise at grazing incidence. Even so, the tolerable limits of peak SPL suggested by Coles *et al.* and by CHABA are 7 dB higher than the data for the same criterion and conditions (see fig. 7.48).

Consistent with the TTS_2 data, a doubling of the number of impulses is equivalent to a 6-dB change in sound pressure level. However, to project this prediction procedure for TTS_2 to NIPTS, allowances must be made for the recovery processes, as described in reference 26 and in the present paper.

The temporal spacing of sessions of gunfire is difficult to generalize, but sessions are probably widely separated in a day of military service or recreational shooting. Accordingly, the effective daily exposure for NIPTS may often be accidentally proportional to that of the equal-energy procedure proposed by Coles *et*

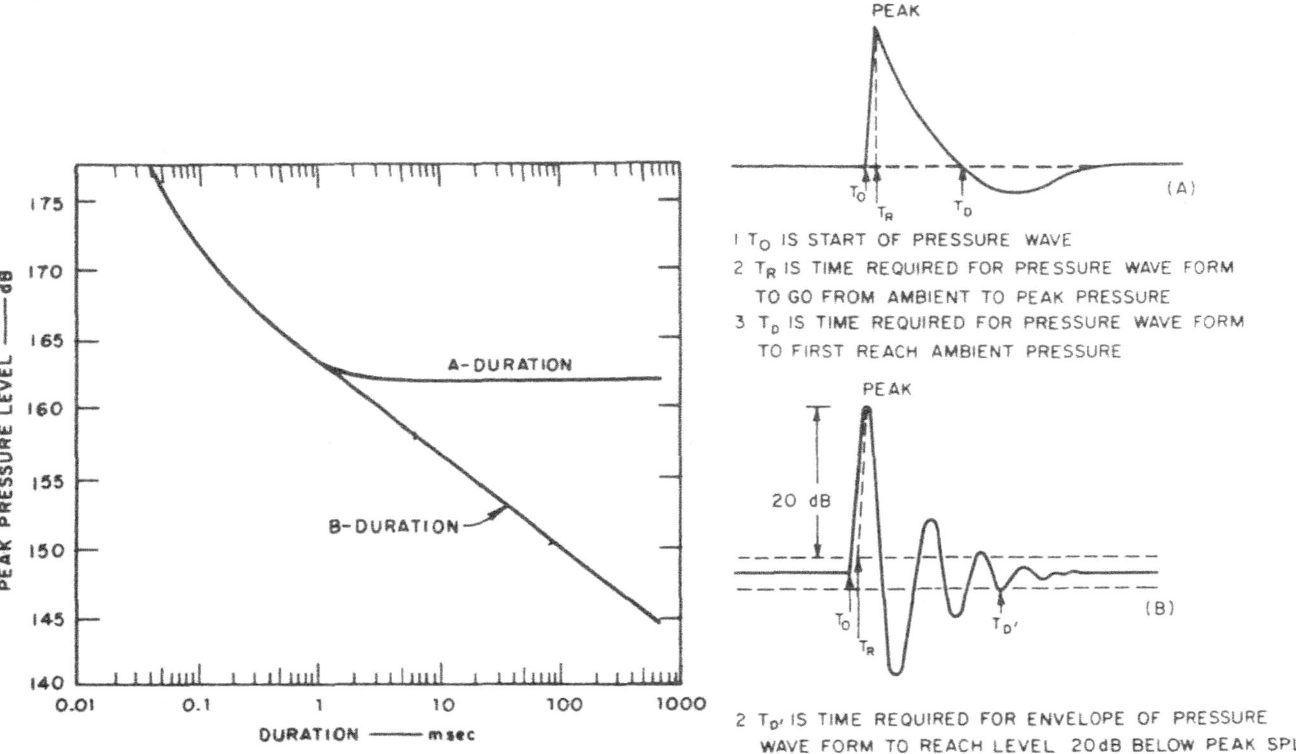

FIGURE 7.47. Peak pressure level and duration limits for 100 impulses daily having near-instantaneous rise times that will not produce more than 14 dB NIPTS in more than 25 percent of the people exposed. (From ref. 141.)

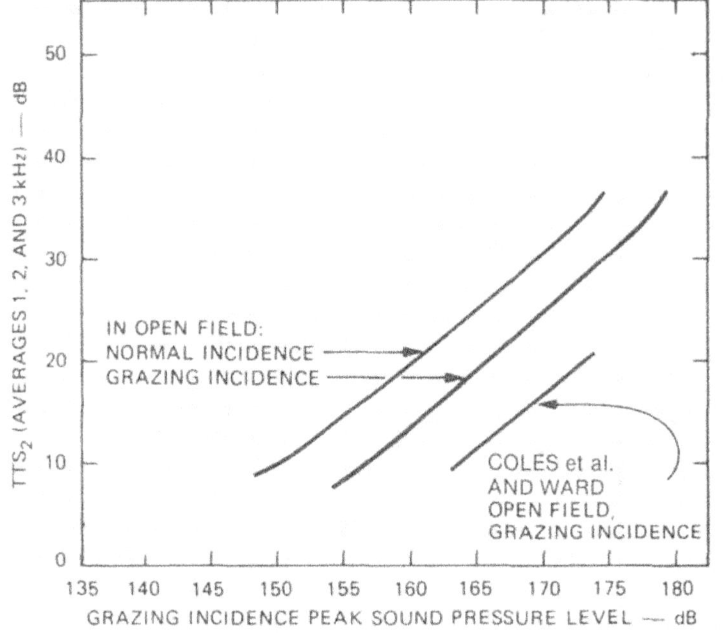

FIGURE 7.48. Tolerable exposure levels for a given TTS_2 (average of 1, 2, and 3 kHz). Short line gives tolerable limits for open field grazing incidence of gunfire impulses (100 per day) as prescribed by Coles *et al.* (ref. 141) and Ward (ref. 144).

al. and by CHABA; that is, the underestimation of TTS_2 from increased shooting time is offset by the lack of allowance for excess recovery between sessions of firing within a day.

Kryter (refs. 2 and 26) compared the percentages of noise-exposed workers whose HL's exceeded 15 and 25 dB (refs. 6 and 7) with the percentage of "better ear" males whose HL's also exceeded 15 and 25 dB (ref. 15) for the same frequencies and ages. One deficiency of this analysis was that of not recognizing and making allowances for greater sociocusis in factory workers than in the general population, as was later found. (See chapter 6 and fig. 7.1.) Also, the HL data for the general population were plotted in a way that tended to underestimate the thresholds of the population having less sensitive ears.

A second mistake was the linear extrapolation of functions relating NIPTS and TTS exceeding 10 dB to levels less than about 10 dB. Later data and analyses show that the rate of decline (with decreased noise) of NIPTS and TTS_2 below about 10 dB is about one-half of that projected from the higher levels of exposure. It is clear that these factors would lead to overestimation of extrapolated NIPTS and TTS, especially for lower values (below 10 dB).

EPA (1973 to 1974)

Johnson (ref. 145) prepared a report for the U.S. Environmental Protection Agency (EPA) on the prediction of NIPTS from exposure to continuous noise. This information was used as the basis for the identification of the maximum noise exposures that would not be harmful to the ear, with "an adequate margin of safety." This was defined as a lifetime exposure equivalent to 8 hours per workday of steady-state noise at a level of 75 dBA. As deduced by Johnson and maintained in the EPA documents, continuous noise at a level of 73 to 75 dBA for 8 hours per workday and a work career of 45 years (or the energy equivalent) will protect virtually the whole population from having more than 5 dB NIPTS at 4000 Hz. The EPA conclusions are based only on, and are presumably applicable only to, the assessment of nonimpulsive noise.

What was actually found by Johnson from the data and the formulations of Burns and Robinson (ref. 8) and of Passchier-Vermeer (refs. 138 and 139) was that after 40 years of exposure to this 75-dBA noise, 10 percent of the population would have about 8.5 dB NIPTS or greater at 4000 Hz, and the remaining 90 percent would have from about 1 to 8 dB NIPTS. (NIPTS for the median was about 2.5 dB; see ref. 145.) Realistically, an 8-hour noise exposure of 75 dBA was the maximum which should cause a loss in hearing of 8 dB or more in at least 10 percent of the population after 45 years of exposure. Johnson arrived at the somewhat lower value (5 dB) reported in the EPA documents by averaging, with the Burns and Robinson and the Passchier-Vermeer conclusions, Baughn's data for workers in 78 dBA noise as being in the non-noise-exposed category.

TABLE 7.11

Estimation of Risk of Having Impaired Hearing in Non-Noise-Exposed Populations

Data source	Total with impaired hearing[a], percent, for age (exposure), years, of—									
	20(0)	25(5)	30(10)	35(15)	40(20)	45(25)	50(30)	55(35)	60(40)	65(45)
ISO 1999[b]	1	2	3	5	7	10	14	21	33	50
U.S. Public Health Service[c]	0	0	0	0	5	8	10	13	19	20
Robinson and Sutton[d]	0	0	0	0	0	0	0	2	6	10

[a] *Impaired hearing defined as average HL at 0.5, 1, and 2 kHz of 26 dB or greater.*
[b] *Reference 146.*
[c] *References 14 and 15.*
[d] *Reference 11.*

ISO (1975)

The International Organization for Standardization (ISO) published a standard for assessing occupational noise exposure for hearing conservation purposes (ISO 1999, ref. 146). The technical information base for this standard is not identified. It is intended for the assessment of both steady-state and impulsive noise.

The standard recommends, among other things, (1) an equal-energy procedure for the calculation of equivalent continuous sound level, (2) a 10-dBA allowance for quasi-stable impulses of less than 1-sec duration, (3) a fence of 25 dB for the average HL at 0.5, 1, and 2 kHz as a start of hearing impairment for speech, and (4) a table for the estimation of the "risk" percentages of noise-exposed and non-noise-exposed people who will exceed that fence when exposed to noise of different sound levels. Given in table 7.11 is a portion of the ISO risk percentages along with comparable risk data based on the HL's of the general, unscreened population in the United States (U.S. Public Health Service, refs. 14 and 15) and from an analysis by Robinson and Sutton (ref. 11) of the HL's of otologically normal and non-noise-exposed males or females.

As shown in table 7.11, ISO 1999 claims that there are much larger percentages of the non-noise-exposed population who are above the fence than are found in the Robinson and Sutton analysis or in the general population. For example, by the age of 65 years, or 45 years of exposure, ISO 1999 indicates that 50 percent of the non-noise-exposed people are above the fence, whereas the U.S. Public Health Service studies show only 20 percent and Robinson and Sutton show only 10 percent above the fence. Accordingly, the increase in risk of hearing impairment with noise exposure is probably much greater than that predicted by the ISO 1999 risk table. For this reason, ISO 1999 is undergoing reevaluation and possibly modifications.

Appendix—Formulas and Tables for E_A and DL Procedures

TABLE 7.A1

Corrections to $L_{A,eq}$ to Calculate DL and E_A

Exposure, years	Value, dB, subtracted to calculate DL	Value, dB, added to calculate E_A[b]	Exposure, years	Value, dB, subtracted to calculate DL	Value, dB, added to calculate E_A	Exposure, years	Value, dB, subtracted to calculate DL	Value, dB, added to calculate E_A
0.25	23.0	−6.0	17	4.7	12.3	36	1.4	15.6
.5	20.0	−3.0	18	4.4	12.6	37	1.3	15.7
1	17.0	0	19	4.2	12.8	38	1.2	15.8
1.5	15.2	1.8	20	4.0	13.0	39	1.1	15.9
2	14.0	3.0	21	3.8	13.2	40	1.0	16.0
2.5	13.0	4.0	22	3.6	13.4	41	.9	16.1
3	12.2	4.8	23	3.4	13.6	42	.8	16.2
4	11.0	6.0	24	3.2	13.8	43	.7	16.3
5	10.0	7.0	25	3.0	14.0	44	.6	16.4
6	9.2	7.8	26	2.9	14.1	45	.5	16.5
7	8.5	8.5	27	2.7	14.3	46	.4	16.6
8	8.0	9.0	28	2.5	14.5	47	.3	16.7
9	7.5	9.5	29	2.4	14.6	48	.2	16.8
10	7.0	10.0	30	2.2	14.8	49	.1	16.9
11	6.6	10.4	31	2.1	14.9	50	0	17.0
12	6.2	10.8	32	1.9	15.1			
13	5.9	11.1	33	1.8	15.2			
14	5.5	11.5	34	1.7	15.3			
15	5.2	11.8	35	1.6	15.3			
16	5.0	12.0						

[a] *DL* = L_A − *20 log*(T − t) − *Exposure factor (see above), where* T *is 28 000 sec (8 hours) and t is duration of* L_A *in seconds.*

[b] E_A = $L_{A,eq}$ + *Exposure factor (see above) + sex factor (1.5 dB for males, −1.5 dB for females, and 0 dB for mixed sexes).*

TABLE 7.A2

Expected Hearing Loss for Percentages of Otologically Normal Population

[Percentages indicate population having expected or less hearing loss; values only to be used with age corrections in table 7.A3; data from ref. 12 but with reverse percentage designation]

Noise emission level, E_A	Hearing loss, dB, at 500 Hz for—				Hearing loss, dB, at 1000 Hz for—				Hearing loss, dB, at 2000 Hz for—				Hearing loss, dB, at 3000 Hz for—				Hearing loss, dB, at 4000 Hz for—			
	10% of population	50% of population	90% of population	Mean	10% of population	50% of population	90% of population	Mean	10% of population	50% of population	90% of population	Mean	10% of population	50% of population	90% of population	Mean	10% of population	50% of population	90% of population	Mean
80	−7.7	0.1	7.9	0.1	−7.6	0.1	8.0	0.2	−7.6	0.3	8.4	0.4	−7.5	0.5	9.2	0.7	−7.4	0.7	9.6	0.9
81	−7.7	0.1	7.9	0.1	−7.6	0.1	8.0	0.2	−7.6	0.3	8.5	0.4	−7.5	0.6	9.4	0.8	−7.4	0.8	9.9	1.1
82	−7.7	0.1	7.9	0.1	−7.6	0.1	8.1	0.2	−7.6	0.3	8.6	0.5	−7.4	0.7	9.6	0.9	−7.4	0.9	10.2	1.2
83	−7.7	0.1	8.0	0.1	−7.6	0.2	8.1	0.2	−7.5	0.4	8.8	0.5	−7.4	0.8	9.9	1.1	−7.3	1.1	10.5	1.4
84	−7.6	0.1	8.0	0.2	−7.6	0.2	8.2	0.3	−7.5	0.4	8.9	0.6	−7.4	0.9	10.2	1.2	−7.3	1.2	10.9	1.6
85	−7.6	0.1	8.1	0.2	−7.6	0.2	8.3	0.3	−7.5	0.5	9.1	0.7	−7.3	1.1	10.5	1.4	−7.2	1.4	11.3	1.8
86	−7.6	0.2	8.1	0.2	−7.6	0.2	8.4	0.3	−7.5	0.6	9.3	0.8	−7.3	1.2	10.9	1.6	−7.1	1.6	11.8	2.0
87	−7.6	0.2	8.2	0.2	−7.6	0.3	8.5	0.4	−7.4	0.7	9.5	0.9	−7.2	1.4	11.3	1.8	−7.0	1.8	12.4	2.3
88	−7.6	0.2	8.3	0.3	−7.6	0.3	8.6	0.4	−7.4	0.8	9.8	1.0	−7.1	1.6	11.8	2.0	−6.9	2.0	13.0	2.6
89	−7.6	0.2	8.3	0.3	−7.6	0.4	8.7	0.5	−7.4	0.9	10.0	1.1	−7.0	1.8	12.4	2.3	−6.8	2.3	13.6	3.0
90	−7.6	0.3	8.4	0.4	−7.5	0.4	8.8	0.6	−7.3	1.0	10.4	1.3	−6.9	2.0	13.0	2.6	−6.7	2.6	14.4	3.3
91	−7.6	0.3	8.5	0.4	−7.5	0.5	9.0	0.6	−7.3	1.1	10.7	1.5	−6.8	2.3	13.6	3.0	−6.6	3.0	15.2	3.8
92	−7.6	0.3	8.6	0.5	−7.5	0.5	9.2	0.7	−7.2	1.3	11.1	1.7	−6.7	2.6	14.4	3.3	−6.4	3.4	16.1	4.2
93	−7.5	0.4	8.8	0.5	−7.5	0.6	9.4	0.8	−7.2	1.5	11.6	1.9	−6.6	3.0	15.2	3.8	−6.3	3.8	17.1	4.8
94	−7.5	0.4	8.9	0.6	−7.4	0.7	9.6	0.9	−7.1	1.7	12.1	2.2	−6.4	3.4	16.1	4.2	−6.1	4.3	18.2	5.3
95	−7.5	0.5	9.1	0.7	−7.4	0.8	9.9	1.1	−7.0	1.9	12.7	2.5	−6.3	3.8	17.1	4.8	−5.8	4.9	19.4	6.0
96	−7.5	0.6	9.3	0.8	−7.4	0.9	10.2	1.2	−6.9	2.2	13.3	2.8	−6.1	4.3	18.2	5.3	−5.6	5.5	20.7	6.7
97	−7.4	0.7	9.5	0.9	−7.3	1.1	10.5	1.4	−6.8	2.4	14.0	3.1	−5.8	4.9	19.4	6.0	−5.3	6.2	22.0	7.4

98	−7.4	0.8	9.8	1.0	−7.3	1.2	10.9	1.6	−6.7	2.8	14.8	3.5	−5.6	5.5	20.7	6.7	−5.0	7.0	23.5	8.3
99	−7.4	0.9	10.0	1.1	−7.2	1.4	11.3	1.8	−6.5	3.2	15.7	4.0	−5.3	6.2	22.0	7.4	−4.6	7.8	25.0	9.2
100	−7.3	1.0	10.4	1.3	−7.1	1.6	11.8	2.0	−6.4	3.6	16.6	4.5	−5.0	7.0	23.5	8.3	−4.2	8.7	26.7	10.1
101	−7.3	1.1	10.7	1.5	−7.0	1.8	12.4	2.3	−6.2	4.0	17.7	5.0	−4.6	7.8	25.0	9.2	−3.7	9.8	28.3	11.2
102	−7.2	1.3	11.1	1.7	−6.9	2.0	13.0	2.6	−6.0	4.6	18.8	5.6	−4.2	8.7	26.7	10.1	−3.2	10.9	30.1	12.3
103	−7.2	1.5	11.6	1.9	−6.8	2.3	13.6	3.0	−5.7	5.2	20.0	6.3	−3.7	9.8	28.3	11.2	−2.6	12.1	31.9	13.5
104	−7.1	1.7	12.1	2.2	−6.7	2.6	14.4	3.3	−5.4	5.8	21.3	7.0	−3.2	10.9	30.1	12.3	−2.0	13.4	33.7	14.8
105	−7.0	1.9	12.7	2.5	−6.6	3.0	15.2	3.8	−5.1	6.6	22.8	7.8	−2.6	12.1	31.9	13.5	−1.3	14.8	35.5	16.1
106	−6.9	2.2	13.3	2.8	−6.4	3.4	16.1	4.2	−4.8	7.4	24.3	8.7	−2.0	13.4	33.7	14.8	−0.5	16.3	37.4	17.5
107	−6.8	2.4	14.0	3.1	−6.3	3.8	17.1	4.8	−4.4	8.3	25.8	9.6	−1.3	14.8	35.5	16.1	0.4	17.8	39.2	18.9
108	−6.7	2.8	14.8	3.5	−6.1	4.3	18.2	5.3	−4.0	9.2	27.5	10.7	−0.5	16.3	37.4	17.5	1.4	19.5	40.9	20.4
109	−6.5	3.2	15.7	4.0	−5.8	4.9	19.4	6.0	−3.5	10.3	29.2	11.7	0.4	17.8	39.2	18.9	2.4	21.2	42.7	21.9
110	−6.4	3.6	16.6	4.5	−5.6	5.5	20.7	6.7	−2.9	11.5	31.0	12.9	1.4	19.5	40.9	20.4	3.6	23.0	44.3	23.5
111	−6.2	4.0	17.7	5.0	−5.3	6.2	22.0	7.4	−2.3	12.7	32.8	14.1	2.4	21.2	42.7	21.9	4.8	24.8	45.9	25.1
112	−6.0	4.6	18.8	5.6	−5.0	7.0	23.5	8.3	−1.6	14.1	34.6	15.4	3.6	23.0	44.3	23.5	6.1	26.6	47.4	26.7
113	−5.7	5.2	20.0	6.3	−4.6	7.8	25.0	9.2	−0.9	15.5	36.5	16.8	4.8	24.8	45.9	25.1	7.6	28.4	48.9	28.3
114	−5.4	5.8	21.3	7.0	−4.2	8.7	26.7	10.1	−0.1	17.1	38.3	18.2	6.1	26.6	47.4	26.7	9.1	30.2	50.2	29.9
115	−5.1	6.6	22.8	7.8	−3.7	9.8	28.3	11.2	0.9	18.7	40.1	19.6	7.6	28.4	48.9	28.3	10.7	32.0	51.4	31.5
116	−4.8	7.4	24.3	8.7	−3.2	10.9	30.1	12.3	1.9	20.3	41.8	21.2	9.1	30.2	50.2	29.9	12.[illegible]	33.8	52.6	33.1
117	−4.4	8.3	25.8	9.6	−2.6	12.1	31.9	13.5	3.0	22.1	43.5	22.7	10.7	32.0	51.4	31.5	14.1	35.5	53.6	34.6
118	−4.0	9.2	27.5	10.7	−2.0	13.4	33.7	14.8	4.2	23.9	45.1	24.3	12.3	33.8	52.6	33.1	15.8	37.2	54.6	36.1
119	−3.5	10.3	29.2	11.7	−1.3	14.8	35.5	16.1	5.5	25.7	46.7	25.9	14.1	35.5	53.6	34.6	17.6	38.7	55.5	37.5
120	−2.9	11.5	31.0	12.9	−0.5	16.3	37.4	17.5	6.8	27.5	48.2	27.5	15.8	37.2	54.6	36.1	19.5	40.2	56.3	38.9
121	−2.3	12.7	32.8	14.1	0.4	17.8	39.2	18.9	8.3	29.3	49.5	29.1	17.6	38.7	55.5	37.5	21.3	41.6	57.0	40.2
122	−1.6	14.1	34.6	15.4	1.4	19.5	40.9	20.4	9.9	31.1	50.8	30.7	19.5	40.2	56.3	38.9	23.1	42.9	57.6	41.5
123	−0.9	15.5	36.5	16.8	2.4	21.2	42.7	21.9	11.5	32.9	52.0	32.3	21.3	41.6	57.0	40.2	24.9	44.1	58.2	42.7
124	−0.1	17.1	38.3	18.2	3.6	23.0	44.3	23.5	13.2	34.7	53.1	33.8	23.1	42.9	57.6	41.5	26.7	45.2	58.7	43.8
125	0.9	18.7	40.1	19.6	4.8	24.8	45.9	25.1	14.9	36.3	54.1	35.4	24.9	44.1	58.2	42.7	28.3	46.3	59.2	44.9
126	1.9	20.3	41.8	21.2	6.1	26.6	47.4	26.7	16.7	37.9	55.1	36.8	26.7	45.2	58.7	43.8	30.0	47.2	59.6	45.8
127	3.0	22.1	43.5	22.7	7.6	28.4	48.9	28.3	18.5	39.5	55.9	38.2	28.3	46.3	59.2	44.9	31.5	48.0	60.0	46.7
128	4.2	23.9	45.1	24.3	9.1	30.2	50.2	29.9	20.4	40.9	56.6	39.6	30.0	47.2	59.6	45.8	33.0	48.8	60.3	47.6
129	5.5	25.7	46.7	25.9	10.7	32.0	51.4	31.5	22.2	42.3	57.3	40.9	31.5	48.0	60.0	46.7	34.3	49.5	60.6	48.3
130	6.8	27.5	48.2	27.5	12.3	33.8	52.6	33.1	24.0	43.5	57.9	42.1	33.0	48.8	60.3	47.6	35.6	50.1	60.8	49.0

TABLE 7.A3

Age Correction for Audiometric Test Frequencies

[Data from ref. 12]

Age, years	Correction, dB, for audiometric frequency, kHz, of—									
	0.5	1	2	3	4	6	Avg. of 0.5, 1, and 2	Avg. of 1, 2, and 3	Avg. of 1, 2, and 4	Avg. of 3, 4, and 6
20	0	0	0	0	0	0	0	0	0	0
21	0	0	0	0	0	0	0	0	0	0
22	0	0	0	0	0	.1	0	0	0	0
23	0	0	.1	.1	.1	.1	0	.1	.1	.1
24	.1	.1	.1	.1	.2	.2	.1	.1	.1	.2
25	.1	.1	.2	.2	.3	.4	.1	.2	.2	.3
26	.1	.2	.2	.3	.4	.5	.2	.2	.3	.4
27	.2	.2	.3	.4	.6	.7	.2	.3	.4	.6
28	.3	.3	.4	.5	.8	.9	.3	.4	.5	.7
29	.3	.3	.5	.6	1.0	1.1	.4	.5	.6	.9
30	.4	.4	.6	.8	1.2	1.4	.5	.6	.7	1.1
31	.5	.5	.7	1.0	1.5	1.7	.6	.7	.9	1.4
32	.6	.6	.9	1.2	1.7	2.0	.7	.9	1.1	1.6
33	.7	.7	1.0	1.4	2.0	2.4	.8	1.0	1.3	1.9
34	.8	.8	1.2	1.6	2.4	2.7	.9	1.2	1.5	2.2
35	.9	1.0	1.3	1.8	2.7	3.2	1.1	1.4	1.7	2.5
36	1.0	1.1	1.5	2.0	3.1	3.6	1.2	1.6	1.9	2.9
37	1.2	1.2	1.7	2.3	3.5	4.0	1.4	1.8	2.1	3.3
38	1.3	1.4	1.9	2.6	3.9	4.5	1.5	2.0	2.4	3.7
39	1.4	1.6	2.2	2.9	4.3	5.1	1.7	2.2	2.7	4.1
40	1.6	1.7	2.4	3.2	4.8	5.6	1.9	2.4	3.0	4.5
41	1.8	1.9	2.6	3.5	5.3	6.2	2.1	2.7	3.3	5.0
42	1.9	2.1	2.9	3.9	5.8	6.8	2.3	3.0	3.6	5.5
43	2.1	2.3	3.2	4.2	6.3	7.4	2.5	3.2	3.9	6.0
44	2.3	2.5	3.5	4.6	6.9	8.1	2.7	3.5	4.3	6.5
45	2.5	2.7	3.7	5.0	7.5	8.7	3.0	3.8	4.6	7.1
46	2.7	2.9	4.1	5.4	8.1	9.5	3.2	4.1	5.0	7.7
47	2.9	3.1	4.4	5.8	8.7	10.2	3.5	4.4	5.4	8.3
48	3.1	3.4	4.7	6.3	9.4	11.0	3.7	4.8	5.8	8.9
49	3.4	3.6	5.0	6.7	10.1	11.8	4.0	5.1	6.3	9.5
50	3.6	3.9	5.4	7.2	10.8	12.6	4.3	5.5	6.7	10.2
51	3.8	4.1	5.8	7.7	11.5	13.5	4.6	5.9	7.1	10.9
52	4.1	4.4	6.1	8.2	12.3	14.3	4.9	6.2	7.6	11.6
53	4.4	4.7	6.5	8.7	13.1	15.2	5.2	6.6	8.1	12.3
54	4.6	5.0	6.9	9.2	13.9	16.2	5.5	7.1	8.6	13.1
55	4.9	5.3	7.4	9.8	14.7	17.2	5.8	7.5	9.1	13.9
56	5.2	5.6	7.8	10.4	15.6	18.1	6.2	7.9	9.6	14.7
57	5.5	5.9	8.2	11.0	16.4	19.2	6.5	8.4	10.2	15.5
58	5.8	6.2	8.7	11.6	17.3	20.2	6.9	8.8	10.7	16.4

TABLE 7.A3 Concluded

Age, years	Correction, dB, for audiometric frequency, kHz, of—									
	0.5	1	2	3	4	6	Avg. of 0.5, 1, and 2	Avg. of 1, 2, and 3	Avg. of 1, 2, and 4	Avg. of 3, 4, and 6
59	6.1	6.5	9.1	12.2	18.3	21.3	7.3	9.3	11.3	17.2
60	6.4	6.9	9.6	12.8	19.2	22.4	7.6	9.8	11.9	18.1
61	6.7	7.2	10.1	13.4	20.2	23.5	8.0	10.3	12.5	19.1
62	7.1	7.6	10.6	14.1	21.2	24.7	8.4	10.8	13.1	20.0
63	7.4	8.0	11.1	14.8	22.2	25.9	8.8	11.3	13.7	21.0
64	7.7	8.3	11.6	15.5	23.2	27.1	9.2	11.8	14.4	21.9
65	8.1	8.7	12.1	16.2	24.3	28.4	9.7	12.4	15.1	22.9
66	8.5	9.1	12.7	16.9	25.4	29.6	10.1	12.9	15.7	24.0
67	8.8	9.5	13.3	17.7	26.5	30.9	10.5	13.5	16.4	25.0
68	9.2	9.9	13.8	18.4	27.6	32.3	11.0	14.1	17.1	26.1
69	9.6	10.3	14.4	19.2	28.8	33.6	11.4	14.6	17.8	27.2
70	10.0	10.7	15.0	20.0	30.0	35.0	11.9	15.2	18.6	28.3

TABLE 7.A4

NIPTS Expected at Audiometric Test Frequencies

DL, dB, for audiometric frequency, kHZ, of—						NIPTS, dB, at population percent of—			DL, dB, for audiometric frequency, kHZ, of—						NIPTS, dB, at population percent of—		
0.5	1 and avg. of 0.5, 1, and 2	Avg. of 1, 2, and 3	2 and avg. of 1, 2, and 4	3	4	10	50	90	0.5	1 and avg. of 0.5, 1, and 2	Avg. of 1, 2, and 3	2 and avg. of 1, 2, and 4	3	4	10	50	90
75	70	66	65	62	59			0	102	97	93	92	89	86	13	15	17
76	71	67	66	63	60			.5	103	98	94	93	90	87	14	16	18
77	72	68	67	64	61		0	1.0	104	99	95	94	91	88	15	17	19
78	73	69	68	65	62		.5	1.5	105	100	96	95	92	89	16	18	20
79	74	70	69	66	63	0	1.0	2.0	106	101	97	96	93	90	17	19	21
80	75	71	70	67	64	.5	1.5	2.5	107	102	98	97	94	91	18	20	22
81	76	72	71	68	65	1.0	2.0	3.0	108	103	99	98	95	92	19	21	23
82	77	73	72	69	66	1.5	2.5	3.5	109	104	100	99	96	93	20	22	24
83	78	74	73	70	67	2.0	3.0	4.0	110	105	101	100	97	94	21	23	25
84	79	75	74	71	68	2.5	3.5	4.5	111	106	102	101	98	95	22	24	26
85	80	76	75	72	69	3.0	4.0	5.0	112	107	103	102	99	96	23	25	27
86	81	77	76	73	70	3.5	4.5	5.5	113	108	104	103	100	97	24	26	28
87	82	78	77	74	71	4.0	5.0	6.0	114	109	105	104	101	98	25	27	29
88	83	79	78	75	72	4.5	5.5	6.5	115	110	106	105	102	99	26	28	30
89	84	80	79	76	73	5.0	6.0	7.0	116	111	107	106	103	100	27	29	31
90	85	81	80	77	74	5.5	6.5	7.5	117	112	108	107	104	101	28	30	32
91	86	82	81	78	75	6.0	7.0	8.0	118	113	109	108	105	102	29	31	33
92	87	83	82	79	76	6.5	7.5	8.5	119	114	110	109	106	103	30	32	34
93	88	84	83	80	77	7.0	8.0	9.0	120	115	111	110	107	104	31	33	35
94	89	85	84	81	78	7.5	8.5	9.5	121	116	112	111	108	105	32	34	36
95	90	86	85	82	79	8.0	9.0	10.0	122	117	113	112	109	106	33	35	37
96	91	87	86	83	80	8.5	9.5	11.0	123	118	114	113	110	107	34	36	38
97	92	88	87	84	81	9.0	10.0	12	124	119	115	114	111	108	35	37	39
98	93	89	88	85	82	9.5	11	13	125	120	116	115	112	109	36	38	40
99	94	90	89	86	83	10	12	14	126	121	117	116	113	110	37	39	
100	95	91	90	87	84	11	13	15	127	122	118	117	114	111	38	40	
101	96	92	91	88	85	12	14	16	128	123	119	118	115	112	39		
									129	124	120	119	116	113	40		

References

1. Botsford, James H.: Prevalence of Impaired Hearing and Sound Levels at Work. J. Acoust. Soc. America, vol. 45, no. 1, Jan. 1969, pp. 79-82.
2. Kryter, Karl D.: Impairment to Hearing From Exposure to Noise. J. Acoust. Soc. America, vol. 53, no. 5, May 1973, pp. 1211-1234.
3. Burns, W.; Robinson, D. W.; Shipton, M. S.; and Sinclair, A.: Hearing Hazard From Occupational Noise: Observations on a Population From Heavy Industry. NPL Acoustics Rep. Ac 80, British A.R.C., Jan. 1977.
4. Sulkowski, Wieslaw: Some Epidemiological Data on Noise-Induced Hearing Loss in Poland, Its Prophylaxis and Diagnosis. Proceedings of the International Congress on Noise as a Public Health Problem, W. Dixon Ward, ed., 550/9-73-008, U.S. Environ. Prot. Agency, May 1973, pp. 139-155.
5. Howell, R. W.: A Seven-Year Review of Measured Hearing Levels in Male Manual Steelworkers With High Initial Thresholds. British J. Ind. Med., vol. 35, no. 1, Feb. 1978, pp. 27-31.
6. Baughn, W. L.: Noise Control—Percent of Population Protected. Int. Audiology, vol. V, no. 3, Sept. 1966, pp. 331-338.
7. Baughn, William L.: Relation Between Daily Noise Exposure and Hearing Loss Based on the Evaluation of 6,835 Industrial Noise Exposure Cases. Rep. AMRL-TR-7353, U.S. Air Force, June 1973. (Available from DTIC as AD 767 204.)
8. Burns, W.; and Robinson, D. W.: Hearing and Noise in Industry. Her Majesty's Stationery Office, 1970.
9. Raber, A.: The Incidence of Impaired Hearing in Relation to Years of Exposure and Continuous Sound Level (Preliminary Analysis of 26,179 Cases). Proceedings of the International Congress on Noise as a Public Health Problem, W. Dixon Ward, ed., 550/9-73-008, U.S. Environ. Prot. Agency, May 1973, pp. 115-138.
10. Royster, L. H.; and Thomas, W. G.: Age Effect Hearing Levels for a White Nonindustrial Noise Exposed Population (NINEP) and Their Use in Evaluating Industrial Hearing Conservation Programs. J. American Ind. Hyg. Assoc., vol. 40, no. 6, June 1979, pp. 504-511.
11. Robinson, D. W.; and Sutton, G. J.: A Comparative Analysis of Data on the Relation of Pure-Tone Audiometric Thresholds to Age. NPL Acoustics Rep. Ac 84, British A.R.C., Apr. 1978.
12. Robinson, D. W.; and Shipton, M. S.: Tables for the Estimation of Noise-Induced Hearing Loss. NPL Acoustics Rep. Ac 61 (Second ed.), British A.R.C., June 1977.
13. Shipton, M.S.: Tables Relating Pure-Tone Audiometric Threshold to Age. NPL Acoustics Rep. Ac 94, British A.R.C., Nov. 1979.
14. Glorig, Aram; and Roberts, Jean: Hearing Levels of Adults by Age and

Sex: United States, 1960–1962. Rep. PHS-PUB-1000-SER-11-11, Nat. Cent. Health Serv. Res. and Dev., Oct. 1965. (Available from NTIS as PB 267 177/4.)

15. Rowland, Michael: Basic Data on Hearing Levels of Adults 25–74 Years. DHEW Publ. No. (PHS) 80-1663, Series 11, No. 215, U.S. Dep. Health, Educ., & Welfare, Jan. 1980.
16. Rop, Ilse; and Raber, Alfred: Application of a Linear Logistic Model To Describe Hearing Impairment as a Function of Noise Exposure and Age. Noise as a Public Health Problem, Jerry V. Tobias, Gerd Jansen, and W. Dixon Ward, eds., ASHA Rep. 10, American Speech-Language-Hearing Assoc., Apr. 1980, pp. 119–123.
17. Macrae, J. H.: Noise-Induced Hearing Loss and Presbyacusis. Audiology, vol. 10, 1971, pp. 323–333.
18. Loeb, M.; and Fletcher, J. L.: Temporary Threshold Shift for "Normal" Subjects as a Function of Age and Sex. J. Auditory Res., vol. 3, 1963, pp. 65–72.
19. Hetu, R.; and Parrot, J.: A Field Evaluation of Noise-Induced Temporary Threshold Shift. J. American Ind. Hyg. Assoc., vol. 39, no. 4, Apr. 1978, pp. 301–311.
20. Kryter, K. D.: Damage Risk Criterion and Contours Based on Permanent and Temporary Hearing Loss Data. J. American Ind. Hyg. Assoc., vol. 26, Jan.-Feb. 1965, pp. 34–44.
21. Kraak, W.; Ertel, H.; Fuder, G.; and Kracht, L.: Risk of Hearing Damage Caused by Steady-State and Impulsive Noise. J. Sound & Vib., vol. 36, no. 3, Oct. 8, 1974, pp. 347–359.
22. Kryter, Karl D.: Exposure to Steady-State Noise and Impairment of Hearing. J. Acoust. Soc. America, vol. 35, no. 10, Oct. 1963, pp. 1515–1525.
23. Cohen, Alexander; Anticaglia, Joseph R.; and Carpenter, Paul L.: Temporary Threshold Shift in Hearing From Exposure to Different Noise Spectra at Equal dBA Levels. J. Acoust. Soc. America, vol. 51, no. 2, pt. 2, Feb. 1972, pp. 503–507.
24. Robinson, D. W.: Characteristics of Occupational Noise-Induced Hearing Loss. Effects of Noise on Hearing, Donald Henderson, Roger P. Hamernik, Darshan S. Dosanjh, and John H. Mills, eds., Raven Press, c.1976, pp. 383–406.
25. Martin, Alan: The Equal Energy Concept Applied to Impulse Noise. Effects of Noise on Hearing, Donald Henderson, Roger P. Hamernik, Darshan S. Dosanjh, and John H. Mills, eds., Raven Press, c.1976, pp. 421–456.
26. Kryter, Karl D.: The Effects of Noise on Man. Academic Press, Inc., 1970.
27. Atherley, G. R.; and Martin, A. M.: Equivalent-Continuous Noise Level as a Measure of Injury From Impact and Impulse Noise. Ann. Occup. Hyg., vol. 14, no. 1, Mar. 1971, pp. 11–23.
28. Carter, N. L.; and Kryter, K. D.: Studies of Temporary Threshold Shift

Caused by High Intensity Noise. Rep. 949, Bolt Beranek & Newman, 1962.

29. Hecker, Michael H. L.; and Kryter, Karl D.: A Study of Auditory Fatigue Caused by High-Intensity Acoustic Transients. Rep. No. 1158 (Contract No. DA-49-007-MD-985), Bolt Beranek and Newman, Inc., Oct. 15, 1964. (Available from DTIC as AD 450 707.)
30. Kryter, K. D.; and Garinther, G. R.: Auditory Effects of Acoustic Impulses From Firearms. Acta Oto-Laryngol., Suppl. 211, 1965, pp. 1-22.
31. Fletcher, J. L.; and Loeb, M.: The Effect of Pulse Duration on TTS Produced by Impulse Noise. J. Auditory Res., vol. 7, 1967, pp. 163-167.
32. Ward, W. Dixon: Effect of Temporal Spacing on Temporary Threshold Shift From Impulses. J. Acoust. Soc. America, vol. 34, no. 9, pt. 1, Sept. 1962, pp. 1230-1232.
33. Brownsey, C. M.: Aspects of Temporary Hearing Loss Resulting From a 'Short Duration or Occasional Noise Exposure.' INTER-NOISE 73 Proceedings, O. Juhl Pedersen, ed., 1973, pp. 37-46.
34. Schori, Thomas R.: Evaluation of Guidelines for Safe Exposure to Continuous Noise of Moderate and High Intensity. Percept. & Mot. Skills, vol. 44, 1977, pp. 307-321.
35. Dieroff, H.-G.: Some Remarks About Differences in Mechanisms of Damage Following Exposure to Impulse and Continuous Noise. Noise as a Public Health Problem, Jerry V. Tobias, Gerd Jansen, and W. Dixon Ward, eds., ASHA Rep. 10, American Speech-Language-Hearing Assoc., Apr. 1980, pp. 86-91.
36. Henderson, Donald; Hamernik, Roger P.; Dosanjh, Darshan S.; and Mills, John H., eds.: Effects of Noise on Hearing. Raven Press, c.1976.
37. Hamernik, Roger P.; Henderson, Donald; and Salvi, Richard, eds.: New Perspectives on Noise-Induced Hearing Loss. Raven Press, c.1982.
38. Coles, R. R. A.; and Rice, C. G.: High-Intensity Noise Problems in the Royal Navy and Royal Marines. J. Roy. Nav. Med. Serv., vol. 51, 1965, pp. 184-192.
39. Elwood, M. A.: Brasher, P. F.; and Croton, L. M.: A Preliminary Study of Sensitivity to Impulsive Noise in Terms of Temporary Threshold Shifts. Paper presented at British Acoustical Society Meeting on Impulse Noise, Southampton, 1966.
40. Acton, W. I.; Coles, R. R. A.; and Forrest, M. R.: Hearing Hazard From Small-Bore Rifles. Rifleman, vol. 74, 1966, pp. 9-12.
41. Murray, N. E.; and Reid, G.: Temporary Deafness Due to Gunfire. J. Laryngol. Otol., vol. 61, 1946, pp. 95-130.
42. Goldstone, G.; and Smith, M. G.: A Pilot Study of Temporary Threshold Shifts Resulting From Exposure to High-Intensity Impulse Noise. Rep. TM-19-61, U.S. Army, Sept. 1961. (Available from DTIC as AD 269 043.)
43. Smoorenburg, Guido F.: Damage Risk Criteria for Impulse Noise. New

Perspectives on Noise-Induced Hearing Loss, Roger P. Hamernik, Donald Henderson, and Richard Salvi, eds., Raven Press, c.1982, pp. 471-490.

44. Shoji, H.; Yamamoto, T.; and Takagi, K.: Studies on TTS Due to Exposure to Octave-Band Noise. J. Acoust. Soc. Japan, vol. 22, 1966, pp. 340-349.

45. Glorig, A.; Ward, W. D.; and Nixon, J.: Damage Risk Criteria and Noise-Induced Hearing Loss. Arch. Otolaryngol., vol. 74, 1961, pp. 413-425.

46. Ward, W. Dixon; Glorig, Aram; and Sklar, Diane L.: Temporary Threshold Shift Produced by Intermittent Exposure to Noise. J. Acoust. Soc. America, vol. 31, no. 6, June 1959, pp. 791-794.

47. Spieth, Walter; and Trittipoe, W. J.: Intensity and Duration of Noise Exposure and Temporary Threshold Shifts. J. Acoust. Soc. America, vol. 30, no. 8, Aug. 1958, pp. 710-713.

48. Allen, Clayton H.; Jackson, Francis J.; and Kryter, Karl D.: Hearing Threshold Shift Produced by High Level Tone Bursts. 5e Congrès International d'Acoustique, Daniel E. Commins, ed., Liège, 1965, Paper B31.

49. Miller, James D.: Temporary Threshold Shift and Masking for Noise of Uniform Spectrum Level. J. Acoust. Soc. America, vol. 30, no. 6, June 1958, pp. 517-522.

50. Davis, H.; Morgan, C. T.; Hawkins, J. E.; Galambos, R.; and Smith, F.: Temporary Deafness Following Exposure to Loud Tones and Noise. Contract OEMcmr-194, Comm. on Med. Res., OSRD, Harvard Med. School, 1943.

51. Miller, J. D.: Temporary Hearing Loss at 4000 cps as a Function of a Three-Minute Exposure to a Noise of Uniform Spectrum Level. Laryngoscope, vol. 68, 1958, pp. 660-671.

52. Ward, W. Dixon: Temporary Threshold Shift in Males and Females. J. Acoust. Soc. America, vol. 40, no. 2, Aug. 1966, pp. 478-485.

53. Ward, W. Dixon; Selters, Weldon; and Glorig, Aram: Exploratory Studies on Temporary Threshold Shift From Impulses. J. Acoust. Soc. America, vol. 33, no. 6, June 1963, pp. 781-793.

54. Kryter, K. D.; Ward, W. Dixon; Miller, James D.; and Eldredge, Donald H.: Hazardous Exposure to Intermittent and Steady-State Noise. J. Acoust. Soc. America, vol. 39, no. 3, Mar. 1966, pp. 451-464.

55. Ward, W. Dixon; Cushing, E. Marion; and Burns, Edward M.: Effective Quiet and Moderate TTS: Implications for Noise Exposure Standards. J. Acoust. Soc. America, vol. 59, no. 1, Jan. 1976, pp. 160-165.

56. Ward, W. D.; Glorig, A.; and Sklar, D. L.: Dependence of Temporary Threshold Shift at 4 kc on Intensity and Time. J. Acoust. Soc. America, vol. 30, no. 10, Oct. 1958, pp. 944-954.

57. Kylin, B.: Temporary Threshold Shift and Auditory Trauma Following

Exposures to Steady-State Noise. An Experimental and Field Study. Acta Oto-Laryngol., Suppl. 152, 1960, pp. 1-93.
58. Rossi, G.; Scevola, M.; and Magliano, C.: Temporary Threshold Shift (TTS) Due to Exposure to Urban Traffic Noise. Acta Oto-Laryngol., Suppl. 339, 1976, pp. 10-13.
59. Hetu, R.; Laliberte, L.; Filon, J.; and St-Cyr, J.: Ambient Sound Level and Recovery From TTS: Conflicting Results. J. Acoust. Soc. America, vol. 64, suppl. 1, Fall 1978, p. 10.
60. Schmidek, Mark; Henderson, Terry; and Margolis, Bruce: Evaluation of Proposed Limits for Intermittent Noise Exposures With Temporary Threshold Shift as a Criterion. J. American Ind. Hyg. Assoc., vol. 32, 1972, pp. 543-546.
61. Lipscomb, D. M.: Considerations of Community Noise Impact on Hearing Health. Proceedings of the ASTM Conference on Community Noise, May 24-26, 1978.
62. Cohen, A.; Kylin, B.; and LaBenz, P.J.: Temporary Threshold Shifts in Hearing From Exposure to Combined Impact/Steady-State Noise Conditions. J. Acoust. Soc. America, vol. 40, no. 6, Dec. 1966, pp. 1371-1380.
63. Ward, W. Dixon: A Comparison of the Effects of Continuous, Intermittent, and Impulse Noise. Effects of Noise on Hearing, Donald Henderson, Roger P. Hamernik, Darshan S. Dosanjh, and John H. Mills, eds., Raven Press, c.1976, pp. 407-420.
64. Johansson, B.; Kylin, B.; and Reopstorff, S.: Evaluation of the Hearing Damage Risk From Intermittent Noise According to the ISO Recommendations. Proceedings of the International Congress on Noise as a Public Health Problem, W. Dixon Ward, ed., 550/9-73-008, U.S. Environ. Prot. Agency, May 1973, pp. 201-210.
65. Ceypek, Tadeusz; Kuźniarz, Jerzy J.; and Lipowczan, Adam: Hearing Loss Due to Impulse Noise—A Field Study. Proceedings of the International Congress on Noise as a Public Health Problem, W. Dixon Ward, ed., 550/9-73-008, U.S. Environ. Prot. Agency, May 1973, pp. 219-228.
66. Henderson, D.; and Hamernik, R. P.: Impulse Noise-Induced Hearing Loss: An Overview. Noise and Audiology, David M. Lipscomb, ed., Univ. Park Press, 1978, pp. 143-166.
67. Sulkowski, Wieslaw J.; Lipowczan, Adam; and Latkowski, Bozydar: Field Study on Effects of Industrial Impulse Noise Upon Permanent Threshold Shift. Proceedings of the International Congress on Noise as a Public Health Problem, W. Dixon Ward, ed., 550/9-73-008, U.S. Environ. Prot. Agency, May 1973, pp. 129-136.
68. Guberan, E.; Fernandez, J.; Cardinet, J.; and Terrier, G.: Hazardous Exposure to Industrial Impact Noise: Persistent Effect on Hearing. Ann. Occup. Hyg., vol. 14, no. 4, Dec. 1971, pp. 345-350.

69. Carder, Henry M.; and Miller, James D.: Temporary Threshold Shifts From Prolonged Exposure to Noise. J. Speech & Hearing Res., vol. 15, no. 3, Sept. 1972, pp. 603-623.
70. Eldredge, Donald H.; Miller, James D.; Mills, John H.; and Bohne, Barbara A.: Behavioral, Physiological and Anatomical Studies of Threshold Shifts in Animals. Proceedings of the International Congress on Noise as a Public Health Problem, W. Dixon Ward, ed., 550/9-73-008, U.S. Environ. Prot. Agency, May 1973, pp. 237-255.
71. Melnick, William: Human Asymptotic Threshold Shift. Effects of Noise on Hearing, Donald Henderson, Roger P. Hamernik, Darshan S. Dosanjh, and John H. Mills, eds., Raven Press, c.1976, pp. 277-290.
72. Mills, J. H.; Gilbert, R. M.; and Adkins, W. Y.: Temporary Threshold Shifts in Humans Exposed to Octave Bands of Noise for 16-24 Hours. J. Acoust. Soc. America, vol. 65, no. 5, May 1979, pp. 1238-1248.
73. Ward, W. Dixon: Noninteraction of Temporary Threshold Shifts. J. Acoust. Soc. America, vol. 33, no. 4, Apr. 1961, pp. 512-513.
74. Cohen, Alexander: Temporary Hearing Losses for Protected and Unprotected Ears as a Function of Exposure Time to Continuous and Impulse Noise. Tech. Rep. EP-151, U.S. Army, June 1961. (Available from DTIC as AD 262 722.)
75. Harris, J. Donald: Relations Among Aftereffects of Acoustic Stimulation. J. Acoust. Soc. America, vol. 42, no. 6, Dec. 1967, pp. 1306-1324.
76. Ward, W. Dixon: Recovery From High Values of Temporary Threshold Shift. J. Acoust. Soc. America, vol. 32, no. 4, Apr. 1960, pp. 497-500.
77. Guignard, J. C.; and Coles, R. R. A.: Effects of Infrasonic Vibration on the Hearing. 5e Congrès International d'Acoustique, Daniel E. Commins, ed., Liège, 1965, Paper B57.
78. Okada, Akira; Miyake, Hirotsugu; Yamamura, Kotaro; and Minami, Masayasu: Temporary Hearing Loss Induced by Noise and Vibration. J. Acoust Soc. America, vol. 51, no. 4, pt. 2, Apr. 1972, pp. 1240-1248.
79. Yokoyama, T.; Osako, S.; and Yamamoto, K.: Temporary Threshold Shifts Produced by Exposure to Vibration, Noise, and Vibration-Plus-Noise. Acta Oto-Laryngol., vol. 78, no. 3-4, Sept.-Oct. 1974, pp. 207-212.
80. Kile, Jack E.; and Wurzbach, William F.: Temporary Threshold Shifts Induced by Vibratory Stimulation. Sound & Vib., vol. 14, no. 5, May 1980, pp. 26-29.
81. Hamernik, R. P.; Henderson, D.; and Hynson, K.: Impulsive Noise and Synergistic Effects Aggravate Hearing Loss. Occup. Health & Saf., vol. 47, no. 1, Jan.-Feb. 1978, pp. 50-58.
82. Johnson, Daniel L.; and Farina, Edward R.: Description of the Measurement of an Individual's Continuous Sound Exposure During a 31-Day Period. J. Acoust. Soc. America, vol. 62, no. 6, Dec. 1977, pp. 1431-1435.

83. Fairman, T. M.; and Johnson, D. L.: Noise Dosimeter Measurements in the Air Force. Rep. AMRL-TR-79-52, U.S. Air Force, Nov. 1979. (Available from DTIC as AD A081 284.)
84. Schori, Thomas R.; and McGatha, Edward A.: A Real-World Assessment of Noise Exposure. Sound & Vib., vol. 12, no. 9, Sept. 1978, pp. 24-30.
85. Nimura, Tadamoto; and Kono, Shunichi: Personal Noise Exposure and Estimation of Population Distributed by Leq(24). Tenth Int. Congress on Acoustics—Volume 2, Contributed Papers (Part One), Australian Acoustical Soc., May 1980, Paper 7.7.
86. Martin, R. H.; Gibson, E. S.; and Lockington, J. N.: Occupational Hearing Loss Between 85 and 90 dBA. J. Occup. Med., vol. 17, no. 1, Jan. 1975, pp. 13-18.
87. Yerg, R. A.; Sataloff, J.; Glorig, A.; and Menduke, H.: Inter Industry Noise Study—The Effects Upon Hearing of Steady State Noise Between 82 and 92 dBA. J. Occup. Med., vol. 20, no. 5, May 1978, pp. 351-358.
88. Schori, T. R.; and Johnson, D. L.: Inter-Industry Noise Study: Another Look. Sound & Vib., vol. 13, no. 1, Jan. 1979, pp. 16-21.
89. Passchier-Vermeer, W.: Noise-Induced Hearing Loss From Exposure to Intermittent and Varying Noise. Proceedings of the International Congress on Noise as a Public Health Problem, W. Dixon Ward, ed., 550/9-73-008, U.S. Environ. Prot. Agency, May 1973, pp. 169-200.
90. Cohen, Alexander; Anticaglia, Joseph R.; and Jones, Herbert H.: Noise Induced Hearing Loss—Exposures to Steady-State Noise. Sixth Congress on Environmental Health, American Med. Assoc., Apr. 1969.
91. Bray, Don E.: Noise Environments in Public Transportation. Sound & Vib., vol. 8, no. 4, Apr. 1974, pp. 16-20.
92. Tobias, Jerry V.: Cockpit Noise Intensity: Fifteen Single-Engine Light Aircraft. Aerosp. Med., vol. 40, no. 9, Sept. 1969, pp. 963-966.
93. Campbell, Richard A.: A Survey of Noise Levels on Board Pleasure Boats. Sound & Vib., vol. 6, no. 2, Feb. 1972, pp. 28-29.
94. Curtis, Jack; and Sauer, Richard C.: An Analysis of Recreational Snowmobile Noise. Sound & Vib., vol. 7, no. 5, May 1973, pp. 49-50.
95. Keim, R. J.: Impulse Noise and Neurosensory Hearing Loss—Relationship to Small Arms Fire. California Med., vol. 113, no. 3, Sept. 1970, pp. 16-19.
96. Ward, W. D.; and Glorig, A.: A Case of Firecracker-Induced Hearing Loss. Laryngoscope, vol. 71, 1961, pp. 1590-1596.
97. Gjaevenes, K.; Moseng, J.; and Nordahl, T.: Hearing Loss in Children Caused by the Impulsive Noise of Chinese Crackers. Scandinavian Audiol., vol. 3, 1974, pp. 153-156.
98. Chaney, R. B., Jr.; McClain, S. C.; and Harrison, R.: Relation of Noise Measurements to Temporary Threshold Shift in Snowmobile Users. J. Acoust. Soc. America, vol. 54, no. 5, Nov. 1973, pp. 1219-1223.
99. Kabuto, Michinori; and Suzuki, Shosuke: Temporary Threshold Shift

From Transportation Noise. J. Acoust. Soc. America, vol. 66, no. 1, July 1979, pp. 170-175.

100. Shirreffs, J. H.: Recreational Noise: Implications for Potential Hearing Loss to Participants. J. Sch. Health, vol. 44, no. 10, Dec. 1974, pp. 548-550.

101. Ayley, Julie; Bartlett, Betty; Bedford, Wendy; Gregory, Wendy; and Hallum, Gillian: Pilot Study on the Effects of Pop Group Music on Hearing. I.S.V.R. Memo. No. 266, Univ. Southampton, 1968.

102. Lebo, C. P.; and Oliphant, K. S.: Music as a Source of Acoustic Trauma. Laryngoscope, vol. 78, no. 7, 1968, pp. 1211-1218.

103. Jansson, E.; and Karlsson, K.: Sound Levels and Long-Time Spectra Recorded Within the Symphony Orchestra and Risk Criteria for Hearing Loss. STL-QPSR 1/1982, Dept. of Speech and Music Acoustics, Royal Institute of Technology, Stockholm, Sweden, 1982.

104. Ulrich, R. F.; and Pinheiro, M. L.: Temporary Hearing Losses in Teen-Agers Attending Repeated Rock-and-Roll Sessions. Acta Oto-Laryngol., vol. 77, no. 1, Jan.-Feb. 1974, pp. 51-55.

105. Abrol, B. M.; Nath, L. M.; and Sahai, A. N.: Noise and Acoustic Trauma Noise Levels in Discotheques in Delhi. Indian J. Med. Res., vol. 58, Dec. 12, 1970, pp. 1758-1763.

106. Axelsson, A.; and Lindgren, F.: Factors Increasing the Risk for Hearing Loss in 'Pop' Musicians. Scandinavian Audiol., vol. 6, no. 3, 1977, pp. 127-131.

107. Axelsson, A.; and Lindgren, F.: Hearing in Pop Musicians. Acta Oto-Laryngol., vol. 85, no. 3-4, Mar.-Apr. 1978, pp. 225-231.

108. Barry, J. P.; and Thomas, I. B.: A Clinical Study To Evaluate Rock Music, Symphonic Music, and Noise as Sources of Acoustic Trauma. J. Audio Eng. Soc., vol. 20, no. 4, 1972, pp. 271-274.

109. Bohne, B. A.; Ward, P. H.; and Fernandez, C.: Rock Music and Inner Ear Damage. American Fam. Physician, vol. 15, no. 5, May 1977, pp. 117-118.

110. Chüden, H.; and Strauss, P.: Gibt es eine Larmschwerhorigkeit bei Disc-jockeys? Mschr Ohrenheilkd. vol. 108, 1974, p. 377.

111. Cohen, Alexander; Anticaglia, Joseph; and Jones, Herbert H.: "Sociocusis"—Hearing Loss From Non-Occupational Noise Exposure. Sound & Vib., vol. 4, no. 11, Nov. 1970, pp. 12-20.

112. Darcy, F. J.: Noise Exposure of Live Music Groups and Other Employees in Night Clubs. J. American Ind. Hyg. Assoc., vol. 38, no. 8, Aug. 1977, pp. 410-412.

113. Ewertsen, H. W.: Beat Music and Damage to Hearing. Ugeskr. Laeg., vol. 133, no. 19, May 1971, pp. 959-961.

114. Fearn, R. W.: Noise Levels in Youth Clubs. J. Sound & Vib., vol. 22, no. 1, May 8, 1972, pp. 127-128.

115. Fearn, R. W.: Pop Music and Hearing Damage. J. Sound & Vib., vol. 29, no. 3, Aug. 8, 1973, pp. 396-397.
116. Flottorp, G.: Music—A Noise Hazard? Acta Oto-Laryngol., vol. 75, no. 4, Apr. 1973, pp. 345-347.
117. Flugrath, James M.: Modern-Day Rock-and-Roll Music and Damage-Risk Criteria. J. Acoust. Soc. America, vol. 45, no. 3, Mar. 1969, pp. 704-711.
118. Flugrath, James M.; Irwin, John A.; Wolfe, Basil N., Jr.; Krone, Betty; and Parnell, Mike: Temporary Threshold Shift and Rock-and-Roll Music. J. Aud. Res., vol. 11, no. 4, Oct. 1971, pp. 291-293.
119. Fluur, E.: Pop Music as Noise Trauma. Lakartidningen, vol. 64, no. 8, Feb. 22, 1967, pp. 794-796.
120. Gryczỳnska, D.; and Czyzewski, I.: Damaging Effect of Music on the Hearing Organ in Musicians. Otolaryngol. Polska, vol. 31, no. 5, 1977, pp. 527-531.
121. Hanson, D. R.; and Fearn, R. W.: Hearing Acuity in Young People Exposed to Pop Music and Other Noise. Lancet, vol. 2, no. 7927, Aug. 2, 1975, pp. 203-205.
122. Hickling, S.: Noise-Induced Hearing Loss and Pop Music. New Zealand Med. J., vol. 71, no. 453, Feb. 1970, pp. 94-96.
123. Jatho, K.; and Hellmann, H.: The Problem of Acoustic Trauma in Orchestra Musicians. HNO, vol. 20, no. 1, Jan. 1972, pp. 21-29.
124. Jerger, J.; and Jerger, S.: Temporary Threshold Shift in Rock-and-Roll Musicians. J. Speech Hear. Res., vol. 13, no. 1, 1970, pp. 221-224.
125. Kowalczuk, H.: "Big-Beat" Music and Acoustic Traumas. Otolaryngol. Polska, vol. 21, no. 2, 1967, pp. 161-167.
126. Lebo, Charles P.; Oliphant, Kenward S.; and Garrett, John: Acoustic Trauma From Rock-and-Roll Music. California Med., vol. 107, no. 5, Nov. 1967, pp. 378-380.
127. Lipscomb, D. M.: Ear Damage From Exposure to Rock and Roll Music. Arch. Otolaryngol., vol. 90, no. 5, Nov. 1969, pp. 545-555.
128. Reddell, R. C.; and Lebo, C. P.: Ototraumatic Effects of Hard Rock Music. California Med., vol. 116, no. 1, Jan. 1972, pp. 1-4.
129. Rintelmann, W. F.; and Borus, J. F.: Noise-Induced Hearing Loss and Rock and Roll Music. Arch. Otolaryngol., vol. 88, no. 4, Oct. 1968, pp. 377-385.
130. Rintelmann, William F.: A Review of Research Concerning Rock and Roll Music and Noise-Induced Hearing Loss. Audiological Library Ser., Volume VIII, Rep. Seven, Maico Hearing Inst., 1970, pp. 24-27.
131. Rintelmann, William F.; Lindberg, Robert F.; and Smitley, Ellen K.: Temporary Threshold Shift and Recovery Patterns From Two Types of Rock and Roll Music Presentation. J. Acoust. Soc. America, vol. 51, no. 4, pt. 2, Apr. 1972, pp. 1249-1255.

132. Rupp, R. R.; and Koch, L. J.: Effects of Too-Loud Music on Human Ears—But, Mother, Rock'n Roll has To be Loud. Clin. Pediatr., vol. 8, no. 2, Feb. 1969, pp. 60–62.
133. Smitley, E. K.; and Rintelmann, W. F.: Continuous Versus Intermittent Exposure to Rock and Roll Music—Effect on Temporary Threshold Shift. Arch. Environ. Health, vol. 22, no. 4, Apr. 1971, pp. 413–420.
134. Speaks, C.; Nelson, D.; and Ward, W. D.: Hearing Loss in Rock-and-Roll Musicians. J. Occup. Med., vol. 12, no. 6, June 1970, pp. 216–219.
135. Whittle, L. S.; and Robinson, D. W.: Discotheques and Pop Music as a Source of Noise-Induced Hearing Loss—A Review and Bibliography. NPL Acoustics Rep. Ac 66, British A.R.C., Mar. 1974.
136. Ward, W. Dixon; Glorig, Aram; and Sklar, Diane L.: Temporary Threshold Shift From Octave-Band Noise: Applications to Damage-Risk Criteria. J. Acoust. Soc. America, vol. 31, no. 4, Apr. 1959, pp. 522–528.
137. Robinson, D. W.: The Relationships Between Hearing Loss and Noise Exposure. NPL Aero Rep. Ac 32, British A.R.C., July 1968.
138. Passchier-Vermeer, W.: Hearing Loss Due to Exposure to Steady-State Broadband Noise. Rep. 35, Res. Inst. for Public Health Eng., (Delft, The Netherlands), Apr. 1968.
139. Passchier-Vermeer, W.: Hearing Loss Due to Continuous Exposure to Steady-State Broad-Band Noise. J. Acoust. Soc. America, vol. 56, no. 5, Nov. 1974, pp. 1585–1593.
140. Glorig, A.; and Nixon, J.: Distribution of Hearing Loss in Various Populations. Ann. Otol., Rhinol., & Laryngol., vol. 69, no. 2, 1960, pp. 497–516.
141. Coles, R. Ross A.; Garinther, George R.; Hodge, David C.; and Rice, Christopher G.: Hazardous Exposure to Impulse Noise. J. Acoust. Soc. America, vol. 43, no. 2, Feb. 1968, pp. 336–343.
142. Golden, Pamela M.; and Clare, R.: The Hazards to the Human Ear From Shock Waves Produced by High Energy Electrical Discharges. AWRE Rep. No. E1/65, United Kingdom At. Energy Auth., [1966].
143. Hodge, D. C.; Gates, H. W.; Soderholm, R. B.; Helm, C. P., Jr.; and Blackner, R. F.: Preliminary Studies of the Impulse Noise Effects on Human Hearing. TM 15-64, Aberdeen Proving Ground, 1964. (Available from DTIC as AD 618 327.)
144. Ward, W. Dixon, ed.: Proposed Damage-Risk Criterion for Impulse Noise (Gunfire). Nat. Acad. Sci.-Nat. Res. Counc., July 1968. (Available from DTIC as AD 673 223.)
145. Johnson, Daniel L.: Prediction of NIPTS Due to Continuous Noise Exposure. Rep. No. AMRL-TR-73-91, U.S. Air Force, July 1973. (Available from DTIC as AD 767 205.)
146. Acoustics—Assessment of Occupational Noise Exposure for Hearing Conservation Purposes. ISO 1999-1975 (E), Int. Organ. Stand.

Chapter 8
Noise-Induced Hearing Impairment and Handicap

Introduction

A permanent, noise-induced hearing loss has a doubly harmful effect on speech communications. First, the elevation in the threshold of hearing means that many speech sounds are too weak to be heard, and second, very intense speech sounds may appear to be distorted.

The whole question of the impact of noise-induced hearing loss upon the impairments and handicaps experienced by people with such hearing losses has been somewhat controversial partly because of the economic aspects of related practical noise control and workmen's compensation. A considerable portion of the discussion which follows is from reference 1. Of probable interest to some readers are comments on that article from several researchers in this field (refs. 2, 3, 4, and 5) and a rebuttal (ref. 6).

Impairment and Handicap

Physiological impairment is usually defined as a lessening in the ability of a bodily organ or structure to perform with normal full capacity. Thus, if because of some injury, a limb can be rotated less than its normal arc, or the size of the visual field is less than normal, impairments are said to have occurred.

The sense of hearing, however, has not been treated in a similar fashion. Rather, in line with the guidelines issued by the Committee on Conservation of Hearing of the American Academy of Ophthalmology and Otolaryngology

(AAOO) (refs. 7 and 8), impairment is defined primarily with respect to the opinions of some otologists about the relations between pure-tone audiograms and the ability to understand speech communications. In that regard, it was proposed in 1959 by the AAOO (ref. 7) that the ability to understand everyday speech in the quiet is not impaired until the average of the pure-tone hearing levels at 500, 1000, and 2000 Hz exceeds 25 dB re ISO (International Standards Organization) audiometric zero. This average hearing level (HL) of 25 dB is commonly referred to as the AAOO "fence", or threshold of hearing impairment according to the AAOO. The AAOO Committee on Conservation of Hearing is considering changing its formula by adding the HL at 3000 Hz to those at 500, 1000, and 2000 Hz before averaging the HL's.

AAOO method

Some general relations are shown in figure 8.1 between HL and (1) speech intelligibility, as calculated according to the articulation index (AI) in percent of speech items correctly heard (the right-hand ordinate); (2) percent impairment for "everyday speech", (the left-hand ordinate); and (3) degrees of handicap as proposed by the AAOO. In addition to exhibiting losses in understanding speech at normal levels of intensity as the result of elevated thresholds of hearing, persons with noise-induced hearing losses have additional degradation in intelligibility test scores if poor enunciation and other distortions found in everyday speech

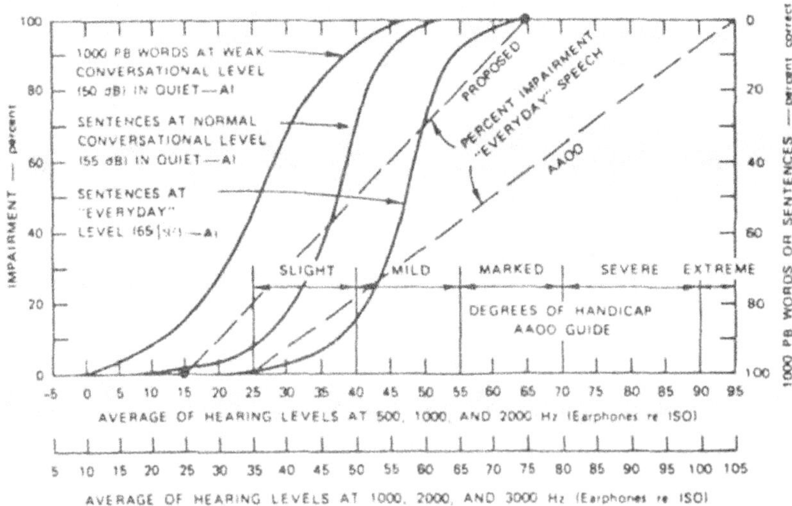

FIGURE 8.1. Relationship between HL and measures of speech understanding as calculated according to the articulation index (AI) (right-hand ordinate), impairment (left-hand ordinate), and handicap. Speech level measured 1 m from talker. (From ref. 1.)

are present, even though the speech signal is of greater than normal intensity (ref. 9).

Speech intelligibility as a function of hearing level

Kryter *et al.* (refs. 10 and 11) found degradation in speech intelligibility test scores with HL's much lower than the AAOO 25-dB fence, especially when the speech was heard in the presence of some noise or frequency distortion (see fig. 8.2). Also, somewhat better prediction of the intelligibility scores was obtained with HL's averaged at 1000, 2000, and 3000 Hz than with HL's averaged at 500, 1000, and 2000 Hz. Although not shown in this figure, HL's averaged at 1000, 2000, 3000, and 4000 Hz also were better predictors of intelligibility scores than were HL's averaged at 500, 1000, and 2000 Hz. These findings were later confirmed by Harris (ref. 12), Kuźniarz (ref. 13), and Suter (ref. 14).

AAOO method compared with speech intelligibility

The descriptions of impairment and handicap proposed by the AAOO are not consistent with the percentage speech scores as measured by tests or calculated from the articulation index. For example, at the 25-dB HL AAOO fence, a person could correctly understand only about 90 percent of the sentences and 50 percent of the monosyllabic phonetically balanced (PB) words uttered at a conversational level of effort by a person 1 m from him in the quiet. (See fig. 8.1.) As another example, with an average HL at 500, 1000, and 2000 Hz of 54 dB, a per-

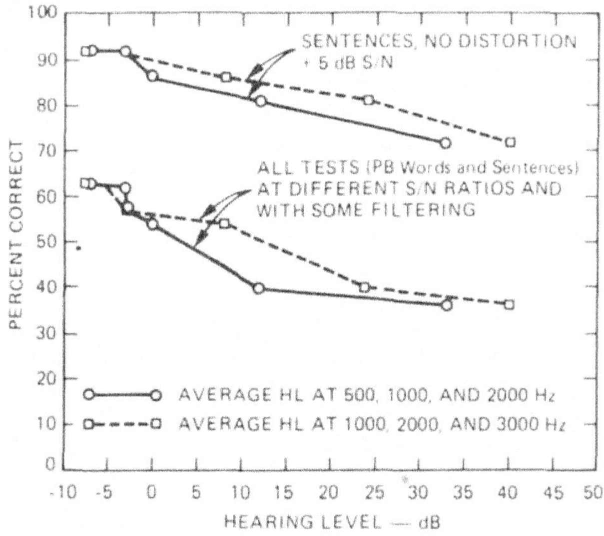

FIGURE 8.2. Intelligibility test scores as a function of average HL. (From ref. 11.)

son would be rated as having a "mild" handicap and a 42 percent impairment in his hearing by the AAOO method, even though he would not be able to understand any words or simple, unpracticed sentences uttered 1 m from him by a talker using a conversational level of effort.

Davis (ref. 15, p. 82) explains this type of apparent discrepancy as follows: "The criterion that was accepted in the Committee on Conservation of Hearing was the ability to understand everyday speech adequately. This does not mean monosyllables in the audiometric discrimination test, nor does it mean nonsense syllables in the psychoacoustic laboratory; the concept is everyday speech 'as she is spoke', and this implies the value of contextual cues and also the careless way that people speak. There is a great deal of redundancy if we are talking about everyday speech and not about the unexpected message, the unfamiliar proper name or the important telephone number."

Although the AAOO does not provide any evidence concerning the consistency of redundancy and contextual clues in "everyday speech," it is clear that such factors do contribute to speech communication, but in varying degrees from time to time. On the other hand, it seems arbitrary to exclude from consideration in the medical assessment of hearing impairment and handicap, the ability to understand speech other than the redundant, relatively high-level speech specified by the AAOO. For example, it is not obvious why the noise-deafened ear should not be considered impaired nor the individual handicapped when losing the ability the normal person has to understand: (a) individual words, the unexpected message, the unfamiliar name, or the important telephone number; (b) the weaker-than-normal-intensity speech that can occur because the talker drops his voice level, or because the distance between the talker and listener is greater than about 1 m and the talker is using a conversational level of effort; or (c) speech in the presence of everyday noise, such as that which occurs at a party or conference when several people are talking at once.

Discrimination capacity of the ear

Perhaps a more basic way to show the impact of threshold shifts typical for an ear with an average shift, or fence, at 500, 1000, and 2000 Hz of 25 dB ("no impairment" by AAOO standards) is by means of figure 8.3 (refs. 16, 17, and 18). The squares in figure 8.3 give the number of differences in pitch and loudness the average ear is capable of perceiving in that area of sound frequencies and intensities. These differences were derived from studies in which the frequency or intensity of a pure tone was changed until the average listener detected a change in pitch or loudness—the so-called just noticeable difference (JND). For example, in the frequency region of 1000 to 1500 Hz and the intensity region of 120 to 130 dB, the ear can detect about 3600 such JND's.

The curves labelled 1, 2, and 3 in figure 8.3 represent the audiograms (the sound intensities required before different frequencies can be heard) of persons with hearing thresholds at several levels of hearing impairment defined by the

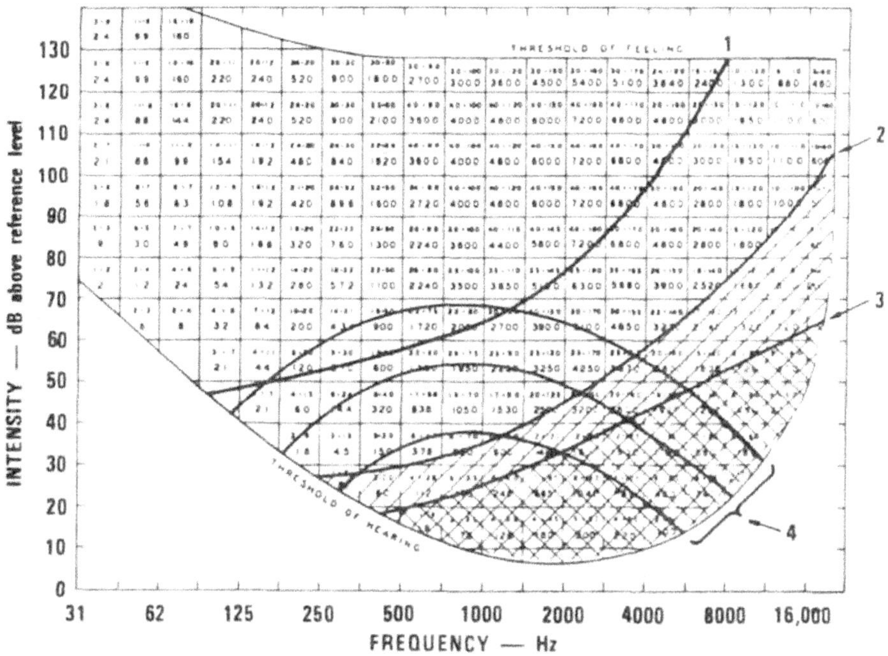

1 Audiogram for a person at upper bound of a "mild handicap" according to AAOO (HL of 55 dB averaged at 500, 1000, and 2000 Hz). Lost auditory capacity for all audible tones at this HL = 44 percent (150 000 out of 350 000 units). Lost discriminable units of everyday speech = 96 percent (41 293 out of 43 093 units).

2 Audiogram for a person at AAOO fence of "no impairment for everyday speech" (HL of 25 dB averaged at 500, 1000, and 2000 Hz). Lost auditory capacity for all audible tones at this HL = 13 percent (44 082 out of 350 000 units). Lost discriminable units of everyday speech = 31 percent (13 500 out of 43 093 units).

3 Audiogram for a person at proposed fence of "no impairment for everyday speech" (HL of 15 dB averaged at 500, 1000, and 2000 Hz). Lost auditory capacity for all audible tones at this HL = 4 percent (13 915 out of 350 000 units). Lost discriminable units of everyday speech = 16 percent (7020 out of 43 093 units).

4 Mean and 90-percent range of intensities of critical bands of speech frequencies present during "everyday" speech. Includes about 43 093 JND units.

FIGURE 8.3. Audiograms and speech spectra as indicated. In each square, the upper left number is the number of JND units for loudness (intensity), and the upper right number is the number of JND units for pitch (frequency). Lower number in each square is total number of JND units (loudness times pitch) for that square. Total area of hearing includes approximately 350 000 JND units. (From refs. 16, 17, and 18.)

AAOO. Sound frequencies at intensities that fall below the intensities represented by the respective curves would be inaudible to a person with that audiogram. Accordingly, the area of the graph below a given audiogram curve represents the number of units of sound discrimination JND's lost to the person with the given audiogram.

The curves labelled 4 in figure 8.3 show the range of frequencies and intensities covered by typical everyday speech signals 1 m from a talker. The squares of the graph between the upper and lower curves labelled 4 include the JND units of frequency and intensity used by the normal ear for distinguishing and understanding the sounds of everyday speech.

As shown in the legends of figure 8.3, some 31 percent of the possible frequency-intensity discriminations available for the perception of everyday speech in the normal ear are lost by the ear that just meets the AAOO 25-dB fence. Within the limits of what the AAOO calls a "mild handicap", 95 percent of the frequency-intensity discriminations available for the perception of everyday speech and 44 percent of total normal auditory capacity for discriminations among all sounds are lost to the hearer.

Ratings of Effects in Real Life

Perhaps this question of impairment and handicap can best be answered by the results of some studies of the consequences of impaired hearing in "real life." Nett *et al.* (ref. 19) found in a study of critical incidents of hearing handicap in a population of 378 hard-of-hearing persons that half of the people had hearing loss of 34 percent or less as estimated by AAOO procedures, but 60 percent of the group estimated their loss as being more than a 60 percent loss in hearing ability. Many of the incidents of handicap occurred when the talker was 10 to 21 ft from the hearing-impaired person. Twenty-two percent of the group had conductive losses, 32 percent had sensorineural losses, and 46 percent had mixed losses.

A more recent and directly relevant investigation of the social effects of noise-induced deafness was reported by Kell *et al.* (ref. 20). The subjects in this study were 96 female weavers whose HL averaged 39 dB re ISO audiometric zero at 500, 1000, and 2000 Hz, and 96 controls, matched for age, whose average HL at these three frequencies was 16 dB. The mean ages of the groups were 64.1 and 64.5 years, respectively. The weavers worked 8 hours per day in a noise level of about 100 dBA. The results of a questionnaire survey of the subjects are summarized by the authors of the study:

> The social consequences of this impaired hearing ability were:
>
> (a) difficulty at public meetings (weavers 72%, controls 6%)
> (b) difficulty talking with strangers (weavers 80%, controls 16%)
> (c) difficulty talking with friends (weavers 77%, controls 15%)
> (d) difficulty understanding telephone conversation (weavers 64%, controls 5%)

(e) 81% of all weavers considered that their hearing was impaired (5% controls)
(f) 9% of weavers and no controls owned hearing aids
(g) 53% of weavers and no controls used a form of lip-reading.

The speech tests given the subjects revealed that the speech level required to achieve equal understandability (50 percent of the items correct) was about 5 to 20 dB greater for the weavers than for the controls (fig. 8.4). This would seem to constitute a severe decline in ability to perceive speech signals compared with that of the control subjects. Also worth mentioning is that the "50 percent correct" score for determining speech threshold underestimates the relative impairment to speech understanding that is imposed on the noise-deafened individual.

As Davis (ref. 21) has shown, the person with losses as severe as those found for the average weaver would be unable to achieve more than about 50 to 60 percent of test words correct no matter how intense the speech signal. This loss in ability to discriminate among speech sounds, regardless of speech level, typifies noise-induced deafness and, unfortunately, is not helped by hearing aids (hence, no doubt, one reason so few of the noise-deafened weavers used hearing aids). According to the AAOO procedures, the mean impairment for the weavers would be rated at 20 percent and considered to be a "slight" handicap (defined as "diffi-

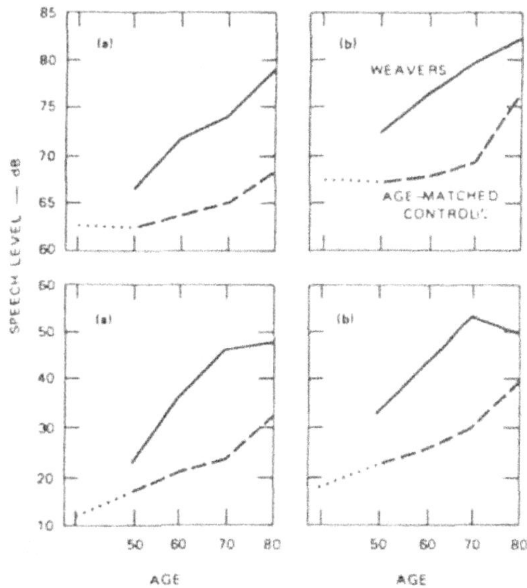

FIGURE 8.4. Mean level at which 50 percent score was achieved for phonemes (a) and words (b) in weavers and age-matched controls. Dotted lines indicate levels obtained for a younger control group. Upper figures (a) and (b) are mean speech audiograms with masking noise; lower figures (a) and (b) are mean speech audiograms with earphones. (From ref. 20.)

culty only with faint speech"). Such a rating is not consistent with the findings of the social studies or the clinical speech tests.

Self-ratings and real-life experiences of hearing impairment have more face validity than pure-tone audiograms and also, but to a lesser extent, than most speech intelligibility tests. They have, however, some drawbacks in that both underrating and overrating of hearing difficulties are a possibility by individuals for both conscious and nonconscious personal reasons. Nonetheless, the proper interpretation and validation of pure-tone thresholds and speech intelligibility tests require consideration of these real-life data and experiences from people who suffer sensorineural, noise-induced hearing losses. (See ref. 22.)

Tinnitus

A somewhat overlooked condition in some cases of permanent noise-induced deafness is that of tinnitus. Tinnitus is a persistent "ringing" or noisy sound in the ear that is often very disturbing and bothersome to those suffering from this condition. Although it is experienced in mild forms from time to time by people with normal hearing, it is associated in its acute forms with sensorineural, noise-induced hearing loss, as well as with some other disorders of the ear.

Tinnitus and these other possible effects from noise-induced hearing loss have not received to date the attention and measurement that pure-tone threshold shifts (NIPTS) and impairment to speech understanding have received. Nonetheless, they, especially tinnitus, are considered severe handicaps by some people. (See Hinchcliffe and Gordon (ref. 23), Sataloff (ref. 24), and Vernon (ref. 25).)

Criterion for raised level speech

It is suggested that a practical criterion for general application is that the start of measurable impairment of hearing speech in the workaday world is said to occur when the listener would be just able to understand all simple, unrehearsed sentences presented to him in the quiet at an average intensity level at the listener's position of about 65 dBA. (See Chapter 4 for a discussion of measuring speech levels.) The 65-dBA level is about 10 dB above that of normal conversational levels, 1 m from the talker, in a quiet home, conference room, or office. This level of 65 dBA has sometimes been called an "everyday" level—that used in a noisy office or department store, in a crowd, or when lecturing.

As shown in figure 8.1, this criterion can be met by a person having an HL that averages about 15 dB at 500, 1000, and 2000 Hz, or 25 dB at 1000, 2000, and 3000 Hz. With this criterion hearing level, the listener would miss about 2 percent of sentences heard at a level of intensity normal for conversation 1 m from a talker in the quiet, and 15 percent of monosyllabic (PB) words at a "weak" level of effort. (See fig. 8.1.) Thus, a person with the amount of impairment specified in the suggested criterion would be handicapped to the extent that he would be unable to correctly perceive some sentences and individual words at speech intensity levels often present in typical quiet living conditions. This person would have a more severe impairment to the understanding of distorted or slurred speech, speech in the presence of noise, or speech in a situation where several persons are talking.

Attention is drawn to the fact that the suggested criterion is concerned solely with hearing at and below frequencies carrying much useful auditory information, including some of that in speech and music. However, as shown in a number of studies (refs. 10, 12, 13, and 14), there is a strong correlation between shifts in audiometric thresholds at these lower frequencies especially 1000, 2000, and 3000 Hz and those present at frequencies above 3000 Hz so that the HL's measured at frequencies below 3000 Hz are a fair index to those above 3000 Hz.

Degrees of impairment for understanding speech

In order to quantify, in a simple way, degrees of hearing impairment for speech with respect to HL's, a straight-line relationship has been drawn in figure 8.1 between percent measured speech impairment and average HL. This line starts at 15 dB for 500, 1000, and 2000 Hz (25 dB at 1000, 2000, and 3000 Hz) and goes to an average HL of 65 dB for 500, 1000, and 2000 Hz (75 dB at 1000, 2000, and 3000 Hz). At this upper point on the proposed curve there is no understanding of sentences presented at a raised, "everyday" level of speech intensity 1 m from the talker. This linear percent rating scheme is similar to that introduced by the AAOO for rating impairment except for the start and termination points. (See fig. 8.1.)

This suggested criterion is purely in terms of a measurable loss in ability to understand speech as described, and not in terms of a threshold or yardstick for specifying a social or behavioral handicap. Arguments could be made that the suggested quantification for measurable speech communication hearing loss underestimates the social or behavioral handicaps that would be present with a given shift in threshold, as specified, because it is based on a raised level of speech rather than on the normal conversational level of speech that will often be found in the real world. However, the hard economics and politics of the workaday world, both private and government, have led to what must, in the author's opinion, be considered pure-tone audiometric assessment procedures that greatly underestimate the social and behavioral handicaps likely to ensue from noise-induced hearing loss. These procedures, concerned primarily with workmen's compensation for damage to hearing, are summarized in various publications (for example, refs. 26 and 27).

References

1. Kryter, Karl D.: Impairment to Hearing From Exposure to Noise. J. Acoust. Soc. America, vol. 53, no. 5, May 1973, pp. 1211-1234.

2. Cohen, Alexander: Some General Reactions to Kryter's Paper "Impairment to Hearing From Exposure to Noise." J. Acoust. Soc. America, vol. 53, no. 5, May 1973, pp. 1235-1236.
3. Davis, Hallowell: Some Comments on "Impairment to Hearing From Exposure to Noise" by K. D. Kryter. J. Acoust. Soc. America, vol. 53, no. 5, May 1973, pp. 1237-1239.
4. Lempert, Barry L.: Technical Aspects of Dr. Kryter's Paper "Impairment to Hearing From Exposure to Noise" With Respect to the NIOSH Statistics. J. Acoust. Soc. America, vol. 53, no. 5, May 1973, pp. 1240-1241.
5. Ward, W. Dixon: Comments on "Impairment to Hearing From Exposure to Noise" by K. D. Kryter. J. Acoust. Soc. America, vol. 53, no. 5, May 1973, pp. 1242-1243.
6. Kryter, Karl D.: Reply to the Critiques of A. Cohen, H. Davis, B. L. Lempert, and W. D. Ward of the Paper "Impairment to Hearing From Exposure to Noise." J. Acoust. Soc. America, vol. 53, no. 5, May 1973, pp. 1244-1252.
7. Guide for the Evaluation of Hearing Impairment. Report of the Committee on Conservation of Hearing. Trans. American Acad. Ophthalmol. & Otolaryngol., vol. 63, 1959, pp. 236-238.
8. Guide for Conservation of Hearing in Noise, Revised ed. 1973. Suppl. Trans. American Academy Ophthalmology & Otolaryngology, c.1973.
9. Harris, J. D.: Combinations of Distortion in Speech. Arch. Otolaryngol., vol. 72, 1960, pp. 227-232.
10. Kryter, Karl D.; Williams, Carl; and Green, David M.: Auditory Acuity and the Perception of Speech. J. Acoust. Soc. America, vol. 34, no. 9, pt. 1, Sept. 1962, pp. 1217-1223.
11. Kryter, K. D.: Hearing Impairment for Speech. Arch. Otolaryngol., vol. 77, June 1963, pp. 598-602.
12. Harris, J. D.: Pure-Tone Acuity and the Intelligibility of Everyday Speech. J. Acoust. Soc. America, vol. 37, no. 5, May 1965, pp. 824-830.
13. Kuźniarz, Jerzy J.: Hearing Loss and Speech Intelligibility in Noise. Proceedings of the International Congress on Noise as a Public Health Problem, W. Dixon Ward, ed., 550/9-73-008, U.S. Environ. Prot. Agency, May 1973, pp. 57-71.
14. Suter, Alice H.: The Ability of Mildly Hearing-Impaired Individuals To Discriminate Speech in Noise. EPA-550/9-78-100, U.S. Environ. Prot. Agency, Jan. 1978.
15. Robinson, D. W., ed.: Occupational Hearing Loss. Academic Press, Inc., 1971.
16. Stevens, Stanley Smith; and Davis, Hallowell: Hearing—Its Psychology and Physiology. John Wiley & Sons, Inc., 1938.
17. Shower, E. G.; and Biddulph, R.: Differential Pitch Sensitivity of the Ear. J. Acoust. Soc. America, vol. III, suppl. no. I, pt. 2, July 1931, pp. 275-287.

18. Riesz, R. R.: Differential Intensity Sensitivity of the Ear for Pure Tones. Phys. Rev., second ser., vol. 31, no. 5, May 1928, pp. 867-875.
19. Nett, Emily; Doerfler, Leo G.; and Matthews, Jack: Summary of the Results of Project SP-167—The Relationship Between Audiological Measures and Actual Social-Psychological-Vocational Disability. U.S. Dep. Health, Educ., & Welfare.
20. Kell, R. L.; Pearson, J. C. G.; Acton, W. I.; and Taylor, W.: Social Effects of Hearing Loss Due to Weaving Noise. Occupational Hearing Loss, D. W. Robinson, ed., Academic Press, Inc., 1971, pp. 179-191.
21. Davis, H.: The Articulation Area and the Social Adequacy Index for Hearing. Laryngoscope, vol. 58, 1948, pp. 761-768.
22. Noble, William G.: Assessment of Impaired Hearing—A Critique and a New Method. Academic Press, Inc., 1978.
23. Hinchcliffe, Ronald; and Gordon, A.: Subjective Magnitude of Symptoms and Handicaps Related to Hearing Impairment. Noise as a Public Health Problem, Jerry V. Tobias, Gerd Jansen, and W. Dixon Ward, eds., ASHA Rep. 10, American Speech-Language-Hearing Assoc., Apr. 1980, pp. 144-146.
24. Sataloff, Joseph: Hearing Loss. J. B. Lippincott Co. c.1966.
25. Vernon, Jack A.: The Other Noise Damage: Tinnitus. Sound & Vib., vol. 12, no. 5, May 1978, p. 26.
26. Ginnold, Richard E.: Occupational Hearing Loss—Workers Compensation Under State and Federal Programs. EPA 550/9-79-101, U.S. Environ. Prot. Agency, Sept. 1979.
27. Sulkowski, Wieslaw J.: Industrial Noise Pollution and Hearing Impairment—Problems of Prevention, Diagnosis and Certification Criteria. TT 76-54047, Natl. Library Med., U.S. Dep. Health, Educ., & Welfare, c.1980. (Available from NTIS.)

Chapter 9
Mental and Psychomotor Task Performance in Noise

Introduction

The possible adverse effects of noise on mental and psychomotor task performance have been a matter of practical concern for centuries. They have also been and continue to be a matter of scientific controversy. With varying degrees of certainty, it has been proposed that noise

1. Masks, or interferes with the perception of, auditory signals such as speech that are needed or helpful to the performance of a specific task

2. Masks irrelevant auditory signals that could distract the worker (the noise then would improve work performance and has been called acoustic perfume

3. Competes for the psychological attention of the worker and thereby distracts and interferes with work performance

4. Creates a monotonous condition which masks normal changes in acoustic environment and thus leads to underarousal, or even sleep

5. Physiologically arouses a worker from a low state of alertness and thereby increases alertness and work performance

6. Physiologically overarouses the worker and thereby decreases work performance

7. Conveys meanings that are not directly required for performance of a particular task but that create feelings of annoyance affecting its performance (*e.g.*, noises from machinery that could cause bodily injury or cause distracting feelings of apprehension)

8. Creates feelings of annoyance because the workers feel the noise is damaging the ear or interfering with the hearing of sounds they wish to hear

9. Creates feelings of annoyance or anger because the worker feels helpless or unable to control his environment

10. Neurologically competes for and somehow preempts the functioning of nonauditory neural pathways or centers involved in the performance of particular nonauditory tasks

11. Preempts the use of some auditory neural pathways or centers that are involved in the internal enunciation and rehearsal of words related to the memory of words required in some mental tasks

A review of the above list indicates that except for the first possible effect (the masking or interference with the hearing of sounds needed to perform a given task), noise does not necessarily interfere with work performance. However, because of difficulties in the experimental control of some of these possible effects, the results of research on work performance in noise have been inconsistent and difficult to encompass in any simple theoretical construct. Indeed, reviews of research in this area over the past 30 years, Kryter (refs. 1 and 2), Gulian (ref. 3), and Loeb (ref. 4) conclude that simple generalizations about possible effects of noise on work performance cannot be made. Nevertheless, several general theories have been put forth.

General theories

Laboratory research has been concerned primarily with three different, but not necessarily independent, general theories:

1. Noise above a certain level of intensity causes general physiological overarousal of the worker.

2. Noise inherently is psychologically aversive, annoying, or distracting because of its loudness.

3. Noise competes with information in nonauditory signals for nonauditory neural pathways and mechanisms involved in cognition and work performance.

In these theories the noises presented to the subjects convey no significant meaning and no feedback information related to task performance. Furthermore, presumably the noise does not mask or interfere with information involved in task performance; that is, the task is entirely nonauditory. In brief, researchers justifiably presumed, or should have, that noise that masks auditory signals helpful to the performance of mental or psychomotor tasks and noise that conveys meanings indicative of possible danger or punishment can degrade task performance. This degradation, if any, would be independent of possible effects postulated in the above three general theories.

Unfortunately, some studies involved unrecognized auditory aspects of the tasks (which make the task auditory and subject to direct masking effects). These auditory aspects complicate the interpretation of research findings in this area, as discussed later in some detail. Some studies also involved noises having emotional meanings to some of the subjects (which means the noise is to some extent informational sound). The unresolved, intriguing, and seemingly elusive research question is whether noise can have adverse effects on the quality or quantity of nonauditory work performance because of some inherent physiological or psychological effects.

Theory of masking artifacts

A major source of inconsistency among the results of research studies has been the role of task-related acoustic cues heard by the subjects when performing the task. These cues can be the slight sounds (clicks or the like) made, for example, by switches, electronic relays, the contact of tracking styluses against metal plates, and the sound of motors that occur when stimuli are presented or when responses are made. These clicks and sounds can give subjects feedback information as to the presence of stimuli or to the accuracy, quickness, and level of performance in ways that tend to improve performance on psychomotor tasks. When these cues are masked by noise, the results show an apparent decline in the subjects' ability to perform the task. The presence of such acoustic cues in what were considered definitive studies of the effects of noise on the monitoring and detection of visual signals (refs. 5 to 18) was uncovered by Poulton (refs. 19 to 24). The implications of these discoveries are discussed and debated by Broadbent (refs. 25 to 27).

Poulton's initial concern about the possibility of masking artifacts in some of these studies arose over a five-choice visual serial-reaction task in which the subjects were to tap as rapidly as possible one of five small brass disks with a steel-tipped metal stylus; the disk to be tapped depended on which of five neon lamps was turned on at a given moment. Poulton and Edwards (ref. 18) found that a low-frequency noise did not degrade task performance relative to that in quiet at a level of 102 dBC, although Hartley (refs. 12 and 13) and Hartley and Carpenter (ref. 14) had earlier found that a flat spectrum noise of but 95 dBC degraded performance on the same task. As Poulton discovered, a plausible reason for these results was that the stylus hitting the brass disks (indicating, in conjunction with a change of lights, a correct response) made a high-pitched click that was not audible in the noise with the flat spectrum, but could be heard in the low-frequency spectrum noise. Thus the subjects received the same auditory response feedback in the low-frequency noise that was available to them in quiet, but not in the flat spectrum noise.

Poulton next examined the equipment used in Broadbent's so-called "20-dial test" for studies of the effects of noise on vigilance, the results of which served as a base for various generally accepted theories about perceptual task performance in noise (refs. 5 to 11). Poulton found that when the subject taking this test turned a control knob in the correct, but not the incorrect direction, a microswitch made a slight click that was audible in the quiet condition (noise at 70 dB) and inaudible in the noisy condition (noise at 100 dB). The experimenters, and probably the subjects, were not aware at the time of the studies of these acoustic feedback cues that were being masked in some conditions of noise and not in others. These data from the 20-dial and the five-choice serial reaction tests served to support the long-held theory of "blinks" that noise above a level of 70 to 80 dB periodically attracts attention from other sensory perceptions for about 1-sec intervals.

Theory of masking of internal speech

To help explain some of the apparent adverse effects of noise on work performance, Poulton (refs. 19, 21, and 23) hypothesized that noise can interfere with internally generated behavior related to the auditory system. According to this theory, the performance of certain types of ostensibly nonauditory work by some subjects may be affected by the noise internally "masking," or interfering with, internally articulated or rehearsed verbalization of words or numbers involved in the tasks. According to this concept, the noise interferes somewhat with the listener's echoic articulation of words or numbers presented for memorization.

The direct masking by noise of auditory cues helpful to the performance of nonauditory work plus Poulton's theory of the role of internal speech interference by noise explain most of the research findings in this area (see Poulton, ref. 24). There are, however, some research results showing effects of noise on the performance of nonauditory work that do not appear to be accounted for by Poulton's composite theory of arousal (beneficial effect), internal speech masking (interfering effect), and direct masking of performance feedback cues. Loeb (ref. 4) suggests that the effects of noise on task performance will be uncovered only if the research done to date is redone with systematic manipulation of task characteristics, multiple-task prioritization, and other variables.

Theory of construed concern

This author suggests that perhaps the effects of meaningless noise on non-auditory work performance which are not explained by the factors of arousal and masking (including internal speech) can be attributed to concepts that subjects may have about the nature and purpose of the experimental procedures the experimenter has concocted for them. This explanation might be called the theory of construed concern. In particular, subjects may believe (whether or not so instructed) that when the loud noise or sound is present, it is supposed to affect them or that it is a signal of importance of some kind. Such attitudes could have at least an initial or transient effect upon task performance, especially in studies wherein the subjects are exposed to the noise conditions for relatively brief periods (*e.g.*, 15 minutes or 1 hour). Some subjects may have a concern (justified or not) that the noise may be harmful to their hearing or that the experimental conditions may become overbearing; this may cause some physiological stress.

These two theories (masking of internal speech and construed concern) are obviously directed to possible auditory and noise-meaning factors not included in the three general theories presented earlier. These factors were presumed to not be present in most research on the effects of noise on mental and psychomotor task performance.

So much by way of introduction for the general concepts and status of research problems on the effects of noise on mental and psychomotor task performance. A number of studies published since 1970 or so in this area of research are discussed below under the following major subject headings:

1. Measured on Controlled Physiological States
2. Task Variables
3. Aftereffects on Insoluble Puzzles and Proofreading Tests
4. Information Overload
5. Industrial Field Studies

Measured or Controlled Physiological States

As discussed in chapter 10, there is no question but that intense sound particularly when it comes as a change in the acoustic environment, can have a general physiological alerting-arousal effect on an organism, for example, increased respiration rate, blood pressure, heart rate, and adrenal gland secretions. As discussed in chapter 10, if the noise is meaningless as far as work performance is concerned (as was uniformly assumed in the research studies to be discussed in this chapter), physiological adaptation or habituation can be expected to take place with repetitions of such noises.

Insofar as task performance depends on alertness, noise might be predicted to have some beneficial effect on work output, at least initially. On the other hand, this physiological arousal could lead to fatigue as well or cause some sensory, perceptual, or motor overactivity that degrades work performance. While this hypothesis is a reasonable one for explaining some variations in work performance in noise, the physiological state of arousal is difficult to measure, as shown in more detail in chapter 10.

EEG state

Using a special analysis of EEG activity as an index of central nervous system arousal, Gulian (ref. 28) divided subjects into two categories, hyperreactive and hyporeactive. These subjects were exposed to intermittent and continuous noises at various levels of intensity for a number of 1 1/2 hour sessions during which they performed auditory signal detection tasks. The tasks consisted of making discriminations among tonal signals. Some of her findings are shown in figure 9.1.

As seen in figure 9.1, while the subjects could be ranked as hyporeactive or hyperreactive on the basis of their EEG state in quiet, in the various noise conditions their states of arousal were not very different from each other. However, relative to their state in quiet, the noise reduced the state of arousal for the hyperreactive and increased it for the hyporeactive. This may be a function of the so-called "law of initial level" (physiological states tend toward a mean level).

Clearly, while the hyperreactive group in general had fewer correct detections than the hyporeactive group, the task performance of neither group was affected by the different noise conditions in any systematic way; however, when in quiet, the hyperreactive group performed exceedingly poorly and were inferior in task performance to the hyporeactive group either in quiet or in noise. An argument could be made for this particular experiment that in quiet the hyperreactive subjects were overaroused and for this reason performed relatively poorly. From these data, noise—intermittent or continuous, weak or intense—does not appear to have had any obvious detrimental effect on task performance or state of arousal as measured by EEG.

Cardiovascular and adrenal states

A major program of research on the effects of noise on physiological stress reactions and task performance was published by Glass and Singer (refs. 29 and 30). Some of the performance test results are discussed later, but note here that the skin conductance (GSR), vasoconstriction and muscle tension measures of physiological reaction to noise showed complete habituation to noise presented during the performance of a variety of mental and motor work tasks.

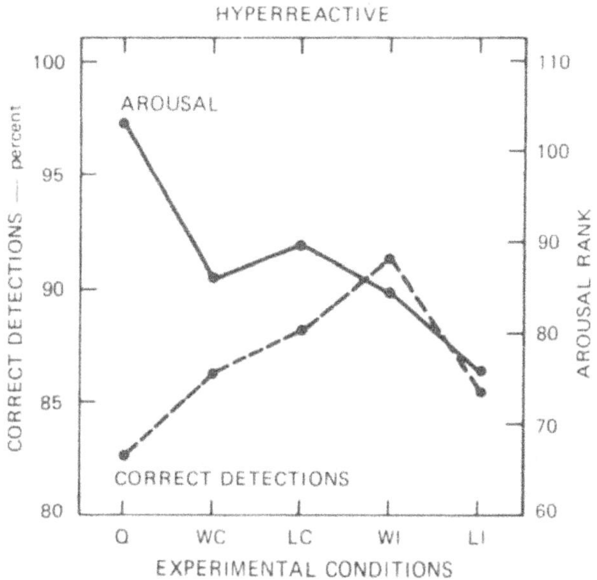

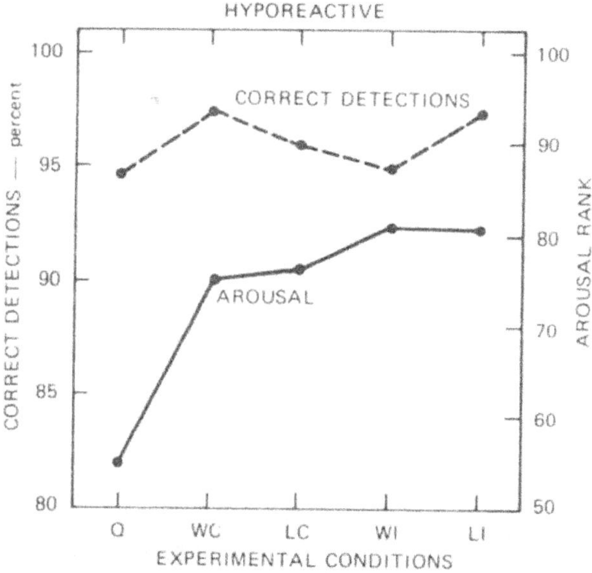

FIGURE 9.1. State of arousal and performance of auditory signal detection tasks during various noise conditions. Q = Quiet, WC = Weak continuous noise (3 tones: 100, 300, 400 Hz at 78 dB), LC = Loud continuous noise (3 tones at 90 dB), WI = Weak intermittent noise (3 tones at 71 dB), and LI = Loud intermittent (3 tones at 92 dB). (From ref. 28.)

Conrad (ref. 31) studied the effect of 93-dBA broadband noise on a serial decoding task. Four panel lights were located above a display window on which a 4-digit code appeared. The four lights corresponded to the four digit positions in the code. Thus, the light that was illuminated indicated a specific digit in the 4-digit code that was to be decoded. The subject responded by using the indicated digit to depress one of the four buttons located under the display window. A trial was given every second. Various physiological measures of muscle tension and vascular changes were taken and the subjects self-rated their sensitivity to noise. Although no effects on task performance were found, there was a significant relation between self-ratings of sensitivity to noise and the vasoconstrictive response. Becker *et al.* (ref. 32) also found that subjects who rated themselves especially sensitive to noise in real life were more psychologically but not more physiologically responsive to noise exposure in laboratory tests than those subjects who rated themselves less sensitive to noise in real life.

Frankenhaeuser and Lundberg (ref. 33) found that aperiodic noise from 65 to 85 dBA did not interfere with the performance of mental arithmetic, nor affect adrenaline excretion (fig. 9.2(a)). After the noise exposure, there appeared to be some relative increase in adrenaline excretion due to the noise compared to the quiet exposure condition, but heart rate showed a relative decline (fig. 9.2(b)).

In another investigation, Lundberg and Frankenhaeuser (ref. 34) asked subjects to choose the highest level of white noise between 70 and 108 dBA that they were willing to endure for 10 min. A second set of subjects was paired to the first set for a subsequent 10-min exposure to the noise level chosen by the first subject of each pair. Comparison of the adrenaline and cortical excretions for quiet control sessions with those for the noise-exposed sessions revealed that the subjects who had selected the noise level were less aroused by the noise than the paired subjects. Further, the self-ratings of discomfort were correlated with the physiological measures. Performance on an arithmetic performance test, however, was not affected by whether the subject had selected the noise level.

Hawel (ref. 35) tested 10 male college students for one 4-hour session per week for 10 weeks. Pulse rate, adrenal gland excretion (catecholamine), and psychological state (anxiety) were monitored. Except for two control sessions (5th and 10th week), the Kraepline-Pauli (simple computations) tests were administered. The noise was interrupted (5 sec on, 12 sec off) pink noise at a level of 90 dB. There were no effects of the noise on performance, but the noise increased pulse rate and cathecholamine excretion and decreased subjective ratings of anxiety. In general, the noise increased general physiological arousal, relative to the control days (when the subjects were in quiet and took no tests), but the state reached did not constitute overarousal detrimental to task performance or to subjective anxiety.

Basow (ref. 36) measured anxiety level (muscle tension) and performance on various work and mathematical tasks. There was no average systematic relation between anxiety level, as measured, and task errors.

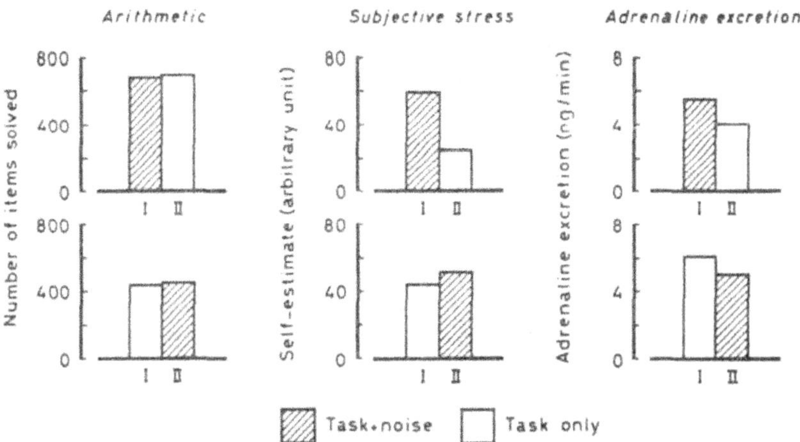

(a) Mean values for performance in arithmetic task, self-estimates of stress, and adrenaline excretion during 80-min noise and no-noise periods in two sessions.

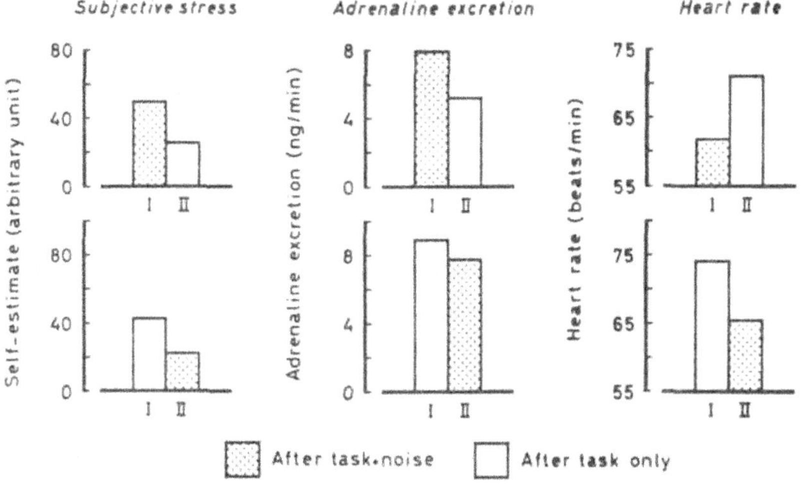

(b) Delayed effects of noise represented by mean values for self-estimated stress, adrenaline excretion, and heart rate, 80 to 90 min after noise and no-noise periods in two sessions.

FIGURE 9.2. Effect during and after noise and no-noise periods. One group of subjects (upper diagrams) had been exposed to noise in session I, and one group (lower diagrams) in session II. (From ref. 33.)

Cohen *et al.* (ref. 37) found that subjects taking a fast-paced, very complex test involving visual signal decoding and response made a greater number of errors in 100 dBA of synthesized office machine noise than in quiet. When in the noise, the subjects also revealed the following physiological conditions: "Profuse sweating (particularly in the palmar and armpit regions), muscle tension (back of the neck and shoulders), hand and finger cramps, blanching of the hand and fingers, and feelings of finger coolness or numbness" (p. 30, ref. 37). Cohen *et al.* concluded that the noise in this study when combined with the fast-paced task caused physiological overarousal that interfered with task performance.

However, in a study similar with respect to task and noise conditions, Kryter and Poza (ref. 38) found no reduction in task performance because of the noise (see lower panel, fig. 9.3). Further, no physiological reactions of the sort reported by Cohen *et al.* were observed in the subjects, nor were there any significant changes in heart rate or blood volume measured as a function of noise condition. However, the typical pulse volume response to broadband noise at an intensity of 100 dBA occurred (see fig. 9.3).

Because of their contrary findings, as well as the usual fact of physiological habituation to intense noise (to be discussed in chapter 10), Kryter and Poza suggested that the reason for the findings of Cohen *et al.* was possibly overarousal from intense motivation and false anxiety about the effects of the noise on performance and not because of any direct physiological overarousal by the noise *per se*.

Mosskov and Ettema (refs. 39 and 40) conducted a series of studies on the effects of noise on task performance and physiological arousal (to be discussed in chapter 10). In these studies young men were exposed to a variety of noises at levels from 40 to 100 dBA and durations from a few minutes to 3 hours. Mosskov and Ettema (ref. 40) conclude that the handling of visual information and mental capacity were impaired by the noise, so that the number of mistakes increased and the subjects needed more time to handle information. Their consistently negative results (adverse effects) seem at odds with some of the studies discussed above and with most of those to be discussed later. As discussed in chapter 10, the physiological overarousal effects of the noise found by these investigators (and concomitant with the adverse effects of the noise on mental and other task performance) are also inconsistent with the majority of other laboratory and field studies on physiological arousal from noise. The most obvious explanation is that for some reason the subjects in the Mosskov and Ettema tests did not show the habituation that normally occurs to such noises. Perhaps habituation did not occur because of the relatively short periods of exposure to different noise conditions.

Sleep and drug states

A number of studies have been done in which the general physiological state of arousal of the subjects was modified by the administration of various substances or by sleep deprivation prior to tests of psychomotor task performance in

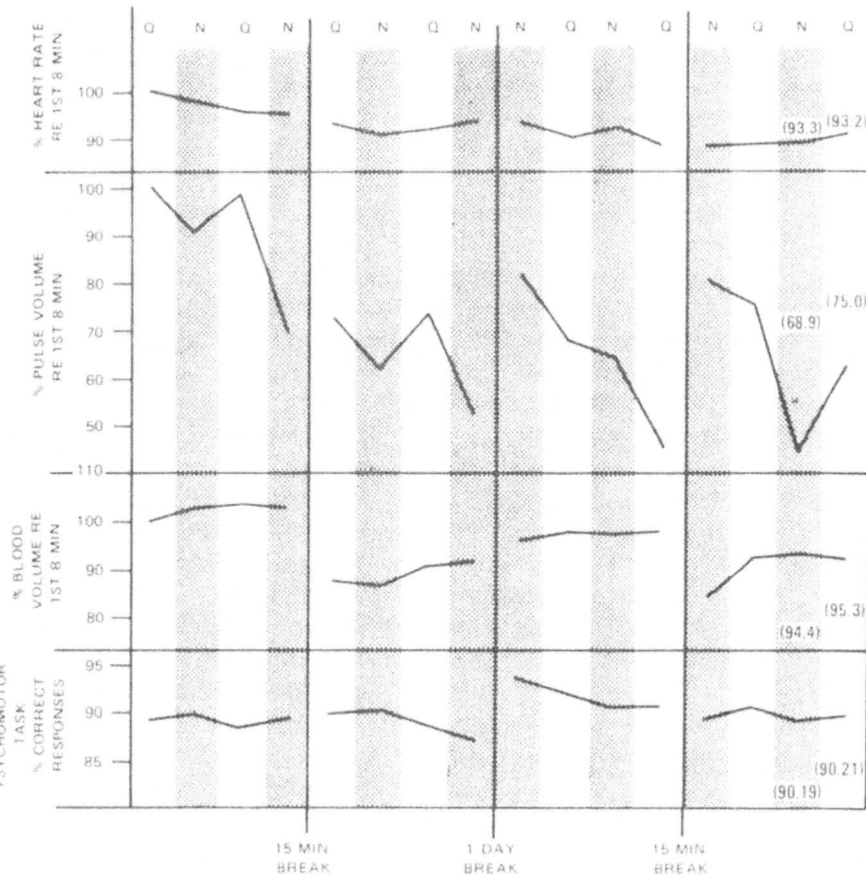

Q - QUIET (50 dBA STIMULUS NOISE) ON FOR 8 MINUTE SEGMENTS, FOLLOWED BY 2 MINUTES OF NO NOISE

N - NOISE (100 dBA) ON FOR 8 MINUTES, FOLLOWED BY 2 MINUTES OF NO NOISE

FIGURE 9.3. Averaged physiological and task performance results in alternate periods of quiet and noise. The numbers in parentheses in the last N and Q segments are the averages over all test segments for the respective N and Q conditions. (From ref. 38.)

noise, the hypothesis being that there could be an interaction between the altered state and noise stimulation that could influence task performance.

Strasser (ref. 41) found that noise had no apparent effect on the learning of, or performance on, a visual tracking task but that it did increase heart rate. However, with a sedative medication, the noise had no appreciable effect on either performance or heart rate.

Hartley *et al.* (ref. 42) exposed male college students for 53-min test periods to white noise at levels of 70 dBA and 95 dBA. Each subject was exposed to two test periods on each of 3 days. The various test conditions were counterbalanced and placebos used as a control for the administration of chlorpromazine (a sedative). No main effects of the drug or noise on a visual target detection (vigilance) task were found. The experimenters also asked the subjects to rate how certain they were of the correctness of their responses on the performance test. The subjects were somewhat more certain of their responses when in noise or when drugged, but these effects were counteracted when both conditions prevailed. All in all, the results indicated no undue arousal from 95-dBA white noise compared with 70-dBA noise, and possibly a slight countereffect to the depressive drug.

Hartley and Shirley (ref. 43) tested the hypothesis that the effects of noise and sleep loss on task performance would be counteractive. The task was the detection of slight changes in luminance of lights, and the noise was white noise at a level of 95 dB and, as control, 70 dB. Sleep conditions were 8 hours at night, 4 hours at night, or 4 hours in the afternoon. The subjects were tested for 1 hour three times per day. In addition to errors in detection, confidence ratings of responses were obtained from the subjects. The authors concluded that noise and sleep loss had mutually antagonistic effects (noise decreased performance after sleep, but increased it with sleep loss; noise made subjects more cautious in their responsiveness, and sleep made them less cautious). However, the variety of conditions and test schedules were complex, and overall there were no main effects of noise or sleep schedules on task (discrimination) performance. In any event, no general overarousal from the noise that would adversely affect task performance was evident.

Colquhoun and Edwards (ref. 44) found that white noise (100 dB) significantly decreased speed of responding to a five-choice serial response test and increased errors. Doses of alcohol caused no change in speed of responding but increased error rate when in quiet and decreased error rate when in noise. However, note that the noise effects data *per se* are suspect because as noted by Poulton (ref. 24) and discussed above, a click made by a response stylus contacting a metal plate could be heard in the control quiet (70-dB noise), but not in the 100-dB noise.

Simpson *et al.* (ref. 45) compared the performance on a tracking task (subjects held a stylus on a rotating visual target) in 50-dBA noise and in 80-dBA noise when the subjects had drunk a solution that may or may not have contained glucose. The results showed that the 80-dBA noise decreased time on target by 20 percent or so (statistically significant) and that the glucose reduced performance by about 10 percent for the 50-dBA exposure condition. These findings seem difficult to relate to arousal vs. counterarousal effects of noise or glucose (and indeed the investigators do not suggest that this is the case). Possibly because of the use of a tracking stylus, there may have been an audible feedback cue present in this study.

Task Variables

The studies presented in the previous section were those in which a major effort was made to measure or control the physiological state of the subjects. The following material is organized, insofar as possible, in terms of (1) possible interactions between task complexity and its performance in quiet and during and after exposure to noise and (2) possible effects on task performance in noise of the subjects' psychological aversion to the noise.

Signal detection tasks

Numerous simple reaction time studies have been conducted which measure the time required for a subject to press a key or switch whenever a light or sound occurs. Intense noise in the task environment has been found to cause the reaction time to increase, decrease, or remain the same. A decrease in reaction time has generally been attributed to some arousal effect of the noise, and an increase to a physiological or psychological stress effect of the noise. However, these explanations of the effects of noise on reaction time are not very convincing because sometimes, even within the same experiment (same subjects and general procedures), a mixture of all three types of results are obtained.

An example of such varied and conflicting results is shown in figure 9.4 from Franszczuk (ref. 46). Relative to reaction time in quiet tones, one-third-octave and wideband noise at center frequencies of 250 Hz and 1000 Hz either had no special effect or shortened reaction time. However, tones and bands of noise with a center frequency of 4000 Hz and white noise increased reaction time by about 0.02 sec. The tones and bands of noise were presented at levels ranging from 80 to 90 dB. There are at least three possible explanations for the variations in these results: (1) the shortened reaction times in the bands of noise at 1000 Hz are due to some beneficial arousal effect of the noises compared with quiet; (2) because higher frequency noise and tones are generally louder at a given sound pressure level than the lower frequency sounds, the subjects are more stressed by the higher frequency than by the lower frequency sounds; and (3) the increased reaction times in the white noise and 4000-Hz sounds are due to some artifact. With respect to the third possibility, the response key may have provided a high-frequency detent click or sound related to the amount of pressure placed on the key by the subject, which was masked by the white noise and the 4000-Hz sounds but not by the other noises. The procedure followed by most subjects in tests of this sort is to complete between signals as much of the switch activation response as possible (pushing the key as far as possible without the final activation) so that the response signal merely serves as a trigger to complete the response act. Possibly when in the masking noise, the subject refrained, for fear of making a false reaction, from starting to press the response key until the test signal occurred. The average 0.02-sec difference involved suggests that this might have been the case.

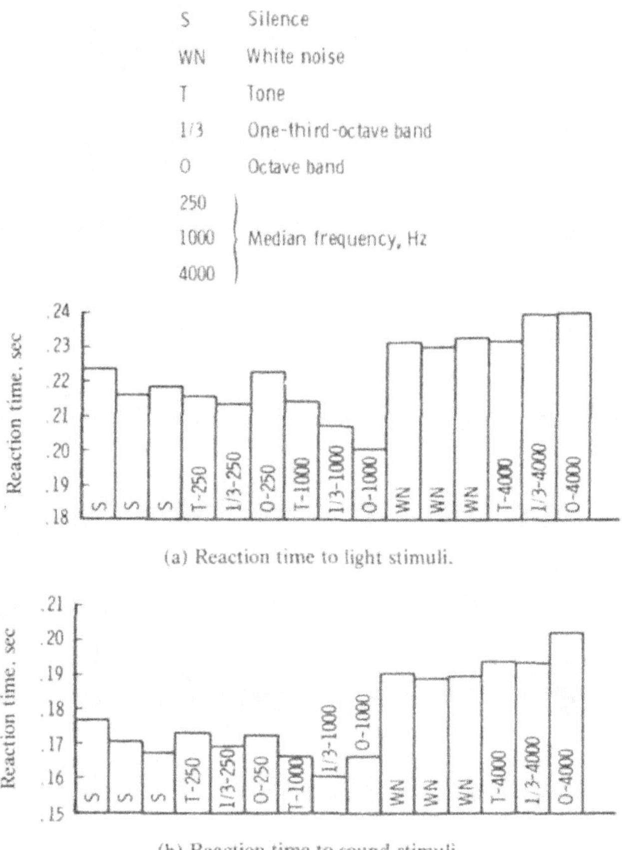

(a) Reaction time to light stimuli.

(b) Reaction time to sound stimuli.

FIGURE 9.4. Comparison of simple reaction time under different acoustical conditions and in silence. (From ref. 46.)

That such an artifact might have been present in Franszczuk's study is pure conjecture, but it is a condition found in the activation of some switches or keys used in some reaction time tests. Interestingly, and perhaps in support of this conjecture, Franszczuk reported that in other psychological tests and tests of choice-reaction times (where the subject must choose which of two keys to touch before responding (thereby obviating the possible partial switch activation artifact), no differences in reaction times were found between performance in noise and quiet.

Osada *et al.* (ref. 47) studied the influence of aircraft, train, and pink noise on both reaction time and time estimation. In the study of reaction time the noises were presented in bursts at levels from 50 to 80 dBA when the subjects were responding (by pushing different switches) to different colored signal lamps, one of which flashed every 5 sec. The noise somewhat reduced the time required for, and the variability of, the reactions. In the studies concerned with the estimation of elapsed time, the subjects pushed a switch when they believed that 10 sec had elapsed after an electric signal light was turned on. No effect on time estimation was observed for the noises except for the "rattling train" noise for which the time estimate was shorter than that given in the quiet listening condition.

Ando (ref. 48) exposed children (boys 13 to 14 years old) in a classroom to recorded aircraft noise. The noise was of about 60-sec duration and repeated 30 times with 60-sec pauses between exposures. He found that when the peak level of the noise was around 75 dBA, time seemed to pass more quickly than during a no-noise session, but when the aircraft noise had peak levels of 90 dBA, the time seemed to pass slower. He also found that subjects who came from an aircraft-noise-impacted area tended to judge the test periods to be shorter than did subjects from quiet neighborhoods. The general conclusions from these studies might be that time filled with noise, at least at moderate levels, passes more quickly than it does in quiet and that this effect is stronger for people not accustomed to living in a noisy environment.

Warner (ref. 49) investigated the effect of intermittent white noise at 80, 90, and 100 dB on the time required to detect the presence of an odd (different) letter from a background of 16 homogeneous letters flashed on a viewing screen. This is obviously a more complex target detection task than the task in a simple signal reaction time test. However, as seen in figure 9.5, the noise had no apparent systematic overall effect on target detection time. Note in figure 9.5 that at 12 and 15 min of time at task, the control (quiet) condition resulted in the fastest detection times. This might be interpreted to mean that the sometimes beneficial (shorter

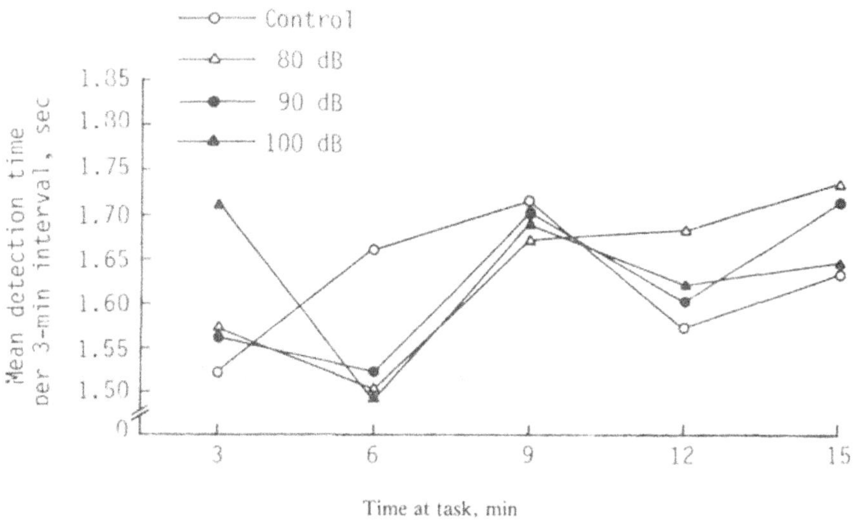

FIGURE 9.5. Mean detection time as a function of time at the task. (From ref. 49.)

detection times) effects of noise at 6 and 9 min of time at task could not be sustained by the subjects for longer periods because of increased fatigue. It has been suggested that performance in quiet may have been better because of a learning effect, whereas performance in noise may have been better earlier in the session because of an initial arousal effect. These conjectures seem unjustified, however, in view of the general variability of the results and the fact that for the longest time at task, 15 min, the detection times for the quiet and the most intense noise, 100 dB, were almost identical.

In subsequent studies in which the number of letter targets was increased from 8 to 16 and to 32, Warner and Heimstra (refs. 50 and 51) found that the mean detection time was greatly increased in quiet and generally increased in continuous noise at levels of 80, 90, and 100 dB. However, the noise improved performance (reduced detection time) for the more difficult, 32-letter task, relative to that in quiet, but had no such improvement effect on the performance of the 8- or 16-letter tasks (see fig. 9.6).

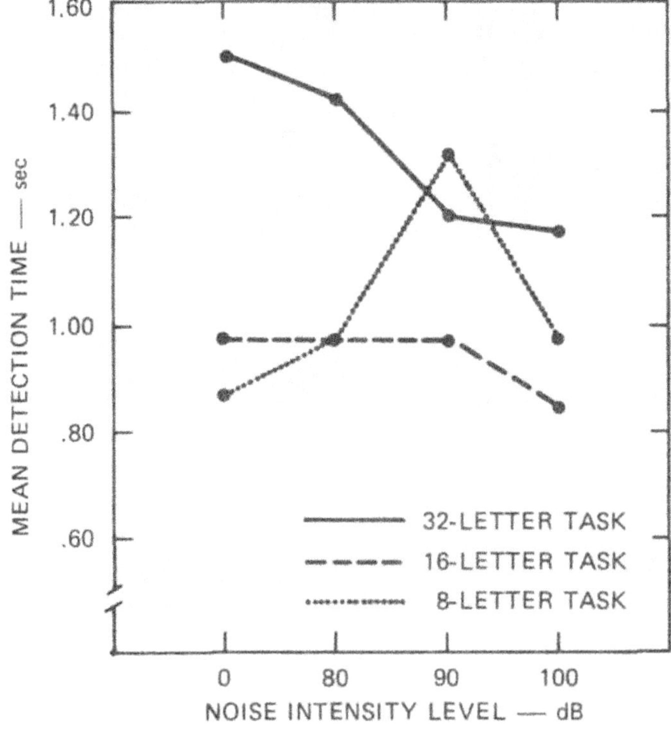

FIGURE 9.6. Detection time as a function of noise intensity and task difficulty. (From ref. 51.)

Psychomotor tasks

So-called psychomotor target tracking tests require the subject to manipulate a control level or grip in coordination with the movements of a visual target or display of lights. As reviewed by Eschenbrenner (ref. 52), studies up to 1971 had shown no effect of high intensity intermittent noise on the ability of subjects to track a moving target on an oscilloscope screen. Some initial decremental effects of noise found in some studies ceased with continued tracking.

Eschenbrenner (ref. 52) found that continuous regular periodic and aperiodic noise all reduced performance time on a complex visual tracking task compared with the performance time of a separate control group of six subjects working in quiet. Also, the more intense the noise (levels of 50, 70, and 90 dB of white noise were used), the greater the decrement. Aperiodic noise had more adverse effects than regular periodic or continuous noise. However, as Eschenbrenner notes, a subject received only twenty 40-sec trials (a total of about 6.5 min) while exposed to any one noise condition, and perhaps habituation to the noise would occur with more exposure.

Kaltsounis (ref. 53), Davies *et al.* (ref. 54), and Hartley and Williams (ref. 55) compared the performance on various visual tasks (figure completion and target detection) when subjects were exposed to noise and to music at comparable levels (75 to 95 dBA), and, by Kaltsounis, to speech and to quiet. Generally, no significant differences were found in task performance for subjects exposed to noise, music, speech, or quiet, although Davies *et al.* found an increase in latency in detection on a difficult task in noise but not in music.

Kunitake *et al.* (ref. 56) assessed the effect of acclimation to aircraft noise upon performance of various visual and time judgment tests by students (average age of 17.8 years). Half of the subjects came from a residential area exposed to intense aircraft noise and half from a quiet neighborhood. The subjects were exposed in the laboratory for 1 3/4 hours to recorded aircraft noises (one every 2 minutes) at a peak level of 95 dBA and for 1 3/4 hours to quiet as a control. The investigators found the following: (1) the acclimated subjects, from the noisy residential area, performed as well in the noise as in quiet on the mental tests; (2) the nonacclimated subjects had greater output in noise than the acclimated, and (3) there were no significant differences between the two groups in terms of fatigue or overall performance.

In addition to previously mentioned physiological measurements during a serial decoding task, Conrad (ref. 31) obtained subjective ratings from his subjects regarding their sensitivity to noise. He found, as shown in table 9.1, that those subjects who rated themselves highly sensitive to noise made fewer errors on the task than did those subjects who rated themselves less sensitive to noise during all test conditions, noise and quiet. There were, however, no significant interactions between overall performances in quiet and in the different noise conditions. As shown more precisely in the table, the main effect of noise condi-

TABLE 9.1

Mean Error Scores in Various Noise Conditions and Subjective Noise Annoyance Sensitivity

[From ref. 31]

Subjective noise annoyance sensitivity	Noise condition				
	Quiet	Continuous	Periodic	Aperiodic	Mean
High............	−7.44	−8.88	−4.97	−7.62	−7.23
Low	−14.09	−13.19	−14.98	−12.93	−13.80
Mean	−10.77	−11.03	−9.98	−10.28	

tion was not significant (F = 0.35) and the main effect of noise annoyance sensitivity was also nonsignificant (F = 1.53). No significant interaction effect was found.

Word and number search and memory tasks

In the period from 1970 to the present, an increasing number of research studies were concerned with the effects of noise on the performance of more intellectual tasks, *e.g.*, tasks involving memory of words and numbers and multiple simultaneous tasks.

Harris and Filson (refs. 57 and 58) used a difficult search task wherein the subjects scanned pairs of numbers (*e.g.*, 12-61, 41-47, and 56-45) printed on sheets of paper and wrote down those number pairs that were related in specified serial relations (*e.g.*, their sums were an odd number). While taking this serial search task, one group of subjects was exposed to 105-dB broadband noise and a second group to quiet. Some of the subjects received 3-min rest breaks during 36-min test periods and other subjects did not receive rest breaks. The results with the rest group indicated an adverse effect of noise on the first 12 min but no effect on the last 24 min of the test period, a result somewhat contrary to the arousal theory of noise. On the other hand, the rest group performed worse in noise than in quiet and performed worse on the last 2 days of testing in the noise than on the first day; these results are consistent with the arousal theory and could also possibly be related to Poulton's theory of noise interfering with internal, repetitious verbalization of the numbers.

However, Harris (ref. 59) later reported no negative effects of bursts of predictable and unpredictable noise, modulated and unmodulated at certain interruption rates at levels from 85 to 106 dBA, upon arithmetic serial search (as described above) and proofreading. In these studies, Harris was concerned with both immediate and aftereffects of the noise. (As will be discussed later, some investigators have found an adverse effect on perseverance on unsolvable puzzles given after exposure to noise.) Harris' conclusions (which differ somewhat from his earlier position that noise can have an adverse effect on the performance of serial search tasks (ref. 57) are as follows (ref. 59, pp. 347–348):

> The results of these series of experiments demonstrate that adverse effects of sound on human performance is not a foregone conclusion. An attempt to degrade performance deliberately in a short time period was unsuccessful. One should not be surprised by this because the literature contains many studies with similar results, and how many unreported studies there are that have failed to find adverse effects is anyone's guess. The effects of noise are so inextricably connected with the motivation of the subjects, the experimental task, and the experimental design, that it will still be a number of years before we begin to understand all of the variables involved.

As mentioned previously, Glass and Singer (ref. 29) found that physiological habituation occurred to their noise with bandwidth from 200 to 5000 Hz (a conglomerate of several people talking in Spanish and English, a mimeograph machine, a mechanical calculator, and a typewriter) at levels as high as 110 dB. This habituation occurred whether the time of occurrence of the noise was predictable or not and whether or not the subjects perceived it as being under their control. The predictable noise consisted of 9-sec bursts occurring regularly once every minute, the unpredictable noise consisted of 3- to 15-sec bursts occurring randomly in different quarters of 1-min periods (the total amount of noise received during 23-min test periods was the same for the predictable and unpredictable presentations). The subjects' attitudes of having or not having control of the noise were achieved by telling the "control" group that they could choose to terminate the noise at any time by pushing a switch on their chairs, but that the experimenter preferred that they not do so. In these investigations it was found that the noise, both predictable and unpredictable, did not significantly affect proofreading and numerical task performance during exposure to the noise. On the proofreading task given after the termination of the noises, the subjects who were exposed to intense unpredictable noise and felt that they had no control over the noise made more proofreading errors than did the other groups. However, Glass and Singer in other experiments and other experimenters, to be discussed later, did not find similar aftereffects of the noise on proofreading errors.

Wittersheim and Salame (ref. 60) studied the effect of 95-dB pink noise on the learning of a series of six digits displayed one at a time for 500 msec, separated by 140 msec. Overt and covert rehearsal was allowed. They found that the noise reduced memory task performance most when the noise was present during the times of acquisition ("information being taken in") and practice ("edited in storage") than in the reproduction stage of the task. Wittersheim and Salame note that earlier studies on memory tasks performed in noise are highly variable and inconclusive. Their findings that noise has a detrimental effect during the learning of number sequences could perhaps also be explained by Poulton's theory of noise interfering with internal speech rehearsal.

Osada *et al.* (ref. 47), in addition to the previously discussed reaction time and time estimation tests, studied the effect of aircraft and train noise (levels of 60 to 90 dB) on performance of a test in which the subjects counted dots flashed on a screen. The noise was found to markedly impair figure counting, a task also possibly involving internal speech.

Noise interference with internal speech or verbalizations of words involved in task performance may possibly have a role in performance on the Stroop color interference test. In this test, subjects must name or sort cards according to the hue of the ink used to spell the name of different colors (*e.g.*, the word "red" might appear in blue ink). Hartley and Adams (ref. 61) found small adverse effects (1 to 3 percent) on task performance in broadband random noise at 100 dB compared with performance in 70-dB noise. Brief exposure to noise was beneficial to performance but long (30-min) exposure increased interference with task performance.

However, O'Malley and Gallas (ref. 62) found that broadband random noise at levels of 75 dB and 100 dB did not degrade performance on the Stroop test from that achieved in quiet, and that noise at 85 dB improved task performance. O'Malley and Gallas also found no effects of the noise on a "rod and frame" test (the subject must set a rod into a vertical position in a nonvertical framework) and a "pathway test" (the subject must trace connections between letters and numbers in some specified pattern).

Weinstein (ref. 63) found that a tape recording of radio news items at 68 dBA, designated as a noise condition, had mixed effects, compared with quiet, on a proofreading task. The noise impaired the detection of grammatical errors, but did not affect speed or detection of spelling errors. Initially in the noise, proofreading was more accurate than in quiet.

Three studies of the effects of broadband noise at a level of about 85 dB on various word and language tests, including the Stroop word-color naming test, were reported in 1980 at a congress on the effects of noise (Smith, ref. 64; Dornic, ref. 65; and Broadbent, ref. 66).

The results of the Smith study are given in table 9.2. In the first test week, slightly fewer (18.0 vs. 18.7) words were recalled in the noise than in quiet, but in the second week this was reversed (17.9 in the noise and 17.5 in the quiet). On a C score (showing "clustering", defined by the formula given in table 9.2), there was less clustering in the noise than in the quiet. The significance of this clustering measure to task behavior is not clear to the present author.

Dornic (ref. 65) studied the effects of combined white noise with real-life recorded noise on the recall of word lists given in the dominant and subdominant language for each subject. Various types of memory tasks for the two languages were administered. Representative findings in figure 9.7 show that compared with quiet, the noise has a negligible adverse effect on performance of the dominant language tasks and some adverse effect on the performance of the subdominant language tasks. Dornic concludes that the greater noise effects with the

TABLE 9.2

Results on Recall of Words on Stroop Test in Conditions of Quiet and Noise

[From ref. 64]

(a) Mean number of words recalled in quiet and noise for each session

Week 1		Week 2	
Quiet	Noise	Quiet	Noise
18.7	18.0	17.5	17.9

(b) Calculation of the Dalrymple-Alford C score

$$C = \frac{R - \text{Min R}}{\text{Max R} - \text{Min R}}$$

R = Number recalled − Observed number of runs

Max R = Number recalled − Number of categories

Min R = 0 (Number recalled + 1 ≥ 2 × Largest category recalled

= 2 × (Largest category recalled − 1) (Number recalled < 2 × (Largest category recalled − 1)

(c) Mean arcsine transformed C scores in quiet and noise for both sessions

[High scores represent greater clustering]

Week 1		Week 2	
Quiet	Noise	Quiet	Noise
2.45	2.19	2.62	2.49

subdominant language are related to covert pronunciation difficulties, a conclusion consistent with Poulton's theory.

Broadbent (ref. 66) studied the time taken to read 100 color names printed in black (W), the time taken to name meaningless patches of colored ink (C), the time to name the ink color in which irrelevant color names were printed (CI), and the time to read color names printed in different irrelevant colors (WI). The performance was measured after, rather than during, exposure to noise. Broadbent's findings are given in table 9.3. The differences between the data after noise and after quiet seem small and variable. However, Broadbent found the C/W ratio after noise to be significantly lower statistically than the ratio after

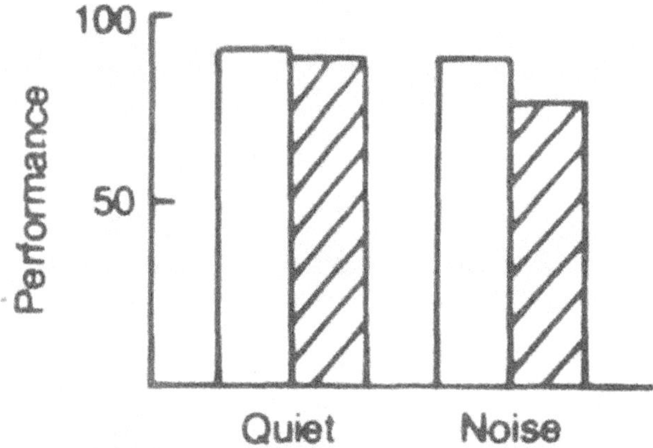

FIGURE 9.7. Performance (combined measure of accuracy and speed) on language tasks in dominant language (open columns) and subdominant language (hatched columns). (From ref. 65.)

TABLE 9.3

Mean of all Subjects Performance After Noise and After Quiet

[From ref. 66]

Condition	C/W	C	W	CI	WI
After noise	1.27	49.76	39.40	73.54	42.89
After quiet	1.31	50.34	38.85	74.11	42.21

quiet, 1.27 vs. 1.31 (actually from the given data the ratios are 1.263 vs. 1.296, respectively). He concludes (pp. 363-364):

> Thus, after noise, people name colored inks relatively faster than they read printed color names. It seems very plausible that effects in the interference condition are secondary to this change; on most theories of the Stroop, interference will be maximal at a particular value of C/W. It may therefore increase or decrease as C/W decreases, depending which side of the maximum one is; or interference may be unchanged as in our results. Notice however the following points: (1) this effect on speech occurs after exposure and cannot be caused by something like masking; (2) the effect is on performance with only one task, not on the interference between two tasks—the effect cannot therefore be included under theories of allocation of attention between stimulus sources; and (3) the effect is not a suppression of speech in general, but of speech in response to one kind of stimulus rather than another—it does not fit any generalization that use of speech as such, is changed by noise.

Aftereffects on Insoluble Puzzles and Proofreading Tests

As discussed above, the effects of noise on proofreading tasks of various kinds appear to be negligible, or at least inconsistent, when measured during the presence of the noise. Glass and Singer (ref. 29) also measured possible aftereffects from the exposure to noise on insoluble puzzles and proofreading tests. The insoluble puzzle task, which except possibly for Broadbent's rather tenuous results with certain color naming tests appears to be of unique importance to the detection of an aftereffect of noise exposure, is described by Glass and Singer as follows (pp. 48–49):

> The postnoise task measuring frustration tolerance was adapted from one used by Feather (1961), and consisted of four line diagrams printed on 5 × 7-inch cards arranged in four piles in front of the subject. . . . Each pile was about 1 inch high and contained only one kind of puzzle. Cards were face down, so a subject was unable to see the puzzle until he began work on that particular pile of cards. The task was to trace over all of the lines of a diagram without tracing any line twice and without lifting the pencil from the figure. He was informed that he could take as many trials at a given item as he wished. However, he was also told that there was a time limit on how long he could work on a given trial, and the experimenter would inform him when his time was up over a loudspeaker. It was emphasized that such notification did not mean he had to go on to the next pile. That was his decision. It simply meant that he must decide whether to take another card from the same pile or move on to the next item. If the subject wanted another trial, he discarded his unsuccessful card into a bin and took another copy of the same item. If he went on to the next pile, however, he could not go back to the previously unsolved item. The subject immediately went on to the next pile following a successful solution.
>
> Two of the line diagrams were mathematically insoluble but sufficiently complex so that subjects were unable to see this. (Postexperimental interviews revealed that most subjects believed the insoluble puzzles were potentially soluble.) The puzzles were arranged in front of the subject so that the first pile always consisted of the same insoluble puzzle, the second the same soluble puzzle, the third the same insoluble puzzle, and the fourth the same soluble puzzle.

Glass and Singer found that perseverance in attempts to complete insoluble puzzles in a 10-min period or so after exposure to some of the noises was less than after quiet. This effect was the greatest when the time of occurrence of the noise bursts was unpredictable and the subjects believed the noises not to be under their control. Figure 9.8 shows that the loud noise (108 dB) had about the same aftereffect on number of attempts to solve insoluble puzzles as soft noise (56 dB) or no noise when the occurrence of the noise was predictable. When the noise occurrences were unpredictable, the subjects attempted fewer puzzles. Figure 9.8 shows that noise with no perceived control had an adverse aftereffect on perseverance compared with noise with the perceived control. Glass and Singer suggest that unpredictable noise conditions with no control levy a psychic cost that results in lower tolerance for frustration after the exposure. Glass and Singer called this condition a state of learned helplessness.

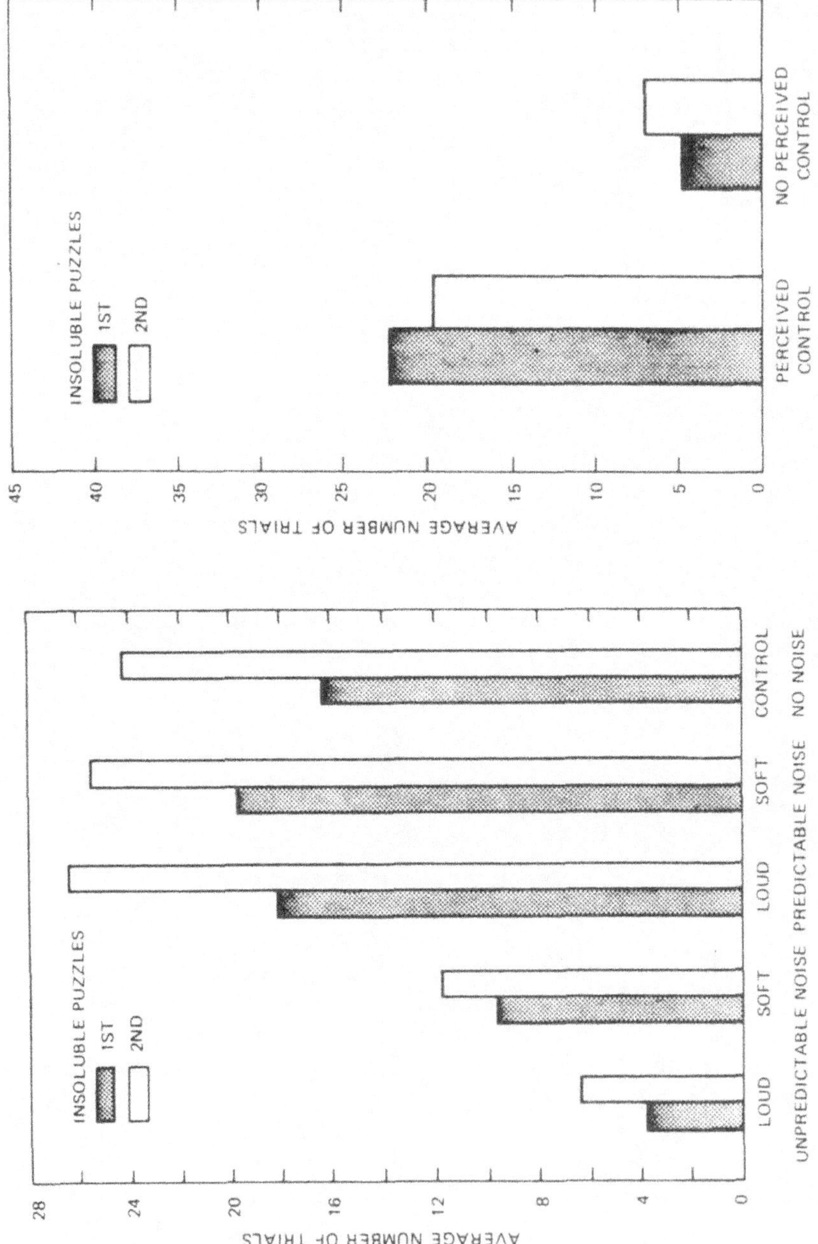

FIGURE 9.8. Average number of trials on the insoluble puzzles. (From ref. 30.)

Wohlwill *et al.* (ref. 67) found results consistent with Glass and Singer's findings, that performance on a dial-monitoring task was unaffected by the presence of noise, but that after exposure to noise, subjects showed less persistence on insoluble puzzles than did the no-noise group. On the other hand, Moran and Loeb (ref. 68), studying the performance of nonauditory and auditory tasks in quiet and in noise, were not able to find any aftereffects on perseverance on insoluble puzzles such as those found by Glass and Singer, even after noise that interfered with the auditory tasks. Moran and Loeb, in a second experiment that more closely replicated Glass and Singer's procedures, again found no aftereffects of the noise conditions on persistence on the unsoluble puzzles. However, Moran and Loeb used recorded aircraft noises and not the conglomerate noise used by Glass and Singer.

Percival and Loeb (ref. 69) conducted two experiments aimed at replicating and extending the studies of Glass and Singer on the aftereffects of noise on performance. In one, subjects were exposed to 24 min of the conglomerate noise (95 dBA) used by Glass and Singer either of fixed schedule (predictable) or of random schedule (unpredictable), plus a control quiet condition (46-dBA background noise). During the test sessions the subjects worked on number comparison, addition, and letter checking tasks. After exposure, the subjects worked on the insoluble puzzles and proofreading tasks. Persistence on insoluble puzzles was least after the unpredictable noise and greatest after the control condition (see table 9.4). However, contrary to some of the results of Glass and Singer, there was no evidence in this study of an aftereffect on the proofreading test. Consistent with Glass and Singer, there were no adverse effects of the noises on tasks performed while the noise was present.

In a second experiment, Percival and Loeb (ref. 69) used essentially the same tasks and procedures as outlined above, but added some new noise conditions. They used 95-dBA intermittent white noise, recorded normal aircraft flyover noises, recorded combinations of the peaks of aircraft flyover noises, Glass and Singer's conglomerate noise, and a 46-dBA background noise (control condition). As in Percival and Loeb's first experiment, the noises had no effect compared with the control condition on any of the performance tasks either during or after noise exposure. In addition, only Glass and Singer's conglomerate noise (GS) and the peaks of aircraft noise (AC), both presented at a random (unpredictable) schedule, significantly affected the number of attempts made on the insoluble puzzles given after exposures (see table 9.5). Percival and Loeb suggest that the sudden abruptness of the peaks of aircraft noise possibly provided the acoustic characteristics that induced the aftereffect of fewer attempts being made on the insoluble puzzles (an average for both puzzles of about 11 attempts compared with about 19 for the control condition). Glass and Singer's conglomerate noise (GS) probably also contained some sudden, irregular peak levels, which by the same token could account for the average of only a few attempts, about 8, being made after that noise.

TABLE 9.4

Mean and Standard Deviation (in parentheses) of the Number of Attempts on Insoluble Puzzles After 24 Min of Exposure to Glass and Singer's Conglomerate Noise

[From ref. 69]

Noise condition	Sex	Insoluble puzzle 1	Insoluble puzzle 2
Random Schedule	Male	7.71 (6.02)	10.71 (5.91)
	Female	7.57 (3.51)	11.14 (7.17)
Fixed Schedule	Male	13.00 (7.30)	14.00 (8.06)
	Female	11.00 (4.80)	12.57 (6.13)
Control	Male	14.00 (7.19)	16.29 (4.72)
	Female	15.86 (4.38)	20.00 (0.00)

TABLE 9.5

Mean and Standard Deviation (in parentheses) of Number of Attempts on the Insoluble Puzzles in Second Experiment

[From ref. 69]

Insoluble puzzle	Noise conditions[a]				
	GS	WN	AC	NA	C
1	8.25 (4.88)	17.67 (7.46)	10.33 (9.27)	16.42 (7.50)	18.83 (6.72)
2	7.60 (4.78)	19.50 (6.71)	11.75 (6.86)	18.08 (7.46)	19.67 (4.31)

[a] *GS* = *Glass and Singer's conglomerate noise*
WN = *White noise*
AC = *Peaks of aircraft noises*
NA = *Normal aircraft flyover noises*
C = *Control (46-dBA background noise)*

Rotton *et al.* (ref. 70) reported that background noise plus meaningful speech (both at a level of 80 dB) presented with tasks requiring memory recall of words reduced an individual's ability to persevere on the insoluble puzzle test. The speech plus noise is somewhat like Glass and Singer's conglomerate noise.

The consistent findings of an apparent reduction in the number of attempts made to solve insoluble puzzles after exposure to some noises irregular in time and spectra attest to the existence of a special noise-task interaction. The possible meaning and significance of this effect, however, is a matter of conjecture.

Glass and Singer's interpretation that the effect represents a reduction of tolerance for frustration and is relatable to learned helplessness is not borne out by the subjective ratings given by the subjects of the irritating, distracting, and unpleasant qualities of the noise. Glass and Singer (p. 54, ref. 29) found that these subjective ratings were more strongly correlated with the intensity of a noise than with its predictability or unpredictability.

Percival and Loeb also found no relation between subjective ratings, including ability to cope with the noise, and the aftereffects of the noise. Similar negative findings were cited earlier with respect to subjective ratings of irritation from noise and performance on signal detection, tracking, and other such tasks. That a person's trying fewer times to solve an insoluble puzzle before going on to work on another puzzle, represents a degradation in performance due to a "psychic cost" is debatable. On the face of it, stopping work on an insoluble puzzle seems to be sensible instead of being a reduction in tolerance for frustration as proposed by Glass and Singer.

For some unexplained reason, the insoluble puzzle test was administered only after exposure to noise, whereas the other performance tasks were given both during and after the noise exposures. Conceivably fewer insoluble puzzles might be attempted during and after exposure to the unpredictable and irregular noise than during and after exposure to quiet or to the other noises tested. But even if this was found, why only this particular type of noise and puzzle task should interact in this way would remain in open question.

In addition to the psychological theories developed by Glass and Singer, a possible explanation could be that judgments of elapsed time are affected (seems shortened) by this type of noise but not by the other noises. Recall that the subjects were instructed to complete as many puzzles as possible during a 10-min period. Also recall that Osada *et al.* (ref. 47) found that during exposure to a "rattling train" noise, but not to other more regular noises, subjects judged the time elapsing after a signal to be somewhat shorter than it was. Could it be that irregular intense noise makes time seem short and that this effect, not a more complex psychological aspect of frustration tolerance, learned helplessness, or psychic cost, is responsible or involved in the aftereffect data in question? But why this effect, if indeed it occurs, should persist for 10-min or so (the estimated period in which the insoluble puzzle test was taken) after exposure to this particular type of noise is another mystery.

Information Overload

Nearly all the mental and psychomotor tasks that have been employed in laboratory studies of the effects of noise on task performance have required the subject to work at a fast pace. Further, the tasks demanded concentrated attention and repeated mental and psychomotor activities. In brief, the subjects were heavily loaded with a work task on the usually tacit assumption that the noise would constitute an additional physiological and/or perceptual load that would somehow interfere with task performance. In addition to increasing the information work load on the subjects, experimenters have taken performance data from subjects on two simultaneous tasks in noise on the theory that the noise might interfere with the ability to concentrate attention on one or the other of the two tasks.

Hockey and Hamilton (ref. 71) and Davies and Jones (ref. 72) required subjects to remember words that were projected in rapid sequence on one of four corners of a screen when in quiet (55-dBA noise) and when in 95-dBA noise. They found that relative to quiet the noise increased the recall of words in their order of appearance (73.75 percent correct in quiet and 80 percent in noise, see table 9.6). The subjects were asked after the word-recall test to try to remember at which corner of the screen the recalled words had been projected. This was designated as an irrelevant task. As seen in table 9.6, after the noise the subjects correctly recalled fewer of the locations of the words than after quiet (33.33 percent vs. 60.12 percent). The experimenters concluded that their data support the theory that noise forces the attention of the subjects towards high priority tasks and away from low priority tasks or irrelevant information. The improvement in performance on the main tasks could also be attributed to a beneficial arousal effect of the noise over the quiet control condition.

TABLE 9.6

Mean Percentage Recall of Relevant and Irrelevant Items (20 subjects in each case)

[Hockey and Hamilton's (ref. 71) results are shown in parentheses for comparison purposes; for their control condition there were 36 subjects and for their noise condition there were 32 subjects. Data from ref. 72.]

	Relevant task, Percentage of words recalled in correct order	Irrelevant task, Percentage of locations recalled
Control (55 dBA)	73.75 (73.12)	60.12 (48.50)
Noise (95 dBA)	80.00 (69.00)	33.33 (32.00)

A study by O'Malley and Poplawsky (ref. 73) supports the concept that noise causes a subject to focus attention on a primary task. The subjects' task was to anticipate words projected on the center of a screen rather than words which were projected onto the periphery of the screen. Fewer peripherally located words (unmentioned to the subjects) were given in free recall after task performance in noise than in quiet.

Loeb *et al.* (refs. 74 and 75) made a direct study of possible interactions of noise (recorded continuous industrial noise at 105 dBA and impact noise at 136 dB) and quiet (white noise at 75 dBA) with performance on two simultaneous tasks. One task consisted of tracking a moving target in the middle of the visual field and the second task consisted of responding to the onset of lights appearing at the periphery of the visual field. By means of instructions, different groups of subjects were made to give different biases (priorities) and degrees of attention to the two different tasks. The results indicated that regardless of priority instructions, the noise conditions impaired, to a small extent, tracking performance but not peripheral light monitoring, a result seemingly contrary to the results cited above from the studies of O'Malley and Poplawsky (ref. 73), Hockey and Hamilton (ref. 71), and Davies and Jones (ref. 72). Poulton suggested that the decrement in the tracking task in noise in the Loeb and Jones study was due to the noise masking feedback cues that were audible in the relative quiet. Loeb and Jones listened for such cues and concluded that this was not a factor, however. In any event the lack of an interaction between noise and performance of two tasks of differing priorities, and presumably given different degrees of attention, was not found in the studies of Loeb *et al.* (refs. 74 and 75).

Finkelman *et al.* (refs. 76 to 80) conducted studies in which, in addition to a no-noise condition, the subjects were presented via earphones with bursts of white noise separated by periods of silence. Predictable noise usually consisted of 9-sec bursts of white noise separated by 3-sec periods of silence; and unpredictable noise consisted of bursts of white noise of random duration (1 to 9 sec) interrupted with random durations (1 to 3 sec) of silence. The subjects performed a primary visual tracking task and also a subsidiary task in which the subjects said out loud a digit previously announced (remembered) over the earphone upon the aural presentation of another randomly selected digit. A digit was presented to the subject every 2 sec. Compared with the quiet condition, neither the regularly nor the randomly presented noise, at levels of 80 dB, was found to have any significant effect on the performance of the tracking task in the Finkelman and Glass study (ref. 76). However, on the subsidiary auditory task, the unpredictable noise had a more degrading effect than did the predictable noise; the mean number of incorrectly repeated digits, out of 60 digits, was 0.6 in the quiet, 4.0 in the predictable noise, and 8.0 in the unpredictable noise. Similar findings were reported by Finkelman *et al.* (ref. 79) in experiments in which the noise was increased to 93 dBA. These investigators interpreted the results to mean that the information handling capacity of the subjects was exceeded when in addition to performing a primary task in noise (especially in unpredictable noise), they must also perform a subsidiary task.

Another possible interpretation of these findings as well as those reported by Finkelman *et al.* (ref. 80) is that the noise masked the digits read to the subjects less in the predictable noise (refs. 76 and 77) than in the unpredictable noise, possibly because the regular 3-sec periods of silence in the predictable noise allowed at least one of the digits coming at 2-sec intervals to occur in silence. The probability of the digits occurring in silence would be less in the unpredictable noise where the periods of silence varied from 1 to 3 sec. These investigators believed the speech to be perfectly audible in the presence of the noise, but errors in speech understanding due to slight masking effects can occur without the listener being necessarily aware of the presence of masking. The high levels of noise, 80 to 93 dBA, would suggest that components of the speech signals (probably presented at about a comparable level) were masked but seemed intelligible because of the small message set (digits only).

That this was perhaps the case is shown by the Zeitlin and Finkelman (ref. 78) study in which as a subsidiary task, the subjects called out digits, from 0 to 9, in random order at a self-paced rate (the subjects were not presented with any aural signals). Task performance was evaluated by determining the randomness of the digits announced by the subjects. There were no interactions between the noise and performance on this nonauditory subsidiary task.

Similar auditory masking, as well as other artifacts, may have also been involved in a study by Bell (ref. 81). In this study a psychomotor pursuit rotor task (performed with the dominant hand) was considered the primary task and a subsidiary, concomitant task (performed by the nondominant hand) consisted of the subject's tapping a telegraph key once if a two digit number was numerically lower than an immediately previous number and twice if it was higher. The numbers were heard by the subjects via earphones at a speech level of about 95 dBA (personal communication from Bell). White noise at 95 dBA of randomly determined duration (1 to 9 sec) and intervals of quiet (55 dBA background noise of 1- to 9-sec duration) and three temperature conditions were present during the performance of the tasks. The results are shown in table 9.7. Performance on the audible number task was degraded by the 95-dBA noise compared with the 55-dBA background noise, but performance on the pursuit tracking was not significantly affected by the 95-dBA noise. The temperature effects followed a similar and additive pattern. Bell ascribes the findings to an information overload that forces greater attention on the primary task. However, the audible signals involved in the subsidiary task, as well as feedback clicks from the telegraph key used in that task, might have been masked by the noise. The effects of temperature differences are indeed interesting, but their relation to the information overload theory is, as Bell notes, a matter of further research.

Grether (ref. 82) and Grether *et al.* (ref. 83) conducted studies in which broadband random noise at a level of 105 dB, heat of 120° F, and vibration of 5 Hz

TABLE 9.7

Mean ($\bar{X}$) and Standard Deviations (σ) for Primary and Subsidiary Task Performance for Three Levels of Temperature and Two Levels of Noise

[From ref. 81]

(a) Primary pursuit rotor task—time on target (seconds)

Noise level	Temperature 22°C	29°C	35°C
55 dB(A)	$\bar{X}$ = 398.21 σ = 151.63	$\bar{X}$ = 392.03 σ = 172.30	$\bar{X}$ = 389.48 σ = 176.16
95 dB(A)	$\bar{X}$ = 383.78 σ = 165.16	$\bar{X}$ = 357.12 σ = 214.10	$\bar{X}$ = 341.40 σ = 179.34

(b) Subsidiary number task—number of errors

Noise level	Temperature 22°C	29°C	35°C
55 dB(A)	$\bar{X}$ = 27.38 σ = 18.39	$\bar{X}$ = 44.54 σ = 42.49	$\bar{X}$ = 52.96 σ = 64.19
95 dB(A)	$\bar{X}$ = 53.21 σ = 24.03	$\bar{X}$ = 55.38 σ = 28.97	$\bar{X}$ = 78.13 σ = 48.83

and 0.30g peak acceleration were imposed singly and in combinations on human subjects. Control, no-stress conditions were also studied. Tasks performed under these conditions included two-dimensional compensatory tracking, choice reaction time, a voice communication test of logical alternatives, mental arithmetic, visual acuity, and subjective ratings of stress conditions.

The subjects rated the subjective severity, but not the intrusiveness, of stress as the number of stressors was increased. However, there were no additive adverse stress interactions with respect to performance of the various tasks. Indeed, on the tracking tack, performance was slightly improved by the addition of noise to the heat and vibration stresses. This was perhaps due to the aforenoted ability of noise to mask distracting sounds in the environment and thereby improve concentration on, and performance of, the tracking task.

Perhaps the feebleness of the theories that sound or noise has inherent adverse effects on nonauditory task performance is illustrated by the results of a recent study by Gawron (ref. 84). Five hypotheses were tested. (Tasks included classifying into three categories, immediate and delayed canceling of visually displayed digit pairs, and joy stick tracking of a moving target.) The following is a summarization concerning four hypotheses in that study.

1. The theory that psychological set is a major determinant of the effect of noise on the performance was not supported by the data.

2. Contrary to hypothesis that greater decrements as a result of noise would occur in dual-task rather than in single-task performances, performance at the single-task level and the lowest noise intensity was comparable to dual-task performance at the highest noise intensity. Also, although there was a significant dual-task decrement, there were no significant differences among performance levels at the three noise intensities.

3. The third hypothesis is based on the theory that noise is a distractor to which subjects habituate over time. Thus, performance decrements in the presence of noise would be expected to occur early in the session and to decrease over time. In this test the opposite effect was found.

4. Arousal theory states that performance in the presence of noise should be a function of intensity and that overarousal should reduce performance. For tracking, and perhaps for percentage-correct scores on delayed digit canceling, there was a linear facilitation of performance associated with increasing noise intensity. The lack of support for this theory may be attributed to the restricted range of noise intensities (55- to 85-dBA white noise), and performance decrements may occur at intensities beyond those investigated.

Gawron concludes that further data must be collected to overcome "dearth of experimental data" and to determine what types of performance are susceptible to noise, what kinds of noise must be controlled, and how such effects depend on the age, physical condition, and personality of the subjects. However, a more realistic conclusion could be that the dearth of data showing adverse effects of noise on nonauditory task performance is due to the plethora of data, including Gawron's, showing that such adverse effects do not occur because of the noise *per se*.

Industrial Field Studies

There are many practical reasons that prohibit, or at least make difficult, the conduct of research investigations on the possible effects of noise in the workplace upon work output and efficiency. Nevertheless, over the years some studies have been published on this question. The following discussion is taken from the author's earlier reviews of these studies (refs. 1 and 2); the author is aware of no new studies published since 1970 in this area of research.

Office and factory noise

Kornhauser (ref. 85) attempted to determine whether typists working in a relatively quiet office do more work and feel less fatigue than those working in a noisy office. In this and some of the experiments described subsequently, "noise" is described only qualitatively, making comparison of results among experiments difficult. Record was kept of two typists who spent the first 2 days working in a quiet office and then 2 days in the noisy office, and two other typists who worked in the reverse order of noise and quiet. The results showed that wasted lineage was 23 percent greater in the quiet room than in the noisy room and also that 1.5 percent more lines were written under noisy conditions. Rating scales revealed that the typists felt that they were working harder in the quiet than in the noise. The differences between the two conditions cannot be accepted as necessarily significant, because of the small number of subjects. But the most important criticism against this study, and the one that can be leveled at nearly all the studies conducted on this problem under working conditions, is that there could have been many differences between the two work offices other than noise level, such as lighting or ventilation, that might account for the results.

There are some reports that purport to show a deleterious effect from noise. It is claimed (refs. 86 to 89) that "moving the assembly department of a regulator company from adjoining a noisy boiler shop to a quiet room resulted in lowering rejections at inspection from 75% to a low figure of 7%." The conclusion drawn was that the reduction in noise level caused the increased efficiency. Changes in lighting, temperature, and facilities were again ignored as possible contributing factors. In spite of the numerous articles published on these results, there are no references to the original study other than the statement that it was done by a "Dr. Sachsenberg" in Germany (ref. 87).

In Great Britain, Weston and Adams (refs. 90 and 91) studied the effect of noise on the work performance of weavers. The looms in weaving sheds generally create noise at a level of about 96 dB. Weston and Adams did three experiments: (1) they had 10 weavers wear earplugs, which reduced the intensity of the noise at the eardrum by 10 to 15 dB, on alternate weeks and recorded their output over a 26-week period; (2) they equated two groups of weavers, 10 in each group, with regard to past efficiency, and then one group wore earplugs while working for a 6-month period while the second group served as a control, working without earplugs; and (3) they repeated the second experiment, using some different subjects, but extended this experiment over a period of 1 year. The results of all three experiments were roughly the same, about a 12 percent average increase in efficiency for those who wore earplugs with respect to those who did not. The gain amounted to a 1 percent increase in the amount of material produced. The results of the first experiment were considered suspect by Weston and Adams, however, because of a difference in humidity between the weeks worked with earplugs and the weeks when earplugs were not worn. Figure 9.9 shows the results of the third

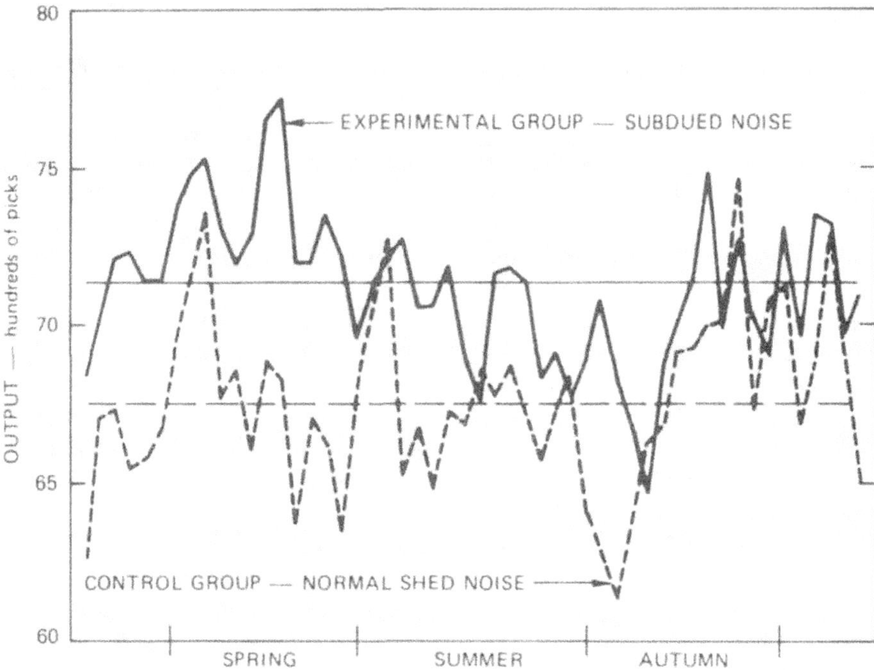

FIGURE 9.9. Weekly variation of output of weavers during 1-year experiment. The experimental group wore earplugs. (From ref. 90.)

experiment. Towards the end of the year the experimental and the control groups were coming closer together with regard to work output. This suggests that an initial difference in motivation between the two groups might have helped make the experimental group superior and the control group inferior. But as the experiment wore on, it might be surmised, the added motivation or interest from being a subject began to wane, bringing the control and experimental groups closer together with regard to the work output. Indeed, the subjective reports of some of the subjects indicate an approval of experiments and attempts to help the worker. Such attitudes alone are known to result in significant changes in work output, in the presence of noise (ref. 92). Another critical point has been made by Berrien (ref. 93), who noted that the equality of the control and the experimental groups in the first and second experiments was never demonstrated.

A major American field study was conducted by the Aetna Life Insurance Company in its own offices (ref. 94). Apparently the study has never been published in its entirety, although sample results have appeared in several places. Semimonthly bonus records (reflecting work productivity) for typists, clerical checkers, punch-card and comptometer operators were compared for a year before, and a year after, sound absorbing material was installed in all offices. As a result of the quieting, calculating machine operators' errors were reduced 52 percent, typists' errors were reduced 29 percent, health improved 37.5 percent, and employee turnover was reduced 47 percent, a truly remarkable achievement for absorbent wall board. The sound level was reported at about 41 dB prior to the sound treatment and 35 dB after.

In view of other studies, these claims are fantastic when credited, as they are, to "adjustment of the noise factor alone." Because of a paucity of relevant facts, to criticize the study is difficult, but one obvious factor undoubtedly contributing to the differences between the variables recorded was the lapse of time. Two years elapsed in which the workers may have improved their efficiency through learning, the ill and nonadept may have changed jobs, *etc.* One control check was made a year later which should, by itself, provide ample data for negating the conclusion concerning the effects of noise. For this check the sound-absorbing walls were covered with gypsum board, thereby raising the sound level by 6 dB. The bonus efficiency dropped to some extent (not as low as the first year), but within 2 months was as high as the level of the quiet year.

Felton and Spencer (ref. 95) comment that ego involvement in a high-status occupation offsets concern about noise (94 to 119 dB). Ganguli and Rao (ref. 96) believe, but present no data, that productivity in most workers is not affected by noise of 100 dB or lower. However, De Almeida (ref. 97) found absenteeism from the work room dropped when noise level was reduced.

Broadbent and Little (ref. 98) reported that the reduction in the noise level from 99 to 89 dB in a factory work space (bay) resulted in fewer numbers of broken rolls of film and equipment shutdowns than were experienced by the same workers when they worked in an untreated bay (the workers moved from one bay to another during the workday). The work performance improved in both the sound-treated and nontreated bays after some of the bays were treated, apparently due to generally improved morale. (See table 9.8.) Broadbent and Little propose that these findings support the findings from the serial search laboratory tests discussed earlier that showed an apparent decline in test performance in noise. However, there may have been an auditory component (as was later found to be the case in some of the laboratory studies) to the work (threading film on spools) that aided the workers threading the film in detecting films slipping from sprockets or malfunctions in the machinery. If this was the case, the reduction of the noise should have lead to improved work performance.

Kourigin and Mikheyen (ref. 99) found that increasing the level of noise (given via loudspeakers) in the room used by postal letter sorters decreased the number of letters correctly sorted (see fig. 9.10). The decrease in performance was systematic with increase in noise level. These results cannot be taken to mean necessarily that the noise *per se* caused the decrease in performance because of some basic physiological or psychological distractive effect; they could be due to personnel viewing the noise as aversive because it bothered their hearing and/or represented a degradation in the concern of management with their comfort and

TABLE 9.8

Comparison of Performance in Acoustically Treated and Untreated Work Bays Before and After Treatment Was Carried Out

[The treatment was applied to the "treated bays" at end of 1957; therefore, the 1956-1957 data for those bays are prior to treatment; the "untreated bays" remained unchanged from 1956-1958; the workers moved from one bay to the other during normal work procedures. From ref. 98.]

	Treated bays		Untreated bays	
	1956/7	1957/8	1956/7	1957/8
Broken rolls (attributed to operator)	75	5	25	22
Other shutdowns (attributed to operator)	158	31	75	56
Calls for maintenance (excluding first 6-week period in each year)	746	597	516	468
Point hour	84.5	89.6	91.2	95.25
Absenteeism (time as % of possible hours worked)	5.18	4.43	2.72	1.556
Labor turnover (mean per 6 weeks)	1956/7 = 6.2%		1957/8 = 0	

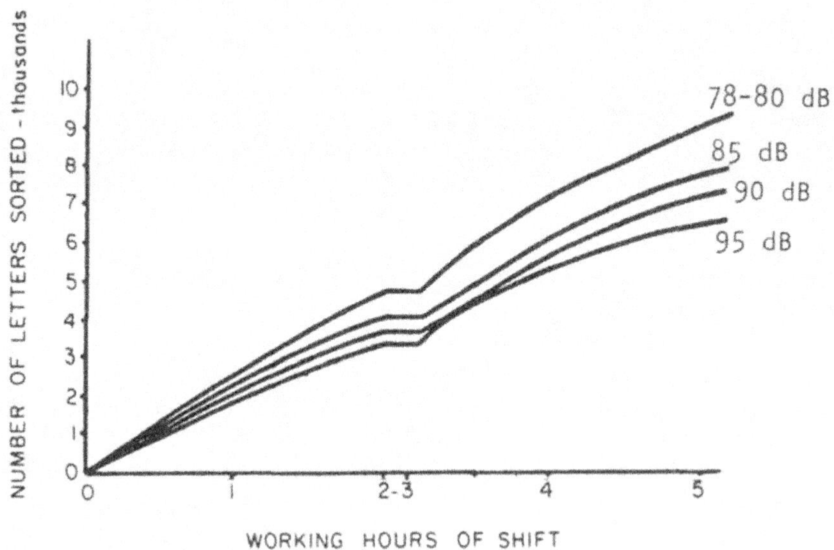

FIGURE 9.10. Effect of change in noise level on working efficiency of postal sorters. (From ref. 99.)

well-being. Also, the measured effects quite possibly could have disappeared with continued exposure or been due to some masking of sounds helpful to the sorting tasks.

Music

As mentioned earlier, a sound or noise may on occasion mask other sounds or noises that can disturb or distract a worker and thereby reduce productivity. For some purposes, in generally quiet surroundings, a low-level broadband random noise may be introduced to increase a sense of privacy with some possible beneficial effects. Music has also been used in work situations not so much perhaps to mask other sounds as to provide some pleasant stimuli to persons doing nonauditory work. The presumptions have been that work output will be increased because of improved morale or that people are kept more aroused and alert than they otherwise would be in monotonous jobs. Figure 9.11 shows some data obtained in one study (ref. 100) on this matter. There appears to be some, but no consistent, relation between the presence of music and work output. The clearly cyclic characteristics of the work output makes firm interpretation of the data difficult.

The purpose of this document is not to review research on the effects of music on work performance. However, note that the beneficial effects of music in industry have been difficult to quantify for some of the same reasons that any detrimental effects of noise on work output are hard to show:

1. In many cases the effects are transitory and related to temporary changes in worker morale.

2. There are no beneficial effects, or perhaps there are adverse effects.

3. There are beneficial effects that are relatively small compared with other task and motivational factors in the situation.

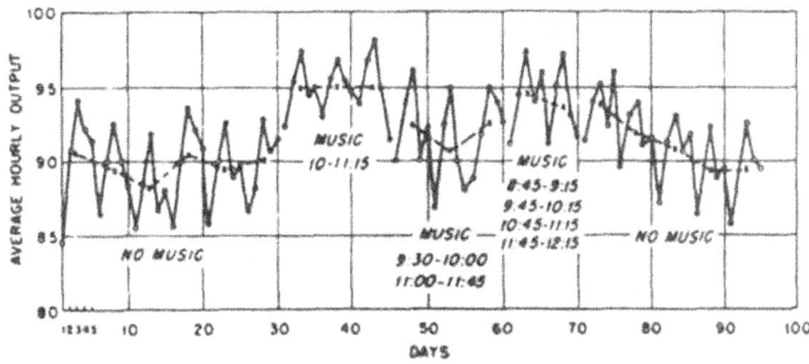

FIGURE 9.11 Output in a light manual task (rolling paper novelties) under various conditions of music presentation. (From ref. 100.)

Summary

The fundamental theories behind most research on the effects of noise on the performance of nonauditory mental and psychomotor tasks are that noise (1) can cause general physiological arousal; (2) is psychologically aversive because of its loudness; and/or (3) engages neural mechanisms involved in nonauditory mental functions (cognition, attention, time judgment, *etc.*). Various investigators have focused these three basic theories into particularized expressions for describing certain experimental results. Examples related to each of the above theories are as follows: (1) as the intensity of noise is increased, arousal occurs and improves task performance up to a point, but beyond that level of intensity, overarousal sufficient to degrade task performance occurs; (2) psychological aversion, inherent to the noise, causes anxiety, learned helplessness, or psychic cost that degrades task performance and/or has adverse aftereffects; and (3) noise causes perceptual blinks, narrowed attention, or information overload that can degrade task performance.

However, it is concluded here, as it was in previous reviews by the author published in 1950 (ref. 1) and 1970 (ref. 2), that noise can have a positive effect, no effect, or a negative effect on performance of nonauditory mental and psychomotor tasks. Further, almost without exception, statistically significant adverse effects found in one experiment have not been found in repeat experiments by the same or by other experimenters. These findings are consistent with (and perhaps are the same as) those to be discussed in chapter 10. Namely, there are no significant inherent adverse nonauditory physiological or psychological responses to sound or noise. Experimental errors, direct and internal masking by noise of task-related auditory signals, and conscious and unconscious construed concerns of the subjects regarding experimental procedures and conditions are variables possibly responsible for the occasionally observed adverse effects of noise on the performance of nonauditory mental and psychomotor tasks.

Some of the factors and distinctions made for theoretical and research purposes between sound and noise may seem academic. Seldom in the real world is mental or psychomotor work strictly nonauditory, and the masking by noise of any audible aspects of a work task (including perhaps inner speech) can be inimical to the performance of that task. Further, the sounds or noises from sources in real life can have emotion-arousing meanings that create psychological and physiological stresses that can have adverse effects on work performance.

References

1. Kryter, Karl D.: The Effects of Noise on Man. J. Speech and Hearing Disorders, Monograph Supp. 1, Sept. 1950.
2. Kryter, Karl D.: The Effects of Noise on Man. Academic Press, Inc., 1970.

3. Gulian, Edith: Psychological Consequences of Exposure to Noise, Facts and Explanations. Proceedings of the International Congress on Noise as a Public Health Problem, W. Dixon Ward, ed., 550/9-73-008, U.S. Environ. Prot. Agency, May 1973, pp. 363-378.
4. Loeb, Michel: Noise and Performance: Do We Know More Now? Noise as a Public Health Problem, Jerry V. Tobias, Gerd Jansen, and W. Dixon Ward, eds., ASHA Rep. 10, American Speech-Language-Hearing Assoc., Apr. 1980, pp. 303-321.
5. Broadbent, D. E.: Noise, Paced Performance, and Vigilance Tasks. British J. Psychol., vol. 44, 1953, pp. 295-303.
6. Broadbent, D. E.: Some Effects of Noise on Visual Performance. Q. J. Exp. Psychol., vol. 6, 1954, pp. 1-5.
7. Broadbent, D. E.: Symposium on Noise at Work: Noise: Its Effects on Behaviour. R. Soc. Promot. Health J., vol. 75, Aug. 1955, pp. 541-548.
8. Broadbent, Donald E.: Effects of Noise on Behavior. Handbook of Noise Control, Cyril M. Harris, ed., McGraw-Hill Book Co., Inc., 1957, pp. 10-1-10-34.
9. Broadbent, D. E.: Perception and Communication. Pergamon Press, Inc., c.1958.
10. Broadbent, D. E.; and Gregory, Margaret: Vigilance Considered as a Statistical Decision. British J. Psychol., vol. 54, no. 4, 1963, pp. 309-323.
11. Broadbent, Donald E.; and Gregory, Margaret: Effects of Noise and of Signal Rate Upon Vigilance Analyzed by Means of Decision Theory. Hum. Factors, vol. 7, no. 2, Apr. 1965, pp. 155-162.
12. Hartley, L. R.: Effect of Prior Noise or Prior Performance on Serial Reaction. J. Exp. Psychol., vol. 101, no. 2, Dec. 1973, pp. 255-261.
13. Hartley, L. R.: Performance During Continuous and Intermittent Noise and Wearing Ear Protection. J. Exp. Psychol., vol. 102, no. 3, Mar. 1974, pp. 512-516.
14. Hartley, L. R.; and Carpenter, A.: Comparison of Performance With Headphone and Free-Field Noise. J. Exp. Psychol., vol. 103, no. 2, Aug. 1974, pp. 377-380.
15. Hockey, G. R.: Effect of Loud Noise on Attentional Selectivity. Q. J. Exp. Psychol., vol. 22, no. 1, 1970, pp. 28-36.
16. Hockey, G. R.: Signal Probability and Spatial Location as Possible Bases for Increased Selectivity in Noise. Q. J. Exp. Psychol., vol. 22, no. 1, 1970, pp. 37-42.
17. Hockey, Robert: Changes in Information-Selection Patterns in Multisource Monitoring as a Function of Induced Arousal Shifts. J. Exp. Psychol., vol. 101, no. 1, Nov. 1973, pp. 35-42.
18. Poulton, E. C.; and Edwards, R. S.: Interactions and Range Effects in Experiments on Pairs of Stresses: Mild Heat and Low-Frequency Noise. J. Exp. Psychol., vol. 102, no. 4, Apr. 1974, pp. 621-628.

19. Poulton, E. C.: Continuous Noise Interferes With Work by Masking Auditory Feedback and Inner Speech. Appl. Ergonomics, vol. 7, June 1976, pp. 79-84.
20. Poulton, E. C.: Arousing Environmental Stresses Can Improve Performance, Whatever People Say. Aviat., Space & Environ. Med., vol. 47, no. 11, Nov. 1976, pp. 1193-1204.
21. Poulton, E. C.: Continuous Intense Noise Masks Auditory Feedback and Inner Speech. Psychol. Bull., vol. 84, no. 5, Sept. 1977, pp. 977-1001.
22. Poulton, E. C.: A New Look at the Effects of Noise: A Rejoinder. Psychol. Bull., vol. 85, no. 5, Sept. 1978, pp. 1068-1079.
23. Poulton, E. Christopher: Composite Model for Human Performance in Continuous Noise. Psychol. Rev., vol. 86, no. 4, July 1979, pp. 361-375.
24. Poulton, E. C.: Psychology of the Scientist: XLI. Continuous Noise Can Degrade Performance When Using Badly Designed Equipment: A Case History. Percept. & Mot. Skills, vol. 50, no. 1, Feb. 1980, pp. 319-330.
25. Broadbent, D.: Noise and the Details of Experiments; A Reply to Poulton. Appl. Ergonomics, vol. 7, no. 4, Dec. 1976, pp. 231-235.
26. Broadbent, Donald E.: Letter to the Editor. Aviat., Space & Environ. Med., vol. 48, no. 4, Apr. 1977, p. 382.
27. Broadbent, Donald E.: The Current State of Noise Research: Reply to Poulton. Psychol. Bull., vol. 85, no. 5, Sept. 1978, pp. 1052-1067.
28. Gulian, Edith: Effects of Noise on Arousal Level in Auditory Vigilance. Attention and Performance III, A. F. Sanders, ed., North-Holland Pub. Co., 1970, pp. 381-393.
29. Glass, David C.; and Singer, Jerome E.: Urban Stress—Experiments on Noise and Social Stressors. Academic Press, Inc., 1972.
30. Glass, David C.; and Singer, Jerome E.: Behavioral Effects and Aftereffects of Noise. Proceedings of the International Congress on Noise as a Public Health Problem, W. Dixon Ward, ed., 550/9-73-008, U.S. Environ. Prot. Agency, May 1973, pp. 409-416.
31. Conrad, D. W.: The Effects of Intermittent Noise on Human Serial Decoding Performance and Physiological Response. Ergonomics, vol. 16, no. 6, 1973, pp. 739-747.
32. Becker, R. W.; Kryter, K. D.; and Poza, F.: A Study of Sensitivity to Noise—Final Report. Rept. EQ-71-4 (Contract DOT-FA69WA-2211), Stanford Res. Inst., June 1971.
33. Frankenhaeuser, Marianne; and Lundberg, Ulf: Immediate and Delayed Effects of Noise on Performance and Arousal. Biol. Psychol., vol. 2, no. 2, 1974, pp. 127-133.
34. Lundberg, Ulf; and Frankenhaeuser, Marianne: Psychophysiological Reactions to Noise as Modified by Personal Control Over Noise Intensity. Biol. Psychol., vol. 6, no. 1, Jan. 1978, pp. 51-59.
35. Hawel, Wolfgang: Investigation of Psychological and Psychophysiological

Effects of Repeated Intermittent Pink Noise Lasting 4 Hours. Z. Exp. & Angew. Psychol., vol. 22, no. 4, 1975, pp. 613-629.

36. Basow, Susan A.: Effect of White Noise on Attention as a Function of Manifest Anxiety. Percept. & Mot. Skills, vol. 39, no. 1, Aug. 1974, pp. 655-662.
37. Cohen, H. Harvey; Conrad, Donald W.; O'Brien, John F.; and Pearson, Richard G.: Effects of Noise Upon Human Information Processing. NASA CR-132469, 1974.
38. Kryter, Karl D.; and Poza, Fausto: Autonomic System Activity and Performance on a Psychomotor Task in Noise. J. Acoust. Soc. America, vol. 67, no. 6, June 1980, pp. 2096-2099.
39. Mosskov, J. I.; and Ettema, J. H.: II. Extra-Auditory Effects in Short-Term Exposure to Aircraft and Traffic Noise. Int. Arch. Occup. & Environ. Health, vol. 40, 1977, pp. 165-173.
40. Mosskov, J. I.; and Ettema, J. H.: Experimental Investigations Into Some Extra-Aural Effects of Exposure to Noise. Noise as a Public Health Problem, Jerry V. Tobias, Gerd Jansen, and W. Dixon Ward, eds., ASHA Rep. 10, American Speech-Language-Hearing Assoc., Apr. 1980, pp. 337-342.
41. Strasser, H.: Effects of Noise, Tranquillizer and Increased Delay Time on Tracking Performance and Heart Rate. Pfluegers Arch., vol. 332, Suppl. 332, 1972, p. R82.
42. Hartley, L.; Couper-Smartt, J.; and Henry, T.: Behavioural Antagonism Between Chlorpromazine and Noise in Man. Psychopharmacology, vol. 55, no. 1, Nov. 24, 1977, pp. 97-102.
43. Hartley, L.; and Shirley, E.: Sleep-Loss, Noise and Decisions. Ergonomics, vol. 20, no. 5, 1977, pp. 481-489.
44. Colquhoun, W. P.; and Edwards, R. S.: Interaction of Noise With Alcohol on a Task of Sustained Attention. Ergonomics, vol. 18, no. 1, 1975, pp. 81-87.
45. Simpson, G. C.; Cox, T.; and Rothschild, D. R.: The Effects of Noise Stress on Blood Glucose Level and Skilled Performance. Ergonomics, vol. 17, no. 4, 1974, pp. 481-487.
46. Franszczuk, Irena: The Effect of Annoying Noise on Some Psychological Functions During Work. Proceedings of the International Congress on Noise as a Public Health Problem. W. Dixon Ward, ed., 550/9-73-008, U.S. Environ. Prot. Agency, May 1973, pp. 425-428.
47. Osada, Yasutaka; Hirokawa, Akiko; and Haruta, Kiyoko: Effect of Intermittent Noise on Mental Tasks—Influence of Aircraft—and Train-Noise on Reaction Time, Time Estimation and Figure Counting Test. Bull. Inst. Public Health, Tokyo, vol. 20, no. 3, 1971, pp. 163-169.
48. Ando, Y.: Effects of Noise on Duration Experience. J. Sound & Vib., vol. 55, no. 4, Dec. 22, 1977, pp. 600-603.

49. Warner, Harold D.: Effects of Intermittent Noise on Human Target Detection. Hum. Factors, vol. 11, no. 3, June 1969, pp. 245-250.
50. Warner, Harold D.; and Heimstra, Norman W.: Effects of Noise Intensity on Visual Target-Detection Performance. Hum. Factors, vol. 14, no. 2, Apr. 1972, pp. 181-185.
51. Warner, Harold D.; and Heimstra, Norman W.: Target-Detection Performance as a Function of Noise Intensity and Task Difficulty. Percept. & Mot. Skills, vol. 36, no. 2, Apr. 1973, pp. 439-442.
52. Eschenbrenner, A. John: Effects of Intermittent Noise on the Performance of a Complex Psychomotor Task. Hum. Factors, vol. 13, no. 1, Feb. 1971, pp. 59-63.
53. Kaltsounis, Bill: Effect of Sound on Creative Performance. Psychol. Rep., vol. 33, no. 3, Dec. 1973, pp. 737-738.
54. Davies, D. R.; Lang, Lesley; and Shackleton, V. J.: The Effects of Music and Task Difficulty on Performance at a Visual Vigilance Task. British J. Psychol., vol. 64, no. 3, Aug. 1973, pp. 383-389.
55. Hartley, L. R.; and Williams T.: Steady State Noise and Music and Vigilance. Ergonomics, vol. 20, no. 3, 1977, pp. 277-285.
56. Kunitake, E.; Ishinishi, N.; and Kodama, Y.: Studies on the Effects of Aircraft Noise in the Experimental Exposure to the Acclimatized and the Non-Acclimatized Students to the Noise. Nippon Eiseigaku Zasshi, vol. 32, no. 2, June 1977, pp. 353-365.
57. Harris, C. Stanley; and Filson, George W.: Effects of Noise on Serial Search Performance. Rept. AMRL-TR-71-56, U.S. Air Force, July 1971. (Available from DTIC as AD 731 184.)
58. Harris, C. Stanley: Effects of Intermittent and Continuous Noise on Serial Search Performance. Percept. & Mot. Skills, vol. 35, no. 2, Oct. 1972, pp. 627-634.
59. Harris, C. Stanley: Effects of Predictable and Unpredictable Sound on Human Performance. Noise as a Public Health Problem, Jerry V. Tobias, Gerd Jansen, and W. Dixon Ward, eds., ASHA Rep. 10, American Speech-Language-Hearing Assoc., Apr. 1980, pp. 343-348.
60. Wittersheim, G.; and Salame, P.: Effects of Noise on a Serial Short-Term Memory Process. Proceedings of the International Congress on Noise as a Public Health Problem, W. Dixon Ward, ed., 550/9-73-008, U.S. Environ. Prot. Agency, May 1973, pp. 417-423.
61. Hartley, L. R.; and Adams, R. G.: Effect of Noise on the Stroop Test. J. Exp. Psychol., vol. 102, no. 1, Jan. 1974, pp. 62-66.
62. O'Malley, John J.; and Gallas, John: Noise and Attention Span. Percept. & Mot. Skills, vol. 44, no. 3, June 1977, pp. 919-922.
63. Weinstein, Neil D.: Noise and Intellectual Performance: A Confirmation and Extension. J. Appl. Psychol., vol. 62, no. 1, Feb. 1977, pp. 104-107.
64. Smith, Andrew P.: Low Levels of Noise and Performance. Noise as a Public

Health Problem, Jerry V. Tobias, Gerd Jansen, and W. Dixon Ward, eds., ASHA Rep. 10, American Speech-Language-Hearing Assoc., Apr. 1980, pp. 365–368.

65. Dornic, Stan: Noise and Language Dominance. Noise as a Public Health Problem, Jerry V. Tobias, Gerd Jansen, and W. Dixon Ward, eds., ASHA Rep. 10, American Speech-Language-Hearing Assoc., Apr. 1980, pp. 331–336.
66. Broadbent, Donald E.: Low Levels of Noise and the Naming of Colors. Noise as a Public Health Problem, Jerry V. Tobias, Gerd Jansen, and W. Dixon Ward, eds., ASHA Rep. 10, American Speech-Language-Hearing Assoc., Apr. 1980, pp. 362–364.
67. Wohlwill, Joachim F.; Nasar, Jack L.; DeJoy, David M.; and Foruzani, Hossein H.: Behavioral Effects of a Noisy Environment: Task Involvement vs. Passive Exposure. J. Appl. Psychol., vol. 61, 1976, pp. 67–74.
68. Moran, Sharon L. Vanderhei; and Loeb, Michel: Annoyance and Behavoral Aftereffects Following Interfering and Noninterfering Aircraft Noise. J. Appl. Psychol., vol. 62, no. 5, Oct. 1977, pp. 719–726.
69. Percival, Lynn; and Loeb, Michel: Influence of Noise Characteristics on Behavioral Aftereffects. Hum. Factors, vol. 22, no. 3, June 1980, pp. 341–352.
70. Rotton, James; Olszewski, Donald; Charleton, Marc; and Soler, Edgardo: Loud Speech, Conglomerate Noise, and Behavioral Aftereffects. J. Appl. Psychol., vol. 63, no. 3, June 1978, pp. 360–365.
71. Hockey, G. R. J.; and Hamilton, P.: Arousal and Information Selection in Short-Term Memory. Nature, vol. 226, no. 5248, May 30, 1970, pp. 866–867.
72. Davies, D. R.; and Jones, D. M.: The Effects of Noise and Incentives Upon Attention in Short-Term Memory. British J. Psychol., vol. 66, no. 1, 1975, pp. 61–68.
73. O'Malley, John J.; and Poplawsky, Alex: Noise-Induced Arousal and Breadth of Attention. Percept. & Mot. Skills, vol. 33, no. 3, Dec. 1971, pp. 887–890.
74. Loeb, Michel; Jones, Paul D.; and Cohen, Alexander: Effects of Noise on Non-Auditory Sensory Functions and Performance. HEW Publ. No. (NIOSH) 76-176, U.S. Dep. Health, Educ., & Welfare, Apr. 1976.
75. Loeb, Michel; and Jones, Paul D.: Noise Exposure, Monitoring and Tracking Performance as a Function of Signal Bias and Task Priority. Ergonomics, vol. 21, no. 4, 1978, pp. 265–272.
76. Finkelman, Jay M.; and Glass, David C.: Reappraisal of the Relationship Between Noise and Human Performance by Means of a Subsidiary Task Measure. J. Appl. Psychol., vol. 54, no. 3, 1970, pp. 211–213.
77. Finkelman, Jay M.: Effects of Noise on Human Performance. Sound & Vib., vol. 10, no. 9, Sept. 1975, pp. 26–28.

78. Zeitlin, Lawrence R.; and Finkelman, Jay M.: Research Note: Subsidiary Task Techniques of Digit Generation and Digit Recall as Indirect Measures of Operator Loading. Hum. Factors, vol. 17, no. 2, Apr. 1975, pp. 218-220.
79. Finkelman, Jay M.; Zeitlin, Lawrence R.; Filippi, John A.; and Friend, Michael A.: Noise and Driver Performance. J. Appl. Psychol., vol. 62, no. 6, Dec. 1977, pp. 713-718.
80. Finkelman, Jay M.; Zeitlin, Lawrence R.; Romoff, Richard A.; Friend, Michael A.; and Brown, Louis S.: Conjoint Effect of Physical Stress and Noise Stress on Information Processing Performance and Cardiac Response. Hum. Factors, vol. 21, no. 1, Feb. 1979, pp. 1-6.
81. Bell, Paul A.: Effects of Noise and Heat Stress on Primary and Subsidiary Task Performance. Hum. Factors, vol. 20, no. 6, Dec. 1978, pp. 749-752.
82. Grether, W. F.: Two Experiments on the Effects of Combined Heat, Noise and Vibration Stress. AMRL-TR-71-113, U.S. Air Force, Nov. 1972.
83. Grether, W. F.; Harris, C. S.; Ohlbaum, M.; Sampson, P. A.; and Guignard, J. C.: Further Study of Combined Heat, Noise and Vibration Stress. Aerosp. Med., vol. 43, no. 6, June 1972, pp. 641-645.
84. Gawron, Valerie J.: Performance Effects of Noise Intensity, Psychological Set and Task Type and Complexity. Hum. Factors, vol. 24, no. 2, Apr. 1982, pp. 225-243.
85. Kornhauser, A. W.: The Effect of Noise on Office Output. Ind. Psychol., vol. 2, 1927, pp. 621-622.
86. Increased Production Resulting From Lessening of Noise. Mon. Labor Rev., vol. 27, no. 2, Aug. 1928, pp. 249-250.
87. Laird, D. A.: Measurement of the Effects of Noise on Working Efficiency. J. Ind. Hyg., vol. 9, Oct. 1927, pp. 431-434.
88. Laird, Donald A.: The Effects of Noise: A Summary of Experimental Literature. J. Acoust. Soc. America, vol. 1, no. 2, pt. I, Jan. 1930, pp. 256-262.
89. Lindahl, Robert: Noise in Industry. Ind. Med., vol. 7, no. 11, Nov. 1983, pp. 664-669.
90. Weston, H. C.; and Adams, S.: The Effects of Noise on the Performance of Weavers. Rep. No. 65, British Industrial Health Research Board, 1932, pp. 38-62.
91. Weston, H. C.; and Adams, S.: The Performance of Weavers Under Varying Conditions of Noise. Rep. No. 70, British Industrial Health Research Board, 1935.
92. Baker, K. H.: Pre-Experimental Set in Distraction Experiments. J. Gen. Psychol., vol. 16, 1937, pp. 471-488.
93. Berrien, F. K.: The Effects of Noise. Psychol. Bull., vol. 43, 1946, pp. 141-161.

94. Wilson, A.: Better Concentration Reduces Employee Turnover 47%. Bankers Mon., vol. 59, 1952, pp. 254-255.
95. Felton, J. S.; and Spencer, C.: Morale of Workers Exposed to High Levels of Occupational Noise. Univ. Oklahoma School of Med., 1957.
96. Ganguli, H. C.; and Rao, M. N.: Noise and Industrial Efficiency: A Study of Indian Jute Weavers. Arbeitsphysiology, vol. 15, 1954, pp. 344-354.
97. De Almeida, H. R.: Influence of Electric Punch Card Machines on the Human Ear. Arch. Otolaryngol., vol. 51, 1950, pp. 215-222.
98. Broadbent, D. E.; and Little, E. A. J.: Effects of Noise Reduction in a Work Situation. Occup. Psychol., vol. 34, 1960, pp. 133-140.
99. Kourigin, S. D.; and Mikheyev, A. P.: The Effect of Noise Level on Working Efficiency. Joint Publications Research Service, July 9, 1965.
100. Wyatt, S.; Langdon, J. N.; and Stock, F. G. L.: Fatigue and Boredom in Repetitive Work. Rep. No. 77, British Industrial Health Research Board, Her Majesty's Stationery Office, London, 1937.

Chapter 10
Nonauditory-System Response to Noise and Effects on Health

Introduction

It is generally believed that continued exposure to noise in real life can be a source of physiological stress possibly capable of causing health disorders beyond that of direct damage to the auditory receptor system (refs. 1 and 2). Some theorists hold that some of these effects occur because of innate, reflexive responses to noise that cannot be prevented or, when suppressed, that require some effort that may itself become somewhat debilitating in time. An alternative theory is that the truly nonhabituating reflexive responses to noise are not sufficient in character to cause any ill health, and that those responses to noise that are or could be significant in this regard are not directly the result of exposure to noise but are responses to the emotional meanings conveyed by the sounds.

Obviously, the degree to which noise can lead to harm to nonauditory physiological systems of the body are questions of utmost importance for the assessment of the need for noise control. However, research on these questions is complicated and somewhat controversial, as seen in the discussions to follow.

Functions of Nonauditory Systems

Figure 10.1 (ref. 1) illustrates some direct and indirect connections between the human auditory system and the neural-muscular-glandular systems in the body. Special attention is invited to the autonomic (sympathetic nervous and glandular) system. The primary function of the autonomic system is to control and coordinate the life-support functions and organs of the body—the digestion

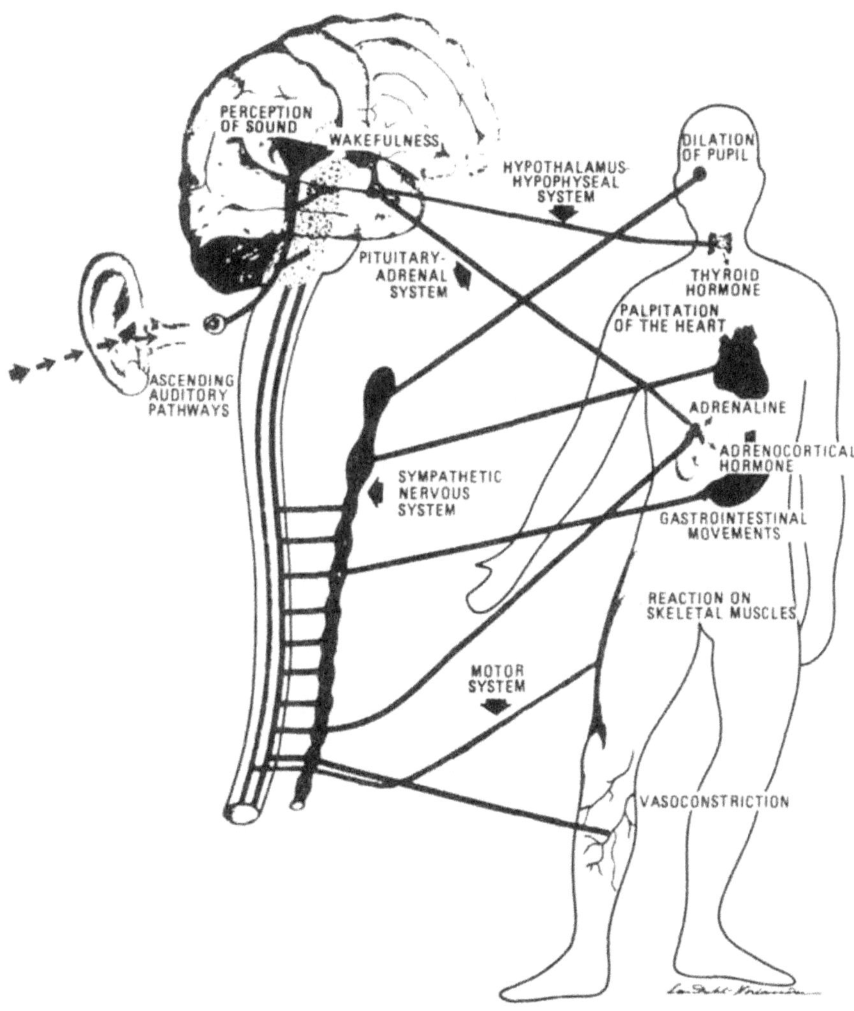

FIGURE 10.1. Major response mechanisms and subsystems of the autonomic system. (Based on ref. 1.)

of food, the cardiovascular blood supply system, the respiration of air, the control of body temperature, *etc*. These functions are carried out automatically and unconsciously. An extensive, recent review of the effects of sound on the non-auditory systems in man and animals is given by Borg (ref. 3).

The auditory system has some direct neural connections with the sympathetic nervous system at levels below the brain proper. It is believed that *via* these connections sounds can cause autonomic-system responses that occur without any conscious "thought" processes as to the meaning or effects of the sounds or noise, and thereby serve as a warning system about the presence of things in the environment. In this regard it can be noted that sounds, unlike light, bend around and, to some extent, go through objects, are omnidirectional, and are generated and transmitted as well at night as in the daytime. In brief, sound (or noise) is generally a more effective warning signal about things, especially moving things, in the environment than is light.

It is considered to be advantageous to have the autonomic system activate the body in ways that best enable it to fight or flee rather than to merely continue to vegetate. However, the primary components of possible direct autonomic-system response to sound (sometimes called the startle response) appear to cease to function, or are suppressed, with repetitions of the sound or noise when it is no longer an unexpected stimulus. As an aid to the interpretation of some experimental data presented, a brief outline of the general physiological responses to sound and major theories that have evolved around these responses are briefly discussed below.

The startle-alerting-arousal response

Since the early 1940's, numerous laboratory measurements have been made of nonauditory-system responses to sudden, intense sounds or noises, both unexpected and repeated or expected. Davis *et al.* (ref. 4) called the resulting complex of responses to these sounds the N-response, herein called the "startle-alerting-arousal" response (alerting response for short). The magnitude of this response generally depends jointly on the suddenness of the onset, the intensity of the noise, and the state of quiescence of the organism. The following physiological responses appeared to be involved:

1. A circulatory response dominated by vasoconstriction of the peripheral blood vessels with other adjustments of blood pressure throughout the body

2. A reduced rate of breathing

3. Galvanic skin response (GSR), a change in the electrical resistance of the skin

4. A brief change in skeletal-muscle tension

To this list can be added, among others, the changes that occur in gastrointestinal motility (refs. 5 and 6) and chemical changes or excretions in the blood and urine from glandular stimulation (refs. 7 to 10).

These various responses are interrelated. As a result, some measure of heart rate, blood circulation (usually blood volume or pulse in the skin of a finger), skeletal-muscle response (such as from the muscles of the forearm), and GSR have been commonly used because they can be readily sensed and recorded by available electronic transducers without greatly inconveniencing the subject. However, the presence of one of these responses does not necessarily mean that any of these other responses have occurred. Further, the lack of response may not mean that an alerting reaction should not have occurred, the reason being that the nonauditory system at the time of the noise stimulation may be somewhat exhausted, overactive, or under the domination of other stimulations. These factors can sometimes make the collection and interpretation of data in this area difficult. (See Surwillo and Arenberg (ref. 11).)

Figures 10.2 to 10.5 (from refs. 4, 12, 13, 14, and 15) show the rather rapid habituation of responsiveness to meaningless sounds or noises that are presented

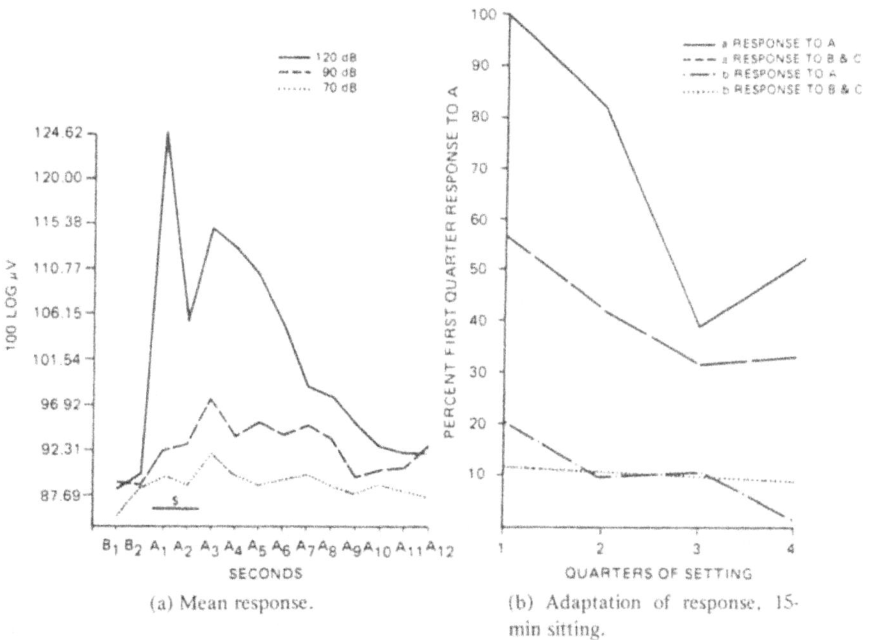

(a) Mean response.

(b) Adaptation of response, 15-min sitting.

FIGURE 10.2. Mean muscle action-potential response and adaptation of response to 1000-Hz tone. Horizontal line indicates stimulus duration; *a* indicates brief latency; *b* indicates long latency. (From ref. 4.)

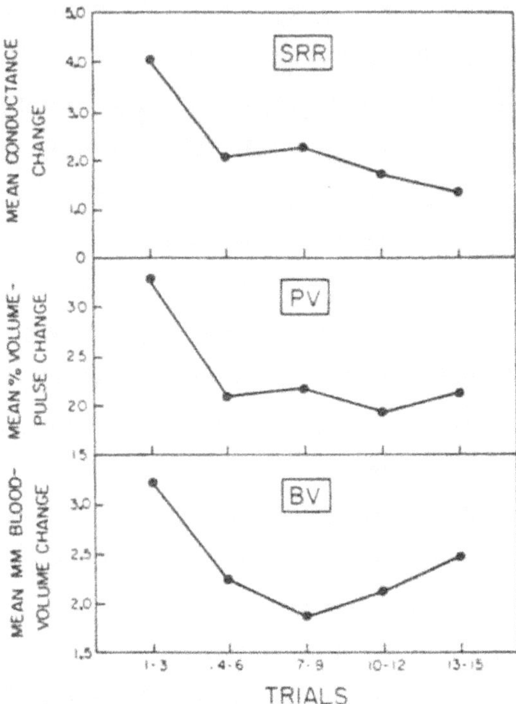

FIGURE 10.3. Mean skin resistance response (SRR), peripheral pulse volume (PV) response, and peripheral blood volume (BV) response to stimulus repetition. (From ref. 12.)

suddenly and unexpectedly to the subjects. It is clear that, although initial responsiveness increases with increases in stimulus intensity, habituation always becomes nearly complete. That this reduction in response with stimulus repetition is not the result of a fatiguing of the mechanism is shown in figure 10.5. It is shown in figure 10.6 that following 15 repetitions (trials) of an 80-dB 200-Hz tone or of a bright light the autonomic responses were much less for the 16th trial when the stimulus was changed from the tone to the light (for half the subjects) or from the light to the tone (for the other subjects). That this habituation should be recognized as some neural "decision" process and not a "fatiguing" of the organism was also shown by Rossi *et al.* (ref. 16), who found that the adaptation of vasoconstriction in subjects exposed to a background noise did not reduce vasoconstriction to superimposed 2000 Hz tones at levels of 80 to 105 dB.

These responses of the autonomic system may not represent any undue stress or health-threatening phenomena, partly because these responses tend to show such rapid adaptation or habituation and partly because the magnitude of the physiological changes that are associated with these responses are rather small in comparison to the range of physiological conditions or states observed in the hu-

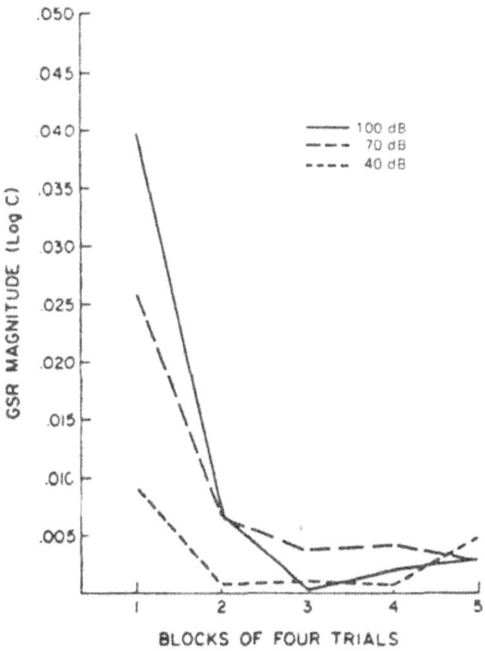

FIGURE 10.4. Galvanic skin response to repetition of 2-sec, 1000-Hz tones of 40, 70, or 100 dB. (From ref. 13.)

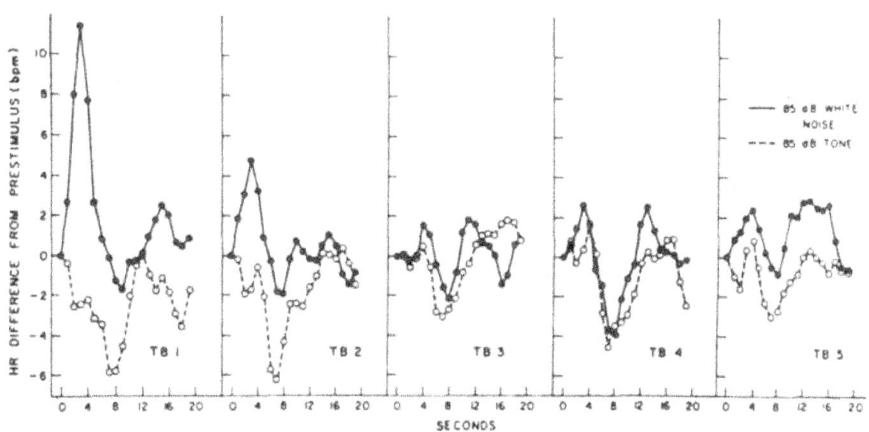

FIGURE 10.5. Heart rate (HR) response to white noise and 1000-Hz tone of 5 sec duration on five trial blocks (TB) of two trials each. (From ref. 15.)

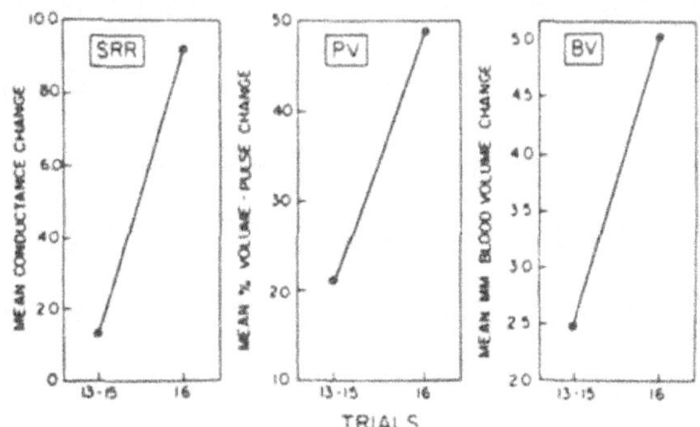

FIGURE 10.6. Mean skin resistance response (SRR), peripheral pulse volume (PV) response, and peripheral blood volume (BV) response to repeated trials and to change trials. (From ref. 12.)

man organism during homeostatic operations of the autonomic system normal to daily living. For example, the greatest heart rate change shown in figure 10.5 is about 11 beats per minute, and this change is for only 1 or 2 beats; the peripheral blood volume changes last only 10 to 20 seconds. Changes much greater than these occur from mild exercise, fright, sudden changes in air temperature, laughter, and so forth.

Reflex theory

Some researchers have proposed that nonauditory-system response to noise can be explained in terms of mechanistic, neural-learning models (ref. 17); that is, innate reflexive responses become conditioned into patterns of behavior which depend on reinforcement of some sort. Some of the research on these concepts has been rather recently reviewed by Graham (ref. 14), Jackson (ref. 18), and Ginsberg and Furedy (ref. 12). This reinforcement, or lack thereof, will cause the responses to be modified, habituated, or inhibited in a way that best adapts the organism to its environment.

Sokolov (ref. 17) postulated that two reflexive types of responses to meaningless sounds are built into humans. One Sokolov calls the orienting response (OR), wherein the autonomic system responds to any sound stimulus in order to alert and make ready the organism for the purposes of receiving and responding as appropriate to this stimulus. This OR is postulated to get stronger as the noise stimuli become weaker, because the organism would require more effort to react to weaker than to more readily observed stimuli.

The second reflex response of the autonomic system to noise postulated by Sokolov is a defensive response (DR) that prepares the organism for fight or flight. This DR becomes stronger as the strength of the noise is increased. These OR's and DR's supposedly occur to meaningless sounds, but as the meaning becomes established through repetition of the noises (*e.g.*, the noise does not warrant either an orienting or a defensive response), the response becomes inhibited or habituated.

Nonreflex cognition theory

An alternative concept is that the auditory system functions solely to bring information represented by the sound waves to the brain. As such, stimulation of the auditory system *per se* by any sounds, including noise, could not be a direct cause of physiological stress in nonauditory systems of the body. According to this view, any association of noise with significant physiological stress responses is because the noise conveys emotion-arousing information to the person or because of emotional reactions arising from the interference effects of the noise with the perception of wanted auditory signals and sleep or rest activities. The significant part of alerting response and defense response to unexpected noises or sounds results from a learned, psychological aspect of the stimulus, that is, "unexpectedness" means potential danger. The dramatic autonomic-system DR-type response is therefore not an innate reflex.

Intuitively, it would seem that the organism would be biologically more successful if, rather than actively inhibiting nonauditory-system OR and DR response to sounds or noises, it responded in these terms only when appropriate, for example, when the change in the sound or noise environment was interpreted through higher brain center memory functions as a stimulus to which some nonauditory-system response was appropriate. Otherwise, in view of the almost constant presence of sounds in the living environment, nonauditory-system organs would be almost constantly in a state of either stress or active inhibition.

In brief, the nonreflex theory holds that what makes sounds into noises is the cognition, consciously or unconsciously, of the potential or actual interference effects of some sounds with the hearing of wanted sounds or with rest and sleep or that the meaning conveyed in the sound is of emotional significance. Accordingly, nonauditory-system responses of a stressful nature come about only after the acts of cognition and are not concomitant with auditory stimulation.

At the same time, it is biologically reasonable that a stimulus (sound) of sufficient intensity to be possibly dangerous to a receptor organ (ear) could elicit a system-wide reflexive response in an organism. Such responses could include some constriction of the peripheral blood vessels, eye blinking, interaural muscle contraction, and so forth. These responses are presumably designed to protect the integrity of the receptor and do not represent significant physiological stress in nonauditory systems of the body. It is possible that this "protective" response and the previously discussed nonauditory-system emotional stress responses could occur together under some circumstances. However, the "protective" reflex response, at least with respect to sound or noise, is not usually consciously perceived and does not, by itself, cause the listener concern about the state of the receptor organ. Research data relevant to these presumptions and factors are presented later.

To summarize, the nonreflex theory holds that there are the following three categories of nonauditory-system responses to sounds or noises:

1. Nonauditory-system stress response based on the cognition of the meanings of the sound or noise or on the interference effects of the noise with sleep, rest, and auditory communications. The degree of response is controlled by the attitudes and knowledge of the individual and by the activities being engaged in by the individual when the sound or noise occurs. Sufficient repetitions of these stress responses could have adverse effects on psychological and physiological health.

2. A startle-alerting-arousal response, possibly stressful, to sudden changes in intensity of the acoustic environment (*i.e.*, unexpected sounds). Unexpectedness is postulated to be a learned psychological attribute of a stimulus. Habituation occurs as the sounds are repeated, so no significant stress from the noise should develop.

3. A brief response of parts of the nonauditory system (*e.g.*, constriction of peripheral blood vessels) occurring with exposure to sounds or noises that are of a spectrum and intensity that constitute some damage risk to or overload of the middle ear or cochlea. These responses may be associated with activation of the aural reflex and presumably do not constitute a condition of stress in nonauditory systems.

The findings of a number of studies on these matters and discussion of some of the inconsistencies among them are presented first, followed by a discussion of studies of health effects of noise in real life. In the discussions to follow, "positive" response or effects refer to measurable responses or effects from exposure to noise and "negative" response or effects refer to no measurable responses or effects from exposure to noise.

Laboratory Studies of People

Positive glandular-cardiovascular response data

Glass and Singer (ref. 19) found that physiological adaptation (GSR and vasoconstriction) invariably occurred in their laboratory studies regardless of the intensity or the unpredictability (in time) of the noise presented to their subjects. However, they found about 4 percent of college students screened for some of their experiments seemed unable to adapt physiologically to any experimental procedures.

Probably more important than the questions of physiological stress responses to impulsive or to unpredictable noises is the question of such stress responses to more continuous or regularly repeated noises in the workplace or living environment. Perhaps other physiological stress responses become manifest during exposures of longer durations.

Table 10.1 (data from Mosskov and Ettema (ref. 20)) shows that either recorded aircraft or street traffic noise caused some statistically significant changes during a 3-hour exposure following a 30-minute rest period. Adding a mental, two-choice task caused an even greater number of statistically significant changes. Further, there is almost no consistent trend in the changes from the first through the third hour of exposure. The authors hold that "these results strongly suggest that long-term exposure to noise is a risk factor for cardiovascular disease in daily living and working conditions." However, one should be cautious in accepting that conclusion. In the first place, the changes noted over the time period could conceivably have occurred with or without the noise being present. Although the authors mention that control data were obtained, these data are not presented to check that possibility. Also, the addition of mental tasks has about the same relation to the rest-period conditions as do the noise data. Again, control data (mental task in the quiet) is required before the significance of these data can be properly assessed.

In addition, only the averages for three 4- to 10-minute segments of a total of some eighteen 10-minute segments of time were examined. Sometimes in experiments of this kind and duration, such segmentation and infrequent sampling can provide an inadequate picture of the true changes taking place in the physiological responses being examined. Di Cantogno *et al.* (ref. 21) also compared various autonomic-system responses of normal persons and persons with heart disorders to recorded road traffic noise for 1 to 10 minutes. The results were variable and difficult to interpret for some of the reasons cited above.

Osada *et al.* (ref. 22) conducted an interesting experiment in which subjects were exposed continuously for 2 or 6 hours to recordings of road traffic noise at levels of 40, 50, and 60 dBA. The noise was presented *via* earphones from a small cassette tape player worn by the subjects. The subjects moved around the laboratory, went out to lunch, and so forth, while wearing the cassette player and earphones. For most of the noise conditions, blood and urinary samples revealed a significant increase (relative to control data) in blood cells and hormones, especially the corticosteroids, which would indicate autonomic-system stress reactions. Osada *et al.* concluded that autonomic-system stress activity is caused by noise levels above about 50 dBA. However, it is suggested that these effects are perhaps related to stress caused by the noise masking the hearing of speech and other wanted environmental sounds useful to the subjects in moving about and not from some direct autonomic system arousal by the noise.

A number of experimenters (*e.g.*, Grandjean (ref. 23), Jansen (refs. 24 and 25), Ohkubo *et al.* (ref. 26), Osada *et al.* (ref. 22), and Kryter and Poza (ref. 27)) have demonstrated that a constriction of peripheral blood vessels occurs from exposures to rather intense, broadband intermittent or impulsive bursts of noise.

TABLE 10.1

Cardiovascular Responses to Noise

[12 subjects; data from ref. 20]

Cardiovascular function	Mean values at condition of—				Significance of—		Noise with mental load[a] affecting—	
	Rest	40th–50th min of exposure	100th–110th min of exposure	160th–170th min of exposure	Difference from rest	Trend for increasing exposure	Mean value	Significance
Aircraft noise[b]								
Heart rate	71	69	64	68			69	
Systolic pressure	117	115	114	115			117	$p < 0.10$
Diastolic pressure	67	71	72	73	$p < 0.05$	$p < 0.10$	75	$p < 0.01$
Pulse pressure	50	44	42	43	$p < 0.05$		42	
Sinus arrhythmia	101	102	111	106			82	$p < 0.01$
Respiratory rate	14	15	16	15			15	$p < 0.01$
Heart rate/respiratory rate	5.0	4.6	3.8	4.5			4.3	$p < 0.02$
Traffic noise[c]								
Heart rate	74	72	68	70			74	
Systolic pressure	117	112	113	113	$p < 0.010$		117	$p < 0.01$
Diastolic pressure	66	71	75	75	$p < 0.005$	$p < 0.05$	75	$p < 0.01$
Pulse pressure	51	41	38	38	$p < 0.005$		41	
Sinus arrhythmia	88	88	102	109		$p < 0.10$	58	$p < 0.01$
Respiratory rate	14	15	15	15	$p < 0.020$		17	$p < 0.01$
Heart rate/respiratory rate	5.3	4.5	4.2	4.5	$p < 0.050$		3.8	$p < 0.02$

[a] *Presented at 55th–60th min of exposure.*
[b] *Twenty flyover noises per hour, with peak levels of 89–100 dBA.*
[c] L_{eq} = *83.5 dBA.*

This vasoconstriction has been explicitly stated by Jansen (ref. 28) as being related to general activations of the autonomic systems that can have deleterious effects on health. However, as discussed later, it appears that constriction of some peripheral blood vessels may occur in response to noises that are not related to the general activation of the autonomic system, or at least not to activation that is stressful to bodily health.

The presumption that this vasoconstriction can be a significant cause of non-auditory-system stress is encouraged by the types of data obtained by Jansen and shown in figure 10.7. Here we see that with each burst of white noise at a level of 95 dB some momentary vasoconstriction (reduction in the amplitude of finger pulse) occurs. Similar results were found by Kryter and Poza (ref. 27) with respect to pulse amplitude and blood volume, but other measures capable of showing activation of the autonomic system showed complete habituation to the noise. (See fig. 10.8.) Similar patterns of vasoconstrictive responses to noise were found by Jansen (ref. 25) and Froehlich (ref. 29). Accordingly, although there may be little or no habituation of the peripheral vasoconstrictive response to intense, broad-band intermittent noises, that response may not always be related to general activation of the autonomic system.

It is shown in figure 10.7 that after 30 minutes in the test room the non-noise control subjects (lowest panel) had as much reduced peripheral blood flow as did the test subjects. Accordingly, it can be questioned whether the subjects had been

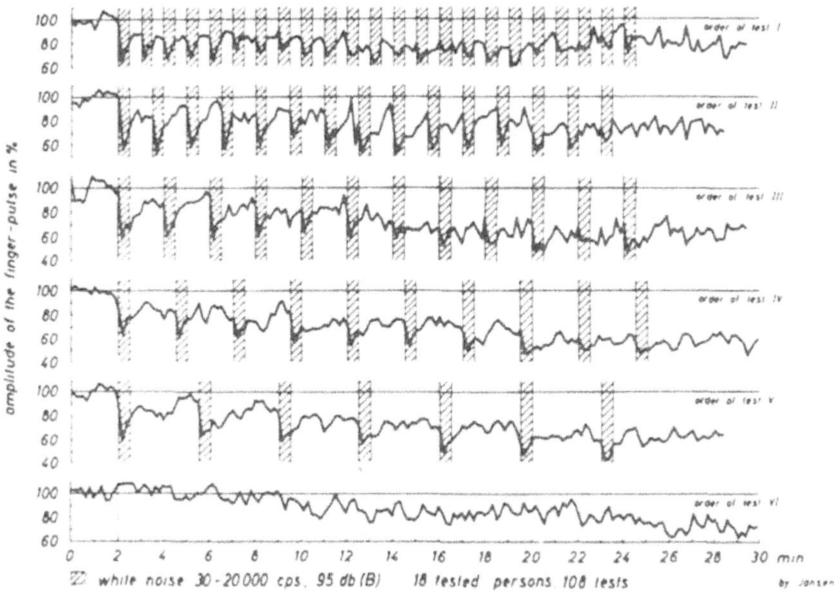

FIGURE 10.7. Autonomic-system reactions during a series of noise exposures. (Data from ref. 25.)

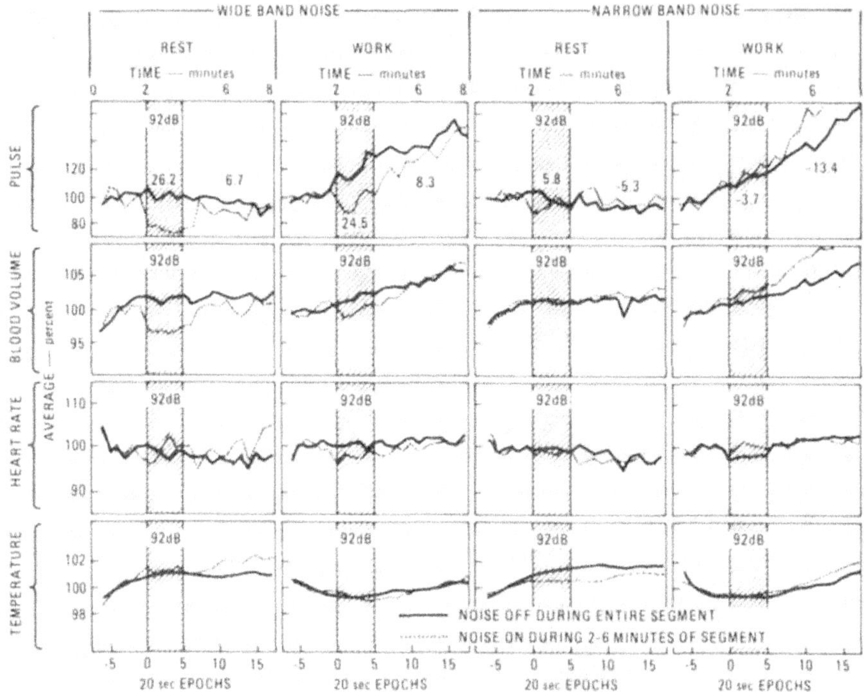

FIGURE 10.8. Pulse amplitude, blood volume, heart rate, and skin temperature changes as a function of noise. High frequency noise, 1/3-octave band with center frequency of 3150 Hz; wideband noise, "pink" random noise sloping 3 dB per octave. (From ref. 27.)

"stressed" by the noise. If we assume that the response could be indicative of a state of stress, it follows that sitting in the test room for 30 minutes with or without the noise was equally stressful.

Jansen (ref. 25) maintains this noise-induced stress response lasts as long as the noise above a certain level of intensity is present. However, as shown in figure 10.9, some of these data indicate that following an initial constrictive response, the peripheral blood circulation takes a state apparently appropriate for either the work or the rest phase of activity. Other data on this point, however, are ambiguous. (See fig. 10.8.)

It is perhaps surprising that there is apparently a complete lack of autonomic-system response to the narrow band of random noise with a center frequency of 3150 Hz (fig. 10.8). At a level of 92 dB this noise sounds very loud and obnoxious to the average listener, more so than the wideband noise at 92 dB. Jansen also reported that the narrow band of noise at 3200 Hz at levels around 92 dB has little effect on pulse amplitude during work, whereas a broadband noise of the same intensity and increased low-frequency energy causes a decrease in pulse ampli-

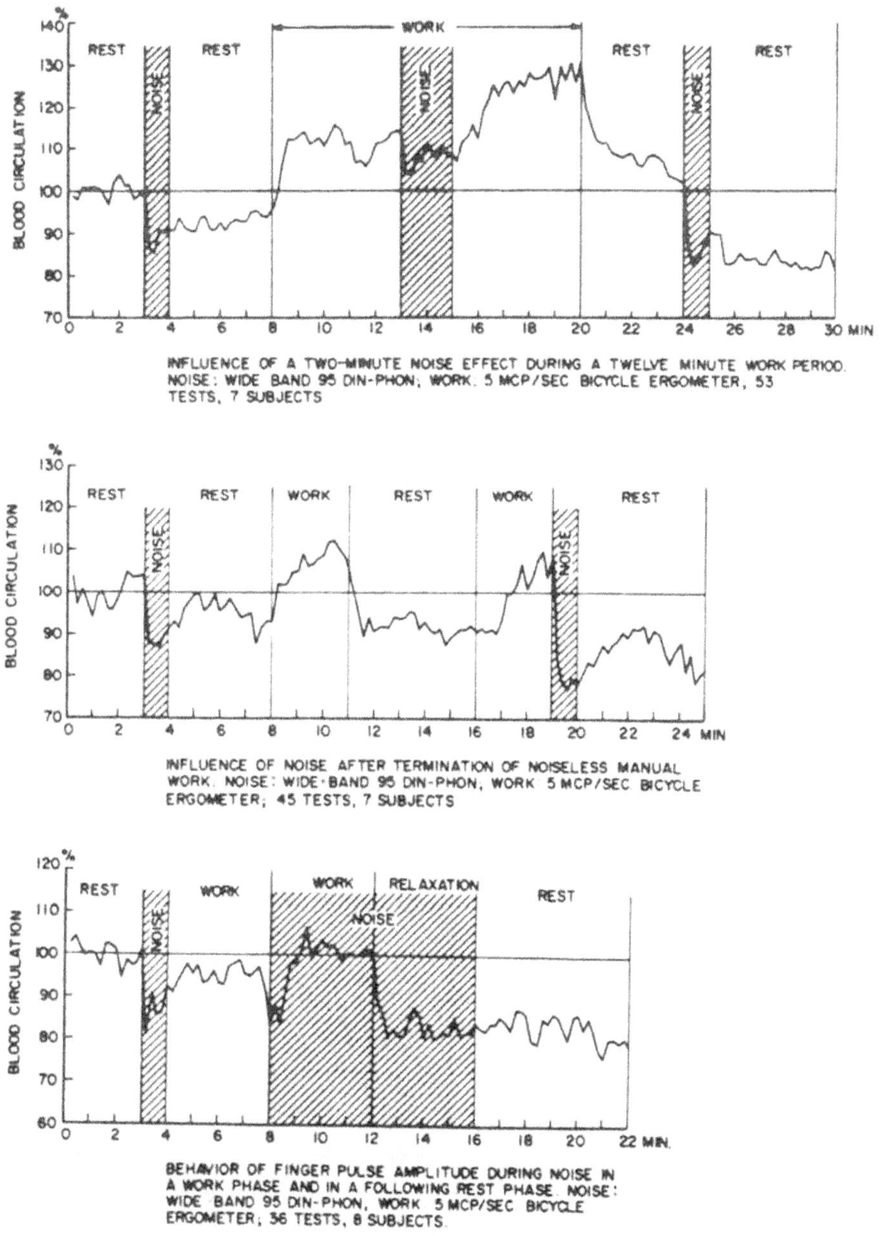

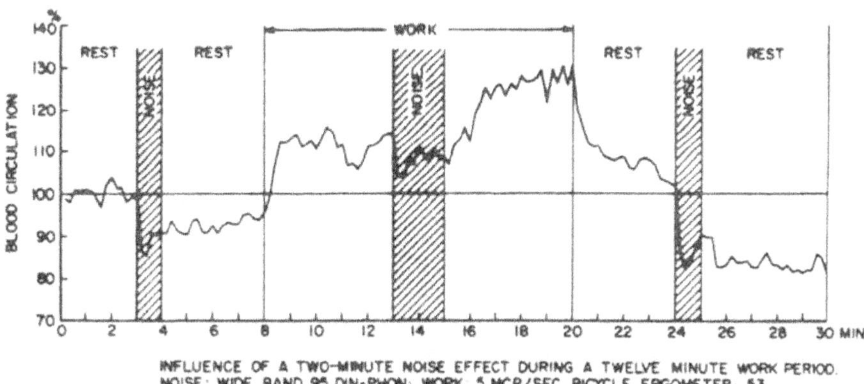

INFLUENCE OF A TWO-MINUTE NOISE EFFECT DURING A TWELVE MINUTE WORK PERIOD. NOISE: WIDE BAND 95 DIN-PHON; WORK: 5 MCP/SEC BICYCLE ERGOMETER, 53 TESTS, 7 SUBJECTS

INFLUENCE OF NOISE AFTER TERMINATION OF NOISELESS MANUAL WORK. NOISE: WIDE-BAND 95 DIN-PHON; WORK 5 MCP/SEC BICYCLE ERGOMETER; 45 TESTS, 7 SUBJECTS

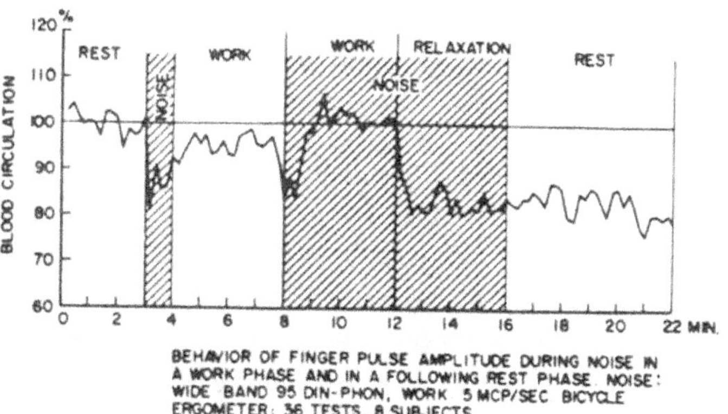

BEHAVIOR OF FINGER PULSE AMPLITUDE DURING NOISE IN A WORK PHASE AND IN A FOLLOWING REST PHASE. NOISE: WIDE BAND 95 DIN-PHON, WORK 5 MCP/SEC BICYCLE ERGOMETER; 36 TESTS, 8 SUBJECTS

FIGURE 10.9. Effects of combinations of work, rest, and noise on blood circulation. Reduced percentage indicates vasoconstriction of the peripheral blood vessels. (From ref. 24.)

tude during work. The thought that this differential response may somehow be related to a reaction of the auditory system to intense noise without necessarily involving the autonomic system in any general way is reinforced by the fact that noises which most readily elicit aural reflex (lower rather than higher frequencies) also most readily cause the peripheral vasoconstrictive response.

Meager or negative glandular response data

Finkle and Poppen (ref. 30) exposed 10 men to 120 dB of jet engine noise for 1 hour per day for 10 days, followed by 5 days of 2-hour exposures. The men wore earmuffs that probably reduced the effective level of noise reaching their ears by 25 dB or so. The men showed complete habituation to the noise following initial responses to the exposures in all the physiological measures (cardiovascular output, basal metabolism, ECG, EEG, urinary functions, kidney functions, and blood tests) taken from them.

Slob *et al.* (ref. 31) conducted a study in which periodically obtained urine samples were analyzed for the presence of corticosteroids, adrenaline, and noradrenaline (glandular secretions associated with activation of the autonomic system). This article also provides a good review of some European studies on this subject. During the hours of 0900 to 1700 urine samples were taken every 4 hours on 2 successive days from 2 groups of 10 male adults. During the second day one group was exposed to a 1/3-octave-band noise centered at 4000 Hz at a level of 80 dBC from 1300 to 1500. Comparison of the morning and afternoon excretions between the 2 days showed that the control group experienced the following: (1) a significant drop in adrenaline excretion on the second day compared with the drop between morning and afternoon on the first day; (2) a slightly greater drop in noradrenaline between afternoon and morning on the second day compared with the difference found on the first day; and (3) a similar difference on both days for the corticosteroids. The experimental group showed similar differences between morning and afternoon excretions of all three hormones on both days. The noise seemingly had no effect.

Slob *et al.* conclude, however, that the noise perhaps had some effect because there was no decrease in adrenaline during the afternoon of the second day for the experimental group, whereas there was a decrease in adrenaline on the second day for the control group. Slob *et al.* suggest that these meager or negative results could be because their noise was presented at a level of only 80 dBC. However, Slob *et al.* seem skeptical and quote Hawel and Starlinger (ref. 32) to the effect that merely taking part in tests produces stress that negates the effects of the agent (in this case, noise) for which the tests are conducted. In such cases the agent may produce no effect at all on the subject.

Brandenberger *et al.* (ref. 33) measured cortisol concentrations in eight human subjects every 10 minutes from 0800 to 1500. In these experiments, 1 day was a control day and on 1 or 2 other days the human subjects were exposed to a broadband pink noise at levels from 96 to 105 dBA and to 1/3-octave-band noises centered at 4000 Hz (84 dBC) and at 8000 Hz (89 dBC). Some of their findings are shown in figures 10.10 and 10.11. Brandenberger *et al.* conclude that noise is not associated with hyperactivity of the pituitary-adrenocortical part of the autonomic system in man. These findings are also consistent with plasma cortisol and EEG data obtained by Favino *et al.* (ref. 34) from human subjects exposed to a band of noise (700 to 1000 Hz) at a level of 90 dB. Fruhstorfer and Hensel (ref. 35) exposed 13 young adults to 16-sec bursts of white noise for 1 hour daily for 10 to 21 days and measured their heart rate, respiration, cutaneous blood flow, and EEG during exposure. They concluded that certain physiological responses adapt to loud noise but that the time needed for this adaptation is somewhat different for different responses.

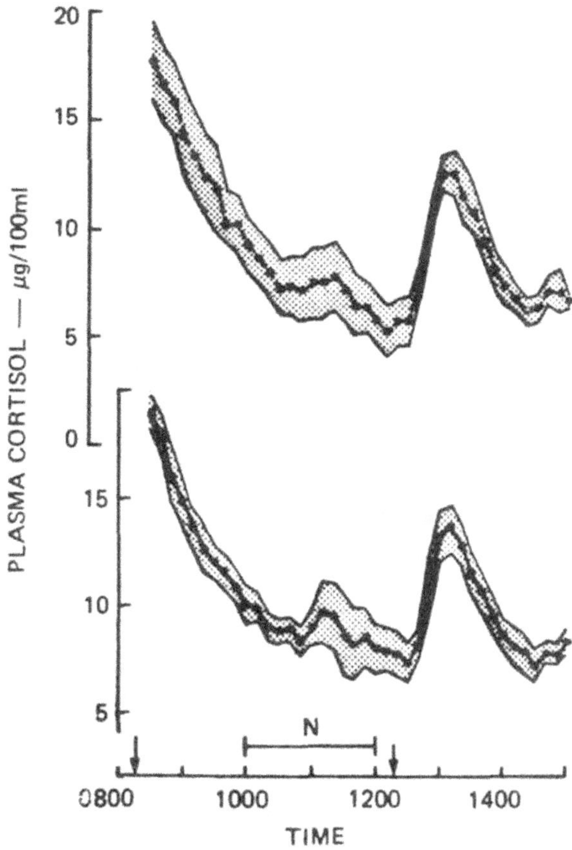

FIGURE 10.10. Mean and standard deviation of plasma cortisol concentration in five subjects during control days (top) and during days with exposure to 96-dBA pink noise (N) of 120 min duration (bottom). (From ref. 33.)

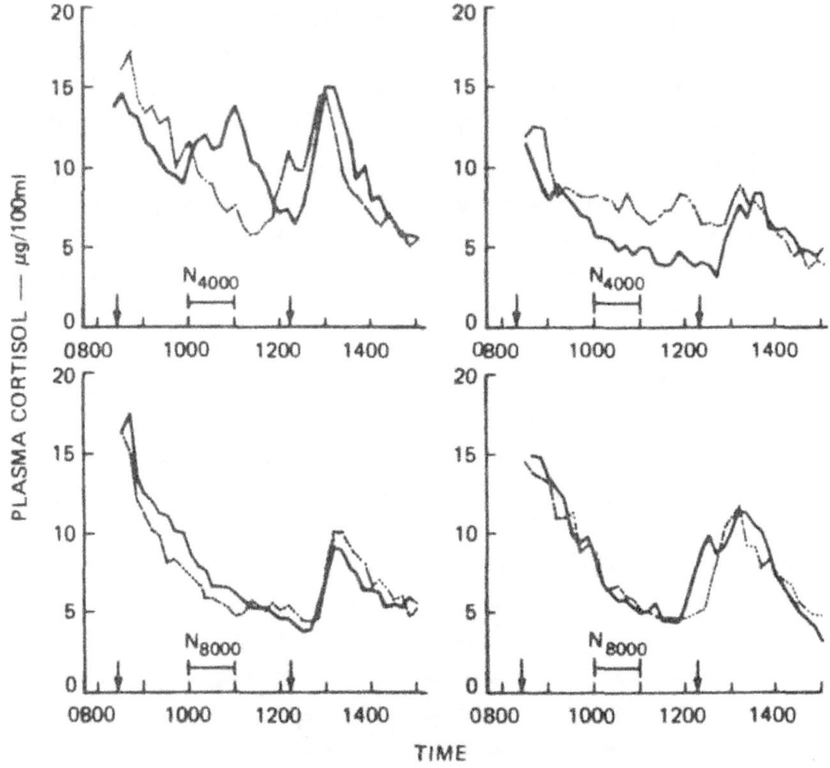

FIGURE 10.11. Plasma cortisol levels during days with exposure to 1/3-octave-band pink noise centered at 4000 Hz or 8000 Hz (continuous line) compared with control day pattern (dotted line). Arrows locate meal intakes. (From ref. 33.)

Negative cardiovascular response data

Diastolic and systolic blood pressure, heart (pulse) rate, and peripheral vascular pulse pressure can all be modified by actions of the autonomic system and have been monitored in man and lower animals to show physiological reactions to "stressful" stimuli. Davis and Van Liere (ref. 36) found habituation in human subjects to repeated loud noises, and Pearsons and Kryter (ref. 37) found similar adaptation of a startle response (heart rate) to simulated sonic booms. Illustrative of the pervasive principle of habituation of organisms to sound is the finding of Bartoshuk (ref. 38) that the acceleration of the heart rate in unborn babies to bursts of acoustic clicks (85-dB level) is adapted out by the end of 40 trials. Also, Ando and Hattori (refs. 39 and 40) report that babies of mothers from neighborhoods subjected to aircraft noise were much less aroused from sleep by aircraft noise than were babies of mothers who lived in quiet neighborhoods during pregnancy.

Cartwright and Thompson (ref. 41) conducted a study in which it was found that 1-hour exposures to white noise at 91 dBA caused no statistically significant changes in any of the cardiovascular responses from the control data obtained when the noise was not present. Some of their findings are shown in figure 10.12. As in the Brandenberger *et al.* study, the need to take frequent samples of data (every 2 minutes by Cartwright and Thompson) is of obvious importance because of the response variabilities unrelated to the noise stimulation.

As was noted earlier, Glass and Singer (ref. 19) found that the peripheral vasoconstriction responses (pulse amplitude) became completely habituated when subjects were exposed to noise at levels of 108 dBA. Also, as shown in figure 10.8, Kryter and Poza (ref. 27) reported that high-frequency, narrowband noise had no effect on heart rate, peripheral pulse pressure or amplitude (vasoconstriction), or peripheral blood volume. However, a broadband noise of the same intensity (92 dB) consistently caused a vasoconstrictive response.

Laboratory studies of Andrén *et al.* (ref. 42) and Andrén (ref. 43) support, in general, the hypothesis that the constriction of peripheral blood vessels that occurs because of broadband noise at levels above 80 dBA or so may be due to a nonstressful reflex response, perhaps associated with the aural reflex or ear protective reaction, rather than being a part of a general, nonspecific stress reaction. Andrén (ref. 43) found that 20-minute exposures of adult males to broadband noise at levels of 95 or 100 dBA caused a significant change in diastolic blood pressure but no changes in systolic blood pressure or heart rate. The increase in diastolic blood pressure was attributed to an increase in constriction of the peripheral blood vessels. No significant changes in cardiovascular behavior were observed from noise at a level of 85 dBA. The subjects rested comfortably on a bed in a recumbent position during the tests.

Important to the above-cited hypothesis is the fact that the noise at all levels caused no significant changes in the blood plasma levels of so-called "stress" hormones (adrenaline, noradrenaline, cortisol, prolactin, and growth hormone) in normotensive subjects. In subjects with mild hypertension there was a significant increase only of norardenaline during noise stimulation. Accordingly, it appears that the broadband noise capable of causing a constriction of the peripheral blood vessels did not cause cardiovascular or autonomic-glandular system responses associated with psychological-physiological stress (ref. 8).

Bättig *et al.* (ref. 44) obtained electromyograms, electrocardiograms, and skin conductance and respiration measures from 33 adults in their homes. At times the subjects performed various written tests and at other times engaged in conversation, rest, and so forth. Bättig *et al.* found that there was no correlation between noise-exposure level and complaint behavior and the various physiological measures, including the electrocardiograms. However, the physiological measures were related to the activities being engaged in. This study is also interesting in that it was conducted in the homes of the subjects rather than under strict laboratory conditions.

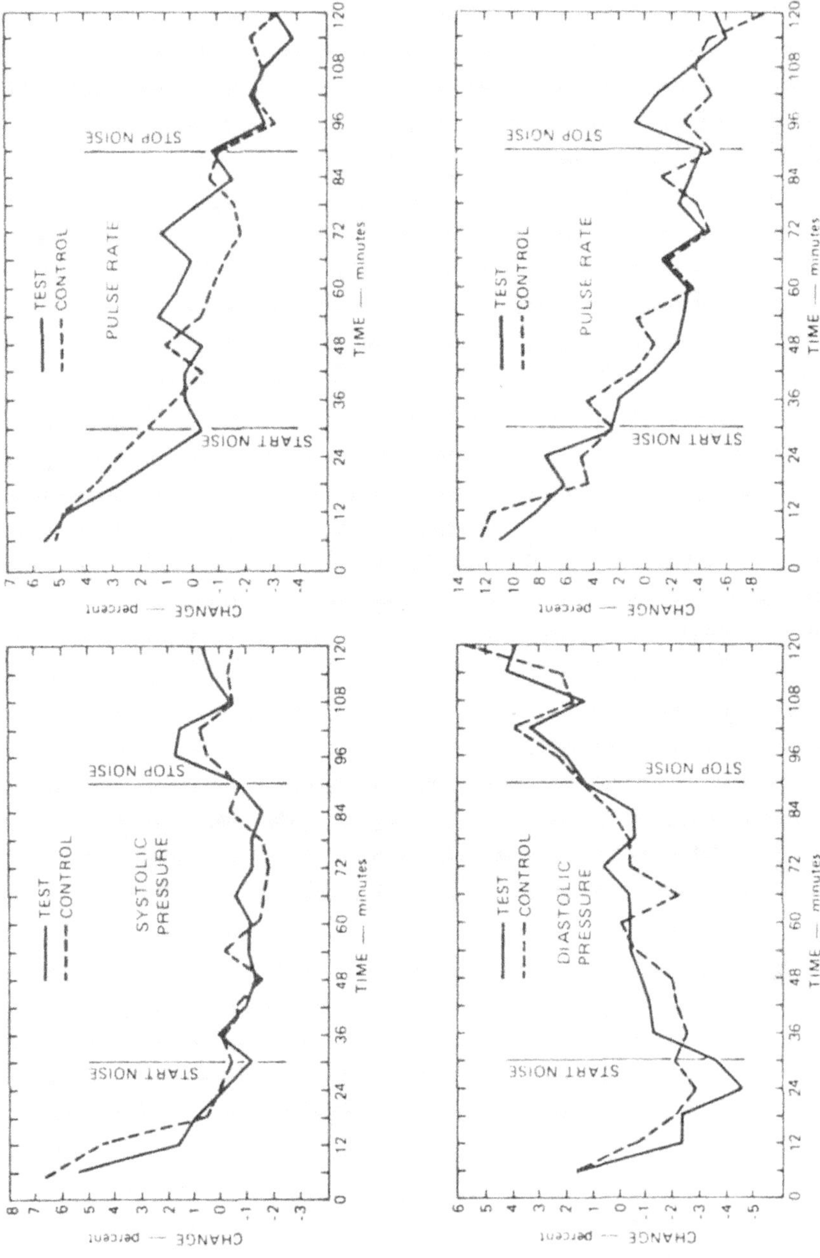

FIGURE 10.12. Change in various physiological activities for controls (quiet) and for test subjects exposed to 91 dBA broadband noise. (From ref. 41.)

Laboratory Studies of Monkeys

Positive and negative cardiovascular response data

A somewhat similar variety of conflicting research results as those found with humans have been obtained with monkeys as subjects. Peterson *et al.* (refs. 45 to 47) have conducted several experiments with monkeys in which cannulas implanted in the thoracic aorta permitted the continuous monitoring (15 sec of each minute) of heart rate and of systolic and diastolic blood pressure. The animals were restrained in chairs within the experimental environment for 24 hours a day for 6 to 9 months prior to the experiments proper and for an additional 30 experimental days for one study and 270 experimental days (9 months) for another study. Although some control animals were used, the analyses by Peterson *et al.* (refs. 45 and 46) to show noise effects were done primarily by comparing the mean of the physiological measures for the experimental animals for a few days before they received any noise to the same measures during and after the periods they received noise.

Figure 10.13 shows some of the data from one of the early experiments of Peterson *et al.* (ref. 45). Peterson *et al.* note that prior to the start of the noise (a recording of street and aircraft noises at L_{eq} = 78 dBA), which started at 0600, the animals showed heart rate and systolic blood pressure increases in anticipation of the noise. Indeed, the levels reached for these two responses prior to the noise were about as great as present at any time during the time the noise was on. The decline in these two physiological activities while the noise is still on shows apparent anticipation of the end of the noise. Accordingly, it would seem logical to conclude that it is the psychological cognitive aspects of the experimental situation and not the noise *per se* that is eliciting the responses shown in figure 10.13.

Somewhat similar deductions can be made from other data published by Peterson *et al.* (refs. 46 and 47). For example, figure 10.14 shows that the experimental animals exhibit a sharp increase in blood pressure on the first day or so of exposure to noise and that the increase stays with the animals for the duration of the experiment and beyond. Figure 10.14, which tracks 9-day averages for control and experimental animals, shows the immediate and the lasting effect of the addition of the noise regime to the environment.

That this elevation of blood pressure may be related to conceptions of anxiety the monkeys have about the experimental procedures and situations that go beyond any direct stress from the noise *per se* is suggested by the upper graphs in figure 10.15. The difference between the lower and the upper graphs in figure 10.15 is that the three curves (A — preexposure days; B — first 12 days of exposure; and C — last 18 days of exposure) in the upper graphs have been normalized with respect to the blood pressure levels for the hours of 12 to 5 a.m., before the noise was turned on, whereas the lower graphs represent the raw data. It is shown in figure 10.15 that the daily pattern of blood pressure changes appeared

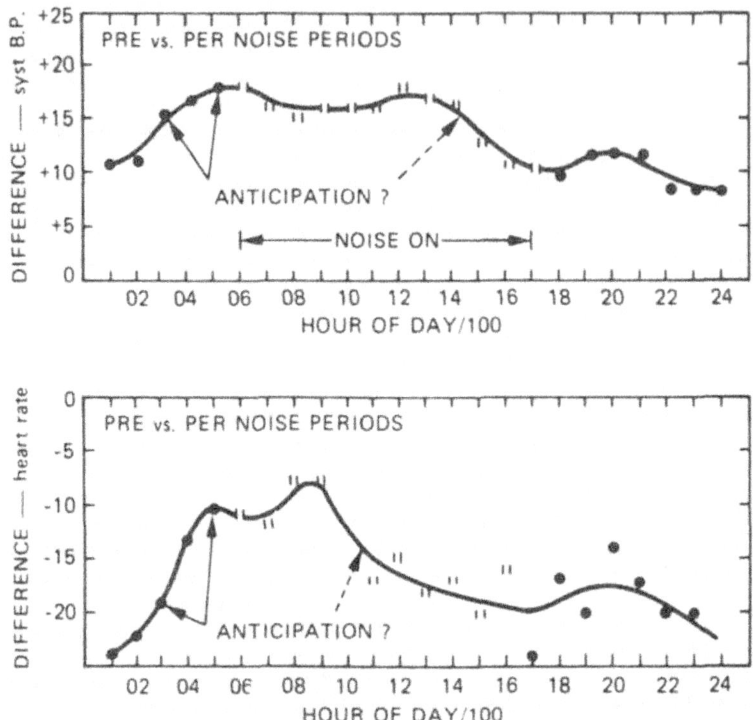

FIGURE 10.13. Hourly effects of noise with diurnal influences removed. "Anticipation?" lines added here. (Data from ref. 45.)

to be more related to the time of day than the condition of noise (*i.e.*, the daily preexposure pattern was about the same as the daily exposure patterns). Even if there are some specific noise relations they are minor in magnitude compared with the overall lack of effect of the introduction of noise.

Peterson (ref. 47) contends that humans exposed to moderately intense noise can experience sustained elevations in blood pressure without also sustaining hearing losses. (Their animals showed no depression in auditory thresholds measured in the brain stem following the noise exposures.) However, the data indicate that psychological factors (factors probably not relevant to noise-exposure conditions and ways of thinking typical for humans) had more impact than the noise on the nonauditory systems of the monkeys. Further, other experiments with both animals and humans make it clear that noise is not necessarily a cause of increased blood pressure or related nonauditory-system stress responses, and that psychological factors make it somewhat risky to extrapolate positive stress effects found in animals to humans.

In support of this interpretation of the data of Peterson *et al.* is the study of Kraft Schreyer and Angelakos (ref. 48) in which monkeys were exposed in the

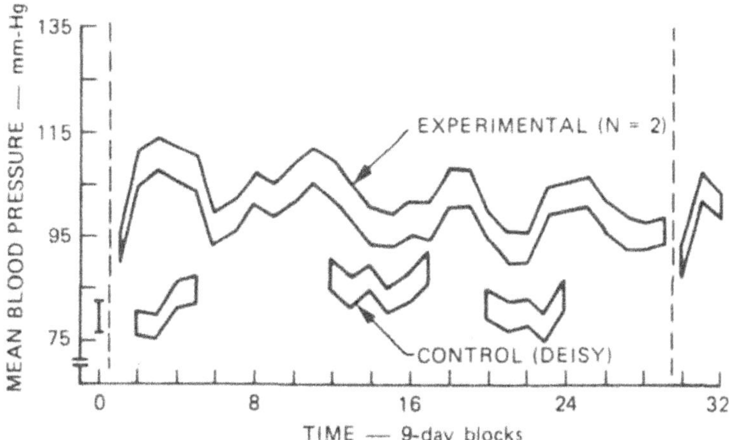

FIGURE 10.14. Comparisons of mean blood pressure for experimental and control animals. Envelopes encompass mean and 95-percent confidence limit. Dashed vertical lines indicate onset and cessation of noise-exposure period. Thick vertical bar at left represents preexposure values for experimental animals. (From ref. 47.)

laboratory to a 400-Hz sound at a level of 100 dB for 4 months. These investigators found no systematic effects on blood pressure in the monkeys from exposure to the intense sound.

Negative glandular response data

An experiment by Hanson *et al.* (ref. 49) would seem to show the predominant role played by psychological aspects of the experimental situation in determining the reaction of an animal to noise. Hanson *et al.* studied the effects on plasma cortisol levels of four 13-minute daily exposure periods to 100 dB of noise for 28 days. Each 13-minute period was separated by 2 minutes of quiet. Twelve monkeys were divided into groups and subjected to the following conditions: (1) no noise and no control over noise; (2) control over noises (at the end of each 13-minute session the animal was presented a lever which, when pushed, turned off the noise); (3) no control over noise (no lever was presented at the end of each 13-minute session); and (4) loss of control (animals who had completed the control-of-noise sessions were given the lever, but pressing the lever now did not turn off the noise). The results showed that blood plasma cortisol levels of animals with control over noise did not differ from animals exposed to no noise at all, and that blood plasma cortisol levels were significantly elevated in animals with no control over noise and in animals experiencing a loss of control over noise. This finding is somewhat reminiscent of the aforementioned habituation findings of Glass and Singer (ref. 19) with humans.

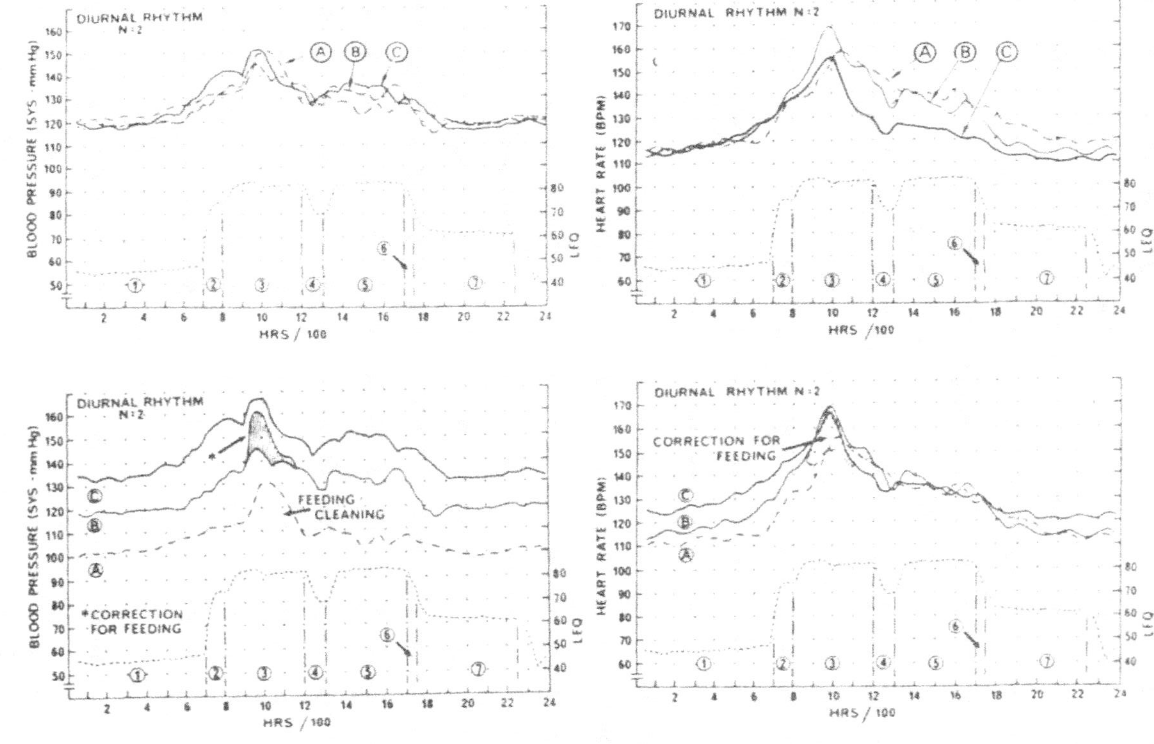

FIGURE 10.15. Diurnal rhythm of cardiovascular responses and noise-exposure sequence. L_{eq24} = 85 dBA; 1—night noise; 2—morning household noise; 3 and 5—work noise; 4—cafeteria noise; 6—transportation noise; 7—evening household noise. (From ref. 46.) See text for data normalization procedure.

Laboratory Studies of Mice and Rats

Audiogenic seizures

A large body of literature is available on the effects of sound and noise on mice and rats. The number of studies done has no doubt been influenced by the ready availability of these animals for laboratory research and by the fact that they exhibit marked responsiveness to some intense sounds or noises. However, conjectures and extrapolations about the effects of noise stress in man from these mice and rat studies have been controversial. (See Falk (ref. 50) and Kryter (ref. 51).) The usefulness of the data from these mice and rat studies is limited for the following reasons:

1. The emotional state of fear engendered in the animals by the hostile experimental conditions must be extremely significant, perhaps more so than any possible acoustical stimulation *per se*. For example, a not atypical condition for research with these animals was used by Buckley and Smookler (ref. 52) to create stress. Rats were placed in a cage on a shaker in a room 4 ft by 4 ft by 6 ft, with loudspeakers in the ceiling and the walls that produced noises such as airblasts, bells, and buzzers at a level of about 100 dB. Spotlights in each corner of the room gave 140 footcandles of light in the cage. Periods of light and darkness were alternated every 1/2 sec. The cage was oscillated at a rate of 140 per minute. There is no question that these conditions should cause stress in the animals, but the interrelationships of psychological and physical stimulation factors can hardly be fathomed.

2. These animals can display stress reactions to some sounds and noises that are peculiar to their species and that are probably pathological within the species. These reactions cover a wide range, from those not detectable from casual observation to convulsive and hebephrenic behavior. These behaviors, sometimes leading to death, are called audiogenic seizures.

The mechanisms involved in audiogenic seizures remain something of a mystery and are beyond the scope of this document to discuss. Certain facts, however, are clear. For one thing, the seizure response depends on several conditions beyond that of an immediate sound or noise stimulation. Examples of some conditions are the following:

1. When certain strains of rats or mice are exposed to intense auditory stimulation at the age of about 15 to 25 days, most will be susceptible to audiogenic seizures to intense noises later in life. This is called priming and the sounds used could conceivably cause damage to the cochlear and vestibular systems (refs. 53 to 55).

2. Animals suffering from middle ear infection (otitis media), which a large number of laboratory animals can have without obvious symptoms, are more

prone to audiogenic seizures than are noninfected animals (refs. 56 and 57). The complexity of audiogenic seizures and interactions among psychological factors, priming, and otitus media is shown by a study done by Niaussat (ref. 58). It was found that in mice genetically nonsusceptible to audiogenic seizures which were primed with exposure to intense noise when 17 days old, 9 percent became seizure prone when raised in colonies and only 3.8 percent became seizure prone when raised in individual, sound-insulated cages. For mice with otitus media, however, 79 percent were seizure prone.

3. The way the animals are handled and the acoustic and other conditions of their living quarters affect their seizure rate (ref. 59).

Whatever the facts might be, the basis, control, and role of audiogenic seizures in rats and mice is very uncertain from present research information. Seldom in studies of audiogenic seizures or of stress from noise in rats or mice are the animals examined for the presence or history of otitus media or for possible damage to the auditory and vestibular receptor systems from exposures to the intense sounds or noises to which the animals have been subjected.

Some investigators hold positive views of the value of experiments with rats and mice exposed to noise. For example, Busnel *et al.* (ref. 60) found that mice dropped in a vat of water to swim for their lives with weights attached to their tails had seizures and submerged more quickly when exposed to tones from 50 to 10 000 Hz at levels ranging from 60 to 115 dB. Busnel *et al.* suggest that the use of these animals avoids the "adaptive" processes that mask the noxious effects of noise in studies with humans. However, it is not clear that their findings are not possibly due to pathological audiogenic seizures rather than normal stress responses to noise, and that in normal mice and rats "adaptive" processes would not, like in humans, be a factor in controlling their behavior to sounds.

A number of studies (refs. 52, 53, and 61 to 71) showing the deleterious effects of audiogenic seizures from exposure to noise on the cardiovascular, reproductive, endocrine, and neurological functions in rats and mice were presented at a symposium on the physiological effects of noise. The concept that these and similar data from rats and mice were indicative of possible deleterious effects of environmental noise upon people was summarized for the symposium by Leake (ref. 72) as follows:

> While we have become conditioned in part to increasing noise in our environment, we still do not realize the extent to which noise can activate subcortical neuronal systems in our brain. Whether continuous or intermittent, noise will modify the pacing by the brain of our cardiovascular, endocrine, metabolic and reproductive functions. While it is true that the cortex may be inhibitory upon lower portions of the brain stem, this is only mildly modulated, and may be related to the extent of conditioning to which we have been subject. Even though we may have learned both behaviourly and physiologically to ignore noise and thus to reduce the intensity of its emotional response, some of the neurological stimulus may spill into the autonomic nervous system, producing cardiovascular-renal disturbance, together with endocrine, metabolic and reproductive abnormalities.

The concepts of neurological function with respect to noise as expressed by Leake and based on many studies done with rats and mice are not consistent with studies of the effects of noise on man or other animals. Indeed, it appears likely that these concepts are applicable strictly to rats and mice that are pathological by birth or that are rendered so by sound or infections causing physical damage to the auditory-vestibular systems.

It should be made clear, at the same time, that psychologically induced stress capable of causing psychosomatic disorders and ill-health in humans can be concomitant with the presence of sound and noise in the environment. But generalizations to humans in these regards from positive data (showing deleterious effects) from studies with rats and mice must be viewed with extreme caution, if not discounted completely. On the other hand, negative results (no effects) could be accepted as being valid, especially if they disprove the existence of a significant reflexive or direct neurological spillover disturbance response to noise.

Relation to epilepsy in people

It is sometimes surmised that audiogenic seizures in mice and rats may be somewhat comparable to epileptic seizures in people. Although both apparently involve pathological-neurological conditions, the events appear to be quite different. Forster (ref. 73) notes that epilepsy affects about 1 percent of the population, that about 6.5 percent of epileptics have their seizures evoked by sensory stimulation, and that the vast majority of those are caused by visual stimuli. Forster indicates that sound elicited seizures can come from a sudden loud noise but that the seizures are usually minor. Some patients will have a seizure only to certain specific sounds (*e.g.*, ringing of a telephone only in their home or a particular song or type of music; also see Rivera (ref. 74)). Forster finds that many of these rare patients are stimulated to seizures from sounds that are not unpleasant or loud to normal people. Apparently sound-induced epileptic seizures are extremely rare and are not necessarily noise related nor similar in etiology to audiogenic seizures in rats or mice.

Negative cardiovascular response data

A different situation exists when mice and rats free from audiogenic seizure symptoms or auditory disease are exposed to noise. As shown by Borg and Moller (ref. 75), such animals show habituation to noises that are meaningless in terms of associations with danger.

Borg and Moller divided 130 normotensive (blood pressure below 125 mm Hg) and spontaneously hypertensive (blood pressure above 125 mm Hg) rats into the following three main experimental groups: (a) exposed to background noise produced by the rats themselves, approximately 50 dBA; (b) exposed to noise at 85 dB; and (c) exposed to noise at 105 dB. The noise was a 1640-Hz-wide band of random noise sweeping from 3 to 30 kHz at a rate of once every 2 sec. It was interrupted randomly seven times per night with rise and decay times of interruption of 5 msec. Except for random interruptions totaling 2 hours during each night, the noise was on continuously for 10 hours each night, which is normal wake time for the rats. After 1 year in the noise the animals in group (b) showed hearing losses of 10 to 15 dB at 6 kHz and those in group (c) showed losses of 40 to 60 dB. Blood pressure was measured with a cuff attached to the tail of each animal.

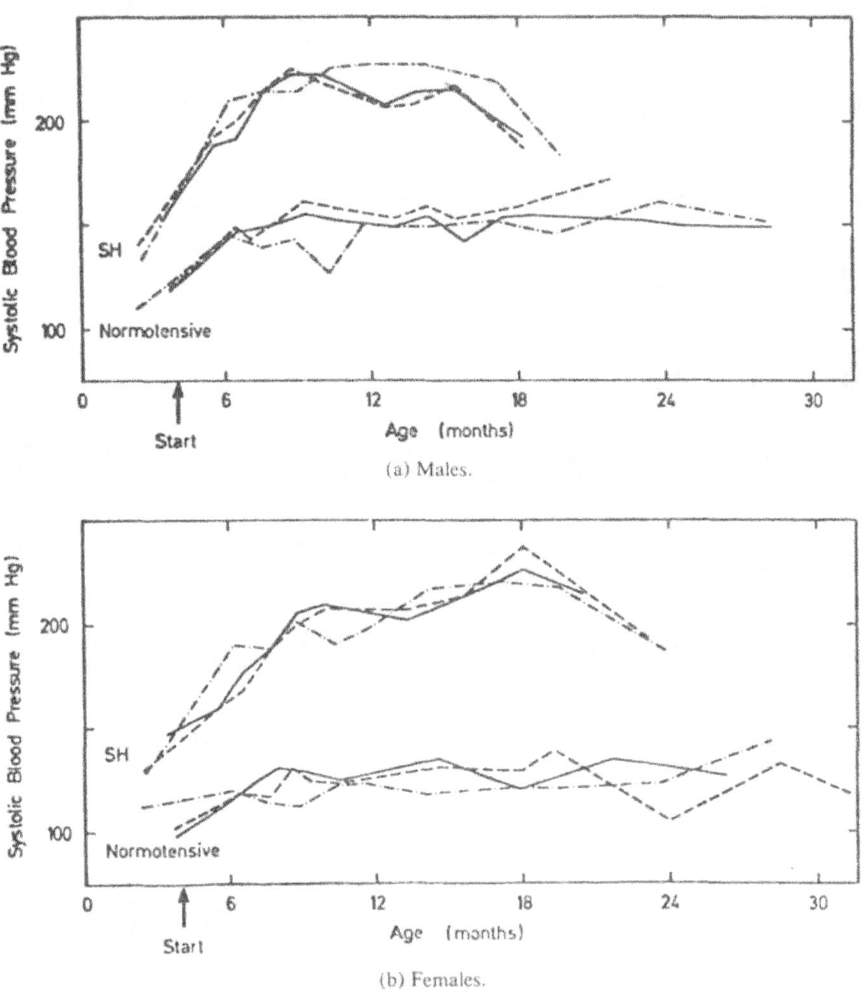

(a) Males.

(b) Females.

FIGURE 10.16. Mean systolic blood pressure measured indirectly from the tail of spontaneously hypertensive (SH) rats and normotensive rats as a function of age for control (solid curve), 85-dB-SPL noise (dashed curve), and 105-dB-SPL noise (dash-dot curve). (From ref. 75.) Average of standard error was not more than 11 mm Hg for any group.

Figure 10.16 shows Borg and Moller's major findings. It is shown that mean systolic pressure in rats over a lifetime of daily exposure to different levels of noise did not differ systematically from that of non-noise-exposed rats. There is some suggestion in figure 10.16 that the hypertensive animals in group (c) showed a somewhat more rapid initial increase in blood pressure than did the rats in the other two groups. To check the reliability of that possibility, an additional 40 hy-

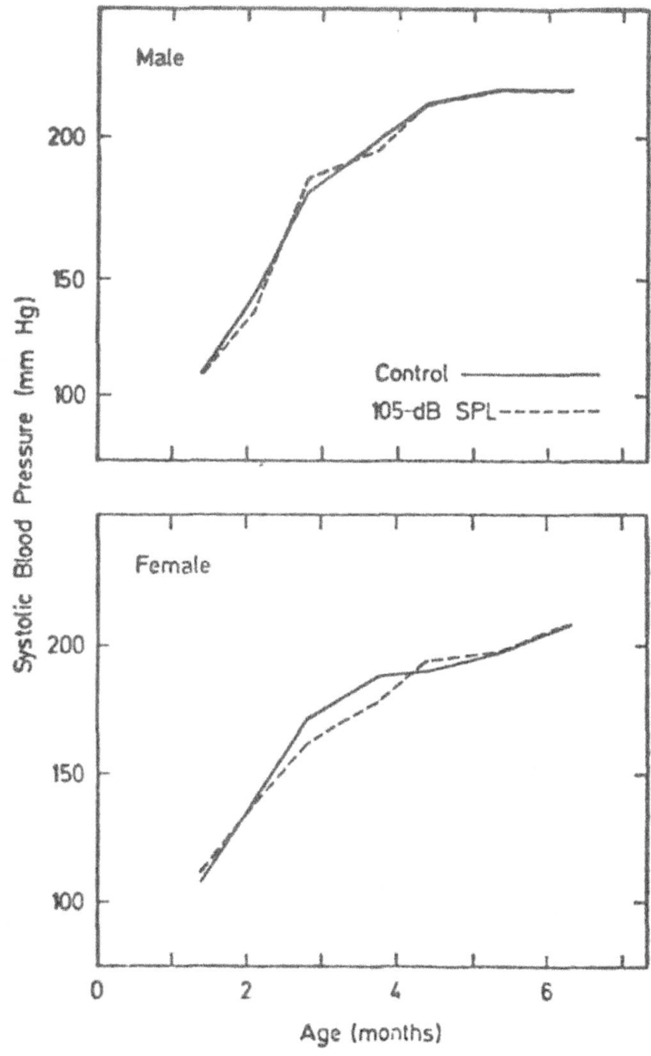

FIGURE 10.17. Average systolic blood pressure for spontaneously hypertensive rats in control environment and 105-dB SPL. (From ref. 75.)

pertensive rats were tested. As shown in figure 10.17 no consistent differences over this initial period were found between the hypertensive animals exposed to the quiet or to the 105-dB-SPL living environments.

Although the noise was sufficiently intense to cause some hearing loss, the losses probably did not reduce the loudness of the noises to the animals because of recruitment. Further, the animals with only a 10- to 15-dB loss exhibited the same general blood pressure changes as those animals with a 40- to 60-dB hearing loss. Borg and Moller concluded that habituation to the noise environments had occurred in the rats, and that lifelong exposure for 10 hours daily to the 85-dB or 105-dB levels of the noise used in their study would not alter blood pressure in normotensive or hypertensive rats.

Field Studies of Animals

Wildlife

Chesser *et al.* (ref. 76) trapped mice from fields near an airport (noise levels from 80 to 120 dB) and from similar fields away from low-flying aircraft (noise levels from 80 to 85 dB). They found that the mice from the fields close to the airport had larger adrenal glands than those from the quieter fields but had the same average gross body weight, and that a group of mice from the quieter fields developed larger adrenal glands than a control group of these mice when exposed to 105 dB aircraft noise for 1 minute every 6 minutes for 2 weeks. The authors indicate that population densities of the mice in the two fields were the same, but that air pollution or other factors in the field near the airport could have contributed to the larger adrenal glands found in the mice from that field. Also, it is possible that the psychological stress or perhaps some incipient audiogenic seizure effects possibly related to ear infections contracted in the laboratory could have contributed to the enlarged adrenal glands in the mice exposed to the aircraft noise in the laboratory experiment.

A possible reason to question the findings of this study is that in another sample of mice taken from the two fields, Pritchett *et al.* (ref. 77) found the control mice to be somewhat heavier, but there were no significant differences in adrenal-pair weights or in adrenal-body weight ratios. Pritchett *et al.* removed the adrenal glands from the mice and incubated them in the presence and absence of ACTH. They found that the adrenal glands from the noise-exposed mice showed somewhat different responsiveness to ACTH than did the glands from the non-noise-exposed mice.

Busnel and Briot (ref. 78) reported that in land zones near commercial airports in France, wildlife of many forms is abundant. Indeed, it is necessary for the government to mount hunting parties to try to control the bird population because they represent a safety hazard to aircraft. Data pertaining to collisions between birds and aircraft and to numbers of animals and birds bagged per year by hunting parties reveal that the animal populations grew independently of the amount of air traffic.

Individual observations show that migratory birds do not hesitate to use airport environs as resting places during migration and do not even necessarily attempt to move because of aircraft noise up to 120 dB. Busnel and Briot conclude that the general absence of humans from the area contributes to the growth of wildlife and that the aircraft noise has little or no effect on the animals. Some habituation no doubt is involved, but it would have to be almost immediate, as evinced the behavior of migratory birds.

Fletcher (ref. 79) reviewed the literature on observations made of wildlife during activities such as the placing of pipelines across wilderness areas and the use of off-the-road recreational vehicles. Noises such as those from helicopters, blasting, earth-moving equipment, and snowmobiles were involved. For the most part, this literature is ambiguous and impossible to interpret with respect to the effects of the noise on the wildlife. The reasons are primarily that the activities were often threatening to animals (destroyed the flora and fauna of some land areas), that people were present (a usually frightening stimulus to wildlife), and that the noise conditions and observations were for such short periods of time that possible habituation to the noise, as well as to the actions of the machinery and the people, was not adequately studied.

In situations where the noise was from fixed structures (*e.g.*, high-voltage lines giving off corona discharge noise levels of up to 63 dBA when wet) in remote wildlife areas, it was found that coyote, sheep, and many other wildlife species were not disturbed by the noise (ref. 79). Animals fed and "played" in its presence. Some birds built nests and raised their young in the towers supporting the high-voltage lines.

Farm animals

A number of studies and observations (*e.g.*, refs. 80 to 83) have been made of the effects of sonic booms and other aircraft noise on farm animals. Table 10.2, from a 2-week study of animals on farms near Edwards Air Force Base (ref. 81), shows that the probability of any significant reaction of any of the animals to sonic booms of the order of 1.0 to 2.0 psf is negligible. Young broilers showed the largest reactions.

The farms involved in these studies had been regularly exposed to about 8 sonic booms per day for several years. This would suggest that habituation had taken place and that previously nonexposed animals could react differently. However, the fact that the farming operations had continued without apparent problems over the years would suggest that this habituation was a rapid phenomenon.

TABLE 10.2

Animal Behavior Under Sonic Booms at Edwards Air Force Base

[Test period from June 6 to 23, 1966; data from ref. 80]

(a) Poultry behavior changes

Parameter	Number of booms	Average effect	0 (a)	1 (b)	2 (c)	3 (d)
Species:						
Broilers	197	1.02	23	158	6	10
Young turkeys	195	.51	100	91	3	1
Adult turkeys	198	.52	95	103	0	0
Young pheasants	85	.81	16	69	0	0
Adult pheasants	125	.96	7	117	0	1
By farm:						
Jones turkeys	187	0.53	90	96	0	1
K-M turkeys	206	0.50	105	98	3	0
Del Mar broilers	106	0.95	9	93	4	0
Ringo broilers	91	1.09	14	65	2	10
Pheasants	210	0.90	23	186	0	1

[a] *Number of booms producing no reaction.*
[b] *Number of booms producing a mild reaction.*
[c] *Number of booms producing a crowding reaction.*
[d] *Number of booms producing pandemonium.*

One interesting observation during the Edwards Air Force Base studies is that turkeys which showed some movement to the sonic booms did so slightly before the airborne acoustic signal reached them. They were apparently responding to the ground wave of the sonic boom, which travels faster than the airborne wave. In short, the turkeys were being stimulated by vibrations from the ground rather than by the audible sound.

Espmark *et al.* (ref. 81) exposed cattle and sheep to 20 sonic booms and 10 subsonic aircraft noises over a period of 4 days. The aircraft noise was at a level of about 94 dBA and the sonic booms at about 3 psf. Espmark *et al.* concluded that the effects of the noises were not unusual and that the animals returned quickly to grazing or other normal activities when interrupted.

Besides reports of young chickens being panicked by sonic booms, there are reports that nesting, farm-raised mink would kill their pups when exposed to sudden, intense noises such as sonic booms. Travis *et al.* (ref. 82) conducted a study in which farm-raised mink were exposed to three real or three simulated sonic booms at a level of 6 psf (very intense booms). A group of control animals were not exposed to booms. The effects of the booms were essentially negligible;

TABLE 10.2 (Continued)

(b) Dairy milking reactions

Date	Number of booms	0 (a)	1 (b)	2 (c)	Average effect
June 6	12	6	6	0	
7	10	10	0	0	
9	12	6	6	0	
13	7	6	1	0	
14	14	13	1	0	
15	1	1	0	0	
20	12	10	2	0	
21	12	11	1	0	
22	13	13	0	0	
23	11	9	2	0	
Totals	104	85	19	0	0.18

[a] *Number of booms producing no reaction.*
[b] *Number of booms producing a mild reaction.*
[c] *Number of booms producing a severe reaction.*

TABLE 10.2 (Concluded)

(c) Percentage changes in animal behavior

Animal	Changed	Returned / Changed	Changed to normal / Changed	Changed to abnormal / Changed	Abnormal / Total	Observations	Total changed	Total number of booms
By species								
Beef	7.68	24.89	73.79	1.31(3)	0.10	2980	229	168
Dairy	3.38	28.92	70.58	0.49(1)	0.01	6032	204	87
Sheep	2.76	0.00	100.00	0.00	0.00	2750	76	99
Horses	4.52	59.25	33.33	7.40(4)	0.33	1193	54	85
By farm								
Beef—1	10.10	25.38	73.09	1.52	0.15	1950	197	65
Beef—10	3.10	21.87	78.12	0.00	0.00	1030	32	103
Horses	4.52	59.25	33.33	7.40	0.33	1193	54	85
Sheep	2.76	0.00	100.00	0.00	0.00	2750	76	99
Dairy	3.38	28.92	70.58	0.49	0.01	6032	204	87

to summarize from their study, no differences ($p > 0.05$) were found among experimental treatments for length of gestation, number of kits born per female whelping, number of kits alive per female at 5 and 10 days of age, weight of kits at 49 days of age, kit pelt value and selling price. A behavioral study showed no evidence that the female mink under observation were sufficiently disturbed by sonic booms to engage in kit packing, kit killing, or to disrupt normal lactation. Results of necropsy examinations showed no mink deaths attributable to real or simulated sonic booms. Likewise, no evidence was found that bacterial disease was induced in the herd following exposure to sonic booms. There were no detectable differences in the overall health of the females at the three sites. The conclusion drawn from these studies was that exposure of farm-raised mink to intense sonic booms during whelping season had no adverse effect on their reproduction or behavior.

Bond (ref. 83) summarized research of the U.S. Department of Agriculture and described the following results: (a) milk production in cows was reduced by noise causing fright reactions, but dairy cows exposed to flyover noises of aircraft and sonic booms in real life showed no such effects; (b) swine exposed to loud noises ranging from 100 to 135 dB showed no changes in growth rate, efficiency of food utilization, or reproduction and showed no detectable microscopic changes in ears or in adrenal or thyroid glands; and (c) mink exposed to sonic booms showed no adverse effects.

Studies of Sleep in the Laboratory

Methods of measuring sleep

The effects of noise on sleep have been qualitatively measured in human subjects primarily in the following three ways:

1. The subject is asked to press a switch when he has been awakened (behavioral awakening).

2. The state of electrical brain activity measured from an electroencephalogram (EEG) is interpreted by the experimenter to show the stage (or change in stage) of sleep of a subject. It is generally held that a change from a "deep" to a lighter stage of sleep (as measured by the EEG) that is correlated with an exposure to a noise is indicative of some arousal. Some other physiological measures (*e.g.*, heart rate and vasoconstriction in the finger) have also been monitored during sleep in noise environments as possible indicators of the arousal effects of noise.

3. Subjects are asked to rate the quality of previous nights of sleep during some of which noises were present in their sleeping environment.

Habituated and nonhabituated responses

It is necessary to allow a number of nights of sleep in laboratory bedrooms for habituation to a new sleeping environment to occur before consistent, normal sleep patterns are established. Also, once tests with noises or sounds are started a further habituation takes place.

Griefahn and Jansen (ref. 84) analyzed the results of a large number of published studies on behavioral awakenings due to noise. They concluded that there was a significant decline in awakenings after the first night of exposure to a noise environment up to about the seventh night. (See fig. 10.18.) After about the seventh night, the effects seemed to remain fairly constant at about 35-percent awakenings. Thiessen (ref. 85) found this same general habituation for behavioral awakenings but found it extended to perhaps 20 nights or so of testing. (See fig. 10.19.) However, in figure 10.19 there was little indication of habituation of the EEG index of arousal (a change in sleep stage) over the durations of the experiments.

This lack of habituation of a change in EEG stages of sleep from exposure to noise has been consistently found. Vasoconstriction in the finger and changes in heart rate because of subsonic aircraft noises, street traffic noises, and sonic

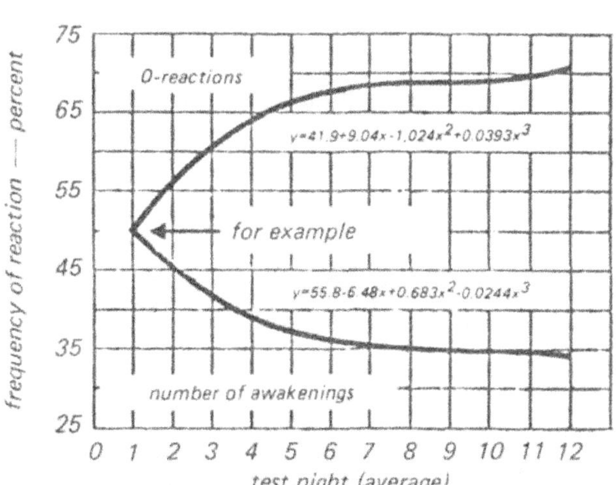

FIGURE 10.18. Sleep disturbances due to noise. (From ref. 84.)

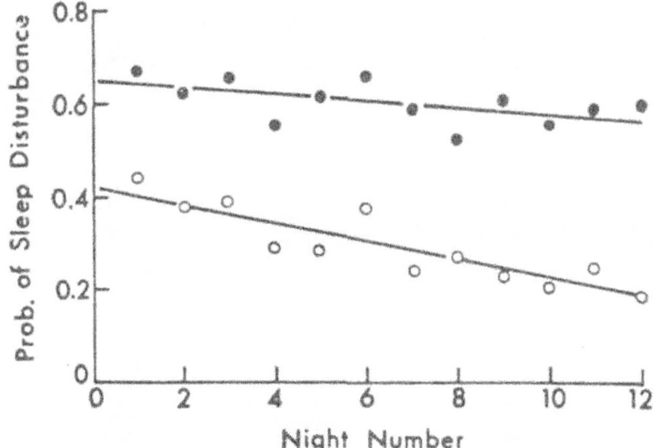

FIGURE 10.19. Laboratory results showing general lack of habituation of shift in EEG level (solid circles) and occurrence of habituation in behavioral awakening (open circles). (From ref. 85.)

booms presented during sleep also show no signs of habituation over many nights (*e.g.*, 53 experimental nights of exposure to sonic booms in ref. 86). In this regard, Muzet and Ehrhart (ref. 87) found in laboratory studies that street traffic noise at a level as low as 45 dBA caused a momentary change in heart rate during sleep that did not habituate over a test period of 15 nights. (See fig. 10.20.)

Kryter and C. E. Williams (unpublished data) found that although there was no habituation of a change in EEG stages from exposure to some noises over a period of 10 test nights, the intensity level of a 5-sec noise had to be higher for deeper stages of sleep than for lighter stages in order to elicit a change in EEG sleep stage. A noise level of about 30 dB relative to a threshold of audibility when awake was needed when the subject was in EEG stage 2 of sleep, about 50 dB was needed when in stage 3, and about 80 dB was needed in stage 4. Williams *et al*. (ref. 88) found that behavioral awakening also required a higher intensity of sound when in stage 4 and in the dream (REM) stage of sleep than when in the lighter stages (2 and 3, as classified by Williams *et al.*). (See fig. 10.21.) Figure 10.22 shows that the finger vasoconstriction response is marginally differentially responsive in different stages of sleep following sleep deprivation; however, for heart rate (fig. 10.20) and EEG (fig. 10.21) the magnitude of change depends somewhat on the intensity level of the noise. Griefahn (ref. 86) also found that about the same amount of vasoconstriction seemed to occur for sonic booms of different levels of intensity.

The results of a number of studies of the effects of noise on sleep in which questionnaires for measuring the opinions of the subjects regarding the quality

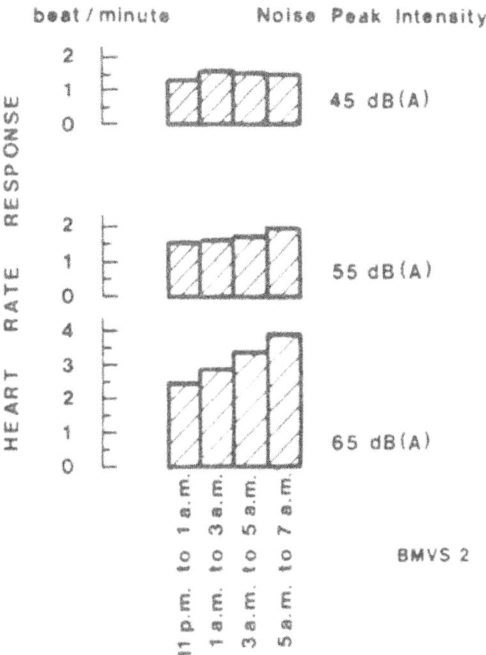

FIGURE 10.20. Changes in heart rate from street traffic noise during sleep. (From ref. 87.)

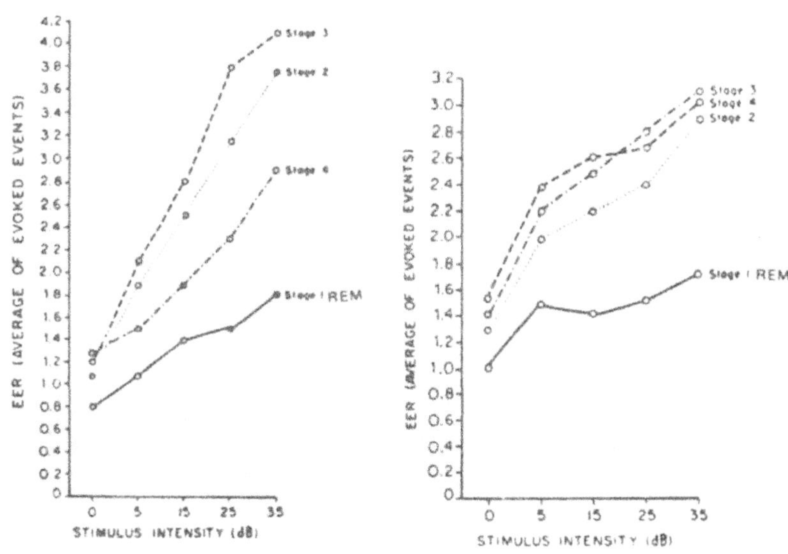

FIGURE 10.21. Effect of stimulus intensity on EER during four stages of sleep after 64 hours of sleep deprivation. (From ref. 88.)

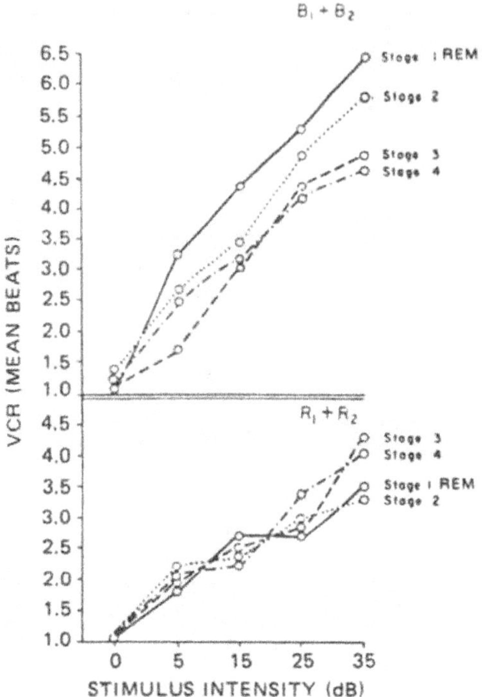

FIGURE 10.22. Effect of stimulus intensity on vasoconstriction response (VCR) during four stages of sleep for baseline nights (B_1 and B_2) and recovery nights (R_1 and R_2) after 64 hours of sleep deprivation. (From ref. 88.)

of the previous night's sleep have been reviewed by Lukas (ref. 89). Figure 10.23 shows how the composite L_{eq} for the sonic booms and aircraft noise predicts rated sleep quality.

Ehrenstein and Müller-Limmroth (ref. 90) have measured mood changes at different times of the day following exposures to awakening noises. Muzet *et al.* (ref. 91) found that subjective ratings of the previous night's sleep were highly correlated with the amount of sleep disturbance observed in the subjects. Öhrström and Rylander (ref. 92) found positive correlations between the levels of exposure to recorded intermittent and continuous traffic noise, sleep arousal (as measured by bodily movements), reduced ratings of sleep quality, and task performance. The continuous noise had a significantly smaller effect than the intermittent noise. The results of this study are difficult to compare with the results of some other similar experiments in that the exposure levels of the different noise conditions are not reported.

LeVere and Davis (ref. 93) found that the correlation between subjective evaluation of subjects regarding their sleep on a previous night and the amount

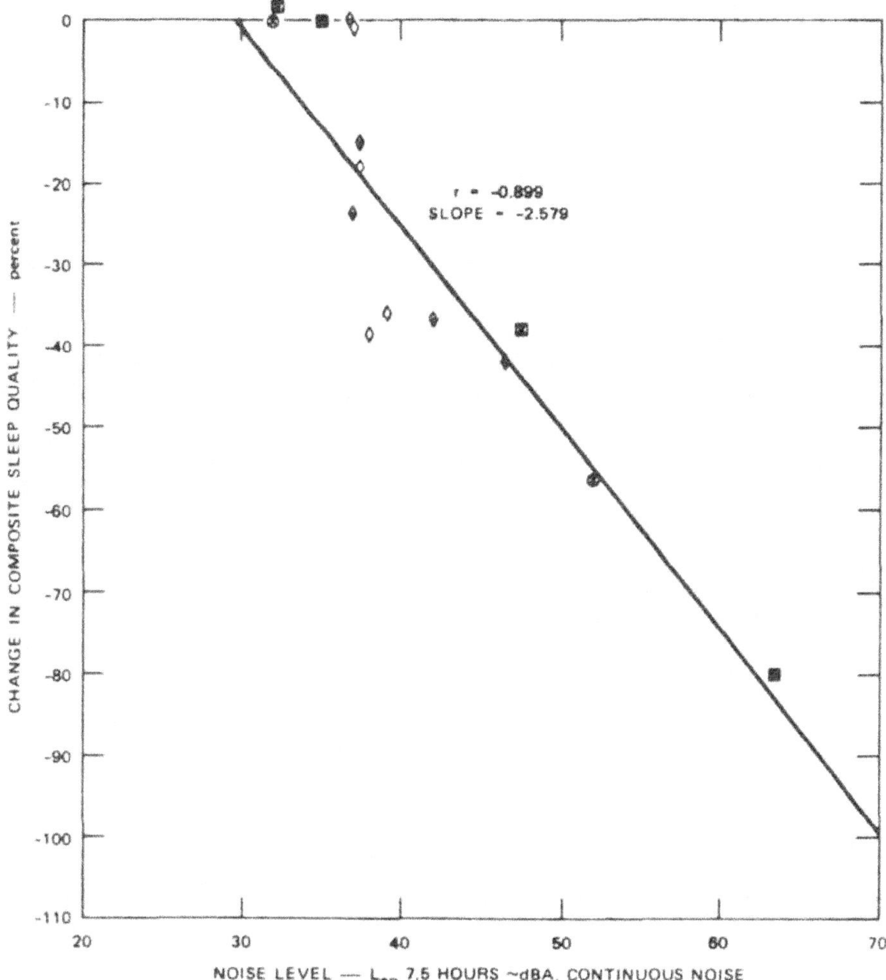

FIGURE 10.23. Relative subjective disturbance of sleep at various total nighttime noise levels calculated in L_{eq}. Solid square is for jet aircraft noise, open diamond is for "clicks," and other symbols are for sonic booms. (From ref. 89.)

of change in EEG activity measured for that night was zero. However, the aircraft noises used by LeVere and Davis did not usually lead to any behavioral awakening. It would thus appear that a change in EEG activity that is short of the pattern related to behavioral awakening is not a disruption in sleep to be noticed subjectively by a person but rather that behavioral awakening is associated with subjective ratings of sleep quality.

Johnson *et al.* (ref. 94) studied heart rate, vasoconstriction in the finger, and EEG in young men exposed to 3- to 4-kHz impulses of 0.75-sec duration every 45 sec for 24 hours per day. The impulses ("pings") were at levels of 80 to 90 dB. In addition to the physiological measures, records of time asleep and performance on some physiological and psychological tests were obtained. Three experiments were conducted: 1 for 15 days with 20 men, 1 for 55 days with 20 men, and 1 for 7 days with 39 men. Figure 10.24 shows that the experimental subjects experienced increased latency in going to sleep as compared with this latency for the control subjects not exposed to pings. Other behavioral effects related to possible sleep disturbance from the pings were not observed.

Of particular interest are the findings that throughout each of these experiments, as well as in other similar studies conducted by Johnson and Lubin (ref. 95) and Cantrell (ref. 96), no extinction occurred in the physiological responses to the pings (vasoconstriction, heart rate, and EEG changes during certain sleep stages), but none of these responses to the pings occurred when the subjects were awake. The lack of these responses to the 3- to 4-kHz pings when the subjects were awake is consistent with all the other studies of the N-response (alerting response, sometimes called the orientation response (OR)), which show a rapid

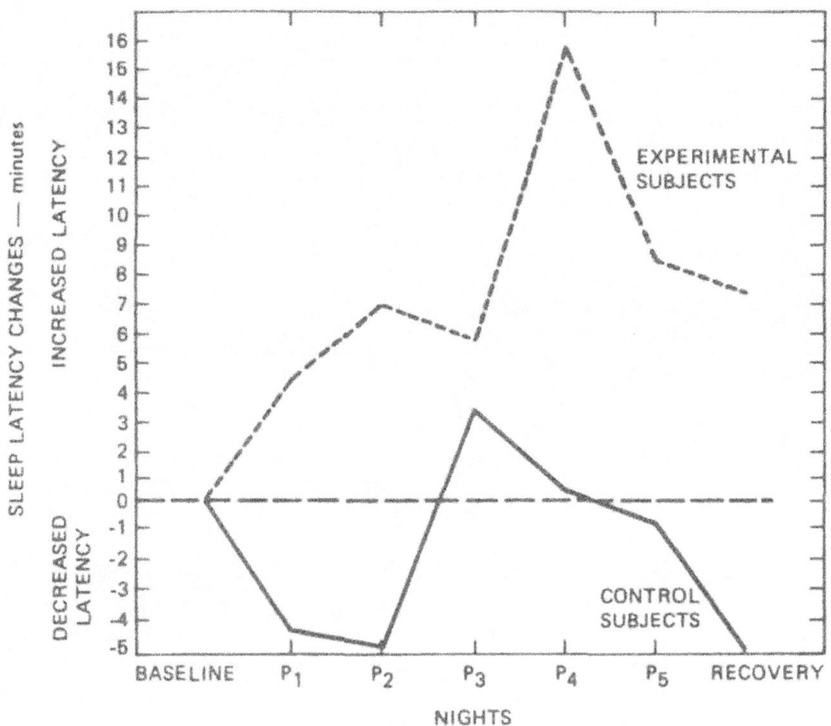

FIGURE 10.24. Changes in sleep latency during 5-day exposure to pings and during recovery compared with sleep latency during baseline. (From ref. 94.)

habituation of these autonomic-system responses to noise in awake subjects. The fact that these responses to the pings did not habituate or adapt when the subjects were asleep would seem to indicate that they are truly reflexive and are not cognitively initiated, but when awake they can be habituated or cognitively (consciously or unconsciously) controlled. The possibility also exists that the alerting response is much more measurable when the subjects are in quiescent sleep than when they are awake. Whether these alerting responses are stressful to the organism is, of course, another consideration.

In brief, from various studies of autonomic-system responses to noise when subjects are awake and asleep, it appears that the following hold:

1. There is an alerting-type reflex response to sudden changes in the sound environment which can become habituated when the subject is awake but not when asleep.

2. There is a vasoconstrictive reflex response probably associated with noises that are also effective in eliciting the aural reflex and which does not become habituated whether the subject is awake or asleep.

It is hypothesized that these two reflexive autonomic-system responses to noise are not sources of stress to the organism but represent normal physiological functionings.

Auditory discrimination during sleep

It is a common observation that one sleeps better in a familiar environment than in an unfamiliar environment containing unfamiliar sounds. Also, people can apparently be instructed to awaken to certain sounds when they are asleep and to ignore others. Oswald *et al.* (ref. 97) found that persons would awaken more readily to the sound of their own names than to other names. However, research data such as that of Johnson and Lubin (ref. 95) suggest the hypothesis that people (a) cannot learn to habituate or not to habituate to a noise when asleep unless they can engage in cognition relative to the input stimulus and (b) cannot engage in cognition unless that stimulus exceeds their threshold of cognitive arousal, that is, unless they are awake. Intuitively, it would seem that a reasonable compromise between the requirements to give the brain a rest and yet maintain some reasonable contact between the organism and the world outside would be for the organism to somehow increase the auditory cognitive arousal threshold by varying amounts during a night of sleep. This hypothesis is consistent with the studies of Emmons and Simon (refs. 98 and 99) who found that so-called "sleep learning" (information recorded on magnetic tape and played *via* an earphone under a person's pillow) only occurred when the listeners exhibited brain wave activity associated with being behaviorally awake. Sometimes the listeners were not consciously aware (particularly the next morning) of the particular times they were awake and capable of cognitive behavior and hearing.

Other studies supporting these hypotheses were conducted by Williams *et al.* (ref. 88) and by Rechtschaffen *et al.* (ref. 100). Williams *et al.* instructed subjects prior to going to sleep that if they did not awaken to a tone to which the subjects were particularly responsive when awake (called a critical stimulus) they would be aroused by a fire alarm and electric shocks to their legs. Further, the subjects were told that they would not be thus awakened if they failed to respond to another tone equally loud but at a different pitch (neutral stimulus). They were instructed not to respond to the neutral stimulus even when awake. The results are shown in figure 10.25, wherein it is shown that the arousal (or awakening) effects of the critical stimulus were much greater than those of the neutral stimulus when in the light stages of sleep but not when in the deep stages of sleep. It is also shown that the relative arousal effects of the stimuli remain relatively unchanged whether or not the failures to respond were punished, except in the REM stage. Rechtschaffen *et al.* likewise found that punishing the subject by shaking him awake when he failed to awaken to a sound did not tend to increase arousal to subsequent exposures to the sound except in the REM stage. Apparently people can be more responsive to sounds that have special meanings when asleep in the brain wave stages 1 and 2 (light stages of sleep), and people can learn this differential responsiveness only when awake or when in the REM stage.

Research data showing that a person in certain stages of sleep can discriminate among auditory stimuli in terms of their meaning are consistent with anec-

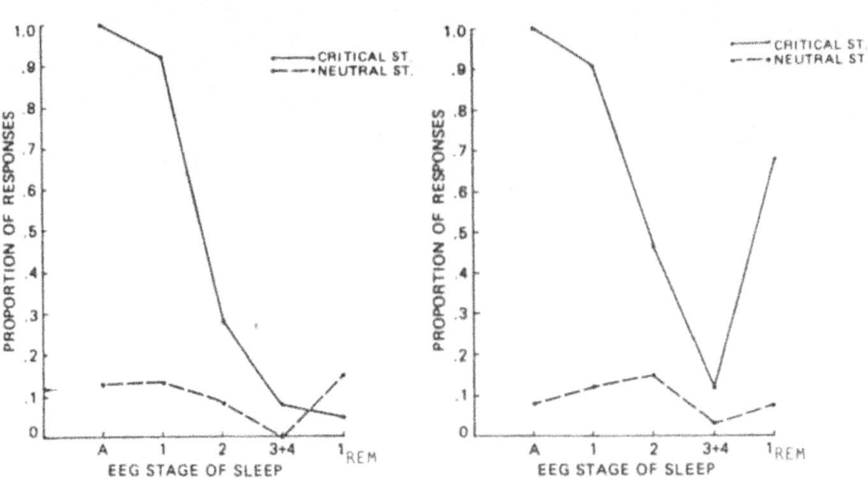

(a) Subjects not punished for failure to respond to critical stimulus.

(b) Subjects punished for failaure to respond to critical stimulus.

FIGURE 10.25. Responsiveness to a critical stimulus and to a neutral stimulus when awake (A) and when in different stages of sleep. (From ref. 88.)

dotes that one can "listen" for certain sounds when asleep and ignore others. This apparently is a form of recognition that is readily learned through previous awake exposure to a noise or to a change in the acoustic environment. Also, it is possible that a person can, to a certain extent, control their general state of arousal so that they spend more time in "light" stages of sleep than in "deep" stages, thereby increasing the probability of hearing sounds because of their lower threshold of auditory arousal.

Age and sex differences

Lukas and Kryter (ref. 101) found in laboratory tests that older persons are much more sensitive, particularly with respect to behavioral awakening, to simulated sonic booms and recorded subsonic aircraft noise than are younger persons. (See fig. 10.26(a).) Indeed, the youngest subjects (ages 7 to 8 years) were not aroused at all by sonic booms more intense than those that awakened the 67- to 72-year-old men nearly 70 percent of the time. It is possible that older people need less deep sleep and are therefore more sensitive to arousal than the younger people, even though their hearing is less acute. Roth *et al.* (ref. 102), using various noises, also found older people more easily aroused from sleep than younger ones. Griefahn and Jansen (ref. 84) derived the functions shown in figure 10.26 on the basis of a number of sleep studies in which age was a variable. The probability of a 70-year old awakening is shown to be about twice that expected for a 20-year old.

It is probable that males and females as a group are equally sensitive to being awakened from sleep by noise. In one study (presented in refs. 103 and 104), Lukas and Dobbs found females to be somewhat more easily awakened by aircraft noise, whereas Muzet *et al.* (ref. 91) found the opposite.

Noise-induced sleep

As mentioned earlier, a more or less steady level of broadband background sound can mask the distraction of intermittent sounds. It is thought by some that sound can induce sleep, either because it prevents other distractions or it serves as a monotonous, hypnotic focus for one's attention. Little experimental research seems to have been done on this phenomenon, although Olsen and Nelson (ref. 105) claim a tone of 320 to 350 Hz calms crying babies and puts them to sleep, and some devices for making soothing sounds to induce sleep are available on the commercial market.

Studies of Sleep in the Home

Aircraft noise

There are obvious reasons to be concerned about using research data on the effects of noise on sleep collected in the laboratory as the basis for a quantitative

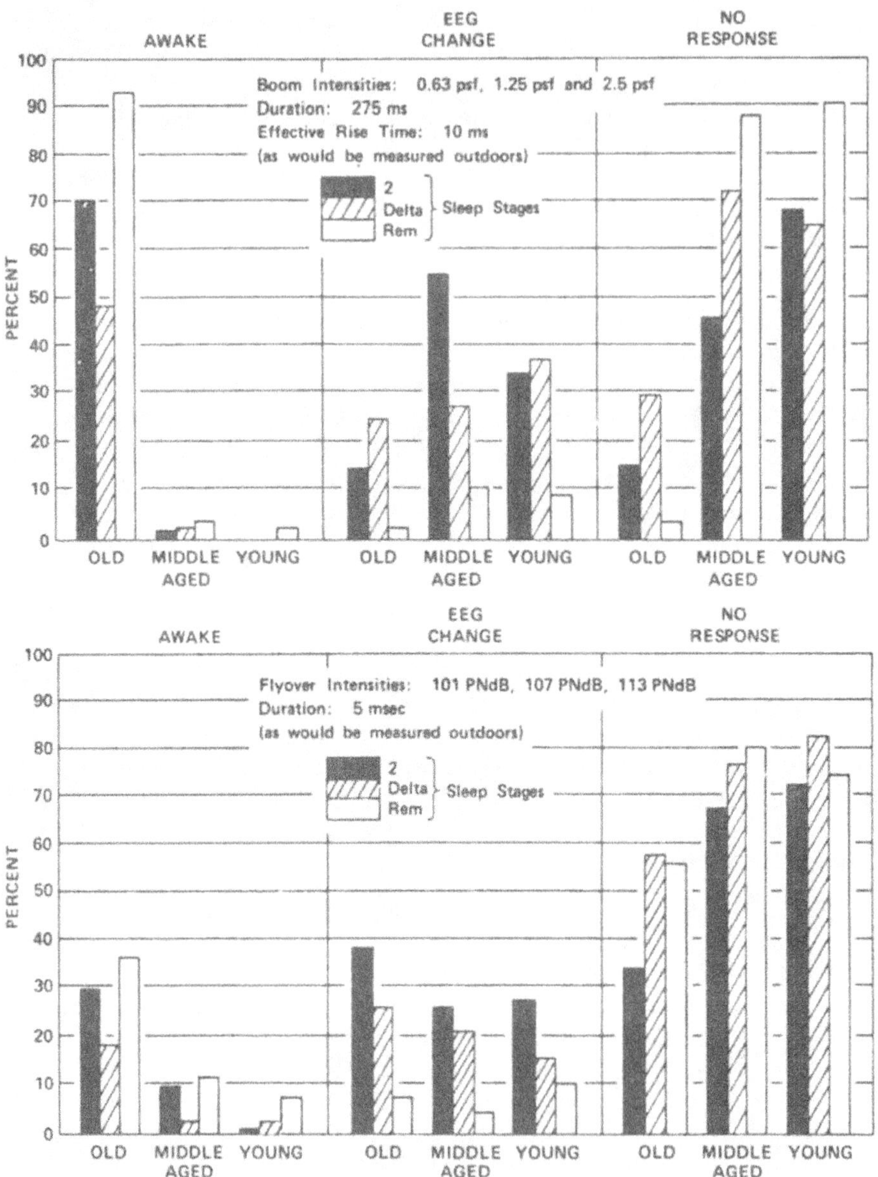

(a) Response to simulated sonic booms and recorded jet aircraft noise. (From ref. 101.)

FIGURE 10.26. Effects of noise on sleep.

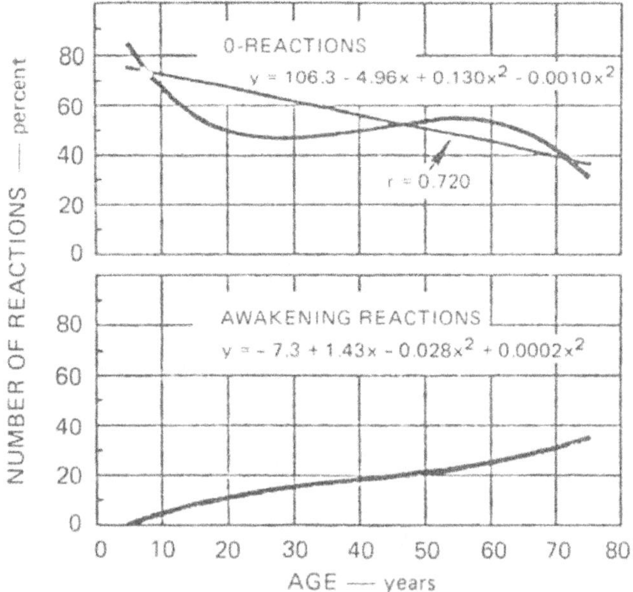

0-reactions = reactions less than a change of one sleep stage from 3 publications, 26 subjects, 368 nights, 4428 stimuli (68-85 dB(A), sonic booms and aircraft noises)

(b) Sleep disturbances by noise. (From ref. 84.)

FIGURE 10.26. Concluded.

description or prediction of what effects are to be expected in the home. One factor is that of "habituation" to the general environment. Another is the disruption of sleep caused by experimental equipment. For example, Johns and Doré (ref. 106) reported that even after 12 nights of sleeping in the laboratory subjects reported about twice as many spontaneous awakenings as occurred in their homes. The major reason was the EEG electrodes pulled on their scalps in the laboratory.

There have been several studies conducted on subjects sleeping in their own homes in which EEG's (and occasionally other electrograms) were obtained (*e.g.*, Globus *et al.* (ref. 107) and Vallet *et al.* (ref. 108)). In the Globus *et al.* study, six middle-aged couples living in the vicinity of the Los Angeles airport and five control couples living several miles from the airport in similar socioeconomic neighborhoods not exposed to aircraft noise had electrodes for EEG's and for electrooculograms (EOG's) attached to their foreheads upon retiring for five consecutive nights. EEG, EOG, and acoustic-noise signals were recorded and later analyzed. In the homes exposed to the noise from aircraft approaching the airport the mean noise level was 77 dBA compared with a mean level of 57 dBA in control homes. Some of the results of this study are shown in figure 10.27. It is

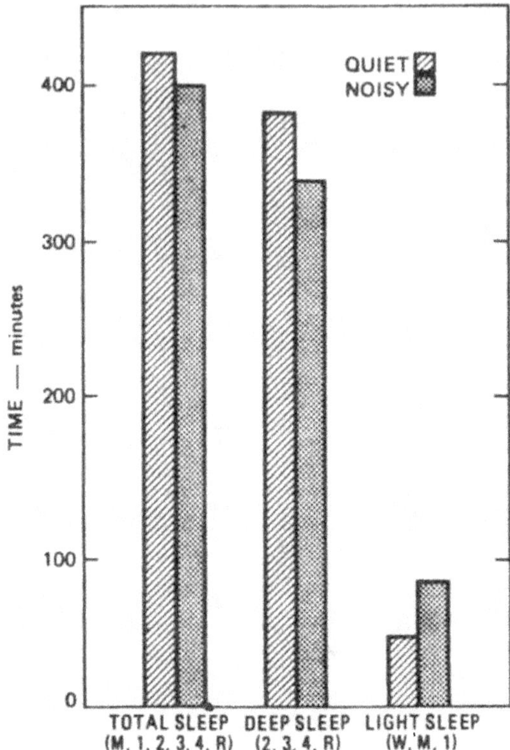

FIGURE 10.27. Average amount of sleeping time in quiet and noisy areas. (From ref. 107.)

shown that more minutes of sleep were experienced in the stages of "useful" sleep (EEG stages 2, 3, and 4) and less in "wasted" sleep (waking (W), movement (M), and stage 1) by the couples in the quiet neighborhood than in the noisier one.

Vallet *et al*. conducted a similar study near Roissy Paris Airport in which 40 men 20 to 55 years old who had been exposed for about 1 year to about 12 aircraft flyovers per night were monitored when sleeping in their homes. Figure 10.28 shows the percentage of EEG responses that indicated awakening. It is shown that for peak noise levels of about 45 to 65 dBA there is about a 20 percent probability of an awakening occurring. Above 65 dBA, the results of Vallet *et al*. show some decline in awakenings, whereas data collected in the laboratory as well as the results of attitude surveys show awakenings increase as the noise level increases, as might be expected.

The Directorate of Operational Research and Analysis (DORA) aircraft noise study (ref. 109) is a particularly important study of sleep disturbance from aircraft noise. Some 4400 people were administered an extensive questionnaire

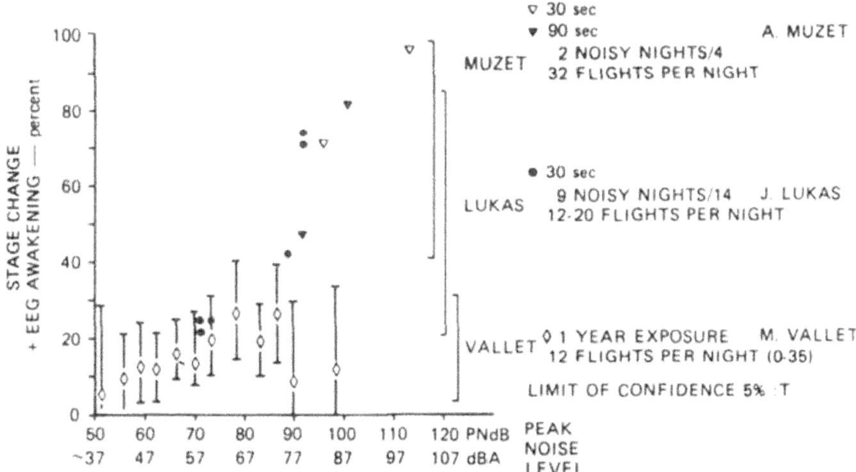

FIGURE 10.28. Comparison of laboratory (Muzet and Lukas) and field studies percent of subjects aroused by noise. (From ref. 108.)

about sleep disturbance over a 4-month period (June to September, 1979) in areas near London's Heathrow and Gatwick Airports.

At the Heathrow and Gatwick Airports, commercial aircraft that create levels of noise above about 90 dBA in residential areas are not allowed to operate from 11:30 p.m. to 6:00 a.m., as shown in figure. 10.29. Also shown in figure 10.29 are the percentages of people going to bed and arising at particular times. Although these temporal patterns may be the same in other residential areas, they may be somewhat influenced by the reduced aircraft noise during the hours of 11:30 p.m. to 6:00 a.m.

The extent to which double glazing is present in the houses involved in this study is variable (apparently from about 20 to 80 percent in different areas). The extent to which bedroom windows were generally kept closed because of aircraft noise and kept closed on a designated night during the period of the study is shown in figure 10.30. Some of the variability in the sleep disturbance data shown below is due to variability in the amount of aircraft noise reaching the beds of individual respondents. (Physical factors affecting noise intrusiveness and attitude survey data are discussed more fully in chapter 11.)

By applying correlation techniques to the data, relations were determined between various physical measures of the aircraft noise and responses to the following three major questions of the survey: (1) percentages of people (respondents) awoken (2) percentages of people having difficulty getting to sleep, and (3) percentages of people feeling "tired" or "very tired." It was found that an energy type of measure (*i.e.*, A-weighted L_{eq}) of aircraft noise correlated best with sleep disturbance from aircraft noise, although some improvement in correlations was

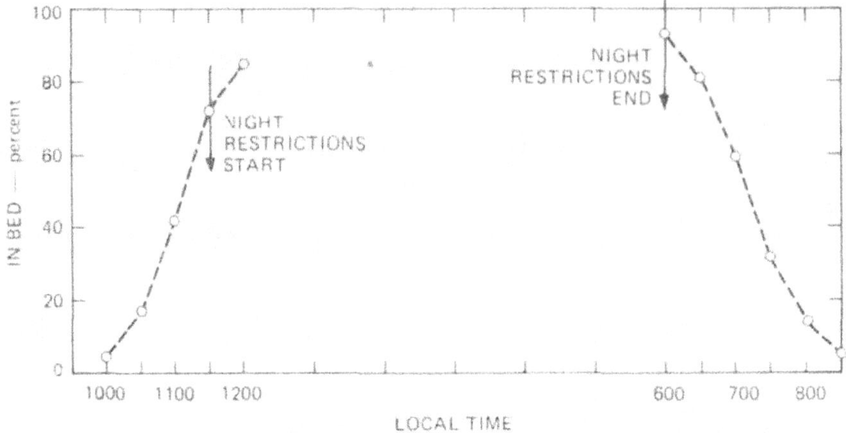

FIGURE 10.29. Times of going to bed and getting up. Noisiest aircraft restricted from operations *during times indicated. (From ref. 109.)*

achieved when information about maximum A-weighted noise levels was included. In support of the appropriateness of the L_{eq} measure of the aircraft noise is the finding in this study that some respondents could not accurately recall the time association between a specific flyover noise event and an arousal from sleep. Although such recollections can be expected to be vague, this finding also suggests the possibility that the energy of the noise events over time increased the general state of arousability from sleep.

The respondents were asked to recollect arousals from sleep and difficulties getting to sleep as a matter of general experience over the past several months and on a recent designated night. Shown in figure 10.31 are percentages of the respondents having difficulty in getting to sleep when living in various amounts of L_{eq} of aircraft noise for the hours 10 to 12 p.m. It is shown that at an L_{eq} of 60 dB or so, about 5 percent of the respondents on the designated night and about 25 percent as a matter of general experience had difficulty in going to sleep because of aircraft noise.

In figure 10.32, the percentages of people awoken by aircraft noise from general experience and on the designated night are shown as functions of L_{eq} from 11 p.m. to 7 a.m. Also shown in figure 10.32 are the percentages of people awoken for all reasons. It is shown in the figure that at an L_{eq} of about 55 dB, about 12 percent of the respondents were awoken by aircraft noise on the designated night and about 25 percent were awoken by aircraft noise more than once per week as a matter of general experience. Unfortunately, the number of sleep arousals per night were not reported. It seems likely that the relative amounts of awakenings

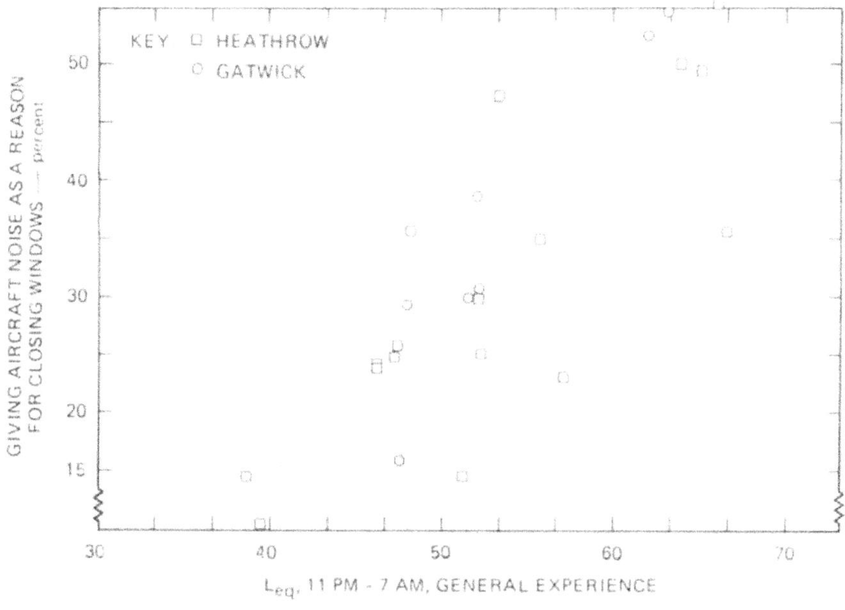

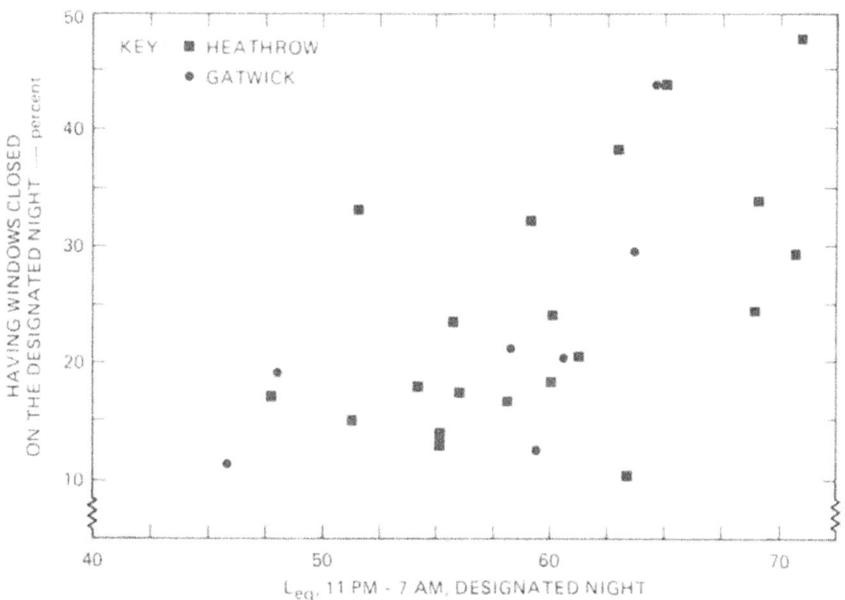

FIGURE 10.30. Respondents giving aircraft noise as a reason for closed windows and sleeping with bedroom windows closed on designated night as functions of L_{eq} from 11 p.m. to 7 a.m. (From ref. 109.)

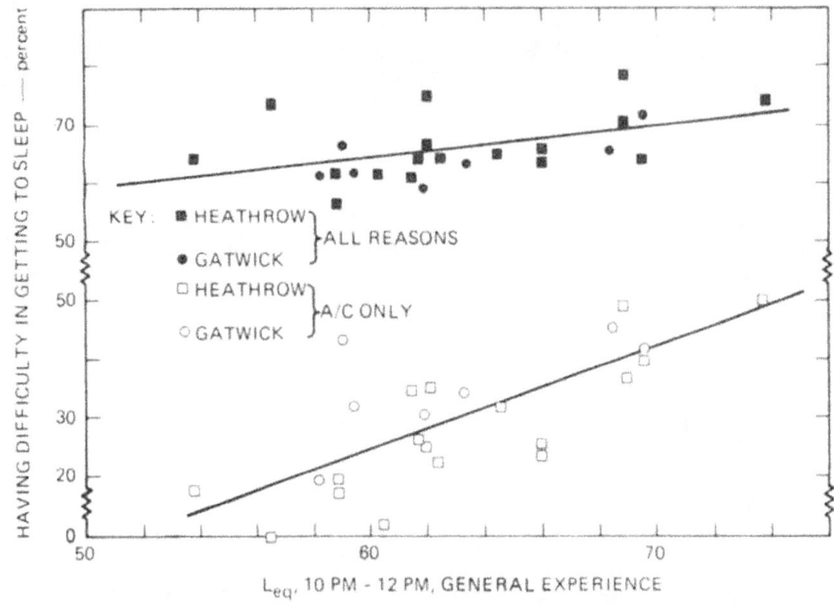

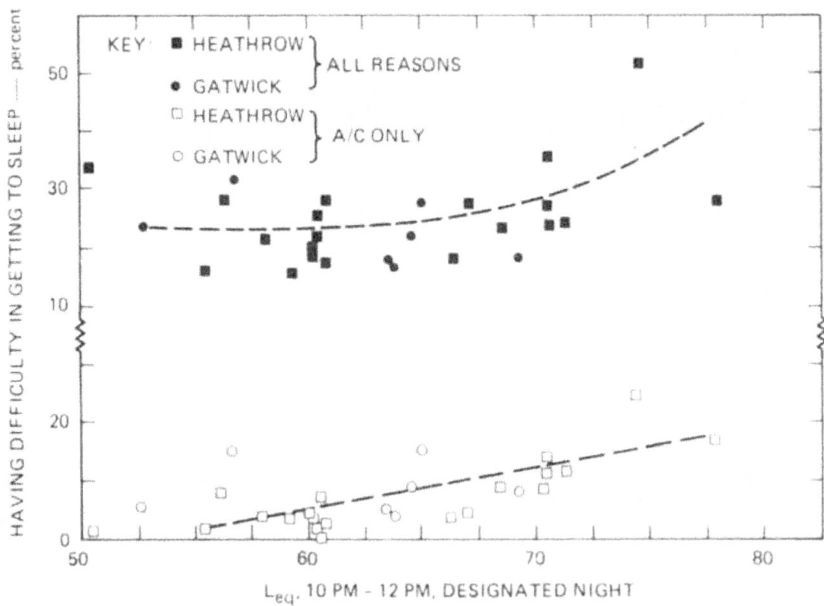

FIGURE 10.31. Respondents having difficulty in getting to sleep according to general experience and on a designated night as a function of L_{eq} from 10 to 12 p.m. Dashed trend curves added for this report. (From ref. 109.)

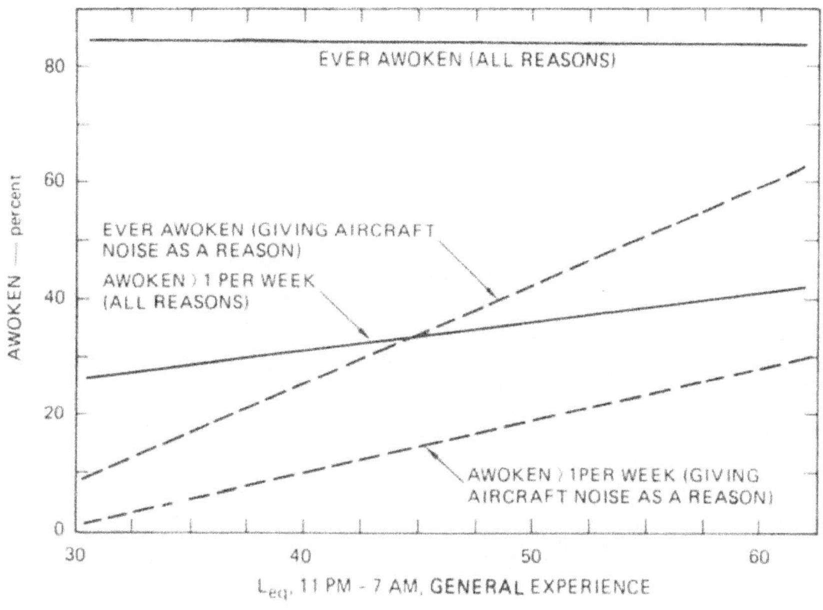

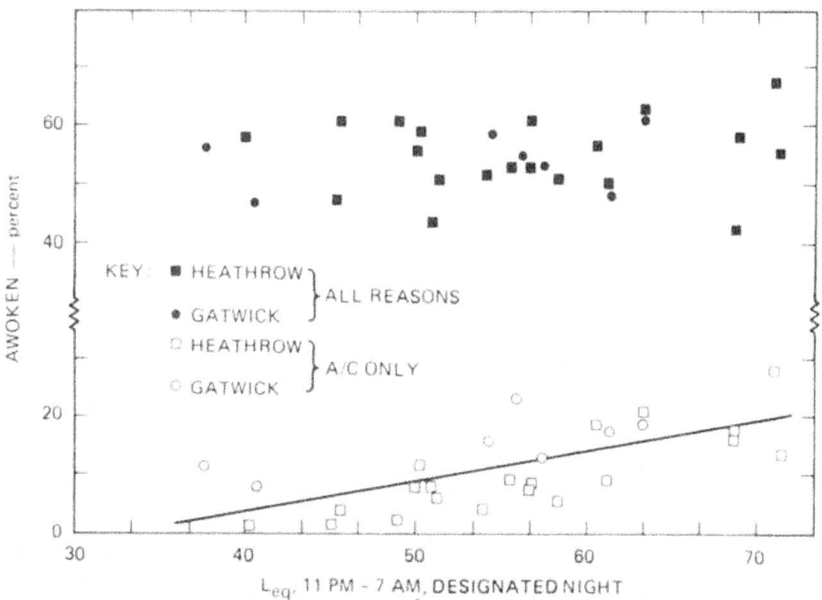

FIGURE 10.32. Respondents awoken according to general experience and on a designated night as a function of L_{eq} from 11 p.m. to 7 a.m. (From ref. 109.)

because of aircraft noise compared with the number of awakenings for all reasons would increase if the actual numbers in a shorter time frame, such as the number per night, were counted. In the DORA report a person awoken "more than once per week" was counted the same whether awoken by aircraft noise three times per night or two times per week. Although the sleep disturbances for other reasons were scored the same way, it seems likely that the potential for multiple disturbances per night would be greater because of aircraft noise than for other causes.

An implicit (and possibly unjustified) assumption in the way the data are analyzed in the DORA report is that sleep disturbances, no matter what the cause, are to be considered equal as causes of annoyance. For example, in reference 109 it is concluded that until noise levels of around L_{eq} = 65 dBA are reached aircraft noise does not result in extra disturbance, or that at least people can adjust to levels below about 65 dBA. However, sleep disturbance for reasons of health or going to the toilet (found in the study to be one of the major causes) would not necessarily reduce the annoyance to be felt from, or increase the "adjustment" to, disturbance the same night because of aircraft noise. Indeed, the annoyance of aircraft noise could even be enhanced with the presence of the additional sleep disturbances from other causes. That these implicit assumptions in the evaluation of sleep disturbance from aircraft noise in the DORA report are questionable is supported by data for the judged annoyance of aircraft noise. Figure 10.33 shows that at an L_{eq} of 60 dB about 35 percent of the respondents were "very much" or "quite a lot" bothered by the aircraft noise, both before going to bed (10 to 12 p.m.) or when in bed during the night (11 p.m. to 7 a.m.), hardly indicative of "adjustment" to L_{eq} below 65 dB.

Figure 10.34 shows the data obtained in response to the question of "tiredness" plotted as a function of L_{eq} for 11 p.m. to 7 a.m. On the basis of the solid and dashed curves, the DORA report concluded that tiredness from sleep disturbance from aircraft noise does not exceed the amount generally found until the L_{eq} goes above about 65 dB. Although the relations between sleep disturbance from aircraft noise, "tiredness" (as measured in the DORA report), and general health are hardly known, the results in figure 10.34 are not inconsistent with research findings presented later regarding relations between exposure to aircraft noise and health disorders.

In the DORA report, the Civil Aviation Authority posed and answered two "Policy-related questions" as follows (pp. 28 and 29 of ref. 109):

Question: What is the level of aircraft noise which will disturb a sleeping person?

Answer: The study indicates that there is a discernible increase in disturbance for both general experience and designated night results when night-time noise *exposure* is about 65 L_{eq}. For the number of aircraft operating at present at night around Heathrow and Gatwick noise exposure at this level can only be ob-

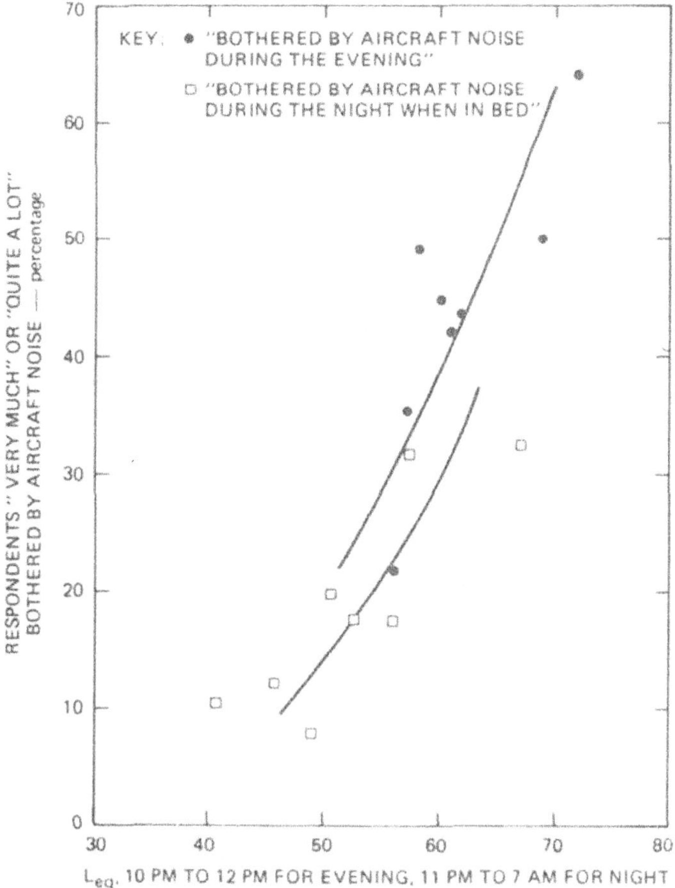

FIGURE 10.33. Respondents bothered by aircraft noise as function of measured L_{eq} for specified time periods. (Data from fig. 12 and tables 6c and 6d of ref. 109.)

tained when some aircraft each night produce noise levels in excess of around 90 dBA on the ground (90 dBA = 103 PNdB approximately). . . .

Question: What level of aircraft noise prevents people from getting to sleep?

Answer: The proportion of people who report difficulty in getting to sleep for all reasons shows a discernible increase for aircraft noise exposure for sites at about 70 L_{eq}, (measured between 1000 and 1200) although aircraft noise is increasingly quoted as a cause of difficulty as aircraft L_{eq} increases.

For a multitude of physical, psychological, and social factors, not to mention the complexity of sleep and related behavior, it is well recognized that it is

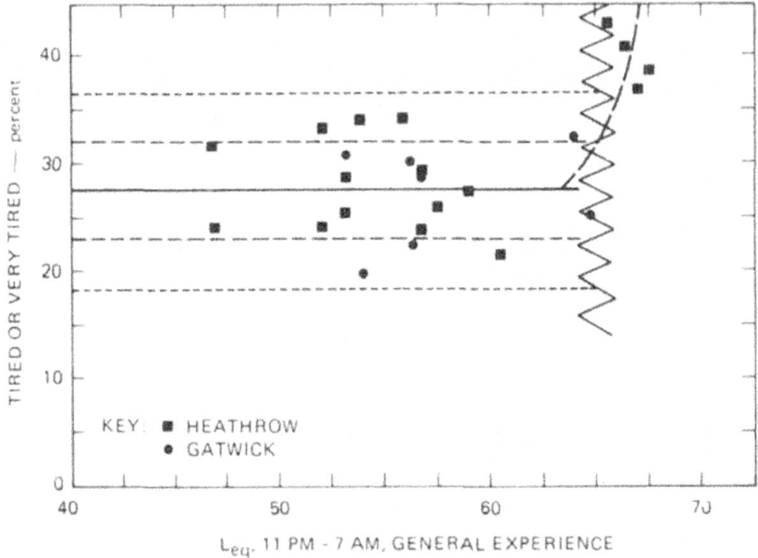

FIGURE 10.34. Respondents "tired" or "very tired" according to general exprience as a function of L_{eq} from 11 p.m. to 7 a.m. (From ref. 109.)

the trends and not the individual data points of such survey data that are most meaningful. In view of the trend curves in figures 10.31 and 10.32, it seems that the aircraft noise levels cited above of an L_{eq} of 65 dB being required for the start of a discernible increase in disturbing sleep and of an L_{eq} of 70 dB for the start of a discernible increase in causing difficulty in going to sleep are about 10 dB too high. These values of L_{eq}, when translated to equivalent L_{dn} or to maximum A-weighted levels for representative flight operations, are also about 10 dB too high to be consistent with other data, such as that shown in figure 10.28 and relevant data presented below in this chapter and in chapter 11.

Fidell and Jones (ref. 110) reported on an attitude survey carried out before and after a 1-month period (May 1973) during which normal nighttime (11 p.m. to 6 a.m.) aircraft landings over a low-economic residential area were suspended. It was found that this reduction in nighttime landings (50 out of an average of 687 landings per 24 hours) did not affect the amount of annoyance measured by the attitude surveys before and after the 1 month of suspended operations. The results were explained as follows: (1) the decrease in the 24-hour exposure level as a result of landing reductions is relatively small; and (2) a 1-month period is not sufficiently long for the people to integrate the annoyance reactions, especially in that there are always periods when nighttime operations are reduced because of such things as weather conditions. Another explanation could be that the "before" annoyance survey was conducted in mid-April, so the sleep-arousal attitudes were probably based on February and March when the weather conditions likely resulted in closed windows, which would tend to reduce the relative amount of noise experienced and thus reduce the overall annoyance experienced.

Non-aircraft noise

A similar relation of increasing disturbances to sleep with an increase in noise-exposure level found with aircraft (figs. 10.31 and 10.32) was also found with road traffic noise, as shown in figure 10.35 (from ref. 111). However, because exposure level for the road traffic noise is expressed as A-weighted level exceeded 10 percent of the time during a 24-hour day, the results cannot be directly compared, numerically, with the L_{eq} levels for the aircraft noise in figures 10.31 and 10.32.

The methodologies used by Horonjeff *et al.* (ref. 112) for studying sleep arousal in the home from non-aircraft noises would seem to avoid some of the problems of previous in-the-home and laboratory studies. In this study 14 subjects slept with a small switch, or response button, on a stand beside their beds and were instructed to push the button whenever they were awakened for any reason. The experimenter was able to present noise into the bedroom over a small loudspeaker and to record the awakenings and noise level in the room from a microphone placed near the bed. Some of their results are shown in figures 10.36(a) and (b).

It is shown in figure 10.36(a) that the different noises had reasonably similar probabilities of awakening as a function of the level of steady-state noise (15-minute durations), but for transient presentations (the noises had a growth and decay rate in level of 2 dB/sec) the data for the different noises show somewhat more divergence. However, figure 10.36(b) shows that when the data for the steady-state conditions for each noise are compared with the data for their tran-

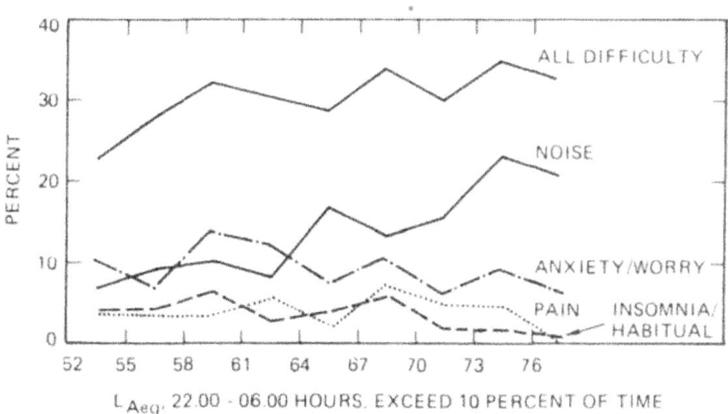

FIGURE 10.35. Reasons given for sleep disturbance compared with level of road traffic noise. (From ref. 111.)

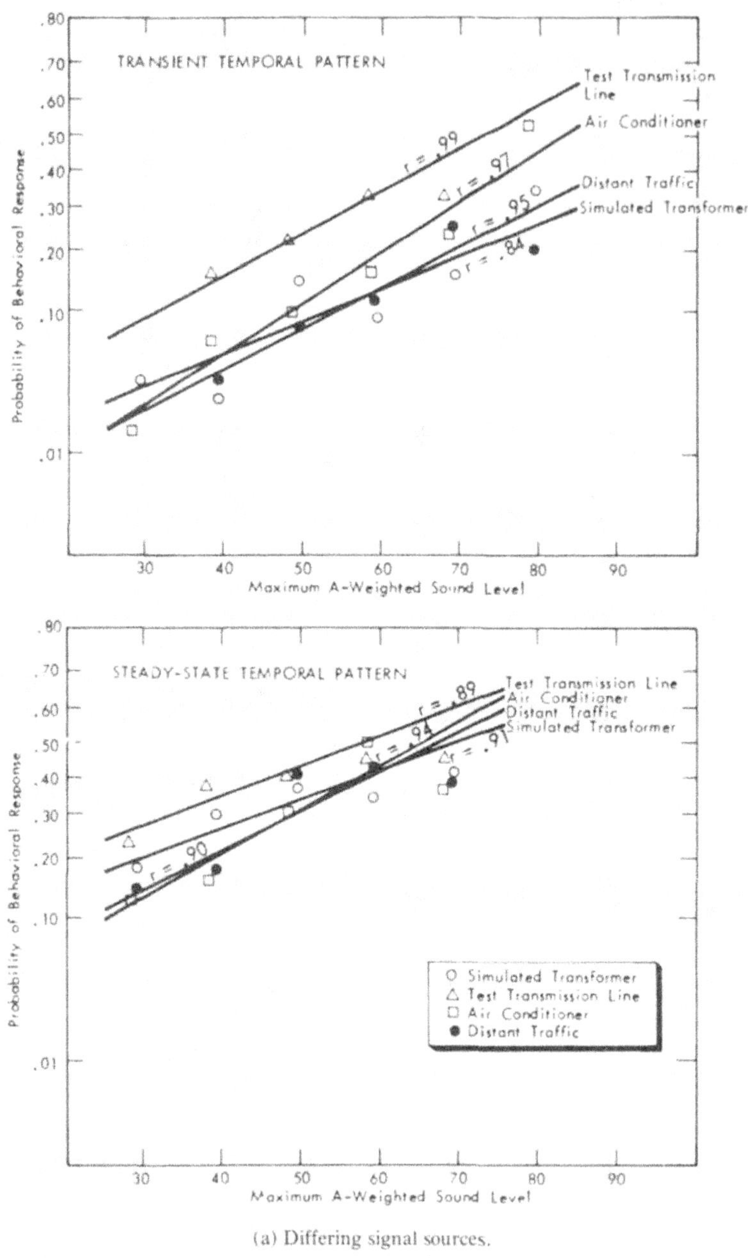

(a) Differing signal sources.

FIGURE 10.36. Observed psychometric functions showing effects on a behavioral awake response. (Data from ref. 112.)

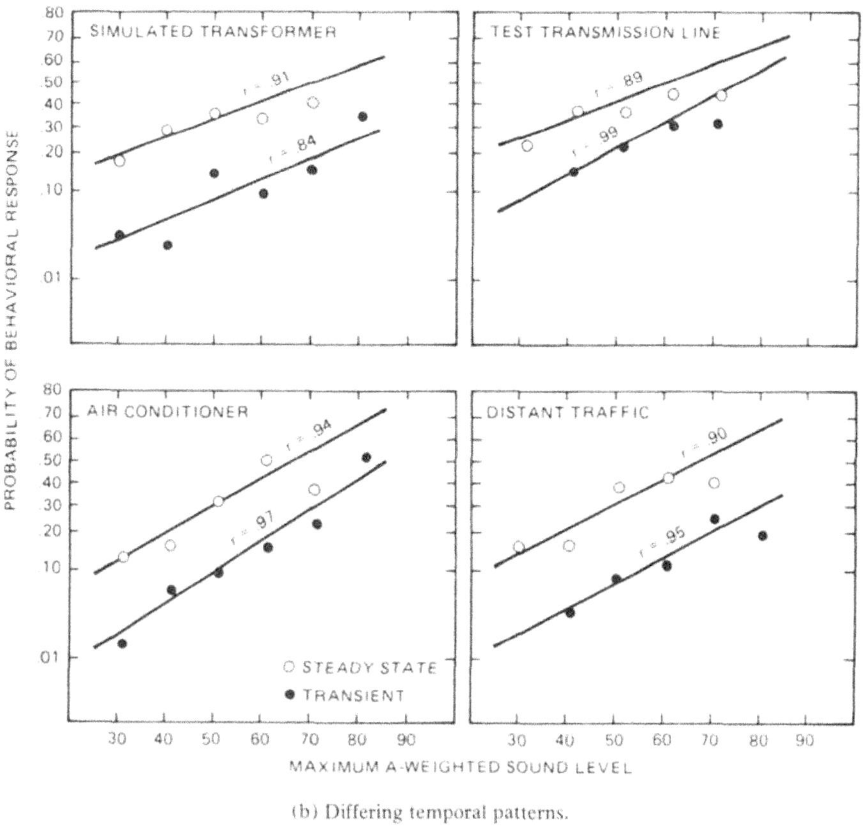

(b) Differing temporal patterns.

FIGURE 10.36. Concluded.

sient presentations at the same maximum A-weighted sound level, the transient presentations had much lower probabilities of causing awakening.

Comparison of home and laboratory sleep data

Steady-state noises have more energy than transient noises of equal maximum sound level and spectrum. Accordingly, the greater arousal effect of steady-state noise compared with transient noise shown in figure 10.36 suggests that a measure of the energy in an event, such as EPNL or SENEL, would be a more accurate index to the sleep-arousal effects than the maximum levels reached during the event. Horonjeff *et al.* converted their noises into EPNL in decibels and Lukas (ref. 89) calculated EPNL in decibels for a number of noises used in laboratory research, as shown in figure 10.37. (See also fig. 10.23.) Also plotted in figure 10.37 are some data from Thiessen (ref. 85) and suggested trend curves for

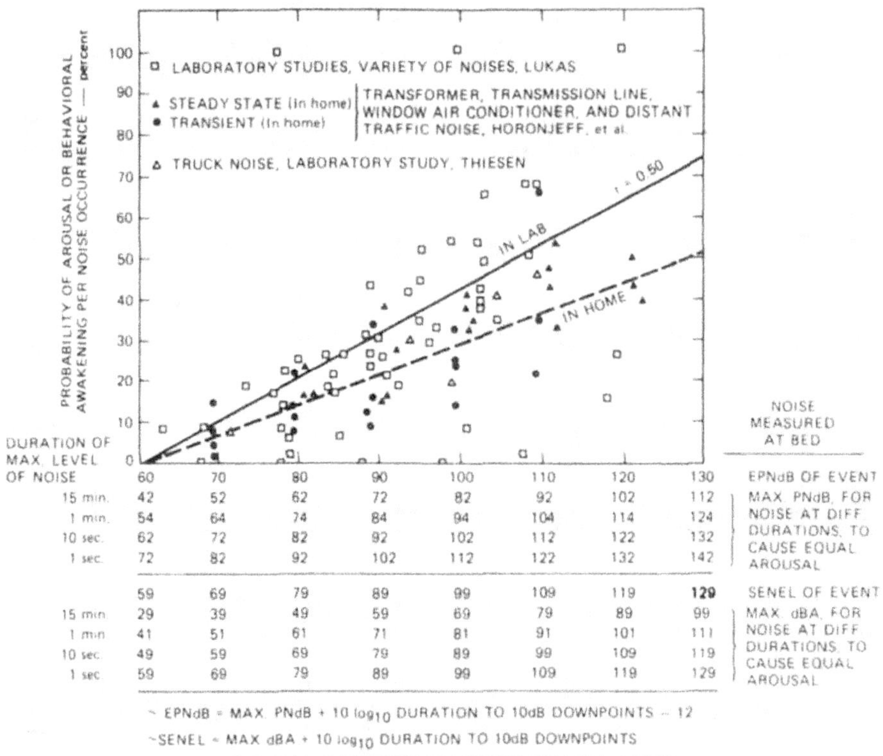

DURATION OF MAX. LEVEL OF NOISE									
	60	70	80	90	100	110	120	130	EPNdB OF EVENT
15 min.	42	52	62	72	82	92	102	112	MAX. PNdB, FOR NOISE AT DIFF. DURATIONS, TO CAUSE EQUAL AROUSAL
1 min.	54	64	74	84	94	104	114	124	
10 sec.	62	72	82	92	102	112	122	132	
1 sec.	72	82	92	102	112	122	132	142	
	59	69	79	89	99	109	119	**129**	SENEL OF EVENT
15 min.	29	39	49	59	69	79	89	99	MAX. dBA, FOR NOISE AT DIFF. DURATIONS, TO CAUSE EQUAL AROUSAL
1 min.	41	51	61	71	81	91	101	111	
10 sec.	49	59	69	79	89	99	109	119	
1 sec.	59	69	79	89	99	109	119	129	

~ EPNdB = MAX. PNdB + 10 $\log_{10}$ DURATION TO 10dB DOWNPOINTS − 12

~SENEL = MAX dBA + 10 $\log_{10}$ DURATION TO 10dB DOWNPOINTS

(DURATION IS FOR MORE OR LESS STEADY LEVEL NOISES. FOR AIRCRAFT OR OTHER "TRIANGULAR" SHAPED NOISES DURATION-LEVELS, DIVIDE THEIR DURATION BY 2 FOR MAKING CALCULATION)

FIGURE 10.37. Comparison of probabilities of arousal as found in a number of laboratory and in home studies. For most of the studies, about 10 noises were presented per night. (Data from Lukas (ref. 89), Horonjeff *et al.* (ref. 112), and Thiessen (ref. 85).)

the laboratory data and for the home data showing probabilities of awakening as a function of EPNL and SENEL. Also, there are some representative values of maximum PNL in decibels or maximum A-weighted SPL in decibels of durations that presumably cause equal arousal.

It is not possible to completely generalize the relations between PNL and A-weighted SPL because of somewhat different spectral weightings given by these two noise measurements; it is also not possible to generalize the relations between a peak level and an energy-exposure counterpart (max PNL vs EPNL or max A-weighted SPL vs SENEL) because of durational differences among the variety of noises involved in figure 10.37. Even so, the trend curves as drawn probably represent the best generalizations about the effects of noise on sleep that can be expected in view of the complexity of the sleep behavior and physiological and psychological factors involved. The difference between the results for laboratory and home sleep studies (somewhat less arousal in the homes for about equal levels of exposure to noise) is probably to be expected because of greater familiarity of the subjects with their home environment.

As noted, Horonjeff *et al.* used steady-state noise of a 15-minute duration. Whether the energy calculations of EPNL or SENEL could be extended, for equal arousal, to noises of longer duration has not been tested. (Nearly all the noises in the aforementioned studies summarized by Lukas were of shorter duration.) As shown in figure 10.23, when impulsive noises such as sonic booms are expressed in terms of their energy, these effects on sleep appear to be comparable to the effects of longer duration noises.

Sleep and health

Richter (ref. 113) observed that a sleeping subject exhibited EEG and vasoconstrictive reactions about every 30 sec because of cars, trains, and motorcycles passing the test room, even though the person slept quietly and upon awakening had no recollection of any disturbances. Richter viewed these responses as indicating that gastrointestinal activities, which are controlled by the autonomic system, are withdrawn from the recovery process of sleep and further believes this withdrawal is detrimental to normal health.

There is no question that people become irritable when deprived of sleep and may show some irrational behavior (ref. 114). However, it is possible that the autonomic-system responses observed by Richter are not indicative of any sleep deprivation to the higher nervous systems and to skeletal muscles. It might also be presumed that activity of the autonomic system is required more or less continuously 24 hours per day, and the small reflexive responses of the system noted by Richter are not likely to be harmful.

As discussed earlier, Johnson *et al.* (ref. 94) found that some increase in the time required to go to sleep was the only significant effect on the sleep behavior of young men exposed to as much as 55 days, 24 hours per day, of intense impulsive "pings" about 1 minute apart. Research findings of Williams *et al.* (ref. 88) suggest that people might develop sleep patterns that would provide some protection for physiological health. It is shown in figures 10.38 and 10.39 that when subjects were deprived of sleep they spent more time in the stages of sleep identified as deep and resistant to arousal than when not deprived. However, as shown by data collected in the home from people exposed to regularly occurring noises in a community (discussed in chapter 11) as well as in the laboratory and field studies of sleep just discussed, noise above certain levels of intensity can delay the onset of sleep and will continue to awaken people, causing feelings of annoyance even after long periods of repeated exposures to the noise. It is surmised that the physiological stress responses are more likely to be autonomic-system responses to the annoyance felt because of arousal rather than the autonomic-system reflexive responses to the noise that may occur prior to or during behavioral awakening.

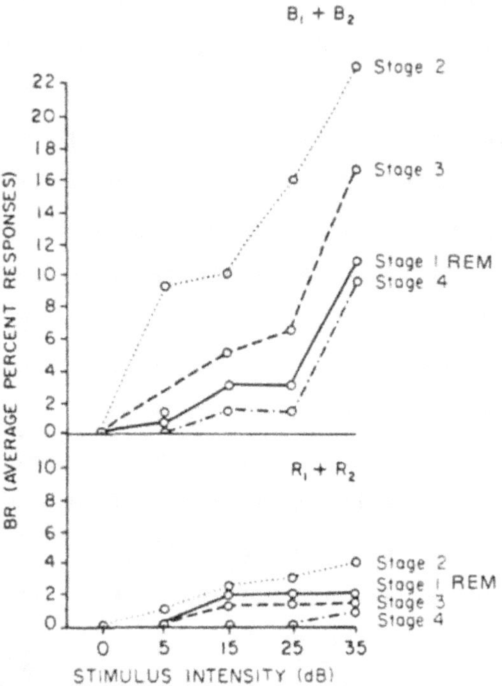

FIGURE 10.38. Effect of stimulus intensity on behavioral response (BR) during four stages of sleep for baseline nights B_1 and B_2 and recovery nights R_1 and R_2 after 64 hours of sleep deprivation. (From ref. 88.)

Effects on Other Senses and Organs

Pain

Gardner and Licklider (ref. 115) developed a device that permitted a dental patient to listen, *via* earphones, to stereophonic music or filtered random noise. The typical procedure was for the patient to relax by listening to the music and to switch to the noise when he felt any dental pain, increasing the intensity of the noise as necessary to "kill" the pain. The dosage of noise had to be controlled so as not to cause undue fatigue or stress to the ear (refs. 116 and 117).

Laboratory experiments conducted by the inventors of the device and by others (see refs. 118 to 122) revealed that pain from things such as heat and cold applied to the hand, electrical stimulation ("tingle" threshold) applied to the teeth, and pressure applied to the arm were not suppressed by the presence of intense random noise, although a reduction in the sensation level of deep muscle pressure was found. Some observations and tests have reportedly been made of

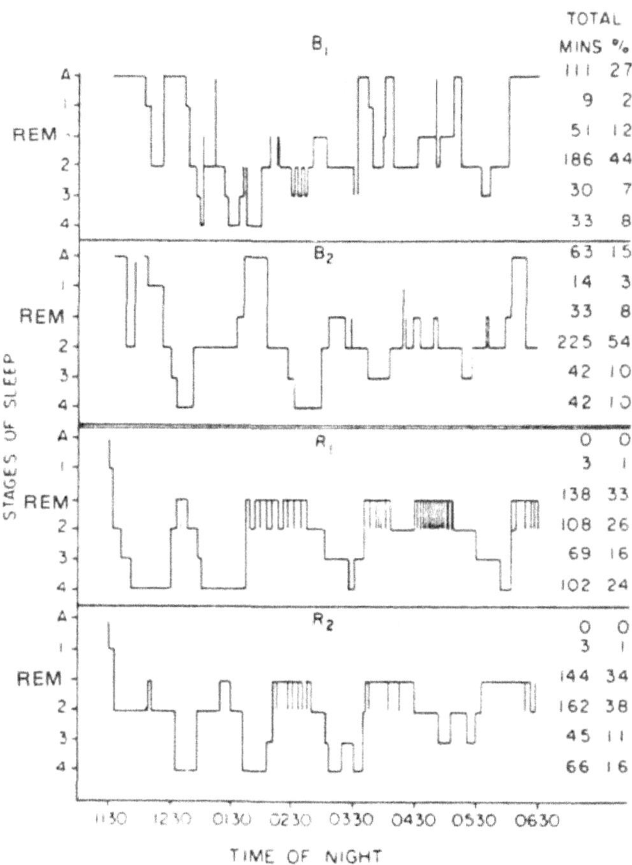

FIGURE 10.39. Effect of sleep loss on the distribution of EEG stages of sleep for one subject. B_1 and B_2 are baseline nights when subject is not deprived of sleep and R_1 and R_2 are nights after sleep deprivation of 64 hours. (From ref. 88.)

possible pain suppression with noise in people during childbirth, surgical operations (other than dental), and diseased conditions with mixed success.

It is established in reference 123 that hypnotic suggestion can be effective in suppressing pain in some people. However, the psychological and neurological mechanisms are not well understood. Also, the effectiveness of morphine in the relief of postoperative pain is statistically about the same as the apparent effectiveness of audio analgesia in clinical dentistry; a third of the patients get relief from morphine that is greater than relief from a placebo, about a third get as much relief from a placebo as from the morphine, and about a third get no relief from either the morphine or a placebo (Beecher (ref. 124)). The results from laboratory experiments on audio analgesia notwithstanding, it is not justifiable to conclude that, in the clinical situation, the audio analgesia may not involve physiological as well as psychological mechanisms in the suppression of pain during dentistry. Also, Beecher has pointed out that morphine and other analgesic agents that do not give consistent suppression of pain at its threshold do provide consistent relief at suprathreshold levels.

It can be concluded that when audio analgesia is effective, one or more of the following explanations are valid:

1. Suggestion, enhanced by attention to the music or noise, gives distraction from any pain.

2. In certain cases no pain would actually be felt except that due to the anxiety of the patient, and the music or noise relaxes the patient and reduces anxiety.

3. Neural impulses from the auditory system preempt, to some extent, the activity of centers in the reticular formation of the brain stem that are involved in the processing of pain impulses to the higher nerve centers. The neurological evidence and theory for this type of activity is meager but not without some substance.

Cutaneous sensations from the ear

Ades *et al.* (ref. 125) and Plutchik (ref. 126) have evoked sensations, other than auditory, from deaf ears exposed to intense noise. They obtained the results for "feeling," "tickle," and "pain" shown in figure 10.40. Figure 10.40 is from an experiment conducted by Ades *et al.* in which persons who were totally deaf were exposed to very intense tones and noise. Because of their deafness, the subjects could be exposed to levels that would be harmful to the normal ear. Also, Plutchik reported that 3 pulses per second were rated as more unpleasant than slower or faster rates.

Ades *et al.* found that persons without eardrums reported no pain sensations with levels up to 170 dB. It seems likely that the discomfort or pain thresholds reported for the deaf and normal ear are attributable to stimulation of the eardrum or some of the middle ear receptors.

Vestibular system

Connected to the cochlea of the inner ear are the sacculus, the utricle, and the semicircular canals. These structures, called the vestibular system, share certain fluids with the cochlea, and their innervations are closely connected. (See fig. 3.1.) This vestibular system is involved in maintaining body balance and orientation in space. When stimulated in certain ways, a person may lose their sense of balance, they may become dizzy, their eyes may show nystagmus movements (a fast movement back and forth of the eyeballs), and, under extreme conditions, they may become nauseated.

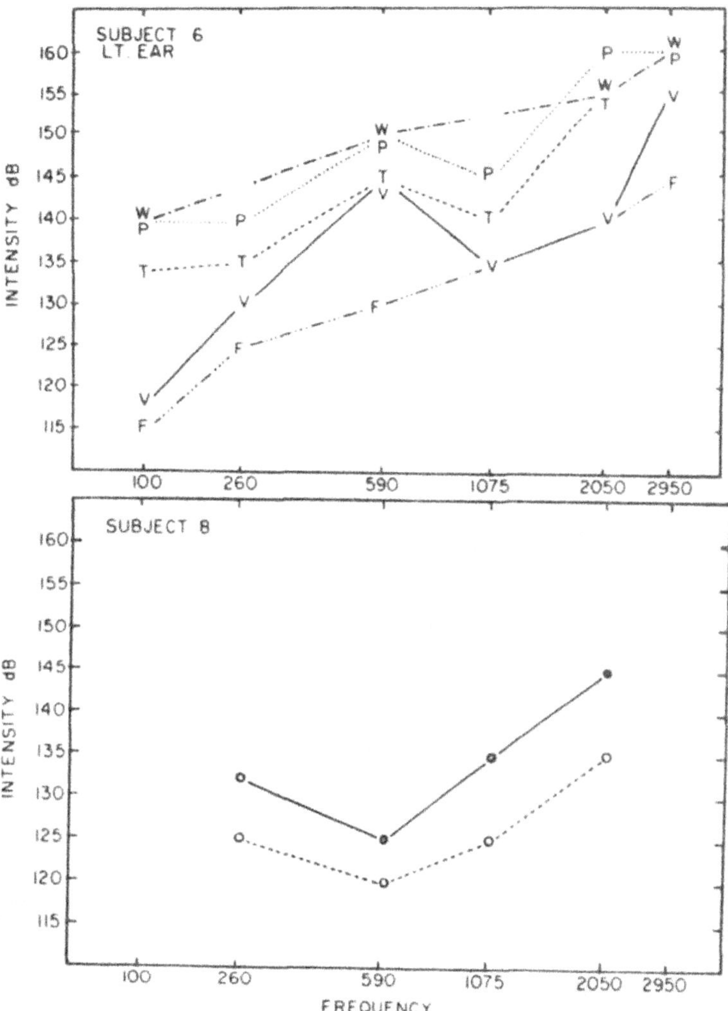

FIGURE 10.40. Threshold curves for vibration (V), tickle (T), pain (P), warmth (W), and feeling (F) for the left ear of subject 6, and threshold curves for small eye movements (broken line) and marked nystagmus (solid line) for subject 8. (From ref. 125.)

Because of their close proximity and fluidic connections, it is not surprising to find that intense sounds affect both the cochlea and the vestibular system. Figure 10.40 shows the intensity required for various tones to reach the threshold of nystagmus in one subject. Dickson and Chadwick (ref. 127) report that for jet aircraft noise over 140 dB or so, a person may feel a sense of disturbance in their equilibrium. Roggevsen and Van Dishoeck (ref. 128) note that in persons who experience vestibular reactions to sounds of relatively weak intensity (considerably less than those shown in fig. 10.40), there are usually lesions present in the bony walls of the vestibular system.

If intense noise can cause a general physiological overarousal, then it is possible that this could affect the ability of a person to maintain balance when standing on a narrow rail. Harris and Von Gierke (ref. 129) exposed subjects wearing earmuffs to white noise at levels of 120, 130, and 140 dB while standing balanced on a rail about 1-1/4 in. wide with eyes open and on a rail about 2-1/4 in. wide with eyes closed. They found that with both ears exposed to the same level of noise, only at 140 dB was there any impairment to balance performance. At the lower levels there was a reduction in balance performance when the noise was louder in one ear than in the other.

Harris (ref. 130) found that an intermittent 1000-Hz tone at a level of 105 dB presented monaurally to an unprotected ear impaired the ability to balance on the rails described above compared with performance with the tone at 65 dB. However, Vanderhei and Loeb (ref. 131), in a later repetition of the experiment, found no effect of the 105-dB, 1000-Hz, monaurally presented tone on the same balance test. These investigators concluded that noise and sound under general field conditions are unlikely to affect equilibrium.

All these results are interpreted as indicating that such impairment as occurred was probably because of a direct effect of the noise on the vestibular system and is not indicative of a general arousal effect by intense noise.

Kinesthesia

Stimulation of the receptors in the muscles and joints of the body by vibrations imposed primarily through contact of the body with physical structures can create sensations of discomfort in the body. Thresholds for these perceptions from body vibrations at different frequencies are shown in figure 10.41 (ref. 132).

Simultaneous exposure to body vibrations and intense airborne noise is found in a number of types of transportation vehicles, but especially in aircraft. Dempsey *et al.* (ref. 133) have conducted experiments that show the contributions of vibration and noise separately and together to subjective discomfort. They developed a model (see fig. 10.42) for the contribution of noise and vibration that shows the trade-offs available in the design or operation of a vehicle so as not to exceed a given level of ride discomfort. Interestingly, the magnitude of the noise-discomfort component is dependent upon the level of vibration present in the combined environment.

Vision

Noise has been thought to influence visual acuity and field, color vision, and critical flicker frequency (CFF). The last phenomenon refers to the fact that alternating dark and light visual fields will become blurred (cease to flicker) at some frequency of alternation.

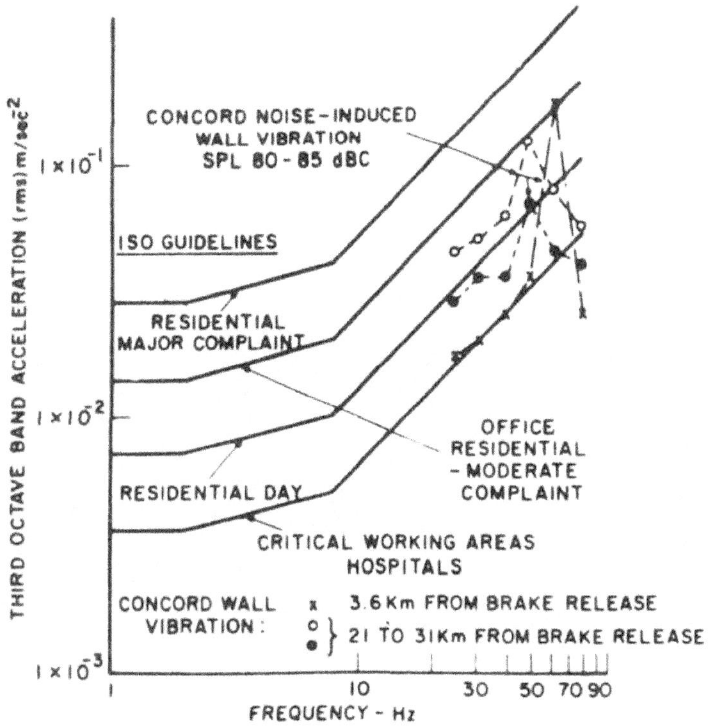

FIGURE 10.41. Acceptability of building vibration to inhabitants. (From ref. 132.)

Visual contrast thresholds (bright target on less bright field) and minimum visual acuity for lines and discs (see fig. 10.43) are apparently not affected by noise levels up to 140 dB or so (refs. 134 and 135), although Dorfman and Zajonc (ref. 136) found some effect of sound level on perception of background brightness but not on size estimation of objects. Loeb (ref. 137) found that broadband noise at a level of 115 dB had no effect on visual acuity; however, Rubenstein (ref. 138) reported adverse effects from noise at 75 to 100 dB, and Chandler (ref. 139) reported a shift of verticality of a visual line away from the ear stimulated with noise. Benko (refs. 140 and 141) reported a narrowing of the visual field with noise.

As discussed earlier in the chapter, noises that invariably cause a dilation of the pupil of the eye along with a vasoconstriction of peripheral blood vessels sometimes cause other responses related to autonomic-system activity. However, Jones *et al.* (ref. 142) found that although exposure to either continuous or intermittent industrial noise increases pupil size, visual acuity is not adversely affected.

Hermann *et al.* (ref. 143) studied the effect on visual depth perception of several minutes of exposure to broadband white noise at levels ranging from 70 dBA to 115 dBA. They found these exposures did not produce any significant changes in stereoscopic depth perception.

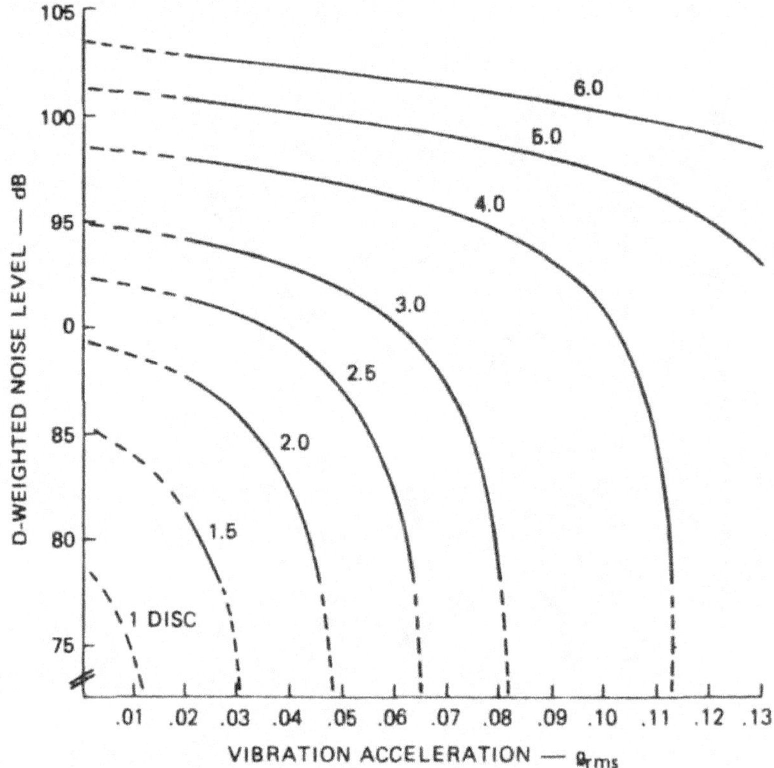

FIGURE 10.42. D-weighted noise level required to produce successive constant discomfort curves as a function of vibration acceleration g_{rms}. (From ref. 133.)

McCroskey (ref. 144) reported that random noise at levels from 85 to 115 dB reduced the CFF from 25 to 23 sec^{-1}. Ogilvie (ref. 145) found no change in CFF with steady-state random noise of 80 to 90 dB, an increase in CFF with noise "fluttered" out of phase with the visual flicker, and a decrease in CFF with noise fluttered in phase with the flicker. Walker and Sawyer (ref. 146), however, were not able to duplicate Ogilvie's findings and got negative results except for a small difference in CFF between steady-state and inphase noise. (See table 10.3.)

The effect of steady-state noise on CFF when the color of the light was varied has also been studied, but the results are very inconsistent. For example, Maier *et al.* (ref. 147) found that CFF decreased with increased loudness when the light was orange-red, but no change occurred with green light. It would appear that noise can sometimes effect about a 10 percent change (usually a decrease) in CFF from the CFF found in quiet, but the exact effects as a function of various noise and light conditions are highly variable and possibly a matter of experimental chance and error.

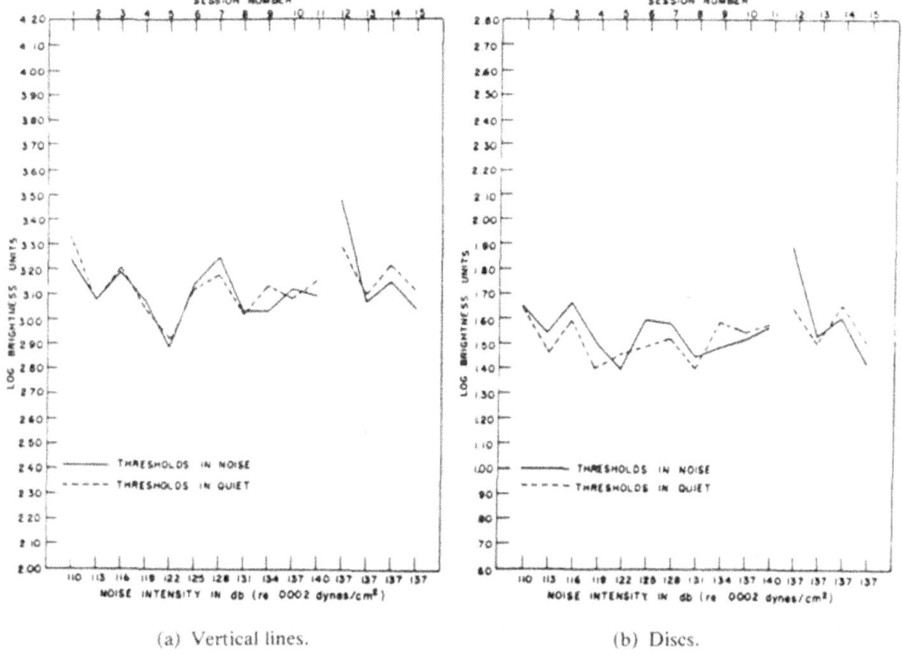

(a) Vertical lines.

(b) Discs.

FIGURE 10.43. Threshold of visual acuity in quiet and in noise. (From ref. 135.)

TABLE 10.3

CFF for Four Conditions

[Data from ref. 146]

Condition	Sample size, N	Mean CFF, sec^{-1}	Standard deviation of CFF, sec^{-1}
With artificial pupil			
No noise	13	32.62	6.20
Steady-state noise	↓	31.94	4.75
Inphase noise	↓	32.34	4.78
Out-of-phase noise	↓	32.78	5.44
Without artificial pupil			
No noise	13	38.12	3.43
Steady-state noise	↓	38.01	3.48
Inphase noise	↓	38.42	3.28
Out-of-phase noise	↓	38.72	3.73

The converse effects of light on auditory threshold are small and are perhaps fortuitous. For example, O'Hare (ref. 148) claimed some colors of light caused a 1- to 2-dB increase whereas some caused a 1- to 2-dB decrease in auditory threshold.

Low- and infrasonic-frequency sounds

It is conceivable that intense airborne low-frequency sound and acoustic energy (frequencies below about 20 Hz are generally called "infrasound") could have particularly adverse effects on man. In addition to possible stimulation of the vestibular system and pain in the ear, low-frequency sound can cause resonant vibration in the chest, throat, and nose cavities of the body.

The effects from vibrations at low frequencies when imposed directly on the body through mechanical contact are fairly well known. (See figs. 10.41 and 10.42.) Because of the impedance mismatch between airborne acoustic energy and the body, acoustic energy has little or no effects on parts of the body other than the ear until the levels become quite intense.

The effects of very intense airborne sound on the body were determined in a series of tests conducted by U.S. Air Force and NASA research personnel and reported by Mohr *et al.* (ref. 149). The stimuli used in these tests are shown in figure 10.44. No nonauditory effects were noted until the spectrum levels exceeded approximately 125 dB. At various higher levels, decrements in visual acuity, some vestibular-system reactions, and chest, nose, and throat responses occurred, and if no ear protection was worn, ear pain and middle-ear fullness were felt. The observations of the subjects for tests 5 to 16 are given in table 10.4. The results of tests 1 to 4 conducted in levels lower than 125 dB revealed no significant effects of the noise on the subjects.

It is clear from tests 5 to 14 that the nonauditory and, to some extent, auditory effects of airborne low-frequency and infrasonic-frequency sound become significant only at spectrum levels (single frequency) in excess of 130 dB. Except in the vicinity of unusual sources of noise, such as near heavy rocket engines or special test sirens, one seldom finds steady-state low frequency acoustic energy at these intensities.

Nixon and Johnson (ref. 150) reviewed research findings on infrasonic-frequency sound with respect to its effect on shifts in thresholds of audibility. Except for one study involving whole-body vibration, little temporary threshold shift was observed from exposure to levels of infrasonic-frequency sound up to 150 dB. (See table 10.5.) Figure 10.45 shows the limiting values for exposure to airborne infrasonic-frequency sound.

In regard to these nonauditory-systems effects, smooth muscles will show reduced tension when exposed to very intense low-frequency sound, as shown by Döring *et al.* (ref. 151) for *in vitro* tests with rabbit and pig smooth muscles. Gerd Jansen (private communication, Johannes Gutenbërg University, Mainz,

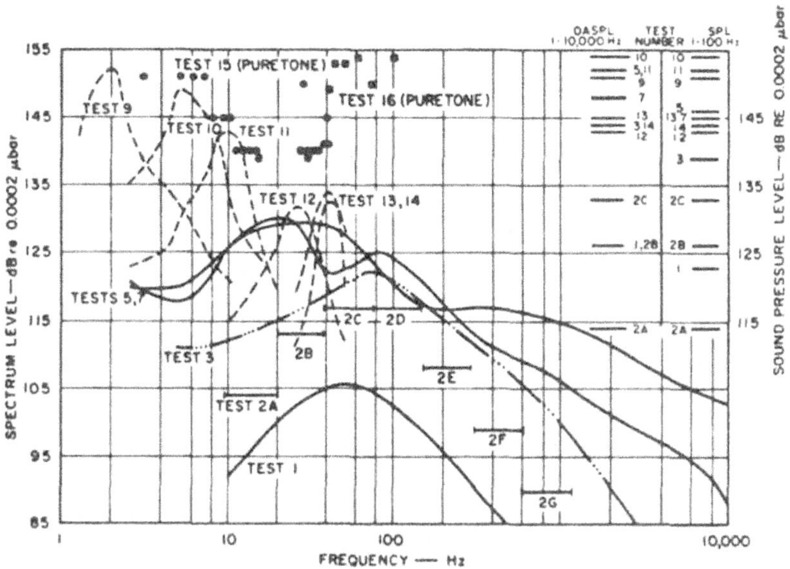

FIGURE 10.44. Summary of test environments for effects of sound on the human body. (From ref. 149.)

TABLE 10.4

Summary of Effects of Acoustic Stimuli on a Group of Five Persons

[Exposure durations of 1 to 2 minutes; ear protective devices usually worn; from ref. 149]

Test 5 – ***Broadband noise. Peak spectrum level 128 dB at 50 Hz.*** The speech signals recorded were completely masked despite the noise reduction provided by microphone and shield. Pulse rates were increased 10 to 40 percent over resting levels. Two subjects reported mild chest wall vibration; two others noted mild nasal cavity vibration; and one of these perceptible throat fullness.

Tests 6, 7, and 8 – *Broadband noise at about same levels as Test 5 with relatively less energy in the higher frequencies.* All subjects considered the exposures tolerable for the short durations involved. Speech signals were completely masked, nevertheless, except those of one subject who was stationed inside a vehicle which afforded appreciable attenuation of the high frequencies. His speech was definitely modulated but the poor intelligibility achieved was attributed to the masking. All subjects reported mild to moderate chest wall vibration; two subjects noted throat pressure; three subjects experienced perceptible though tolerable interference with the normal respiratory rhythm. Pulse rates measured during Test 7 exhibited no significant changes during the exposure.

Throughout these tests visual acuity, hand coordination and spatial orientation were subjectively normal.

TABLE 10.4 Concluded

Tests 9-11 – *Narrow bands, center frequencies, 2 to 10 Hz, spectrum level 142-153 dB.* The most prominent effects attributable to the infrasonic noise spectra (Tests 9-11) occurred during exposure without ear protection. An uncomfortable sensation reflecting pressure build-up in the middle ear was elicited which required frequent Valsalva swallowing to relieve. This effect was almost entirely absent when insert earplugs were used. Earmuffs alone helped prevent the middle ear pressure changes. Three subjects described an occasional tympanic membrane tickle sensation during these exposures without protection and one subject observed marked nostril vibration. Another noted mild abdominal wall vibration during exposure to the test 10 spectrum (5-10 Hz). No shifts in hearing threshold were detectable one hour following these exposures. When ear protectors were worn to lessen the middle ear pressure changes, exposures to infrasound of these levels were judged well within tolerance.

Tests 12-14 – *Narrow bands, center frequencies 15 to 50 Hz, spectrum level at 140 dB.* The maximum intensity low sonic exposures produced moderate chest wall vibration, a sensation of hypopharyngeal fullness (gagging) and perceptible visual field vibration in all subjects. Two subjects experienced mild middle ear pain during brief periods without ear protection but a third had no sensation of tickle or pain. Recorded speech sounds exhibited audible modulation. Post-exposure fatigue was generally present after a day of repeated testing. The exposures as a group were not considered pleasant; however, all subjects concurred that the environments experienced were within the tolerance range.

Test 15 – *Pure tones 3 to 40 Hz, spectrum level 145-153 dB.* Exposures to 24 discrete frequency noise fields showed both objective and subjective responses qualitatively similar to those elicited by the corresponding narrow band spectra. Pressure build-up in the middle ear was not a factor at 30 Hz and above, but the gag sensation was magnified for at least one subject. Although all exposures were judged tolerable, it was noted that the subjective sensations rose to intensity very rapidly as sound pressure levels were increased above 145 db.

Test 16 – *Pure tones 40 to 100 Hz, spectrum level 150-155 dB.* Voluntary tolerance of the subjects was reached at 50 Hz (153 dB), 60 Hz (154 dB), 73 Hz (150 dB), and 100 Hz (153 dB). The decision to stop exposures at these levels was based on the following subjectively alarming responses: mild nausea, giddiness, subcostal discomfort, cutaneous flushing and tingling occurred at 100 Hz; coughing, severe substernal pressure, choking respiration, salivation, pain on swallowing, hypopharyngeal discomfort and giddiness were observed at 60 Hz and 73 Hz. One subject developed a transient headache at 50 Hz; another developed both headache and testicular aching during the 73 Hz exposure.

A significant visual acuity decrement (both subjective and objective) occurred for all subjects during the 43, 50, and 73 Hz exposures. Speech sounds were perceptibly modulated during all exposures. All subjects complained of marked post-exposure fatigue. No shifts in hearing threshold were measurable two minutes post exposure; the earplug and muff combinations worn are known to provide sufficient protection against the higher harmonics of the noise fields and were apparently effective to an appreciable degree in attenuating the fundamental tones. Recovery from most of the symptoms was complete upon cessation of the noise. One subject continued to cough for 20 minutes, and one retained some cutaneous flushing for approximately four hours post exposure. Fatigue was resolved by a night's sleep.

West Germany) has found similar effects in humans when exposed to low-frequency engine noise at levels exceeding 170 dB.

It has been reported (ref. 152) that some people in intense infrasonic-frequency sound at levels sometimes found in automobiles, in subways, and during high winds in some high-rise office buildings have feelings of unpleasantness and even loss of balance. To explore these observations the researchers vibrated flexible walls of a small office to produce intense infrasonic-frequency sound in the room. Some of their results are shown in figure 10.46. Clearly, some nonauditory sensations can be created by infrasonic-frequency sound in a narrow region around 10 Hz with fairly prolonged exposures at relatively high intensities. The prevalence in real life of infrasonic-frequency sound at the intensities and durations required to create these sensations is rather small.

Ultrasonic-frequency sound

Acoustic energy in the frequency region above 20 000 Hz is called ultrasonic because it is inaudible to man. Actually, for most adults, acoustic energy above 15 000 Hz is ultrasonic. As noted in a review of the effects of ultrasound by Parrack (ref. 153), the advent of the jet aircraft engine, high-speed dental drills, and ultrasonic cleaners provided relatively common sources of high-intensity ultrasounds. Tables 10.6 to 10.8 show the spectra of the noise from representative samples of these devices.

Although there is considerable energy in the bands above 20 000 Hz for each of the spectra, there is energy in the audible frequency region that often exceeds the damage risk values specified as tolerable for long exposures. (See chapter 7.) For this reason, the tinnitus, dizziness, headache, nausea, and fullness of the ears often reported by some persons exposed to these noises are probably not because of ultrasonic frequencies but are probably because of audible frequencies in the noise.

In Parrack's opinion, the reactions of tinnitus, dizziness, nausea, and headache listed above are psychosomatic and are engendered by unwarranted apprehension. Although the energy above 20 000 Hz (or possibly at any frequency that is inaudible to a given individual) is not the source of the tinnitus, dizziness, nausea, and headache experienced by persons exposed to such sound energy, the reactions in question are perhaps not "psychosomatic" in origin either. At the same time, the belief of Parrack and others (see refs. 154 and 155) that any acoustic energy at high frequencies that significantly affects man does so only through his inner ear, appears to be true. The physical arguments against ultrasound entering or stimulating man except through normal stimulation of the inner ear are as follows:

1. The absorption coefficient of the skin for sound above 20 000 Hz is less than 0.1 percent and levels at these frequencies that would cause any slightly

TABLE 10.5

Summary of Studies of Temporary Hearing Loss Following Exposure to Infrasonic-Frequency Sound

[From ref. 150]

Investigator	Exposure	Hearing response	Recovery
Tonndorf	Submarine diesel room; 10–20 Hz; no level given	Depression of upper limits of hearing as measured by number of seconds a tuning fork was heard; no conversion to maximum audible pressure	Recovery in few hours outside of diesel room
Mohr *et al.*	Discrete tones; narrowband *noise in* 10–20 Hz region; 150–154 dB exposures of about 2 min	No *change in hearing sensitivity* reported by subjects; no TTS measured about one hour post exposure	
Jerger *et al.*	Successive 3-min whole-body exposures; 7–12 Hz; 119–144 dB	TTS in 3000–6000 Hz range for 11 of 19 subjects (TTS of 10–22 db)	Recovery within hours
Nixon	Pistonphone coupled to ear *via* earmuff; 18 Hz at 135 dB; series of 6 5-min exposures rapid in succession	Average TTS of 0–15 dB after 30-minute exposures	Recovery within 30 min

Nixon	Pistonphone coupled to ear *via* ear-muff; 14 Hz at 140 dB; 6 individual exposures of 5, 10, 15, 20, 25, and 30 min	Three experienced subjects; no TTS in 1; slight TTS in 1; 20–25 dB TTS in 1	Recovery within 30 min
Johnson	Ear only: pressure chamber coupled to ear *via* tuned hose and muff:		
	171 dB (1–10 Hz), 26 sec, 1s;	No TTS	
	168 dB (7 Hz), 1 min, 1s;	No TTS	
	155 dB (7 Hz), 5 min, 2s;	No TTS	
	140 dB (4, 7, 12 Hz), 30 min, 1s;	14–17 dB TTS	Recovery within 30 min
	140 dB (4, 7, 12 Hz), 5 min, 8s;	8 dB TTS for 1 subject	
	135 dB (0.6, 1.6, 2.9 Hz), 5 min, 12s;	No TTS	Recovery within 30 min
	126 dB (0.6, 1.6, 2.9 Hz), 16 min, 11s;	No TTS	
	Whole body: All exposures, 2s:		
	8 min at 8 Hz at SPL's of 120, 126, 132, 138	No TTS	
	8 min 15 1, 2, 4, 6, 8, 10 Hz at 144 dB	No TTS	
	8 min at 12, 16, 20 Hz at 135–142 dB	No TTS	

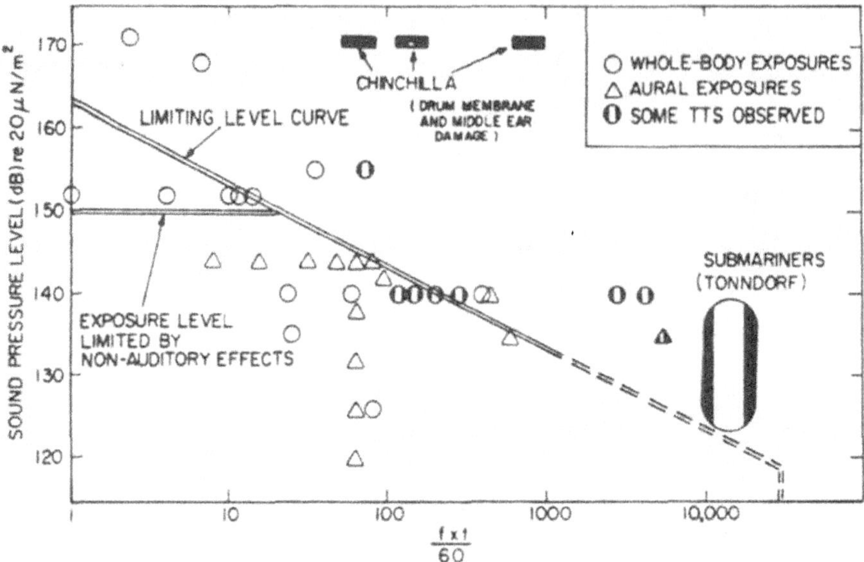

FIGURE 10.45. Various laboratory infrasonic-frequency sound exposures in terms of level and number of cycles ($f \times t/_{60}$) and a limiting sound pressure level curve based on the formulation. Limiting SPL = 10 log(t/8 min) + 10 log(f/10) + 144. (From ref. 150.)

noticeable local heating effects would have to be in excess of 100 dB. (The absorption coefficient of acoustic energy at 20 000 Hz in small, furry animals is of the order of 21 percent, so that lethal heating can occur in these animals at levels of ultrasound that go unnoticed by, or are harmless to, man.)

2. Ultrasonic frequencies generated by crystals (refs. 156 to 158) and applied to bones and tissues of the head resulted (if sufficiently intense) in the person perceiving an audible, high-pitched tone usually around 8000 to 10 000 Hz, depending upon their upper limit of hearing. Deaf subjects in these experiments heard nothing. Crystals have many resonant modes, and the rubbing of the crystal against the surface of the skin could have created an audible subharmonic that was radiated as an acoustic signal to the ear (analogous to the mode of detection of the presumed "electrical" stimulation of the ear mentioned in chapter 3). Also, it is possible that subharmonics falling in the normal frequency range of audibility may be generated in the middle ear when the ear is exposed to intense ultrasonic frequencies.

Acton and Carson (ref. 159) (see also ref. 160) found convincing evidence that unless a person's range of hearing extended to about 17 000 Hz, and unless the energy in that frequency region exceeded 70 dB, no subjective effects (tinnitus, headaches, fatigue, *etc.*) were experienced from exposure to ultrasonic frequencies. Figure 10.47 (lower graph) shows the results of laboratory tests with

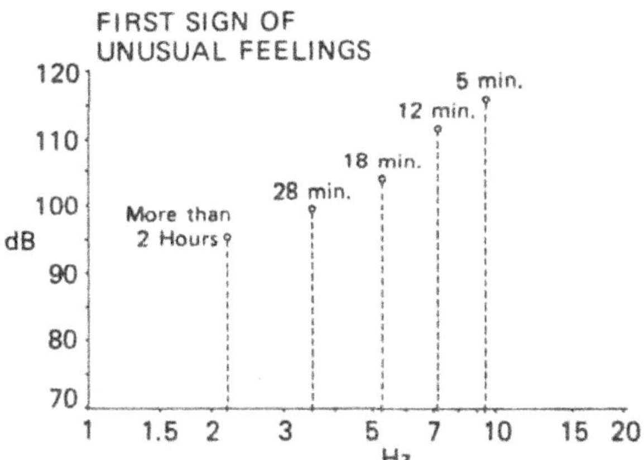

The sensory response of one person to infrasound excitation

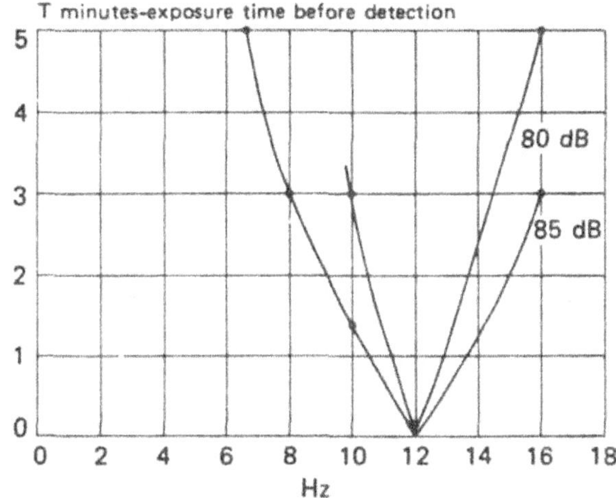

The threshold/exposure time relationship around the most sensitive frequency

FIGURE 10.46. Thresholds of subjective feelings of "unusualness" from exposure to infrasonic-frequency sound in an office space. (From ref. 152.)

ultrasonic signals. The upper curve of the upper graph of figure 10.47 shows the spectrum of noise which did not cause significant complaints. Acton and Carson noted that women had adverse symptoms more often than men and that young men had adverse symptoms more often than older men. This was presumed to be because of the auditory acuity of the people involved and not because of their sex or age *per se*.

TABLE 10.6

Sound Pressure Level Around a Jet Aircraft in 1/3-Octave Bands

[Aircraft is F-102; from ref. 153]

Positions and operating conditions	SPL, dB, at 1/3-octave-band center frequencies, kHz, of—														
	2	2.5	3.15	4	5	6.3	8	10	12.5	16	20	25	31.5	40	50
[a]500 ft; idle	57	67	63	54	53	51	44	38	37	37	35				
500 ft; military	98	95	92	90	88	84	82	77	76	75	74	74			
500 ft; afterburner	105	103	103	100	99	96	95	91	89	87	83	80	78		
[b]100 ft; idle	75	83	82	74	77	77	74	73	70	64	60	57	56	55	59
100 ft; military	111	114	112	111	109	108	106	104	102	100	97	95	92	88	
100 ft; afterburner	123	122	123	121	119	118	118	115	115	113	112	111	110	107	106
[c]25 ft forward; idle	90	96	94	96	94	92	92	90	90	88	86	84	80	76	73
25 ft forward; military	102	107	104	101	102	100	100	97	96	93	91	89	86	83	80
25 ft forward; afterburner	109	110	107	107	105	104	103	101	100	97	95	93	90	87	87
Maintenance[d]; idle	91	95	94	94	93	90	93	85	85	83	79	77	73	70	
Maintenance; military	117	115	114	116	112	112	111	108	108	105	103	101	99	96	
Maintenance; afterburner	121	120	120	123	118	117	117	114	113	111	108	106	103	100	

[a] *On radius located 125° from nose.*
[b] *On radius located 120° from nose.*
[c] *About 30° from nose of aircraft.*
[d] *Just off main landing gear and under fuselage.*

TABLE 10.7

Sound Pressure Levels Measured in Air Around the Weber High-Speed Dental Drill

[Drill suspended in fixed position; drill and microphone 54 in. from floor; from ref. 153]

Measurement position (a)	OASPL, dB	SPL, dB, for center frequencies of octave bands, Hz, measured at—										
		31.5	63	125	250	500	1000	2000	4000	8000	16 000	31 500
1	97	46	44	50	48	52	55	64	87	84	93	95
2	89	44	43	47	43	46	47	54	77	82	83	83
3	82	47	51	43	45	44	45	52	73	76	78	76
4	81	48	54	41	42	44	45	51	74	76	77	74

[a] *Position 1—Location of patient's ear was 6 in. from source.*
Position 2—Location of patient's ear was 20 in. from source.
Position 3—Far field on radius 65 in. from source.
Position 4—Far field on radius 90 percent to position 3—65 in. from source.

TABLE 10.8

Sound Pressure Levels in 1/3-Octave Bands

[Bendix model sec 1825A sonic energy cleaning system; from ref. 153]

Positions	SPL, dB, for 1/3-octave-band center frequencies, kHz, of—																			
	1.0	1.25	1.6	2.0	2.5	3.15	4	5	6.3	8	10	12.5	16	20	25	31.5	40	50	63	80
Operator,[a] cover closed	56	55	55	57	62	68	70	72	73	83	96	83	83	101	85	91	89	86	85	
Operator,[a] cover open		62	63	63	66	70	74	77	81	91	104	91	93	109	93	102	99	95	95	
At desk,[b] cover closed	45	43	42	42	47	51	54	56	57	64	77	64	64	80	63	71	69	64	62	
At desk,[b] cover open	49	48	49	49	52	55	58	61	64	73	86	73	73	89	72	79	78	72	71	
Office,[c] cover closed	45	45	44	43	41	42	42	41	40	44	57	45	45	64	46	45	42	38		
Office,[c] cover open	46	45	44	43	42	42	42	43	43	51	63	50	50	57	49	52	48	42		

[a] *Immediately adjacent to one end of cleaner system.*

[b] *About 15 ft from edge of tank where operator stands in laboratory work area.*

[c] *Adjacent room (door open) about 12 ft from cleaner system; about 13 ft from operator position.*

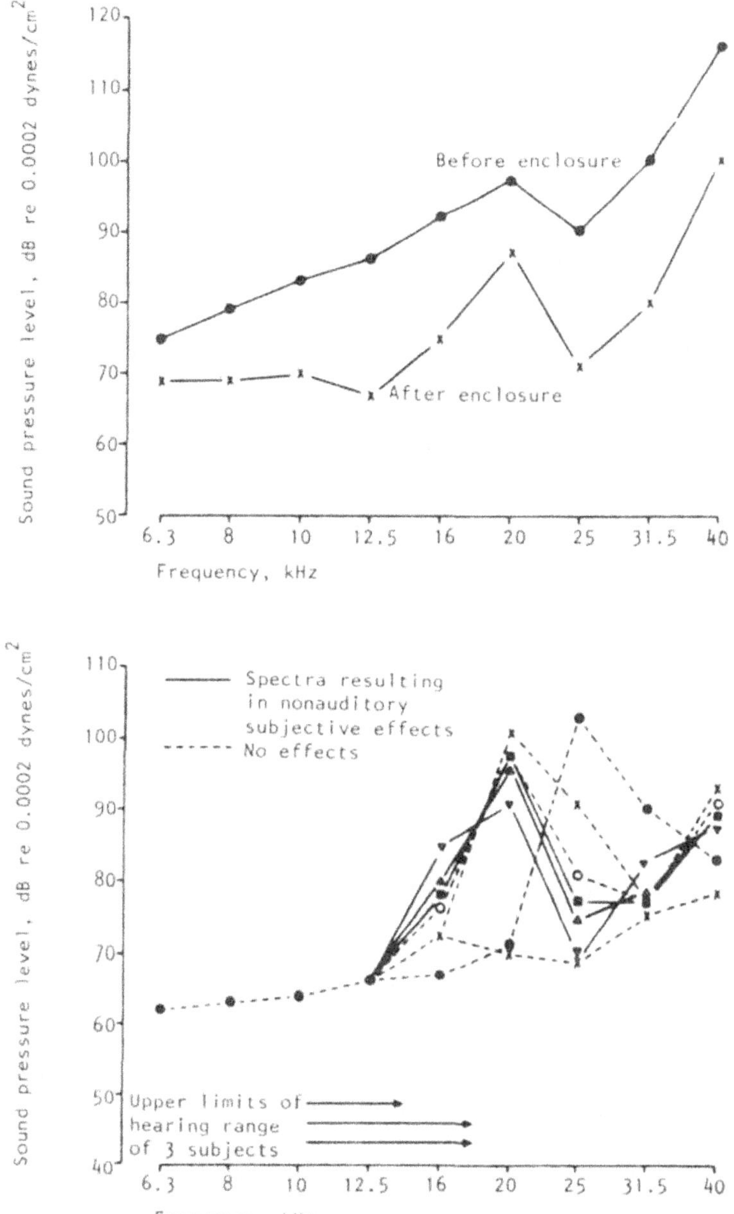

FIGURE 10.47. Band spectra of noise from an ultrasonic washer and spectra of noise and tones that were used in laboratory tests. (From ref. 159.)

It also appears that hearing in the higher frequency regions of most industrial workers sooner or later is reduced by the upward spread of a noise-induced permanent threshold shift from lower frequency noise as much as it is by the acoustic energy above 10 000 Hz. It might be conjectured that the upturn in the normal threshold of audibility from 4000 Hz to, say, 20 000 Hz is severely influenced by everyday noise in the frequency region 2000 to 8000 Hz. In any event, the subjective effects of ultrasonic frequenices are not due to apprehension on the part of the listener but are due to sound that exceeds 78 dB in the frequency region of about 16 000 Hz and that is audible to the listener. Continued exposure to sufficiently intense sound at or below 16 000 Hz results in the elimination of the subjective and audible effects apparently because of noise-induced threshold shifts in those frequency regions.

The question remains as to why these adverse subjective effects are more often noticed from ear-damaging exposures to these higher frequencies than from damaging exposures to noise at lower frequenices, for example, below 2000 Hz. Of course, there may be no fundamental difference since some comments regarding headache, unusual fatigue, and certainly tinnitus are also sometimes reported from initial exposures to lower frequency noise when sufficiently intense and for sufficiently long exposures.

NIOSH studies

A program of studies were conducted for the National Institute for Occupational Safety and Health (NIOSH) on the effects of noise on nonauditory sensory functions and performance (ref. 161). Groups of adult subjects were exposed to continuous noise ranging from 105 to 110 dBA, to impulsive noise at 136 dBA, and to quiet. No effects of the noises, relative to quiet, were found for tests of tactile, thermal, or vestibular functions. A few measures of visual functions showed noise-induced effects, but these could not be altogether replicated on retesting.

Health-Related Effects in Workers

Introduction

The concept of events in everyday life that cause stress responses in the autonomic system contributing to body and mental illness is a generally accepted principal of psychiatry and medicine (Levi (ref. 162) and Kagan (ref. 163)). Kagan states that "There is much evidence showing that psychological stimuli arising from a large variety of social stressors may cause the catecholamine or corticosteriod stress responses and that these in turn cause a large variety of secondary physiological changes which are associated with high risk for a large variety of diseases. For example, such diverse secondary physiological changes as increased heart rate, raised blood pressure, increased peripheral resistance, increased fat metabolism, decreased glucose tolerance, impaired myocardial uptake of oxygen, cardiac arrhythmias, gastroenteric activity, and possibly effects on the histo-immunological system are caused by or highly associated with the endocrine changes." However, data supporting causal relations between these events and health effects are not plentiful because of complexity of the events, individual differences, and other effects. (See discussions by Rabkin and Struening (ref. 164) and Graeven (ref. 165).) Whether stress associated with exposure to noise is mostly the result of indirect psychological effects rather than direct "overstimulation" effects of noise is the major question concerning the causal role of noise in the work environment on ill health. If they are and if these psychological factors can be controlled, it would follow that noise at work not involving auditory communications need not contribute to ill health in the nonauditory systems of people. To what degree these noise conditions exist in real life is an empirical question that may be answerable by retrospective epidemiological research such as that presented below.

Effects from long-term exposures

Studies of the health of people working in noisy industries compared with workers from quieter industries are plagued by the problems of adequately equating different groups of workers with respect to socioeconomic and familial health-status variables. There also appears to be at least two work-related psychological variables that can have strong physiological effects (especially on the autonomic system) and that make difficult the matching, for comparative purposes, of groups of workers in noisy industries with workers in quiet industries:

(1) The work conditions may be unsafe. Often intense industrial noise is indicative of the operation of moving machinery that must be attended to in order that the worker avoid bodily injury. Not only does the noise connote this danger, but it can also mask or interfere with the hearing of acoustic cues and signals that must be perceived in order to avoid the dangers involved. These conditions contribute to psychological emotions which may be reflected as physiological stress.

(2) The work may require the perception of certain sounds (machinery made, speech, or other) in order that the work tasks be properly and quickly performed. The noise may interfere with these perceptions, creating some anxiety which may also be reflected as physiological stress.

There are obviously other environmental factors, such as air pollution and general physical work conditions, that must also be equated for noise-exposed versus non-noise-exposed workers before valid conclusions on the effects of the noise on the health of the workers are drawn. For example, Pilawska *et al.* (ref. 166) cited the following on-the-job factors in shipyards as having decisive significance as causes of illness and absences: (1) improper climatic conditions and (2) excessive intensity of vapors of welding gases, vapors of paints and solvents, dust, noise, vibration, ultraviolet radiations, and drafts. Pilawska *et al.* found in a particular industry 1826 persons worked in noise levels greater than 85 dBA (group A) and 5825 worked in noise levels less than 75 dBA (group B). A comparison of the medical records of these two groups revealed that hearing disorders were 22 times more frequent, stomach and intestinal ulcers were 5 times more frequent, and high blood pressure was about 2 times more frequent in group A than in group B. Similar types of results were found by Jansen (ref. 167) (see fig. 10.48), Shatalov and Murov (ref. 168), and Zvereva *et al.* (ref. 169) among metal workers and other workers, and by Cieslewicz (ref. 170) among workers in weaving mills in noisy work locations.

Most of the studies of the effects of noise on workers have been concerned with abnormal cardiac conditions as measured by the electrocardiogram (ECG) and by the development of high blood pressure, or hypertension. Hypertension is defined as systolic and diastolic blood pressures greater than normative values for a given age group. Major studies of these effects include those of Andriukin (ref. 171), Geller *et al.* (ref. 172), Andrukovich (ref. 173), Kavoussi (ref. 174), Capellini and Maroni (ref. 175), Parvizpoor (ref. 176), Cuesdean *et al.* (ref. 177), Dega and Klajman (ref. 178), Kachnyi (ref. 179), and Friedlander *et al.*

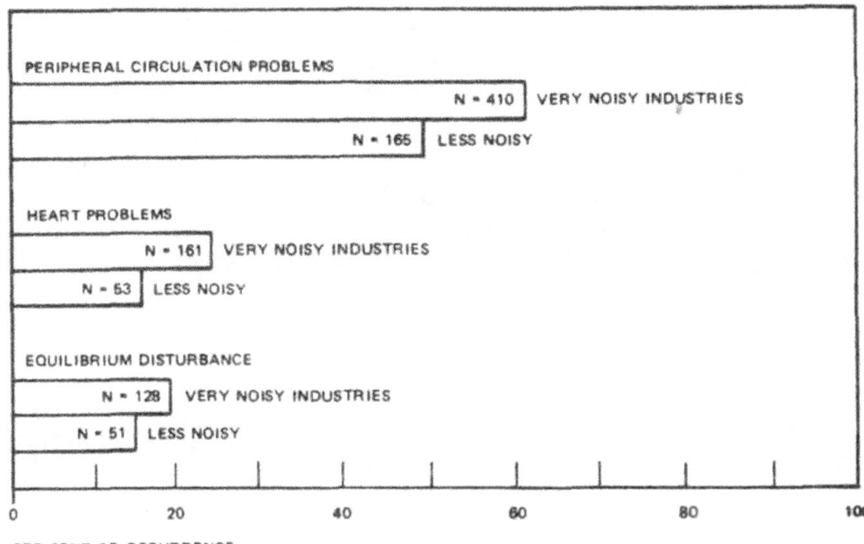

FIGURE 10.48. Differences in percent of occurrence of physiological problems in 1005 German industrial workers. The differences in peripheral circulation and heart problems in the two classes of work environments were statistically significant. (Data from ref. 167.)

(ref. 180). Additional studies in this research area are reviewed in an article by Welch (ref. 181).

The results for these studies appear to Welch to be mostly positive; that is, persons working in intense noise show a significantly greater incidence of hypertension or ECG abnormalities than persons who work in less noise. For example, Parvizpoor found increased hypertension in weavers in textile mills working in noise above 95 dBA compared with a group randomly selected from a population matched socioeconomically to that of the weavers. (See table 10.9.) Parvizpoor notes, however, that physiological stressors in the mills besides the noise include cotton dust (7.5 mg/m^3, a very heavy dust content), high temperatures and humidity, and other adverse environmental factors.

Possible causal relations of cardiovascular effects with noise are not established by these studies, especially the concept that noise *per se* is overstimulating the worker and thereby causes the cardiovascular effects. For example, Kachnyi reports that weavers operating smaller, less noisy looms showed more disorders of arterial pressure than did weavers using fewer, noisier looms. In addition, the data are not consistent in showing any uniform relation between noise intensity and cardiovascular functions. For example, Geller *et al.* found that workers in the Soviet petroleum industry in noise of 115 to 125 dB and exposed to oil and gas fumes suffered no more hypertension or cardiac neurosis (as determined by ECG) than administrators or laborers working in the quiet. However, petroleum workers in 115 to 125 dB noise who were not exposed to oil and gas fumes suffered more hypertension and cardiac neurosis than workers in the quiet. Cuesdean *et al.* found that ECG abnormalities were less for air compressor operators and stokers in 100 to 106 dB noise than for mechanics in 95 to 100 dB noise, and the mechanics showed fewer ECG abnormalities than were found in laboratory assistants in 85 to 95 dB noise.

TABLE 10.9

Effect of Noise on Cardiovascular Function of Textile Mill Weavers

[95 dBA noise level; hypertension, >160/90 mm Hg blood pressure; data from ref. 176]

Age, years	Hypertension, percent, shown by—	
	Weavers	Controls[a]
20 to 29	1	0
30 to 39	7.3	1.2
40 to 49	12.1	6.5
50 to 59	27.1	8.6
Total	8.5	2.4

[a] *Socioeconomically matched to weavers.*

Table 10.10 shows that the incidence of hypertension is not related in an orderly way with noise levels from relative quiet to a range of 115 to 125 dB. For example, the percent of people with hypertension is the same, for the "quiet" and the 115 to 125 dB groups. Obviously these results do not support the notion that intense noise in and of itself is a cause of hypertension or other cardiovascular problems.

Further evidence along these lines comes from a study by Cohen (ref. 182) of health and accident records in two U.S. industries with both high-noise (above 95 dBA) and low-noise (below 80 dBA) work areas. Figure 10.49 shows that in Plant Complex A the number of medical problems was significantly higher in the high-noise group compared with that in the low-noise group, but in Plant Complex B the number of problems was about equal. Also, the lower curve in figure 10.49 shows the number of accidents in Plant Complex A to be much greater than in Plant Complex B in high noise.

Cohen suggests that the differential risk of injury in the high-noise work areas (boiler factory) in Plant Complex A was much greater than in Plant Complex B (electronics and missile parts plant) and that it was the anxiety over danger of injury or accidents and not the high-noise level *per se* that could have been the cause of increased medical problems in the workers in high noise in Plant Complex A. These problems included digestive, respiratory, urological, glandular, and cardiovascular disorders, all suggestive of physiological stress.

Lees *et al.* (ref. 183) found that 70 men who worked in industrial noise levels of 85 dBA or less for a period of 15 years had the same incidence of absenteeism, headaches, and accident rates (called indicators of "stress") as 70 matched cohorts who had worked for 3 to 15 years in noise levels exceeding 90 dBA. Lees *et*

TABLE 10.10

Hypertension as Function of Noise in the Workplace

Groups	Hypertension, percent affected, for noise levels of—			
	Quiet	87 to 102 dB	103 to 120 dB	115 to 125 dB
General population[a]	23	26	17	8
Exposed to little noise[b]	11	12		20
Administrative workers[c]	12			8
Manual workers[c]	8			20
Average	14	19	17	14

[a] *Reference 173.*
[b] *Reference 171.*
[c] *Reference 172.*

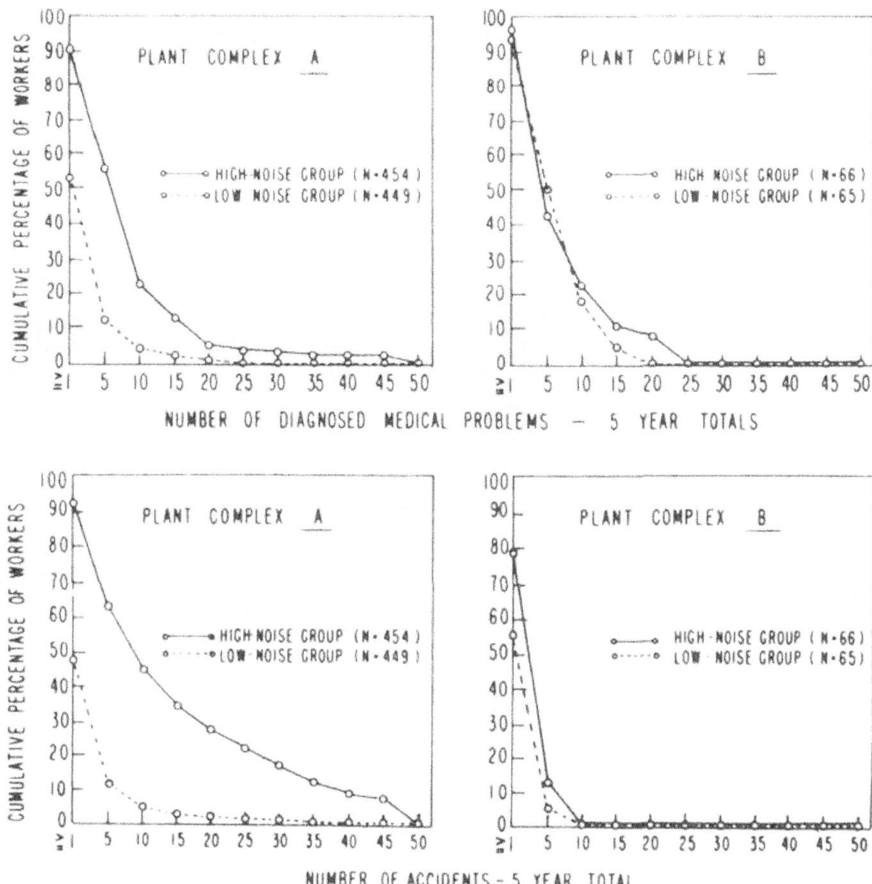

FIGURE 10.49. Cumulative percentages of workers in high- (above 95 dBA) and low- (below 80 dBA) noise groups with specifiable number of diagnosed medical disorders and accidents over the 5-year period 1966 to 1970. (From ref. 182.)

al. concluded that noise-exposure level was not necessarily a cause of stress in industrial work.

Hearing loss and hypertension

Jonsson and Hansson (ref. 184) examined the prevalence of hypertension in industrial workers who had been exposed to noise at the workplace and had suffered elevated hearing thresholds greater than expected for their age. They found that men with noise-induced hearing losses had a significantly greater incidence of hypertension (systolic/diastolic blood pressure above 160/100 mm Hg) than matched groups with normal or nearly normal hearing for their age.

On the other hand, Drettner *et al.* (ref. 185), Hedstrand *et al.* (ref. 186), Takala *et al.* (ref. 187), Brown *et al.* (ref. 188), Cohen *et al.* (ref. 189), and Lees and Roberts (ref. 190) found no correlation between noise-induced hearing loss and hypertension. Drettner *et al.*, using 1000 50-year-old men, also measured other cardiovascular risk factors (serum cholesterol and triglycerides and a glucose tolerance test) with similar results. The results of some of these studies are summarized in table 10.11.

In the Brown *et al.* study, 22 professional airline pilots were compared over a period of 8 years with 29 males of the same age who did not fly. In addition to hypertension, these investigators measured serum cholesterol and glucose. They found that although the pilots developed somewhat more elevated audiometric thresholds compared with the control group (indicating some noise-induced hearing loss from aircraft cockpit noise), there were no significant differences between the groups with respect to the cardiovascular or blood serum tests during the 8-year period.

Figure 10.50 shows the differences in hearing levels for workers with noise-induced hearing loss and for a control group with normal hearing. Table 10.10 compares the systolic and diastolic blood pressures of the NIPTS and control groups found in the various studies discussed above. It is shown in the table that the mean blood pressures and numbers of hypertensives were nearly the same for the NIPTS and control groups in all except the Jonsson and Hansson study.

TABLE 10.11

Summary of Studies on Hearing Loss and Hypertension

Study	Hearing level	Number of subjects	Mean age, years	Mean blood pressure, mm Hg		Number of Hyper-tensives
				Systolic	Diastolic	
Jonsson and Hansson (ref. 184)	Normal HL	74	54	133	81	6
	NIPTS[a]	44	57	[b]145	[b]89	10
Hedstrand *et al.* (ref. 186)	Normal HL	376	50	134	84	24
	NIPTS[c]	393	50	132	84	26
Takala *et al.* (ref. 187)	Normal HL	67	45	151	95	14
	NIPTS[a]	32	47	155	99	11
Cohen *et al.* (ref. 189)	Normal HL	51	34	122	68	7
	NIPTS[a]	51	47	123	70	5

[a] *NIPTS = HL > 65 dB above 3000 Hz.*

[b] *Difference between normal and NIPTS groups statistically significant.*

[c] *History of noise exposures and a "significant" hearing loss.*

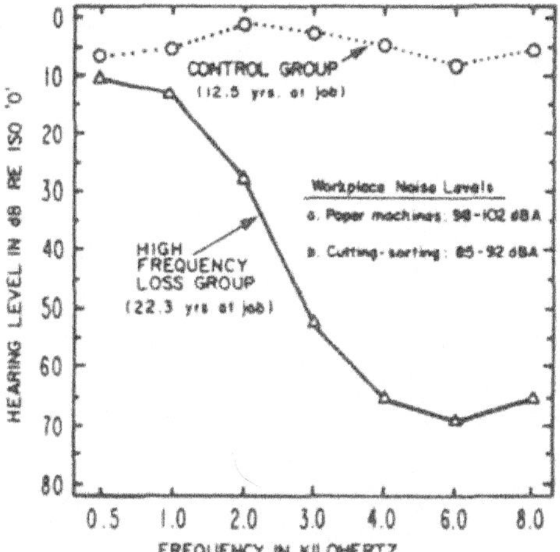

FIGURE 10.50. Mean hearing levels of group with high-frequency hearing loss vs control group with normal hearing. Only right ear data are shown. (From ref. 189.)

Manninen and Aro (ref. 191) divided a sample of 188 male and 92 female engineering industry workers into 3 groups—those with normal hearing, those with moderate hearing losses, and those with severe hearing losses. It was assumed that the hearing losses were noise induced. The average diastolic and systolic blood pressures were essentially the same between the three categories of hearing level in workers below the age of 41 years. The findings for workers between the ages of 41 and 64 years were somewhat inconsistent; those with moderate hearing losses had about a 15-mm Hg elevation in blood pressure above those with normal hearing, but the blood pressure of those with severe hearing losses was only 5 mm Hg above those with normal hearing.

Apparently, exposure to industrial noise sufficient to cause significant amounts of noise-induced hearing loss from many years of working in it does not mean that the workers will exhibit any blood pressure or other cardiovascular blood serum conditions that are indicative of any abnormal stress. It seems likely that in those studies in which workers in intense noise suffered increased incidence of cardiovascular, gastrointestinal, or other disorders related to the autonomic system, there are psychological, personnel selection, and sometimes adverse physical factors other than noise in the work environments that may be responsible for the observed adverse health effects. In some studies done of weaving, metal, and lumber mills and of shipyards, all involving the presence of generally harsh and sometimes hazardous environments, the presence of the subject health problems were observed. It is, of course, possible that an additive effect of all the adverse conditions, including that of noise, takes place. Even so, the role of the psychological and auditory interference aspects of noise cannot be dismissed as an explanation for such adverse nonauditory system health effects that could be attributed to the presence of the noise.

Short-term experiments

Some of the problems of retrospective studies, such as those just discussed, can be avoided in experiments in which the noise environment is controlled or manipulated and the physiological effects on the workers are monitored for any changes. Experiments of this type can be divided into two groups—those reporting adverse nonauditory-system effects of the noise and those showing no adverse effects.

In an experiment conducted by Ortiz *et al.* (ref. 192), 18 jet-engine test-bench workers who had been at that job for at least 3 years were exposed for one 3-hour test period to jet-engine turbine noise at levels from 105 to 115 dB. It is not stated whether or not earplugs were worn. Various analyses of urine, blood content, and blood pressure were made before, during, and after the exposure to the noise. Of the 18 subjects, 13 showed marked elevation in catecholamine excretions and increases in cholesterol, plasma-free fatty acids, blood pressure, and pulse frequency. The average age of this group was 39 years. Six of the subjects, average age 49 years, showed no significant responses to the noise. Ortiz *et al.* concluded that the blood pressure and catecholamine elevations found in some of the workers might be detrimental to people suffering from arteriosclerosis and other forms of vascular pathology.

These data should be viewed with caution because of the small number of subjects, the brief duration of the tests, and the lack of controls (no non-noise-exposed periods). Also contrary to the above findings are those from a study by Paolucci (ref. 193) who measured catecholamine excretion in experienced aircraft ground personnel the day before and the day of exposure to different levels of jet aircraft noise. Ten men were exposed to noise at a level of 120 dB for 1-1/2 hours in the jet-engine test area; 10 other men were exposed to jet aircraft takeoff noise at levels of 80 to 100 dB every 20 minutes for 5 hours. All the men wore earplugs. Neither group of men showed a significant increase in catecholamine excretion for the noise-exposure period of the second day over that found for the same period of the preceding "quiet" day.

If the subjects in the Ortiz *et al.* study did not wear earplugs (whether or not they did was not specified), they received noise at their eardrums some 15 to 20 dB more intense than did the subjects in the Paolucci study. This could explain the difference in the results between the two studies. However, another possible explanation is found in the aforementioned study of Finkle and Poppen (ref. 30), who exposed men wearing earmuffs to daily jet-engine noise at a level of 120 dB in a laboratory setting for a period of over a week. The subjects showed greatly increased secretions of catecholamine and other related hormones for the first days of exposure, but by the end of the week the noise ceased to cause these or related stress responses (*i.e.*, habituation had occurred). It is suggested that the positive findings for some of the subjects in the Ortiz *et al.* study could be due to a concern of these subjects that the noise was potentially harmful or stressful, and this concern was not habituated out during one 3-hour exposure.

Physiological stress responses to noise at the workplace were reported by Ising *et al.* (refs. 194 and 195). Twelve workers in a brewery where the noise was at an average level of 95 dBA worked for 1 week with earplugs and 1 week without earplugs. The earplugs provided an average daily noise reduction of about 13 dB.

It was found that working without earplugs increased the systolic blood pressure by almost 7 mm Hg, increased the excretion of vanillyl mandelic acid in urine by 67 percent, and increased the excretion of noradrenaline by 16 percent over those when working with earplugs. (All differences were statistically significant.) After 1 week of work without earplugs, magnesium concentration in the blood of the 12 workers was 5 percent lower than after 1 week of work with earplugs. The evaluation of the parameters of 26 test subjects showed a negative correlation of −0.52 between the magnesium content of blood sediment and the increase in blood pressure when exposed to noise.

There are, of course, other variables to be mentioned in the interpretation of these findings besides that of the 13-dB reduction in noise exposure reaching the eardrum. One is the well-known Hawthorne effect that workers respond favorably to changes in their environment that are intended to improve the environment. Why this should cause the particular physiological changes found is not obvious, however. Other possible factors are the following: (1) the workers feel some apprehension in the noise because of fear of auditory fatigue, or damage to hearing, and interference effects of the noise with hearing speech or other auditory cues helpful to job performance. The wearing of earplugs would not only relieve the apprehension but could also improve hearing performance (see chapter 4); and (2) the higher level of noise provides a greater continuous degree of physiological arousal than does the lower level.

Ising *et al.* (ref. 195) postulate that the stress effects elicited by noise can be schematically summarized as follows (ref. 196):

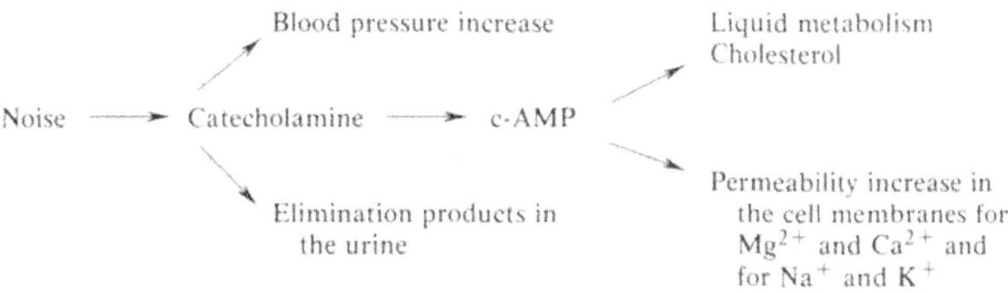

Whether or not these effects are related to psychological factors or to some reflexive arousal mechanism or to both and are sufficient to be pathogenic in some workers is, of course, a matter of conjecture.

A study previously mentioned was conducted by Cantrell (ref. 96) in which workers (U.S. Naval submarine personnel) were exposed to noise (sonar pings) in a simulation of cruising in a submarine. Twenty healthy young male volunteers were evaluated audiometrically, medically, and psychologically. They were then confined to a dormitory for 55 days. During the first 10 days, audiometric, mental and motor performance, and sleep patterns were evaluated. The subjects were then exposed to a pulsed, 10-step tone in the 3000 to 4000-Hz range for 0.66 sec every 22 sec, 24 hours per day for 30 days. The tonal pulses were presented at 80 dB (re 0.0002 dyne/cm^2) for 10 days, 85 dB for 10 days, and at 90 dB for 10 days *via* 118 loudspeakers hung from the ceiling throughout the building.

The most noticeable physiological effect was a statistically significant rise in plasma cortisol and blood cholesterol levels compared with preexposure levels. The levels decreased after cessation of the noise. Figure 10.51 shows the data for these two physiological measures.

The results of this study must be taken as inconclusive in demonstrating any causal relation between the 24-hour-per-day exposure to the intense tonal pulses and any noted physiological or psychological responses, however. The major reason is that the increase in cholesterol and plasma cortisol levels could well be attributed to the prolonged confinement imposed rather than the tonal pulses. Indeed, except for possibly the first period of exposure to the pulses, there was no consistent pattern of these physiological responses as the intensity level (including the cessation altogether) of the pulses was changed. Even the initial increases plotted in figure 10.51 are not necessarily related to the onset of the pulses, as clearly indicated by Cantrell; that is, the first increases plotted could have occurred prior to the onset of the pulses.

It is also obvious that in addition to the psychological aspects of confinement *per se*, with possibly related physiological stress reactions, the subjects could have been apprehensive about the possible effects of the noise on their hearing. Indeed, subjects exhibited varying amounts of audiometric threshold shifts during the study.

In a study made of some personnel involved and not involved in aircraft launch operations aboard a U.S. Navy aircraft carrier (refs. 196 and 197), there was a consistent tendency for those most exposed to noise to perform less well on a variety of physiological and psychological tests. Psychiatric examination of the men revealed that the men most exposed to noise had somewhat greater feelings of anxiety than the others. This anxiety may be because their jobs are inherently more dangerous or difficult than those of the men less exposed to noise. (Aircraft launch operations occasionally result in injuries and death to launch operators.) This seems borne out by data which showed that the men most exposed to the noise did not rate the jet aircraft noise as more disturbing than did other groups, but did express the most anxiety about their jobs.

Carlestam *et al.* (ref. 198) studied 22 young female IBM operators in their usual work situation. Half of the group was exposed to 6-dB increases in noise of

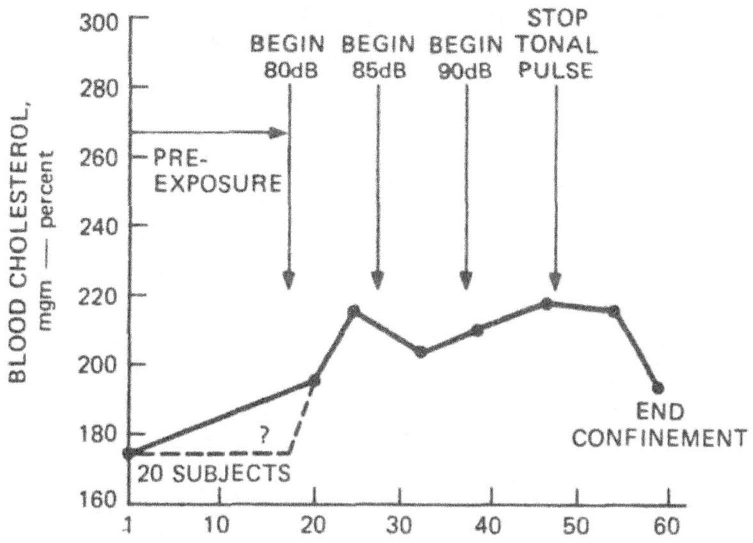

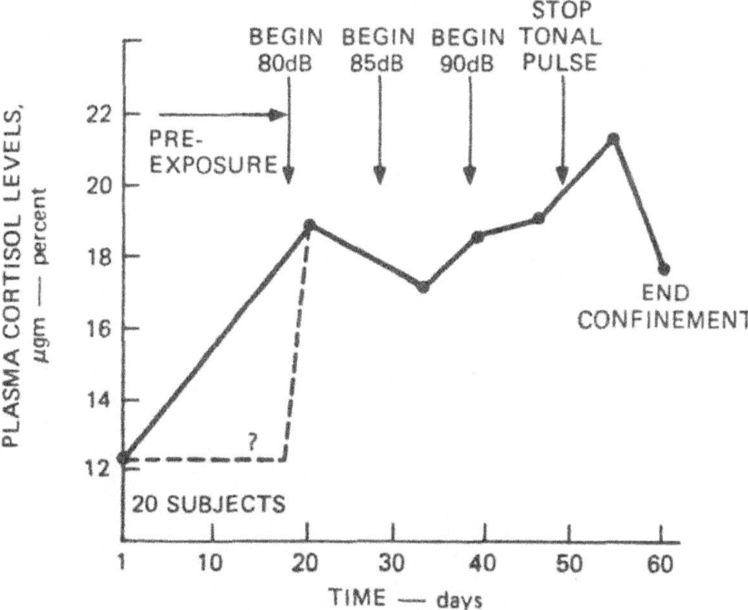

FIGURE 10.51. Mean blood cholesterol levels and plasma cortisol levels before, during, and after exposure to tonal pulses. (From ref. 96.)

their IBM machines for 4 consecutive days. The noise levels used were 76, 82, 88, and 94 dBC. The other half was exposed to the same noise levels but in decreasing order (*i.e.*, 94, 88, 82, and 76 dBC). The normal noise level for the office was 76 dBC. Each work day started with a 2-hour period of rest without noise exposure, followed by exposure to the appropriate noise for three 2-hour work periods.

The subjects experienced only minor increases in fatigue and "distress" with increasing noise. (See fig. 10.52.) However, the rating differences between those for the highest and lowest noise levels were very small.

Adrenaline and noradrenaline excretion levels remained low or moderate (fig. 10.53), and the changes in these levels were not significant for either control to noise periods or from lowest to highest noise levels. Not even the higher levels

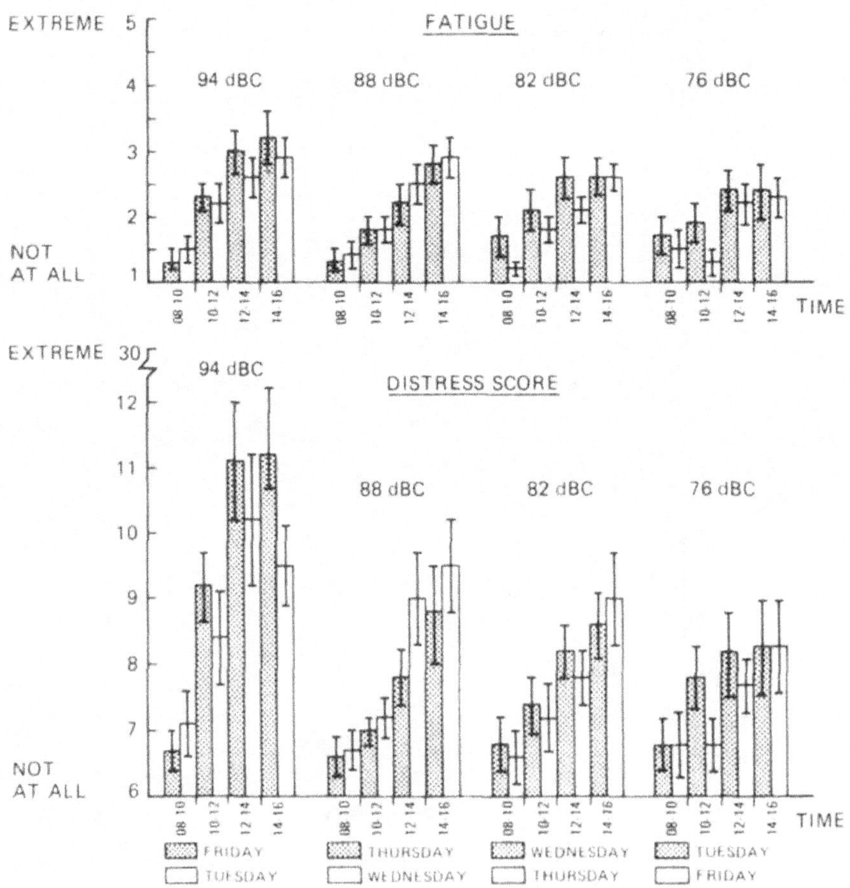

FIGURE 10.52. Self-rated fatigue and "distress" of IBM operators under different noise conditions and during different times of the day. (From ref. 198.)

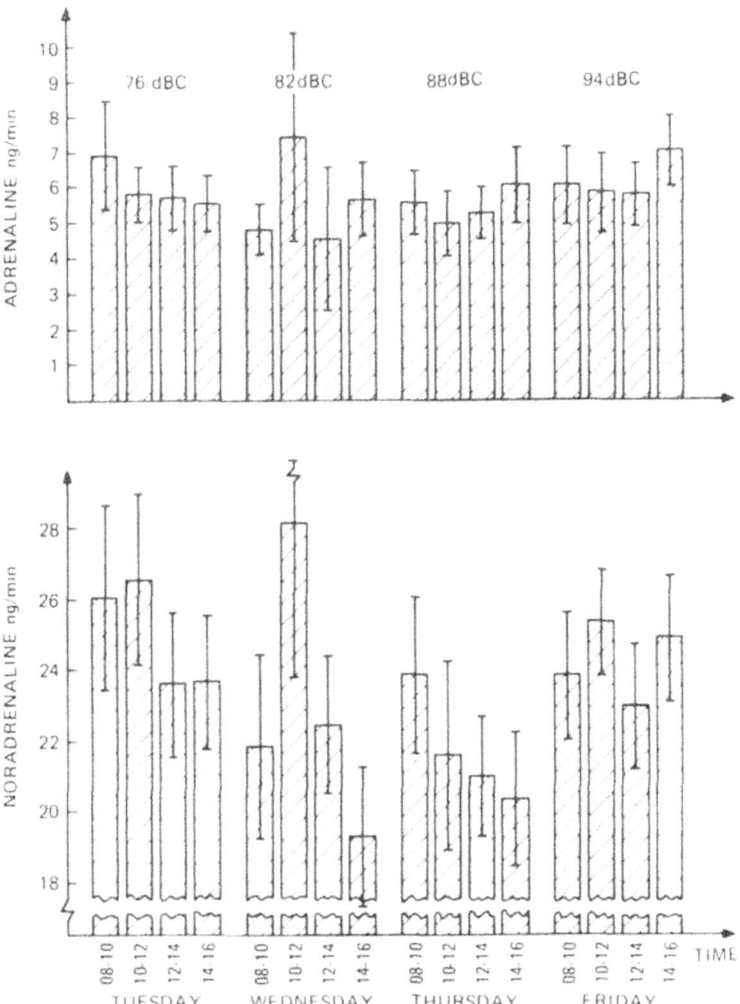

FIGURE 10.53. Urinary excretion of adrenaline and noradrenaline of IBM operators during 4 consecutive days with increasing noise levels. (From ref. 198.)

of noise seemed to be particularly stressful. Although they indicate that the subjects were familiar with the noise and had positive attitudes about their jobs and the experiment, Carlestam *et al.* question the general validity that noise at work is necessarily a pathogenic agent. To this might be added the comment that the trend towards increased fatigue and distress scores, though small with increased noise, could be due to some masking effects of the noise on hearing wanted sounds in the work environment or concern about auditory fatigue.

Health-Related Effects in Residential Communities

In most of the studies of the effects of noise as a "stressor" on workers, it is usually implied or assumed that (1) the effects are due to some reflexive form of over-arousal of the autonomic system, (2) the noise does not interfere with work performance by masking sounds relevant to good job performance, and (3) the noise does not cause feelings of fear or anxiety related to the work situation or concern about the damaging effects of the noise on hearing. From the above review of the effects of noise on autonomic-system stress reactions found in persons working in noisy environments, there are reasons to believe that all these assumptions may be invalid for a number of such studies.

In residential communities exposed to noise, however, it is obvious that the masking of wanted auditory signals and sounds and anxiety about safety and unwanted arousal from rest and sleep are effects of noise that will cause feelings of emotion and annoyance. Further, these emotions can cause autonomic-system stress responses, as previously discussed. In addition, it is doubtful there are any reflexive reactions to sounds or noises sufficient to be considered stressful to non-auditory systems. In any event, the existence of such reactions would not change the possibility that noise in the general living environment can prevent satisfactory living behavior and thereby cause ill health. Extensive reviews of research on noise as a possible factor in mental health, physical health, and behavior are found in papers by McLean and Tarnopolsky (ref. 199) and by Cohen and Weinstein (ref. 200).

Expecting adverse effects on health to occur because of noise in general environments seems unrealistic in view of the laboratory experiments with humans and animals that show complete habituation to noises with their repetition. There is, however, as mentioned before, no habituation possible to some of the interference effects of noise on auditory communications and sleep, and thus no needed relief from the annoyance and autonomic-system stress responses thereto. Also habituation does not necessarily occur for the annoyance experienced from noise-induced house vibrations or for the concern about accidents from the noise maker (*e.g.*, aircraft flying overhead at close distances). If anything, some accumulative annoyance with repetitions of the interference effects of some noises seems to occur.

It can be argued that people in residential communities would (or could) avoid the stressful indirect effects of the noise by putting little or no value on the events disrupted. Indeed, it would seem that this process could help explain individual differences in regard to annoyance from noise in the environment. However, to what degree accommodation to behavior interference effects of noise is to be expected in residential environments can only be found in empirical data obtained from communities of people. These data would provide normative descriptions of the liabilities to mental and physical health imposed by the behavior interference-annoyance effects of the noise.

Mental health

Meecham and Smith (ref. 201) compared admissions to mental hospitals in two groups that were matched to a considerable degree socioeconomically. One group, the control group, came from an area where aircraft noise was considerably below 90 dBA, and the other group, the maximum-noise area (MNA) group, came from an area near Los Angeles Airport where the aircraft noise reached levels of 90 dBA and higher (L_{dn} about 75 dBA). The results indicate a 29-percent increase in mental hospital admissions for the MNA group over the control group. The chi-square test of statistical significance is at the 90-percent level of confidence. These findings are in good agreement with those reported by Abey-Wickrama *et al.* (ref. 202) and Herridge and Chir (ref. 203), although Chowns (ref. 204) questioned the methodology of these last two studies.

Gattoni and Tarnopolsky (ref. 205) examined the admission rates to the same psychiatric hospital (Springfield Hospital) from the same areas near Heathrow Airport as analyzed by Abey-Wickrama *et al.* and by Herridge and Chir but for a later period of time (1966 to 1968 vs 1970 to 1972). Gattoni and Tarnopolsky defined the noise areas somewhat differently and also removed some data (those for "old people's care homes") from the high-noise areas that were included in the earlier studies. Table 10.12, from Gattoni and Tarnopolsky, shows how the original analysis of data from the Springfield Hospital compares with that for the 1970 to 1972 period.

It is shown in table 10.11 that the 1966 to 1968 data indicate a number of statistically significant differences between the admission rates for the maximum-noise areas and the lower noise areas, with the higher rates for the maximum-noise area. However, the 1970 to 1972 data show no statistically significant differences in this regard, but they do show a similar trend in that for 9 of the 16 category comparisons made, the rate of admissions from the high-noise zone was at least 10 percent greater than that from the lower noise area.

Taking another time period (1969 to 1973), the same population source (people near Heathrow Airport), and data from several psychiatric hospitals (including Springfield), Jenkins *et al.* (refs. 206 and 207) and Hand *et al.* (ref. 208) made further analyses of hospital admission data for residential areas that were exposed to levels of aircraft noise of $L_{dn} < 67$ dB, L_{dn} of 67 to 75 dB, and $L_{dn} > 75$ dB. Figure 10.54 shows that the results obtained are somewhat ambiguous—progressively greater hospital admission rates were generally found for the higher noise zones than for the lower noise zones at Holloway and St. Bernard's Hospitals but for Springfield Hospital there was a decrease in admissions as a function of noise zone. Figure 10.55 gives a somewhat more detailed analysis of admission of single household males to Holloway Hospital.

This difference between the data for Springfield Hospital and the two other hospitals is as perplexing as the fact that the low-noise-exposure data for Springfield Hospital are so different from those found for the same hospital in the ear-

TABLE 10.12

Comparison of two studies of admissions to Springfield Hospital

[From ref. 205]

		Abey-Wickrama *et al.* (1966 to 1968)[a]		Gattoni and Tarnopolsky (1970 to 1972)[b]	
Category	Admission	Area with higher admission rates	Statistical significance (c)	Area with higher admission rates (d)	Statistical significance (c)
Both sexes	All	Maximum-noise area[e]	0.005	None	Not significant
Both sexes	First	Maximum-noise area[e]	.01	High-noise zone[e]	
Females	All	Maximum-noise area[e]	.025	None	
Females	First	Maximum-noise area[e]	.10	High-noise zone[e]	
Males	All	None	Not significant	High-noise zone[e]	
Males	First	None	Not significant	High-noise zone[e]	
Females, age > 45	All	Maximum-noise area[e]	.005	None	
Females, age > 45	First	Maximum-noise area[e]	.0005	High-noise zone[e]	
Females, married	All	None	Not significant	None	
Females, married	First	None	Not significant	High-noise zone[e]	
Females, other	All	Maximum-noise area[e]	.01	High-noise zone[e]	
Females, other	First	Maximum-noise area[e]	.01	High-noise zone[e]	
Females, neurotic	All	Maximum-noise area[e]	.05	None	
Females, neurotic	First	None	Not significant	None	
Females, organic	All	Maximum-noise area[e]	.005	High-noise zone[e]	
Females, organic	First	Maximum-noise area[e]	.0005	High-noise zone[e]	

[a] *Reference 202.*
[b] *Reference 205.*
[c] *Probability that could be due to chance.*
[d] *An area is chosen for this column if its rate is at least 10 percent greater than the other noise zones.*
[e] *Noise greater than an NNI of 55 ($L_{dn} > 75$ dB).*

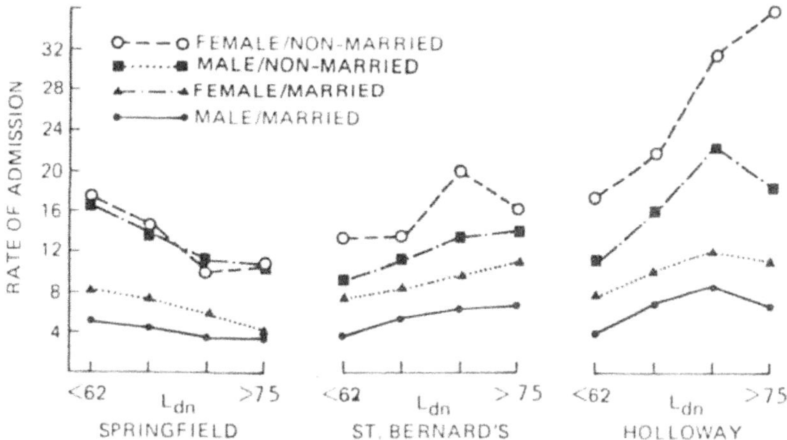

FIGURE 10.54. Age-standardized psychiatric hospital admission rates per 1000 people as a function of noise, sex, and marital status. (From ref. 207.)

lier studies. It is shown in table 10.13 that the admission rate per 1000 at risk, taken over the 4-year period, was 15.09 in areas exposed to aircraft noise of $L_{dn} < 67$ dB, 6.09 in L_{dn} of 67 to 75 dB, and 5.619 in $L_{dn} > 75$. The average of the rates for $L_{dn} < 67$ dB and L_{dn} of 67 to 75 dB is taken as being representative of the same areas as those for $L_{dn} < 75$ dB in the Abey-Wickrama *et al.* and the Gattoni and Tarnopolsky analyses. This average rate of 10.6 per 1000 for the 4-year period of the Jenkins *et al.* analysis is divided by 2 to give a 2-year rate of 5.3 per 1000 that should be comparable to those rates found in the 2-year periods analyzed by Abey-Wickrama *et al.* and by Gattoni and Tarnopolsky (rates of 3.471 and 3.210 per 1000, respectively).

Abey-Wickrama *et al.* found a significant difference between admission rates for low-noise and high-noise areas; more hospital admissions were found in high-noise areas. (Gattoni and Tarnopolsky's data had a similar, but nonsignificant, trend.) However, Jenkins *et al.* found a highly significant difference in the opposite direction. They found that even higher admission rates occurred in low-noise areas.

Jenkins *et al.* (ref. 207) suggest that these findings may indicate that the trends of a positive relationship between psychiatric hospital admission rates and exposure to aircraft noise found in the other analyses were due to chance.

However, unusual circumstances may have been associated more with data from the Springfield Hospital than with the other data analyzed by Jenkins *et al.*

Jenkins *et al.* obtained socioeconomic information (1971 census) about the populations in the catchment areas of the three hospitals (tables 2, 3, and 4 of ref. 207) that provides a basis for better understanding the seemingly deviant Springfield Hospital data and the somewhat neutral results of the earlier Gattoni and Tarnopolsky analysis. For example, the socioeconomic data indicate a nega-

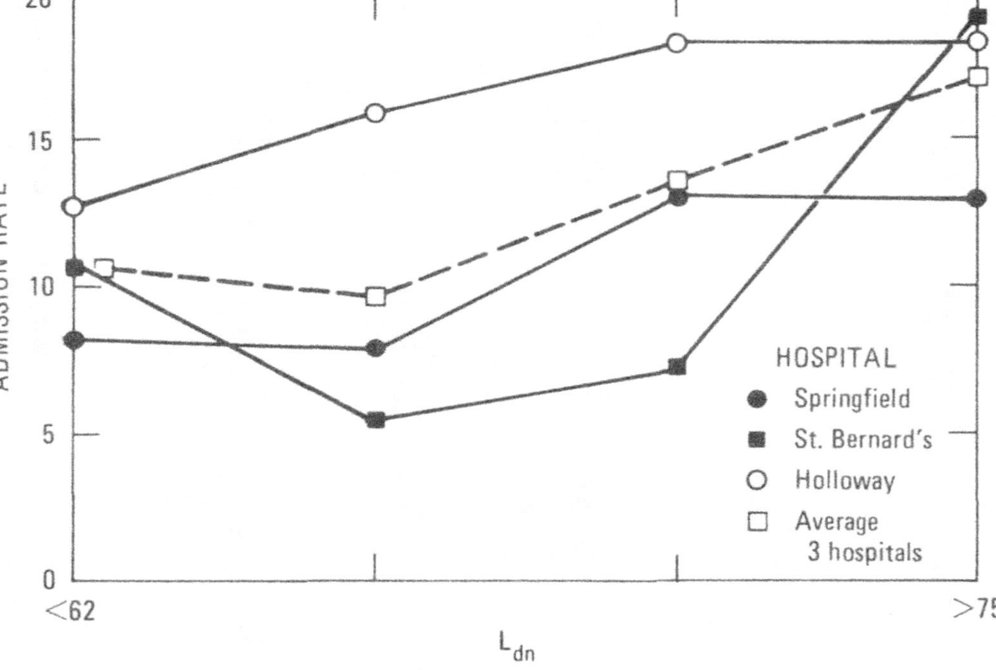

FIGURE 10.55. Age—and affluence/unemployment—standardized psychiatric hospital admission rates per 1000 people regardless of sex or marital status as a function of aircraft-noise-exposure level. (Basic data from ref. 107.)

TABLE 10.13

Comparison of Springfield Hospital Admission Rates

[Rates per 1000 persons; all categories of admissions]

Study	Admission rate for high-noise area[a]	Admission rate for low-noise area[b]
Abey-Wickrama *et al.* (1966 to 1968), ref. 202	[c]4.537	[c]3.471
Gattoni and Tarnopolsky (1970 to 1972), ref. 205	[d]3.462	[d]3.210
Jenkins *et al.* (1969 to 1973), refs. 206 and 207	[e]2.8	[e]5.3

[a] $L_{dn} > 75$ *dB.*
[b] $L_{dn} < 75$ *dB.*
[c] *Statistically significant difference supporting hypothesis that aircraft noise increases hospital admissions.*
[d] *Difference not statistically significant.*
[e] *Statistically significant difference supporting hypothesis that aircraft noise decreases hospital admissions. The Jenkins et al. study is for a 4-year period, and the rates they report must be divided by a fact of of 2 in order to be roughly compared with the earlier 2-year studies.*

tive relation is the Springfield catchment area between aircraft-noise-exposure level and several socioeconomic characteristics (percentage of one person households, percentage of people unemployed, and general level of affluence) that are factors probably positively correlated with psychiatric disorders.

I have undertaken a further analysis of the hospital admission rate, socioeconomic, and aircraft-noise-level data reported by Jenkins *et al.* Some preliminary findings from this analysis are shown in figure 10.55. It is shown in that figure that standardizing the admission rates for the different noise zones of the three hospitals with respect to general level of affluence and unemployment in each area provides a more consistently positive relation between level of exposure to aircraft noise and admission rates to the three hospitals than is depicted in figure 10.54. It might be further noted that the downturn in admission rate at the Holloway Hospital shown in figure 10.55 for the area with an L_{dn} of greater than 75 dB as compared with the next lower level of exposure could be due to data unreliability as a result of the relatively small number of persons at risk in that particular area with $L_{dn} > 75$ dB. Only 1360 people lived in that area of the Holloway Hospital catchment basin, whereas a minimum of 19 000 lived in each of the other areas of the hospitals. (In the analysis just described, "affluence" is based on data with respect to the percentage of people renting from local authorities, percentage of households with exclusive use of an indoor toilet, percentage of houses owners occupied, and percentage of people of professional, managerial, or employer class. The admission rates reported by Jenkins *et al.* were further standardized by multiplying them by the ratio of an affluence and unemployment score for each noise area of the three hospitals to the average affluence score of all the noise areas of the three hospitals.)

Annoyance and mental health

Distinctions can be made between "sensitivity" to exposures to noise and annoyance from noise as measured in attitude surveys of what bothers people in everyday living. For example, the noise occurrences that interfere with speech communication are likely to do so more frequently and cause greater feelings of annoyance in people who engage most actively in such behavior. Thus, although noise annoyance can perhaps be a burden that creates problems for highly sensitive people, it appears that it creates the most annoyance for persons engaging in normal behavior; that is, it occurs most often in normal people. This finding comes from a number of studies (*e.g.*, ref. 209).

The role of aircraft noise as a contributor to annoyance in people with psychiatric illnesses and the role of annoyance as a possible contributor to psychiatric illness were studied in a survey of psychiatric illness and annoyance from aircraft noise in different aircraft-noise-impacted areas near London Heathrow Airport (ref. 210). In this study psychiatric illness was identified by means of a screening questionnaire and not by admission to a psychiatric hospital. The principal results are shown in figures 10.56 and 10.57.

It is shown in figure 10.56 that a greater percentage of psychiatrically morbid people suffered the highest annoyance than suffered any other degree of annoyance, which would seem to suggest that the aircraft noise could contribute to an increased degree of suffering in those who are psychiatrically ill. However, as shown in figure 10.57, there was no general increase in the number of psychiatric "cases" as a function of the level of aircraft noise exposure. Indeed, the increase in annoyance as the aircraft-noise-exposure level increased occurred predominantly in the normal population. Tarnopolsky *et al.* concluded that most of the people who complain of aircraft noise are psychiatrically normal.

The fact that an average of about 22 percent of the population was identified as being psychiatrically morbid in each level of aircraft noise exposure would indicate that there were many types of such illnesses revealed by the questionnaire given and that aircraft noise was not a significant cause, if at all, of the total group of illnesses. At the same time, the data showing an increased degree of annoyance in those psychiatrically ill would be consistent with the trends in increased psychiatric hospitalization admission rate with increased exposure to aircraft noise.

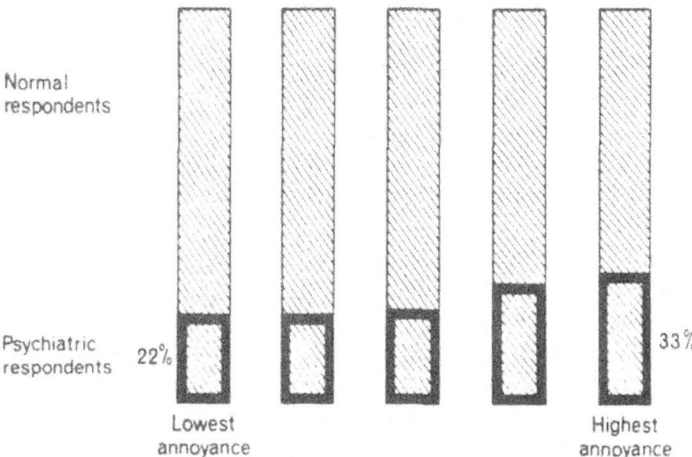

FIGURE 10.56. Proportion of psychiatric cases among persons who report interference with their activities. (From ref. 210.)

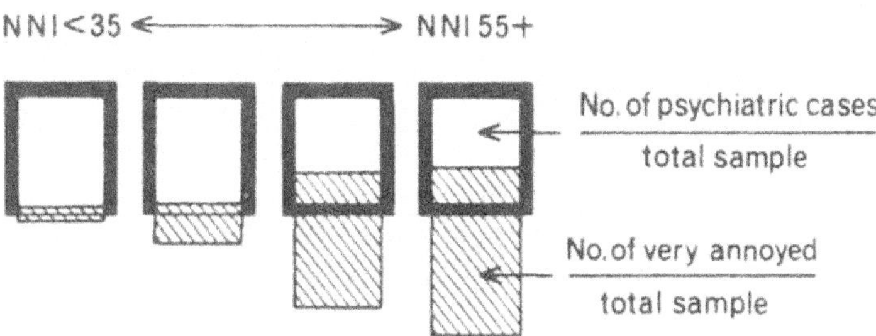

FIGURE 10.57. Relations between noise, psychiatric cases, and extreme annoyance. (From ref. 210.)

In summary, the Los Angeles area data, the 1966 to 1968 and 1970 to 1972 Springfield Hospital data, and the 1969 to 1973 Holloway and St. Bernard Hospital data for the London area support, to varying degrees, the hypothesis that intense aircraft noise increases admission rates to psychiatric hospitals. The 1969 to 1973 data for the Springfield Hospital indicate an opposite trend, that less aircraft noise increases the hospital admission rate. However, the pattern of admissions to the Springfield Hospital from the "quieter" areas in the 1969 to 1973 period suggests these particular data were possibly influenced by strong population-social factors not comparable with those present in the other studies cited. Also, McLean and Tarnopolsky (ref. 199) make the point that mental disorders sufficient to warrant admission to a psychiatric hospital represent but the "tip of the iceberg of mental health problems."

Stress- versus non-stress-related health disorders

It is possible that the apparent stressful effects of environmental noise could be reflected in more common physical and mental ailments than pathological mental disorders requiring hospitalization. Several studies have been published reporting such effects of aircraft noise on people.

Karagodina *et al.* (ref. 211) conducted a large-scale study of the effects of aircraft noise on (among other things) the physical health of the adult population (over 15 years of age) living near nine Soviet airports. Health statistics (145 000 medical diagnostic records were analyzed) revealed that the populations within about 3.7 statute miles (6 km) had 2 to 4 times the amount of otorhinolaryngological diseases (otitis and auricular neuritis), cardiovascular diseases (hypertension and hypotension), nervous diseases (neuritis and asthenic states), and gastrointestinal diseases (gastric and duodenal ulcers and gastritis) as those outside that perimeter. These investigators concluded from these and related laboratory and survey studies that the maximum permissible exterior level of aircraft noise should be set at 85 dBA during the daytime and 75 dBA at night.

Unfortunately, any trend in the incidence of these diseases as a function of specific levels of exposure to aircraft noise were not given in the Karagodina *et al.* paper. Nonetheless, this study, published in 1969, is important particularly in that its findings are strongly supported by a later investigation (described below) in which the relations between exposure to L_{dn} and certain diseases are derived.

The results of an interdisciplinary study of aircraft noise effects in the area of Munich airport were published in 1974 (ref. 212). As part of that study, 192 men and 200 women were selected from 32 areas in which aircraft noise exposure was measured. These persons were given certain clinical medical tests. It was concluded that there were no major clinical disorders related to the aircraft noise present in the more or less healthy population selected, but that there were disorders of sleep and blood pressure reactions to the aircraft noise with an increased risk of hypertension in those more heavily exposed to noise.

Koszarny *et al.* (ref. 213) administered a health questionnaire to 256 residents in an area where the aircraft noise levels exceeded 100 dBA and to 255 residents in somewhat quieter areas (80 to 90 dBA peak noise). No statistically significant differences in complaints about ailments were found in groups of men living in the two areas. However, significantly greater numbers of complaints related to the cardiovascular system, the digestive system, frequency of taking medication for heart problems or headaches, and nervousness were found in women living in the noisier area than in women who lived in the lower noise level area. The authors note that the men worked in acoustically unfavorable environments outside their residential area. The results suggest that the health of the women in the noisier area was relatively more adversely affected than the health of the women in the less noisy residential areas, but the health of the men from the two residential areas did not differ because both groups worked in noisy industries and spent relatively less time in their homes than did the women.

Grandjean *et al.* (ref. 214) found a progressive increase in the reported use of tranquilizers and sleeping pills as the exposure level to aircraft noise increased in different neighborhoods. (See table 10.14.) The data collected by Grandjean *et al.* that are shown in the table and other data of theirs discussed later indicate that the increased use of these sedatives in higher levels of exposure to aircraft noise was due to the interference effects of the noise with sleep and speech communications.

The most substantial research data showing a probably causal relation between exposure to aircraft noise in residential neighborhoods and adverse health effects are those obtained by Knipschild (refs. 215 to 217) and Knipschild and Oudshoorn (ref. 218). In one of these studies (ref. 215) about 6000 people were given medical examinations (heart X rays, ECG, blood pressure, height, and weight) and World Health Organization standard questionnaires for angina pectoris, for medical treatment and drugs for heart trouble, and for smoking habits. The 6000 people represented about 42 percent of the people who were invited to participate in the cardiovascular screening study from eight villages near the Amsterdam Schiphol Airport. (See fig. 10.58.) All the subjects were screened by the same staff, equipment, and methods.

The main results of this community cardiovascular screening survey are shown in table 10.15 and figure 10.59. Table 10.15 shows that the incidence of medical treatment of heart disease, use of cardiovascular drugs, pathological heart shape, and hypertension were all significantly greater (statistically) in people from the higher aircraft noise areas than in the lower aircraft noise areas. The differences were in the same direction with respect to angina pectoris and pathological ECG but were not statistically significant. The most significant statistical

TABLE 10.14

Effects of Aircraft Noise on Behavioral Reactions

[Hours from 0600 to 1800; data from ref. 214]

Behavioral reaction	Percent of people interviewed[a] responding affirmatively for L_{dn} of—						
	<47 (223)	47 (485)	55 (460)	63 (1066)	71 (1065)	77 (540)	87 (73)
Use ear protection	2	0.5	2	4	4	7	22
Use tranquilizers + sleeping pills	1	0	2	5	5	10	20
Close windows	2	.4	13	20	28	46	55
Remain less outdoors	1	0	4	4	11	22	34
Wish to move away	.4	0	2	9	17	35	33

[a] *Number of people interviewed given in parentheses.*

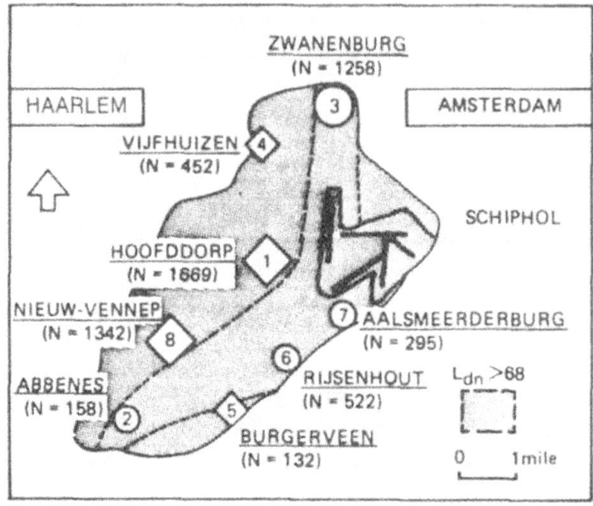

(a) Villages in cardiovascular survey. (Data from ref. 215.)

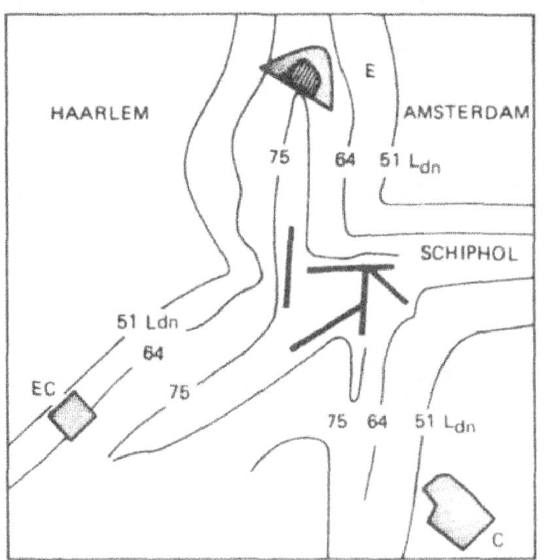

(b) Village areas for general practice survey. (Data from ref. 216.)

FIGURE 10.58. Location of villages and village areas for surveys.

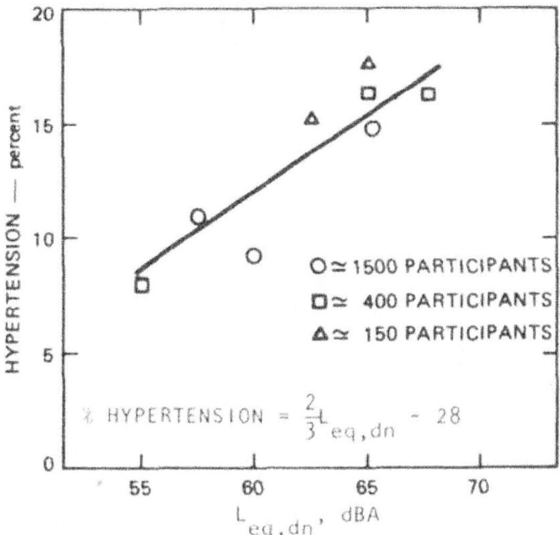

FIGURE 10.59. Aircraft noise and the prevalence rate of hypertension. (From ref. 217.)

TABLE 10.15

Main Results of the Community Cardiovascular Survey

[Data from ref. 217]

Cardiovascular condition	Participants[a], percent, affected by L_{dn}, dBA, of—		Fisher's test for significance
	<62.5 (3595)	>62.5 (2233)	
Angina pectoris	2.8	3.0	Not significant
Medical treatment of heart disease	1.8	2.4	0.04
Use of cardiovascular drugs	5.6	7.4	.003
Pathological ECG	4.5	5.0	Not significant
Pathological heart shape	1.6	2.4	0.01
Hypertension[b]	10.1	15.2	<.001

[a] *Number of participants given in parentheses.*
[b] *Blood pressure >175/100 mm Hg or use of antihypertensive drugs or both.*

difference is that for hypertension (systolic/diastolic blood pressure greater than 175/100 mm Hg or the taking of antihypertensive drugs or both). As shown in figure 10.59 and the table, the prevalence rate increased from about 10 to 15 percent.

One possible confounding factor in such a study is that the people from the less noisy areas may be socioeconomically advantaged (and perhaps in somewhat better health) than those from the areas with more aircraft noise. It can be argued from some research data that the more affluent and better educated people are more annoyed by aircraft and road traffic noise (and presumably would suffer more stress-related health effects) than the less affluent and less educated groups of people who more generally live in noise-blighted areas. For example, Ko and Wong (ref. 219) concluded that people of higher education, higher income level, and with managerial jobs are more annoyed by road traffic noise. Bradley and Jonah (ref. 220) also found a similar socioeconomic influence on annoyance from traffic noise. Knipschild (refs. 215 and 216) was of the opinion that there were some indications that although those in the lower noise levels were more affluent, the difference was not sufficient to explain the differences in health disorders found between the two groups of people. Age, sex, smoking habits, weight of the subjects, and the size of the villages in which they resided did not account for the differences.

Possibly the people who volunteered for the screening examinations were not representative of the general population, even though they constituted 42 percent of it. It is likely that mostly those people concerned about possible cardiovascular problems would volunteer for such testing. However, there is no obvious reason that this would be a greater motivation in the noisier area than in the quieter areas, other than the possibility that there were more cardiovascular problems present in the noisier areas because of the noise.

A second study by Knipschild (ref. 216) demonstrated that the apparent increase in the incidence of cardiovascular health problems was almost certainly not because of some non-health-related differences in volunteer motivations of people from the noisier and quieter areas. In the study, the records of general medical practitioners serving local villages were examined to determine the physician contacts that had been made for health problems during a 1-week period (March 13-18, 1974). The location of the villages studied are shown in figure 10.58(b). The respective number of general practice physicians and the size of the populations at risk were 9 and 17 500 in area C, 4 and 5650 in area EC, and 6 and 12 000 in area E.

The main results are shown in tables 10.16 and 10.17 and figure 10.60. The data in table 10.16 are consistent with the cardiovascular survey data in table 10.15 in that the incidence of hypertension is less in the lower L_{dn} areas than in the higher areas (an increase of about 24 percent in the physician contact rate for hypertension between $L_{dn} < 60$ and L_{dn} of 60 to 65, 39 percent between L_{dn} of 60 to 65 and $L_{dn} > 65$, and 72 percent between $L_{dn} < 60$ and $L_{dn} > 65$).

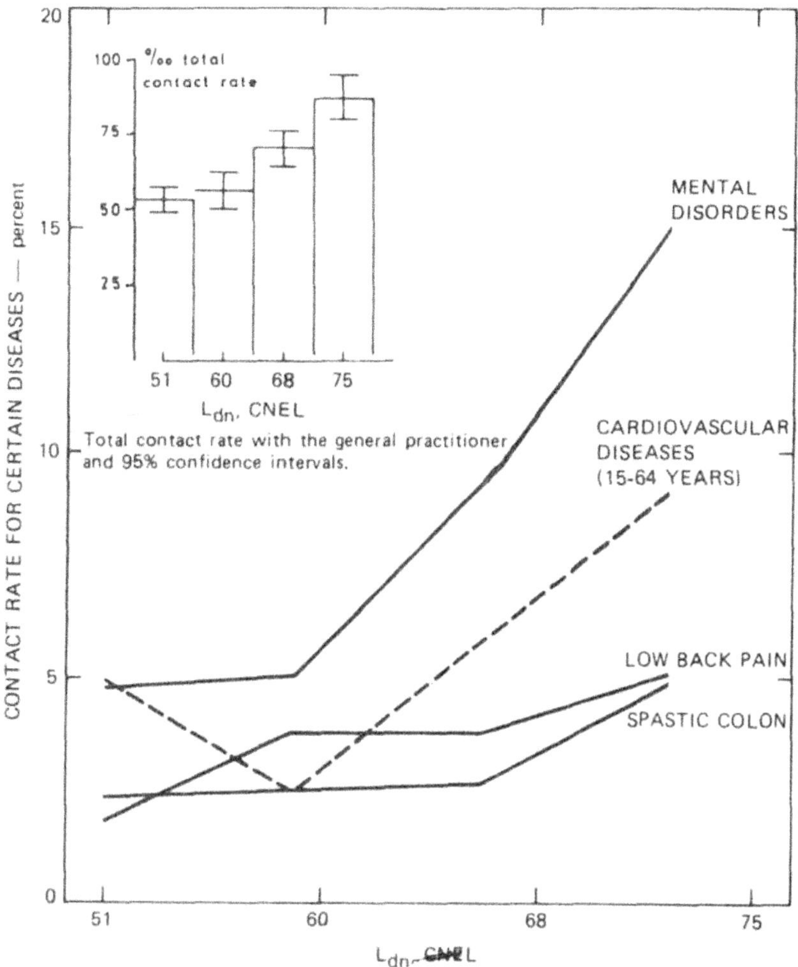

FIGURE 10.60. Physician contact rate for certain diseases in areas with more and less aircraft noise. (Data from ref. 216.)

For the purpose of further analysis, the four disorders identified in table 10.16 as psychological problems, psychosomatic problems, cardiovascular disease, and hypertension are grouped together as representing the disorders most likely due to stress from, among other things, exposure to aircraft noise. The remaining physician contacts are, for the present discussion, assumed to be more likely due to non-stress-related causes.

Table 10.17 shows that as the noise level increases (*e.g.*, $L_{dn} < 60$ vs $L_{dn} =$ 60 to 65), there is an increase in the percent of physician contacts for the non-

TAELE 10.16

Main Results of General Practice Survey

[Data from ref. 217]

Reason for physician contact	Population[a], percent, contacting physician for L_{dn}, dBA, of—			χ^2 test for linear trend
	<60 (14625)	60-65 (4050)	>65 (3650)	
Psychological problems	0.65	1.13	1.75	<0.001
Psychosomatic problems[b]	1.12	1.54	1.69	.001
Cardiovascular disease	.46	.60	.82	.004
Hypertension	.25	.31	.43	.03
Total, stress effects	2.48	3.58	4.69	
Total, contacts	5.71	7.97	9.34	<0.001

[a] *Population at risk given in parentheses.*

[b] *Consist of low back pain, spastic colon, stomach complaints, allergic diseases, tinnitus, dizziness, and headache.*

TABLE 10.17

Percentages and Percent Increases of Physician Contact for Stress- and Non-Stress-Related Health Disorders

[See table 10.16 for basic data]

Description of parameter	Physician contacts, percent, for L_{dn}, dBA, of—			Increase in contacts, percent, for L_{dn}, dBA, comparison of—		
	<60	60-65	>65	<60 vs 60-65	60-65 vs >65	<60 vs >65
Sum of 4 stress[a] categories of health disorders	2.48	3.58	4.69	44	31	89
Total contacts, all disorders, minus sum of 4 stress categories	3.23	4.39	4.65	36	6	44

[a] *Stress categories are psychological problems, psychosomatic problems, cardiovascular disease, and hypertension.*

stress-related health disorders of 35, 6, and 44 percent. Accordingly, there is some indication that the people living in $L_{dn} > 60$ suffer from more health disorders in general than do people in $L_{dn} < 60$.

Table 10.17 also shows that there are greater increases in stress-related disorders than in non-stress-related disorders as noise levels increase (41, 31, and 89 percent compared with 35, 6 and 44 percent). This greater increase in stress-related health disorders than in non-stress-related disorders suggests that increasing the level of aircraft noise exposures is at least partially responsible for increases in stress-related disorders.

It is also possible that stress-related disorders increase with noise-exposure level at a greater rate than other types of health disorders because of a stronger correlation between socioeconomic status and stress-related disorders than with other types of health. On the other hand, it is also possible that some of the non-stress-related health disorders were also related to aircraft noise exposure and that the socioeconomic status differences noted by Knipschild were not large enough to account for differences in physician contacts.

Longitudinal study

Supportive evidence that aircraft noise was in fact a causal factor in the stress-related health disorders found in the community cardiovascular survey and the physician contact study was obtained by Knipschild and Oudshoorn (ref. 218). In this study, records of the purchase of certain drugs for the period of 1967 to 1974 were obtained from pharmacists in village areas C and E. (See fig. 10.58(b).) Area C had essentially no exposure to aircraft noise during this period whereas area E had essentially no aircraft noise until 1969, after which time, because of the opening of a new runway at Schiphol Airport, the noise was of the order of $L_{dn} > 64$. The authors state that there were essentially no changes in age distribution or socioeconomic status within each of the areas during the 1967 to 1974 period.

Figure 10.61 shows the number of certain drugs obtained per adult per year in the two areas. Clearly, the use of most of the drugs shows a steady increase after 1967 in area E (the aircraft-noise area) and no systematic change in area C (the no-aircraft-noise area). An exception appears to be for sedatives, which showed a decrease in area E after 1972 from a previously increasing trend in their use. The authors note that in 1973 nighttime aircraft operations were shifted to the daytime because of a curfew placed on nighttime aircraft operations. This possibly reduced the need for such sedatives to induce sleep during the night.

The trend curves in figure 10.61 seem to show a possible causal relation between exposure to aircraft noise and stress-related health conditions, particularly hypertension requiring medication. This relation cannot be readily attributed to any socioeconomic or other such factors in that the health disorders requiring certain medications steadily increased in a population following the onset and continuation of exposure to aircraft noise, whereas in a nearby control popula-

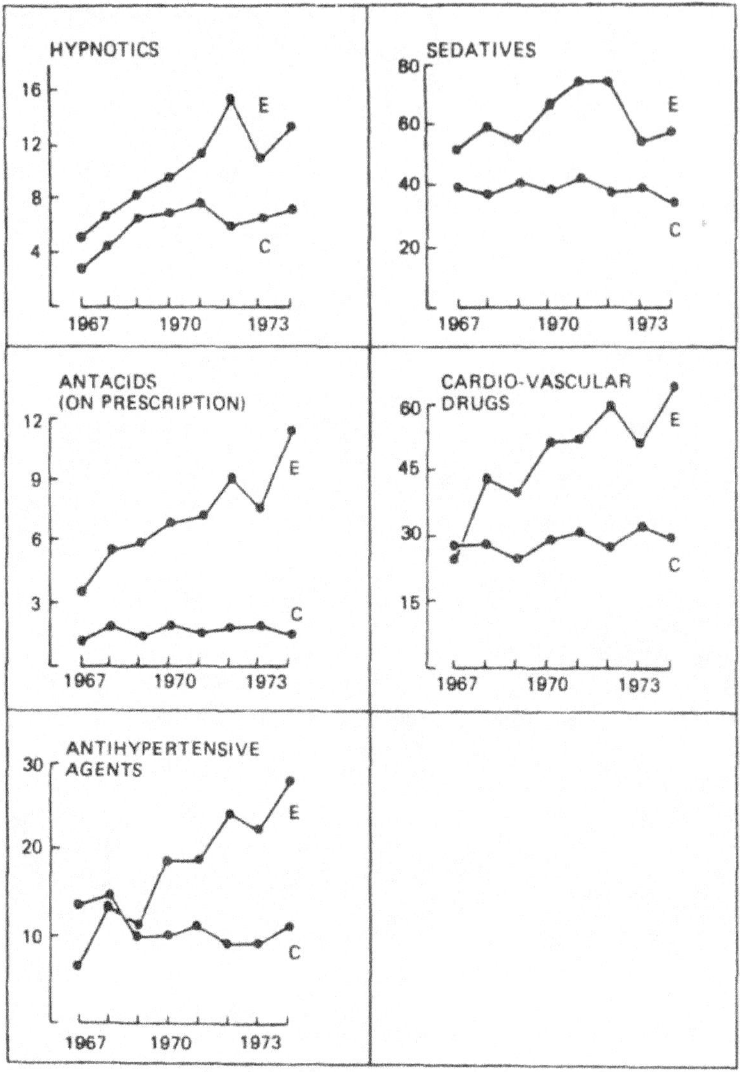

FIGURE 10.61. Number of certain drugs per adult per year. Area C—$L_{dn} \leq 51$; area E—$L_{dn} \leq 51$ (1967–1969) and $L_{dn} \geq 64$ (1969–1974). (From ref. 218.)

tion not exposed to aircraft noise, studies over the same period showed no such increase.

It might be suggested that healthier (*i.e.*, wealthier) people moved from the higher aircraft noise area over time and were replaced by generally less healthy people. Alternatively, it would seem more likely that there would be a trend for those most bothered by the noise to move to quieter areas and bias the results in the opposite direction. More importantly, the investigators maintain that the populations involved remained stable with respect to socioeconomic average distributions over the period of the study.

The Knipschild and Oudshoorn study is unique among studies of the impact of community noise on the health condition of a large population *before* exposure to an aircraft noise environment as well as during exposure for a number of subsequent years. Further, the study was "double blind" in that neither the populations under study nor the experimenters or pharmacists knew before or during the period of noise exposure that the impact of the aircraft noise on drug usage was to be investigated.

Knipschild's studies are of particular importance in that they demonstrate a quantitative relation between health disorders over a relatively wide range of aircraft noise dosages measured in L_{dn}. Other studies merely compare health conditions of groups of people from areas with some unspecified amount of "high" aircraft noise exposure with groups from areas with "low" levels of aircraft noise exposure.

Meecham and Shaw (ref. 221) examined mortality rates and causes in aircraft noise-impacted areas versus quiet areas in Los Angeles. They reported greater death rates due to cirrhosis of the liver and strokes in the noisier areas than in the quieter areas. However, a reanalysis of the same data, taking into account some differences between the two areas with respect to age, race, and sex, showed no such differences in mortality rates (ref. 222). Meecham and Shaw (ref. 223) and Frerichs *et al.* (ref. 224) have published additional, inconclusive data and comments on this question.

The use of mortality rates as a measure of a possible increase in deaths due to diseases supposedly indirectly related to stress because of aircraft noise (stress causing alcoholism causing cirrhosis of the liver and stress causing cardiovascular disorders causing strokes) provide data that are perhaps too weak to be meaningful. Not only are these diseases responsible for relatively small numbers of deaths, they are presumably related to a number of other more direct causes, thus making the data from even large populations of people insensitive to the possible impact of indirect causes upon the subject mortality rates.

A reasonable hypothesis would be that stress of the magnitude induced by aircraft noise would not be sufficient to cause a measurable change in the pattern of mortality rates for different diseases. Rates of admission to psychiatric hospitals, contacts with physicians for stress-related diseases, use of drugs, and, as discussed later, changes in fetal development are all probably more sensitive methods of detecting possible adverse effects of aircraft or other environmental noise on health than is a study of relative mortality rates for different diseases.

Street traffic noise

It is shown in chapter 11 that for a given level of L_{dn}, aircraft noise causes greater annoyance than does street traffic noise (by a factor of about 10 dB). Not inconsistent with this finding is the result of a study by Knipschild and Sallé (ref. 225) in which it was found that women living in areas with high levels of street noise did not have significantly more cardiovascular disorders than women living in quieter neighborhoods. Also, Mroz *et al.* (ref. 226) reported that a survey of 1901 Warsaw residents found that about 1/3 were annoyed to some degree because of street traffic noise but that there were no obvious health-related effects. (As noted previously, Koszarny *et al.* (ref. 213) found women living in Warsaw in high aircraft noise areas suffered more health disorders than did women living in less noisy areas.)

In brief, it would appear that street traffic noise in residential areas is generally not sufficiently intense to cause cardiovascular stresses in adults, but aircraft noise can be. As mentioned previously, this is consistent with attitude surveys of annoyance from these two kinds of environmental noise.

Health in children

Karagodina *et al.* (ref. 211) reported the results of a clinical examination of school children 9 to 13 years old carried out in 1965 and 1967 in settlements adjacent to Moscow airports and in a control settlement remote from the airport. It was found that those children living near the airports had functional changes in the cardiovascular system and in the nervous system consisting of increased fatigue, blood pressure abnormalities, higher pulse lability, and cardiac insufficiency (revealed by functional loads and autonomic vascular changes).

Figure 10.62 (from Karsdorf and Klappach (ref. 227)) shows blood pressure levels in high-school children exposed in the classroom to intruding street traffic noise of different intensities. Figure 10.63 (from Cohen *et al.* (ref. 228)) shows blood pressure for students in elementary school in Los Angeles that were matched socioeconomically but differed in that three of the schools were in areas impacted by aircraft noise (peak levels of 95 dBA) and three were in quiet neighborhoods.

It is shown in both figures that the students in the higher levels of noise had elevated diastolic and systolic blood pressures compared with the students in quieter areas. Cohen *et al.* (ref. 229) suggested that children are possibly somewhat more susceptible to increased blood pressure from environmental noise than adults. However, tests made 1 year later revealed that the students then in the noisier schools had blood pressure levels similar to those in the quieter schools because the students with higher blood pressure had moved from the area. Also, they found no difference in height/weight ratios (a supposed indirect measure of health) between the students from the quiet and from the noisy areas.

Cohen *et al.* (ref. 230) have reviewed their own findings and those of other investigators concerned with cardiovascular and behavioral effects of community noise. The apparent deleterious effects of aircraft noise in areas around schools and around the home upon learning achievement in the school are discussed in chapter 11.

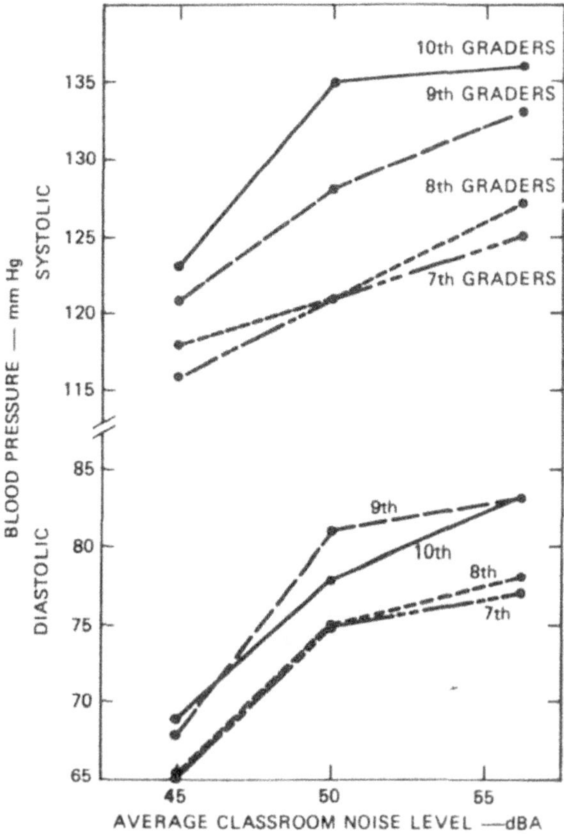

FIGURE 10.62. Blood pressure of students as a function of noise level of intruding street traffic noise. (Data from ref. 227.)

Health in fetuses

Ando and Hattori (ref. 231) and Knipschild *et al.* (ref. 232) found that infants born to mothers living in areas exposed to intense aircraft noise had lower average weights at birth than did infants born to mothers living in quiet neighborhoods. Ando and Hattori found that as the level of the aircraft noise increased the incidence of babies with low birth weights increased. In a later study, Ando and Hattori (ref. 233) found human placental lactogen (HPL) to be lower in the serum of expectant mothers in areas of high aircraft noise than in areas of low noise. (See fig. 10.64.) It is highly likely that these effects, if valid, are due to feelings of fear and annoyance engendered in the mothers and are not due to any acoustic energy transmitted to the fetus through tissues or fluids of the mother's body.

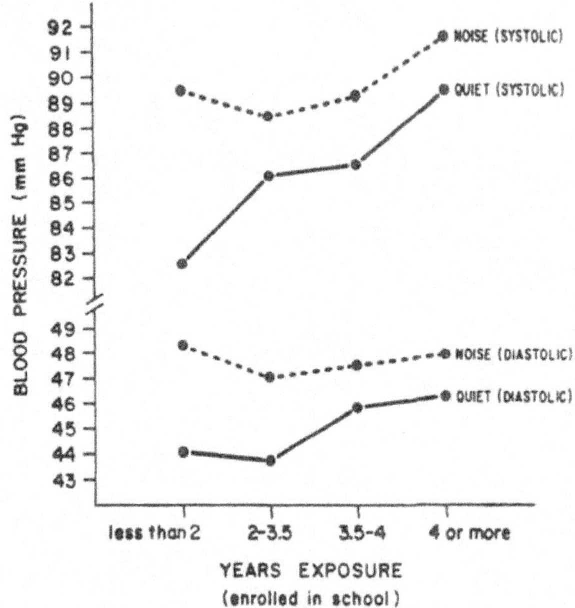

FIGURE 10.63. Systolic and diastolic blood pressure as a function of school noise level and duration of exposure. Each period on the years-exposure coordinate represents approximately one quarter of the sample. (From ref. 228.)

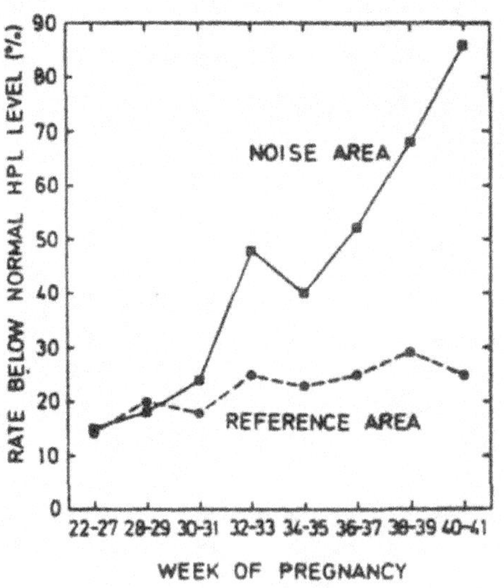

FIGURE 10.64. Percentage of subjects with HPL levels more than one standard deviation below the mean by stage of pregnancy. (From ref. 233.)

In the Ando and Hattori studies the socioeconomic status and the environmental factors (other than the noise) of those exposed and not exposed to the noise were felt to be reasonably similar. In the analyses of birth weight data from hospitals in six villages near the Amsterdam airport, Knipschild *et al.* (ref. 232) attempted to control the following factors: family income, birth order, and sex of the infant. Only single birth infants (no twins) were considered. Table 10.18 shows that significantly more female infants had birth weights below 3000 g whose mothers' lived in high levels of aircraft noise (L_{dn} = 65 to 75) than in low levels of noise (L_{dn} < 65). A further division of L_{dn} = 65 to 75 into L_{dn} of 65 to 70 and 70 to 75 found 23 and 29 percent, respectively, of all infants with birth weights below 3000 g. As shown in the table the percentage was 18.1 percent for all infants from areas with L_{dn} < 65.

Rehn, according to Knipschild *et al.* (ref. 232), did not find a significant relation between birth weight and aircraft noise in the city of Dusseldorf. Knipschild *et al.* suggest that the differential age and socioeconomic factors favoring the higher noise areas (an unusual condition for most cities) may have contributed to Rehn's findings.

Schell (ref. 234) studied the effect on the length of gestation and on the birth weight of infants of mothers who were exposed to aircraft noise while pregnant. A multiple-correlation analysis technique was used to partial out the effects of the mother's height, weight, age, and smoking habits, and the father's weight

TABLE 10.18

Birth Weight Statistics as a Function of Noise Level

[Data from ref. 232]

Socioeconomic category and sex of infant	Infants, percent, with low birth weight[a] for L_{dn}, dBA, of— (b)		
	< 65	65 to 75	< 75
Low income:			
Female	17.0 (112)	30.3 (175)	25.1 (287)
Male	21.8 (119)	22.7 (150)	22.3 (269)
High income:			
Female	17.9 (84)	24.4 (82)	21.1 (166)
Male	14.6 (89)	14.3 (91)	14.4 (180)
All infants	18.1 (404)	24.1 (498)	21.4 (902)
Adjusted[c]	18.1	23.8	

[a] *Birth weights less than 3000 g.*
[b] *Number of infants tested given in parentheses.*
[c] *Combination of the two populations used as standard population.*

and education. A statistically significant negative correlation between aircraft noise exposure level and gestation length was found for female infants but not for male. A larger negative correlation was also found between noise exposure level and birth weights of female than male infants, but neither were statistically significant.

A seemingly startling finding was that birth defects were more prevalent in infants born to mothers from neighborhoods with high levels of aircraft noise than in infants born to mothers from areas with lower levels of noise in Los Angeles (Jones and Tauscher (ref. 235).) Jones (ref. 236) suggests that the indirect effects of the noise (interference with communications, sleep, *etc.*) causes women to be psychologically stressed, resulting in more than normal use of drugs, alcohol, and so forth. However, Bader (ref. 237) points out the variability among hospitals with respect to birth defects is so large that the comparisons made by Jones and Tauscher are not statistically meaningful. In addition, Edmonds *et al.* (ref. 238) found no differences in birth defects among children from areas in Atlanta having differing levels of aircraft noise exposure. Also, the role of genetic factors as a dominating cause in birth defects (compared with environmental stress on the mother), as well as the relatively small absolute numbers of infants with nongenetically determined birth defects, probably makes birth defect data of questionable import to the understanding of possible adverse physiological health effects from exposure to environmental noise (ref. 239).

EPA summary and research plan for effects of noise on health

Recent EPA documents (refs. 240 to 244) report on the basic physiological-biochemical mechanisms involved in the pathological functioning of the cardiovascular system when activated by the autonomic-endocrine system following exposure to "stressful stimuli." Major obstacles that have hindered adequate demonstration of the relationships between noise dose and nonauditory-system stress response and the etiology of cardiovascular disease are discussed and general research studies that should be undertaken to overcome these obstacles are outlined. However, the discussions and plans for research of these reports appear to be based on the implicit presumption that noise is inherently a stressful stimulus.

Included in references 240 and 241 are reviews and assessments of the research literature on the effects of noise on the cardiovascular system. The author (ref. 241) had a group of scientists assign numerical rating scores of adequacy (0 to 9, lowest to highest) of a number of published research studies with respect to each of the following: noise exposure, health effects, and epidemiologic methodology.

Unfortunately, at least some of the given subjective ratings need to be challenged. For example, a paper by Cohen *et al.* (ref. 228) was rated in all aspects as being superior to a paper by Knipschild (ref. 216). In the Cohen *et al.* study, the amounts of aircraft noise present at the four different schools involved were not identified other than that the schools were located under a flight corridor and that the peak levels reached 95 dBA at one of the schools. On the other hand, Knipschild's aircraft noise dosages were stratified for the exposed populations according to noise contours by the local airport authorities. Cohen *et al.* tested 142 children in schools exposed to aircraft noise and 120 children in schools not exposed. As discussed earlier, Knipschild obtained records of all physician contacts made during a 1-week period from a population of 17 500 people exposed to aircraft noise and from 12 000 people living in nearby areas not exposed. Surprisingly, the Cohen *et al.* study was rated as being superior to the Knipschild study with respect to noise exposures and methodology even though Knipschild's study is quantitatively and qualitatively more impressive by normal standards.

Most startling of the relative ratings given for these two studies is in regard to the health effects. The health effects data obtained by Cohen *et al.* from each of their subjects consisted of measurements of blood pressure, weight and height, and school attendance record. No significant differences were found, as noted earlier, between the subjects from the noisy school versus the quiet school with respect to height or weight, but children from the noisy schools had a statistically significant higher attendance record than the children from the quiet school. Some of their blood pressure data indicated that the children in the noisy school had higher blood pressures, but other of their data showed no such differences. It is very questionable that any of the data collected in this study have any interpretable meaning as showing noise-health relations in the children.

In contrast, the health data obtained by Knipschild consisted of physician-diagnosed health disorders. (See fig. 10.60.) These data show a monotonic increase of physician contact rate for certain psychological and cardiovascular-hypertension diseases as a function of aircraft noise exposure level. The EPA ratings on the health effects of these two studies was 9 for Cohen *et al.* and 0 for the Knipschild study. These relative ratings are completely inconsistent with the relative merits of the two studies.

It should be made clear that Cohen *et al.* themselves did not attempt or claim to find definitive relations between aircraft noise dosage and health. Indeed, the main concern of their extensive and carefully performed study was the possible effects of aircraft noise on cognition and motivation in school children.

Summary

Reflexive responses

Sounds or noises of a spectrum and intensity that could, over time, damage the ear can elicit nonauditory-system responses such as constriction of peripheral blood vessels and dilation of the pupils. It is likely that these responses are not a source of stress to the organism because of their nature and small size.

Indirect effects

Many sounds (or noises) can indirectly (nonreflexively) cause autonomic-system reactions that are deemed physiologically stressful. These are sounds which create feelings of emotion (startle, fear, anger, frustration, *etc.*) in the listener because of unexpectedness or other meanings the sounds convey or because of the annoyance caused by interference with sleep, with rest, or with the hearing of wanted sounds, or both.

Experimental evidence demonstrates that autonomic-system responses that are probably stressful occur only after conscious or unconscious cognitive processes are completed. That is to say, sound or noises are not inherently aversive or a cause of physiological stress except to the ear. These findings indicate that autonomic-system responses that might be physiologically stressful and appear to be associated with noise are due to psychological factors related to the situation or to the experiment.

Health effects

Research studies done to date indicate that $L_{dn} \approx 65$ or greater of aircraft noise in residential areas is correlated with:

1. An increase in the number of people with psychological and physiological health problems requiring the increased use of certain types of drugs and visits to physicians

2. An increase in the incidence of female infants with a reduced gestation period and body weight at time of birth

3. An increase in the number of adults requiring admission to psychiatric hospitals

It appears that these effects are due to physiological stress induced by annoyance and emotions arising from interferences of the noise with normal auditory functions and sleep behavior and from fear by associating the noise from close-by aircraft with possible crashes.

Although real-life data are by their nature complex, these findings (except possibly that of admission rates to psychiatric hospitals) are not now contradicted by conflicting epidemiological data. The degrees of prevalence of these objectively measured adverse effects as a function of the level of exposure to aircraft and ground vehicle noise are consistent with the subjective attitudes, feelings, and complaint behavior of communities of people exposed to different levels of aircraft and ground vehicle noise, as is discussed in the next chapter.

Increased hypertension (or other nonauditory-system health disorders) in workers exposed to high levels of industrial noise are not consistently found. When they are found, they are probably due to psychological factors associated with the general work environment and with danger of injury from the noise source and are not from exposure to the noise *per se.*

Effects on animals

There are no consistent data from which to conclude that under real-life conditions (or laboratory conditions, for that matter), wild or domesticated lower animals react much differently to sound or noise than do people. When normal animals (excluding those exhibiting audiogenic seizures or symptoms) do react in a stressful manner to "noise," it is probably for the same reasons as in people, namely, the psychological aspects related to the information conveyed by the noise.

References

1. Moller, Aage R.: Occupational Noise as a Health Hazard. Scandinavian J. Work, Environ. & Health, vol. 3, 1977, pp. 73–79.
2. Levi, L.: Sympatho-Adrenomedullary and Related Biochemical Reactions During Experimentally Induced Emotional Stress. Endocrinology and Human Behaviour, R. P. Michael, ed., Oxford Univ. Press, 1968.
3. Borg, E.: Physiological and Pathogenic Effects of Sound. Acta Oto-Laryngol., Suppl. 381, 1981, pp. 1-68.
4. Davis, R. C.; Buchwald, A. M.; and Frankman, R. W.: Autonomic and Muscular Responses and Their Relation to Simple Stimuli. Psychol. Monogr., vol. 69, no. 405, 1955.
5. Davis, R. C.; and Berry, T.: Gastrointestinal Reactions to Response-Contingent Stimulation. Psychol. Rep., vol. 15, 1964, pp. 95–113.
6. Stern, R. M.: Effects of Variation in Visual and Auditory Stimulation on Gastrointestinal Motility. Psychol. Rep., vol. 14, 1964, pp. 799–802.
7. Hale, H. B.: Adrenalcortical Activity Associated With Exposure to Low Frequency Sounds. American J. Physiol., vol. 171, 1959, p. 732.
8. Levi, Lennart: Sympathoadrenomedullary Responses to "Pleasant" and "Unpleasant" Psychosocial Stimuli. Stress and Distress in Response to Psychosocial Stimuli, Lennart Levi, ed., Pergamon Press, Inc., c.1972, pp. 55–73.
9. Arguelles, A. E.; Ibeas, D.; Ottone, J. Pomes; and Chekherdemian, M.: Pituitary-Adrenal Stimulation by Sound of Different Frequencies. J. Clin. Endocrinol., vol. 22, 1962, pp. 846–852.
10. Arguelles, A. E.; Martinez, M. A.; Pucciarelli, Eva; and Disisto, Maria V.: Endoctrine and Metabolic Effects of Noise in Normal, Hypertensive and Psychotic Subjects. Physiological Effects of Noise, Bruce L. Welch and Annemarie S. Welch, eds., Plenum Press, 1970, pp. 43–55.
11. Surwillo, Walter W.; and Arenberg, David L.: On the Law of Initial Value and the Measurement of Change. Psychophysiology, vol. 1, no. 4, Apr. 1965, pp. 368–370.

12. Ginsberg, Stanley; and Furedy, John J.: Stimulus Repetition, Change, and Assessments of Sensitivities of and Relationships Among an Electrodermal and Two Plethysmographic Components of the Orienting Reaction. Psychophysiology, vol. 11, no. 1, Jan. 1974, pp. 35-43.
13. Harper, M. M.: The Effects of Signal Property and Stimulus Intensity on the Skin Resistance Response to Repetition of Auditory Stimuli. M.A. Thesis, Univ. of Wisconsin (Madison), 1968.
14. Graham, Frances K.: Habituation and Dishabituation of Responses Innervated by the Autonomic Nervous System. Habituation, Volume 1—Behavioral Studies, Harman V. S. Peeke and Michael J. Herz, eds., Academic Press, Inc., c.1973, pp. 163-218.
15. Graham, Frances K.; and Slaby, Diana Arezzo: Differential Heart Rate Changes to Equally Intense White Noise and Tone. Psychophysiology, vol. 10, no. 4, July 1973, pp. 347-362.
16. Rossi, L.; Oppliger, G.; and Grandjean, E.: Neurovegetative Effects of Man of Noises Superimposed on a Background Noise. Med. Lav., vol. 50, 1959, pp. 332-377.
17. Sokolov, Evgeni Nikolaevich (Stefan W. Waydenfeld, transl.): Perception and the Conditioned Reflex. Pergamon Press, Inc., 1963.
18. Jackson, Jan C.: Amplitude and Habituation of the Orienting Reflex as a Function of Stimulus Intensity. Psychophysiology, vol. 11, no. 6, Nov. 1974, pp. 647-659.
19. Glass, David C.; and Singer, Jerome E.: Urban Stress—Experiments on Noise and Social Stressors. Academic Press, Inc., 1972.
20. Mosskov, J. I.; and Ettema, J. H.: IV. Extra-Auditory Effects in Long-Term Exposure to Aircraft and Traffic Noise. Int. Arch. Occup. & Environ. Health, vol. 40, no. 3, Nov. 29, 1977, pp. 177-184.
21. Di Cantogno, L. Verdun; Dallerba, R.; Teagno, P. S.; and Cocola, L.: Urban Traffic Noise, Cardiocirculatory Activity and Coronary Risk Factors. Acta Oto-Laryngol., Suppl. 339, 1976, pp. 55-63.
22. Osada, Y.; Ogawa, S.; Hirokawa, A.; and Haruta, K.: Physiological Effects of Long-Term Exposure to Low-Level Noise. Bull. Inst. Public Health, Tokyo, vol. 22, no. 2, 1973, pp. 61-67.
23. Grandjean, E.: Biological Effects of Noise. Paper presented at Fourth International Congress on Acoustics, Copenhagen, 1962.
24. Jansen, Gerd: The Influence of Noise at Manual Work. Int. J. Appl. Physiol., vol. 20, 1964, pp. 233-239.
25. Jansen, Gerd: Extra-Auditory Effects of Noise. Transl. Beltone Inst. Hearing Res., no. 26, Mar. 1972.
26. Ohkubo, Chiyoji; Miyazaki, Kuratoshi; and Osada, Yasutaka: Response of Finger Pulse Amplitude to Intermittent Noise. Bull. Inst. Public Health, Tokyo, vol. 25, no. 1, 1976, pp. 1-8.

27. Kryter, Karl D.; and Poza, Fausto: Effects of Noise on Some Autonomic System Activities. J. Acoust. Soc. America, vol. 67, no. 6, June 1980, pp. 2036–2044.
28. Jansen, Gerd: Non-Auditory Effects of Noise. Physiological and Psychological Reactions in Man. Proceedings of the International Congress on Noise as a Public Health Problem, W. Dixon Ward, ed., 550/9-73-008, U.S. Environ. Prot. Agency, May 1973, pp. 431–439.
29. Froehlich, G. R.: The Effects of Ear Protectors on Some Autonomic Responses to Aircraft and Impulsive Noise. Effects of Long Duration Noise Exposure on Hearing and Health, Milton A. Whitcomb, ed., AGARD-CP-171, Nov. 1975, pp. C8-1–C8-5.
30. Finkle, A. L.; and Poppen, J. R.: Clinical Effects of Noise and Mechanical Vibrations of a Turbo-Jet Engine on Man. J. Appl. Physiol., vol. 1, 1948, pp. 183-204.
31. Slob, A.; Wink, A.; and Radder, J. J.: The Effect of Acute Noise Exposure on the Excretion of Corticosteroids, Adrenalin and Noreadrenalin in Man. Int. Arch. Arbeitsmed., vol. 31, no. 3, July 10, 1973, pp. 225-235.
32. Hawel, W.; and Starlinger, H.: Effect of Repeated 4-Hour Intermittent (So-Called) Pink Noise on Catecholaminar Separation (In Urine) and Pulse Frequency. Int. Z. Angew. Physiol. Einschol. Arbeitsphysiol., vol. 24, 1967, pp. 351-362.
33. Brandenberger, G.; Follenius, M.; and Trémolières, C.: Failure of Noise Exposure To Modify Temporal Patterns of Plasma Cortisol in Man. European J. Appl. Physiol., vol. 36, no. 4, May 10, 1977, pp. 239-246.
34. Favino, A.; Maugeri, U.; Kauchtschischvili, G.; Robustelli, Della Cuna G.; and Nappi, G.: Radioimmunoassay Measurements of Serum Cortisol, Thyroxine, Growth Hormone and Luteinizing Hormone With Simultaneous Electroencephalographic Changes During Continuous Noise in Man. J. Nucl. Biol. & Med., vol. 17, no. 3, July-Sept. 1973, pp. 119–122.
35. Fruhstorfer, Beate; and Hensel, Herbert: Extra-Auditory Responses to Long-Term Intermittent Noise Stimulation in Humans. J. Appl. Physiol., vol. 49, no. 6, Dec. 1980, pp. 985–993.
36. Davis, R. C.; and Van Liere, D. W.: Adaptation of the Muscular Tension Response to Gunfire. J. Exp. Psychol., vol. 39, 1949, pp. 114–117.
37. Pearsons, K. S.; and Kryter, K. D.: Laboratory Tests of Subjective Reactions to Sonic Booms. NASA CR-187, [1965].
38. Bartoshuk, A. K.: Response Decrement With Repeated Elicitation of Human Neonatal Cardiac Acceleration to Sound. J. Comp. Physiol. Psychol., vol. 55, 1962, pp. 9–13.
39. Ando, Y.; and Hattori, H.: Effects of Intense Noise During Fetal Life

Upon Postnatal Adaptability (Statistical Study of the Reactions of Babies to Aircraft Noise). J. Acoust. Soc. America, vol. 47, no. 4, pt. 2, Apr. 1970, pp. 1128-1130.

40. Ando, Y.; and Hattori, H.: Effects of Noise on Sleep of Babies. J. Acoust. Soc. America, vol. 62, no. 1, July 1977, pp. 199-204.

41. Cartwright, Lyle B.; and Thompson, Robert N.; The Effects of Broadband Noise on the Cardiovascular System in Normal Resting Adults. J. American Ind. Hyg. Assoc., Sept. 1975, pp. 653-658.

42. Andrén, Lennart; Hansson, Lennart; Björkman, Martin; and Jonsson, Anders: Noise as a Contributory Factor in the Development of Elevated Arterial Pressure. Acta Med. Scand., vol. 207, 1980, pp. 493-498.

43. Andrén, Lennart: Cardiovascular Effects of Noise. Acta Med. Scand., Suppl. 657, 1982.

44. Bättig, K.; Zeier, H.; Müller, R.; and Buzzi, R.: A Field Study on Vegetative Effects of Aircraft Noise. Arch. Environ. Health, vol. 35, no. 4, July/Aug. 1980, pp. 228-235.

45. Peterson, E. A.; Augenstein, J. S.; Hosek, R. S.; Klose, K. J.; Manas, K.; Bloom, J.; Lovett, S.; and Greenberg, D. A.: Noise and Cardiovascular Function in Rhesus Monkeys. J. Aud. Res., vol. 15, no. 4, Oct. 1975, pp. 234-251.

46. Peterson, E. A.; Tanis, D. C.; Augenstein, J. S.; Seifert, R. A.; and Bromley, H. R.: Noise and Cardiovascular Function in Rhesus Monkeys: II. Noise as a Public Health Problem, Jerry V. Tobias, Gerd Jansen, and W. Dixon Ward, eds., ASHA Rep. 10, American Speech-Language-Hearing Assoc., Apr. 1980, pp. 246-253.

47. Peterson, E. A.; Augenstein, J. S.; Tanis, D. C.; and Augenstein, D. G.: Noise Raises Blood Pressure Without Impairing Auditory Sensitivity. Science, vol. 211, no. 4489, Mar. 27, 1981, pp. 1450-1452.

48. Kraft Schreyer, Nancy; and Angelakos, E. T.: Effects of Sound Stress on Norepinephrine Responsiveness and Blood Pressure. Fed. Proc., vol. 38, no. 3, pt. II, Mar. 1, 1979, p. 883.

49. Hanson, John D.; Larson, Mark E.; and Snowdon, Charles T.: The Effects of Control Over High Intensity Noise on Plasma Cortisol Levels in Rhesus Monkeys. Behav. Biol., vol. 16, no. 3, Mar. 1976, pp. 333-340.

50. Falk, Stephen A.: Environmental Noise. American J. Public Health, vol. 63, no. 10, 1973, pp. 833-834.

51. Kryter, Karl D.: Reply to Dr. Falk. American J. Public Health, vol. 63, no. 10, 1973, pp. 834-836.

52. Buckley, Joseph P.; and Smookler, Harold H.: Cardiovascular and Biochemical Effects of Chronic Intermittent Neurogenic Stimulation. Physiological Effects of Noise, Bruce L. Welch and Annemarie S. Welch, eds., Plenum Press, 1970, pp. 75-84.

53. Henry, Kenneth R.; and Bowman, Robert E.: Acoustic Priming of Audio-

genic Seizures in Mice. Physiological Effects of Noise, Bruce L. Welch and Annemarie S. Welch, eds., Plenum Press, 1970, pp. 185-201.

54. Saunders, James C.; Bock, Gregory R.; Chen, Chia-Shong; and Gates, G. Richard: The Effects of Priming for Audiogenic Seizures on Cochlear and Behavioral Responses in BALB/c Mice. Exp. Neurol., vol. 36, no. 3, Sept. 1972, pp. 426-436.
55. Chen, Chia-Shong: Acoustic Trauma-Induced Developmental Change in the Acoustic Startle Response and Audiogenic Seizures in Mice. Exp. Neurol., vol. 60, no. 2, June 1978, pp. 400-403.
56. Patton, R. A.: Purulent Otitis Media in Albino Rats Susceptible to Sound Induced Seizures. J. Psychol., vol. 24, 1947, pp. 313-317.
57. Kenshalo, D.; and Kryter, K. D.: Middle Ear Infection and Sound Induced Seizures in Rats. J. Comp. Physiol. Psychol., vol. 42, 1949, pp. 328-331.
58. Niaussat, M. M.: Experimentally Induced Otitis and Audiogenic Seizure in the Mouse. Experientia, vol. 33, no. 4, Apr. 15, 1977, pp. 473-474.
59. Pfaff, J.: Noise as an Environmental Problem in the Animal House. Lab. Anim., vol. 8, no. 3, Sept. 1974, pp. 347-354.
60. Busnel, R. G.; Busnel, M. C.; and Lehmann, A. G.: Synergic Effects of Noise and Stress on General Behavior. Life Sci., vol. 16, no. 1, Jan. 1975, pp. 131-137.
61. Jensen, Marcus M.; and Rasmussen, A. F., Jr.: Audiogenic Stress and Susceptibility to Infection. Physiological Effects of Noise, Bruce L. Welch and Annemarie S. Welch, eds., Plenum Press, 1970, pp. 7-19.
62. Lockett, Mary F.: Effects of Sound on Endocrine Function and Electrolyte Excretion. Physiological Effects of Noise, Bruce L. Welch and Annemarie S. Welch, eds., Plenum Press, 1970, pp. 21-41.
63. Geber, William F.: Cardiovascular and Teratogenic Effects of Chronic Intermittent Noise Stress. Physiological Effects of Noise, Bruce L. Welch and Annemarie S. Welch, eds., Plenum Press, 1970, pp. 85-90.
64. Árvay, A.: Effect of Noise During Pregnancy Upon Foetal Viability and Development. Physiological Effects of Noise, Bruce L. Welch and Annemarie S. Welch, eds., Plenum Press, 1970, pp. 91-115.
65. Tamari, I.: Audiogenic Stimulation and Reproductive Function. Physiological Effects of Noise, Bruce L. Welch and Annemarie S. Welch, eds., Plenum Press, 1970, pp. 117-130.
66. Sontag, Lester W.: Effect of Noise During Pregnancy Upon Foetal and Subsequent Adult Behavior. Physiological Effects of Noise, Bruce L. Welch and Annemarie S. Welch, eds., Plenum Press, 1970, pp. 131-141.
67. Krushinsky, L. V.; Molodkina, L. N.; Fless, D. A.; Dobrokhotova, L. P.; Steshenko, A. P.; Semiokhina, A. F.; Zorina, Z. A.; and Romanova, L. G.: The Functional State of the Brain During Sonic Stimulation. Physiological Effects of Noise, Bruce L. Welch and Annemarie S. Welch, eds., Plenum Press, 1970, pp. 159-183.

68. Fuller, John L.; and Collins, Robert L.: Genetic and Temporal Characteristics of Audiogenic Seizures in Mice. Physiological Effects of Noise, Bruce L. Welch and Annemarie S. Welch, eds., Plenum Press, 1970, pp. 203-210.
69. Fink, Gregory B.; and Iturrian, W. B.: Influence of Age, Auditory Conditioning, and Environmental Noise on Sound-Induced Seizures and Seizure Threshold in Mice. Physiological Effects of Noise, Bruce L. Welch and Annemarie S. Welch, eds., Plenum Press, 1970, pp. 211-226.
70. Lehmann, Alice G.: Psychopharmacology of the Response to Noise, With Special Reference to Audiogenic Seizure in Mice. Physiological Effects of Noise, Bruce L. Welch and Annemarie S. Welch, eds., Plenum Press, 1970, pp. 227-257.
71. Sze, Paul Y.: Neurochemical Factors in Auditory Stimulation and Development of Susceptibility to Audiogenic Seizures. Physiological Effects of Noise, Bruce L. Welch and Annemarie S. Welch, eds., Plenum Press, 1970, pp. 259-269.
72. Leake, Chauncey D.: Summary of the Symposium. Physiological Effects of Noise, Bruce L. Welch and Annemarie S. Welch, eds., Plenum Press, 1970, pp. 337-340.
73. Forster, Francis M.: Human Studies of Epileptic Seizures Induced by Sound and Their Conditioned Extinction. Physiological Effects of Noise, Bruce L. Welch and Annemarie S. Welch, eds., Plenum Press, 1970, pp. 151-158.
74. Rivera, Reyes L.: Musicogenic Epilepsy. Bol. Asoc. Med. Puerto Rico, vol. 70, no. 5, May 1978, pp. 143-145.
75. Borg, Erik; and Moller, Aage R.: Noise and Blood Pressure: Effect of Lifelong Exposure in the Rat. Acta Physiol. Scandinavica, vol. 103, 1978, pp. 340-342.
76. Chesser, Ronald K.; Caldwell, Ronald S.; and Harvey, Michael J.: Effects of Noise on Feral Populations of *Mus Musculus*. Physiol. Zool., vol. 48, no. 4, Oct. 1975, pp. 323-325.
77. Pritchett, John F.; Caldwell, Ronald S.; Chesser, Ronald K.; and Sartin, James L.: Effect of Jet Aircraft Noise Upon *In Vitro* Adrenocortical Response to ACTH in Feral *Mus Musculus*. Life Sci., vol. 18, no. 4, 1976, pp. 391-396.
78. Busnel, Rene-Guy; and Briot, Jean-Lue: Wildlife and Airfield Noise in France. Noise as a Public Health Problem, Jerry V. Tobias, Gerd Jansen, and W. Dixon Ward, eds., ASHA Rep. 10, American Speech-Language-Hearing Assoc., Apr. 1980, pp. 621-631.
79. Fletcher, John L.: Effects of Noise on Wildlife: A Review of Relevant Literature 1971-1978. Noise as a Public Health Problem, Jerry V. Tobias, Gerd

Jansen, and W. Ward Dixon, eds., ASHA Rep. 10, American Speech-Language-Hearing Assoc., Apr. 1980, pp. 611-620.

80. Sonic Boom Experiments at Edwards Air Force Base. NSBEO-1-67 (Contract AF 49(638)-1758), CFSTI, U.S. Dep. Com., July 28, 1967.

81. Espmark, Yngue; Falt, Lars; and Falt, Birgitta: Behavioural Responses in Cattle and Sheep Exposed to Sonic Booms and Low-Altitude Subsonic Flight Noise. Vet. Rec., vol. 94, no. 6, Feb. 9, 1974, pp. 106-113.

82. Travis, H. F.; Bond, J.; Wilson, R. L.; Leekley, J. R.; Menear, J. R.; Curran, C. R.; Robinson, F. R.; Brewer, W. E.; Huttenhauer, G. A.; and Henson, J. B.: An Interdisciplinary Study of the Effects of Real and Simulated Sonic Booms on Farm-Raised Mink (Mustela Vison). FAA-EQ-72-2, Aug. 1972.

83. Bond, James: Effects of Noise on the Physiology and Behavior of Farm-Raised Animals. Physiological Effects of Noise, Bruce L. Welch and Annemarie S. Welch, eds., Plenum Press, 1970, pp. 295-304.

84. Griefahn, Barbara; and Jansen, Gerd: EEG-Responses Caused by Environmental Noise During Sleep—Their Relationships to Exogenic and Endogenic Influences. Sci. Total Environ., vol. 10, 1978, pp. 187-199.

85. Thiessen, George J.: Habituation of Behavioral Awaking and EEG Measures of Response to Noise. Noise as a Public Health Problem, Jerry V. Tobias, Gerd Jansen, and W. Dixon Ward, eds., ASHA Rep. 10, American Speech-Language-Hearing Assoc., Apr. 1980, pp. 397-400.

86. Griefahn, Barbara: Effects of Sonic Booms on Fingerpulse Amplitudes During Sleep. Int. Arch. Occup. Environ. Health, vol. 36, no. 1, Nov. 4, 1975, pp. 57-66.

87. Muzet, Alain; and Ehrhart, Jean: Habituation of Heart Rate and Finger Pulse Responses to Noise in Sleep. Noise as a Public Health Problem, Jerry V. Tobias, Gerd Jansen, and W. Dixon Ward, eds., ASHA Rep. 10, American Speech-Language-Hearing Assoc., Apr. 1980, pp. 401-404.

88. Williams, H. L.; Hammack, J. T.; Daly, R. L.; Dement, W. C.; and Lubin, A.: Responses to Auditory Stimulation, Sleep Loss, and the EEG Stages of Sleep. Electroencephalogr. & Clin. Neurophysiol., vol. 16, 1964, pp. 269-279.

89. Lukas, Jerome S.: Measures of Noise Level: Their Relative Accuracy in Predicting Objective and Subjective Responses to Noise During Sleep. EPA-600/1-77-010, U.S. Environ. Prot. Agency, Feb. 1977.

90. Ehrenstein, Wolfgang; and Müller-Limmroth, Wolf: Laboratory Investigations Into Effects of Noise on Human Sleep. Noise as a Public Health Problem, Jerry V. Tobias, Gerd Jansen, and W. Dixon Ward, eds., ASHA Rep. 10, American Speech-Language-Hearing Assoc., Apr. 1980, pp. 433-441.

91. Muzet, A.; Schieber, J. P.; Olivier-Martin, N.; Ehrhart, J.; and Metz, B.:

Relationship Between Subjective and Physiological Assessments of Noise-Disturbed Sleep. Proceedings of the International Congress on Noise as a Public Health Problem, W. Dixon Ward, ed., 550/9-73-008, U.S. Environ. Prot. Agency, May 1973, pp. 575-586.

92. Öhrström, E.; and Rylander, R.: Sleep Disturbance Effects of Traffic Noise—A Laboratory Study on After Effects. J. Sound & Vib., vol. 84, no. 1, Sept. 8, 1982, pp. 87-103.
93. LeVere, T. E.; and Davis, N.: Arousal From Sleep: The Physiological and Subjective Effects of a 15 dB(A) Reduction in Aircraft Flyover Noise. Aviat., Space & Environ. Med., vol. 48, no. 7, July 1977, pp. 607-611.
94. Johnson, Laverne C.; Townsend, Richard E.; Naitoh, Paul; and Muzet, Alain G.: Prolonged Exposure to Noise as a Sleep Problem. Proceedings of the International Congress on Noise as a Public Health Problem, W. Dixon Ward, ed., 550/9-73-008, U.S. Environ. Prot. Agency, May 1973, pp. 559-574.
95. Johnson, Laverne C.; and Lubin, Ardie: The Orienting Reflex During Waking and Sleeping. Electroencephalogr. & Clin. Neurophysiol., vol. 22, no. 1, 1967, pp. 11-21.
96. Cantrell, Robert W.: Prolonged Exposure to Intermittent Noise: Audiometric, Biochemical, Motor, Psychological and Sleep Effects. Laryngoscope, suppl. 1, vol. LXXXIV, no. 10, pt. 2, Oct. 1974, pp. 1-55.
97. Oswald, I.; Taylor, A. M.; and Treisman, M.: Discriminative Responses to Stimulation During Human Sleep. Brain, vol. 83, 1960, pp. 440-453.
98. Emmons, W. H.; and Simon, C. W.: The Non-Recall of Material Presented During Sleep. Rev. Rep. P-619, The Rand Corp., 1955.
99. Simon, Charles W.; and Emmons, William H.: A Critical Review of the "Learn-While-You-Sleep" Studies. Rep. P-534, The Rand Corp., June 14, 1954. (Rev. Oct. 14, 1954.)
100. Rechtschaffen, A.; Hauri, P.; and Zeitlin, M.: Auditory Awakening Thresholds in REM and NREM Sleep Stages. Percept. & Mot. Skills, vol. 22, 1966, pp. 927-942.
101. Lukas, Jerome S.; and Kryter, Karl D.: Awakening Effects of Simulated Sonic Booms and Subsonic Aircraft Noise. Physiological Effects of Noise, Bruce L. Welch and Annemarie S. Welch, eds., Plenum Press, 1970, pp. 283-293.
102. Roth, T.; Kramer, M.; and Trinder, J.: The Effect of Noise During Sleep on the Sleep Patterns of Different Age Groups. Canadian Psychiatr. Assoc. J., vol. 17, suppl. 2, 1972, pp. 197-201.
103. Lukas, J. S.; and Dobbs, M. E.: Effects of Aircraft Noises on the Sleep of Women. NASA CR-2041, 1972.
104. Lukas, J. S.: Awakening Effects of Simulated Sonic Booms and Aircraft Noise on Men and Women. J. Sound & Vib., vol. 20, no. 4, Feb. 22, 1972, pp. 457-466.

105. Olsen, J.; and Nelson, E. N.: Calming the Irritable Infant With a Simple Device. Minnesota Med., vol. 44, 1961, pp. 527-529.
106. Johns, M. W.; and Doré, Caroline: Sleep at Home and in the Sleep Laboratory: Disturbance by Recording Procedures. Ergonomics, vol. 21, no. 5, 1978, pp. 325-330.
107. Globus, Gordon; Friedmann, Joyce; Cohen, Harry; Pearsons, Karl S.; and Fidell, Sanford: The Effects of Aircraft Noise on Sleep Electrophysiology as Recorded in the Home. Proceedings of the International Congress on Noise as a Public Health Problem, W. Dixon Ward, ed., 550/9-73-008, U.S. Environ. Prot. Agency, May 1973, pp. 587-591.
108. Vallet, Michel; Gagneux, J. M.; and Simonnet, F.: Effects of Aircraft Noise on Sleep: An *In Situ* Experience. Noise as a Public Health Problem, Jerry V. Tobias, Gerd Jansen, and W. Dixon Ward, eds., ASHA Rep. 10, American Speech-Language-Hearing Assoc., Apr. 1980, pp. 391-396.
109. Aircraft Noise and Sleep Disturbance: Final Report. DORA Rep. 8008, Civil Aviation Authority, London, 1980.
110. Fidell, S.; and Jones, G.: Effects of Cessation of Late-Night Flights on an Airport Community. J. Sound & Vib., vol. 42, no. 4, 1975, pp. 411-427.
111. Langdon, F. J.; and Buller, I. B.: Road Traffic Noise and Disturbance to Sleep. J. Sound & Vib., vol. 50, no. 1, Jan. 8, 1977, pp. 13-28.
112. Horonjeff, R. D.; Fidell, S.; Teffeteller, S. R.; and Green, D. M.: Behavioral Awakening as Functions of Duration and Detectability of Noise Intrusions in the Home. J. Sound & Vib., vol. 84, no. 3, 1982, pp. 327-336.
113. Richter, R.: Sleep Disturbances Which We are Not Aware of Caused by Traffic Noise. EEG Station, Neurolog. Univ. Clinic, Basel.
114. West, L. J.: Psychopathology Produced by Sleep Deprivation. Proceedings of the Association for Research in Nervous and Mental Disease, Sleep and Altered States of Consciousness, S. S. Ketty, ed., Williams & Wilkins, 1967.
115. Gardner, W.; and Licklider, J. C. R.: Auditory Analgesia in Dental Operations. J. American Dental Assoc., vol. 59, 1959, pp. 1144-1149.
116. Davis, H.; and Glorig, A.: Audio Analgesia: A New Problem for Otologists. Arch. Otolaryngol., vol. 75, 1962, pp. 498-501.
117. Kryter, K. D.; Weisz, A. Z.; and Wiener, F. M.: Auditory Fatigue From Audio Analgesia. J. Acoust. Soc. America, vol. 34, no. 4, Apr. 1962, pp. 383-391.
118. Gardner, Wallace J.; Licklider, J. C. R.; and Weisz, A. Z.: Suppression of Pain by Sound. Science, vol. 132, no. 3418, July 1, 1960, pp. 32-33.
119. Melzack, Ronald: Perception of Pain. Sci. American, vol. 204, no. 2, Feb. 1961, pp. 41-49.
120. Carlin, Sidney; Ward, W. Dixon; Gershon, Arthur; and Ingraham, Rex:

Sound Stimulation and Its Effect on Dental Sensation Threshold. Science, vol. 138, no. 3546, Dec. 14, 1962, pp. 1258-1259.

121. Camp, Walter; Martin, Robert; and Chapman, Loring F.: Pain Threshold and Discrimination of Pain Intensity During Brief Exposure to Intense Noise. Science, vol. 135, no. 3506, Mar. 9, 1962, pp. 788-789.

122. Robson, J. G.; and Davenport, H. T.: The Effects of White Sound and Music Upon the Superficial Pain Threshold. Canadian Anesthesiol. Soc. J., vol. 9, 1962, pp. 105-108.

123. Hilgard, E. R.: Pain as a Puzzle for Psychology and Physiology. American Psychol., vol. 24, 1969, pp. 103-113.

124. Beecher, Henry K.: Pain: One Mystery Solved. Science, vol. 152, no. 3711, Feb. 11, 1966, pp. 840-841.

125. Ades, H. W.; Graybiel, A.; Morrill, S.; Tolhurst, G.; and Niven, J.: Non-Auditory Effects of High Intensity Sound Stimulation on Deaf Human Subjects. Joint Rep. 5, Univ. Texas Southwestern Med. School and U.S. Naval School of Aviation Med., 1958.

126. Plutchik, R.: Physiological Responses to High Intensity Intermittent Sound. Psychol. Rec., vol. 13, 1963, pp. 141-148.

127. Dickson, E. D. D.; and Chadwick, D. L.: Observations on Disturbances of Equilibrium and Other Symptoms Induced by Jet Engine Noise. J. Laryngol. Otol., vol. 65, 1951, pp. 154-165.

128. Roggevsen, L. S.; and Van Dishoeck, H. A.: Vestibular Reactions as a Result of Acoustic Stimulation. Pract. Oto-Rhino-Laryngol., vol. 18, 1956, pp. 205-213.

129. Harris, C. Stanley; and Von Gierke, Henning E.: The Effects of High Intensity Noise on Human Equilibrium. Rep. AMRL-TR-6761, U.S. Air Force, Dec. 1971. (Available from DTIC as AD 737 826.)

130. Harris, C. S.: Effects of Increasing Intensity Levels of Intermittent and Continuous 1000-Hz Tones on Human Equilibrium. Percept. Mot. Skills, vol. 35, no. 2, Oct. 1972, pp. 395-405.

131. Vanderhei, Sharon L.; and Loeb, Michel: Effects of Bilateral and Unilateral Continuous and Impact Noise on Equilibrium as Measured by the Rail Test. J. Appl. Psychol., vol. 61, no. 1, 1976, pp. 123-126.

132. Von Gierke, Henning E.: Exposure to Combined Noise and Vibration Environments. Noise as a Public Health Problem, Jerry V. Tobias, Gerd Jansen, and W. Dixon Ward, eds., ASHA Rep. 10, American Speech-Language-Hearing Assoc., Apr. 1980, pp. 649-656.

133. Dempsey, Thomas K.; Leatherwood, Jack D.; and Clevenson, Sherman A.: Development of Noise and Vibration Ride Comfort Criteria. J. Acoust. Soc. America, vol. 65, no. 1, Jan. 1979, pp. 124-132.

134. Broussard, I. G.; Walker, R. Y.; and Roberts, E. E., Jr.: The Influence of Noise on the Visual Contrast Threshold. Rep. 101, U.S. Army Med. Res. Labs., 1952.

135. Krauskopf, J.; and Coleman, P. D.: The Effect of Noise on Eye Movements. U.S. Army Med. Res. Labs., 1956.
136. Dorfman, D. D.; and Zajonc, R. B.: Some Effects of Sound, Background Brightness, and Economic Status of the Perceived Size of Coins and Discs. J. Abnormal Social Psychol., vol. 66, 1963, pp. 87-90.
137. Loeb, M.: A Further Investigation of the Influence of Whole-Body Vibration and Noise on Tremor and Visual Acuity. Rep. 165, U.S. Army Med. Res. Labs., 1954.
138. Rubenstein, M. K.: Interaction Between Vision and Audition. Rep. 151, U.S. Army Med. Res. Labs., 1954.
139. Chandler, Kenneth A.: The Effect of Monaural and Binaural Tones of Different Intensities on the Visual Perception of Verticality. American J. Psychol., vol. 74, no. 2, June 1961, pp. 260-265.
140. Benko, E.: Objekt-und Farbengeisichtsfeldeinengung bei Chronischen Larmschaden. Ophthalmologica, vol. 138, 1959, pp. 449-456.
141. Benko, E.: Further Information About the Narrowing of the Visual Fields Caused by Noise Damage. Ophthalmologica, vol. 140, 1962, pp. 76-80.
142. Jones, P. D.; Loeb, M.; and Cohen, A.: Effects of Intense Continuous- and Impact-Type Noise on Pupil Size and Visual Acuity. J. American Audiol. Soc., vol. 2, no. 6, May-June 1977, pp. 202-207.
143. Hermann, Edward R.; Hesse, Carolyn S.; Hoyle, E. Robinson; and Leopold, Anne C.: Effects of Noise and Hearing Acuity Upon Visual Depth Perception and Safety Among Humans. Rep. IIEQ-77/10, Illinois Inst. Environ. Qual., Apr. 1978. (Available from NTIS as PB 280 365.)
144. McCroskey, R. L., Jr.: The Effect of Specified Levels of White Noise Upon Flicker Fusion Frequency. Rep. 80, U.S. Naval School Aviat. Med., Bur. Med. Surgery, 1958. (Available from DTIC as AD 211 759.)
145. Ogilvie, J. C.: Effect of Auditory Flutter on the Visual Critical Flicker Frequency. Canadian J. Psychol., vol. 10, 1956, pp. 61-68.
146. Walker, E. L.; and Sawyer, T. M., Jr.: The Interaction Between Critical Flicker Frequency and Acoustic Stimulation. Psychol. Rec., vol. 11, 1961, pp. 187-191.
147. Maier, Barbara; Bevan, William; and Behar, Isaac: The Effect of Auditory Stimulation Upon the Critical Flicker Frequency for Different Regions of the Visible Spectrum. American J. Psychol., vol. 74, no. 1, Mar. 1961, pp. 67-73.
148. O'Hare, J.: Intersensory Effects of Visual Stimuli on the Minimum Audible Threshold. J. Gen. Psychol., vol. 54, 1956, pp. 167-170.
149. Mohr, George C.; Cole, John N.; Guild, Elizabeth; and Von Gierke, Henning E.: Effects of Low Frequency and Infrasonic Noise on Man. Aerosp. Med., vol. 36, no. 9, Sept. 1965, pp. 817-824.
150. Nixon, Charles W.; and Johnson, Daniel L.: Infrasound and Hearing. Proceedings of the International Congress on Noise as a Public Health Prob-

lem, W. Dixon Ward, ed., 550/9-73-008, U.S. Environ. Prot. Agency, May 1973, pp. 329-347.

151. Döring, Hans J.; Hauf, Gerhard; and Seiberling, Michael: Effects of High-Intensity Sound on the Contractile Function of the Isolated Ileum of Guinea Pigs and Rabbits. Noise as a Public Health Problem, Jerry V. Tobias, Gerd Jansen, and W. Dixon Ward, eds., ASHA Rep. 10, American Speech-Language-Hearing Assoc., Apr. 1980, pp. 288-293.

152. Bruel, Per V.; and Olesen, P.: Infrasonic Measurements. B & K Instr., Inc. Tech. Rev., no. 3, 1973, pp. 14-25.

153. Parrack, H. O.: Effect of Air-Borne Ultrasound on Humans. Intl. Aud., vol. 5, 1966, pp. 294-308.

154. Davis, Hallowell: Biological and Psychological Effects of Ultrasonics. J. Acoust. Soc. America, vol. 20, no. 5, Sept. 1948, pp. 605-607.

155. Davis, H.; Parrack, H. O.; and Eldredge, D. H.: Hazards of Intense Sound and Ultrasound. Ann. Otol. Rhinol. Laryngol., vol. 58, 1949, pp. 732-738.

156. Haeff, Andrew V.; and Knox, Cameron: Perception of Ultrasound. Science, vol. 139, no. 3555, Feb. 15, 1963, pp. 590-592.

157. Bellucci, R. J.; and Schneider, D. E.: Some Observations on Ultrasonic Perception in Man. Ann. Otol. Rhinol. Laryngol., vol. 71, 1962, pp. 719-726.

158. Deatherage, Bruce H.; Jeffress, Lloyd A.; and Blodgett, Hugh C.: A Note on the Audibility of Intense Ultrasonic Sound. J. Acoust. Soc. America, vol. 26, no. 4, July 1954, p. 582.

159. Acton, W. I.; and Carson, M. B.: Auditory and Subjective Effects of Airborne Noise From Industrial Ultrasonic Sources. British J. Ind. Med., vol. 24, no. 4, Oct. 1967, pp. 297-304.

160. Skillern, C. P.: Human Response to Measured Sound Pressure Levels From Ultrasonic Devices. J. American Ind. Hyg. Assoc., vol. 26, 1965, pp. 132-136.

161. Loeb, Michel; Jones, Paul D.; and Cohen, Alexander: Effects of Noise on Non-Auditory Sensory Functions and Performance. HEW Publ. No. (NIOSH) 76-176, U.S. Dep. Health, Educ., & Welfare, Apr. 1976.

162. Levi, Lennart: Stress, Distress and Psychosocial Stimuli. Occup. Mental Health, vol. 3, no. 3, Mar. 1973, pp. 2-10.

163. Kagan, Aubrey: Stress and Noise Principles of Research. Noise as a Public Health Problem, Jerry V. Tobias, Gerd Jansen, and W. Dixon Ward, eds., ASHA Rep. 10, American Speech-Language-Hearing Assoc., Apr. 1980, pp. 237-240.

164. Rabkin, Judith G.; and Struening, Elmer L.: Live Events, Stress, and Illness. Science, vol. 194, no. 4269, Dec. 3, 1976, pp. 1013-1020.

165. Graeven, David B.: The Effects of Airplane Noise on Health: An Examina-

tion of Three Hypotheses. J. Health & Soc. Behav., vol. 15, no. 4, 1974, pp. 336-343.

166. Pilawska, H.; Mikulski, T.; Rusin, J.; Soroka, M.; and Wysocki, K.: Effect of Acoustic Microclimate Prevailing in Shipyards on the Health of Workers. Med. Pr., vol. 28, no. 5, 1977, pp. 441-447.
167. Jansen, G.: Adverse Effects of Noise on Iron and Steel Workers. Stahl. Eisen., vol. 81, 1961, pp. 217-220.
168. Shatalov, N. N.; and Murov, M. A.: The Influence of Intensive Noise and Neuropsyhic Tension on the Level of Arterial Pressure and the Incidence of Hypertensive Vascular Disease. Klin. Med. (Moscow), vol. 48, Mar. 1970, pp. 70-73.
169. Zvereva, G. S.; Ratner, M. V.; and Kolganov, A. V.: Noise of a Rolling Mill and Its Influence on the Bodies of Workers. Gig. Sanit., vol. 11, Nov. 1975, pp. 104-105.
170. Cieslewicz, J.: Attempt To Evaluate the Extra-Auditory Impact of Noise Upon the Workers of a Weaving Mill in the Cotton Industry. Med. Pr., vol. 22, no. 4, 1971, pp. 447-459.
171. Andriukin, A. A.: Influence of Sound Stimulation on the Development of Hypertension—Clinical and Experimental Results. Cor. Vassa., vol. 3, no. 4, 1961, pp. 285-293.
172. Geller, L. I.; Sakaeva, S. Z.; Musina, S. S.; Kogan, Ia. D.; Belomytseva, L. A.; Ostrovskaia, R. S.; Volokhov, Ia. P.; Lukianova, E. S.; Popova, P. M.; and Moskatel'nikova, E. V.: The Influence of Noise on Arterial Blood Pressure (On the Etiology of Arterial Hypertension). Ter. Arkh., vol. 35, no. 7, July 1963, pp. 83-86.
173. Andrukovich, A. I.: The Effect of Industrial Noise in Winding and Weaving Factories on the Arterial Pressure of Operators. Gig. Tr. Prof. Zabol., vol. 9, no. 12, Dec. 1965, pp. 39-42.
174. Kavoussi, N.: The Relationship Between the Length of Exposure to Noise and the Incidence of Hypertension at a Silo in Terran. Med. Lav., vol. 64, no. 7-8, July-Aug. 1973, pp. 292-295.
175. Capellini, A.; and Maroni, M.: Clinical Studies of Arterial Hypertension and Coronary Disease and Their Possible Relations to the Work Environment in Chemical Industry Workers. Med. Lav., vol. 65, no. 7-8, July-Aug. 1974, pp. 297-305.
176. Parvizpoor, D.: Noise Exposure and Prevalence of High Blood Pressure Among Weavers in Iran. J. Occup. Med., vol. 18, no. 11, Nov. 1976, pp. 730-731.
177. Cuesdean, L.; Teganeanu, S.; Tutu, C.; Raiciu, M.; Carp, C.; and Coatu, S.: Study of Cardiovascular and Auditory Pathophysiological Implications in a Group of Operatives Working in Noisy Industrial Surroundings. Physiologie, vol. 14, no. 1, Jan.-Mar. 1977, pp. 53-61.

178. Dega, K.; and Klajman, S.: The Effect of Noise on Some Indexes of the Circulatory System Efficiency of Shipyard Grinders. Bull. Inst. Marit. Trop. Med. Gdynia, vol. 28, no. 3-4, 1977, pp. 143-150.
179. Kachnyi, G. G.: Blood Pressure in Weavers Under Conditions of Industrial Noise. Vrach. Delo, vol. 4. Apr. 1977, pp. 107-109.
180. Friedlander, Barry; Greberman, Mel; Wathen, George; and Zeidler, William H.: An Analysis of Noise and Its Relationship to Blood Pressure in an Industrial Population. Maryland State Department of Health and Mental Hygiene (undated).
181. Welch, Bruce L.: Extra-Auditory Health Effects of Industrial Noise—Survey of Foreign Literature. Contract No. 16-BB-7, Welch Assoc., Dec. 1, 1978.
182. Cohen, Alexander: Industrial Noise and Medical, Absence, and Accident Record Data on Exposed Workers. Proceedings of the International Congress on Noise as a Public Health Problem, W. Dixon Ward, ed., 550/9-73-008, U.S. Environ. Prot. Agency, May 1973, pp. 441-453.
183. Lees, Ronald E. M.; Romeril, Carol Smith; and Wetherall, Lee D.: A Study of Stress Indicators in Workers Exposed to Industrial Noise. Canadian J. Public Health, vol. 71, no. 4, July-Aug. 1980, pp. 261-265.
184. Jansson, A.; and Hansson, L.: Prolonged Exposure to a Stressful Stimulus (Noise) as a Cause of Raised Blood-Pressure in Man. Lancet, vol. 1, no. 8002, Jan. 8, 1977, pp. 86-87.
185. Drettner, B.; Hedstrand, H.; Klockhoff, I.; and Svedberg, A.: Cardiovascular Risk Factors and Hearing Loss. A Study of 1000 Fifty-Year-Old Men. Acta Oto-Laryngol., vol. 79, no. 5-6, May-June 1975, pp. 366-371.
186. Hedstrand, H.; Drettner, B.; Klockhoff, I.; and Svedberg, A.: Noise and Blood Pressure. Lancet, vol. 2, no. 8051, Dec. 17, 1977, p. 1291.
187. Takala, J.; Varke, S.; Vaheri, E.; and Sievers, K.: Noise and Blood Pressure. Lancet, vol. 2, no. 8045, Nov. 5, 1977, pp. 974-975.
188. Brown, J. E., III; Thompson, R. N.; and Folk, E. D.: Certain NonAuditory Physiological Responses to Noises. J. American Ind. Hyg. Assoc., vol. 36, no. 4, 1975, pp. 285-291.
189. Cohen, Alex; Taylor, William; and Tubbs, Randy: Occupational Exposures to Noise Hearing Loss, and Blood Pressure. Noise as a Public Health Problem, Jerry V. Tobias, Gerd Jansen, and W. Dixon Ward, eds., ASHA Rep. 10, American Speech-Language-Hearing Assoc., Apr. 1980, pp. 322-326.
190. Lees, R. E. M.; and Roberts, J. Hatcher: Noice-Induced Hearing Loss and Blood Pressure. Canadian Med. Assoc. J., vol. 120, no. 9, May 5, 1979, pp. 1082-1084.
191. Manninen, O.; and Aro, S.: Noise-Induced Hearing Loss and Blood Pressure. Int. Arch. Occup. Environ. Health, vol. 42, no. 3-4, Jan. 15, 1979, pp. 251-256.

192. Ortiz, G. A.; Arguelles, A. E.; Crespin, H. A.; Sposari, Griselda; and Villafane, Carmen T.: Modifications of Epinephrine, Norepinephrine, Blood Lipid Fractions and the Cardiovascular System Produced by Noise in an Industrial Medium. Horm. Res., vol. 5, no. 1, Jan. 1974, pp. 57-64.
193. Paolucci, G.: Influence of the Noise on Catecholamine Excretion Effects of Long Duration Noise Exposure on Hearing and Health. Milton A. Whitcomb, ed., AGARD-CP-171, Nov. 1975, pp. C9-1-C9-2.
194. Ising, H.; and Melchert, H.-U.: Endocrine and Cardiovascular Effects of Noise. Noise as a Public Health Problem, Jerry V. Tobias, Gerd Jansen, and W. Dixon Ward, eds., ASHA Rep. 10, American Speech-Language-Hearing Assoc., Apr. 1980, pp. 241-245.
195. Ising, H.; Gunther, T.; Havestadt, T.; Krause, Ch.; Markert, B.; Melchert, H. U.; Schoknecht, G.; Thefeld, W.; and Teitze, K. W.: Study on the Quantification of Risk for the Heart and Circulatory System Associated With Noise Workers. EPA Transl. TR-79-0857, 1979.
196. Davis, H.: Project ANEHIN: Auditory and Non-Auditory Effects of High Intensity Noise. Joint Rep. 7, Central Inst. for the Deaf and U.S. Naval School Aviat. Med., 1958.
197. Davis, H.: Effects of High Intensity Noise on Naval Personnel. U.S. Armed Forces Med. J., vol. 9, 1958, pp. 1027-1048.
198. Carlestam, Gösta; Karlsson, Claes-Göran; and Levi, Lennart: Stress and Disease in Response to Exposure to Noise—A Review. Proceedings of the International Congress on Noise as a Public Health Problem, W. Dixon Ward, ed., 550/9-73-008, U.S. Environ. Prot. Agency, May 1973, pp. 479-486.
199. McLean, E. K.; and Tarnopolsky, A.: Noise, Discomfort and Mental Health. A Review of the Socio-Medical Implications of Disturbance by Noise. Psychol. Med., vol. 7, no. 1, Feb. 1977, pp. 19-62.
200. Cohen, Sheldon; and Weinstein, Neil: Nonauditory Effects of Noise on Behavior and Health. J. Social Issues, vol. 37, no. 1, 1981, pp. 36-70.
201. Meecham, W. C.; and Smith, H. G.: Effects of Jet Aircraft Noise on Mental Hospital Admissions. British J. Audiol., vol. 11, no. 3, Aug. 1977, pp. 81-85.
202. Abey-Wickrama, I.; A'Brook, M. F.; Gattoni, F. E. G.; and Herridge, C. F.: Mental-Hospital Admissions and Aircraft Noise. Lancet, vol. 2, no. 633, Dec. 13, 1969, pp. 1275-1277.
203. Herridge, C. F.; and Chir, B.: Aircraft Noise and Mental Hospital Admission. Sound, vol. 6, 1972, p. 32-36.
204. Chowns, R. H.: Mental-Hospital Admissions and Aircraft Noise. Lancet, vol. 1, Feb. 28, 1970, p. 467.
205. Gattoni, F.; and Tarnopolsky, A.: Aircraft Noise and Psychiatric Morbidity. Psychol. Med., vol. 3, no. 4, Nov. 1973, pp. 516-520.
206. Jenkins, L. M.; Tarnopolsky, A.; Hand, D. J.; and Barker, S. M.: Com-

parison of Three Studies of Aircraft Noise and Psychiatric Hospital Admissions Conducted in the Same Area. Psychol. Med., vol. 9, no. 4, Nov. 1979, pp. 681-693.

207. Jenkins, Linda; Tarnopolsky, Alex; and Hand, David: Psychiatric Admissions and Aircraft Noise From London Airport: Four-Year, Three-Hospitals' Study. Psychol. Med., vol. 11, no. 4, Nov. 1981, pp. 765-782.

208. Hand, David J.; Tarnopolsky, Alex; Barker, Sandra M.; and Jenkins, Linda M.: Relationships Between Psychiatric Hospital Admissions and Aircraft Noise: A New Study. Noise as a Public Health Problem, Jerry V. Tobias, Gerd Jansen, and W. Dixon Ward, eds., ASHA Rep. 10, American Speech-Language-Hearing Assoc., Apr. 1980, pp. 277-282.

209. Broadbent, D. E.: Individual Differences in Annoyance by Noise. Sound, vol. 6, 1972, pp. 56-61.

210. Tarnopolsky, Alex; Hand, David J.; Barker, Sandra M.; and Jenkins, Linda M.: Aircraft Noise, Annoyance, and Mental Health: A Psychiatric Viewpoint. Noise as a Public Health Problem, Jerry V. Tobias, Gerd Jansen, and W. Dixon Ward, eds., ASHA Rep. 10, American Speech-Language-Hearing Assoc., Apr. 1980, pp. 588-593.

211. Karagodina, I. L.; Soldatkina, S. A.; Vinokur, I. L.; and Klimukhin, A. A.: Effect of Aircraft Noise on the Population Near Airports. Hyg. & Sanit. (USSR), vol. 34, nos. 4-6, Apr.-June 1969, pp. 182-187.

212. Fluglärmwirkungen—Eine Interdisziplinare Untersuchung uber die Auswirkungen des Fluglärms auf den Menschen. Hauptbericht. Harald Boldt Verlag KG (West Germany), c.1974. (Available as NASA TM-75819.)

213. Koszarny, Zbigniew; Maziarka, Stefan; and Szata, Wanda: The Effect of Airplane Noise on the Inhabitants of Areas Near the Okecie Airport in Warsaw. NASA TM-75879, 1981.

214. Grandjean, Etienne; Graf, P.; Lauber, A.; Meier, H. P.; and Muller, R.: Survey on the Effects of Aircraft Noise Around Three Civil Airports in Switzerland. INTER-NOISE 76 Proceedings, Roger L. Kerlin, ed., c.1976, pp. 85-90.

215. Knipschild, Paul: Medical Effects of Aircraft Noise: Community Cardiovascular Survey. Int. Arch. Occup. Environ. Health, vol. 40, no. 3, Nov. 29, 1977, pp. 185-190.

216. Knipschild, Paul: Medical Effects of Aircraft Noise: General Practice Survey. Int. Arch. Occup. Environ. Health, vol. 40, no. 3, Nov. 29, 1977, pp. 191-196.

217. Knipschild, Paul.: Aircraft Noise and Hypertension. Noise as a Public Health Problem, Jerry V. Tobias, Gerd Jansen, and W. Dixon Ward, eds., ASHA Rep. 10, American Speech-Language-Hearing Assoc., Apr. 1980, pp. 283-293.

218. Knipschild, Paul; and Oudshoorn, Nelly: Medical Effects of Aircraft Noise:

Drug Survey. Int. Arch. Occup. Environ. Health, vol. 40, No. 3, Nov. 29, 1977, pp. 197-200.

219. Ko, N. W. M.; and Wong, V. L. P.: Responses to Road Traffic Noise: A Socio-Economic Approach. J. Sound & Vib., vol. 68, no. 1, Jan. 8, 1980, pp. 147-152.

220. Bradley, J. S.; and Jonah, B. A.: The Effects of Site Selected Variables on Human Responses to Traffic Noise. Part II: Road Type by Socio-Economic Status by Traffic Noise Level. J. Sound & Vib., vol. 67, no. 3, Dec. 8, 1979, pp. 395-407.

221. Meecham, W. C.; and Shaw, N.; Effects of Jet Noise on Mortality Rates. British J. Audiol., vol. 13, no. 3, Aug. 1979, pp. 77-80.

222. Frerichs, Ralph R.; Beeman, Barbara L.; and Coulson, Anne H.: Los Angeles Airport Noise and Mortality—Faulty Analysis and Public Policy. American J. Public Health, vol. 70, no. 4, Apr. 1980, pp. 357-362.

223. Meecham, William C.; and Shaw, Neil: Comments on 'Los Angeles Airport Noise and Mortality—Faulty Analysis and Public Policy.' American J. Public Health, vol. 70, no. 5, May 1980, p. 543.

224. Frerichs, Ralph R.; and Coulson, Anne Hersey: Frerichs, *et al.*, Respond. American J. Public Health, vol. 70, no. 5, May 1980, pp. 543-544.

225. Knipschild, Paul; and Sallé, Herman: Road Traffic Noise and Cardiovascular Disease. A Population Study in The Netherlands. Int. Arch. Occup. Environ. Health, vol. 44, no. 1, Aug. 1979, pp. 55-59.

226. Mroz, E.; Kopczynski, J.; Szudrowicz, B.; and Sadowski, J.: Health Sequelae of Community Noise. Rocz. Panstw. Zakl. Hig., vol. 29, no. 2, 1978, pp. 219-227.

227. Karsdorf, G.; and Klappach, H.: Einflüsse des Verkehrslärms auf Gesundheit und Leistung bei Oberschulerneiner Grossstadt. Z. Gesamte Hyg., vol. 14, Jan. 1968, pp. 52-54.

228. Cohen, Sheldon; Evans, Gary W.; Krantz, David S.; and Stokols, Daniel: Physiological, Motivational, and Cognitive Effects of Aircraft Noise on Children: Moving From the Laboratory to the Field. American Psychol., vol. 35, no. 3, Mar. 1980, pp. 231-243.

229. Cohen, Sheldon; Krantz, David S.; Evans, Gary W.; Stokols, Daniel; and Kelly, Sheryl: Aircraft Noise and Children: Longitudinal and Cross-Sectional Evidence on Adaptation to Noise and the Effectiveness of Noise Abatement. J. Pers. & Soc. Psychol., vol. 40, no. 2, Feb. 1981, pp. 331-345.

230. Cohen, Sheldon; Krantz, David S.; Evans, Gary W.; and Stokols, Daniel: Cardiovascular and Behavioral Effects of Community Noise. American Sci., vol. 69, no. 5, Sept.-Oct. 1981, pp. 528-535.

231. Ando, Y.; and Hattori, H.: Reaction of Infants to Aircraft Noise and Effects of the Noise on Human Fetal Life. Pract. Otol. Kyoto, vol. 67, 1974, pp. 129-136.

232. Knipschild, Paul; Meijer, Hans; and Sallé, Herman: Aircraft Noise and Birth Weight. Int. Arch. Occup. Environ. Health, vol. 48, 1981, pp. 131-136.

233. Ando, Y.; and Hattori, H.: Effects of Noise on Human Placental Lactogen (HPL) Levels in Maternal Plasma. British J. Obstet. & Gynaecol., vol. 84, Feb. 1977, pp. 115-118.

234. Schell, Lawrence M.: Environmental Noise and Human Prenatal Growth. American J. Phys. Anthropol., vol. 56, no. 1, 1981, pp. 63-70.

235. Jones, F. Nowell; and Tauscher, Judy: Residence Under an Airport Landing Pattern as a Factor to Teratism. Arch. Environ. Health, vol. 33, no. 1, Jan.-Feb. 1978, pp. 10-12.

236. Jones, F. Nowell: Nonauditory Effects of Noise on Fetal Life. Noise as a Public Health Problem, Jerry V. Tobias, Gerd Jansen, and W. Dixon Ward, eds., ASHA Rep. 10, American Speech-Language-Hearing Assoc., Apr. 1980, pp. 274-276.

237. Bader, Max: Residence Under an Airport Landing Pattern as a Factor in Teratism. Arch. Environ. Health, vol. 33, no. 4, July-Aug. 1978, p. 214.

238. Edmonds, Larry D.; Layde, Peter M.; and Erickson, J. David: Airport Noise and Teratogenesis. Arch. Environ. Health, vol. 34, no. 4, July-Aug. 1979, pp. 243-247.

239. Kryter, Karl D.: Physiological Acoustics and Health. J. Acoust. Soc. America, vol. 68, no. 1, July 1980, pp. 10-14.

240. Hattis, Dale; Richardson, Barbara; and Ashford, Nicholas A.: Noise, General Stress Responses, and Cardiovascular Disease Processes: Review and Reassessment of Hypothesized Relationships. Rep. EPA 550/9-80-101, June 1980.

241. Thompson, Shirley: Epidemiology Feasibility Study: Effects of Noise on the Cardiovascular System. Rep. EPA 550/9-81-103, Sept. 1981. (Available from NTIS as PB82 147 752.)

242. Thompson, Shirley J.: Epidemiology Feasibility Study: Effects of Noise on the Cardiovascular System. Appendix B: Annotated Bibliography. Rep. EPA 550/9-81-103B, July 1981. (Available from NTIS as PB82 147 760.)

243. Keil, Julian E.; and Propert, David M.: Epidemiology Feasibility Study: Effects of Noise on the Cardiovascular System. Appendix C: Review of Non-Noise Related Research of Cardiovascular Disease. Rep. EPA 550/9-81-103C, July 1981. (Available from NTIS as PB82 147 778.)

244. Five-Year Plan for Effects of Noise on Health. Rep. EPA 550/9-82-101, Dec. 1981. (Available from NTIS as PB82 168 972.)

Chapter 11
Reactions to Community Noise

Introduction

Most of the research on reactions of people to community noise can be divided into studies of annoyance as measured by attitude surveys and annoyance as measured by complaint behavior, including legal actions. This research has provided means of testing the concept promulgated over 20 years ago that the average amount of noise energy from significant sources that intrudes daily into houses and living areas can be used to predict the impact of the noise on people in a community. However, as is discussed below, research data on annoyance and complaint behavior collected over the past 10 to 20 years have shown that there are significant limitations and variables that must be considered in the fair application of the noise energy concept in its simplest form.

Relations Between Noise Exposure Level and Annoyance

Attitude surveys are the primary means used for measuring the amount of annoyance felt in a community from noises intruding into homes from outdoors. A variety of fairly synonymous terms have been used in attitude survey questions to elicit responses that show an attitude of annoyance because of noise. These terms include "annoyance," "bothersomeness," "unacceptability," and "disturbance." The responses to these questions have usually been scored on scales ranging from "not at all annoyed" (or the like) to "extremely annoyed" (or the like). Responses within that range have been divided, in different surveys, into steps of different size on 3-, 4-, 7-, and 9-point scales. Langdon and Griffiths (ref. 1) found that more reliable results were obtained with a 7-point scale than with a 4-point scale and that the word "dissatisfaction" used in a survey questionnaire was interpreted somewhat differently than the word "bother" in the same general question concerning noise.

In order to compare the results of different attitude surveys, it is helpful to express the response data in terms of a common verbal and quantitative scale. Figure 11.1 shows such a common scaling procedure. The basis (refs. 2 to 5) for figure 11.1 will be presented later.

Most of the research using attitude surveys over the past 40 years or so has been concerned with the effects of the noise from vehicles of transportation, especially aircraft, upon people in their homes. As seen in table 11.1 (from preliminary results prepared for the Department of Aviation, Dallas, Texas, under the

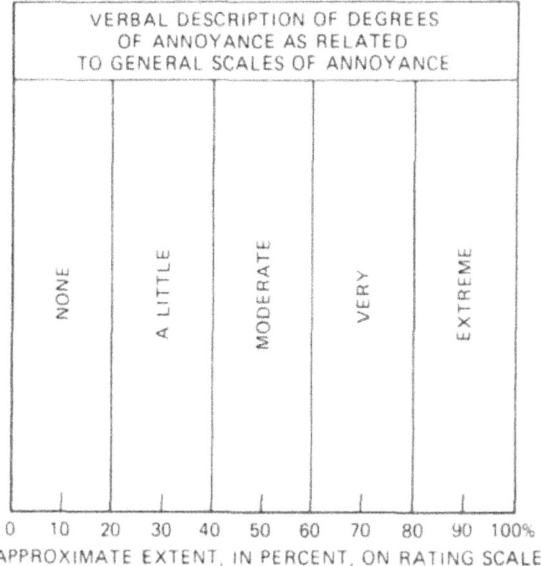

FIGURE 11.1. Proposed quantification of degrees of annoyance and percent of typical unipolar scale for rating noise environments. (Data from ref. 2.)

Noise Control Program) the noises from aircraft and secondarily from street traffic are the primary causes of high degrees of disturbance to people from noise in a community. In the residential areas involved, some 80 to 90 percent of the people were not at all disturbed by the variety of typical neighborhood noises, only 65 percent were not disturbed by noise from autos, and only 9 percent were not disturbed by the noise from aircraft. Indeed, it is seen that the percentage of people moderately, very, or extremely disturbed by aircraft noise day-night average sound levels (L_{dn}) as low as 55 dB exceeded the percentages thus disturbed by noise from any other source, including autos, buses, or trucks.

Two surveys conducted near Heathrow airport in London are among the most informative studies of annoyance from aircraft noise (refs. 6 and 7). Further, the similarity between these results and those of other aircraft noise surveys, discussed later, indicates that findings drawn from the Heathrow data have general validity. The division of the scale extent and the verbal descriptive labeling of these divisions in figure 11.1 utilize the fractional scale points and verbal applications developed in the Heathrow studies. In the Heathrow studies, the answers to several questions concerning annoyance and disturbance from aircraft noise were summed in certain ways to arrive at the overall annoyance score. This is described on pages 205 to 206 in reference 6 as follows:

> It is, of course, possible to ask informants to rate the degree of annoyance that they feel with aircraft, and such ratings were in fact obtained in answers to the question

TABLE 11.1

Subjective Ratings of Disturbance

(a) Environmental noises

Noise source	Percent of people rating disturbance as—			
	None	A little	Moderately	Very and extremely
Loud neighbors	88	3	3	5
Loud radios	89	3	3	4
Barking dogs	81	5	9	5
Children	94	2	1	4
Lawnmowers	95	3	1	1
Motorcycles	80	6	7	5
Industries	90	3	4	2
Autos	65	11	16	7
Buses	95	2	1	3
Trucks	86	3	6	5
Aircraft	9	16	31	41

(b) Aircraft noise in different zones

Aircraft zone, L_{dn}, dB	Percent of people rating disturbance as—			
	None	A little	Moderately	Very and extremely
Less than 55	15	38	38	9
55-60	20	35	35	10
60-65	10	10	53	27
65-70	17	17	39	27
70-75	14	14	18	54
75-80	0	0	50	50

13B, which was 'Does the noise of aircraft bother you very much, moderately, a little, not at all?' . . . the scale finally used was formed from the answers to the question 13B, and the five questions 'Does the noise of aircraft ever (a) wake you up, (b) interfere with listening to T.V. or radio, (c) make the house vibrate or shake, (d) interfere with conversation, (e) interfere with or disturb any other activity, or bother, annoy or disturb you in any other way?' An informant scored one if he rated himself at least a little annoyed by aircraft in question 13B, and an additional point for each kind of disturbance from aircraft—sleep, television or radio, house vibrating or interference with conversation— which he said annoyed him when it occurred. . . .

The scale gave annoyance ratings from 0 to 6, but the number of people scoring 6 was so small that for most purposes they could be combined with those scoring 5. The answers to question 13B showed that the scale points 0, 2, 3 and 4 correspond approximately to the verbal categories 'not at all,' 'a little,' 'moderately' and 'very much' annoyed.

L_{dn}, sometimes designated as DNL, is the composite daily average unit of noise developed by the Environmental Protection Agency (EPA) (refs. 8 and 9). As noted in chapter 2, the EPA specified two methods of the calculation of the exposure experienced from an occurrence of the fly-by noise from an aircraft. The preferred method is that of integrating the sound energy present during the entire noise event. A second procedure, commonly used in the 1961 to 1972 period when a number of attitude surveys were conducted, is to estimate the energy in an aircraft noise event by adding to the maximum level reached during about 0.5-second intervals of the event, a correction factor based on the number of seconds the noise is above some reference level. These two procedures can give somewhat different values for the same aircraft noise event, as discussed below. Unfortunately, it is not always possible to be sure which method of measurement was used when reading of L_{dn} values in the literature on aircraft noise.

In recent years, after about 1972, the integrated sound energy has been generally employed. In this book, L_{dn} is used to represent the preferred integrated energy metric. Average day-night sound level based on the maximum, or peak, level during a fly-by noise plus a duration correction procedure is identified, when known, as $L_{dn(m)}$. Specific formulae relevant to these two types of L_{dn} values are presented later.

Figure 11.2 (from refs. 6, 7, 10, and 11) shows as a function of $L_{dn(m)}$ the percentage of people who rated themselves as annoyed as, or more annoyed than, some degree of annoyance represented on the scale. Also shown, to indicate further the consistency among the results for different aircraft noise attitude surveys, is the trend curve given by Alexandre based on the combined results of several surveys, including the first Heathrow survey. Grandjean (refs. 12 and 13); a related study (ref. 14); Schultz (ref. 15); and Kryter (ref. 2) also showed the commonality of the results from a number of different aircraft noise attitude surveys.

Figure 11.3 shows interpolations from the curves of figure 11.2. In figure 11.3, the abscissa represents the degree of annoyance reported, and the noise exposure in $L_{dn(m)}$ is the parameter.

Normalized annoyance functions

Figure 11.3 indicates that as the noise is reduced to sufficiently low values, the annoyance curves become asymptotic. For example, it appears that 4 percent of the people will rate the aircraft noise environment as "extremely annoying" no matter how far it falls below an $L_{dn(m)}$ of 50 dB, and that even at an extrapolated $L_{dn(m)}$ of 35 dB, 30 percent of the people will say they are "a little" annoyed. For example, five daily aircraft noise occurrences at an outdoor peak level of about 71 dBA, and a 10-second duration above 61 dBA, would have an $L_{dn(m)}$ of 35 dB. Such noise occurrences conceivably could be sources, on occasion, of some disturbance and annoyance to speech communication or sleep. However, the asymptotic slope of these curves would seem to suggest the likelihood that there were

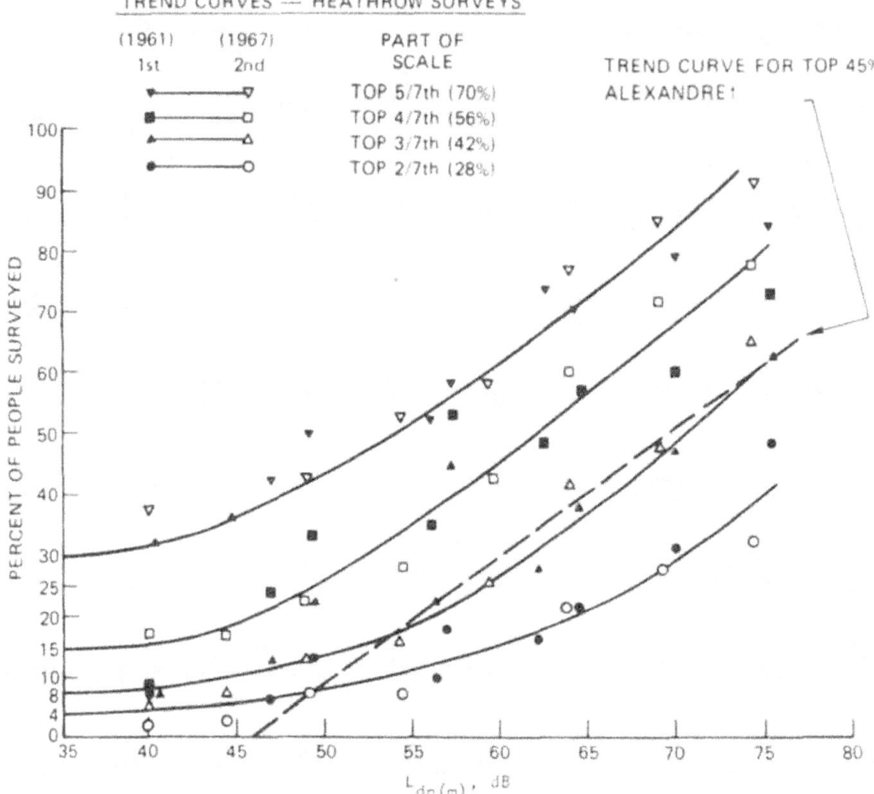

FIGURE 11.2. Percentages of people rated at or above various points of the annoyance scale in the Heathrow studies (refs. 6 and 7) as a function of $L_{dn(m)}$ as calculated or estimated in present analysis. Also shown are data from Alexandre (refs. 10 and 11).

people responding to aircraft noise independently of its intensity or effects on sleep, hearing, or house vibration.

The author suggests that these asymptotic percentages, for the given degrees of annoyance, represent people who are either supersensitive to noise, people who are actually exposed to aircraft noise events not included in the calculated or measured L_{dn}, or people whose attitudes of annoyance because of noise are unduly influenced by biases against aircraft operations in general.

It is proposed that the percentages of persons reporting given degrees of annoyance be corrected for the atypical responses due to one or more of the reasons just cited. The remainder of the responses, for want of a better term, would be called "normal-sensitives" and would represent feelings of annoyance primarily

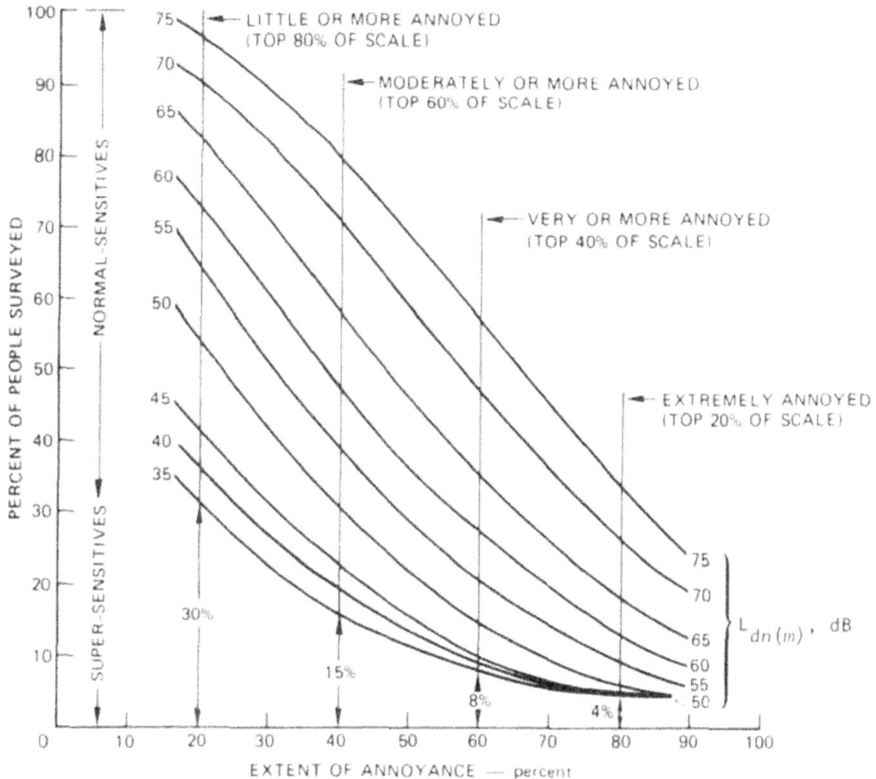

FIGURE 11.3. Percentage of people rating noise at or above a given point on a scale of annoyance. Based on Heathrow aircraft noise studies (refs. 6 and 7). (See figure 11.2.) Parameter in $L_{dn(m)}$ in decibels for aircraft noise.

related to the measured aircraft noise exposure. The unbiased percentage of people annoyed is found by subtracting from the total percentage of people rating a noise environment at or above a given degree of annoyance, the percentage of responses that can be expected regardless, apparently, of how weak the intensity of the noise exposure. Figure 11.4(a), based on figure 11.3, shows the percentages of all people found to have the different degrees of annoyance indicated as a function of L_{dn}, and figure 11.4(b) shows the same for the normal-sensitive population as determined by the procedure just described. This interpretation of these attitude survey data is speculative and perhaps only of academic interest. However, it does address the question of why aircraft noise at apparently inaudible levels in typical residential environments appears from attitude surveys to be a cause of feelings of extreme annoyance to some individuals.

In laboratory tests it is found that different "personality" types (*e.g.*, introvert, extrovert, or neurotic) identified by means of psychological tests respond, to

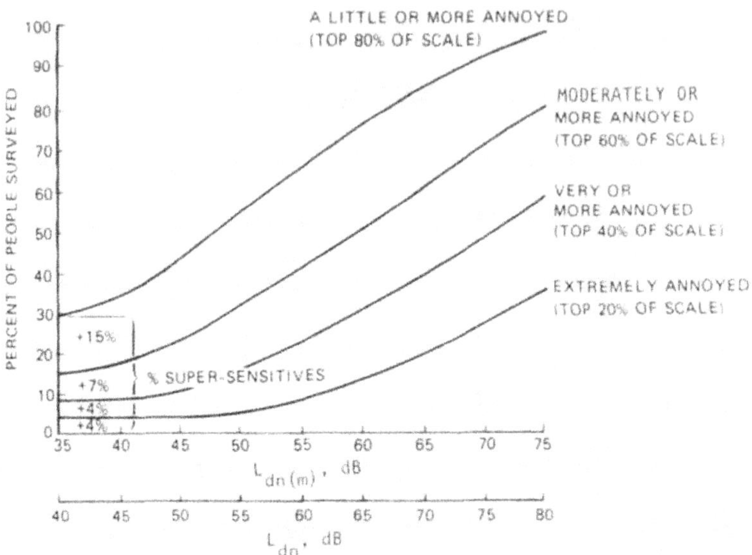

(a) Total survey.

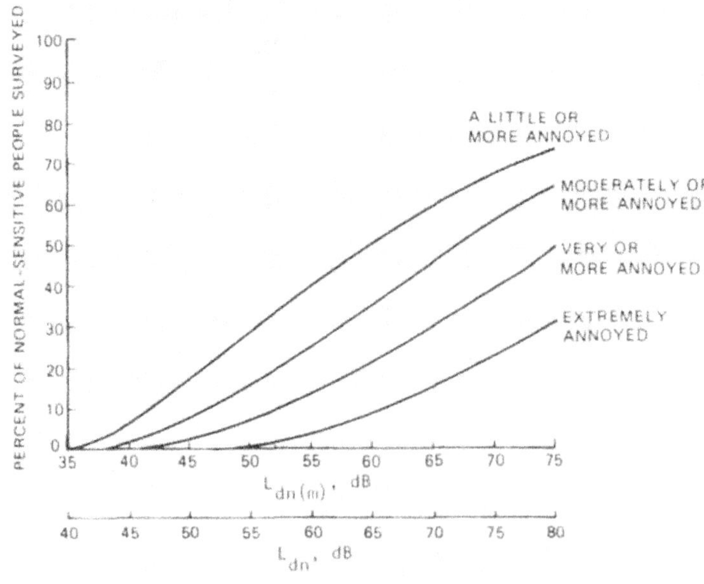

(b) Normal-sensitive people. Total survey minus "super-sensitive" respondents given "normal-sensitive" population. (See figure 11.3.)

FIGURE 11.4. Percentage of people surveyed rating aircraft noise at or above given points on scale of annoyance. (See figure 11.3.)

a small degree, differently to expected recordings of aircraft noises presented under varying conditions of room illumination (refs. 16 to 19). Such factors as personality-emotionality and the context of other environmental conditions undoubtedly contribute to the prevalence of "atypical" respondents in attitude surveys; however, from a practical point of view, as is discussed in more detail later, these personality and other environmental factors presumably do not distort the results of the surveys because they are distributed similarly among groups of people in different levels of exposure to aircraft noise.

Noise effects causing annoyance

In addition to rating the annoyance, or the like, due to a type of noise, people have also been asked to state why they felt annoyed. A sample of these findings is presented in figure 11.5(a). It is to be noted that the terminology used in questions concerned with disturbances from the noise is so different from study to study (*e.g.*, "ever disturbed" to "strongly disturbed") that a close clustering of the results for the various studies is not to be expected.

Nevertheless, a comparison of figure 11.4 with figure 11.5 shows that annoyance and disturbances (speech communications, sleep, and house vibration) increase in a somewhat similar way as the level of exposure increases. Disturbances from house vibration and vibration-induced rattles include cessation of auditory communication or interference with sleep. Some concern and annoyance may also be expressed about possible damage to windows and plaster in the house because of vibration.

It is to be expected that relative amounts of different effects from aircraft noise near different airports will depend upon the pattern of daily aircraft operations. For example, Taylor *et al.* (ref. 20) found that the overall annoyance from aircraft noise in communities with equal L_{dn} exposure was about equal for major airports in Canada and in Japan. The relative percentages of people highly annoyed by the noise because of sleep interference compared with speech interference differed between the two cities in a way commensurate with differences in the number of high level nighttime operations.

One of the major correlates of disturbance from aircraft noise is that of the "fear" people feel when the noise becomes sufficiently intense. The apparent reason is that the aircraft making the noise is so close that an aircraft accident or flight failure could cause damage to residents and structures in their neighborhood. It is seen in figure 11.6 (from refs. 21 and 22) that fear becomes increasingly prevalent as the aircraft noise exposure exceeds an L_{dn} of about 55 dB. The fear expressed by people in attitude surveys is generally not considered in the assessment of the impact of aircraft noise on people because the sound energy *per se* is not the source of the felt annoyance, but rather it is the meaning (approaching danger) to which subjects are reacting. While this distinction is logical in terms of the assessment and control of acoustical factors in the noise, the physiological stress reactions (discussed in chapter 10) to emotions, such as fear, and possible

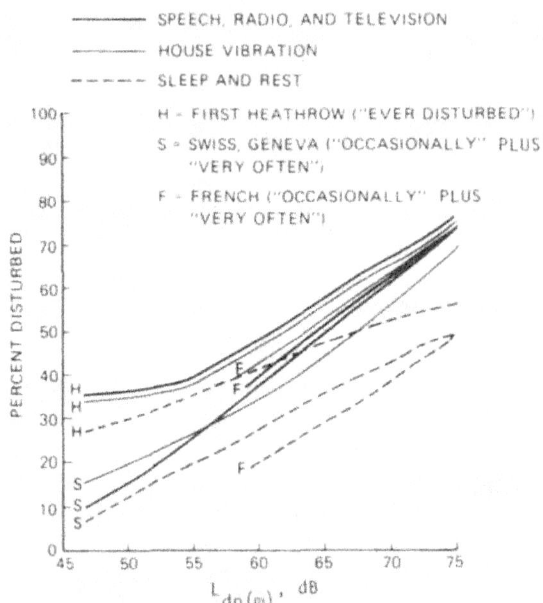

(a) Smoothed curves of data from first Heathrow (ref. 6), Geneva (ref. 14), and French (ref. 11) surveys.

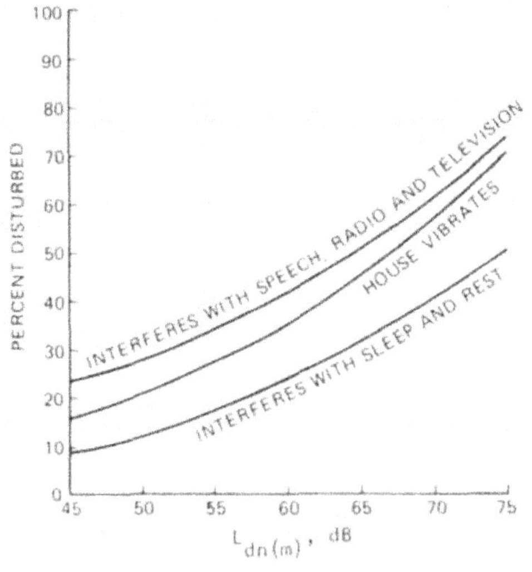

(b) Smoothed averages of curves in (a).

FIGURE 11.5. Disturbance from aircraft noise.

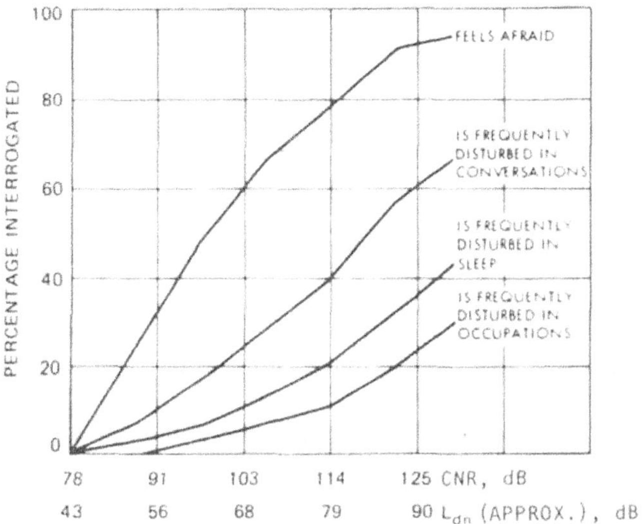

FIGURE 11.6. Results of community survey in the Netherlands on effects of aircraft noise. (From refs. 21 and 22.)

subsequent effects of that stress on mental and physical health should not be overlooked.

It appears from figure 11.6 that both annoyance from interference effects with normal behavior and fear similarly increase with increased intensity of aircraft noise. However, although this is found to be consistently true for the annoyance from the interference effects of the noise on auditory communications and sleep, Gunn *et al.* (ref. 23) found that ratings of general annoyance in a laboratory setting were greater when the subjects perceived that noise above a level of about 80 dBA came from aircraft flying directly overhead than from aircraft flying to one side, even though the noises were of equal intensity. Gunn *et al.* suggest that this finding is due to a greater fear from aircraft flying on a track taking the aircraft directly overhead than from one to the side of an observer, and that this factor probably operates in real-life situations. Hall *et al.* (ref. 24) deduced from an attitude survey conducted near the Toronto, Canada airport that nonacoustical factors such as nearness to flight paths and fear of crashes do affect annoyance responses but not activity interferences by aircraft noise.

Acoustical factors influencing individual response to noise

It has generally been held (refs. 25 and 26) that personality characteristics, differences in individual attitudes of misfeasance toward airport and aircraft operators, and fear of aircraft crashes influence annoyance reactions to aircraft noise and are at the root of the low correlations (usually less than 0.50) between noise exposure and annoyance in individuals. While there is little reason to doubt that such idiosyncratic factors influence the subjectively felt degrees of annoyance, it is possible that part of this relatively low correlation is attributable to the fact that the L_{dn} values for aircraft noise (which are typically measured or estimated as being homogeneous over a given neighborhood) do not reflect large differences in noise dosage received by individuals inside different houses. Perhaps differences in attitudes of misfeasance and fear found in persons from a given area are *caused* by differences in actual noise dosage; *i.e.*, the person who suspects more misfeasance or has more fear than his neighbor may actually be exposed inside his house to levels of aircraft noise that are considerably higher than those experienced by his neighbor inside his house, although both houses are in the same outdoor L_{dn} zone. In addition, differences in actual, at-the-person exposures occur because of differences in life-styles. For example, one person may stay at home more or use the outdoor areas of his home more than a neighbor exposed to supposedly the same L_{dn} from aircraft noise.

Cheifetz and Borsky (ref. 27) found that acoustical factors rather than presumed individual differences in noise sensitivity may often be the basis of differences among some people in reported annoyance from aircraft noise. These investigators selected people taken from comparable, in terms of L_{dn}, aircraft noise neighborhoods, but who differed greatly with respect to how annoying they found the aircraft noise to be in their homes. The subjects did not necessarily differ in their sensitivities to other noises. For this experiment, people with greatly different feelings of annoyance, but from similar L_{dn} aircraft noise environments, were brought to a laboratory "living room" with TV and in which recordings of aircraft noises were played. The people were asked to consider the noises as though they were in their own homes.

It was found that in the laboratory people similarly rated the annoyance level of different aircraft noises regardless of their differences in sensitivity to aircraft noise in their homes. These results suggest that the levels of the aircraft noises may have actually differed inside different homes (although the levels were about equal outdoors) and were perhaps a significant factor underlying the in-the-home feelings of annoyance.

It is possible that the subjects in the laboratory situation were unable to follow the instructions to respond to the noises as if they were in their homes. While this is undoubtedly true to some extent, there has been generally reasonable agreement between laboratory and field research findings in this area of research. For example, Becker *et al.* (ref. 28) found that differences in psychological sensitivity to noises in general in the home were also present in the same people when tested in the laboratory.

Correlations between L_{dn} and percentage of people annoyed

The factors influencing individual noise dosages and responses thereto do not significantly detract from the validity and utility of effective L_{dn} as a means for understanding and predicting the annoyance response of neighborhoods or groups of people to transportation noise. Indeed, as shown in table 11.2, the correlations between percentages of people annoyed "a little or more" and aircraft noise dosage in L_{dn} are of the order of 0.90 to 0.95 for the Heathrow studies. Clearly, those factors which contribute to variability in individual responses to the noise, be they personality differences, acoustical factors, or both, are rather evenly distributed among different groups of people exposed to given outdoor aircraft noise strata in the Heathrow studies. Hall *et al.* (ref. 29) found similarly high correlations (0.90 or higher) between L_{dn} and attitudes of annoyance from exposure to aircraft or to street traffic noise when the attitude scores were averaged over neighborhoods exposed to given levels of exposure.

It is also worth noting in table 11.2 that as the people who are included in the "annoyed" group are progressively restricted to include only the more severely annoyed, the correlations tend to become smaller but are still quite significant, of the order of 0.85 to 0.90. Restricting the range of noise strata over which the annoyance data are considered from a total range of 35 dB (an L_{dn} of 40.0 to 75.0 dB) to the lower 23 dB (an L_{dn} of 40.0 to 62.6 dB) or the upper 20 dB (an L_{dn} of 55.0 to 75.0 dB) does not greatly reduce the magnitude of the correlations.

The general reliability of attitude questionnaire surveys is shown in figure 11.7. These data were obtained from random sample surveys (over 3000 different households) taken repeatedly from the same general neighborhood areas (refs. 30 and 31). These surveys were administered several times to people in aircraft noise environments with L_{dn} values ranging from about 58 to 69 dB, as shown by the dots in the upper graphs of figure 11.7. The surveys were given about 2 weeks apart, covering a period of about 2 months. The subjects were asked to rate (on a 5-point scale from "not at all" to "extremely") their annoyance because of street traffic or aircraft noise during the past week, and for the noise from large airliners, during the past year. The percentage of people highly annoyed in this study is highly correlated (coefficient of about 0.90) with the L_{dn} exposure. The annoyance ratings for the past week were roughly the same as they were for remembrances of the past year (middle row compared with bottom row of graphs). In this investigation, "highly" annoyed referred to those people who rated their annoyance as "very" or "extreme," the top two categories or top 40 percent of the 5-point annoyance scale employed. As discussed later, the phrase "highly annoyed" was used by Schultz (ref. 15) to describe the top 27 to 29 percent of annoyance scales.

Griffiths and Delauzun (ref. 32) found that much of the variability in predictions of individual annoyance from noise exposure appears to rest upon the unreliability of the questionnaires rather than with individual differences in sensitivity to noise. Griffiths *et al.* (ref. 33) found that although the correlation coefficient was of the order of 0.4 between individual ratings of dissatisfaction with traffic noise and outdoor noise exposure level, the correlation became about 0.9 when the median of four dissatisfaction ratings (obtained over a period of a num-

TABLE 11.2

Correlation Coefficients Between $L_{dn(m)}$ and Cumulative Percentages of People Having Different Degrees of Annoyance from Aircraft Noise in the Heathrow Studies

[Data from refs. 6 and 7]

Study	$L_{dn(m)}$ range, dB	Correlation coefficient			
		"A little or more annoyed" (scale categories 2, 3, 4, 5, and 6)	"Moderately or more annoyed" (scale categories 3, 4, 5, and 6)	"Very much or more annoyed" (scale categories 4, 5, and 6)	"Extremely annoyed" (scale categories 5 and 6)
First Heathrow	47.7-74.2	0.94	0.89	0.88	0.87
Second Heathrow	40.0-75.0	.97	.96	.97	.95
First and second combined	40.0-75.0	.95	.93	.93	.88
	40.0-62.6	.91	.86	.79	.77
	55.0-75.0	.90	.82	.86	.87
Average (first and second)		.93	.89	.89	.87

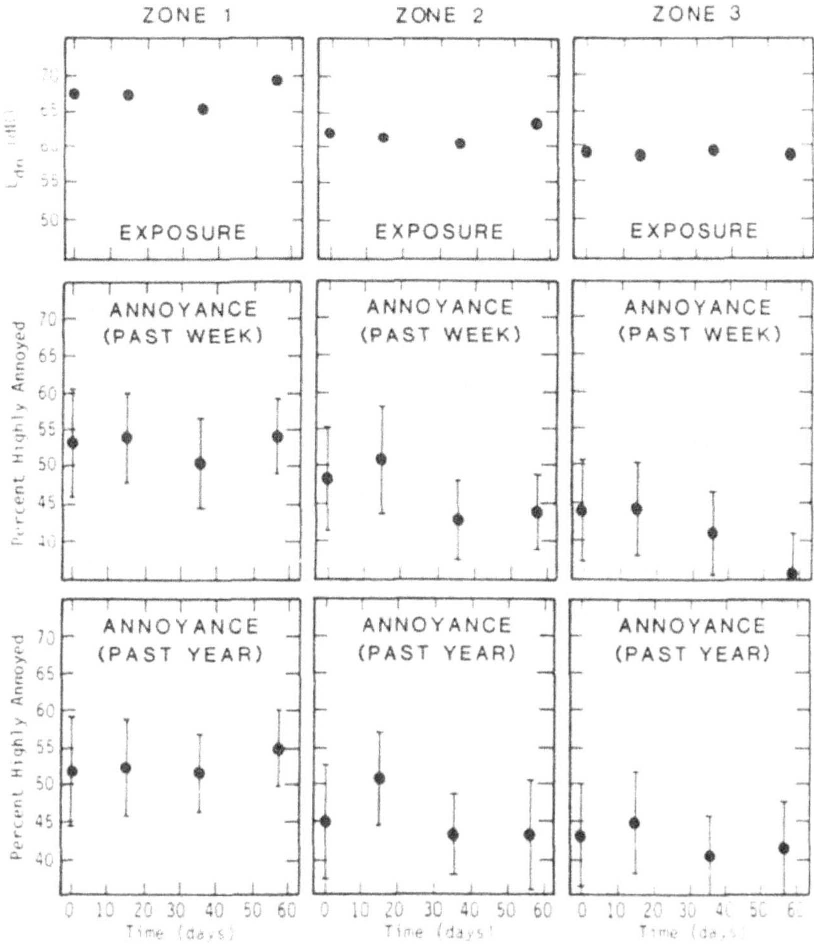

FIGURE 11.7. History of exposure to aircraft noise and annoyance. (From ref. 30.)

ber of months) for each individual respondent (222) was related to noise exposure level.

It seems likely that the ratings obtained at any one time are influenced by the activities the individuals were engaged in around the period of the interview. The typical annoyance for an individual requires several assessments perhaps because of acoustical and seasonal activity factors, rather than idiosyncratic personality factors. In any event, the reliability of attitude survey data appears to be greater than originally thought (refs. 25 and 26).

Analysis of aircraft and ground vehicle noise surveys

An important finding of surveys of annoyance from aircraft and ground vehicle noises is that nationalistic differences appear to be minimal factors in determining the impact of noise on feelings of annoyance. Presumably, this being the case, procedures for the assessment and regulation of such noises could be standardized on an international basis.

A major effort was made by Schultz (ref. 15) to compare the results of a number of attitude surveys of annoyance from aircraft and ground vehicle noise conducted in a number of countries. A similar analysis of the findings of the same surveys was made by Kryter (ref. 2). Although the end result of these two analyses was approximately the same for the aircraft noise surveys, there were disagreements about certain fundamental matters, and the conclusions reached with regard to the impact of ground vehicular noise differed significantly (refs. 2, 3, 4, and 5).

Six major attitude surveys of aircraft noise were selected by Schultz (ref. 15) and later by Kryter (ref. 2) for detailed comparative analyses. In some of these studies, the aircraft-noise-exposure levels were based on the peak or maximum level of intensity (L_A or PNL), reached at a given point on the ground during each flyover noise event. From the daily number of aircraft noise events of given peak levels, either a noise and number index (NNI) or an L_{dn} can be calculated. The basic formulae for L_{dn} are given in chapter 2. The formula for NNI was developed in Great Britain for measuring aircraft noise (ref. 6).

$$\text{NNI} = \text{Peak PNL} + (15 \log_{10} N) - 80$$

or,

$$\text{NNI} = \text{Peak } L_A + 13 + (15 \log_{10} N) - 80$$

where N is the average number of flyover noise events per 24-hour day.

However, to calculate (or rather to estimate) L_{dn} from peak levels, a duration correction must be added in order to estimate the "energy" in the noise event. In the discussions to follow, the exposure level of a noise event found from energy-type measures is called L_{ex}, and when estimated from a peak level plus a duration correction the result will be labeled as $L_{ex(m)}$. The formula used for estimating L_{ex} from peak levels is

$$L_{ex(m)} = \text{Peak } L_A + [10 \log_{10}(t/2)]$$

or,

$$L_{ex(m)} = \text{Peak PNL} - 13 + [10 \log_{10}(t/2)]$$

where t is the time in seconds between the 10-dB downpoints from the peak level. It is, of course, necessary to add a nighttime penalty for noise events occurring from 10 p.m. to 7 a.m. in the calculation of L_{dn}. Figures 11.8(a), (b), and (c)

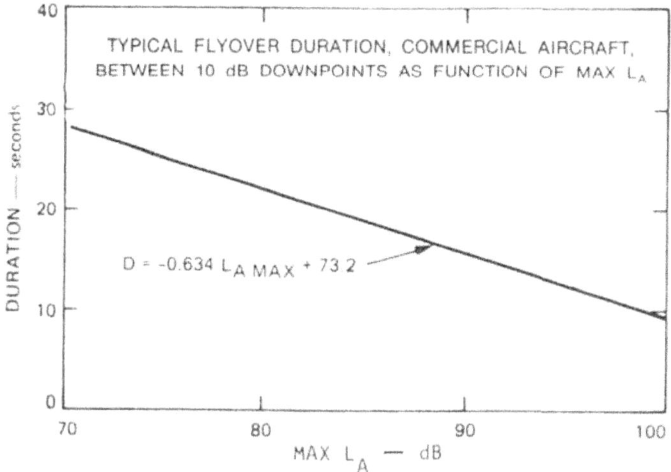

(a) Duration of typical commercial aircraft flyover noise between 10-dB downpoints as a function of maximum L_A on ground under aircraft. (Data from ref. 15, fig. 32.)

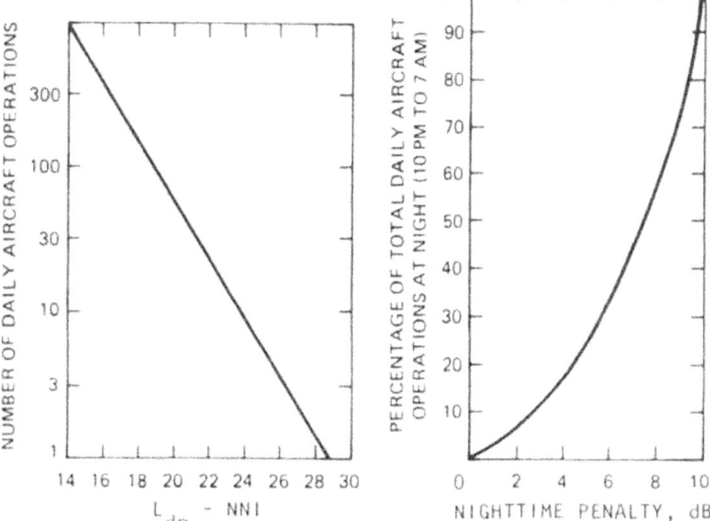

(b) Difference between L_{dn} or $L_{dn(m)}$ and NNI as a function of number of daily aircraft operations. (Note: 24-hr L_{eq} should be corrected for nighttime penalty in accordance with fig. 8(c).

(c) Number of decibels to be added to 24-hr $L_{eq(m)}$ of L_{eq} as a function of percentage of total daily aircraft operations occurring between the hours of 10 p.m. and 7 a.m. The result is $L_{dn(m)}$ or L_{dn}, as appropriate.

FIGURE 11.8. Graphs for converting NNI to L_{dn}.

present graphs useful for converting NNI to L_{dn}, given a specific number of aircraft operations (noise events), and for finding the night penalty to be assigned with a given percentage of the noise events occurring between the hours of 10 p.m. and 7 a.m. As noted previously, L_{dn} values obtained from energy-type L_{ex} are labeled L_{dn}. L_{dn} values based on $L_{ex(m)}$ are labeled as $L_{dn(m)}$.

Energy-type L_{eq} (equivalent continuous sound level) measures of the total daily aircraft noise exposure were obtained in some of the selected surveys. In these cases, L_{dn} is found by adding an estimated nighttime penalty (see fig. 11.8(c)) to the reported L_{eq} values.

British and Swedish surveys

In several important aircraft noise surveys the daily composite noise unit reported by the original investigators was NNI (refs. 6, 7, and 34). In the Heathrow (British, refs. 6 and 7) and Swedish (ref. 34) studies, the number of daily operations for various noise levels (maximum PNL or maximum L_A) are provided in published reports. However, in these studies, energy-type L_{ex} values or daily L_{eq} values were not reported. Accordingly, for these three surveys, $L_{dn(m)}$, but not L_{dn}, can be found.

French, Swiss, and German surveys

Three additional major surveys of reactions to aircraft noise that provide somewhat comparable data on the percentage of people "highly annoyed" are the so-called Swiss (ref. 14), Munich (refs. 35 and 36), and French (ref. 10) studies. The French, Swiss, and Munich surveys do not provide the detailed information on numbers of daily aircraft operations as a function of different strata of noise level, as is required to calculate directly $L_{dn(m)}$. These studies do, however, give NNI (presumably calculated from peak or maximum flyover noise levels) and measured L_{eq} of daily aircraft noise dosage.

In figure 11.9 are data points from the Swiss and Munich surveys, and curve thereto, for NNI plotted against L_{dn}, where the NNI values are those calculated by the investigators from numbers of aircraft operations and maximum L_A levels (as would be read on a standard sound level meter set on "fast") of individual flyover noises, and where L_{dn} is the measured integrated L_{eq} plus a nighttime penalty for those noises. Also shown in figure 11.9 are data points from the British and Swedish surveys, and curve thereto, for NNI calculated from maximum levels of individual flyover noises and plotted against L_{dn} calculated from the maximum level plus an energy correction factor for the estimated time between the levels 10 dB below the maximum level. (Unlike the Swiss and Munich surveys, the British and Swedish surveys did not measure L_{eq}, and it is for this reason that $L_{dn(m)}$ and not L_{dn} must be used in order to make a comparative analysis of the findings for these different surveys.)

It is clear in figure 11.9 that there is about a 5-dB difference between NNI versus $L_{dn(m)}$ and NNI versus L_{dn}. The reasons for this difference are not fully

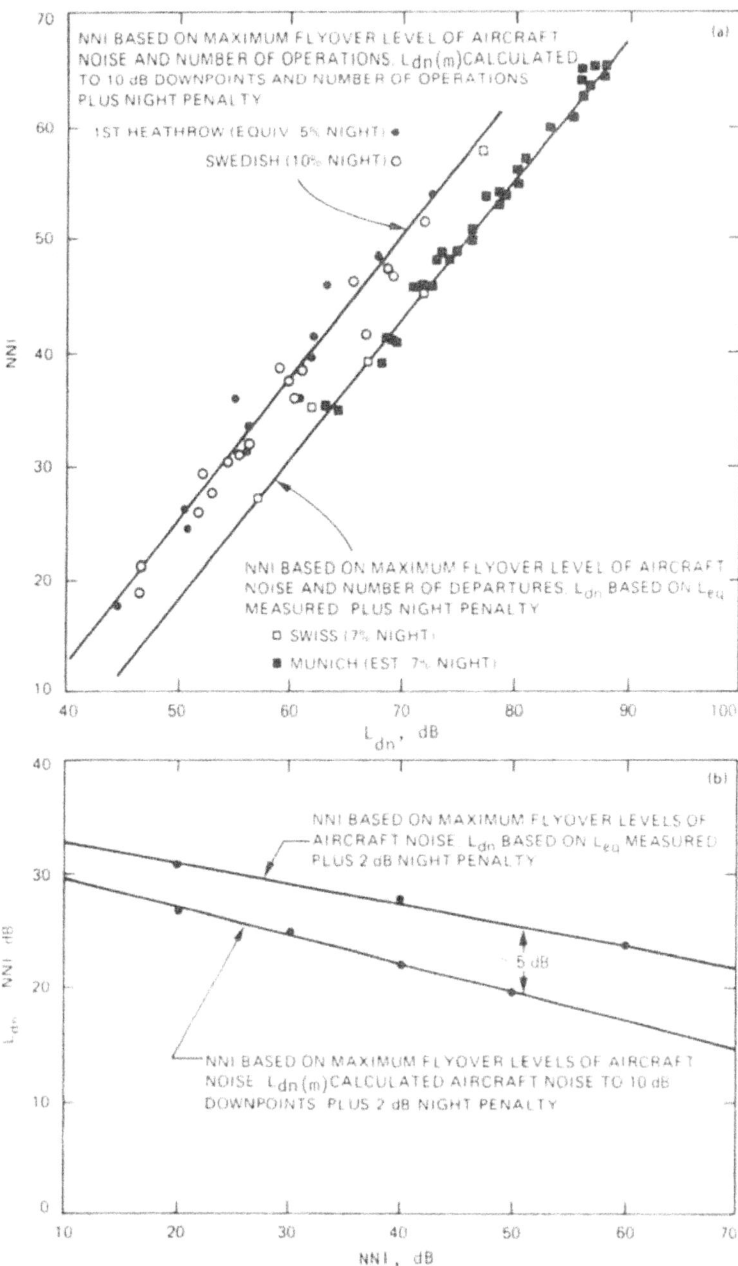

FIGURE 11.9. Upper figure shows NNI as a function of L_{dn} when L_{dn} is based on maximum levels plus calculated duration correction (upper curve) and when L_{dn} is based on energy-measured L_{eq} (lower curve). Lower figure shows as a function of NNI, differences between each form of L_{dn} and NNI. (From ref. 2.)

understood but may to some extent be related to the duration correction procedure. It was suggested (Kryter, ref. 2) that this difference might be due to the fact that the level of the flyover noise was at or near a maximum for a longer duration than 1 second. As a result, more sound energy would be actually present and measured by the computer-aided techniques than was presumed in this development of the simple duration-correction-plus-maximum-level formula. However, calculations of L_{ex} to the 10-dB downpoints for a number of aircraft flyover noises according to the duration correction formula and from integration of the sound energy present showed that no more than 1 dB of the above noted 5-dB difference could be caused by oversimplification of the flyover noise pattern by the formula estimation procedure (J. R. Young, SRI International, Menlo Park, Calif., personal communication).

An additional and more probable cause of the subject 5-dB difference is that the peaks of the levels in the British and Swedish studies were found on A-weighted standard sound level meters, which do not handle such short duration noises well. Whereas in the Munich and Swiss studies, the maximum levels were calculated from a large number of samples integrated over 1-second intervals. As discussed in chapters 5 and 7, for many types of sound signals, standard sound level meters do not give a good measure of the sound energy over 1-second intervals of time. In any event, in order to strictly compare the results of the attitude surveys it is somewhat important to take into account the procedures used for determining the exposure level of aircraft noise.

United States survey

A major program of attitude surveys of aircraft noise was conducted in the United States (U.S.) (refs. 37, 38, and 39). However, the results of this study cannot be readily averaged with the results of the six European surveys because the upper portion of the scale used is too broad to be comparable with the top 27 to 29 percent of the scale used in the analysis of the other surveys. (See Schultz, ref. 15.) Further, the U.S. survey in metropolitan areas comparable with those involved in European surveys was conducted during the summertime, whereas the European surveys were conducted in the spring and fall seasons. As discussed later, these seasonal differences can be significant factors in the amount of annoyance found in such surveys.

Trend curves for aircraft noise

Figure 11.10 shows trend curves for seven (six European and one U.S.) large-city attitude surveys of annoyance from aircraft noise. Except for the U.S. survey, the results fall in a compact range and suggest that the annoyance and disturbance effects of comparable dosages of aircraft noise in residential areas are much the same in all the countries. An average trend curve to represent this relation for the rating of "highly annoyed" is proposed in figure 11.10.

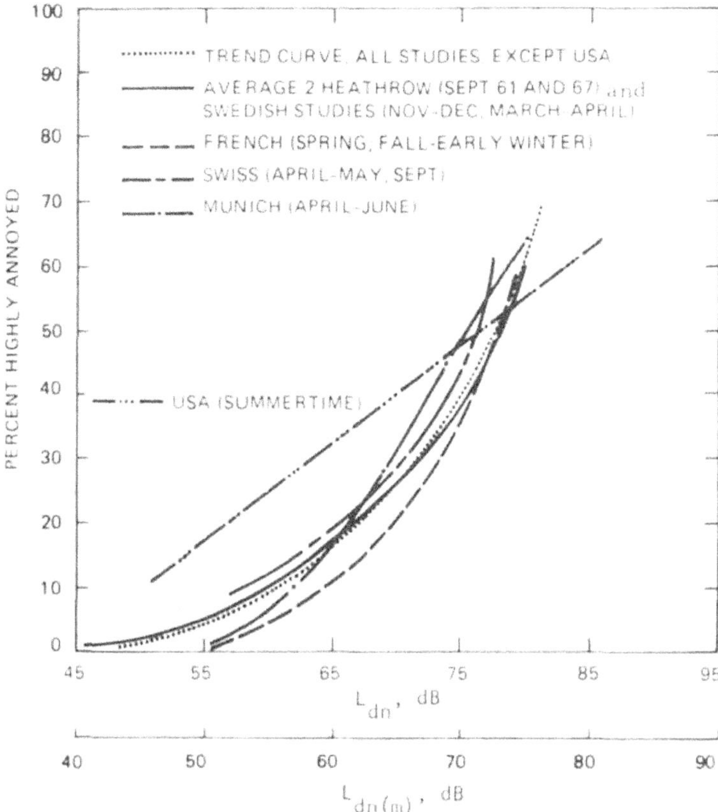

FIGURE 11.10. Results of attitude surveys of aircraft noise. Percent "highly annoyed" is for the top 27 to 29 percent of an annoyance scale. (From ref. 2. Basic data from the Heathrow (British) (refs. 6 and 7), Swedish (ref. 34); French (ref. 10); Swiss (ref. 14); Munich (ref. 35), and U.S. (refs. 37 and 38) studies.)

The somewhat greater amount of annoyance found in the survey conducted in the U.S. is perhaps explained by the aforementioned differences in the annoyance scales used and seasons of administration of the surveys. The effects of these differences could be expected to cause the observed general increase, relative to the European results, in the percentages of people called "highly annoyed" in the U.S. study.

Street, road, and railroad noise surveys

In street, road, and railroad surveys, noise is usually measured directly in time-integrated values (such as L_{50}, the noise level exceeded 50 percent of the time). Table 11.3 presents a compilation of data pertaining to L_{dn}, L_{eq}, and L_{50}

TABLE 11.3

Relation Between L_{dn}, L_{eq}, and L_{50} for U.S. Urban Road-Street Traffic

[Data from ref. 40. L_{eq} and L_{50} are for daytime only]

(a) Single-family dwellings or small apartments

[Average microphone location 8 ft above ground and 25 ft from curb]

City	Site number	Noise metric, dB			
		L_{dn}	L_{eq}	L_{50}	$L_{dn} - L_{50}$
Atlanta	401	63.8	61.1	51.0	12.8
	404	60.7	56.9	49.7	11.0
	406	67.3	63.5	54.6	11.7
Boston	0001	61.2	59.7	53.8	7.4
	0003	59.6	55.4	50.6	9.0
	0004	59.6	56.5	47.7	10.9
	0005	57.0	50.6	46.8	10.2
	0006	67.8	66.3	57.4	10.4
	0007	61.7	59.8	51.7	10.0
	0008	65.2	62.6	56.9	8.3
Chicago	502	71.2	68.0	63.7	7.5
	503	60.6	60.1	50.9	9.7
	505	59.0	46.8	47.0	12.0
	506	64.4	61.8	56.2	8.2
Dallas	1401	65.9	65.8	52.6	7.3
	1402	57.8	54.1	46.8	11.0
	1403	61.1	58.6	51.0	10.1
	1404	61.1	59.4	53.2	8.9
	1405	56.3	52.4	47.1	9.2
	1406	61.6	58.6	51.5	10.1
	Average	62.3	59.1	52.0	10.3

(b) Street corners in nonresidential areas

[Average microphone location about 18 ft above ground and 20 ft from curb]

City	Site number	Noise metric, dB			
		L_{dn}	L_{eq}	L_{50}	$L_{dn} - L_{50}$
Chicago	504	66.9	67.6	57.4	8.5
	508	63.1	61.7	56.2	6.9
	510	68.4	67.3	62.2	6.2
	511	70.7	69.7	65.7	5.0
	Average	67.3	66.6	60.4	6.9

TABLE 11.3 Concluded

(c) Masonry, multistory, contiguous apartments

[Average microphone location about 12 to 30 ft above ground and 12 ft from curb]

City	Site number	Noise metric, dB			
		L_{dn}	L_{eq}	L_{50}	$L_{dn} - L_{50}$
New York	0201	74.0	71.2	66.1	7.9
	2	72.9	68.6	64.9	8.0
	3	68.0	64.5	62.6	5.4
	4	73.6	68.3	65.3	8.0
	5	69.5	67.2	63.9	5.6
	7	71.6	69.4	65.1	5.5
	8	72.2	69.5	67.7	4.5
	9	71.1	69.8	62.6	8.5
	11	71.5	69.6	65.2	6.3
	12	68.0	66.1	63.5	5.5
	13	72.3	70.1	65.7	6.6
	Average	71.4	68.6	65.1	6.3

for urban street traffic noise in the U.S. (ref. 40). There are some systematic differences between L_{dn} and L_{50} as a function of type of neighborhood (or housing) and absolute level of L_{50}. Attention is invited to table 11.4, which shows some differences between European and U.S. data on this matter. Schultz (ref. 15) derived a formula for L_{dn} from L_{50} for the U.S. (See table 11.4.)

A significant difference emerged between analyses made by Schultz and by Kryter of the results of surveys of annoyance from ground vehicle noise in residential areas. Figure 11.11 shows the data used and curves fitted to these data. The synthesis curve of figure 11.11(a) (which was based on a clustering of the results of aircraft and ground vehicle noise surveys) shows a higher percentage of "highly annoyed" for a given L_{dn} than does the trend curve of figure 11.11(b). The data points in figure 11.11(b) are taken from references 39 to 47.

Worthy of note is that the synthesis curve of figure 11.11(a) lies above most of the data points. In addition, there are grounds for questioning the propriety of the inclusion of the upper data points (those for the London and Paris ground vehicle noise surveys) because the annoyance scales were not comparable in certain major respects to those used in the aircraft or street traffic noise surveys. The reasons for questioning and omitting from the present analysis the noted London, Paris, and U.S. data are that a "bipolar" scale of annoyance was used in the London survey, a somewhat ambiguous and unique rating procedure for annoyance was used in the Paris survey, and the annoyance scale was divided into larger sized steps in the U.S. survey than in the other studies.

TABLE 11.4

Relationship Between L_{dn} and L_{50} for Road Traffic

(a) Europe (calculated)

Location	L_{dn}, dB	L_{50}, dB	$L_{dn} - L_{50}$, dB	Schultz equation (ref. 15) used to calculate L_{50} (a)
Paris urban	52.5 62.5 72.5	45.6 56.1 66.6	6.9 6.4 5.9	$L_{dn} = 0.95\,L_{50} + 9.2$
Paris suburb	52.5 62.5 72.5	46.9 56.5 66.2	5.6 6.0 6.3	$L_{dn} = 1.04\,L_{50} + 3.7$
Belgium	52.5 62.5 72.5	43.9 56.0 68.0	8.6 6.5 4.5	$L_{dn} = 0.829\,L_{50} + 16.1$
Sweden	52.5 62.5 72.5	45.5 56.4 67.3	7.0 6.1 5.2	$L_{dn} = 0.92\,L_{50} + 10.6$
Average			6.3	

(b) United States

Location	Measured averages (b)			Calculated from Schultz equation (ref. 15) $L_{dn} = 0.762\,L_{50} + 22.0$		
	L_{dn}, dB	L_{50}, dB	$L_{dn} - L_{50}$, dB	L_{dn}, dB	L_{50}, dB (b)	$L_{dn} - L_{50}$, dB
Single-family dwellings or small apartments	62.3	52.0	10.3	61.6	52.0	9.6
Street corners in nonresidential areas	67.3	60.4	6.9	68.0	60.4	7.6
Masonry, multistory, contiguous apartments	71.4	65.1	6.3	71.6	65.1	6.5

[a] L_{50} *values for about 7 a.m. to 10 p.m.*
[b] *From table 11.3. Data from ref. 40.*

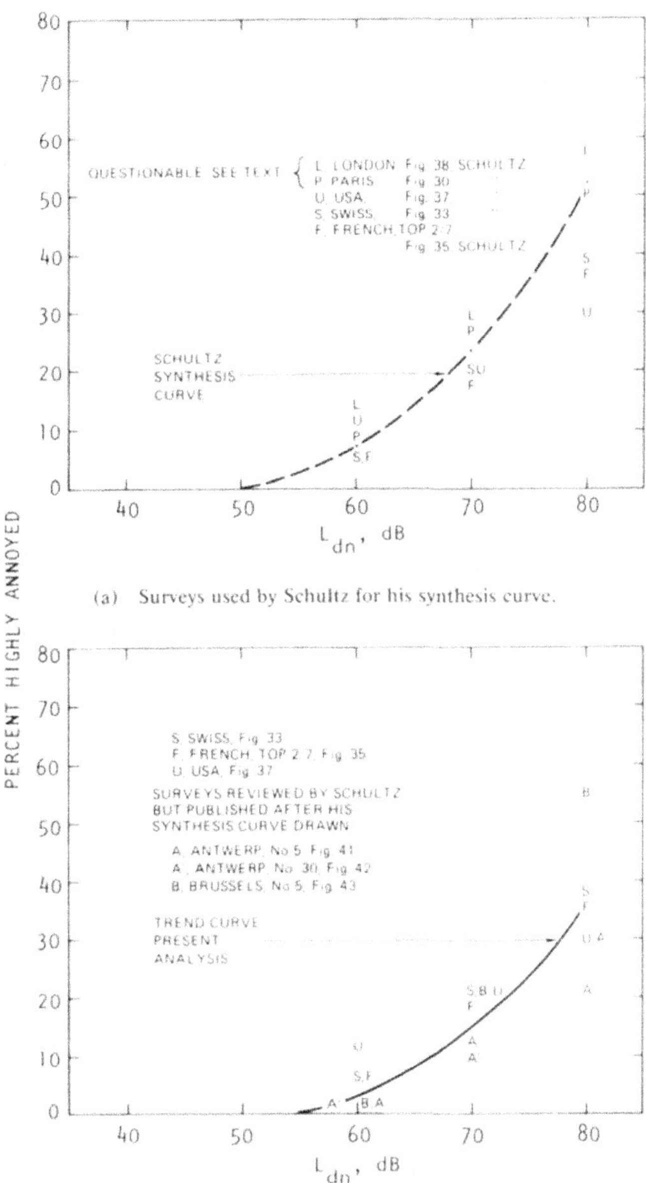

(a) Surveys used by Schultz for his synthesis curve.

(b) Surveys used in present analysis. Plotted data points are taken from respective (refs. 39 to 47) figures cited by Schultz (ref. 15).

FIGURE 11.11. Results of ground vehicle (street, road, and railroad) noise surveys. "Highly annoyed" refers to top 27 to 29 percent of annoyance scale.

Attention is invited to a number of other important attitude surveys of street and railroad traffic noise: Rylander *et al.* (ref. 48), Bruckmayer and Lang (ref. 49), Fog and Jonsson (ref. 50), Lamure and Bacelon (ref. 51), Nimura *et al.* (ref. 52), Hall and Taylor (ref. 53), Relster (ref. 54), and Lang (ref. 55). For a variety of reasons, the results of these surveys could not be readily converted to the annoyance scaling and noise measurements adopted in the cluster analyses of attitude survey data made by Schultz (ref. 15) and Kryter (ref. 2).

Aircraft versus ground vehicle noise trend curves

Figure 11.12(a) shows that there is about a 10-dB difference between the L_{dn} cluster trend curves for equal percentages of persons highly annoyed in different surveys of aircraft and ground vehicle noise. However, some of the difference for figures 11.12(a) (from ref. 2), 11.12(c) (from ref. 56), and 11.12(d) (from ref. 57) could possibly be caused by variations in the groups of people and survey questionnaires involved. These problems are overcome in studies conducted by Hall *et al.* (ref. 29) and Taylor (ref. 58) in which the same attitude questionnaire was administered to people exposed to road vehicle and aircraft noise in their neighborhoods. As seen in figure 11.12(b), Hall *et al.* (ref. 29) found a greater sensitivity to aircraft than to road traffic noise. The difference was equivalent to a difference of 6 to 14 dB in the noise levels measured (at the fronts of the the houses). Similar results, discussed later with respect to effects of noise from multiple sources, were found by Taylor (ref. 58).

The percentages of people "highly annoyed" are somewhat greater in figure 11.12(b) than in figure 11.12(a) for both types of noises. The reason is that the attitude scale used in the Hall *et al.* study, figure 11.12(b), was bipolar and intended for application to sounds in general; *i.e.*, ranging from "extremely agreeable" to "extremely disturbing," and for this reason, as mentioned earlier, biased the response relative to the annoyance response elicited by unipolar scales. However, the relative difference between the aircraft and road vehicle noise found by Hall *et al.* should not be affected by this factor.

Figure 11.13 summarizes the differences found in various studies and analyses of the relation between L_{dn} for ground-based vehicle noise and aircraft noise when the two noises cause equal annoyance or dissatisfaction. It is seen that, except for the Schultz analysis, there is a significant difference in favor of the ground vehicle noise (causes less annoyance) and that this difference, in general, increases as L_{dn} rises above about 60 dB. This progressively greater difference is presumably due to the accelerated growth in annoyance at the higher levels of aircraft noise exposure. The dashed curve in figure 11.13 is suggested to represent the approximate average of the top four curves.

Types of disturbances from road vehicle noise as compared with aircraft noise

The results of studies by Grandjean *et al.* (refs. 12 and 13) and a related study (ref. 14) support the proposition that the noise from ground-based vehicles

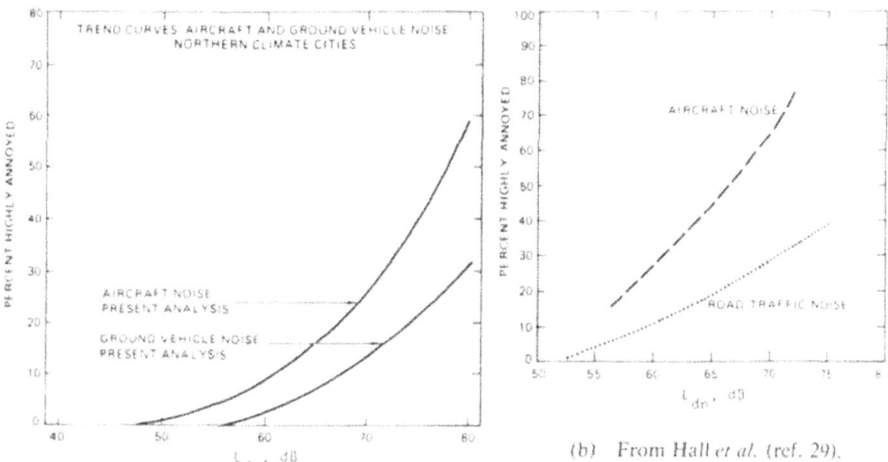

(b) From Hall *et al.* (ref. 29).

(a) From Kryter (ref. 2). Top 27 to 29 percent of scale = Highly annoyed.

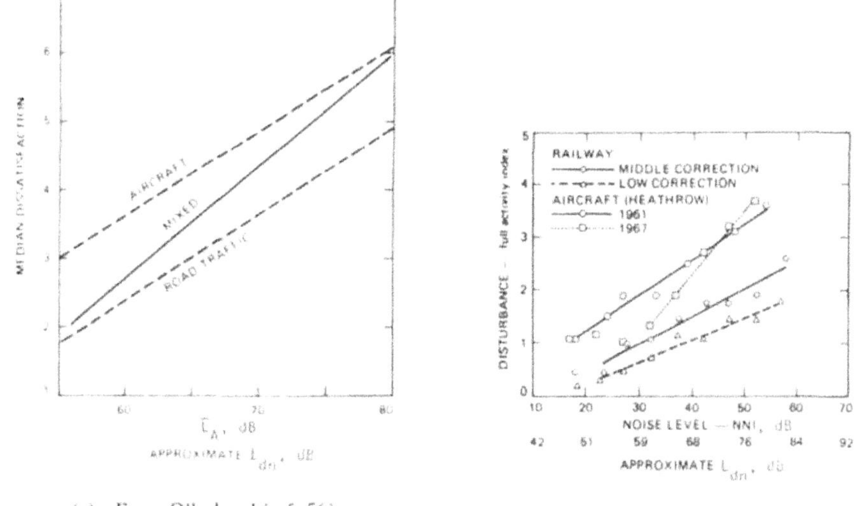

(c) From Ollerhead (ref. 56).

(d) From Fields and Walker (ref. 57).

FIGURE 11.12. Trend curves showing relations between annoyance or dissatisfaction and L_{dn} for noise from aircraft and ground-based vehicles.

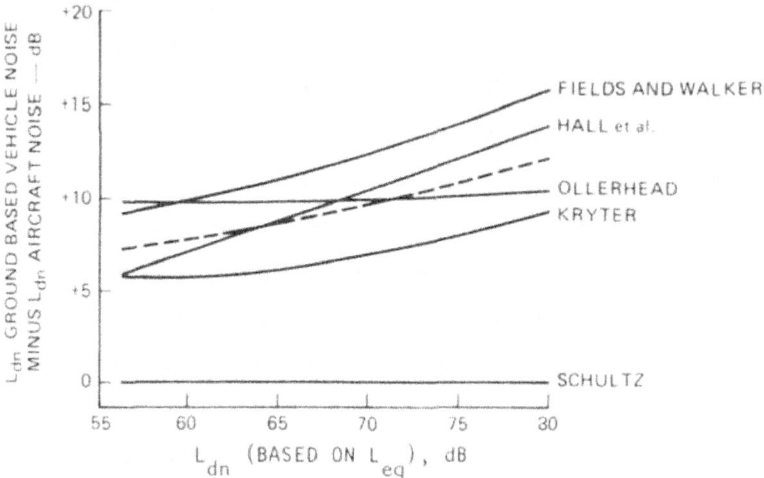

FIGURE 11.13. Number of decibels by which L_{dn} of ground-based vehicle noise must exceed that of aircraft noise to cause equal annoyance in residential areas. The dashed curve is a proposed conversion function for L_{dn} values of equal annoyance. Based on analyses by Fields and Walker (ref. 57), Hall *et al.* (ref. 29), Ollerhead (ref. 56), Kryter (ref. 2), and Schultz (ref. 15).

is less intrusive physically than is aircraft noise in many of the living areas in and around houses. These investigators surveyed nearly 5000 people in the region of Basel, Switzerland, with respect to disturbances caused by both aircraft and by road vehicle noises. Figure 11.14, based on their data, shows that it takes an L_{dn} 5 to 15 dB higher, depending on L_{dn} level, for road traffic noise than for aircraft noise to cause equal disturbance, averaged over all effects. The conversion of L_{50} (the statistic used by Grandjean *et al.* for describing road noise) to L_{dn}, as plotted in figure 11.14, was based on the average difference (≈ 6 dB) for European locations given in table 11.4.

It should be noted, however, that road noise and aircraft noise cause somewhat different patterns of effects. For example, conversation was less disturbed than was sleep by road noise, with the reverse being true for aircraft noise.

It would follow, then, that the data from attitude surveys of annoyance from ground-based vehicular noises should not cluster, as claimed by Schultz (ref. 15), as a function of L_{dn} with similar data for aircraft noise, but should be separate. This is not to say that a noise event at the listeners' ears of comparable energy and spectrum will cause more speech interference or sleep arousal because it is from an airplane instead of, say, an automobile. Rather, it is deduced that these two noises will generally be significantly different in level at the listeners' ears in many living areas of a home when the noises are reported as being equal in level according to measurements made outdoors. Possible explanations

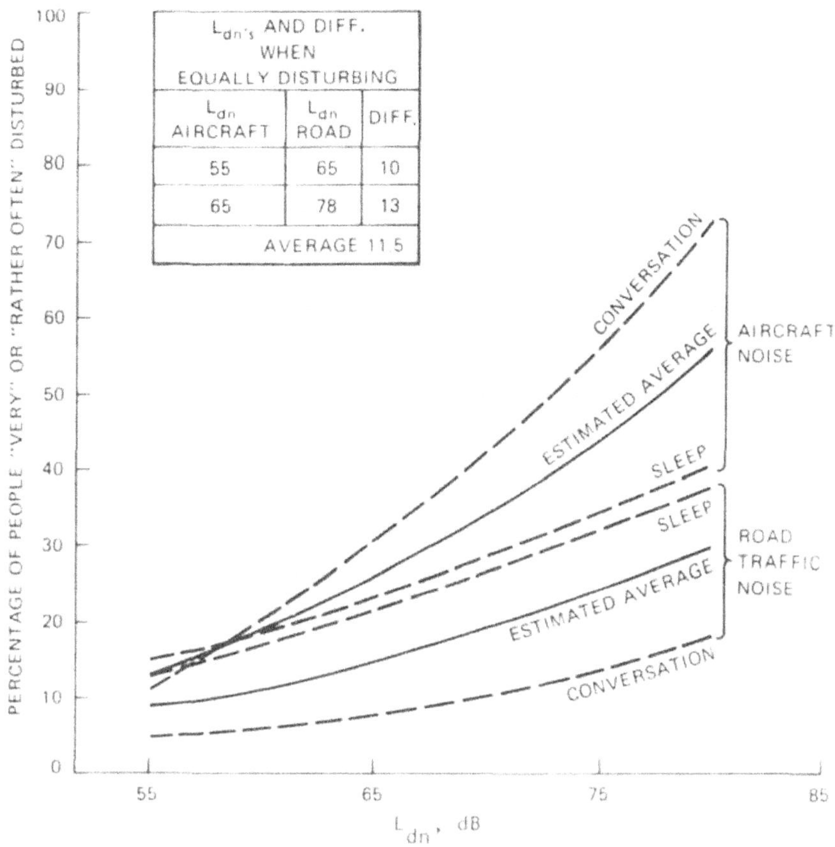

$L_{dn(m)}$ for aircraft = NNI + conversion factor. (See fig. 11.9.) L_{dn} for road traffic = L_{50} + conversion factor. (See table 11.4, average for European locations.) Curves for road traffic noise are extrapolations above about L_{dn} = 75 dB. Aircraft noise data are for Geneva, and road traffic noise data are for Basel.

FIGURE 11.14. Average percentage of people disturbed from sleep and conversation by aircraft and by road traffic noise. (Based on refs. 12 and 14.)

for these rather large apparent differences in the effectiveness of outdoor aircraft and ground vehicle noise in causing annoyance are given later in this chapter.

Additional studies (Nemecek *et al.* (ref. 59), Fields and Walker (refs. 60 and 61), and Ahrlin and Rylander (ref. 62)) conducted, for the most part, since the above discussed Schultz and Kryter analyses report that variations in demographics, attitudes, research methodologies, and types of disturbances (*e.g.*, sleep or speech interference) appear to contribute some uncertainties in the results and reliability of attitude surveys. It might be noted that these referenced studies do not attempt to assess the possible contributions to this variability by acoustical variables intervening between the place of the exterior noise measurements and the noise as received in the houses of the people annoyed.

Öhrström *et al.* (ref. 63) found that judgments in the laboratory of recorded noises from aircraft, lorries, trains, and mopeds were generally predictable from A-weighted L_{eq} measured at the position of the listener. However, the noise from lorries was somewhat less disturbing at equal L_{eq} than the other noises. The authors suggest that some "irregularity of the noise or individual experience" could be responsible for this finding. An additional factor to be mentioned is the previously discussed shortcoming of A-weighting with respect to the subjective loudness of low and mid-high sound frequencies. It is possible that some of this apparent variability in uniform predictability of the annoyance from different sources would be reduced were a more appropriate weighting for loudness than A used in the assessment of environmental noises.

Sonic booms

A special type of noise from aircraft is that of booms created when aircraft fly at supersonic speeds (generally greater than about 700 miles per hour). Some military aircraft and the commercial Concorde are capable of such flight. A great deal of research on the impact of sonic booms on people, animals, and structures was conducted during the 1960's with the advent of the Concorde and other proposed commercial supersonic transports.

In order to establish the equivalent effectiveness of sonic booms and typical subsonic aircraft flyover noise as causes of annoyance, a series of tests were conducted with subsonic and supersonic jet aircraft (ref. 64). Aircraft were flown over people inside and outside one-story and two-story air-conditioned frame houses. Over 100 of the people lived in residences at Edwards Air Force Base and had been exposed to five or so sonic booms per day for an average of 2 years. Some 200 subjects who had no prior experience with sonic booms were also tested. The major findings with respect to the effect of vibration on judged annoyance are discussed in chapter 5.

Figure 11.15 is based on figure 5.21 and shows the $L_{A,ex}$ value of subsonic aircraft noise judged to be equally as acceptable or unacceptable as sonic booms to people "accommodated" to both sonic booms and subsonic aircraft noise. Using these relations, it is possible to calculate an equivalent (to subsonic aircraft noise) 24-hour L_{eq} or L_{dn} for sonic booms of different levels and numbers of daily occurrences. To calculate L_{dn}, a 10-dB penalty would be added to the equivalent $L_{A,ex}$ values of the booms occurring between the hours of 10 p.m. to 7 a.m.

Calculations made for the proposed operations of supersonic transports over the U.S. indicated that some 50 million people per day would be exposed to an equivalent L_{dn} from sonic booms of about 70 dB or greater (ref. 65). Ancillary laboratory experiments and real-life experiences in cities substantiated, in gen-

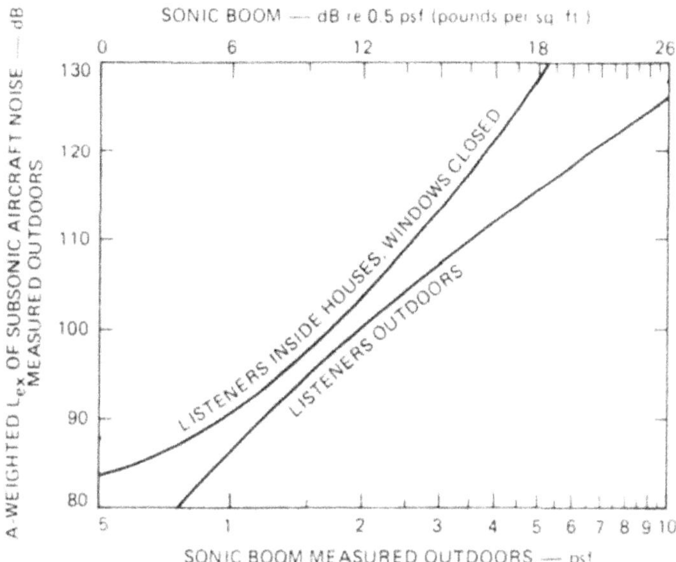

FIGURE 11.15. Graph for converting sonic booms judged equivalent $L_{A,ex}$ of subsonic jet aircraft flyover noise. Boom and noise measured outdoors; listeners indoors and outdoors as indicated. Based on figure 5.21.

eral, this use of equivalent $L_{A,ex}$ or L_{dn} for assessing exposure to sonic booms (ref. 65).

Attention is invited to the more rapid growth in annoyance ("unacceptability" rating) to sonic booms as their level is increased compared with ratings for the noise from subsonic aircraft. For example, for the listener indoors, a sonic boom of 1 psf would be equivalent to a subsonic aircraft noise at an $L_{A,ex}$ of about 90 dB. With 50 booms during the hours of 7 a.m. to 10 p.m., the equivalent L_{dn} would be 58 dB. However, for a sonic boom of 2 psf (an increase of but 6 dB in psf), the equivalent $L_{A,ex}$ is about 103 dB (an increase of 13 dB in $L_{A,ex}$), and the L_{dn} with 50 daily daytime booms would be 71 dB. As shown earlier in this chapter, an L_{dn} of 71 dB for regular aircraft noise represents a very significant source of annoyance in residential areas. It seems likely that the accelerated growth of annoyance from sonic booms relative to subsonic aircraft noise at higher boom levels shown in figure 11.15 is due to noticeably more house vibration and window rattles at the higher levels.

Most of the energy in sonic booms, as well as booms from cannon fire and other explosive sources, is generally in the sound frequencies below about 100 Hz. In addition to direct auditory effects of the booms, sound energy in these lower frequencies can cause vibrations and rattles that are a source of annoyance to residents. The relation of the spectra of such booms to judged annoyance is discussed in chapter 5.

Electric transformer and transmission line noise

A study was conducted by Fidell *et al.* (ref. 66) on feelings of annoyance in residents exposed to the "hum" from transformers in ground-based substations in urban areas and to the so-called "corona" discharge noise, a "crackling, frying" sound that comes from high voltage power lines when wet from rain, snow, or dense fog. The power lines are usually suspended a hundred or more feet in the air and several hundred feet from backyards of homes along electrical rights-of-way in suburban areas. Those respondents who self-rated themselves as "very" or "extremely" annoyed on a 5-point scale (top 40 percent of the scale) by either the transformer or corona noise were called "highly annoyed" by Fidell *et al.* These ratings correspond to "very or more annoyed" used in this book for the top 40 percent of the noise annoyance scale. The percentages who were "very or more annoyed" are shown in figure 11.16.

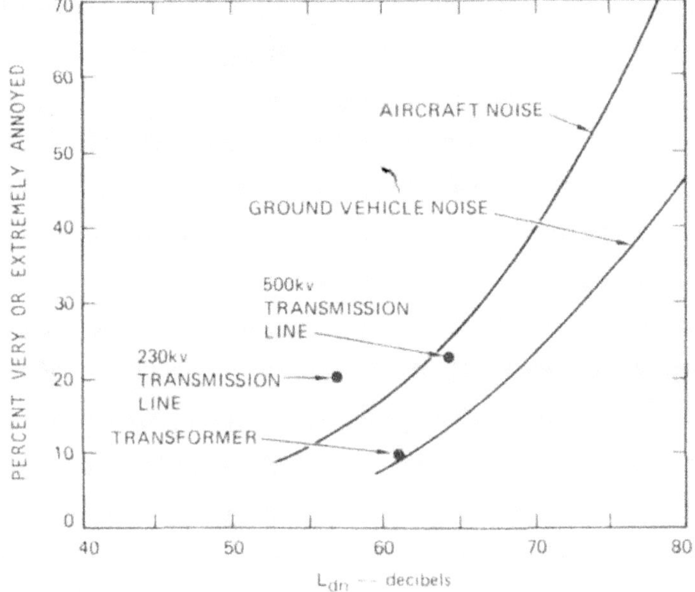

FIGURE 11.16. Electric transformer and transmission line noise annoyance data collected from 17 sites in southern California (from ref. 66) and aircraft and ground vehicle noise annoyance data collected in northern climate cities. "Very or extremely annoyed" refers to the top 40 percent of annoyance scale, (not the top 27 to 29 percent as used in the percent "highly annoyed" trend curve analysis, see figs. 11.10 and 11.11). The conversion from the 27 to 29 percent scale point to the 40 percent scale for the aircraft and ground vehicle noise trend curves was accomplished by the application of relations in figure 11.3 to trend curves given in figure 11.12(a).

It is evident in figure 11.16 that the transformer noise (from ground-based structures about 20 to 160 ft from residences) is less effective, in terms of L_{dn}, than the noise from 230-kV and 500-kV transmission lines. It is estimated that the transmission lines were 200 to 500 ft from residences and at an altitude of about 100 ft. It is perhaps reasonable to presume that this difference in the effectiveness, as a function of L_{dn}, of these two noises to cause annoyance is related to differences in acoustical factors that are somewhat similar to the differences in these regards for the noises from aircraft and ground vehicles. However, the considerable differences in the character and temporal patterns of exposure to electrical noises may also be contributive factors. In any event, it appears that some adjustment, such as that required for equating ground vehicle and aircraft noise with respect to annoyance is also needed to give L_{dn} values for ground-based electrical noise sources that are comparable for predicting annoyance to the L_{dn} values for more distant and higher altitude electrical noise sources.

Effective Exposure Level of Noises for Annoyance

Acoustical factors in noise intrusiveness

In interpreting attitude survey studies, it is important to note that the noise measurements or estimates for these various studies generally refer to the noise present near the front facade of houses, whether the noise is from street, or road, traffic or from aircraft. This being the case, some probable explanations for the difference shown in figures 11.13 and 11.14 between aircraft noise and ground vehicular noise in effectiveness as a source of annoyance are as follows:

1. The sound power of the aircraft noise is greater than that of the ground traffic noise. Typically, in residential areas close to a commercial airport, the aircraft would be at a slant-range distance of 500 to 2500 ft (depending on whether a landing or take-off operation is occurring) from the noise monitor or measurement microphone at a residence, whereas the street or road traffic would be, for many urban areas, only 25 to 50 ft away from the measurement microphone. Accordingly, the difference in noise levels in the front and backyard of a house will be significantly different for street noise solely because of an approximate doubling in the distance travelled by the noise from the street to the backyard, as compared with the distance travelled by the noise from the street to the front yard. Doubling the distance results in an air attenuation of about 6 dB. Whereas, for aircraft noise, the difference in the transmission distance from the aircraft to the front yard or to the backyard is negligible.

2. The aircraft noise tends to fall equally over the entire roof structure and, at times, on the sides of the house. Because of the usual altitude of the aircraft over residential areas, shielding of the noise by interfering structures, including

the same house, tends to be relatively small except when the aircraft is at very low altitudes and rather long distances to the side. Aircraft noise measurement data in 104 rooms in residences (ref. 67, appendix N) revealed the following: "Rooms with windows facing the source (*i.e.*, aircraft in flight) had about 3 dB less noise reduction than for rooms with windows facing 90 degrees to the direction of the source and about 4 dB less noise reduction than rooms which had all windows facing away from the source (*e.g.*, shielded by the house)."

3. The relatively close distance between the front facade of the house and the street or road traffic (coupled with the depth of the house and the relatively small distances between houses—10 to 20 ft in compact areas) makes for acoustical-barrier shielding of the ground traffic noises from the rear facade and backyard-patio-porch areas of a house (on the order of 14 dB, not including a typical air attenuation of about 6 dB).

4. It is common practice that people do not open the windows of their homes that face the street and do much of their "outdoor living" in the backyards or patios of their homes rather than in the front yards.

Some of the physical variables involved are illustrated in figure 11.17.

It is readily calculated (ref. 68) that because of differences in sound attenuation due to barriers and atmospheric attenuation, the noise intruding into the backyard and rooms of a house would typically be at least 16 dB less from road traffic than from aircraft flyover operations when the noises are measured as having equal sound pressure levels at a point somewhere in the front yards of most urban houses. To obtain an empirical check on this predication, Ortega and Kryter (ref. 69) made measurements of the noise from aircraft flying overhead and from cars and trucks driving past the same houses. Simultaneous measurements were made from microphones placed in the front yard (10 ft from the front facade) and in the backyard patio (10 ft from the rear facade). Table 11.5, which summarizes the findings, shows an average difference in noise level between the front yard and backyard of 19 dB for the road traffic noise and 0.3 dB for the aircraft noise.

Whether it is with respect to aircraft noise alone or ground traffic noise alone, it is possible, as aforementioned, that the variability observed in attitude surveys among people exposed to apparently equal dosages of noise (as measured in their front yards) may be caused more by real differences in the noise dosage actually reaching the listeners than by "personality" differences among the people. Further evidence that it is not the source of the noises but the exposure level of the noises reaching the listeners' ears that primarily determines how annoying they are, is shown by field studies reported in appendices IV, VIII, and X of reference 6 and by laboratory studies of Stephens and Powell (ref. 70). In all these investigations it was found that, except at very high L_{ex} values, aircraft and ground traffic noises were judged as being about equally annoying when of equal L_{ex} measured at the position of the ears of the subjects.

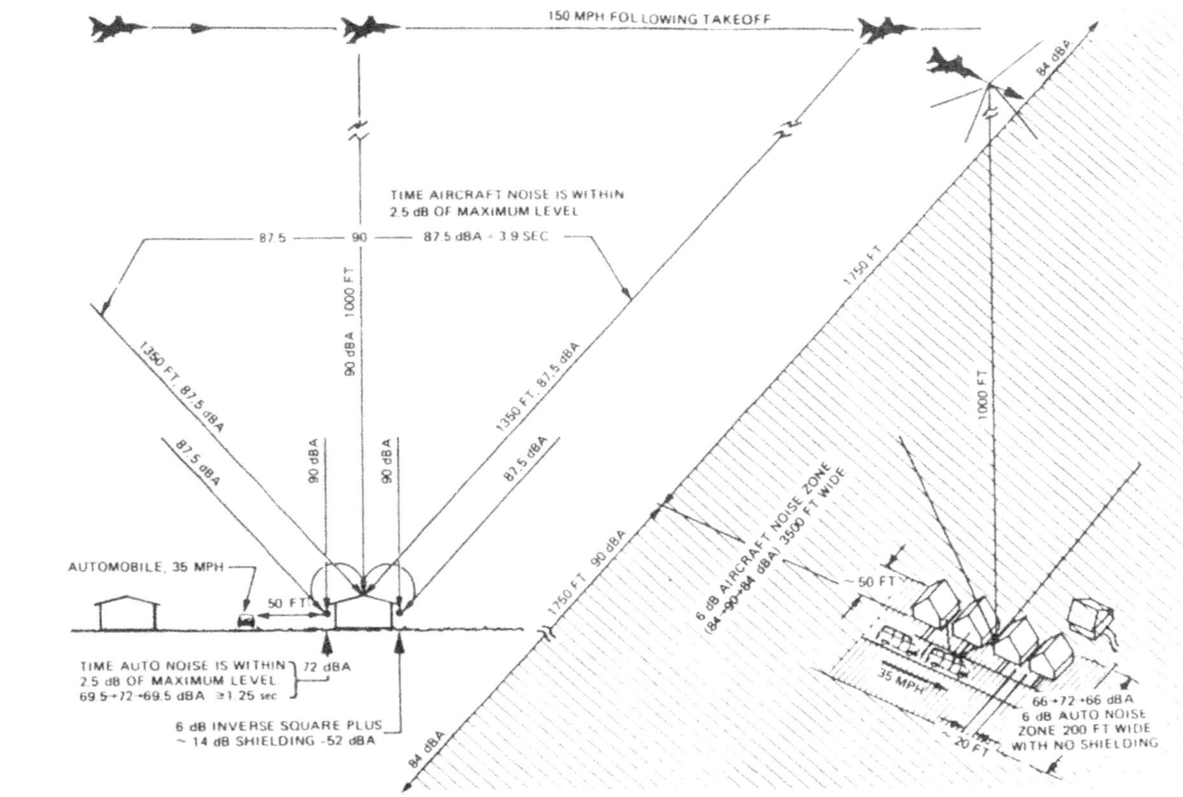

FIGURE 11.17. Schematic diagram of acoustical factors that affect noise levels from aircraft and ground vehicles measured in and around houses. (From ref. 2.)

TABLE 11.5

Peak L_A Levels of Noises Measured at Two Residences Located Near a General Aviation Airport in Southern California

[From ref. 69]

Source and date	Residence number	Events/time period	L_A, dB				
			Front yard		Backyard		Avg. front yard − Avg. backyard
			Average	Range	Average	Range	
Aircraft (general aviation, following take-off)							
27 February 1980	1	31/hr	77.4	68–91	77.0	68–90	0.4 } 0.3
26 February 1980	2	24/hr	76.7	63–91	76.5	60–90	0.2 }
Street vehicle							
27 February 1980 (city trash truck)	1	5/10 min[a]	71	66–75	54	51–57	17.0 } 19.0
26 February 1980 (1975 4-cylinder car, manual transmission)	2	11/10 min[a]	72	68–76	50.9	48–54	21.1 }

[a] *Time period is approximate.*

Effectiveness factor for ground vehicle noise

It is proposed that the dashed curve in figure 11.13 is a reasonable representation of the average of the upper four curves. The slope of the dashed curve is such that for each 5-dB increase in L_{dn} there is an additional 1-dB increase in the difference between the effectiveness of aircraft noise as compared with ground-based vehicle noise as a cause of annoyance. The decibel corrections to be applied to measured ground-based vehicle noise (the vertical ordinate of fig. 11.13) are given in table 11.6.

Kryter (ref. 2) suggested earlier, before the more definitive data of Hall *et al.* (ref. 29) and Taylor (ref. 58) were available, that regardless of level, 10 dB be added to L_{dn} ground vehicle noise to correct to effective L_{dn} re aircraft noise. The varying corrections as a function of L_{dn} level shown by the dashed curve in figure 11.13 and table 11.6 now seem more realistic.

Practical threshold levels of annoying noise

The EPA conducted surveys of noise exposure in different areas and types of neighborhoods in the U.S. (ref. 9). From these surveys, estimates have been made of the numbers of people living in various levels of noise exposure (primarily from vehicles of transportation) as measured outdoors, near the fronts of houses.

It is relevant to note that noise interference with low-level conversational speech is, especially for intermittent noise, likely to become noticeable when the noise level exceeds about 50 dBA at the listener's ears (ref. 8). Inasmuch as a house with windows open will offer, on the average, 15 dB attenuation of outdoor noise (see tables 11.7 and 11.8), it follows that annoyance effects will start (during open-window conditions) only when the outdoor noise exceeds about 65 dBA for

TABLE 11.6

Ground Vehicle/Aircraft Noise Correction Factor

[Decibel correction to be added to measured L_{dn} or L_{eq} of ground-based vehicle noise to obtain an L_{dn} or L_{eq} of equivalent aircraft noise annoyance]

L_{dn} or L_{eq}, dB	Correction factor, dB
50	6
55	7
60	8
65	9
70	10
75	11
80	12

TABLE 11.7

Measured Reduction in A-Weighted Noise Levels for Residential Structures

[Single-family detached dwellings except for nine rooms in apartments in New York City (data from ref. 67)]

Climate	Cities	Windows open			Windows closed		
		Number of rooms	ΔL_A, dB (a)	S.D. (b)	Number of rooms	ΔL_A, dB (a)	S.D. (b)
Warm	Miami, Wallops (VA), Los Angeles	14	11.1	4.6	132	26.2	4.8
Cold	Winthrop (MA), Boston, New York	33	18.4	5.1	27	27.7	5.2
	Overall Average		14.8	4.85		27.0	5.0

[a] ΔL_A = *Average difference between A-weighted noise level outside and inside room. The number of measurements per room varies widely from 1 to 46 with an average of 6.0. Almost all sources were jet aircraft.*

[b] *S.D. = standard deviation of average ΔL_A over number of rooms measured.*

speech interference. As seen in tables 11.7 and 11.8, for winter or closed-window conditions, this outdoor threshold can be about 10 dB higher.

Unfortunately, the EPA and others include in noise surveys outdoor noise energy that is as low as 40 dBA. Such a procedure can give an exaggerated statement of the amount of outdoor community noise that is capable of causing any appreciable annoyance. These threshold differences may account for, interestingly, the results shown in figure 11.14. It was seen there that disturbance of speech conversation from street and road traffic is less than disturbance to sleep,

TABLE 11.8

Typical Sound Level Reductions of Buildings

[From ref. 9]

Climate	Sound level reduction, dB	
	Windows open	Windows closed
Warm	12	24
Cold	17	27
Approximate national average	15	25

but that the reverse is true for aircraft noise. It is speculated that this would follow if the traffic noise is more often below the threshold for speech interference than for sleep arousal, whereas aircraft noise is likely to be above both thresholds. This would indeed be the case, in general, for these two types of noises.

The need for application of a relatively high threshold level for estimating annoyance from physical measures of noise has been demonstrated by Gjestland (ref. 71) and Gjestland and Oftedal (refs. 72 and 73). These investigators found in some laboratory tests that this threshold at the listener's ears was probably of the order of 40 or 50 dBA, depending on whether quiet or busy daytime activities were involved. This would translate to outdoor noise levels of approximately 55 to 65 dBA for open-window houses.

Difference in peak L_A levels of aircraft and automobile noise

It is worth noting that an L_{dn} of 55 from aircraft noise could be indicative of the presence, 50 or so times during the daytime, of aircraft noise having a peak level of 81 dBA in the front and backyards of a house (peak levels of about 66 dBA inside a typical house with windows partly open and 56 dBA with the windows closed). Such a noise can interfere with sleep (levels above 40 dBA) and with some speech (noise levels above 50 dBA) and other auditory communications during each noise occurrence. This condition would not be unusual at about 1500 ft from the ends of a 5000-ft-long runway at a general aviation airport with the operation of moderate-sized general aviation aircraft.

On the other hand, an L_{dn} of 55 from automobile noise would be indicative of the passing during the daytime of 1000 or so automobiles that make noise having a peak level of 72 dBA in the front yard of a house (peak levels of 52 dBA in the backyard, 57 dBA in rooms facing the street with windows partly open, and 42 dBA in rooms facing the street with windows closed). The peak levels in rooms facing the backyard would be about 37 dBA with windows open and about 22 dBA with windows closed. These conditions would be typical for a house 35 ft or so from the street and for automobiles traveling at a speed of about 35 miles per hour, and they would have no disturbing effect on speech conversation or sleep in the backyard or rear rooms of a house.

U.S. population exposed to aircraft and ground vehicle noise

In any event, as discussed earlier, it appears that because of acoustical reasons the L_{dn} values of street and road traffic noise should be adjusted when estimating the percentage of the population exposed to street and road noise that is comparable with aircraft noise in effectiveness as a cause of community annoyance. A 10-dB reduction in the measured L_{dn} values reported for street and road noise was made to achieve the effective L_{dn} values for urban traffic plotted in figure 11.18.

It is seen in figure 11.18 that at an effective L_{dn} of 55 dB, about 18 percent of the urban population is exposed to outdoor noise from aircraft or ground vehicles,

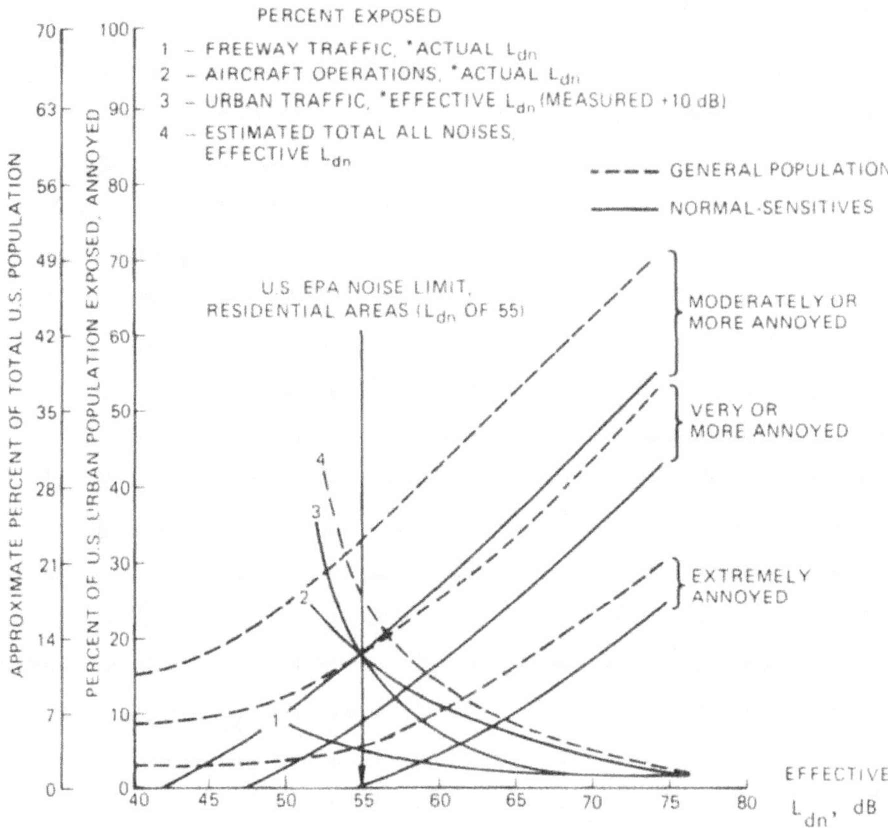

* BASED ON FIGURE 8, EPA REPORT, Ref. 9

† ASSUMES NIGHTTIME NOISES CONTRIBUTE ABOUT 2 dB TO TOTAL VALUE OF L_{dn}

FIGURE 11.18. Percentages of U.S. total and urban populations exposed to a given actual and effective L_{dn}, and the levels of annoyance felt by normal-sensitive people thus exposed. The effective level of urban traffic noise relative to aircraft noise is actual (measured) L_{dn} minus 10 dB.

and that of that part of the general population, about 33 percent are "moderately or more annoyed," 17 percent are "very or more annoyed," and 5 percent are "extremely annoyed." These percentages are reduced somewhat (solid curves) when the data are normalized for atypical responses. Although not shown in figure 11.18, at an L_{dn} of 55 dB, 55 percent of the general population and 30 percent of the normal-sensitives are "a little or more annoyed."

The curves for annoyance are based on attitude surveys conducted in the spring and fall in northern climate cities. The curves could be adjusted upwards somewhat (equivalent to 5 dB of exposure) when projected to those expected in warm climate cities.

Relations Between Noise Exposure and Complaints and Costs

Complaint activity

Complaints, or rather lack of complaints, to authorities and legal and political actions by individuals and citizens are not as reliable or consistent measures of the effects of environmental noise on people as are attitude surveys of annoyance (ref. 74). The reasons for such variability include a variety of social, individual, experiential, and legal factors and constraints. However, some meaningful information can be obtained from records of complaints.

In an attitude survey conducted in the U.S. (refs. 37 and 39), people were asked, among other things, how annoyed they were by aircraft noise and whether they had complained to any authorities about the noise. The results, shown in figure 11.19, indicate that some complaints can be expected when 5 to 10 percent

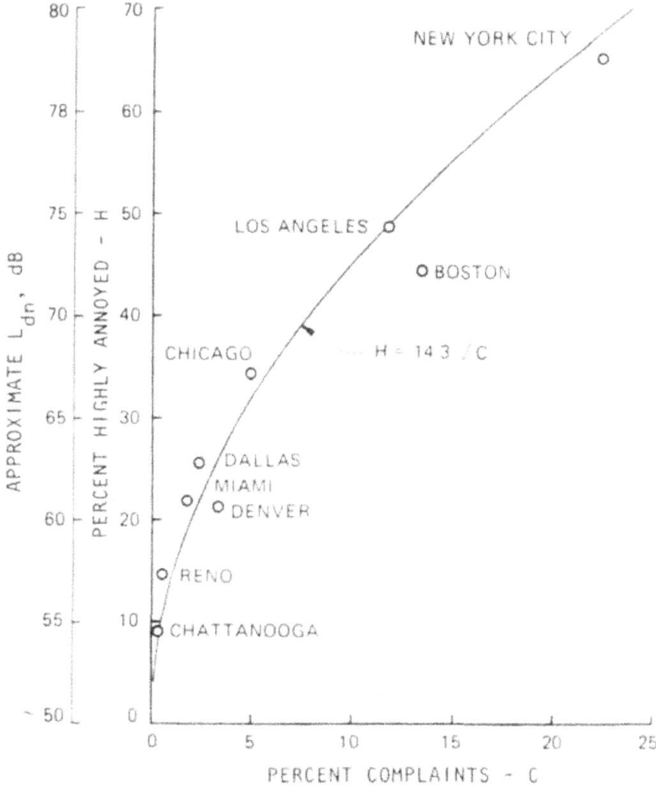

FIGURE 11.19. Percent of people who complained about aircraft noise as a function of percent highly annoyed and approximate L_{dn}. (From ref. 39.)

of the people are highly annoyed by the noise, a percentage typically associated with an L_{dn} of 55 dB. Interestingly, McKennell in Great Britain (ref. 75) found that when 1 percent of the people had complained about noise, 10 percent felt like complaining, and when the noise was such that 10 percent had complained, 40 percent felt like complaining.

Additional difficulties to the quantitative interpretation of complaints about aircraft noise are exemplified in table 11.9 (from refs. 76, 77, and 78). In that table, it is shown that in 1976, 8 percent of the households exposed to aircraft noise called in complaints to the airport operator (a not atypical annual average), yet 37 percent of the households interviewed during January of 1977 indicated they had at some time called or written protests about, primarily, the aircraft noise. Obviously, the percentage could be higher for those interviewed because their protests could have occurred in a year previous to 1976, but does this mean that they had grown tired of protesting or had adapted to the noise? The results of attitude surveys made in the same area, to be presented and discussed more fully later in regard to figure 11.34, revealed continued high annoyance and that the people had not adapted significantly to the aircraft noise. In brief, while complaint records have some research applications, the numbers and derived percentages of complaints do not necessarily reflect the numbers and percentages of the noise-exposed people who are adversely affected by the noise.

Generalized functions betwen noise exposure and complaints

The validity of L_{dn} for assessing annoyance and complaints because of outdoor noises in residential areas from sources other than aircraft and ground vehicles is not well established. Because these other noises are less widespread and sometimes present on only an irregular basis, they have not been the subject of large scale attitude surveys of annoyance. Nevertheless, L_{dn} or similar noise dosage measures appear to be generally useful in this regard provided certain adjustments or corrections are made to the measurements.

In order to assess the generality of L_{dn}, a special study was undertaken by the EPA (ref. 79). In this study, 55 case histories of community noise environments from the literature and the files of acoustical consultants were examined. The 55 noise cases, associated L_{dn} values, and complaint behavior are presented in table 11.10. A wide variety of noise sources are included in the 55 cases. Figure 11.20, upper graph, shows that while there is a general monotonic relation between L_{dn} and severity of complaint activity, there is a range in clustered L_{dn} values of 15 to 20 dB among the different cases at a given degree of community reaction. However, as seen in figure 11.20, lower graph, this variability is reduced to about 10 dB when certain corrections are applied to measured L_{dn}. These corrections, given in table 11.11, are essentially those developed earlier for the aforementioned composite noise rating (CNR) (ref. 80). Note that these corrections are no longer used by the EPA or other government agencies in the measurement or calculation of L_{dn} or similar community noise exposures (ref. 8).

TABLE 11.9

Complaint Activity About Aircraft Noise at Orange County Airport During 1976

$\sim L_{dn}$, dB	Approximate no. of households in $\sim L_{dn}$ zones (a)	Percent interviewed in Jan. 1977 who have called or written protests about airport, primarily aircraft noise (b)	Total no. of complaints received in 1976 by airport operator (c)
>65	1333	55.6	
60	4000	31.2	433, all L_{dn} areas
<60		7.5	
Total 5333		Average, 37.0	Total, 433 (8 percent)

[a]*Estimated from tables IV-1, IV-3, and map in ref. 76.*
[b]*From ref. 77.*
[c]*From ref. 78.*

Legal actions

A special type of complaint that has been used to assess the tolerability of aircraft and other noise exposure conditions is that of lawsuits filed for relief or damages from noise. An implication of this approach is that if lawsuits are not filed, the noise environment is to be considered compatible with those being exposed. Such a position is hardly justified in view of the complexities, restraints, and costs involved in private citizens' taking such action. However, this criteria does have an upper-bounds quality that has a practical appeal. It places the final assessment of the meaning of scientific data and facts concerning the effects of noise on people and communities in a judge or jury adversary situation.

Figure 11.21 illustrates the range of L_{dn} levels over which threats of or actual legal actions occurred in case studies of civic problems involving aircraft noise. In general, the legal action consisted of lawsuits to prohibit or reduce the level of aircraft noise in specific residential areas and/or to pay compensation to residents for the taking of property. It is seen that in relatively warm climate areas, some legal action started at L_{dn} levels of about 60 dB.

Costs of lost time

As mentioned in chapter 1, a seemingly objective measure of the impact of noise on people is that of assessing the cost to the people adversely affected. One method of assessment is to assign a "dollar" cost for the time lost by the exposed parties because of the noise. For example, it is noted from studies of classroom

TABLE 11.10

Summary of Data for 55 Community Noise Reaction Cases

[Data from ref. 79]

Type of reaction	Case no.	Noise description	$\sim L_{dn}$, dB
Vigorous	A-1	Rocket testing	63
	A-2	Wind tunnel	70
	A-3	Aircraft landing	82
	A-4	Aircraft take-off	71
	A-5	Circuit breaker testing	73
	A-6	Auto race track	77
	A-7	Aircraft take-off	69
	A-8	Aircraft landing	84
Threats of legal action	B-1	Rocket testing	57
	B-2	Aircraft ground runup	72
	B-3	Wind tunnel	61
	B-4	Freeway	86
	B-5	Aircraft overflight	57
	B-6	Plant blower	67
	B-7	Asphalt quarry	64
	B-8	Glass bead plant blower	62
	B-9	Plastics plant	61
	B-10	Target shooting range	54
	B-11	Residential air conditioning	57
	B-12	Unloading newsprint	76
	B-13	Auto body shop	70
	B-14	Motorcycle raceway	65
Widespread complaints	C-1	Transformer substation	49
	C-2	Cement plant	64
	C-3	Aircraft landing	62
	C-4	Paperboard plant cyclone	50
	C-5	Oil refinery	61
	C-6	Milling & grinding metal	66
	C-7	Chemical plant material handling	58
	C-8	Residential air conditioning	56
	C-9	Transformer substation	62
	C-10	Rail car shaker	57
	C-11	Transformer substation	52
	C-12	Positive displacement blower	55
	C-13	Aircraft take-off	63
	C-14	Glass manufacturing plant	67

TABLE 11.10 Concluded

Type of reaction	Case no.	Noise description	~L_{dn}, dB
Sporadic complaints	D-1	Factory air pump	71
	D-2	Manufacturing plant	58
	D-3	Chemical plant	56
	D-4	Local automobile traffic	56
	D-5	Plastics plant	51
	D-6	Power station	69
No observed reaction	E-1	Transformer substation	40
	E-2	Aircraft runup	46
	E-3	Asphalt tile shaker	59
	E-4	Asphalt tile reddler	40
	E-5	Power plant	47
	E-6	Aircraft overflight	53
	E-7	Aircraft landing	60
	E-8	City traffic	56
	E-9	Aircraft log and take-off	57
	E-10	Local traffic	59
	E-11	Auto assembly plant	66
	E-12	Can manufacturing	67
	E-13	Oil refinery	69

behavior during lectures, studying, and similar activities and from laboratory studies of sleep activities that an aircraft flyover noise of sufficient intensity will typically cause a disruptive loss of about 1 minute in these activities in people acclimated to the noise, although the noise itself may have a duration of only about 20 seconds.

An example of the application of this method of assessment is as follows. About 25 000 000 U.S. citizens are exposed to a calculated L_{dn} from aircraft noise of 55 dB or higher (ref. 9), and about 15 percent of the normal-sensitive people are very or more annoyed by this noise. An L_{dn} of 55 dB for aircraft noise could be the result of, typically, 60 daily flyover noises during the daytime at peak levels of about 80 dBA, a level capable of causing the cited disruptive-annoyance effects. Thus, 1 hour per day, from 60 overflights, is lost by 15 percent of the people. Expressed in another way, the probabilities are such that every time one of these flyover noises occurs, 15 percent of the exposed people are engaged in conversation, listening to TV or radio, using telephones, or sleeping in ways that are significantly interfered with by the noise. Assigning an average value of $10 per hour for a person's time, the costs are $37 500 000 per day or $13 687 500 000 per year to the citizens of the U.S.

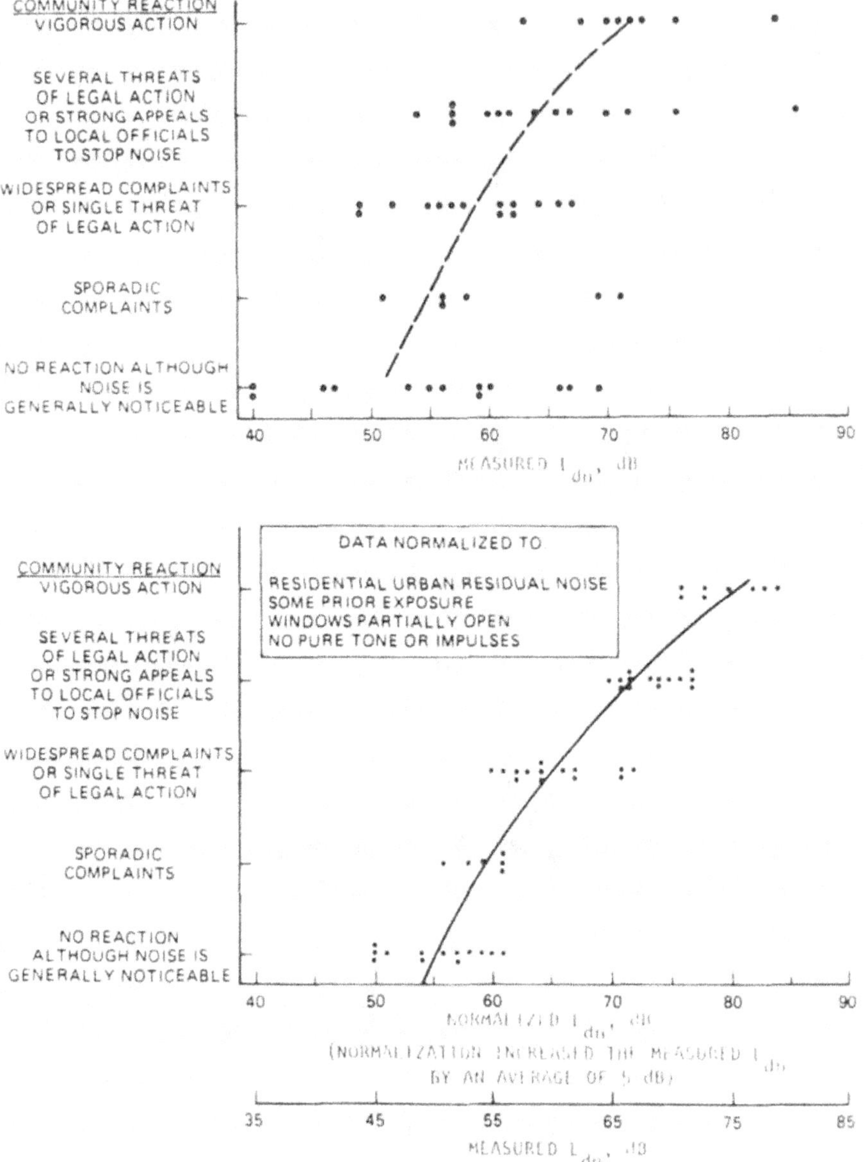

FIGURE 11.20. Upper graph: community reaction to intensive noise as a function of outdoor L_{dn} of the intruding noise. Types of noise situations for the 55 data points are given in table 11.10. Lower graph: data points normalized to equivalent effectiveness in accordance with corrections in table 11.11. (Data from ref. 79.)

TABLE 11.11

Correction to L_{dn} for Outdoor Noise in Residential Areas

[Data from ref. 79]

Noise description	Correction, dB
Noise only in winter or windows always closed	−5
Quiet suburban	+10
Normal suburban (no industry)	+5
Suburban near busy roads or industry	−5
Very noisy urban	−10
No prior experience with the noise	+5
Experienced with noise and good relations with maker	−5
Noise is necessary and temporary	−10
Pure tone or impulsive noise	+5

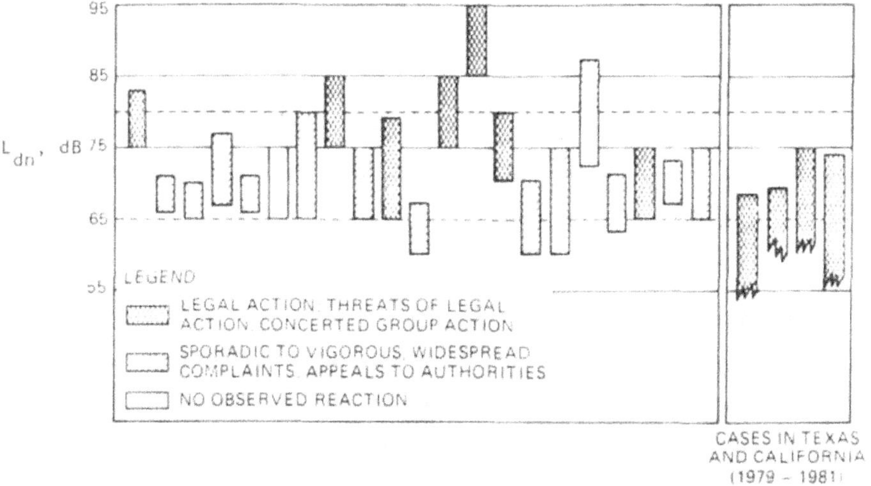

FIGURE 11.21. Reactions of people in communities exposed to aircraft noise environments. The height of the bars represents the approximate range of L_{dn} values found over a given neighborhood. Left-hand panel after Galloway and Von Gierke (ref. 80).

House depreciation

One of the obvious consequences of the presence of noise that is sufficiently annoying is that people do not consider that environment as desirable for residential living. A study of the possible relations between aircraft noise and housing values was undertaken by the British Government (ref. 81). In addition to examination of records of sales and appraisals, questionnaires were administered to panels of real estate experts regarding samples of houses around Heathrow and Gatwick airports in London. Some of the findings are given in table 11.12.

The results of the British study and studies conducted in the U.S. (ref. 82) and Canada (ref. 83) are shown in figure 11.22. Of perhaps special interest is the finding in the British study that the rate of depreciation as a function of exposure to aircraft noise was significantly related to the general class of house—the higher priced the dwellings, the greater was the percentage depreciation. The U.S. and Canadian data were not stratified with respect to the cost of the houses.

Also shown in figure 11.22 are estimated linear rates of depreciation for the three price ranges of houses, namely, 1 percent per decibel (L_{dn}) for the high priced houses, and 0.75 percent per decibel for medium and low cost houses. An analysis, conducted for the Federal Aviation Administration (FAA), of the effects of aircraft noise on residential property values in seven cities indicated that there was about a 0.5 percent decrease in value for each 1 dB increase in L_{dn} (ref. 84).

TABLE 11.12

Estimated Percent Depreciation in Housing Values

[From ref. 81]

Location	Housing price range	Percent depreciation		
		$L_{dn} \approx 68$ dB	$L_{dn} \approx 75$ dB	$L_{dn} \approx 83$ dB
Around Heathrow airport	Low	0	2.9	5.0
	Medium	2.6	6.3	10.5
	High	3.3	13.3	22.5
Around Gatwick airport	Low	4.5	10.3	[a]15
	Medium	9.4	16.5	[a]22
	High	16.4	29.0	[a]39
Average	Low	2.2	6.6	10.0
	Medium	6.0	11.4	14.0
	High	10.0	21.2	31.0

[a] *Extrapolation.*

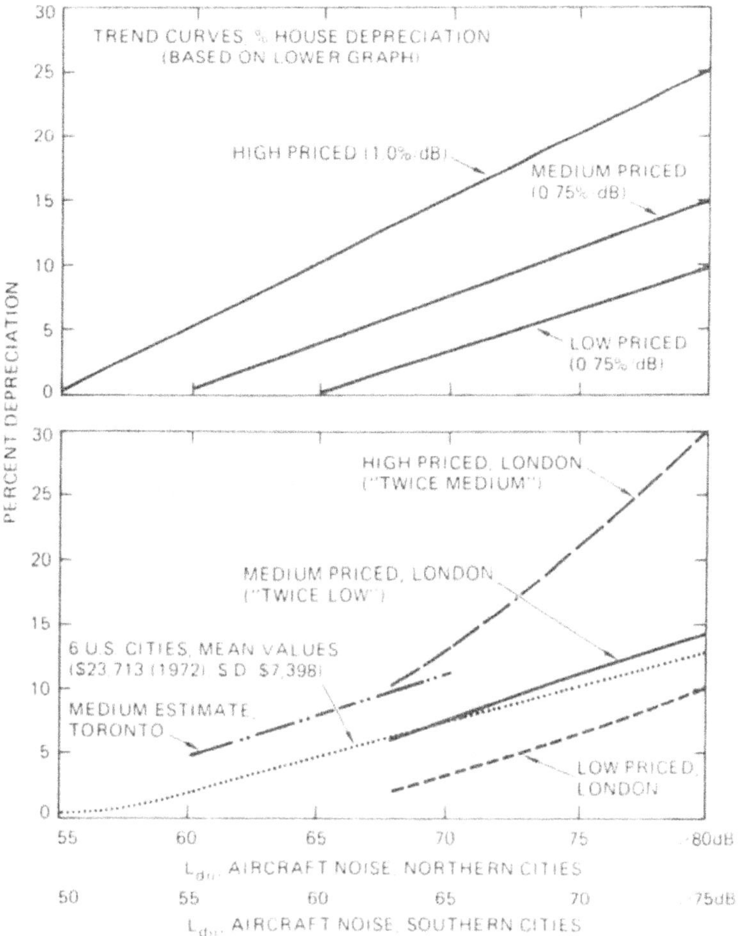

FIGURE 11.22. Depreciation of house values as function of aircraft noise near London (ref. 81); near U.S. cities (San Francisco, St. Louis, Cleveland, New Orleans, San Diego, and Buffalo) (ref. 82); and near Toronto, Canada (ref. 83).

Taylor *et al.* (ref. 85) studied the effect of arterial and expressway road traffic noise on house prices in Ontario, Canada. They found that there was a depreciation of approximately 0.5 percent per decibel of ground traffic noise in the value of medium priced houses (average cost of $60 000) at a 24-hour L_{eq} of about 70 dB (an L_{dn} of about 72 dB). This compares roughly with the effect of aircraft noise on house costs at an L_{dn} of about 60 dB as shown in figure 11.22. This relative difference between the impact of the two types of noises on house costs is obviously consistent with 10 dB or so difference found in the effectiveness between these two noises as a cause of annoyance, and for the same acoustical reasons as discussed earlier.

On the other hand, the L_{dn} levels at which these house depreciation effects occur are presumably somewhat lower in warm climate areas as compared with the relatively cold climate, for the most part, cities involved in these various studies just discussed. The difference in the effectiveness of aircraft noise, in L_{dn}, as a cause of annoyance and complaints in warm as compared with colder climate residential areas is discussed in a later section of this chapter.

There seems little reason to question the relations shown in figure 11.22 regarding the impact of aircraft noise on housing values in general. There is at least one factor, however, pertaining to closeness to an airport and housing values that for certain individuals may have a compensating, positive effect on housing values that offsets to varying degrees the negative effect on those values because of the aircraft noise. The factor is the reduction in cost and time to reach the airport for those people working at or near the airport, or who use the airport on a very frequent basis. In addition, the demand by certain industries to be near an airport may also increase the number of workers who wish to live nearby.

Among other variables, De Vany (ref. 86) studied the relations between distance from an airport, aircraft noise level, and housing costs. He found an interactive effect between distance from an airport (Love Field, Dallas, Texas), aircraft noise level, and house and land values. Figure 11.23 is after figure 1 from De Vany's paper. Certain liberties have been taken in preparing figure 11.23, especially in regard to quantifying the distance dimensions on the basis of ancillary information regarding typical L_{dn} values.

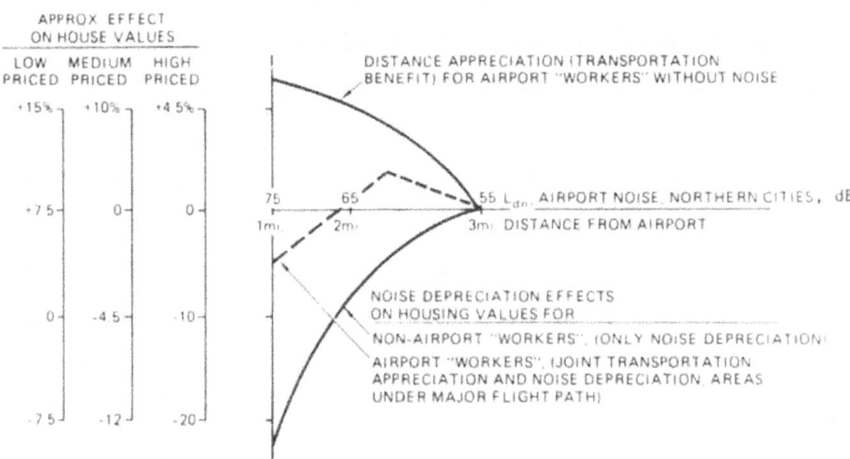

FIGURE 11.23. Model for effects of aircraft noise and transportation distance from airport on approximate housing values. Airport workers are those whose employment location is at or near the airport. (Data from ref. 86.)

Figure 11.23 can be used to interpret (a) depreciation as a function of L_{dn} for "non-airport workers" and (b) appreciation as a function of closeness and depreciation with L_{dn} for "airport workers" living in a band under the major aircraft flight path from the airport. Figure 11.23 would seem to indicate that at a distance of about 2 miles from the airport (under aircraft flight paths), the interaction between noise costs to residential living and the benefits of close-by ground transportation to airport workplace results in a net increase in house value to those people who work at or near the airport. It is to be emphasized that figure 11.23 represents a purely statistical model and simplification of available data. There are, of course, a number of other factors that would influence possible depreciation and/or appreciation in the value of a given house.

Also, the functions in figure 11.23 are not generalizable to all locations around an airport because the relation between distance from an airport and L_{dn} is quite different from that found under the main take-off flight paths. A typical example of the L_{dn} levels at different distances from an airport is given in figure 11.24 (from preliminary results prepared for the Department of Aviation, Dallas, Texas, under the Noise Control Program), where it is seen that at a distance of only 1 mile to the side of an airport, the aircraft noise is negligible (L_{dn} less than 55). However, at the end of the runway (on the flight path), an L_{dn} of 55 dB is not reached until about 6 miles from the end of the runway.

Payment of damages to individuals

Monetary compensation for adverse effects of aircraft noise on health and house values has been awarded in lawsuits (ref. 87). Studies of the effects of aircraft noise on health are presented in chapter 10, and some further discussion of the general question of the zoning of land usage with respect to the protection of people from the impact of environmental noise appears in chapter 12.

Noise effects in schools

Some of the external costs of aircraft noise have been identified in the development of lawsuits brought by city school authorities against city airport authorities for damages to the educational process in schools heavily impacted by aircraft noise (ref. 88). An example of the effects of aircraft noise having peak levels of about 87 dBA outdoors and L_{dn} levels of about 70 dB upon various school activities as observed by the teachers in some schools in San Diego, California, is shown in figure 11.25. Comparable data were obtained from teachers in London (ref. 89) and Hong Kong (ref. 90). Data very similar to those shown in figure 11.25 were also obtained from the students in the San Diego schools. It is obvious from knowing the masking and other basic effects of noise, that these activity interferences from aircraft noise are to be expected.

Figure 11.26 shows estimates made by the teachers in the San Diego study of the frequency and duration of interference effects of the aircraft noise. Other

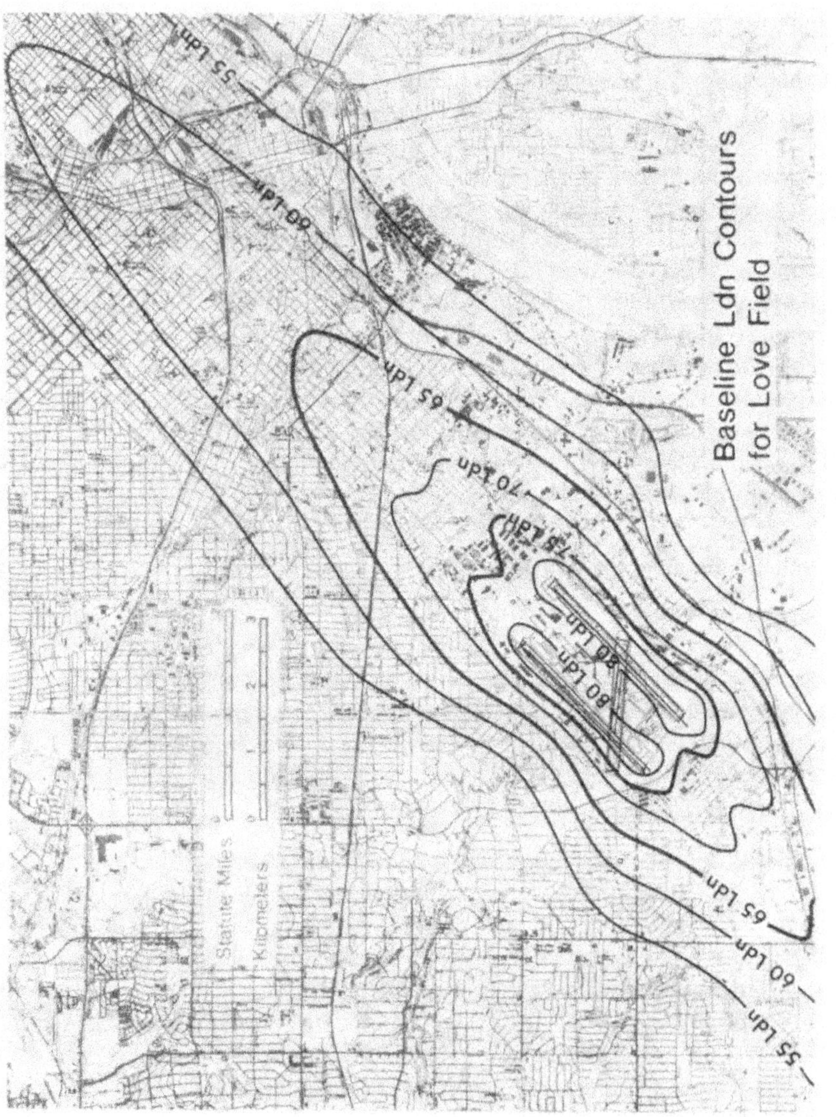

FIGURE 11.24. L_{dn} contours around Love Field, Dallas, Texas.

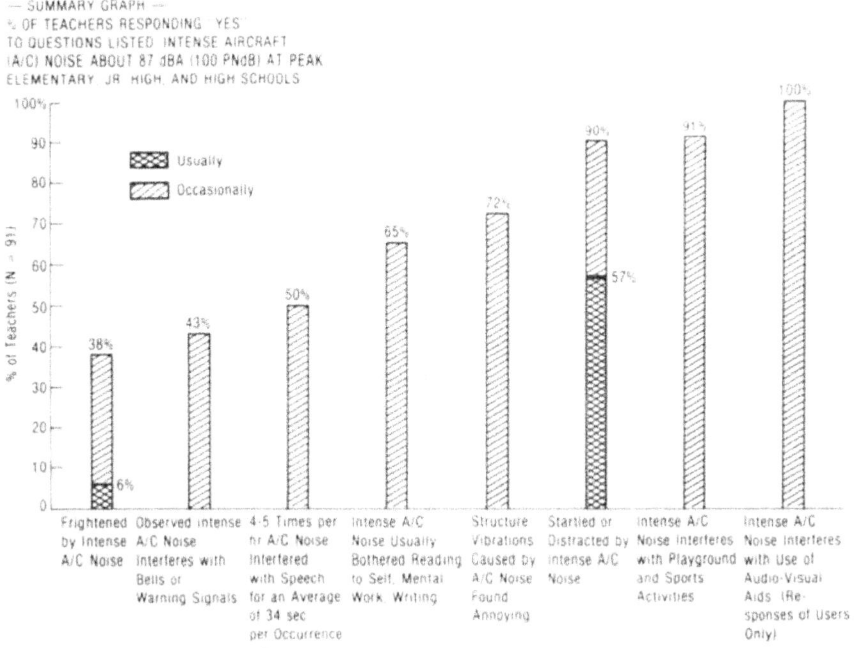

FIGURE 11.25. Results from questionnaire on aircraft noise effects in some San Diego, California, school areas. L_{dn} approximately 70 dB. (From an unpublished report by Karl D. Kryter for the City Schools Attorney, San Diego, California, 1978.)

studies (unpublished) conducted by the Highline School District (near Seattle, Washington) revealed somewhat longer duration effects from aircraft noise in school rooms than are shown in figure 11.26.

In one Highline School District study (ref. 91), an analysis of the school achievement test scores (standardized tests administered to the students each school year) over the school grades of 3 to 7 and 5 to 10 revealed that high academic aptitude students in schools exposed to aircraft noise did as well as those in quiet schools. However, middle and, especially, low academic aptitude students in the noisy schools showed progressive deterioration in school achievement tests with continued school attendance as compared with the achievement of cohorts of equal aptitude in quiet schools (ref. 91). The results of this study are shown in figure 11.27. The differences between the test scores of students in the lower third of academic aptitude in the noisy and in the quiet schools were statistically significant for the seventh and tenth grades.

This apparent cumulative effect of the aircraft noise on the less capable students seems intuitively reasonable. The socioeconomic characteristics of the students from the noisy and quiet schools were similar. The only significant difference between the schools was the fact that the "noisy" schools were exposed 50

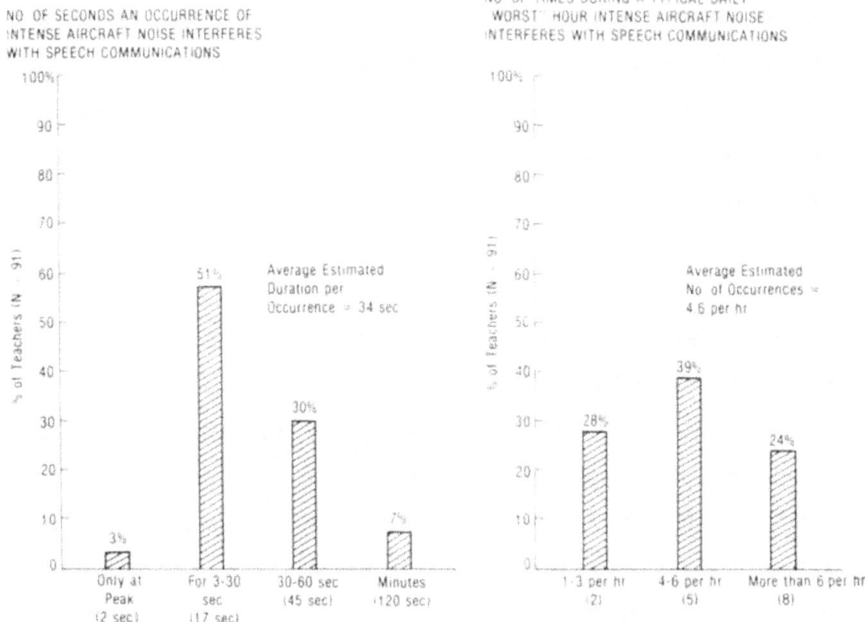

FIGURE 11.26. Estimates made by teachers of duration and frequency of occurrences of speech-interfering aircraft noise in some San Diego, California, school rooms. L_{dn} approximately 70 dB. (From an unpublished report by Karl D. Kryter for the City Schools Attorney, San Diego, California, 1978.)

times or so on a near daily basis to aircraft noise reaching peak levels of about 90 dBA (L_{dn} about 70 dB).

It should be noted in the San Diego and Seattle studies just discussed, the buildings were typical masonry school structures, but they were not air-conditioned or specially soundproofed. The related lawsuits were settled by the aviation interests paying for the costs of soundproofing certain school structures and/or building new structures in quieter areas.

Cohen *et al.* (ref. 92) also found a generally adverse effect of intense aircraft noise (peak levels as high as 95 dBA outdoors) upon reading and math achievement in grades 3 and 4. (See table 11.13.) In addition, Cohen *et al.* measured some increased blood pressure levels in the students in the noisier schools in comparison with those of students from the quieter schools, as discussed in chapter 10.

Additional evidence of a cumulative adverse effect of freeway noise is seen in figure 11.28, from a study by Lukas *et al.* (ref. 93). Some apparent degradation in reading achievement occurred with increased classroom noise in third graders. This effect was accelerated by the sixth grade. Bronzaft and McCarthy (ref. 94)

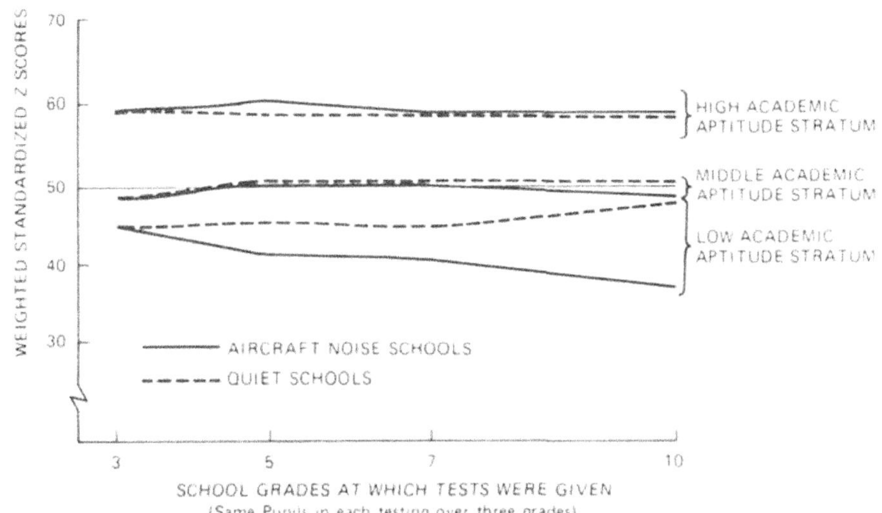

FIGURE 11.27. Average weighted standardized total battery scores on the CTBS (Comprehensive Test of Basic Skills) for testing from Fall 1970 to Fall 1976. Grades 3, 5, 7, and 10; academic group scores set for each group to be the same for the initial test grade, Grade 3; 269 students from aircraft noise schools (L_{dn} about 70 dB) and 370 students from quiet schools (no aircraft noise) near Seattle, Washington. (Data from ref. 91.)

TABLE 11.13

Mean (Adjusted) School Achievement Percentiles as a Function of Classroom Noise Abatement and Grade

[From ref. 92]

Classroom condition	Mean achievement percentile			
	Reading		Math	
	3rd grade	4th grade	3rd grade	4th grade
Noise	30.30	35.96	34.35	39.35
Noise-abated	47.36	37.90	56.24	37.54
Quiet	37.85	39.09	36.96	42.76

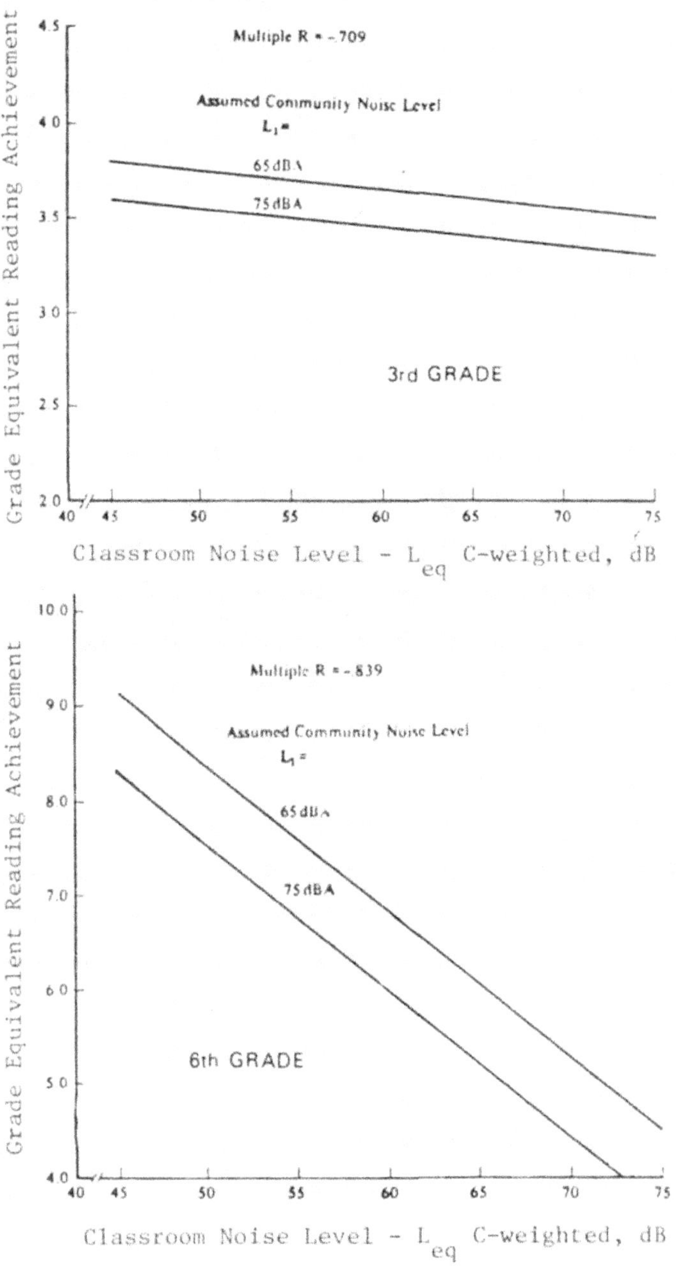

FIGURE 11.28. Relationship between achievement of third and sixth graders and classroom noise levels given different noise levels in the community. (Data from ref. 93.)

and Bronzaft (ref. 95) reported that the noise from elevated trains going by grade schools also had more pronounced adverse effects on the reading achievement scores of sixth than third graders.

It is to be noted that Cohen *et al.* (ref. 92) found that the amount of aircraft noise around the residences of the children, as well as that at their schools, contributed to a lowering of school achievement scores. This is consistent with an earlier study by Cohen *et al.* (ref. 96) of the effects of freeway noise at high-rise apartments upon the reading ability of children measured in the school.

Review of Concept of Day-Night Average Noise Exposure Dose (L_{dn})

As discussed above, L_{dn} and similar measures of the amount of noise received by groups of people (usually neighborhoods) during a typical day have been developed for those environments where nearly day after day the noise exposures follow similar patterns. Accordingly, it is presumed that people develop reliable impressions and attitudes about the average daily or seasonal impact of those particular noises upon them. The basic model and concepts underlying L_{dn} are those developed by Rosenblith *et al.* (ref. 97) in 1953. (Also see refs. 80 and 21.) The unit of noise measurement then proposed was called the CNR (composite noise rating).

Borsky (ref. 98) found that while there may be an initial period of getting used to noises in a new environment, the annoyance reported for aircraft and street traffic noise does not diminish; if anything, it tends to increase with continued years of exposure. This increased annoyance could perhaps be attributed to some degree to the increase in sensitivity to sound as people get older. (See chapter 10.) Vallet *et al.* (ref. 99) surveyed the annoyance of people living near expressways in 10 French towns. These investigators found in a second survey conducted after a lapse of 2 years that annoyance had not declined during this period and that no habituation had taken place.

This annoyance appears to be caused by the effects of the noise on sleep, speech communications, and the like, and not by the unexpectedness of the noise. Relevant to this are comments obtained during attitude surveys that many respondents become used to the aircraft or street noise and hardly notice it except when it occurs during a conversation or party, arouses them from sleep, or shakes the house, at which times they are annoyed.

Several important, and to some extent quantified, factors to be used or considered in the calculation of L_{dn}-type energy measures have been identified. Among them are (a) the role of ambient and background noises, (b) the combined effects of noise from multiple sources, (c) seasonal weather or climate differences, (d) the effects of nighttime as compared with daytime exposures to noises, (e) A-weighted measures of narrowband as compared with wide-frequency-band noises, and (f) additional effects of impulsive and low-frequency noises compared with more typical noise events.

Ambient, background, and identified noises

In its broadest form, the L_{dn} concept could be used to estimate the annoyance-disturbance impact on people of the total daily dosage of noise regularly received from all sources in a living environment. In this context, the words "ambient" and "background" noise are sometimes used as labels to distinguish such noise from noise identified as coming from sources such as aircraft, automobiles, or factories.

Exact definitions of ambient and background noise are difficult to make, at least with respect to the calculation of a total L_{dn}. Ambient is generally used to mean the sounds that typify a given area (*e.g.*, the sound of birds, insects, wind, weak intermittent music, and people's voices). Realistically, ambient noise is probably a misnomer in that what is meant by ambient noises are sounds that are too weak to interfere with speech or other normal behavioral activities. Ostensibly such sound is not unwanted sound, or noise, and accordingly should not be used in calculating the total L_{dn} of an environment.

Background noise is usually used to describe a continuous condition (*e.g.*, the hum of distant traffic or factory noises) that is less intense than intermittent noises from closer aircraft, street traffic, or motors. Background noise qualifies for inclusion in estimating a total L_{dn}, provided it is of sufficient intensity to cause some annoyance or disturbance.

In most living environments where noise is a problem, noise occurrences that are readily identified as coming from a particular source determine the value of L_{dn}. Background noise, unless of near equal intensity, will not contribute sufficient sound energy to increase the value of L_{dn}. It is sometimes surmised that a high background noise level makes an identified intruding noise less annoying because the increase in noise level may seem relatively small compared with the change caused when the identified noise occurs in the presence of a low ambient or background level. However, a high level of background noise may in its own right be bothersome. Some laboratory research on this question is discussed in chapter 5 and below.

Laboratory tests and attitude surveys of the effects of continuous background noise level on judged annoyance from aircraft noise have been somewhat inconsistent. Powell and Rice (ref. 100) and Johnston and Haasz (ref. 101) conducted laboratory tests in which subjects judged the annoyance of recorded aircraft noise heard in different levels of recorded background (road traffic noise). It was found that the subjective rating of the aircraft noise decreased when the level of the background noise approached the level of the aircraft noise. However, Taylor *et al.* (ref. 102) found in an attitude survey conducted near Toronto Airport that the effect of background noise level was generally not significant.

It is to be noted that the subjects in these studies were asked to judge the annoyance of the aircraft noise and not the annoyance of the total environment (background and aircraft noise). Accordingly, the findings of reduced aircraft noise annoyance in high levels of background noise could be expected as the result of the background noise masking to some extent the loudness of the aircraft noise: at that point the background noise would presumably be the primary cause of annoyance from noise. The results of both laboratory and field tests of the effect of continuous background noise are consistent with the phenomenon of loudness recruitment discussed in chapter 3; namely, that the loudness of the noise (aircraft) to be masked is not appreciably lessened until the loudness of the masking noise (the continuous background) approaches that of the noise being masked.

Aperiodic noise

The annoyance felt by people from extremely aperiodic noises (*e.g.*, those occurring once a week or once a month) may not be predictable from L_{dn} values taken on an annualized (average day during a 1-year period) exposure level. For the most part, these aperiodic conditions are not readily available for research study, although the aforementioned electric line corona noise, which occurs only when there is heavy moisture in the air, may qualify for such an aperiodic exposure condition. Obviously, cases such as one burst of noise per day (*e.g.*, $L_{A,ex}$ of 140 dB) are too atypical to cause predictably equal annoyance, even though both may have L_{dn} levels of 65 dB. Thus, L_{dn} is not a good metric to use for predicting annoyance from regularly occurring intermittent, predominantly daytime noises such as one might find in typical neighborhoods.

The energy-trading relation presumed for the L_{dn} measure between the intensity level of individual aircraft noise events and the number of daily occurrences was examined in the Heathrow noise surveys discussed earlier. This was done by comparing the annoyance scores for people exposed to differing average levels of intensity and numbers of daily events. The noise exposures were converted to $L_{dn(m)}$ values and related to the percentages of people for each noise condition who reported various degrees of annoyance, as shown earlier in figure 11.2.

It is seen in figure 11.2 that for a given part (percent) of the annoyance scale, the data points for a given L_{dn} fall reasonably close together, even though the L_{dn} represents different combinations of numbers of aircraft noise events and levels of intensity. There were three average numbers of aircraft per day (5.75, 22.5, and 81) for each of four ranges of peak noise levels (84 to 90, 91 to 96, 97 to 102, and 103 to 108 PNdB). This result obviously supports the equal energy concept of L_{dn}.

Analyses by Rylander *et al.* (refs. 103, 104, and 105) of some aircraft noise attitude surveys conducted in Scandinavia and France indicate that the energy contribution of large numbers (above 50 or so) of noise occurrences was not as important as the energy in a few high level noise events. Schultz (ref. 106) challenged this conclusion as not being adequately supported by at least some of the data involved.

Robinson (ref. 107) developed and proposed a modification to the L_{eq} or L_{dn} procedure to include the contributions of the deviations in the levels of different noise events during a typical day to the overall annoyance. Accordingly, a standard deviation statistic was incorporated into what Robinson called the NPL (noise pollution level). However, because of its relative complexity and lack of significant verifications (ref. 56), the NPL has not been widely used.

Multiple source noises

Another research question facing the L_{eq} or L_{dn} noise energy summation concept is whether the annoyance from different sources (*e.g.*, aircraft and street traffic) is predictable from the total L_{eq} for the two noises. Unlike continuous background noise, the noises may not occur at the same time so that masking does not necessarily, or perhaps generally, occur. As discussed above, it appears that the annoyance attributable to either source alone is fairly well predicted from the L_{eq} of different numbers of different intensities of aircraft noise.

Taylor (ref. 58) asked people living in residential areas in Toronto, Canada, to rate on a scale from 0 ("not at all disturbing") to 10 ("unbearably disturbing") the noise in their neighborhood from, among other things, aircraft, main road traffic, and lastly (after the ratings to individual noises were given) to "overall" noise. The exposures to the noises were estimated for, or measured at, positions in the fronts of houses facing the street.

The findings obtained at 17 different sites are plotted in figure 11.29(a). There it is seen that the overall noise at a given L_{eq} (combined noises) is rated somewhat higher than road traffic noise at the same L_{eq}. However, aircraft noise of a given L_{eq} was judged to be much more disturbing than road traffic noise at the same L_{eq}. For equal annoyance, road traffic noise is about 8 dB higher than aircraft noise and 2 to 3 dB higher than overall noise. This finding is consistent with the previously discussed differences associated with outdoor noise measurements of aircraft as compared with ground-based vehicle noise.

A more interesting research finding is the large difference (equivalent to -5 to -6 dB in exposure levels) in the ratings for overall noise and for aircraft. A possible exception to this occurs at the highest levels of exposure from both sources. The puzzling result—that annoyance from noise overall is considerably less than the annoyance from a single but dominant source—is discussed later.

Powell (ref. 108) asked subjects seated in a simulated living room and engaged in reading or knitting to rate their annoyance reaction to 15-minute sessions during which various exposure levels of recorded aircraft and road traffic noises were presented separately or in combinations. The ratings were made on a scale of 0 ("not at all annoyed") to 9 ("extremely annoyed"). Figure 11.29(b) shows the data obtained by Powell.

As is seen in figure 11.29(b), about the same amount of annoyance was reported for either the aircraft or the road traffic noises when the noises were of equal L_{eq} and when only one type of noise was heard during the test period. This is consistent with other laboratory and field data, as discussed earlier, showing that when common noises from different sources are of about equal, A-weighted energy $L_{A,ex}$ at the position of the listener, they are judged to be equally annoying.

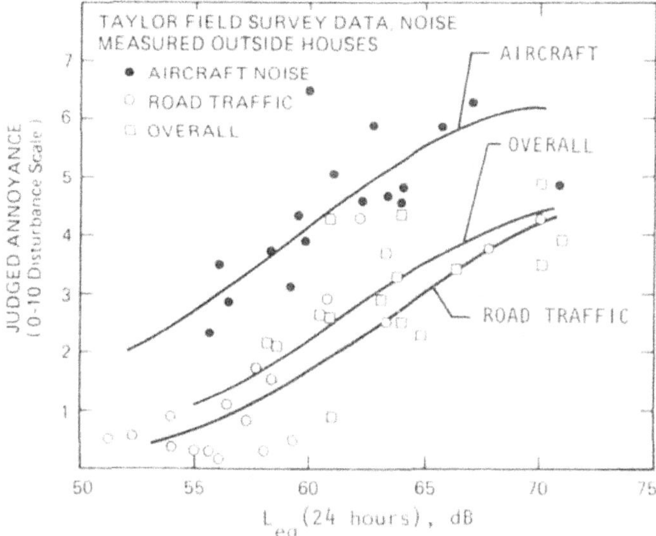

(a) Based on data collected by Taylor (ref. 58) in attitude survey conducted in Toronto, Canada, with noise measured or estimated for a position outdoors, near the fronts of houses.

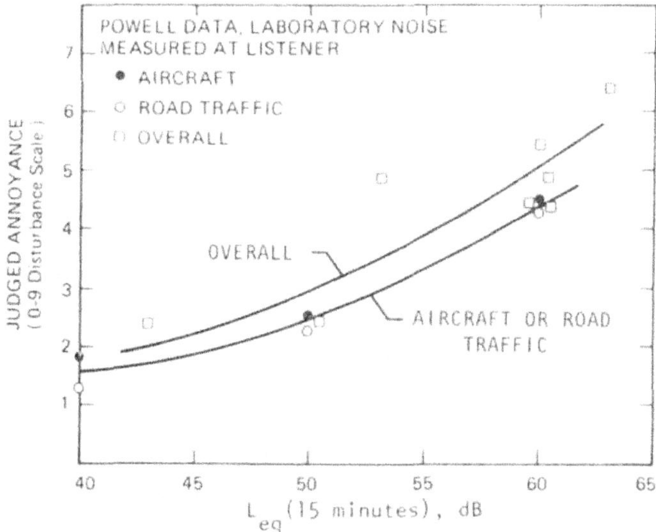

(b) Based on data collected by Powell (ref. 108) from recordings presented under laboratory conditions in 15-minute test sessions with noise measured at position of listeners.

FIGURE 11.29. Annoyance ratings given to separate sources (aircraft or road traffic noise) and overall annoyance from noise.

Figure 11.29(b) also shows that when the two types of noises were presented together, the annoyance ratings at a given L_{eq} were generally higher than would be the ratings given when only one or the other of the noises was present at the same L_{eq}. The difference for equal annoyance between the two trend curves is about 3 dB in terms of L_{eq}.

Powell's data indicate that this enhanced, or at the least additive, annoyance effect is not strongly, if at all, related to the magnitude of the difference in exposure level between the aircraft and road traffic noises. Table 11.14 shows that the average scale point differences between the overall and the average of individual noise ratings were +1.67 for 0-dB L_{eq} differences between the two noises (both had L_{eq} values of 40, 50, or 60 dB), +0.78 for 10-dB differences, and +1.57 for 20-dB differences. Certainly there is no consistent "level difference" effect on the annoyance. It is to be noted, however, that the exposure levels in the Powell study were quite high, probably more typical to outdoor than to indoor, living room listening. In the Powell study, the 15-minute L_{eq} values (40, 50, and 60 dB) correspond to 24-hour L_{eq} values as given in the Taylor study of about 58, 68, and 78 dB. Further, these levels were at the position of the listener, whereas the levels reported in the Taylor study were outdoor levels, but the annoyance

TABLE 11.14

Increase in Overall Annoyance Rating Above Average of Annoyance Ratings Given Separate Noise Sources

[Recorded aircraft and road traffic noise. Data from ref. 108]

L_{eq} (15 minutes), dB				Annoyance rating response (scale 0-9)				
Aircraft	Road	Both	Aircraft - Road	Aircraft	Road	Average	Overall	Overall - Average
40	40	43	0	1.88	1.23	1.56	2.56	+1.00
50	50	53	0	2.51	2.35	2.43	4.29	+1.86
60	60	63	0	4.51	4.24	4.38	6.52	+2.14
								Avg. +1.67
40	50	50.4	−10	1.88	2.35	2.11	2.29	+ .18
50	40	50.4	10	2.51	1.23	1.87	2.42	+ .55
50	60	60.4	−10	2.51	4.24	3.37	4.93	+1.56
60	50	60.4	10	4.51	2.35	3.43	4.26	+ .83
								Avg. +0.78
40	60	60	−20	1.88	4.24	3.06	5.59	+2.53
60	40	60	20	4.51	1.23	2.87	4.47	+1.60
								Avg. +2.07

reported by the respondents was presumably as much, if not more, from the noise experienced when they were indoors than when outdoors.

There are some striking differences between figures 11.29(a) and (b). The fact that the aircraft noise and road traffic noise at a given L_{eq} were rated the same in the Powell but not in the Taylor study, as discussed above, can be attributed to the differences in the noise measurement procedures (at the listener's position by Powell and outdoors by Taylor) and the effective levels of the two types of noises. A more difficult to explain difference between the Powell and Taylor results is with regard to the "overall" ratings. Taylor's data indicate that the overall annoyance rating was considerably less than that given the aircraft noise, even though the aircraft noise was the dominant noise in the environment being judged. Powell's data, on the other hand, indicate that not only does the annoyance from noises from multiple sources summate on an exposure energy basis, but also there is some enhancement to the overall annoyance.

It is speculated that the explanation lies in the differences in the questioning procedures used in the two studies. In the Powell study, the subjects rated their annoyance after independent test sessions during which only aircraft noise, only road traffic noise, or a combination of both, were present. For the latter type of session, the subjects did not rate their feelings of annoyance separately for the two noises. As noted above, in the Taylor study, the subjects rated their annoyances first for each type of noise present, presumably, during each typical day and, after giving those ratings, for "overall" noise.

It is tempting to think that the respondents in the Taylor study reported for "overall" noise annoyance their impression of the annoyances averaged. For example, if a person in the Taylor survey rated for his neighborhood the aircraft noise as, say, a "7," and the road traffic noise as a "1," and was then asked to rate the noise "overall," the respondents would consider the appropriate response to be somewhere between the two ratings. Figure 11.30 (ref. 58) shows that overall annoyance in the Taylor study about equals the average judged annoyance for aircraft and for road traffic noises. Annoyance ratings are based on a scale of 0 (not at all disturbing) to 10 (unbearably disturbing). This area of research is obviously challenging because of methodological problems and the very nature of the annoyance judgment. Also sometimes confounding the results of surveys in real life are acoustical factors differentially affecting the transmission of noise from outdoor sources to listeners.

Seasonal effects

Although the assumptions that the annualized L_{dn} (especially when expressed in terms of effective at-the-ear L_{dn} values) may be valid for predicting annoyance from aperiodic noises of practical importance, it cannot be presumed that differences in climates do not have significant influence upon the amount of annoyance experienced as a function of L_{dn}. The obvious reason is that the effective, at-the-ear amount of aircraft noise will be different in a typical northern-moderate

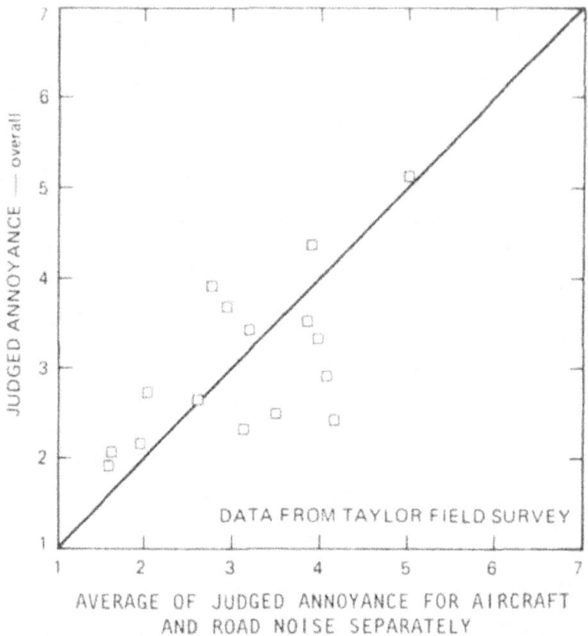

FIGURE 11.30. Relation between judged annoyance overall and the average of judged annoyance for aircraft and road noises given separately. (Data from ref. 58.)

climate (such as New York, London, Geneva, or Amsterdam) where the open window, out-of-doors living season may average fewer months of the year than is the case for a warmer climate (such as in southern cities in the U.S.).

Figure 11.31 from Beranek *et al.* (ref. 109) shows that during the warmer months higher degrees of annoyance are reported than during colder months. The number of airline operations at these airports does not increase sufficiently during the warmer months or in a pattern which would explain this increase in annoyance. In figure 11.32 from Patterson and Connor (ref. 38), it is seen that higher percentages of people were annoyed at a given L_{dn} in the summertime than in the winter months. It is conceivable that the data in figure 11.32 are somehow confounded with a city size factor, although such a factor has not been explored or identified in previous research. Also, the "small" cities involved have populations of around 60 000 (Reno) and 150 000 (Chattanooga), and their residential neighborhoods are not necessarily different from those in the "large" cities.

Figure 11.33 shows that surveys in Dallas, Texas (July) and in southern California (Sept.-Nov.) (ref. 30) revealed appreciably more annoyance than that found in the northern-moderate-climate cities involved in the major attitude surveys included in the previously discussed cluster analyses. Note in figure 11.33

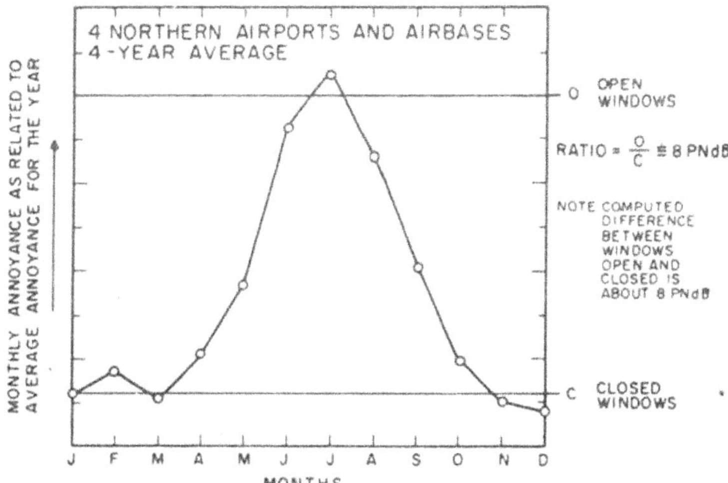

FIGURE 11.31. Total monthly annoyance, as measured by number of complaints registered with airports, as a function of season of the year. Based on 4-year average of four airports in northeastern U.S. (From ref. 109.)

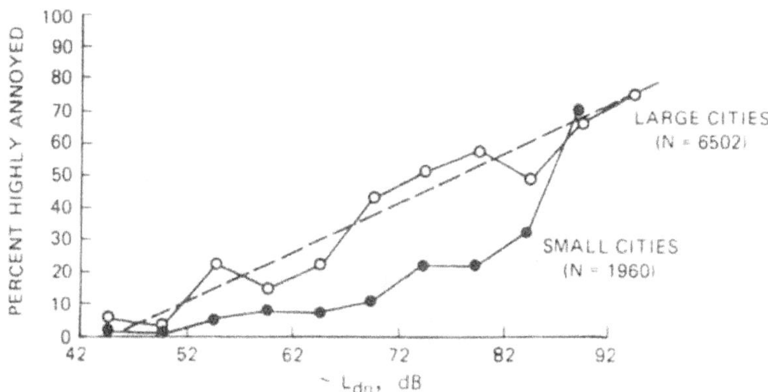

FIGURE 11.32. Percentage of people highly annoyed by aircraft noise. Small cities (Chattanooga, Tennessee, and Reno, Nevada) surveyed in winter; large cities (Chicago, Dallas, Denver, Los Angeles, Boston, Miami, and New York) surveyed in summer. The dashed curve is a visual straight-line fit to the data for the large cities. (Data from ref. 38.)

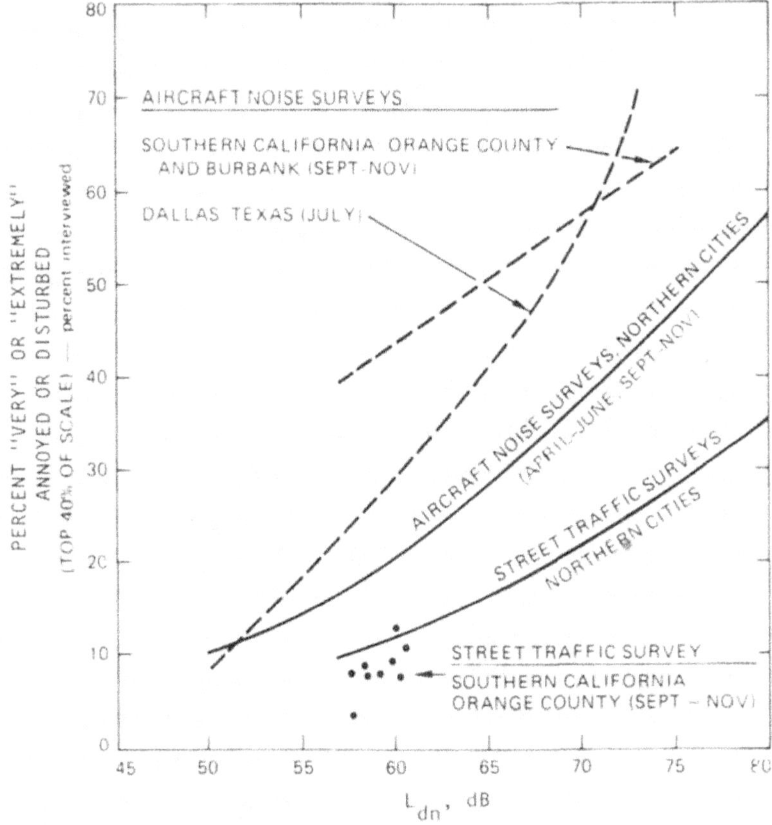

FIGURE 11.33. Comparison of percentages of people very or extremely annoyed (top 40 percent of annoyance scale) by aircraft noise and street noise in northern climate cities (see fig. 11.12(a)), in Dallas, Texas (from preliminary results prepared for the Department of Aviation, Dallas, Texas, under the Noise Control Program), and in Orange County, California (ref. 30). Note all data are for top 40 percent of annoyance scale, although identified as "highly annoyed" in reference 30.

that it does not appear that ground vehicle noise in a warm climate is a greater source of annoyance than it is in colder climates.

Regardless of the measured level of street traffic noise below an L_{dn} of about 58 dB, 5 to 10 percent of people will still report "very much or more annoyance." Indeed, it is seen in figure 11.4(a) that 10 percent of people exposed to aircraft noise L_{dn} levels of 50 dB or less are "very much or more annoyed." As discussed above, ground vehicle noise with an L_{dn} level of 56 dB would be as effective as aircraft noise with an L_{dn} level of 50 dB.

Griffiths *et al.* (ref. 33) found that annoyance ratings of road traffic noise in some London residential areas did not vary significantly for different times of the year even though the proportion of open windows did. It is possible that this apparent lack of a seasonal effect was due to the relatively low effective level of the road traffic noise, as discussed above with respect to figure 11.33. As already mentioned, it is also possible that people tend to keep the windows facing streets, especially heavily traveled streets, closed, which if true, would tend to make the effective levels reaching the people indoors somewhat the same for both the warmer and colder climates.

At stake, because of the disparity between these colder and warmer climate data, is probably not the utility of the basic L_{dn} concept as a general means of predicting annoyance. Rather, these data show that, as with aircraft versus ground vehicle noise, in order to predict the amount of annoyance to be expected from L_{dn} values in communities having appreciably different climates, it is necessary to make appropriate corrections to measured L_{dn} to reflect effective indoor levels.

Also in support of this notion are the findings from a survey of the attitudes and beliefs of people living near a southern California (Orange County) airport (ref. 76). Figure 11.34 shows that with an L_{dn} of 60 to 65 dB, about 50 percent of the people believed the aircraft noise was a serious problem causing decreases in residential property values and danger to health. These percentages, although not directly comparable with aircraft noise data from the northern climate surveys because of differences in the wording of the questions asked, indicate greater adverse effects than would be expected on the basis of the annoyance data obtained in the northern climate surveys discussed.

On the basis of the data that are available, it is perhaps not unreasonable to suggest that 5 dB be added to L_{dn} values for aircraft noise in warm climate areas to provide comparability with L_{dn} values in appreciably colder climates. It is of interest to note that the original CNR procedure briefly discussed earlier in this chapter incorporated a 5-dB correction for a windows-closed condition compared with a windows-open environment. If one wished to preserve the annualized L_{dn} concept, it could be argued that the increase from, for example, 3 open-window, outdoor-living months in the colder climates, to for example, 9 months in the warmer climates, represents a factor of about 5 dB in increased "at-the-ear" noise energy over the year.

Clearly, some further research or analysis of differences in climate (such as average number of days or months per year with temperatures at or above certain levels) is needed to implement a better, more practical procedure for expressing climate-effective L_{dn} values with northern city climates as a reference condition. It is to be noted that the data and discussions on this point have been with respect to the prediction of the effects of aircraft and ground vehicle noise on people living in typical single- or two-family houses. Air conditioning and acoustic shielding afforded in masonry or soundproofed houses, multi-family dwellings, and office-type buildings tend to make climate less of a variable factor. Some estimates of the effective L_{dn} values appropriate when these latter types of structures are of interest are presented in chapter 12.

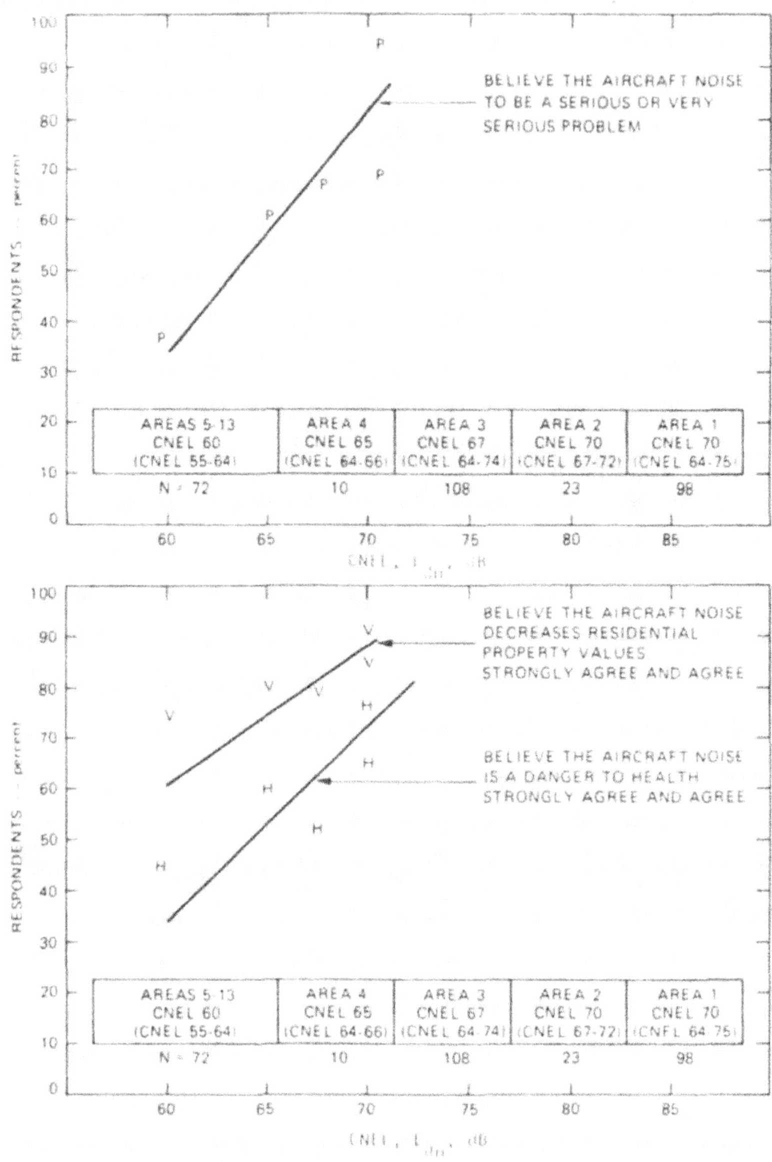

FIGURE 11.34. Percentage of samples of people living near John Wayne Airport, Orange County, California, perceiving aircraft noise as a serious or very serious problem, as decreasing the value of their houses, and as a danger to their health. (Based on data from ref. 76, chapter X and appendix A3.)

Night and evening time penalties

As noted earlier, the L_{dn} noise measurement procedures involve the addition of a 10-dB penalty to noises occurring during typical hours of sleep, 10 p.m. to 7 a.m., and the CNEL procedure involves, in addition, a 5-dB penalty to be added to aircraft noise exposures occurring between 7 p.m. and 10 p.m. These penalties are based on a mixture of laboratory and field research and the general experience of acoustical consultants working on community noise problems. As a practical matter, for areas around typical commercial airports, application of each of these penalties will increase the equivalent continuous sound level (L_{eq}) for 24 hours by 1 to 2 dB. For example, it can be deduced from basic laboratory data that the *threshold level for noise that will start to interfere with quiet conversational speech* (about 45 dBA, see chapter 4) is about 10 dB higher than the level that will start to interfere with sleep (about 35 dBA, see chapter 10). Also, as shown in figure 11.35, the records of a telephone complaint bureau maintained by the operator of several major airports in the northeastern U.S. revealed that for a given number of aircraft operations, the complaints for nighttime operations were greater than for daytime operations (ref. 109).

Borsky (ref. 110), in a review of research on this and other questions pertaining to the measurement of annoyance from aircraft noise, notes that some unpublished studies by him and work by Ollerhead (ref. 56) indicate that this 10-dB penalty may be somewhat too large. However, as shown in the analyses of surveys of annoyance and disturbance from aircraft noise discussed above, the L_{dn} measure with the 10-dB nighttime penalty seemed to properly assess overall annoyance. The general experience of airport operators is that compared with daytime flight operations, nighttime flight operations are an inordinate source of complaints. Indeed, in a number of major European cities and to some extent in the U.S., operations of noisy commercial aircraft are prohibited during late nighttime hours. The proceedings of a conference on concepts and some research related to the 10-dB nighttime penalty are found in reference 111.

The 5-dB evening time penalty used in the CNEL noise assessment procedure was proposed on the grounds that more social, speech, and other auditory communications behavior that can be disturbed by noise occurs from about 7 p.m. to 10 p.m. than during any other 3-hour period from 7 a.m. to 7 p.m.

A-weighted measures of narrowband noises

A-weighting, which is used in the measurement of noise for purposes of calculating L_{dn}, ostensibly serves to equate the contribution of different parts of the frequency spectrum content of a noise to its loudness. As discussed in chapter 5, A-weighting does not agree well with judgments of the relative loudness or noisiness of very low- and high-frequency bands of sound or noise, although with most broad spectrum noises these inaccuracies tend to balance each other and become relatively insignificant.

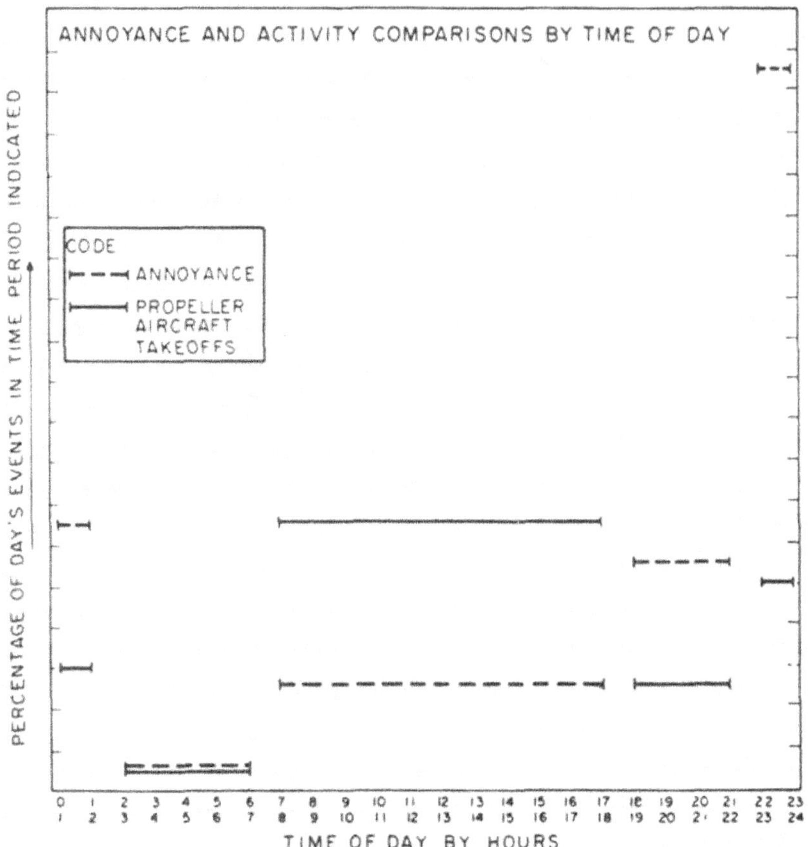

FIGURE 11.35. Annoyance, as measured by number of complaints registered at a major airport, and aircraft take-off activity at that airport as a function of time of day. Data for 9 months averaged. Complaints and aircraft activity are expressed as the percentage of day's complaint or aircraft activity events occurring in the hour indicated. (From ref. 109.)

However, there occasionally are found in communities sounds or noises that have most of their energy in relatively narrow frequency bands. In these cases A-weighted measurements may not be valid for assessing relative loudness, noisiness, or annoyance to people in the community. For these noises, D-weighting, or PNdB, should be used. (See chapters 2 and 5.)

Impulsive and low-frequency noises

The L_{dn} procedure can be used with impulsive-type noises when their measured levels can be converted to equivalent, effective $L_{A,ex}$ values and, if intense low-frequency components are involved, corrected for house-vibration factors. At the present time, the conversion of impulsive sound measurements to equivalent A-weighted decibel energy values can best be done through judgment tests, such as described earlier for sonic booms and in chapter 5 for sonic booms and artillery-firing noise. Procedures for correcting $L_{A,ex}$ or equivalent $L_{A,ex}$ values to take into account vibration-caused annoyance from noises with intense low frequencies are described in chapter 5.

The impulses from hand-held and shoulder guns have such a rapid rise-time and short duration that the energy in the lower frequency regions is generally not sufficient to cause house vibrations. However, the sound energy in the more audible frequencies and the impulsive nature of small arms fire can be a source of annoyance. (See chapter 7 for information on the spectra of such impulses and their damage risk to hearing.)

Bullen and Hede (ref. 112) measured the noise in a residential area near a rifle range and administered an annoyance attitude survey to a sample of the residents. They found only low correlations between various noise measurements (including A-weighted sound pressure levels) and annoyance. The investigators noted that the noise was very directional and variable from day to day. It was concluded that A-weighted energy or L_{eq} was probably the most useful unit of measurement, even though it did not predict annoyance well. Some of the variables and problems involved in the physical measurement and subjective measurement of such impulses are discussed in chapter 5.

The noise from helicopters is noted for its relatively intense impulsive frequency components, as briefly discussed in chapter 5. Recent studies (Powell and McCurdy (ref. 113), Ollerhead (ref. 114), and a review by Molino (ref. 115) of numerous related studies) indicate the A-frequency weighting is not as appropriate as D-weighting or PNdB for assessing judged loudness and noisiness. When L_D or PNL is used as a measure of helicopter noise, it is generally found that no impulse correction is needed. However, when helicopter noise is simulated electronically under laboratory conditions, it appears in some studies that an impulsive "beating" sometimes present in helicopter noise increases its noisiness to the extent that some correction to exposure PNL is justified. Some of this effect may be due to the introduction of higher frequency components to the noise in laboratory simulations than are present in the field from helicopters.

Summary

1. There is a remarkable consistency in the results of most attitude surveys conducted in different countries that show the percentage of groups of people who feel various degrees of annoyance and types of disturbances when exposed to equal day-night average sound levels (L_{dn}) of aircraft noise. Correlation coefficients of 0.90 to 0.95 are found between the percentages of people annoyed and the level of exposure to aircraft noise.

2. There are strong correlations between the subjective effects of aircraft noise as reflected in attitude surveys of annoyance and various measures of adverse effects of aircraft noise on people, such as disturbances because of interference with speech communications and sleep, house vibration, and housing depreciation.

3. The relatively low correlation, of the order of 0.5 or less, between exposure to aircraft noise as measured by L_{dn} and ratings of annoyance by individuals can be attributed to differences in actual noise dosages received at the ears of individuals indoors for equal outdoor levels of L_{dn}, as well as to the effects of individual differences in personalities, changing behavior activities, and general attitudes.

4. The percentages of people feeling various degrees of annoyance are significantly different for exposure to urban ground vehicular traffic noise than for exposure to aircraft noise of equal L_{dn}. This difference is largely due to the acoustical factors that lead to less noise (for given equal sound pressure levels measured outdoors near the fronts of houses) reaching the insides of houses and backyard living areas from street traffic than from aircraft operations. Also, ground vehicle noise, generally having a much lower peak level of intensity than aircraft noise, will more often fall below the threshold levels capable of causing annoyance in and around houses, than will aircraft noise. Approximately 10 dB should be subtracted from L_{dn} values for street and road traffic noise measured at the fronts of houses in order to compare the annoyance impact therefrom to the annoyance from aircraft flyover noise.

5. "Energy" corrections to achieve an effective annualized L_{dn} appear to be in order when estimates are made of the adverse effects to be expected in residential areas in different climates from exposure to exterior aircraft noises. It appears that the annualized L_{dn}, or the like, must be about 5 dB less in a warm climate such as southern California than in a northern-moderate climate such as New York City or London in order to have about equal noise impact in the two areas.

6. Physical measures of impulsive and low-frequency noise exposures must be corrected for the effects of impulsiveness and house vibration upon annoyance in order to calculate their equivalent A-weighted exposure ($L_{A,ex}$) levels and L_{dn} values.

7. L_{dn}, which measures the effective noise energy combined from all sources, will either accurately predict or slightly underestimate community annoyance from the total noise environment. Only effective noise energies above the threshold of annoyance are to be included in such an assessment.

8. Night (10 p.m. to 7 a.m.) and evening (7 p.m. to 10 p.m.) penalties appear to be appropriate for noise events that occur during those periods. The night penalty is presumably due to the greater sensitivity to annoyance from sleep disturbance, for a given noise event, than from other disturbing effects of the noise.

The evening penalty is presumably due to the generally greater amount of speech communication and social behavior during the evening hours than during the daytime. As a practical matter, for areas around typical commercial airports, application of each of these penalties will increase the equivalent continuous sound level (L_{eq}) for 24 hours by 1 to 2 dB.

References

1. Langdon, F. J.; and Griffiths, I. D.: Subjective Effects of Traffic Noise Exposure. II: Comparisons of Noise Indices, Response Scales, and the Effects of Changes in Noise Levels. J. Sound & Vib., vol. 83, no. 2, July 22, 1982, pp. 171-180.
2. Kryter, Karl D.: Community Annoyance From Aircraft and Ground Vehicle Noise. J. Acoust. Soc. America, vol. 72, no. 4, Oct. 1982, pp. 1222-1242.
3. Schultz, Theodore J.: Comments on K. D. Kryter's Paper, "Community Annoyance From Aircraft and Ground Vehicle Noise." J. Acoust. Soc. America, vol. 72, no. 4, Oct. 1982, pp. 1243-1252.
4. Kryter, Karl D.: Rebuttal by Karl D. Kryter to Comments by T. J. Schultz. J. Acoust. Soc. America, vol. 72, no. 4, Oct. 1982, pp. 1253-1257.
5. Kryter, Karl D.: Response of K. D. Kryter to Modified Comments by T. J. Schultz on K. D. Kryter's Paper, "Community Annoyance From Aircraft and Ground Vehicle Noise." J. Acoust. Soc. America, vol. 73, no. 3, Mar. 1983, pp. 1066-1068.
6. Committee on the Problem of Noise: Noise—Final Report. Her Majesty's Stationery Off., July 1963.
7. MIL Research Ltd: Second Survey of Aircraft Noise Annoyance Around London (Heathrow) Airport. Her Majesty's Stationery Off., 1971.
8. Information on Levels of Environmental Noise Requisite To Protect Public Health and Welfare With an Adequate Margin of Safety. Rep. 550/9-74-004, U.S. Environ. Prot. Agency, Mar. 1974. (Available from NTIS as PB 239 429.)
9. Protective Noise Levels: Condensed Version of EPA Levels Document. Rep. EPA-550/9-79-100, Nov. 1978. (Available from NTIS as PB82 138 827.)
10. Alexandre, Ariel: Prevision de la Gene due au Bruit autour des Aeroports et Perspectives sur les Moyens d'y Remedier. Doc. A.A. 28/70, Anthropologie Appliquee, Centre d'Etudes et de Recherches, Apr. 1970.
11. Alexandre, Ariel: Decision Criteria Based on Spatio-Temporal Comparisons of Surveys on Aircraft Noise. Proceedings of the International Congress on Noise as a Public Health Problem, W. Dixon Ward, ed., 550/9-73-008, U.S. Environ. Prot. Agency, May 1973, pp. 619-626.
12. Grandjean, Etienne; Graf, Peter; Lauber, Anselm; Meier, Hans Peter; and Muller, Richard: A Survey of Aircraft Noise in Switzerland. Proceedings

of the International Congress on Noise as a Public Health Problem, W. Dixon Ward, ed., 550/9-73-008, U.S. Environ. Prot. Agency, May 1973, pp. 645-659.

13. Grandjean, Etienne; Graf, P.; Lauber, A.; Meier, H. P.; and Muller, R.: Survey on the Effects of Aircraft Noise Around Three Civil Airports in Switzerland. INTER-NOISE 76 Proceedings, Roger L. Kerlin, ed., c.1976, pp. 85-90.
14. Socio-Psychological Airplane Noise Investigation in the Districts of Three Swiss Airports, Zurich, Basel, Geneva. NASA TM-75787, 1980.
15. Schultz, Theodore, J.: Synthesis of Social Surveys on Noise Annoyance. J. Acoust. Soc. America, vol. 64, no. 2, Aug. 1978, pp. 377-405. (Erratum, vol. 65, no. 3, Mar. 1979, p. 849.)
16. Shigehisa, T.; and Gunn, Walter J.: Annoyance Response to Recorded Aircraft Noise: I. Effect of Intensity of Illumination. J. Aud. Res., vol. 18, no. 3, July 1978, pp. 175-182.
17. Shigehisa, T.; and Gunn, Walter J.: Annoyance Response to Recorded Aircraft Noise: II. Effect of Intensity of Illumination in Relation to Noise Spectrum. J. Aud. Res., vol. 18, no. 3, July 1978, pp. 183-190.
18. Shigehisa, T.; and Gunn, Walter J.: Annoyance Response to Recorded Aircraft Noise: III. In Relation to Personality. J. Aud. Res., vol. 19, no. 1, Jan. 1979, pp. 41-46.
19. Shigehisa, T.; and Gunn, Walter J.: Annoyance Response to Recorded Aircraft Noise: IV. Effect of Intensity of Illumination in Relation to Personality. J. Aud. Res. vol. 19, no. 1, Jan. 1979, pp. 47-58.
20. Taylor, S. M.; Hall, F. L.; and Birnie, S. E.: A Comparison of Community Response to Aircraft Noise at Toronto International and Oshawa Municipal Airports. J. Sound & Vib., vol. 77, no. 2, July 22, 1981, pp. 233-244.
21. Galloway, William J.; and Bishop, Dwight E.: Noise Exposure Forecasts: Evolution, Evaluation, Extensions, and Land Use Interpretations. Rep. FAA-NO-70-9, Aug. 1970.
22. Bitter, C.: Annoyance Due to Aircraft Noise. Paper presented at the Colloquium on Human Demands With Respect to Noise (Paris), Nov. 18-19, 1968.
23. Gunn, Walter J.; Shigehisa, T.; Fletcher, John L.; and Shepherd, William T.: Annoyance Response to Aircraft Noise as a Function of Contextual Effects and Personality Characteristics. J. Aud. Res., vol. 21, 1981, pp. 51-83.
24. Hall, F. L.; Taylor, S. M.; and Birnie, S. E.: Spatial Patterns in Community Response to Aircraft Noise Associated With Non-Noise Factors. J. Sound & Vib., vol. 71, no. 3, Aug. 8, 1980, pp. 361-381.
25. McKennell, Aubrey: Psycho-Social Factors in Aircraft Noise Annoyance. Proceedings of the International Congress on Noise as a Public Health Problem, W. Dixon Ward, ed., 550/9-73-008, U.S. Environ. Prot. Agency, May 1973, pp. 627-644.

26. Borsky, Paul N.; and Leonard, H. Skipton: A New Field Survey-Laboratory Methodology for Studying Human Response to Noise. Proceedings of the International Congress on Noise as a Public Health Problem, W. Dixon Ward, ed., 550/9-73-008, U.S. Environ. Prot. Agency, May 1973, pp. 743-763.
27. Cheifetz, Philip; and Borsky, Paul N.: Laboratory Study of Effects of Acoustic and Nonacoustic Variables on Annoyance With Aircraft Noise. Noise as a Public Health Problem, Jerry V. Tobias, Gerd Jansen, and W. Dixon Ward, eds., ASHA Rep. 10, American Speech-Language-Hearing Assoc., Apr. 1980, pp. 522-528.
28. Becker, R. W.; Kryter, K. D.; and Poza, F.: A Study of Sensitivity to Noise—Final Report. Rep. EQ-71-4 (Contract DOT-FA69WA-2211), Stanford Res. Inst., June 1971.
29. Hall, Fred L.; Birnie, Susan E.; Taylor, S. Martin; and Palmer, John E.: Direct Comparison of Community Response to Road Traffic Noise and to Aircraft Noise. J. Acoust. Soc. America, vol. 70, no. 6, Dec. 1981, pp. 1690-1698.
30. Evaluation of Three Noise Abatement Departure Procedures at John Wayne Airport. U.S. Dep. Transp., Nov. 1981.
31. Fidell, S.; Horonjeff, R.; Teffeteller, S.; and Pearsons, K.: Community Sensitivity to Changes in Aircraft Noise Exposure. NASA CR-3490, 1981.
32. Griffiths, I. D.; and Delauzun, F. R.: Individual Differences in Sensitivity to Traffic Noise: An Empirical Study. J. Sound & Vib., vol. 55, no. 1, Nov. 8, 1977, pp. 93-107.
33. Griffiths, I. D.; Langdon, F. J.; and Swan, M. A.: Subjective Effects of Traffic Noise Exposure: Reliability and Seasonal Effects. J. Sound & Vib., vol. 71, no. 2, July 22, 1980, pp. 227-240.
34. Rylander, R.; Sörensen, S.; and Kajland, A.: Annoyance Reactions From Aircraft Noise Exposure. J. Sound & Vib., vol. 24, no. 4, Oct. 22, 1972, pp. 419-444.
35. Fluglärmwirkungen—Eine Interdisziplinare Untersuchung uber die Auswirkungen des Fluglärms auf den Menschen. Hauptbericht. Harald Boldt Verlag KG (West Germany), c.1974. (Available as NASA TM-75819.)
36. Rohrmann, B.; Schümer, R.; Schümer-Kohrs, A.; Guski, R.; and Finek, H.-O.: An Interdisciplinary Study on the Effects of Aircraft Noise on Man. Proceedings of the International Congress on Noise as a Public Health Problem, W. Dixon Ward, ed., 550/9-73-008, U.S. Environ. Prot. Agency, May 1973, pp. 765-776.
37. TRACOR, Inc.: Community Reaction to Airport Noise.
Volume I. NASA CR-1761, 1971.
Volume II. NASA CR-111316, 1970.
38. Patterson, Harrold P.; and Connor, William K.: Community Responses to Aircraft Noise in Large and Small Cities in the U.S.A. Proceedings of the

International Congress on Noise as a Public Health Problem, W. Dixon Ward, ed., 550/9-73-008, U.S. Environ. Prot. Agency, May 1973, pp. 707-718.

39. Connor, William K.; and Patterson, Harrold P.: Community Reaction to Aircraft Noise Around Smaller City Airports. NASA CR-2104, 1972.
40. Galloway, W. J.; Eldred, K. M.; and Simpson, M. A.: Population Distribution of the United States as a Function of Outdoor Noise Level—Volume 2. Rep. EPA-550/9-74-009-A-VOL-2, June 1974. (Available from NTIS as PB 257 617/1.)
41. Langdon, F. J.: Noise Nuisance Caused by Road Traffic in Residential Areas, Parts I and II. J. Sound & Vib., vol. 47, no. 2, July 22, 1976, pp. 243-263, 265-282.
42. Aubrée, D.: A Study of Annoyance Due to Urban Automobile Traffic. Centre Scientifique et Technique du Bâtiment, Paris, June 1971.
43. Simpson, Myles A.; Pearsons, Karl S.; Fidell, Sanford A.; and Muehlenbeck, Richard H.: Social Survey and Noise Measurement Program To Assess the Effects of Noise on the Urban Environment: Data Acquisition and Presentation. Rep. No. 2753, Bolt Beranek and Newman, Inc., July 1974.
44. Aubrée, D.: Enquête Acoustique et Sociologique Permettant de Definir une Echelle de la Gêne Eprouvée par l'Homme dans son Logement du Fait des Bruits de Train. (Acoustical and Sociological Investigation Permitting the Definition of a Scale of Annoyance Felt by People in Their Dwellings Due to the Noise of Trains.) Centre Scientifique et Technique du Bâtiment, Paris, June 1973. (Available in English translation as Bolt Beranek and Newman Technical Information Report No. 88, August 1973.)
45. Myncke, H.; Cops, A.; Steenackers, P.; Bruyninckx, W.; Gambart, R.; and Verleysen, P.: Studie van het Verkeerslawaai in Steden en de Hinder Ervan voor de Bevolking. (Study of Urban Traffic Noise and the Annoyance Felt by the Population.) 13 Volumes, Lab. for Acoustics & Thermal Conductivity of the Catholic Univ. of Leuven, & Ministry of Public Health and of the Environment of Belgium, 1977.
46. Myncke, H.; Cops, A.; and Steenackers, P.: Traffic Noise Measurements in Antwerp and Brussels. Part I: Physical Measurements. 9th International Congress on Acoustics, Contributed Papers, Volume 1, July 1977, p. 168.
47. Myncke, H.; Cops, A.; and Gambart, R.: Traffic Noise Measurements in Antwerp and Brussels. Part II: Enquiry Concerning Annoyance. 9th International Congress on Acoustics, Contributed Papers, Volume 1, July 1977, p. 169.
48. Rylander, R.; Sörensen, S.; and Kajland, A.: Traffic Noise Exposure and Annoyance Reactions. J. Sound & Vib., vol. 47, no. 2, July 22, 1976, pp. 237-242.

49. Bruckmayer, F.; and Lang, J.: Störung der Befölkerung durch Verkehrslärm. (Annoyance of People by Traffic Noise.) Österreichische Ingenieur-Zeitschrift, vol. 10, nos. 8, 9, 10, 1967, pp. 302-306, 338-344, 376-385.
50. Fog, Hans; and Jonsson, Erland: Traffic Noise in Residential Areas. Rep. No. 36 E, Nat. Swedish Inst. for Building Res., 1968.
51. Lamure, Claude; and Bacelon, Michel: The Nuisance Due to the Noise of Automobile Traffic. An Investigation in the Neighborhoods of Freeways. NASA TM-75812, 1980.
52. Nimura, T.; Sone, T.; and Kono, S.: Some Considerations on Noise Problem of High-Speed Railway in Japan. INTER-NOISE 73 Proceedings, O. Juhl Pedersen, ed., 1973, pp. 298-307.
53. Hall, F. L.; and Taylor, S. M.: Predicting Community Response to Road Traffic Noise. J. Sound & Vib., vol. 52, no. 3, June 8, 1977, pp. 387-399.
54. Relster, Else: Traffic Noise Annoyance. Polyteknisk Forlag, Lyngby, 1975.
55. Lang, Judith: Den Zusamenhang Zwischen Objectiven Messergebnissen und Subjectiv Empfundener Störung von Verkehrslärm. (The Relation Between Objectively Measured Data and Subjectively Expressed Disturbance Due to Traffic Noise.) Rep. No. 3486/WS, Der Physikalisch-Technischen Versuchsanstalt für Wärme-und Schalltechnik, Vienna, Oct. 27, 1975.
56. Ollerhead, John B.: Accounting for Time of Day and Mixed Source Effects in the Assessment of Community Noise Exposure. Noise as a Public Health Problem, Jerry V. Tobias, Gerd Jansen, and W. Dixon Ward, eds., ASHA Rep. 10, American Speech-Language-Hearing Assoc., Apr. 1980, pp. 556-561.
57. Fields, John M.; and Walker, J. G.: Comparing Reactions to Transportation Noises From Different Surveys: A Railway Noise vs. Aircraft and Road Traffic Comparison. Noise as a Public Health Problem, Jerry V. Tobias, Gerd Jansen, and W. Dixon Ward, eds., ASHA Rep. 10, American Speech-Language-Hearing Assoc., Apr. 1980, pp. 580-587.
58. Taylor, S. M.: A Comparison of Models To Predict Annoyance Reactions to Noise From Mixed Sources. J. Sound & Vib., vol. 81, no. 1, Mar. 8, 1982, pp. 123-138.
59. Nemecek, J.; Wehrli, B.; and Turrian, V.: Effects of the Noise on Street Traffic in Switzerland, A Review of Four Surveys. J. Sound & Vib., vol. 78, no. 2, Sept. 22, 1981, pp. 223-234.
60. Fields, J. M.; and Walker, J. G.: Comparing the Relationships Between Noise Level and Annoyance in Different Surveys: A Railway Noise vs. Aircraft and Road Traffic Comparison. J. Sound & Vib., vol. 81, no. 1, Mar. 8, 1982, pp. 51-80.
61. Fields, J. M.; and Walker, J. G.: The Response to Railway Noise in Residential Areas in Great Britain. J. Sound & Vib., vol. 85, no. 2, Nov. 22, 1982, pp. 177-255.

62. Ahrlin, U.; and Rylander, R.: Annoyance Caused by Different Environmental Noises. J. Sound & Vib., vol. 66, no. 3, Oct. 8, 1979, pp. 459-462.
63. Öhrström, E.; Bjorkman, M.; and Rylander, R.: Laboratory Annoyance and Different Traffic Noise Sources. J. Sound & Vib., vol. 70, no. 3, June 8, 1980, pp. 333-341.
64. Kryter, K. D.; Johnson, P. J.; and Young, J. R.: Psychological Experiments on Sonic Booms Conducted at Edwards Air Force Base. Contract AF49(638)-1758, Nat. Sonic Boom Eval. Off., Aug. 1978. (Available from DTIC as AD 689 844.)
65. Kryter, Karl D.: Sonic Booms From Supersonic Transport. Science, vol. 163, no. 3865, Jan. 24, 1969, pp. 359-367.
66. Fidell, S. A.; Teffeteller, S. R.; and Pearsons, K. S.: Initial Study on the Effects of Transformer and Transmission Line Noise on People, Volume 3: Community Reaction. Rep. EPRI-EA-1240-VOL-3, Bolt Beranek and Newman, Inc., Dec. 1979.
67. Sutherland, Louis C.; Braden, Marcia H.; and Colman, Richard: A Program for the Measurement of Environmental Noise in the Community and Its Associated Human Response, Volume I—A Feasibility Test of Measurement Techniques. DOT-TST-74-5, Dec. 1973. (Available from NTIS as PB 228 563.)
68. Piercy, J. E.; and Embleton, Tony F. W.: Sound Propagation in the Open Air. Handbook of Noise Control, Second ed., Cyril M. Harris, ed., McGraw-Hill Book Co., c.1979, pp. 3-1-3-16.
69. Ortega, Jose C.; and Kryter, Karl D.: Comparison of Aircraft and Ground Vehicle Noise Levels in Front and Backyards of Residences. J. Acoust. Soc. America, vol. 71, no. 1, Jan. 1982, pp. 216-217.
70. Stephens, David G.; and Powell, Clemans A.: Laboratory and Community Studies of Aircraft Noise Effects. Noise as a Public Health Problem, Jerry V. Tobias, Gerd Jansen, and W. Dixon Ward, eds., ASHA Rep. 10, American Speech-Language-Hearing Assoc., Apr. 1980, pp. 488-494.
71. Gjestland, T.: The Importance of a Threshold Level When Assessing Noise Annoyance. J. Acoust. Soc. America, vol. 65, suppl. no. 1, Spring 1979, p. S45.
72. Gjestland, Truls; and Oftedal, Gunnhild: Assessment of Noise Annoyance—The Introduction of a Threshold Level in L_{eq} Calculations. Paper presented at the 50th Anniversary Meeting of the Acoustical Society of America (Boston, Massachusetts), June 1979.
73. Gjestland, T.; and Oftedal, G.: Assessment of Noise Annoyance: The Introduction of a Threshold Level in L_{eq} Calculations. J. Sound & Vib., vol. 69, no. 4, 1980, pp. 603-610.
74. Avery, G. C.: Comparison of Telephone Complaints and Survey Measures of Noise Annoyance. J. Sound & Vib., vol. 82, no. 2, 1982, pp. 215-225.

75. McKennell, A. C.: Aircraft Noise Annoyance Around London (Heathrow) Airport. S.S. 337, Cent. Off. Inf. (British), Apr. 1963.
76. VTN Consolidated, Inc.: Airport Master Plan/ANCLUC Plan for John Wayne Airport, Orange County. Volume II: Airport Noise Control and Land Use Compatibility (ANCLUC) Plan Draft Report. General Services Agency (Orange County, California), Oct. 1980.
77. Opinion Research of California: Report of Findings—A Study Among Residents of Orange County Relative to Operations at Orange County Airport. City of Newport Beach [California], Mar. 1977.
78. Appendix AA—Summary of Citizen Complaints Regarding Aircraft Noise at John Wayne Airport, Orange County for Calendar Year 1979 and Complaint History From 1975 Through March 31, 1980. John Wayne Airport Noise Abatement Program Report for the Periods October 1, 1979 Through October 31, 1979 and January 1, 1980 Through March 31, 1980. Airports Div., General Services Agency (County of Orange, Santa Ana, California), 1980.
79. Wyle Labs., Inc.: Community Noise. Rep. NTID300.3 (Contract 68-04-0046), Dec. 31, 1971.
80. Galloway, W. J.; and Von Gierke, H. E.: Individual and Community Reaction to Aircraft Noise; Present Status and Standardization Efforts. INC/C4/P9, Amer. Stand. Assoc., Nov. 1966.
81. Commission on the Third London Airport—Papers and Proceedings, Volume VII. Her Majesty's Stationery Off., 1970.
82. Nelson, Jon P.: Airport Noise, Location Rent, and the Market for Residential Amenities. J. Environ. Econ. & Management., vol. 6, 1979, pp. 320-331.
83. Mieszkowski, Peter; and Saper, Arthur M.: An Estimate of the Effects of Airport Noise on Property Values. J. Urban Econ., vol. 5, 1978, pp. 425-440.
84. Nelson, Jon P.: Aircraft Noise and the Market for Residential Housing: Empirical Results for Seven Selected Airports. Rep. DOT/RSPA/DPB-50/78/24, Sept. 1978. (Available from NTIS as PB 297 681.)
85. Taylor, S. M.; Breston, B. E.; and Hall, F. L.: The Effect of Road Traffic Noise on House Prices. J. Sound & Vib., vol. 80, no. 4, Feb. 22, 1982, pp. 523-541.
86. De Vany, Arthur S.: An Economic Model of Airport Noise Pollution in an Urban Environment. Theory and Measurement of Economic Externalities, S. A. Y. Lin, ed., Academic Press, 1976, pp. 205-214.
87. Greater Westchester Homeowners Assn. *et al.* v. City of Los Angeles, 152 Cal. Rptr. 878, aff'd. in part, rev'd. in part and remanded, 26 Cal. 3d 86, 603 P. 2d 1329, 160 Cal. Rptr. 733 (1979), Cert. Denied, 449 U.S. 820 (1980).

88. Highline School District No. 401, King County v. Port of Seattle, 87 Wash. 2d 6, 548 P. 2d 1085 (1976).
89. Crook, M. A.; and Langdon, F. J.: The Effects of Aircraft Noise in Schools Around London Airport. J. Sound & Vib., vol. 34, no. 2, May 22, 1974, pp. 221-232.
90. Ko, N. W. M.: Responses of Teachers to Aircraft Noise. J. Sound & Vib., vol. 62, no. 2, Jan. 22, 1979, pp. 277-292.
91. Maser, Arthur L.; Sorensen, Philip H.; Kryter, Karl D.; and Lukas, Jerome S.: Effects of Intrusive Sound on Classroom Behavior: Data From a Successful Lawsuit. Western Psychological Assoc., Apr. 1978.
92. Cohen, Sheldon; Krantz, David S.; Evans, Gary W.; Stokols, Daniel; and Kelly, Sheryl: Aircraft Noise and Children: Longitudinal and Cross-Sectional Evidence on Adaptation to Noise and the Effectiveness of Noise Abatement. J. Pers. & Soc. Psychol., vol. 40, no. 2, Feb. 1981, pp. 331-345.
93. Lukas, J. S.; DuPree, R. B.; and Swing, J. W.: Effects of Noise on Academic Achievement and Classroom Behavior. FHWA/CA/DOHS-81/01 (Contract I.A.A. 19-7165), Office of Noise Control, California Dep. Health Services, Sept. 1981. (Available from NTIS.)
94. Bronzaft, Arline L.; and McCarthy, Dennis P.: The Effect of Elevated Train Noise on Reading Ability. Environ. & Behav., vol. 7, no. 4, Dec. 1975, pp. 517-527.
95. Bronzaft, Arline L.: The Effect of Elevated Train Noise on Reading Ability: Followup Report. J. Acoust. Soc. America, vol. 68, suppl. no. 1, Fall 1980, p. S91.
96. Cohen, Sheldon; Glass, David C.; and Singer, Jerome E.: Apartment Noise, Auditory Discrimination and Reading Ability in Children. J. Exp. Soc. Psychol., vol. 9, no. 5, Sept. 1973, pp. 407-422.
97. Rosenblith, Walter A.; Stevens, Kenneth N.; and the Staff of Bolt, Beranek, and Newman: Handbook of Acoustic Noise Control—Volume II. Noise and Man. WADC Tech. Rep. 52-204, U.S. Air Force, June 1953.
98. Borsky, Paul N.: Community Reactions to Air Force Noise. Parts I and II. WADD Tech. Rep. 60-689 (I) and (II), U.S. Air Force, Mar. 1961.
99. Vallet, M.; Maurin, M.; Page, M. A.; Favre, B.; and Pachiaudi, G.: Annoyance From and Habituation to Road Traffic Noise From Urban Expressways. J. Sound & Vib., vol. 60, no. 3, Oct. 8, 1978, pp. 423-440.
100. Powell, C. A.; and Rice, C. G.: Judgments of Aircraft Noise in a Traffic Noise Background. J. Sound & Vib., vol. 38, no. 1, Jan. 8, 1975, pp. 39-50.
101. Johnston, G. W.; and Haasz, A. A.: Traffic Background Level and Signal Duration Effects on Aircraft Noise Judgment. J. Sound & Vib., vol. 63, no. 4, Apr. 22, 1979, pp. 543-560.
102. Taylor, S. M.; Hall, F. L.; and Birnie, S. E.: Effect of Background Levels on

Community Responses to Aircraft Noise. J. Sound & Vib., vol. 71, no. 2, July 22, 1980, pp. 261-270.

103. Rylander, R.; Sorensen, S.; Alexandre, A.; and Gilbert, Ph.: Determinants for Aircraft Noise Annoyance—A Comparison Between French and Scandinavian Data. J. Sound & Vib., vol. 28, no. 1, May 8, 1973, pp. 15-21.

104. Rylander, R.; Sorensen, S.; and Berglund, K.: Re-analysis of Aircraft Noise Annoyance Data Against the dB(A) Peak Concept. J. Sound & Vib., vol. 36, no. 3, Oct. 8, 1974, pp. 399-406.

105. Rylander, R.; Bjorkman, M.; Ahrlin, U.; Sorensen, S.; and Berglund, K.: Aircraft Noise Annoyance Contours: Importance of Overflight Frequency and Noise Level. J. Sound & Vib., vol. 69, no. 4, Apr. 22, 1980, pp. 583-595.

106. Schultz, T. J.: Comments on "Determinants for Aircraft Noise Annoyance—A Comparison Between French and Scandinavian Data." J. Sound & Vib., vol. 33, no. 3, Apr. 8, 1974, pp. 369-371.

107. Robinson, D. W.: The Concept of Noise Pollution Level. NPL Aero Rep. Ac 38, British A.R.C., Mar. 1969.

108. Powell, Clemans A.: A Summation and Inhibition Model of Annoyance Response to Multiple Community Noise Sources. NASA TP-1479, 1979.

109. Beranek, L. L.; Kryter, K. D.; and Muller, L. N.: Reaction of People to Exterior Aircraft Noise. NOISE Contr., vol. 5, no. 5, Sept. 1959, pp. 23-31.

110. Borsky, Paul N.: Review of Community Response to Noise. Noise as a Public Health Problem, Jerry V. Tobias, Gerd Jansen, and W. Dixon Ward, eds., ASHA Rep. 10, American Speech-Language-Hearing Assoc., Apr. 1980, pp. 453-474.

111. Clevenson, Sherman A.; and Shepherd, William T., eds.: Time-of-Day Corrections to Aircraft Noise Metrics. NASA CP-2135, FAA-EE-80-3, 1980.

112. Bullen, R. B.; and Hede, A. J.: Assessment of Community Noise Exposure From Rifle Shooting. J. Sound & Vib., vol. 82, no. 1, May 8, 1982, pp. 29-37.

113. Powell, Clemans A.; and McCurdy, David A.: Effects of Repetition Rate and Impulsiveness of Simulated Helicopter Rotor Noise on Annoyance. NASA TP-1969, 1982.

114. Ollerhead, J. B.: Laboratory Studies of Scales for Measuring Helicopter Noise. NASA CR-3610, 1982.

115. Molino, John A.: Should Helicopter Noise Be Measured Differently From Other Aircraft Noise?—A Review of the Psychoacoustic Literature. NASA CR-3609, 1982.

Chapter 12
Guidelines for Assessment and Control of Noise

Introduction

There is an obvious need for guidelines and standards with respect to what intensities and durations of noise exposure should be considered incompatible with the health and well-being of the people exposed. These guidelines would presumably be useful for the zoning of land areas to avoid overexposure of people to environmental noise and for legislative-judicial adjudication of liabilities for possible damages to individuals and groups from exposure to noise. A number of guidelines for noise exposure have been promulgated over about the past 25 years by various governmental agencies for these purposes.

Partly because of the need for an integrated and consistent program of noise control for all elements of the government, the United States Environmental Protection Agency (EPA) was established in 1972. (See ref. 1.) In addition to previously issuing some specific documents regarding "safe" noise exposure limits, the EPA in 1980 joined several Federal operating and regulatory agencies in issuing guidelines for considering noise in land use planning and control. These government guideline documents, as well as some issued by non-government agencies, are examined in this chapter. Also, newly proposed guidelines for noise in residential areas, and the scientific basis for these guidelines, are presented.

Various Federal, State, and local ordinances and certification procedures have been developed to limit noise from specific pieces of machinery. By and large, although obviously related to various proposed guidelines, these "certification" ordinances and procedures are not concerned with the assessment of the effects of exposure to environmental noise and are not discussed herein.

EPA Protective Levels Documents

The purposes of the "Levels" reports (refs. 2 and 3) prepared by the EPA are the identification of "the levels of environmental noise the attainment and maintenance of which in defined areas under various conditions are requisite to protect the public health and welfare with an adequate margin of safety." (See Foreword, ref. 2.) As prescribed by Congress in the Noise Control Act of 1972 (ref. 1), this information was to be developed by the EPA without regard to the possible economic costs of achieving the noise levels identified, and, as such, does not necessarily constitute regulations or standards appropriate to given living areas in a community. However, these documents of the EPA presumably represent an interpretation of information as to the effects of the noise on people and methods for quantitatively describing noise environments that are fundamental to the preparation of regulations and codes for its control.

What is meant by "health and welfare" in the context of the instructions to the EPA by Congress is defined as follows (page 7 of ref. 2):

> The phrase "health and welfare" as used herein is defined as "complete physical, mental and social well-being and not merely the absence of disease and infirmity." This definition would take into account subclinical and subjective responses (*e.g.*, annoyance or other adverse psychological reactions) of the individual and the public. As will be discussed below, the available data demonstrate that the most serious clinical health and welfare effect caused by noise is interference with the ability to hear. Thus,

as used in this document, the phrase "health and welfare" will necessarily apply to those levels of noise that have been shown to interfere with the ability to hear.

The phrase "health and welfare" also includes personal comfort and well-being and the absence of mental anguish and annoyance. In fact, a considerable portion of the data available on the "health and welfare" effects of noise is expressed in terms of annoyance.

Noise interference

The EPA Levels Document identified for typical residential areas an average daily noise exposure level, over a year, of 55 L_{dn} as the maximum noise exposure permissible in typical residential housing areas for public health and welfare, with an adequate margin of safety. The only factual basis for that conclusion is presented as follows in Appendix D of reference 2.

> The levels of environmental noise . . . clearly identified in terms of the national public health and welfare . . . are the levels which are required to assure that speech communication in the home and outdoors is adequate . . . Lower levels may be desirable and appropriate for specific local situations.
>
> The level identified for the protection of speech communication is 45 dB within the home. Allowing for the 15-dB reduction in sound level between outdoors and indoors, this level becomes an outdoor day-night sound level of 60 dB (re 20 micropascals) for residential areas. For outdoor voice communication, the outdoor day-night level of 60 dB allows normal conversation at distances up to 2 meters with 95% sentence intelligibility.
>
> Although speech interference has been identified as the primary interference of noise with human activities, and as one of the primary reasons for adverse community reactions to noise and long-term annoyance, a margin of safety of 5 dB is applied to the maximum outdoor level to give adequate weight to all of these other adverse effects.
>
> Therefore, the outdoor day-night sound level identified for residential areas is a day-night sound level of 55 dB.

Table D-10 from reference 2 (here table 12.1) is a compilation of recommended maximum noise levels put forth over the years that support the limit set by the EPA. These recommendations are consistent with the data discussed in chapter 4 on masking of speech by noise.

Extrapolation from steady-state to intermittent noise

The limit of L_{dn} for a steady-state (constant level of intensity) noise of 60 dBA outdoors will, of course, generally provide for little or no speech masking inside a typical house with windows partly opened. However, an intermittent noise (such as aircraft noise), which has an outdoor L_{dn} of 60, may not provide a degree of quiet that will not interfere with speech communication in the home. This is true even with a 5 dB margin of safety, as 55 L_{dn}.

For example, 30 overflights of general aviation aircraft, shortly after take-off during the hours of 7 a.m. to 10 p.m., each of which makes a noise on a point on the ground that reaches a maximum level of about 83 dBA (an $L_{A,ex}$ typically of about 90), will give an outdoor L_{dn} of about 60. Allowing 15 dB for partly open-window house attenuation, the maximum level indoors would be about 68 dBA.

The level would be above 58 dBA for about 6 sec and above 45 dBA (about the threshold level of masking quiet, conversational speech) for about 15 to 20 sec. Accordingly, 30 times per day, for 15 to 20 sec each time, there could be some degree of interference with typical conversation or listening in the home or school from the aircraft noise. This is true if $L_{dn} \geq 55$ dBA and if windows are open.

This example and others given subsequently raise questions about the validity of the presumption by EPA. Their presumption is that minimal speech interference indoors exists for a noise of 55 L_{dn}, whether that noise is intermittent or steady state.

Incomplete speech intelligibility data

The EPA Levels Document (ref. 2) also presents information about speech interference from noise as measured by intelligibility test procedures, and estimates thereof, through application of the articulation index. As presented, this information (see figs. 12.1 and 12.2) seems to indicate less noise interference than

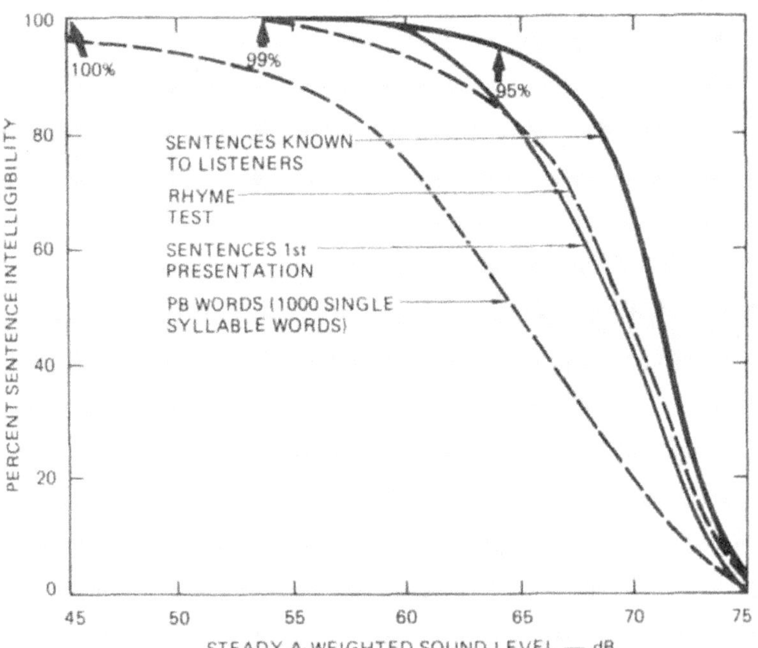

FIGURE 12.1. Speech intelligibility scores for trained talkers and listeners using several types of standard tests.

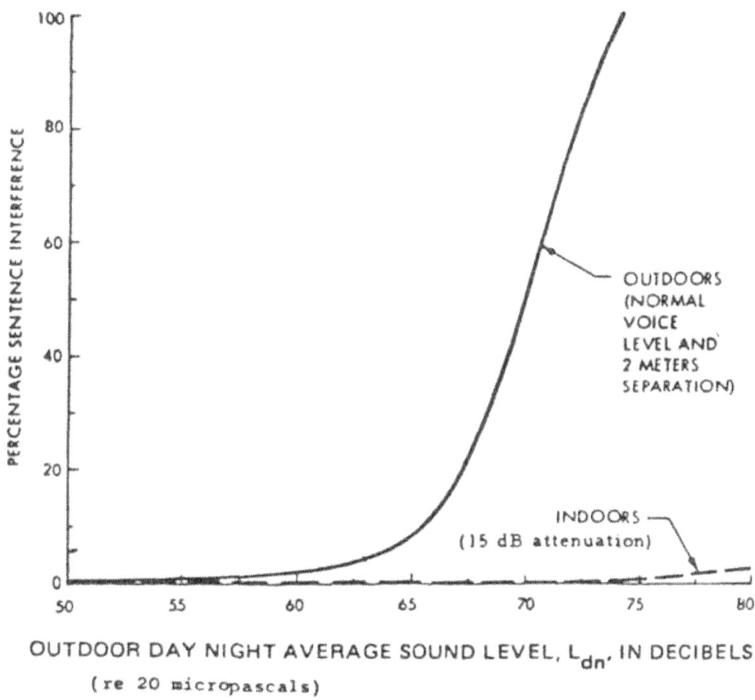

FIGURE 12.2 Maximum percentage interference with sentences as a function of day-night average sound level. (From ref. 2.)

implied by the architectural standards of 35 to 45 dBA shown in table 12.1. Indeed, figure 12.2 shows that not until a steady-state noise level of about 65 dBA is reached at the position of the listener is there a distinctly measurable effect (5 percent or less errors) on sentence intelligibility, when people are talking outdoors, and only 2 percent errors when people are indoors. It is assumed that indoors the listener is about 1 m from the talker, and outdoors about 2 m from the talker.)

For steady-state noise, the results in figure 12.2 can be expected. However, with more realistic, intermittent noise conditions, such as that from aircraft at $L_{dn} = 65$, the intelligibility of sentences at times would be 0 percent rather than the 95 to 98 percent intelligibility shown for $L_{dn} = 65$ in figure 12.2. In fact, 100 occurrences per day (7 a.m. to 10 p.m.) of aircraft noise with a peak outdoor level of 85 dBA typically has an L_{dn} of 65. In real life for this example, 100 times per day, 2 or 3 sentences in most speech conversations would be at or near 0 percent intelligibility (fig. 12.1) and would probably have adverse interruptive effects on the comprehension of even longer parts of the conversation. In brief, sentence intel-

TABLE 12.1

Prior Recommendations of Sound Levels in Various Spaces

[Data from table D-10 of ref. 2]

	Knudsen-Harris 1950 dB(A)	Beranek 1953 dB(A)	Beranek 1962 dB(A)	Lawrence 1962 dB(A)	Kosten-Van Os 1962 dB(A)	Ashrae 1967 dB(A)	Kryter 1970 dB(A)	
Resident:								
Home								
Bedroom	35-45	35	35-45	25	30	25-35	40	
Living Room	35-45	35		40	35	30-40	40	
Apartment	35-45		35-40	30		35-45	38	
Hotel	35-45		35-45	35-40		35-45	38	
	USSR 1971 dB(A)	**Beranek 1971 dB(A)**	**Doelle 1972 dB(A)**	**Wood 1972 dB(A)**	**Rettinger 1973 dB(A)**	**Sweden dB(A)**	**Switzerland 1970 dB(A)**	**Czechoslovakia 1967 dB(A)**
Resident:								
Home								
Bedroom	35	34-47	35-45	35	34-42	25	35-45	40
Living Room	35	38-47		40		25	35-45	40
Apartment		34-47			38-42		35-50	40
Hotel	35	34-47	35-54	30-40	42		35-50	40

ligibility at the time the intermittent aircraft noise occurs can be zero or near zero for values of L_{dn} that, for steady-state noise, would not significantly affect the hearing of normal speech.

In addition, omitted from the discussions of figures 12.1 and 12.2 are some relevant facts that would give the reader a better perspective of the speech intelligibility test data. In the first place, these figures were developed for the condition in which the speech reaching the listeners' ears averaged 65 dBA, with the talker using a constant level of effort. As discussed previously, in real life, much of the speech at the listeners' ears does not reach 65 dBA, whether that speech is from talkers, telephones, TV, or radio.

Also, the sentence intelligibility data in these figures are for trained talkers and listeners using short, familiar sentences. (The intelligibility tests were designed to obtain relative measures of performance with electronic communications equipment, and not as a measure of speech communication in real life.) Further, the EPA Levels Documents fail to mention that in 65-dBA steady-state noise, and with 65-dBA speech at the listeners' ears, even trained listeners and talkers would (as shown in fig. 12.1) misunderstand about 15 percent of multiple choice "rhyme" words, and 48 percent of single-syllable words (speech materials also used for speech intelligibility testing).

Speech communication does not usually consist of the short, known sentences, or the single-syllable tests shown in figure 12.1. However, the hearing of single-syllable words is not uncommon in everyday life, and the curve for the 1000 "PB" word tests is probably no less indicative of some of the interference effects to be expected from noise than is the sentence test-score function in figure 12.1. (See discussion on hearing handicap in chapter 8.)

The proof of what levels of noises under real-life living conditions are bothersome in these regards comes from studies in which people report disturbances and annoyance due to interference effects on speech from noises intruding into the home from outdoors. That an L_{dn} of 55 from intermittent noises, such as aircraft, does not provide complete protection from annoyance from speech communication interference is shown clearly from the attitude survey data discussed in chapter 11, as well as in the EPA Levels Documents themselves.

Condensed version of Levels Document

The purpose of the Condensed Version of the Levels Document (ref. 3) is to serve as an introduction, or a supplement, to the Levels Document (ref. 2). As such, most of the analysis of the Levels Document is applicable to the Condensed Version. In addition, comments follow regarding the somewhat exaggerated adverse picture given in both documents with respect to non-aircraft noises, especially street noise, and implications of that picture for the assessment of noise impact on communities.

Figure 12.3 shows annoyance from aircraft noise. This figure does not appear as such in the Levels Document proper, but is a version of Figure D-13 of the

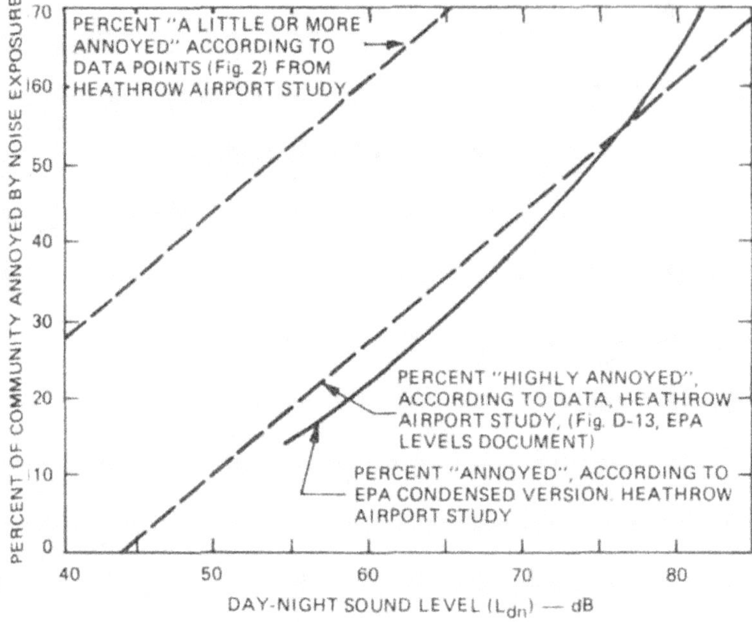

FIGURE 12.3. Percentage of population annoyed by community noise (Heathrow Airport study). Dashed curves are added herein to actual EPA figure 12 (ref. 3).

Levels Document. However, in this new version in the Condensed Levels Document, percent "annoyed" is shifted about 3 dB to the right (indicating less annoyance) in the critical region of 55 to 65 L_{dn}. Of more importance is the use of the word "annoyed" in the Condensed Version in place of "highly annoyed," as properly labeled in the first Levels Document. (See chapter 11.)

Indoor sounds versus noise

Some of the figures in the Levels Document and the discussion in the Condensed Version pertaining to indoor noise can be misleading by suggesting that there is more "noise" present in everyday conditions than is often the case. Probably a good share of what is measured as being "noise" (speech, music, radio, television, etc.) in Figure 2 and Figures B-2 through B-7 of the Levels Document, and figures 7 and 9 of the Condensed Version, is really wanted sounds.

Outdoor noise from different sources

Data are lacking to show that the L_{dn} concept can be applied to any and all noises in a residential area, as is promoted in the EPA Levels documents. To the contrary, as discussed in detail in chapter 11, it is clear that the impact of urban traffic noise upon community annoyance is significantly (by about 10 dB) exaggerated relative to the impact from aircraft noise of the same L_{dn}.

Summary of EPA Levels Documents

In summary, with respect to the assessment of the impact of noise in the general living environment, the EPA Levels Documents significantly underestimate the adverse interference effects of speech. These documents overestimate to some extent the amount of noise present in the general environment and greatly overestimate the impact on people of general ground vehicle noise compared with the impact of aircraft noise on people.

Guidelines for Limits on Environmental Noise

Unlike the EPA documents, which attempted to identify "safe levels" of environmental noise exposure, the guidelines for noise control issued by governmental aviation, industry, and labor agencies identify acceptable levels of noise exposure. These levels are not only considered acceptable to people in actual living and work areas, but are also economical and practical.

FAA-DOD

In 1964, the Federal Aviation Administration (FAA), part of the Department of Transportation (DOT), and the Department of Defense (DOD) published similar documents to be used for land use planning with respect to aircraft noise. (See refs. 4 and 5.) On the basis of case histories involving aircraft noise problems at civilian and military airports, a relationship was deduced between composite noise exposure (equivalent L_{dn}) for aircraft noise environments and the complaint behavior to be expected in typical residential neighborhoods. (See ref. 6.). This relationship is shown in table 12.2. Some of the data on which this deduction is based are shown in figure 11.21.

The data in figure 11.21 show that the basic criterion for the difference between Zone 1 (L_{dn} less than 65) and Zone 2 (L_{dn} above 65) is that complaint behavior in Zone 2 includes legal actions and threats of legal action because of the aircraft noise. In the right-hand segment of figure 11.21 are some additional data on legal actions or threats of legal action that were not available during the preparation of the FAA-DOD guideline.

HUD

In 1971, the Department of Housing and Urban Development (HUD) published noise assessment guidelines for land use planning. (See ref. 7.) Somewhat different terminologies than those employed by FAA-DOD were employed to define noise zones of differing degrees of compatibility for aircraft noise

TABLE 12.2

Chart for Estimating Response of Communities Exposed to Aircraft Noise

[Data from refs. 4 and 5]

Noise rating	Zone	Description of expected response
Less than 65 L_{dn} 100 CNR	1	Essentially no complaints would be expected. The noise may, however, interfere occasionally with certain activities of the residents.
65 to 80 L_{dn} 100 to 115 CNR	2	Individuals may complain, perhaps vigorously. Concerted group action is possible.
Greater than 80 L_{dn} 115 CNR	3	Individual reactions would likely include repeated, vigorous complaints. Concerted group action might be expected.

measured in terms of NEF, CNR, or L_{dn}. (See table 12.3.) Table 12.3 shows that in areas greater than L_{dn} = 65, typical residential usage is considered as "normally unacceptable," and below as "normally acceptable."

The procedures in the 1971 HUD document for the assessment of noise from ground vehicles is more complex than that used for aircraft noise. The ground vehicle procedure does not use a composite noise measurement unit, such as L_{dn}, and includes some on-site listening tests and other analysis procedures. The 1971 HUD procedures were superceded by the publication by HUD of environmental noise criteria and standards in 1979 (ref. 8) that use the L_{dn} unit of measurement for aircraft as well as for ground-vehicle noise. (See table 12.4.) New noise assessment guidelines for HUD were published in 1980 (ref. 9) that provide a means of estimating L_{dn} for ground vehicular noise. These guidelines take various acoustical factors, such as sound barriers and distance to source into account. They also specify certain listening tests.

Joint Federal agencies

In order to put various Federal agency policy and guidance packages into perspective and to encourage local land use that is compatible with various noise levels, a number of Federal agencies have issued a set of environmental noise guidelines for land usage, along with some supporting data and comments. (See ref. 10.) Table 12.5 shows suggested land compatibility for residential noise zones from 55 to 85 L_{dn}. Compatibility guidelines suggested for a variety of nonresidential areas, as proposed in the Federal Interagencies Guidelines, are presented subsequently.

"Compatibility" for noise in residential areas (L_{dn} of 55 to 65) should be interpreted by the user of the guidelines with reservations because some Federal

TABLE 12.3

Site Exposure to Aircraft Noise

[From ref. 7]

Distance from site to the center of the area covered by the principal runways	Acceptability category
Outside the L_{dn} = 65 (NEF-30, CNR-100) contour at a distance greater than or equal to the distance between the contours L_{dn} = 65 and L_{dn} = 75	Clearly acceptable
Outside the L_{dn} = 65 contour, at a distance less than the distance between the L_{dn} = 65 and L_{dn} = 75 contours	Normally acceptable
Between the L_{dn} = 65 and L_{dn} = 75 contours	Normally unacceptable
Within the L_{dn} = 75 contour	Clearly unacceptable

TABLE 12.4

Site Acceptability Standards

[From ref. 8]

	Day-night average sound level, dB	Special approvals and requirements
Acceptable	Not exceeding 65 dB (1)	None
Normally unacceptable	Above 65 dB but not exceeding 75 dB	Special approvals (2) Environmental review (3) Attenuation (4)
Unacceptable	Above 75 dB	Special approvals (2) Environmental review (3) Attenuation (5)

[1] *Acceptable threshold may be shifted to 70 dB in special circumstances pursuant to Section 51.105(a).*

[2] *See Section 51.104(b) for requirements.*

[3] *See Section 51.104(b) for requirements.*

[4] *5-dB additional attenuation required for sites above 65 dB but not exceeding 70 dB, and 10-dB additional attenuation required for sites above 79 dB but not exceeding 75 dB. (See section 51.104(a).)*

[5] *Attenuation measures to be submitted to the Assistant Secretary for CPD for approval on a case-by-case basis.*

TABLE 12.5

Suggested Land Use Compatability for Residential Areas

[Data from ref. 10]

Key

SLUCM	Standard Land Use Coding Manual.
Y (yes)	Land use and related structures are compatible without restrictions.
N (no)	Land use and related structures are not compatible and should be prohibited.
25, 30, or 35	Land use and related structures are generally compatible; measures to achieve NLR of 25, 30, or 35 must be incorporated into design and construction of structure.

Land use		Noise zones/DNL levels in L_{dn}						
SLUCM no.	Name	A 0–55	B 55–65	C-1 65–70	C-2 70–75	D-1 75–80	D-2 80–85	D-3 85+
10	**Residential**							
11	Household units							
11.11	Single units—detached	Y[a]	Y[a]	25[b]	30[b]	N	N	N
11.12	Single units—semidetached	Y	Y[a]	25[b]	30[b]	N	N	N
11.13	Single units—attached row	Y	Y[a]	25[b]	30[b]	N	N	N
11.21	Two units—side-by-side	Y	Y[a]	25[b]	30[b]	N	N	N
11.22	Two units—one above the other	Y	Y[a]	25[b]	30[b]	N	N	N
11.31	Apartments—walk up	Y	Y[a]	25[b]	30[b]	N	N	N
11.32	Apartments—elevator	Y	Y[a]	25[b]	30[b]	N	N	N
12	Group quarters	Y	Y[a]	25[b]	30[b]	N	N	N
13	Residential hotels	Y	Y[a]	25[b]	30[b]	N	N	N
14	Mobile home parks or courts	Y	Y[a]	N	N	N	N	N
15	Transient lodgings	Y	Y[a]	25[b]	30[b]	35[b]	N	N
16	Other residential	Y	Y[a]	25[b]	30[b]	N	N	N

[a] *The designation of these uses as "compatible" in this zone reflects individual Federal agencies' consideration of general cost and feasibility factors as well as past community experiences and program objectives. Localities, when evaluating the application of these guidelines to specific situations, may have different concerns or goals to consider.*

[b] *Although local conditions may require residential use, it is discouraged in C-1 and strongly discouraged in C-2. The absence of viable alternative development options should be determined and an evaluation indicating that a demonstrated community need for residential use would not be met if development were prohibited in these zones should be conducted prior to approvals. Where the community determines that residential uses must be allowed, measures to achieve outdoor to indoor noise level reduction (NLR) of at least 25 dB (Zone C-1) and 30 dB (Zone C-2) should be incorporated into building codes and be considered in individual approvals. Normal construction can be expected to provide a NLR of 20 dB, thus the reduction requirements are often stated as 5, 10, or 15 dB over standard construction and normally assume mechanical ventilation and closed windows year round. Additional consideration should be given to modifying NLR levels based on peak noise levels. NLR criteria will not eliminate outdoor noise problems. However, building location and site planning, design and use of berms and barriers can help mitigate outdoor noise exposure particularly from ground level sources. Measures that reduce noise at a site should be used wherever practical in preference to measures which only protect interior spaces.*

agencies define compatibility not solely with respect to the effects of the noise but also take economic costs and technical feasibility factors of noise control into account (footnote to table 12.5). This table of land use compatibility and a similar table published by the FAA (ref. 11) represent only recommendations to local authorities, and do not relieve the local agencies of responsibilities for determining acceptable and permissible land uses: "The designations contained in this table do not constitute a Federal determination that any use of land covered by the program is acceptable or unacceptable under Federal, State, or local law. This responsibility for determining the acceptable and permissible land uses remains with the local authorities. FAA determinations under part 150 are not intended to substitute federally determined land uses for those determined to be appropriate by local authorities in response to locally determined needs and values in achieving noise-compatible land uses." (Table 2 of ref. 11.)

In the preparation of their guidelines, the Federal agencies utilized some research data and findings concerned with the effects on people of environmental noise in residential areas. These data and findings are presented in table 12.6. Table 12.6 is unfortunately deficient in several respects:

1. It uses concepts pertaining to speech interference effects of noise that underestimate the interference effects for the assessment of intermittent noises such as those from aircraft, autos, etc. (See section entitled "EPA Protective Levels Documents.")

2. It uses a relationship between "percent highly annoyed" and L_{dn} that is somewhat suspect (chapter 11).

3. The "NOTE" in table 12.6 regarding noise as a stress-related health factor does not take cognizance of important published findings on this matter (chapter 10).

4. Important differences in regard to the relationship between aircraft noise and ground vehicle noise measured in L_{dn} and their effects on annoyance are not referred to or considered (chapter 11).

The speech interference shown in table 12.6 is not true for aircraft and other intermittent noises, which interfere more than shown with everyday speech. Also, contrary to "NOTE" in table 12.6, adverse health effects from exposure to aircraft noise have been recently quantified in environments of $L_{dn} = 60$ dB and greater. These health effects and the annoyance, community reaction, and attitudes cited in table 12.6 are for aircraft noise studies in nothern cities, such as London and New York, and are not applicable to noise in warm climates or noise from ground vehicles.

ANSI

In 1980, the American National Standards Institute (ANSI) issued a standard on sound level descriptions for the determination of compatible land use with respect to noise. (See ref. 12.) That document includes, as a matter of information, figure 12.4. Figure 12.4 is basically consistent with the recommendation of the Federal Interagency Guidelines, and presumably (no references to supporting research data are given in the ANSI document) bears the same critical comments given above in regard to the Federal Interagency document.

CHABA

This document from the Committee on Hearing, Bioacoustics, and Biomechanics (CHABA) of the National Research Council (ref. 13) is primarily concerned with steps to be followed in the preparation of so-called environmental impact reports about exposure to noise and vibration. Although no specific recommendations with respect to noise and vibration levels compatible with land usage are given, some information is given in appendices on the effects of noise and vibration on people.

The sections on noise effects are open to some criticism in that the aforementioned specious description of noise interference effects of intermittent noise developed by EPA, and a questionably accurate annoyance response curve for use with either aircraft or ground vehicle noises, are included. Also, except for noise-induced hearing loss, data on the adverse effects of environmental noise on health are not discussed.

California Department of Aeronautics

In 1970, the California Department of Aeronautics (CDA) developed a noise regulation for California airports. (See ref. 14.) In terms of methodology of measuring aircraft noise, this regulation is similar to the L_{dn} procedure used by the EPA and in the recent HUD, Federal Interagency, ANSI, and CHABA guidelines discussed previously. It differs to some extent from these other guidelines in the way in which an annualized, average-composite, daily-noise-exposure unit is calculated. In the CDA procedure, in addition to the 10-dB nighttime penalty, as used with L_{dn}, a 5-dB penalty for aircraft noises occurring between the hours of 7 and 10 p.m. is employed. The result is called the community noise equivalent level (CNEL). For this reason, CNEL has a numerical value of 1 to 2 dB higher than the L_{dn} value for the same aircraft noise environment (assuming about 15 percent of the operations occur from 7 to 10 p.m.).

The CDA regulation states that: (1) It is established that with respect to speech and sleep effects, aircraft noise of 65 CNEL is acceptable to a reasonable person residing in the vicinity of an airport in California. To quote from Section 5005:

> The level of noise acceptable to a reasonable person residing in the vicinity of an airport is established as a community noise equivalent level (CNEL) value of 65 dB for purposes of these regulations. This criterion level has been chosen for reasonable persons residing in urban residential areas where houses are of typical California construc-

TABLE 12.6

Effects of Noise on People In Residential Areas

[From ref. 10]

Effects[1]	Hearing loss	Speech interference		Annoyance[2]		
		Indoor	Outdoor			
Day-night average sound level, dB	Qualitative description	Percent of sentence intelligibility	Distance in meters for 95% sentence intelligibility	Percent of population highly annoyed[3]	Average community reaction[4]	General community attitude towards area
75 and above	May begin to occur	98	0.5	37	Very severe	Noise is likely to be the most important of all adverse aspects of the community environment.
70	Will not likely occur	99	0.9	25	Severe	Noise is one of the most important adverse aspects of the community environment.
65	Will not occur	100	1.5	15	Significant	Noise is one of the important adverse aspects of the community environment.
60	Will not occur	100	2.0	9	Moderate to slight	Noise may be considered an adverse aspect of the community environment.

55 and below	Will not occur	100	3.5	4		Noise considered no more important than various other environmental factors.

[1] *"Speech interference" data are drawn from the following tables in EPA's "Levels Document" (ref. 3): Table 3, Fig. D-1, Fig. D-2, Fig. D-3. All other data from reference 13.*

[2] *Depends on attitudes and other factors.*

[3] *The percentages of people reporting annoyance to lesser extents are higher in each case. An unknown small percentage of people will report being "highly annoyed" even in the quietest surroundings. One reason is the difficulty all people have in integrating annoyance over a very long time.*

[4] *Attitudes or other non-acoustic factors can modify this. Noise at low levels can still be an important problem, particularly when it intrudes into a quiet environment.*

NOTE: *Research implicates noise as a factor producing stress-related health effects such as heart disease, high-blood pressure and stroke, ulcers and other digestive disorders. The relationship between noise and these effects, however, have not as yet been quantified.*

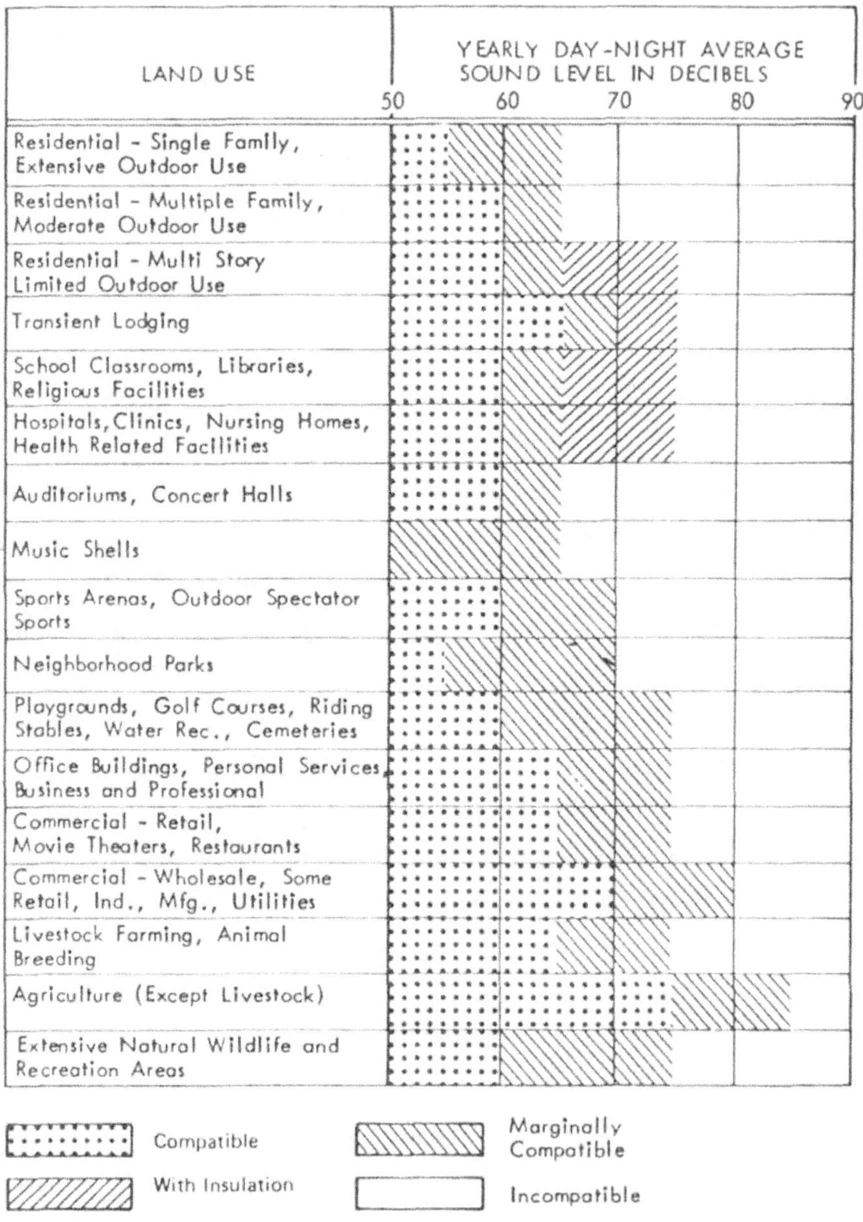

FIGURE 12.4. Land use compatibility with yearly day-night average sound level at a site for buildings as commonly constructed. (From ref. 12.)

tion and may have windows partially open. It has been selected with reference to speech, sleep, and community reaction.

and (2) A residential area with aircraft noise of less than 65 CNEL is to be considered as a "no-noise impact area."

These definitions presented and used by CDA are, however, not consistent with data on the effects of noise on people and communities discussed in a technical report (ref. 15) prepared as a basis for the CDA regulation, and data and analyses thereof since the preparation of the CDA regulation show these definitions to be even more incorrect. For example, in the supporting CDA technical report it is shown that (as also reported in chapter 11) in the Heathrow Airport studies at NNI levels of aircraft noise equivalent to CNEL's of about 65, about 50 percent of the people were disturbed because of speech and TV sound interference, and over 40 percent were disturbed because of arousal from sleep. Further, about 15 percent of the people exposed to an equivalent 65 CNEL were highly annoyed. These data hardly establish that this exposure level of aircraft noise is acceptable to a reasonable person residing in California, or that a CNEL of 64 represents no noise impact.

The technical report for the CDA regulation states (p. 57):

> This decision is made with the realization that further human factors research (when a way is found to establish the effect of flight frequency) may show the CNEL limit of 65 to be somewhat too high. When that occurs, the CNEL limits in the noise standard should be lowered accordingly. We therefore strongly recommend that: The CNEL 65 limit should be periodically reviewed by the State with a view to the possible necessity of reducing the limit in light of any new human factors research results which may become available.
>
> *It is our opinion that these reviews should be conducted at an interval of five (5)* years as a maximum, to take advantage of advances in research in the effects of noise on people and advances in the technology of aircraft noise reduction.

To the best of the author's knowledge, no review of the CDA regulation or supporting technical data has been made by the California Department of Aeronautics since their original publications in 1971.

Summary of reviewed guidelines and their limitations

Table 12.7 summarizes the maximum noise exposure levels (annualized, daily average) commensurate with their acceptability-compatibility in typical residential neighborhoods as promulgated in the guidelines reviewed above. It is clear from the material presented in chapters 10 and 11 that each of the guideline limits cited in table 12.7 needs to be challenged on two major accounts: (1) these guidelines recommend a single L_{dn} or CNEL number for the boundary between compatible and incompatible land usage that is the same regardless of climate or source of noise. Research data show that acoustical factors differentially affect the effective noise dosages received in homes and yards from aircraft noise, compared with street traffic noise, and in warm, as compared with colder climates;

TABLE 12.7

Exposure Level in Annualized Daily Average L_{dn} or CNEL for Acceptability-Compatibility of Exterior Environmental Noise in Typical Residential Houses With Windows and Doors Open

[15-dB exterior sound attenuation with windows and doors partly open, 28 dB when closed]

Agency	Source of noise	Criteria	Level limit
EPA, 1972	Any	Health and well-being, with 5-dB margin of safety	55 L_{dn}
FAA-DOD, 1964	Aircraft	Essentially no complaints	<65 L_{dn}
		Vigorous complaints, group action	>65 L_{dn}
HUD, 1971	Aircraft	Normally acceptable	<65 L_{dn}
	(Ground vehicle treated separately)	Normally unacceptable	>65 L_{dn}
HUD, 1979-1980	Aircraft and ground vehicles	Acceptable	<65 L_{dn}
		Normally unacceptable	>65 L_{dn}
Joint federal agencies, 1980	Aircraft and ground vehicles	Compatible	55 L_{dn}
		Marginally[1] compatible	55 − 65 L_{dn}
		Incompatible	>65 L_{dn}
CDA, 1971	Aircraft	Acceptable	<65 CNEL
		Unacceptable	>65 CNEL
ANSI, 1980	Aircraft and ground vehicles	Compatible	<55 L_{dn}
		Marginally compatible	55 − 65 L_{dn}
		Incompatible	>65 L_{dn}

[1] *Reflects cost and feasibility factors.*

and (2) some research findings pertaining to the impact of aircraft noise on health, property values and annoyance were not adequately considered in the preparation of these guidelines.

It also appears that the recommendations of these Federal guidelines with respect to aircraft noise are sometimes rationalized on the grounds that noise from ground vehicles, or other sources, is generally of the order of $L_{dn} = 65$. Accordingly, it would be unjustified to reduce aircraft noise beyond that level (See refs. 16 and 17.) This argument is not true, however, with respect to noise in residential areas.

Table 12.8 (from a nationwide noise survey made by Galloway *et al.* (ref. 18)) shows that of the 71 sites of single-family or small apartment dwellings studied, only 6 had an L_{dn} from street traffic noise of 65 or greater (65.2, 65.9, 66.1, 67.3, 67.8, and 71.2). The other 65 sites had L_{dn}'s ranging from 51.0, and a median L_{dn} of 61.0. More importantly, as discussed in chapter 11, the impact of this noise on residents is some 10 dB less effective than is the noise from aircraft.

TABLE 12.8

L_{dn} at Single- or Two-Family Dwellings from Urban Street Traffic (Cars, Trucks, Buses) in Various U.S. Cities

[Microphone located avg. 8 ft above ground and 25 ft from street curb; median all sites, 61.0; data from ref. 18]

Site	L_{dn}, dB	Site	L_{dn}, dB	Site	L_{dn}, dB
Atlanta:		Denver—Continued		St. Louis:	
401	63.8	1107	59.0	1201	58.5
404	60.7	1109	59.3	1203	62.9
406	67.3	1110	56.6	1204	56.2
Boston:		1112	61.3	1206	60.0
0001	61.2	Kansas City:		1208	61.0
0003	59.6	1301	60.1	1209	61.5
0004	59.6	1302	53.2	1210	64.2
0005	57.0	1304	61.4	San Francisco:	
0006	67.8	Los Angeles:		1005	61.9
0007	61.7	1601	56.4	1007	60.5
0008	65.2	1602	57.1	1008	61.2
Chicago:		1603	56.1	1010	63.8
502	71.2	1604	61.0	1011	61.1
503	69.6	1605	58.6	1012	61.9
505	59.0	1606	59.2	Seattle:	
506	64.4	1607	58.9	1501	55.6
Dallas:		1608	55.8	1502	55.1
1401	65.9	Miami:		1503	52.4
1402	57.8	0601	66.1	1504	53.2
1403	61.1	0602	58.1	1505	51.0
1404	61.1	0603	63.2	1506	54.7
1405	56.3	0604	64.4	Washington, D.C.:	
1406	61.6	0606	60.9	0102	63.0
Denver:		Pittsburgh:		0103	62.4
1101	62.6	0301	58.1	0104	64.1
1103	58.3	0302	59.1	0106	64.6
1104	60.7	0303	61.9		
1106	57.3	0305	62.2		

Galloway (in his testimony before the U.S. District Court) and others also suggest that aircraft noise limits should not be less than $L_{dn} = 65$ because it is difficult to "verify" lesser levels in real life. This is a specious basis for such a conclusion, in that aircraft noise contours as low as 55 are readily calculated by means of computer programs (ref. 19) developed by the FAA. (See fig. 11.24 for an example.) These methods of calculating aircraft noise around airports have been developed from aircraft noise certification tests (ref. 20) and validated by extensive noise monitoring procedures around airports. Further, the peak sound pressure levels (around 75 to 80 dBA) of the aircraft noise occurrences that contribute significantly to L_{dn} values of 55 or greater in typical residential areas are readily measured by standard sound level measurement procedure (ambient noise levels of around 45 dBA and peak automobile noise of around 70 dBA).

Criteria for Noise Exposure in Residential Areas

In succeeding paragraphs, the relations between amounts of environmental noise and various criteria effects are summarized. On the basis of these criteria effects and noise dosage relations, new guidelines for assessing the compatibility of aircraft and street-traffic noise with residential land usage are proposed.

Annoyance scale criterion

Feelings of annoyance obviously have value as a criterion effect that could be used as a basis for defining the compatibility of noise with residential areas. In addition, annoyance attitude scale data provide a quantitative means of equating the relative impact of different exposure and cultural conditions and should, if properly "calibrated," have value for predicting the magnitudes of other adverse criteria effects under a given condition.

Complaint behavior criterion

As discussed in chapter 11, the quantification of complaint behavior with respect to noise exposure is hindered by the fact that for a variety of socio-economic and other environmental reasons, complaints and legal action, especially the lack of legal action, are highly variable among different communities exposed to similar noise environments. Nevertheless, complaints, including threats of legal action and the like, can be meaningful indicators that a noise environment is having an adverse effect upon people.

Objective effects criteria

The credibility of subjective annoyance scale and complaint behavior data would obviously benefit from supportive noise effects data that are objectively related to noise exposure. Two such objective effects which have significant practical importance are the effects of noise on residential housing values and on health. These data are called objective in that they are not subjective feelings of annoyance on the part of the people exposed to the noise, but are independent data taken from real estate appraisers and records, physicians and medical records, *etc*.

Proposed Guidelines for Noise in Residential Areas

The adjudication of a depreciation effect, if any, of a particular noise environment, on housing values is a matter of specific appraisals. However, the fact that research data shows a progressive increase in housing depreciation with increased exposure, in L_{dn}, to aircraft noise lends some objectivity to the interpretation of annoyance survey data and lends credibility to any proposed guidelines for limiting such noise.

More significant is the quantification in the Knipschild studies (discussed in chapter 10) of the effects of exposure to aircraft noise in terms of L_{dn} upon health. (See refs. 21 to 24.) A particular strength of these studies is that all the observed adverse effects increase more or less monotonically with increases in aircraft L_{dn}, and in ways consistent with all subjective and objective criteria data that have been collected in real-life and laboratory studies of the effects of aircraft noise on people.

The data for the major adverse criterion effects of noise summarized above are plotted on a common set of coordinates in figure 12.5. The effects on health are in terms of hypertension and percentage rate of physician contacts from figures 10.59 and 10.60; the curve showing percent depreciation in house prices is taken from figure 11.22, and the curve for percent of people annoyed is taken from figure 11.4(a). The locations of the lines in figure 12.5, indicating degrees of complaint behavior, are based on figures 11.20 and 11.21.

The generality of the data and relations in figure 12.5 for noises from sources other than aircraft and ground vehicles is difficult to establish. Some case studies (see fig. 11.20 and table 11.10) and general theory suggest that the effects-dose relations shown in figure 12.5 are somewhat similar for other exterior noises (such as that from industry, heavy machinery, gunnery ranges, *etc.*) when the noise is measured in terms of L_{dn} and equated to aircraft or ground vehicle noise with respect to acoustic barriers and house intrusiveness. Exceptions are predominately low frequency noises; such noises are somewhat louder than reflected in the weighting given to low frequencies by the dBA unit used with L_{dn}, and such noises can also cause house vibrations that cause annoyance not predicted by dBA or L_{dn}. (See chapters 5 and 11.)

Thresholds of marginal compatibility and noncompatibility

It is proposed that the limits of exposure to noise to be considered as compatible with residential living be based on the operational definition that the threshold level of marginal compatibility for a given adverse effect of noise is the lowest level at which the effect becomes measurable, or derivable, by extrapolation from results obtained at higher levels of exposure. The exposure level at which the noise becomes practically not compatible with residential living is taken to be 10 dB above the threshold of compatibility. While the matter of prac-

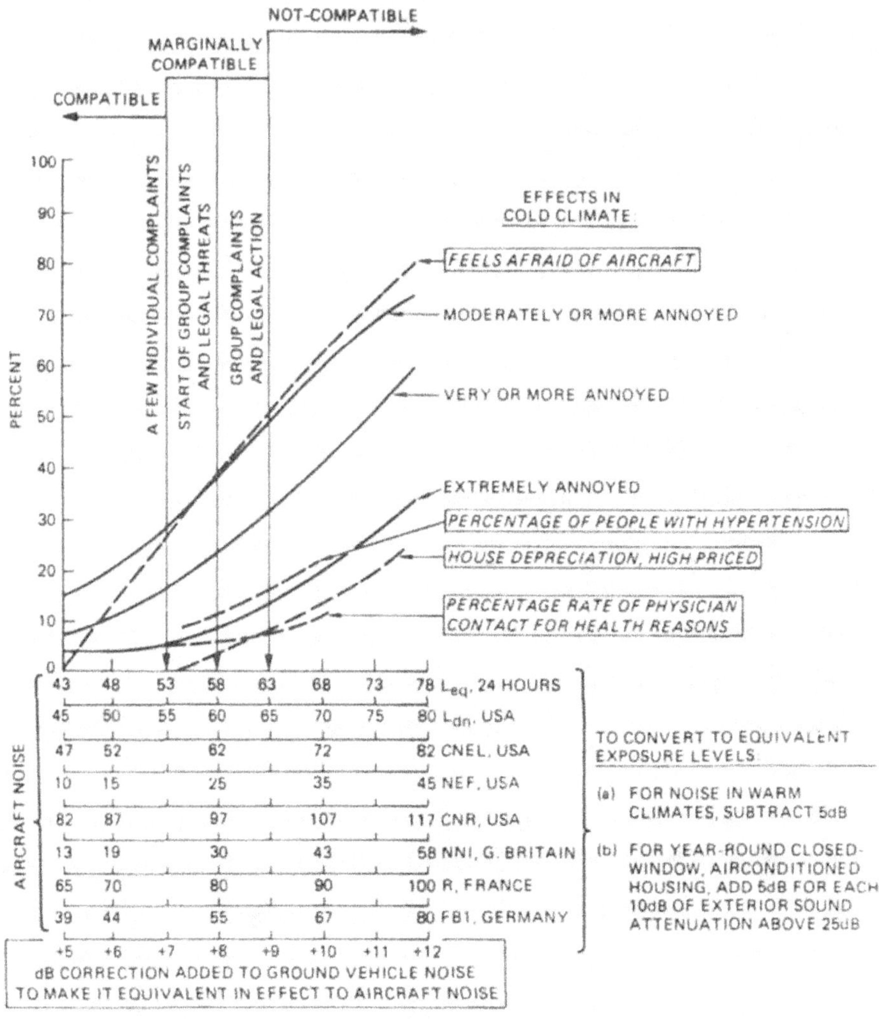

FIGURE 12.5. Graph showing, as a function of exposure level, effects of aircraft noise on annoyance, health, housing values, and complaint activity. Also indicated are suggested compatible and not compatible threshold levels for residential living and corrections to be applied for use with ground vehicle noise and in warmer (*e.g.*, southern U.S.) climates. Noise dosage is expressed in composite daily units used in the U.S. and in some European countries (see chapter 11).

ticality can be a matter of judgment and controversy, the 10-dB difference between the threshold of marginal compatibility and the practical level of noncompatibility is that traditionally used for this purpose in the various Federal and other guidelines discussed previously. (See table 12.7.)

Compatibility as defined by the start of "complaints," "extreme annoyance," "physician contacts for health problems," and "high-priced house depreciation" all occur at $L_{dn} \approx 55$ for aircraft noise in colder climates. However, a number of people experience various degrees of annoyance below extreme annoyance from this noise at $L_{dn} = 55$ and lower. Even so, setting the threshold of marginal compatibility for the aircraft noise specified at $L_{dn} = 55$ is defensible on the grounds that the amount of annoyance at $L_{dn} = 55$ for aircraft noise is perhaps typical for other environmental conditions present in average residential areas. (This subject is discussed more fully in a subsequent section.)

House attenuation factor

It is indicated in figure 12.5 that significantly more exterior noise is considered compatible and causes less adverse effects when the attenuation of exterior sound afforded by a house is greater than that present in typical houses. Indeed, research data (see chapter 11) show that when the windows and doors of houses are kept closed, annoyance and complaints about aircraft noise are reduced. Accordingly, a progressive allowance is generally made, as shown in most of the guidelines discussed previously, in the amount of exterior noise considered compatible as the exterior sound attenuation properties of the house, or other structure, are increased. (See table 12.5 and fig. 12.4.) Indeed, as discussed in chapter 11, it is the noise at the position of the listener that really matters in causing disturbances and annoyance.

However, in many climates, a significant part of residential living is done outdoors in yards, patios, and parks. For this reason it is likely that reducing the exterior sound reaching the person indoors by 10 dB, for example, from that present with windows and doors partly open will not reduce the overall annoyance that is experienced by an amount fully equivalent to a 10-dB reduction of the noise reaching indoors. Rather, it is hypothesized that for present and practical purposes the reduction is equivalent to only 5 dB on a year-round, annualized basis. It of course follows that, for some residential apartments and office buildings, where outdoor activities are minimal, the compatibility of the exterior noise should be almost totally a function of its indoor (structure attenuated) level.

Source differences

One of the major findings of research conducted on annoyance in residential areas from aircraft and from street noises is that aircraft noise of a given L_{dn} as typically measured (in the front yards of houses) is a much greater cause of annoyance (see chapter 11) and apparently of adverse health effects (see chapter 10) than is ground vehicle traffic noise of the same L_{dn} value. This difference in equivalent effective exposure level averages about 10 dB (ranging from 7 dB at L_{dn} = 55 to 12 dB at L_{dn} = 80), as shown by the bottom abscissa on figure 12.5. The differences can be attributed to accelerated growth in these adverse effects as a function of exposure level and to the greater acoustic attenuation, by house structures, of the noise from cars and trucks reaching the backyards and rooms of the houses compared with aircraft noise coming from overhead. In addition to more barrier attenuation, the noise from nearby street vehicles (25-50 feet) is significantly decreased between the front and rear yards of houses, but the noise from aircraft is essentially uniform over an entire house and yard's areas.

Climate factor

In figure 12.5, the increased impact of the noise from aircraft is specified as being equivalent to 5 dB in noise level between "warm" and "cold" climate cities. This 5 dB is an apparent minimum, as shown by the data on figures 11.10, 11.31, 11.32, and 11.33.

The definition of warm versus cold is in terms of general knowledge concerning the locations and climates of the various cities involved in the subject studies (London, Amsterdam, Munich, Zurich, Copenhagen, New York, Chicago, *etc.* versus some cities in California, see fig. 11.33) and between seasons of the year (summer versus spring, fall, and winter, see figs. 11.31 and 11.32). Clearly, it would be desirable to develop a method for better quantifying the underlying variable, such as the probable average number of days per year of "open-window, outdoor" living in a given city or locale.

Relative compatibility of other environmental conditions

Some data are available that show how much, on a comparative basis, annoyance from environmental conditions other than noise is being experienced by people. Comparing the magnitude of the differences in relative disturbance from aircraft noise versus nonaircraft noise factors around London Heathrow Airport with the magnitude of those differences around Orange County Airport in Southern California may be instructive. It is shown in the left-hand graphs of figure 12.6 that aircraft noise is the cause of much disturbance around Heathrow. However, as shown in the upper right-hand graph, it does not become the dominant factor for disliking a neighborhood until the level of exposure reaches about L_{dn} = 70; at L_{dn} = 65, it is the worst factor except one. However, in the 10-mile area around Heathrow Airport, where the surveys were conducted, only about 35 percent of the people liked their neighborhood or found it attractive, and only about 25 percent liked the people living there (ref. 25). That would seem to indicate that the local socioeconomic mix was not generally well liked by the residents.

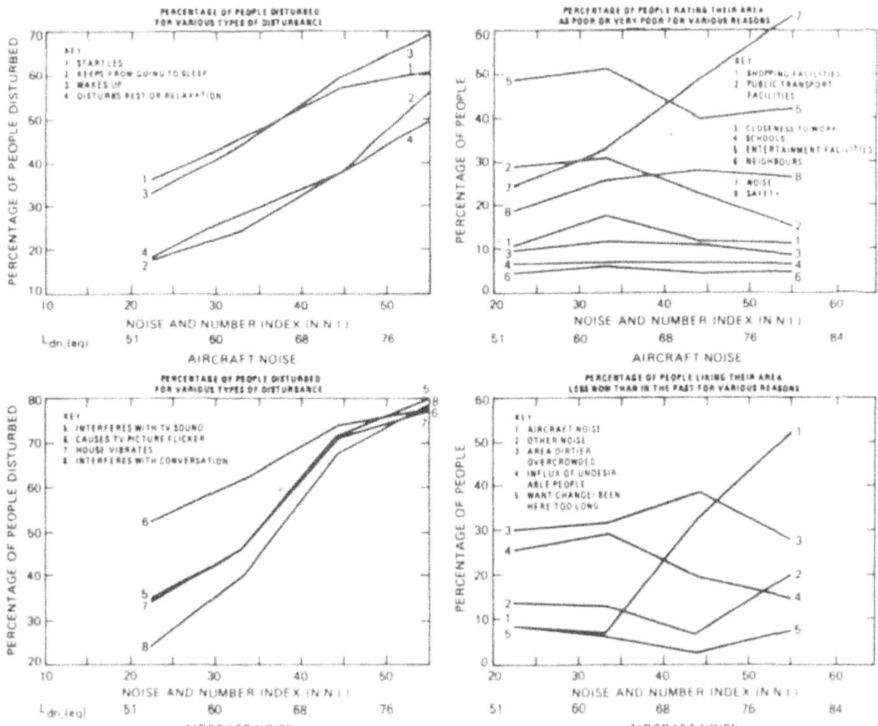

FIGURE 12.6. Graphs on left: Percentage of people disturbed for various reasons by aircraft noise at different values of L_{dn} and NNI. Graphs on right: Percentage of people rating poor or disliking various conditions of neighborhoods with different levels of aircraft noise. (Data from appendix XI of ref. 27.)

On the other hand, table 12.9 (from ref. 26) shows that although the aircraft noise at $L_{dn} = 60$, the Orange County people considered aircraft noise to be a more major problem (25.9 percent) than other local problems (about 1 to 15 percent). This apparently wide difference in what might be called priorities of noise versus other environmental and social conditions within these two residential areas is perhaps caused by the milder climate and more strictly residential and nonindustrial character of Orange County compared with the London Heathrow areas.

In the context of identifying the threshold of noncompatibility of noise for residential neighborhoods, the aircraft noise becomes the most incompatible environmental factor to living at L_{dn} above about 65 when environmental factors other than noise appear to be generally adverse. In a warm climate and more uniformly residential areas, this threshold appears to be 5 to 10 dB lower. This

TABLE 12.9

Major Local Problems and Community Response Around Orange County Airport Among Subsample Groups

[From ref. 26]

(a) Problems

Problems	Total, percent	~55* CNEL,** percent	60 CNEL percent	65 CNEL percent
Aircraft noise	16.5	5.5	25.9	55.6
Traffic congestion	16.5	19.0	15.1	
Crime	11.3	12.5	10.2	
Population growth	8.3	9.5	6.8	
Growth/development	5.8	6.5	5.4	
Taxes in general	4.8	5.0	4.4	
Airport	4.3	3.0	5.9	11.1
Noise	3.8	2.5	4.4	11.1
Housing costs	3.3	4.0	2.4	
Pollution in general	2.5	1.0	3.9	
Air quality/smog	2.3	4.0	0.5	
Inflation	2.0	3.0	1.0	
Don't know	4.5	7.0	1.5	11.1

(b) Community response

Response	~55 CNEL, percent	60 CNEL, percent	65 CNEL, percent
Noise interferes with activities	26	60	67
Does not interfere	74	40	33

**2-mile area outside 60 CNEL, average CNEL less than 55.*
***CNEL* ≈ L_{dn}.

analysis is taken to support the thresholds of noncompatibility identified in figure 12.6 (from ref. 27).

Comparison of proposed guidelines with previous guidelines

At first glance, comparison of the limits 55 and 65 L_{dn} for marginal compatible and noncompatible, typical residential areas, as proposed in table 12.10, may seem fairly consistent with most of the previous guidelines summarized in table 12.7. However, the new guidelines proposed in table 12.10 differ significantly from the previous guidelines for certain specific situations.

TABLE 12.10

Proposed Exposure Levels, in Annualized Daily Average L_{dn}, for Compatibility of Exterior Environmental Noise in Typical Residential Areas and Housing

[Compatibility based on multiple criteria—health effects, housing depreciation, complaint behavior, and attitudes of annoyance (see fig. 12.5)]

	Windows sometimes open				Windows always closed, sound attenuation ≈ 28 dB			
	Aircraft noise, dBA		Ground vehicle noise, dBA		Aircraft noise, dBA		Ground vehicle noise, dBA	
Climate	Cold	Warm	Cold	Warm	Cold	Warm	Cold	Warm
Not compatible	>65	>60	>75	>70	>70	>65	>80	>75
Marginally compatible	55-65	50-60	65-70	60-70	60-70	55-65	70-80	65-75
Compatible	<55	<50	<65	<60	<60	<55	<70	<65

For example, for typical residential housing with windows open, the level limit for the start of "not compatible" is $L_{dn} = 65$ in the Federal Agency guideline for ground vehicle noise, but is $L_{dn} = 75$ in the proposed guidelines for a cold climate. On the other hand, for aircraft noise in warm climates, this limit is $L_{dn} = 60$ for the proposed and $L_{dn} = 65$ for the Federal Agency guidelines.

A common question is whether it is realistic to set a specific number, in dB, of guideline limits in view of uncertainties in some of the response-to-noise data as well as in some of the physical noise factors themselves. This matter is reflected in statements such as: "It has been established that the average person cannot distinguish between noise levels which differ by 3 dB or less." (See p. A3 of ref. 28.)

While an average person may exhibit such variability under real-life conditions, the average of the judgments of a group of people, say from a given homogeneous neighborhood, generally has about a 1-dB standard error. By enlarging the group size and increasing the number of exposures, the predictability of the average response of the group to a given noise condition becomes increasingly accurate. This is especially true when, as is generally the case for environmental noise, the response behavior to be assessed is an attitude or effect created over rather extensive periods of time, such as weeks or months. Whatever error may be involved in predicting the effects of some environmental noises by the use of figure 12.5, or similar methods, the error is just as likely to lead to an underestimation as an overestimation of the effects of the noise on a typical group of people during any one day, week, or other relatively brief period of time.

As shown in figure 12.5, for each dB increase in aircraft noise exposure above $L_{dn} = 65$ (cold climate), there is a 1- to 2-percent increase in the number of people "very or more annoyed," and a 0.5- to 1-percent increase in the number of people with health disorders and in house depreciation. A 1-dB increase at these levels translates into about a 5-percent increase in the actual number of people very or more annoyed, and about a 10-percent increase in the actual number of people with health disorders.

It seems clear that in a cold climate with aircraft noise at or above $L_{dn} = 65$, or the equivalent, these adverse effects (and considerably below $L_{dn} = 65$ for feelings of annoyance) are statistically measurable in large population studies. The practical significance of a given noise exposure level, or a change in 1-dB steps in a given exposure condition, is of course a matter of judgment about the costs to the health, well-being, and property of those exposed to the noise versus the costs, in a broad sense, of controlling the noise.

All the research data underlying the previous and proposed guidelines are based on studies of large samples of people from a rather wide range of noise exposure levels and socio-economic areas. As such, the findings are statistical in nature and can be used to predict the effects of noise exposure in any specific neighborhood or locale only with some degree of uncertainty and variability. However, the variability appears to be just as likely in the direction of underestimating as overestimating the adverse effects of the noise.

Guidelines for Nonresidential Areas

The Federal Agencies Guidelines are the most complete, up-to-date recommendations of noise limits suitable for compatibility in a large variety of nonresidential land and structure usages. The guidelines proposed by the Federal Interagencies for all land usages are given in table 12.11. The recommended exterior noise limits will result, because of exterior noise attenuation properties of the different buildings, in interior noise levels compatible with speech communications, listening to music and lectures, and sleeping. Because of the use of air conditioning and generally reduced open-window and outdoor activities associated with many nonresidential structures, acoustical climate and source factors, cited as being important for typical resident houses, would generally be less important variables in nonresidential areas.

Ordinances for Fixed-Source Noises

A number of municipalities and counties have established ordinances which specify maximum limits for the noise generated by a source located on one piece of property when measured at the boundaries of that piece of property. These specifications presumably provide for a legal control over undesirable noise levels in land areas adjacent to the source of the noise.

TABLE 12.11

Suggested Land Use Compatability Guidelines

[From ref. 10]

Key

SLUCM	Standard Land Use Coding Manual.
Y (yes)	Land use and related structures are compatible without restrictions.
N (no)	Land use and related structures are not compatible and should be prohibited.
25, 30, or 35	Land use and related structures generally are compatible; measures to achieve NLR of 25, 30, or 35 must be incorporated into design and construction of structure.
25*, 30*, or 35*	Land use generally is compatible with NLR; however, measures to achieve an overall noise reduction do not necessarily solve noise difficulties and additional evaluation is warranted.

Land use		Noise zones/DNL levels in L_{dn}						
SLUCM no.	Name	A 0-55	B 55-65	C-1 65-70	C-2 70-75	D-1 75-80	D-2 80-85	D-3 85+
10	**Residential**							
11	Household units							
11.11	Single units—detached	Y	Y[a]	25[b]	30[b]	N	N	N
11.12	Single units—semidetached	Y	Y[a]	25[b]	30[b]	N	N	N
11.13	Single units—attached row	Y	Y[a]	25[b]	30[b]	N	N	N
11.21	Two units—side-by-side	Y	Y[a]	25[b]	30[b]	N	N	N
11.22	Two units—one above the other	Y	Y[a]	25[b]	30[b]	N	N	N
11.31	Apartments—walk up	Y	Y[a]	25[b]	30[b]	N	N	N
11.32	Apartments—elevator	Y	Y[a]	25[b]	30[b]	N	N	N
12	Group quarters	Y	Y[a]	25[b]	30[b]	N	N	N
13	Residential hotels	Y	Y[a]	25[b]	30[b]	N	N	N
14	Mobile home parks or courts	Y	Y[a]	N	N	N	N	N
15	Transient lodgings	Y	Y[a]	25[b]	30[b]	35[b]	N	N
16	Other residential	Y	Y[a]	25[b]	30[b]	N	N	N
20	**Manufacturing**							
21	Food and kindred products—manufacturing	Y	Y	Y	Y[c]	Y[d]	Y[e]	N
22	Textile mill products—manufacturing	Y	Y	Y	Y[c]	Y[d]	Y[e]	N
23	Apparel and other finished products made from fabrics, leather, and similar materials—manufacturing	Y	Y	Y	Y[c]	Y[d]	Y[e]	N
24	Lumber and wood products (except furniture)—manufacturing	Y	Y	Y	Y[c]	Y[d]	Y[e]	N

TABLE 12.11—Continued.

Land use		Noise zones/DNL levels in L_{dn}						
SLUCM no.	Name	A 0-55	B 55-65	C-1 65-70	C-2 70-75	D-1 75-80	D-2 80-85	D-3 85+
25	Furniture and fixtures—manufacturing	Y	Y	Y	Y[c]	Y[d]	Y[e]	N
26	Paper and allied products—manufacturing	Y	Y	Y	Y[c]	Y[d]	Y[e]	N
27	Printing, publishing, and allied industries	Y	Y	Y	Y[c]	Y[d]	Y[e]	N
28	Chemicals and allied products—manufacturing	Y	Y	Y	Y[c]	Y[d]	Y[e]	N
29	Petroleum refining and related industries	Y	Y	Y	Y[c]	Y[d]	Y[e]	N
30	**Manufacturing**							
31	Rubber and misc. plastic products—manufacturing	Y	Y	Y	Y[c]	Y[d]	Y[e]	N
32	Stone, clay, and glass products—manufacturing	Y	Y	Y	Y[c]	Y[d]	Y[e]	N
33	Primary metal industries	Y	Y	Y	Y[c]	Y[d]	Y[e]	N
34	Fabricated metal products—manufacturing	Y	Y	Y	Y[c]	Y[d]	Y[e]	N
35	Professional, scientific, and controlling instruments; photographic and optical goods; watches and clocks—manufacturing	Y	Y	Y	25	30	N	N
39	Miscellaneous manufacturing	Y	Y	Y	Y[c]	Y[d]	Y[e]	
40	**Transportation, communication, and utilities**							
41	Railroad, rapid rail transit and street railway transportation	Y	Y	Y	Y[c]	Y[d]	Y[e]	Y
42	Motor vehicle transportation	Y	Y	Y	Y[c]	Y[d]	Y[e]	Y
43	Aircraft transportation	Y	Y	Y	Y[c]	Y[d]	Y[e]	Y
44	Marine craft transportation	Y	Y	Y	Y[c]	Y[d]	Y[e]	Y
45	Highway and street right-of-way	Y	Y	Y	Y[c]	Y[d]	Y[e]	Y
46	Automobile parking	Y	Y	Y	Y[c]	Y[d]	Y[e]	N
47	Communication	Y	Y	Y	25[f]	30[f]	N	N
48	Utilities	Y	Y	Y	Y[c]	Y[d]	Y[e]	Y
49	Other transportation, communication, and utilities	Y	Y	Y	25[f]	30[f]	N	N
50	**Trade**							
51	Wholesale trade	Y	Y	Y	Y[c]	Y[d]	Y[e]	N
52	Retail trade—building materials, hardware, and farm equipment	Y	Y	Y	Y[c]	Y[d]	Y[e]	N

TABLE 12.11—Continued.

Land use		Noise zones/DNL levels in L_{dn}						
SLUCM no.	Name	A 0-55	B 55-65	C-1 65-70	C-2 70-75	D-1 75-80	D-2 80-85	D-3 85+
53	Retail trade—general merchandise	Y	Y	Y	25	30	N	N
54	Retail trade—food	Y	Y	Y	25	30	N	N
55	Retail trade—automotive, marine craft, aircraft, and accessories	Y	Y	Y	25	30	N	N
56	Retail trade—apparel and accessories	Y	Y	Y	25	30	N	N
57	Retail trade—furniture, home furnishings and equipment	Y	Y	Y	25	30	N	N
58	Retail trade—eating and drinking establishments	Y	Y	Y	25	30	N	N
59	Other retail trade	Y	Y	Y	25	30	N	N
60	**Services**							
61	Finance, insurance, and real estate services	Y	Y	Y	25	30	N	N
62	Personal services	Y	Y	Y	25	30	N	N
62.4	Cemeteries	Y	Y	Y	Y[c]	Y[d]	Y[e,g]	Y[g,h]
63	Business services	Y	Y	Y	25	30	N	N
64	Repair services	Y	Y	Y	Y[c]	Y[d]	Y[c]	N
65	Professional services	Y	Y	Y	25	30	N	N
65.1	Hospitals, nursing homes	Y	Y[a]	25*	30*	N	N	N
65.1	Other medical facilities	Y	Y	Y	25	30	N	N
66	Contract construction services	Y	Y	Y	25	30	N	N
67	Governmental services	Y	Y[a]	Y	25*	30*	N	N
68	Educational services	Y	Y[a]	25*	30*	N	N	N
69	Miscellaneous services	Y	Y	Y	25	30	N	N
70	**Cultural, entertainment, and recreational**							
71	Cultural activities (including churches)	Y	Y[a]	25*	30*	N	N	N
71.2	Nature exhibits	Y	Y[a]	Y[a]	N	N	N	N
72	Public assembly	Y	Y	Y	N	N	N	N
72.1	Auditoriums, concert halls	Y	Y	25	30	N	N	N
72.11	Outdoor music shells, amphitheaters	Y	Y[a]	N	N	N	N	N
72.2	Outdoor sports arenas, spectator sports	Y	Y	Y[i]	Y[i]	N	N	N
73	Amusements	Y	Y	Y	Y	N	N	N
74	Recreational activities (incl. golf courses, riding stables, water recreation)	Y	Y[a]	Y[a]	25*	30*	N	N

TABLE 12.11—Continued.

Land use		Noise zones/DNL levels in L_{dn}						
SLUCM no.	Name	A 0–55	B 55–65	C-1 65–70	C-2 70–75	D-1 75–80	D-2 80–85	D-3 85+
75	Resorts and group camps	Y	Y[a]	Y[a]	Y[a]	N	N	N
76	Parks	Y	Y[a]	Y[a]	Y[a]	N	N	N
79	Other cultural, entertainment, and recreation	Y	Y[a]	Y[a]	Y[a]	N	N	N
80	**Resource production and extraction**							
81	Agriculture (except livestock)	Y	Y	Y[j]	Y[k]	Y[l]	Y[g,l]	Y[g,l]
81.5 to 81.7	Livestock farming and animal breeding	Y	Y	Y[j]	Y[k]	N	N	N
82	Agricultural related activities	Y	Y	Y[j]	Y[k]	Y[l]	Y[g,l]	Y[g,l]
83	Forestry activities and related services	Y	Y	Y[j]	Y[k]	Y[l]	Y[g,l]	Y[g,l]
84	Fishing activities and related services	Y	Y	Y	Y	Y	Y	Y
85	Mining activities and related services	Y	Y	Y	Y	Y	Y	Y
89	Other resource production and extraction	Y	Y	Y	Y	Y	Y	Y

[a] *The designation of these uses as "compatible" in this zone reflects individual Federal agencies consideration of general cost and feasibility factors as well as past community experiences and program objectives. Localities, when evaluating the application of these guidelines to specific situations; may have concerns or goals to consider.*

[b] *Although local conditions may require residential use, it is discouraged in C-1 and strongly discouraged in C-2. The absence of viable alternative development options should be determined and an evaluation indicating that a demonstrated community need for residential use would not be met if development were prohibited in these zones should be conducted prior to approvals. Where the community determines that residential uses must be allowed, measures to achieve outdoor to indoor noise level reduction (NLR) of a least 25 dB (Zone C-1) and 30 dB (Zone C-2) should be incorporated into building codes and be considered in individual approvals. Normal construction can be expected to provide a NLR of 20 dB, thus the reduction requirements are often stated as 5, 10, or 15 dB over standard construction and normally assume mechanical ventilation and closed windows year round. Additional consideration should be given to modifying NLR levels based on peak noise levels. NLR criteria will not eliminate outdoor noise problems. However, building location and site planning, design and use of berms and barriers can help mitigate outdoor noise exposure particularly from ground level sources. Measures that reduce noise at a site should be used wherever practical in preference to measures which only protect interior spaces.*

[c] *Measures to achieve NLR of 25 must be incorporated into the design and construction of portions of these buildings where the public is received, office areas, noise sensitive areas, or where the normal noise level is low.*

[d] *Measures to achieve NLR of 30 must be incorporated into the design and construction of portions of these buildings where the public is received, office areas, noise sensitive areas, or where the normal noise level is low.*

TABLE 12.11 Concluded

[e] *Measures to achieve NLR of 35 must be incorporated into the design and construction of portions of these buildings where the public is received, office areas, noise sensitive areas, or where the normal noise level is low.*

[f] *If noise sensitive, use indicated NLR; if not, use is compatible.*

[g] *Land use not recommended, but if community decides use is necessary, hearing protection devices should be worn by personnel.*

[h] *No buildings.*

[i] *Land use is compatible provided special sound reinforcement systems are installed.*

[j] *Residential buildings require a NLR of 25.*

[k] *Residential buildings require a NLR of 30.*

[l] *Residential buildings not permitted.*

The variation in the tolerable boundary levels specified in these ordinances is quite wide, as illustrated in figures 12.7, 12.8, and 12.9. These figures, taken from a compilation by the National Institute of Municipal Law Officers and the EPA (ref. 29), presumably reflect an uneven interpretation of knowledge about the effects of noise on people and communities as well as differences in the perceived needs and goals of different communities with respect to public health and land usages.

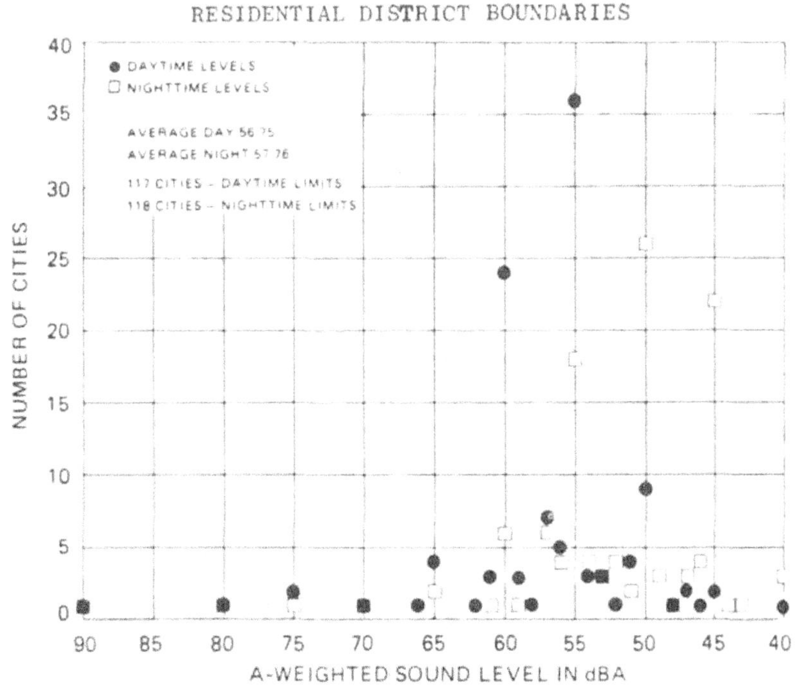

FIGURE 12.7. Fixed source noise levels allowable at residential district boundaries. (From ref. 29.)

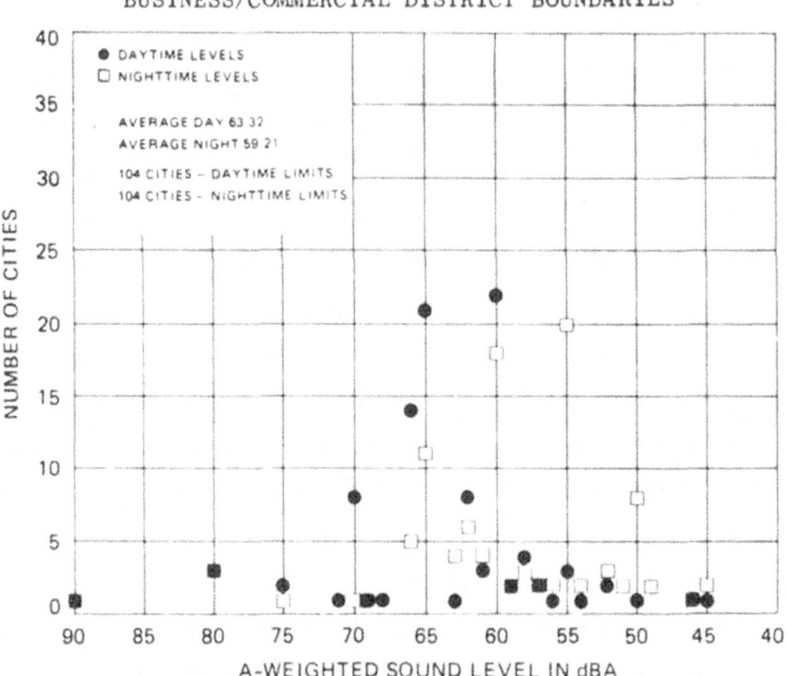

FIGURE 12.8. Fixed source noise levels allowable at business/commercial district boundaries. (From ref. 29.)

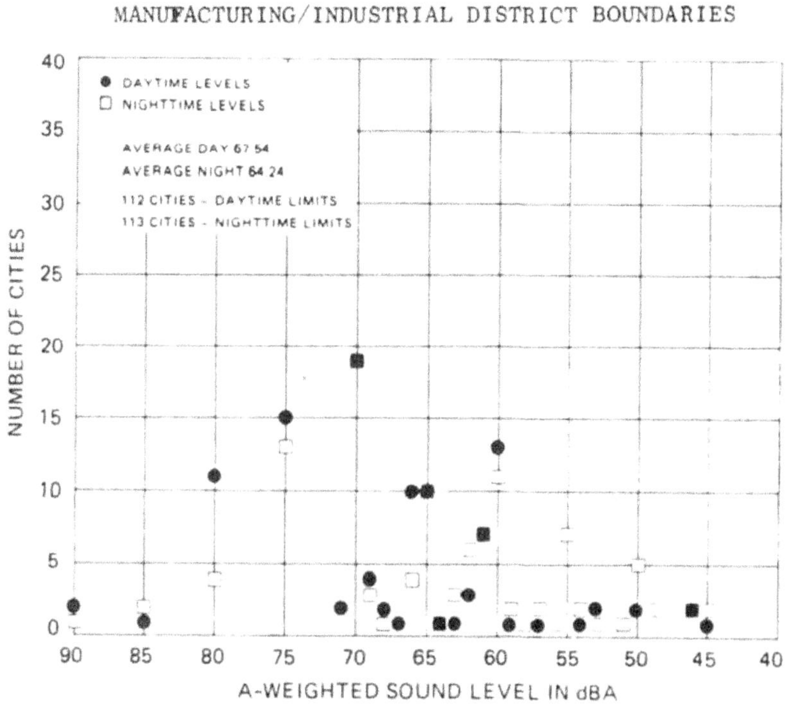

MANUFACTURING/INDUSTRIAL DISTRICT BOUNDARIES

FIGURE 12.9. Fixed source noise levels allowable at manufacturing/industrial boundaries. (From ref. 29.)

References

1. Anon.: Noise Control Act of 1972. Public Law 92-574, Oct. 1972.
2. Information on Levels of Environmental Noise Requisite To Protect Public Health and Welfare With an Adequate Margin of Safety. Rep. 550/9-74-004, Environ. Prot. Agency, Mar. 1974. (Available from NTIS as PB 239 429.)
3. Protective Noise Levels: Condensed Version of EPA Levels Document. Rep. EPA-550/9-79-100, Nov. 1978. (Available from NTIS as PB82 138 827.)
4. Land Use Planning Relating to Aircraft Noise. Tech. Rep., Bolt, Beranek & Newman, Inc., Oct. 1964. (Available from DTIC as AD 615 015.)
5. Land Use Planning With Respect to Aircraft Noise. AFM 86-5, TM 5-365, NAVDOCKS P-98, U.S. Dep. Defense, Oct. 1, 1964. (Available from DTIC as AD 615 015.)
6. Galloway, W. J.; and Von Gierke, H. E.: Individual and Community Reaction to Aircraft Noise; Present Status and Standardization Efforts. INC/C4/P9, American Stand. Assoc., Nov. 1966.

7. Schultz, Theodore J.; and McMahon, Nancy M.: HUD Noise Assessment Guidelines. Rep. No. HUD TE/NA 171, Aug. 1971. (Available from NTIS as PB 210 590.)
8. Environmental Criteria and Standards. Fed. Regist., vol. 44, no. 135, pt. 51, July 12, 1979, pp. 40860-40866.
9. Galloway, W. J.; and Schultz, T. J.: Interim Noise Assessment Guidelines. Rep. HUD-0001934 (Contract No. HUD-H-2243R), Bolt Beranek and Newman, Inc., Oct. 1980. (Available from NTIS as PB82 125 527.)
10. Guidelines for Considering Noise in Land Use Planning and Control. Federal Interagency Committee on Urban Noise, June 1980.
11. Establishment of New Part 150 To Govern the Development and Submission of Airport Operator's Noise Compatibility Planning Programs and the FAA's Administrative Process for Evaluating and Determining the Effects of Those Programs. Fed. Regist., vol. 46, no. 16, pt. 150, Jan. 26, 1981, pp. 8316-8346.
12. Sound Level Descriptors for Determination of Compatible Land Use. ANSI S3.23-1980 (ASA 22-1980), American Nat. Stand. Inst., Inc., 1980.
13. Guidelines for Preparing Environmental Impact Statements on Noise. Report of Working Group 69 on Evaluation of Environmental Impact of Noise, Committee on Hearing, Bioacoustics, and Biomechanics, Contract No. N00014-75-C-0406, Nat. Res. Council, Nat. Academy Sci., 1977. (Available from DTIC as AD A044 384.)
14. California Airport Noise Standards (California Administrative Code, Title 21, Chapter 2.5, Subchapter 6, Articles 1 Through 14; As Amended Through May 26, 1979). Noise Regulation Reporter, 81:3581, Nov. 3, 1980, pp. 87-97.
15. Wyle Labs. Res. Staff: Supporting Information for the Adopted Noise Regulations for California Airports. Rep. No. WCR 70-3(R), Wyle Labs., Jan. 29, 1971.
16. Schultz, Theodore J.: Synthesis of Social Surveys on Noise Annoyance. J. Acoust. Soc. America, vol. 64, no. 2, Aug. 1978, pp. 377-405. (Erratum, vol. 65, no. 3, Mar. 1979, p. 849.)
17. Testimony of W. J. Galloway. Charles Frances Davison *et al.* vs. Department of Defense *et al.*, U.S. District Court, Southern District of Ohio, Eastern Div., Case NO. C-2-80-871, Apr. 17, 1982.
18. Galloway, W. J.; Eldred, K. M.; and Simpson, M. A.: Population Distribution of the United States as a Function of Outdoor Noise Level—Volume 2. Rep. EPA-550/9-74-009-A-VOL-2, June 1974. (Available from NTIS as PB 257 617/1.)
19. FAA Integrated Noise Model Version 1, Basic User's Guide. Rep. FAA-EQ-78-01, Dec. 1977. (Available from DTIC as AD A052 790.)
20. Noise Standards: Aircraft Type Certification. Federal Aviation Regulations, vol. III, pt. 36, FAA, Dec. 1969.

21. Knipschild, Paul; and Oudshoorn, Nelly: Medical Effects of Aircraft Noise: Drug Survey. Int. Arch. Occup. Environ. Health, vol. 40, no. 3, Nov. 29, 1977, pp. 197-200.
22. Knipschild, Paul: Medical Effects of Aircraft Noise: Community Cardiovascular Survey. Int. Arch. Occup. Environ. Health, vol. 40, no. 3, Nov. 29, 1977, pp. 185-190.
23., Knipschild, Paul: Medical Effects of Aircraft Noise: General Practice Survey. Int. Arch. Occup. Environ. Health, vol. 40, no. 3, Nov. 29, 1977, pp. 191-196.
24. Knipschild, Paul: Aircraft Noise and Hypertension. Noise as a Public Health Problem, Jerry V. Tobias, Gerd Jansen, and W. Dixon Ward, eds., ASHA Rep. 10, American Speech-Language-Hearing Assoc., Apr. 1980, pp. 283-293.
25. MIL Research Ltd: Second Survey of Aircraft Noise Annoyance Around London (Heathrow) Airport. Her Majesty's Stationery Off., 1971.
26. Opinion Research of California: Report of Findings—A Study Among Residents of Orange County Relative to Operations at Orange County Airport. City of Newport Beach [California], Mar. 1977.
27. Committee on the Problem of Noise: Noise—Final Report. Her Majesty's Stationery Office, July 1963.
28. Aircraft Noise and Sleep Disturbance: Final Report. DORA Rep. 8008, Civil Aviation Authority, London, 1980.
29. Nat. Inst. Municipal Law Off., U.S. Environ. Prot. Agency: Model Community Noise Control Ordinance. EPA 550/9-76-003, Sept. 1975. (Available from NTIS as PB 262 005/2.)

Index

www.ingramcontent.com/pod-product-compliance
Lightning Source LLC
Chambersburg PA
CBHW081713170526
45167CB00009B/3562